Optical Metrology

NATO ASI Series

Advanced Science Institutes Series

A Series presenting the results of activities sponsored by the NATO Science Committee, which aims at the dissemination of advanced scientific and technological knowledge, with a view to strengthening links between scientific communities.

The Series is published by an international board of publishers in conjunction with the NATO Scientific Affairs Division

A	Life Sciences	Plenum Publishing Corporation
B	Physics	London and New York
C	Mathematical and Physical Sciences	D. Reidel Publishing Company Dordrecht, Boston, Lancaster and Tokyo
D	Behavioural and Social Sciences	Martinus Nijhoff Publishers Boston, Dordrecht and Lancaster
E	Applied Sciences	
F	Computer and Systems Sciences	Springer-Verlag Berlin, Heidelberg, New York
G	Ecological Sciences	London, Paris, Tokyo
H	Cell Biology	

Series E: Applied Sciences – No. 131

Optical Metrology

Coherent and Incoherent Optics for Metrology, Sensing and Control in Science, Industry and Biomedicine

edited by:

Olivério D.D. Soares

Professor of Physics
University of Porto
Porto
Portugal

1987 **Martinus Nijhoff Publishers**
Dordrecht / Boston / Lancaster
Published in cooperation with NATO Scientific Affairs Division

Proceedings of the NATO Advanced Study Institute on "Optical Metrology", Viana do Castelo, Portugal, July 16-27, 1984

Library of Congress Cataloging in Publication Data

NATO Advanced Study Institute on Optical Metrology
   (1984 : Viana do Castelo, Portugal)
   Optical metrology.

   (NATO ASI series. Series E, Applied sciences ;
no. 131)
   "Proceedings of the NATO Advanced Study Institute
on Optical Metrology, Viana do Castelo, Portugal, July
16-27, 1984"--T.p. verso.
   "Published in cooperation with NATO Scientific
Affaris Division."
   Includes index.
   1. Optical measurements--congresses. 2. Physical
measurements--Congresses. I. Soares, O. D. D.
(Olivério D. D.) II. North Atlantic Treaty Organization.
Scientific Affairs Division. III. Title. IV. Series.
QC367.N38 1984   681.2   87-7800

ISBN 978-94-010-8115-3   ISBN 978-94-009-3609-6 (eBook)
DOI 10.1007/978-94-009-3609-6

Distributors for the United States and Canada: Kluwer Academic Publishers, P.O. Box 358, Accord-Station, Hingham, MA 02018-0358, USA

Distributors for the UK and Ireland: Kluwer Academic Publishers, MTP Press Ltd, Falcon House, Queen Square, Lancaster LA1 1RN, UK

Distributors for all other countries: Kluwer Academic Publishers Group, Distribution Center, P.O. Box 322, 3300 AH Dordrecht, The Netherlands

Hardcover edition, printed in 1987 by Martinus Nijhoff Publishers, Dordrecht

PRELIMINARIES

portugal

OPTICAL METROLOGY

PREFACE

Optical Metrology is a rapidly expanding field in both its scientific foundations and technological developments, being of major concern to measurements, quality control, non-destructive testing and in fundamental research.

In order to define the state-of-the-art, and to evaluate present accomplishments, whilst giving an appraisal of how each of the particular topics will evolve the Optical Metrology - an Advanced Study Institute was organized with a concourse of the world's acknowledged experts. Thus, the Institute provided a forum for tutorial reviews blended with topics of current research in the form of a progressive and comprehensive presentation of recent promising developments, leading techniques and instrumentation in incoherent and coherent optics for Metrology, Sensing and Control in Science, Industry and Biomedicine.

Optical Metrology is a very broad field which is highly interdisciplinary in its applications, and in its scientific and technological background. It is related to such diverse disciplines as physical and chemical sciences, engineering, electronics, computer sciences, biological sciences and theoretical sciences, such as statistics.

Although there was an emphasis on photomechanics and industrial applications, a marked diversity was reflected in the different background and interests of the participants. The vitality and viability of the discipline was enhanced not only by the encouraging number of young scientists and industrialists participating and authoring, but also by the remarkably promising prospects found in the practical applications supported by advanced electronic hybridization.

Demonstrations of various techniques involving projections of films, video, and live experiments during lectures contributed significantly to the scope of the Institute. Equipment was also made available especially for manipulation by the participants, complementing the material delivered in the lecture's room. A parallel hologram exhibition displayed holograms of various types including: reflection, embossed, rainbow, multiplex, and light-in-flight holograms.

It was felt that the relevance and continuous development of Optical Metrology deserved a critical examination of the present status of education and training activities in the field. Consequently, curriculum development was studied by a working group that provided recommendations for further action to identify cohesive modules for education and practical training.

The formal and informal discussions often extended into vivid debates, the foremost topics being: research activities, application in industrial environments, and the consequences of scientific and technological development to human society.

Despite the inevitable limitations, a broad and reasonably representative coverage of the field was achieved, great advantage being taken of the very active audience which included a significant number of experts in the field who contributed posters and written papers, some of which are included in the proceedings.

The written material was grouped in eleven major sections:Foundations; Beam Deflection and Image Forming Systems; Interferometry, Velocimetry and Photon Correlation; Image Processing Techniques; Moiré Methods; Holographic Metrology; Speckle Metrology; Optical Fibre Metrology; Picosecond Pulse Metrology; Photoelasticity, and Further Topics. The papers were assembled in a pedagogical manner in order to establish a transition between research and education, and the needs of the industrial community.

It is to be hoped that independently of background, scientific or industrial, those dedicated to Optics, Mechanics and Engineering who wish to work in the field or study it in depth will find the proceedings an important tool for the design of applications, and the development of major ideas and new directions of research. Furthermore, there is an excellent up-to-date source of bibliographical material.

The proceedings are of course, the result of considerable efforts by authors for which they are cordially thanked and acknowledged.

The editor would like to express profuse thanks to the organizations and societies that lent indispensable financial assistance. The Scientific Affairs Division of NATO provided the essential foundations of support to the Institute. Likewise, great support was made available by the Council of Europe. Hewlett Packard is to be thanked both for its kindness in awarding funds used in grants to deserving students who could not have attended otherwise, and for the exhibition and demonstration of essential equipment. Other Institutions providing complementary support are acknowledged with gratitude. In particular, my own Institution, the University of Porto, is thanked for its constant support.

Special thanks are also due to the diverse contributions and assistance that engendered the particularly stimulating social atmosphere that all the one hundred or so participants from 20 countries around the world so much enjoyed, and which contributed decisively to informal exchange of new ideas, the identification of problem areas and laid the roots for future cooperation among participants. This spirit of conviviality which grew out of many informal interactions has not been properly captured in the proceedings but will continue to increase between individuals and foster further inter-disciplinary applications of Optical Metrology.

The invaluable help of the staff working with the organizing committee is gratefully acknowledged, the tireless efforts of Mr. J.S. Fernandes deserves a particular mention.

Last, but not least the editor is also indebted to Mrs. H. Hoogervorst of Martinus Nijhoff for her contagious enthusiasm and devoted work to the publication of the proceedings

Professor O.D.D. Soares
Universidade do Porto
4000 Porto, Portugal

INSTITUTE PROGRAMME

OPTICAL METROLOGY

INTERNATIONAL ADVANCED STUDY INSTITUTE
16-27 July 1984
HOTEL DO PARQUE
VIANA DO CASTELO - PORTUGAL

Progressive and Comprehensive presentation of recent developments
and leading techniques in coherent and incoherent OPTICS for METRO-
LOGY, SENSING and CONTROL in SCIENCE, INDUSTRY and BIOMEDICINE.

MOTIVATION OF THE INSTITUTE

Optical metrology is receiving great attention stimulated partially
by the availability of new light sources (e.g. Lasers), and components (e.g. optical fibres, photodetectors) and partly by develoments,
in general, of the new field of coherent Optics. In consequence, the
range and variety of optical techniques and applications have grown
recently. Applications are being progressively and extensively dis-
seminated in several fields of science, medicine and industry.
The amount of scientific literature is by now vast and much of it
dispersed in interdisciplinary scientific magazines as a consequence
of its relevance to several new directions in science and technology.
Instrumentation, techniques, and components for optical metrology
have grown significantly with considerable diversity and sophistication to deserve an integrated view of later developments.

PROGRAMME

16th JULY - Monday

9.00/10.00 - Openning. Aims and Programme Overview (O. D. D. SOARES)
10.00/11.00 - Fundamentals of Optical Metrology (H.J.CAULFIELD)
11.00/11.30 - Coffee Break
11.30/12.30 - Spectral and Temporal Metrology (H.J. CAULFIELD)
12.30/14.30 - Lunch
14.30/15.30 - Optical Metrology Overview (J.M. BURCH)
15.30/16.30 - Stabilised Lasers and Length Measurement (J. M. BURCH)
16.30/17.00 - Coffee Break
17.00/18.00 - Engineering Applications of Laser-Beam-Deflection Measurement Methods (R.J. PRYPUTNIEWICZ)

17th JULY - Tuesday

9.00/10.00 - Optical Metrology of Fish Scales (S.P. ALMEIDA)
10.00/11.00 - Hybrid Coherent Optical and Electronic Filtering (S.P. ALMEIDA)
11.00/11.30 - Coffee Break
11.30/12.30 - NATO Scientific Affairs Division - Various possibilities of support (V. da CUNHA)
12.30/14.30 - Lunch
14.30/15.30 - Application of Novel Heterodyne Techniques to Metrology (K. WICKRAMASINGHE)
15.30/16.30 - Laser Probes (K. WICKRAMASINGHE)
16.30/17.00 - Coffee Break
17.00/18.20 - Medium Scale and Large Scale Metrology + Film (J.M. BURCH)

18th JULY - Wednesday

 9.00/10.00 - Optical Inspection and Measurement along Optical
             Axis (F.C. RODRIGUES)

10.00/11.00 - 3 - Axis Measurement and Photogrametry (J.M. BURCH)

11.00/11.30 - Coffee Break

11.30/12.30 - Moiré Techniques (J.M. BURCH)

12.30/14.30 - Lunch

14.30/15.30 - Holography - Measuring Tool in Science and Industry
             (N. ABRAMSON)

15.30/16.30 - Holographic Vibration Analysis (N. ABRAMSON)

16.30/17.00 - Coffee Break

17.00/18.00 - Holo-diagram, a Practical Device for the Making
             and Evaluation of Holograms (N. ABRAMSON)

19th JULY - Thursday

 9.00/10.00 - Sandwich Holography (N. ABRAMSON)

10.00/11.00 - Fringe Interpretation in Hologram Interferometry
             by Matrix Methods (K.A. STETSON)

11.00/11.30 - Coffee Break

11.30/12.30 - Prediction of Fringe Patterns in Hologram Interfe-
             rometry (R.J. PRYPUTNIEWICZ)

12.30/14.30 - Lunch

14.30/15.30 - Unification of Laser Holography with Finite Element
             Methods (R.J. PRYPUTNIEWICZ)

15.30/16.30 - Holographic and Numerical Studies of Convective
             Heat Transfer from Horizontal Cylinders (R. J.
             PRYPUTNIEWICZ)

16.30/17.00 - Coffee Break

17.00/18.00 - High-Speed Holographic Motion Pictures of Ultra-
             -fast Phenomena + Film (N. ABRAMSON)

20th JULY - Friday

 9.00/10.00 - Computer Holograms in Metrology (H.J. CAULFIELD)

10.00/11.00 - Automated System for Engineering Applications of
              Holography (R.J. PRYPUTNIEWICZ)

11.00/11.30 - Coffee Break

11.30/12.30 - Automatic Holography NDT (P. MEYREUIS)

12.30/14.30 - Lunch

14.30/15.30 - Optical Metrology, Microprocessor and Microcompu-
              ters (C.M. VEST)

15.30/16.30 - Phase Conjugation Metrology (S.P. ALMEIDA)

16.30/17.00 - Coffee Break

17.00/18.00 - Biomedical Applications of Holography (G. von BALLY)

18.00/19.00 - Biomedical Applications of Holography (G. von BALLY)

21st JULY - Saturday

 9.00/10.00 - Holography Metrology - Past and Future (C.M. VEST)

10.00/11.00 - Laser Velocimetry (E.O. SCHULTZ-DUBOIS)

11.00/11.30 - Coffee Break

11.30/12.30 - Microscopic Particle Analysis (E.O. SCHULTZ-DUBOIS)

12.30/13.00 - Resumé (J.M. BURCH)

13.00/14.30 - Lunch

23rd JULY - Monday

 9.00/10.00 - Electronic Processing of Holographic Interferograms
              (R. DÄNDLIKER)

10.00/11.00 - Electronic Processing of Holographic Interferograms
              (R. DÄNDLIKER)

11.00/11.30 - Coffee Break

11.30/12.30 - Measuring Microscopic Vibrations by Heterodyne
              Interferometry (R. DÄNDLIKER)

12.30/14.30 - Lunch

14.30/15.30 - Microscopic Surface Roughness Metrology (S. P.
              ALMEIDA)

15.30/16.30 - Laser Roughness Measurement (P. MEYREUIS)

16.30/17.00 - Coffee Break

17.00/18.00 - Speckle Interferometry (K.A. STETSON)

24th JULY - Tuesday

9.00/10.00 - Matrix Procedures for Interpretation of Holograms and Specklegrams (R.J. PRYPUTNIEWICZ)

10.00/11.00 - Strain Analysis by Speckle Heterodyne Photogrammetry (K.A. STETSON)

11.00/11.30 - Coffee Break

11.30/12.30 - Computer Aided Analysis of Specklegrams (R.J. PRYPUTNIEWICZ)

12.30/14.30 - Lunch

14.30/15.30 - Biomedical Applications of Holography and Speckle Metrology (R.J. PRYPUTNIEWICZ)

15.30/16.30 - ESPI versus Hologram Interferometry. Fringe Formation, Basic Interferometers (O.J. LØKBERG)

16.30/17.00 - Coffee Break

17.00/18.00 - Opto-mechanical and Electronic Construction of an ESPI system (O.J. LØKBERG)

25th JULY - Wednesday

9.00/10.00 - ESPI Mode Operation. Observation of Dynamical and Static Changes. Contouring (O.J. LØKBERG)

10.00/11.00 - Application of ESPI in Industry and Biology. Future Trends (O.J. LØKBERG)

11.00/11.30 - Coffee Break

11.30/12.30 - Photoelasticity (J.F.S. GOMES)

12.30/14.30 - Lunch

14.30/15.30 - Photoelasticity (J.F.S. GOMES)

15.30/16.30 - Non-Destructive 3-D Photoelasticity (R.DESAILLY)

16.30/17.00 - Coffee

17.00/18.00 - Hybrid Coherent Optical and Electronic Object Recognition (R. DÄNDLIKER)

26th JULY - Thursday

 9.00/10.00 - Visual Inspection (B.G. BATCHELOR)

10.00/11.00 - Visual Inspection (B.G. BATCHELOR)

11.00/11.30 - Coffee Break

11.30/12.30 - Industrial Application of Practical Electro-Optical
       Instrumentation (B.G. BATCHELOR)

12.30/14.30 - Lunch

14.30/15.30 - Thermography LCD and Infra-red (P. MEYREUIS)

15.30/16.30 - Monomode Fiber. Optical Sensors (R. DÄNDLIKER)

16.30/17.00 - Coffee Break

17.00/18.00 - Monomode Fiber. Optical Sensors (R. DÄNDLIKER)

27th JULY - Friday

 9.00/10.00 - Glass Fiber Sensors (E.O. SCHULTZ-DUBOIS)

10.00/11.00 - Picosecond Optical Pulses (C. FROEHLY)

11.00/11.30 - Coffee Break

11.30/12.30 - Metrology by Picosecond Pulses and Nonlinear Optics
       (C. FROEHLY)

12.30/14.30 - Lunch

14.30/15.30 - Metrology by Picosecond Pulses and Temporal Holo-
       graphy (C. FROEHLY)

15.30/16.30 - Fibre Characterization by Picosecond Pulses Metro-
       logy (C. FROEHLY)

16.30/17.00 - Coffee Break

17.00/17.30 - Conclusions and Future Events (C. VEST and O.
       SOARES)

17.30/18.00 - Closing Remark (N. ABRAMSON)

SATELLITE ACTIVITIES

- Posters (J.M. BURCH; O.J. LØKBERG; R.J. PRYPUTNIEWICZ - coordinators.)
- Curriculum Development (K.A. STETSON; S.P. ALMEIDA - coordinators.)

- Informal Speakers Corners - Research and Culture (P: MEYREUIS;
  J.M. SANTOS - coordinators)
  Examples: - Most recent results of Research
            - Computer Graphics
            - Industrial Optical Metrology

INTERNATIONAL COMMITTEES
ORGANIZING

H.J. CAULFIELD, Aerodyne Research, Inc. U.S.A.

P. MEYREUIS, Université Louis Pasteur, FRANCE

R. PRYPUTNIEWICZ, Worcester Polytechnic Institute, U.S.A.

J.C.M. SANTOS, University of Porto, PORTUGAL

O.D.D. SOARES, University of Porto, PORTUGAL

K.A. STETSON, United Technology, U.S.A.

ADVISORY

S.P. ALMEIDA, Virginia Polytechnic Institute, U.S.A.

J.M. BURCH, National Physical Laboratory, U.K.

H.J. CAULFIELD, Aerodyne Research, Inc. U.S.A.

J.W. GOODMAN, Stanford University, U.S.A.

A. LOHMANN, University of Erlangen, F.R.G.

O. LØKBERG, University of Trondheim, NORWAY

K.A. STETSON, United Technologies, U.S.A.

J. TSUJIUCHI, Tokyo Institute of Technology, JAPAN

POSTERS

- Optical Standards and Frequency Synthesis at the Italian Metrology Institute.
    P. Sassi

- Three- Dimensional Inspection of Large Objects.
    M. Maul, G. Bickel and G. Häusler

- Laser Dimensional Metrology
    O.D.D. Soares and A.O.S. Gomes

- Distance Sensing via Spectrographic Coding
    G. Molesini and F. Quercioli

- Velocity Measurement in a Rotating Tank using a L.D.V. System
    M.S. Kilickaya and N. Ekem

- Deformation of Rail Fastening Clips with Automatic Fring Evaluation
    A.E. Ennos

- Moiré interferometry and its Developments in Strain Analysis
    L. Pirodda

- Measurements of Fracture Toughness and Integrity by Moiré Inter-ferometry
    J. Mckelvie, C.A. Walker, P. Mackenzie and T.G.F. Gray

- Phase Measuring Moiré Topography
    G.T. Reid

- Moiré Evaluation with Fringe Patterns of Interferograms,Holograms, Moirégrams and Specklegrams
    O.D.D. Soares, A.L.V.S. Lage and L.M. Bernardo

XXII

1 Abramson,Nils
Industrial Metrology
Royal Institute of Metrology
1044 STOCKOLM - SWEDEN

2 Abrantes,João
Lab. Biomecânica
Instituto Gulbenkian de Ciência
2781 OEIRAS CODEX - PORTUGAL

3 Aktas,Oslan
Bobazici University - Dep. of Physics
BEBEK - ISTANBUL - TURKEY

4 Almeida,Silvério P.
Virginia Polytechnic Institute and
State University, Dep. of Physics
BLACKSBURG, VIRGINIA 24061, U.S.A.

6 Bernardo,Luis Miguel
Lab. de Física-Fac.de Ciências
Univ. do Porto
4000 PORTO - PORTUGAL

7 Boulanger,Raymond
Université de Strasbourg
16 Boulevard de la Marne
67000 STRASBOURG - FRANCE

12 Costa,Carlos Alberto Pereira
Laboratório Nac.de Eng.Civil
Av. do Brasil 101
1800 LISBOA - PORTUGAL

13 Costa,Luis Cadilhon M.
Universidade de Aveiro
Dep. de Física
3800 AVEIRO - PORTUGAL

14 Couto,Paulo Martins
C.A.T.I.M.
Rua do Rio-Edifício Mipercentro
4300 PORTO - PORTUGAL

15 Cruickshank,Robert F.III
Center for Holographic Studies and
Laser Technology, Dep. of Mech.Engrs.
Worcester Polytechnic Institute
WORCESTER,MA 01609 - U.S.A.

16 Cruz, António
Direcção G.Qualidade
Rua Prof. Reinaldo Santos
Lote 1378
1500 LISBOA - PORTUGAL

17 Dandliker,Rene
Institute de Microtechnique
Université de Neuchatel,Maladiere 71
CH 2000 NEUCHATEL - SWITZERLAND

18 Desailly,Roger
CNRS-Lab.du Mech.des Solides
40 Av.du Recteur Pineau
86022 POITIERS - FRANCE

19 Ekem,Naci
Dept. of Physics - Anatolian University
ESKISEHIR - TURKEY

20 ENNOS,Tommy
National Physical Laboratory
Teddington,MIDDLESEX TW1 OLW
UNITED KINGDOM

21 Fabião,Luis Eugénio Bataglia
I.N.D.E.P.
Rua Fernando Palha
1099 LISBOA CODEX - PORTUGAL

22 Fernandes,J.C.Aparicio
Lab. de Física - Fac.Ciências
Praça Gomes Teixeira
4000 PORTO - PORTUGAL

24 Freitas,Diamantino R.de Silva
F.E.U.P. - Rua dos Bragas
4000 PORTO - PORTUGAL

25 Freitas,José António Cabrita
Lab. Nac. Eng.e Tech.Industrial
Estrada Nacional 10
2685 SACAVEM - PORTUGAL

26 Froehly, Claude
Université de Limoges
25,Rue des Sablons
87100 LIMOGES - FRANCE

27 Goncali, Osman Fayzi
Tech. Univ. of Istanbul
GUMUSSUYU - ISTANBUL - TURKEY

29 Gomes,J.F.Silva
Dept.Eng.Mech. - Univ. do Porto
Rua dos Bragas,4000 PORTO-PORTUGAL

31 Grossmann,Michel
Univ.Louis Pasteur
3,Rue de l'Université
67000 STRASBOURG-FRANCE

34 Haywood,Peter Ward
British Robotics Syst.Ltd.
90 Newington Causeway
LONDON SE 1,UNITED KINGDOM

35 Khan, Anan U.
Center for Holographic Studies and
Laser Technology-Dept. of Mech. Engrs.
Worcester Polytechnic Institute
WORCESTER,MA 01609 - U.S.A.

34 Ellichoya,M.S.
Dept. of Physics-Anatolian University
ESKISEHIR - TURKEY

35 Kneull, Hartwig
MTB- Munich
Buchauerstr. 665 - Postfach 500640
D-8000 MUNICH 50, FED.REP.GERMANY

36 Kurtoglu,Asim
Royal Institute of Tech.
Drimellivagen 6A,1
S-100 44 STOCKHOLM
SWEDEN

37 Lofberg,Ole J.
Physics Dept.
N-7034 -N.T.H.
TRONDHEIM - NORWAY

39 Lemozzi,José Joaquin
Univ. Estadual de Campinas
Instituto de Física
13100 CAMPINAS - SP,C.P.6165
CAMPINAS - BRASIL

42 Mckelvie,James
University of Strathclyde
2 Materside Gdns
CARDONALD,GLASGOW G76 9AL
UNITED KINGDOM

44 Montgomery, Paul C.
Mech.Engineering Dept.
Loughborough University
LOUGHBOROUGH - LEICESTER LE 11 3TU
UNITED KINGDOM

45 Morais, António C.Meireles
Secção Aut.de Eng. Mecânica
Fac. Ciências e Tecnologia
Universidade de Coimbra
3000 COIMBRA - PORTUGAL

46 Noson, Brian S.
Center for Holographic Studies and
Laser Technology, Dept.of Mech. Engrs.
Worcester Polytechnic Institute
WORCESTER, MA 01609 - U.S.A.

47 Nunes,Urbano José Correia
Universidade de Coimbra
Dept.Eng.Electrotécnica
3000 COIMBRA - PORTUGAL

49 Ots,Heikki
National Council for Metrology
and Testing
Box 878, S-501 15 BORAS, SWEDEN

55 Reid, Graeme T.
National Eng. Laboratory
EAST KILBRIDE
GLASGOW - UNITED KINGDOM

56 Renaud,Blaise
Institute CERAC
CH - 1074 ECUBLENZ
SWITZERLAND

57 Reynolds, George Frederick
Dept. of Cybernetics - Research Unit.
in Instr. Physics, Univ.of Reading
J.Corley Gate, White Knight
READING, BERKS RG62 AL
UNITED KINGDOM

58 Rodriguez, A. Corrams
Instituto de Optica
SERRANO 121, MADRID 6, SPAIN

59 Rowland, Adrian
Mechanical Engineering Dept.
Loughborough University
LOUGH.LEICESTERSHIRE LE 11 3TU
UNITED KINGDOM

62 Sassi, Maria Paola
Istituto de Metrologia-CNR TORINO
Strada delle Cacce, 73
10135 TORINO - ITALY

64 Seabra,Manuel Joaquim
Direcção Geral da Qualidade
Rua Prof.Reinaldo Santos,Lote 1378
1500 LISBOA - PORTUGAL

65 Shabas-Roctchkinon,Pablo
Instituto de Física, U.N.A.M.
Apdo.Postal 20-364
01000 MEXICO CITY - MEXICO

66 Sobrig,Gordion
Hahn-Meitner-Inst. fur Keraforschung
Clayallee 787
1000 BERLIN 37, FED.REP.GERMANY

67 Silva,Jordaino F.Araújo
Lab.Nac.de Eng.e Tec. Industrial
Estrada Nacional 10
2685 SACAVEM - PORTUGAL

69 Silva, Mário Zenith
Lab. Física-Univ. do Minho
4719 BRAGA - PORTUGAL

70 Soares,Olivério D.D.
Lab. de Física - Fac.Ciências
Praça Gomes Teixeira
4000 PORTO - PORTUGAL

72 Svensson, L. B.Michel
Dept. of Prod.Engineering
Operatörgrand 4
S - 175 69 JAMFALLA - SWEDEN

73 Tamer,Kemal
Fac.Eng.Mechanical Eng.Dept.
Anatolian University
ESKISEHIR - TURKEY

74 Tarantieff,Serge
Hewlett-Packard - Geneve
COSSY 1
1260 NYON - SWITZERLAND

76 Teixeira,Maria do Bidrio E.C.M.
I.N.D.E.P. - Av. do Rovacavão
MORCAVIDE - PORTUGAL

77 Triol, Marcelo Ricardo
Centro de Inv. Opticos
CONICET - UNLP - CIC
C.P.124, 1900 LA PLATA
ARGENTINA

78 Vest, Charles M.
Div. of Michigan
249 Chrysler Center
Ann Arbor,MI 48109 - U.S.A.

79 Vinhas, Maria da Graça
Secção Aut.de Eng. Química
Universidade de Coimbra
Largo Marquês de Pombal
3000 COIMBRA - PORTUGAL

81 Visser, Bert
Philips Research Laboratories
Wag 1, P.O.Box 80 000
5600 JA EINDHOVEN
The NETHERLANDS

82 Wei,Fu-Qiang
Div.Machine,L.N.F.-I.N.2.N.
C.P.13 00046
FRASCATI - ROMA - ITALY

84 Wickramasinghe,Kumar
IBM,T.J.Watson Res.Center
P.O.Box 218
YORKTOWN HEIGHTS
NEW YORK 10598 - U.S.A.

SPONSORS

Scientific Affairs Division - NATO
Higher Education and Research Division - Council of Europe
E.P.A. - European Photonics Association

Calouste Gulbenkian Foundation
HP - Hewlett Packard Laboratories

SPIE - International Society of Optical Engineering
EOC - European Optics Committee
INIC - Instituto Nacional de Investigação Científica
C.A.T.I.M. - Centro de Apoio Tecnológico à Industria Metalome-
            cânica

Reitoria da Universidade do Porto
Faculdade de Ciências da Universidade do Porto
EFACEC - Empresa Fabrìl de Máquinas Eléctricas
CTT-TLP - Correios e Telecomunicações de Portugal
BPA - Banco Português do Atlântico
Lloyds Bank International

CO-SPONSORS

National Science Foundation
ACC - American Cultural Council
The British Council
Service de Cooperation Scientifique et Technique - Ambassade
de France au Portugal

C.C.R.N. - Comissão de Coordenação da Região Norte
Governo Civil de Viana do Castelo
Câmara Municipal de Viana do Castelo
Comissão de Turismo do Alto-Minho (Costa Verde)
Direcção-Geral de Turismo
CDUP - Centro Desportivo Universitário do Porto

Câmara Municipal do Porto

Câmara Municipal de Braga

Câmara Municipal de Barcelos

Instituto do Vinho do Porto

Academia de Música de Viana do Castelo

Comissão de Viticultura da Região dos Vinhos Verdes

RN-Tours

NORTECOPIA

Comercial Laborum

SOGRAPE - Vinhos de Portugal SARL

A VIANENSE - Fábrica de Chocolates

TAP - Air Portugal

TABLE OF CONTENTS

IMAGE PROCESSING TECHNIQUES

MOIRE METHODS

HOLOGRAPHIC METROLOGY

SPECKLE METHODS

OPTICAL FIBRE METROLOGY

PICOSECOND PULSE METROLOGY

PHOTOELASTICITY

FOUNDATIONS

OPTICAL METROLOGY - SOME PERSONAL REFLECTIONS

H.J. Caulfield
Center for Applied Optics
University of Alabama
Huntsville AL 35899, USA

ABSTRACT

Three of several major themes of modern optical metrology are
reviewed briefly: effective use of the system operating curve, clever use of references or standards, and spectral and temporal measurement of spatial effects. Despite the exciting progress observed,
a major non-scientific problem is noted: the widespread failure of
most academic institutions to treat and teach optical metrology as
a coherent discipline. Some reasons for that failure are suggested.

I. INTRODUCTION

Optics has profound advantages over all other methods in metro-
logy. Among those advantages (Not all achievable at once) are
- High resolution (small wavelength),
- "Non contact" capability (of course photons do contact solid obje-
  cts as laser drilling and welding attest.),
- High scanning speed (faster than the speed of light if need be),
- Low mass for scanning or deflecting (individual photons have very
  low mass and usually the number of photons is not large enough to
  require or generate large forces), and
- Convenient components (lasers, lenses, detectors, scanners, etc.).
Thus, it is not surprising that optics dominates metrology.
  In the last few years some major clear trends in optical metro-
logy have become clear. In this paper, I comment on a few of these.
No attempt at completeness has been made. The message is in the
"big picture" not in the detail.

II. EFFECTIVE USE OF THE SYSTEM OPERATING CURVE

Since wavelengths are not getting smaller, resolution improvement must come from clever use of light. Most of this improvement comes from what is best termed good use of the system operating curve (SOC). We explore that basic concept here.

A measurement system produces a signal S which depends for its value on the parameter of interest x (usually distance or length in optical metrology) as well as on uncontrolled (noise) factors u. Thus we can write

$$S = f(x,u) \quad . \tag{1}$$

More often we supress the u dependence and write

$$S = <f(x,u)>_u$$

$$= g(x) \quad . \tag{2}$$

This is what we call the "response curve". In metrology we measure S and infer x. For a monotonic g(x) and no "noise", the accuracy with which we can measure x depends solely on (but is seldom equal to) the accuracy with which we can measure S. Symbolically (because an analytic inversion of g(x) is seldom done), we write

$$x = g^{-1}(S) = h(S) \tag{3}$$

The sensitivity to S is

$$\left|\frac{dS}{dx}\right| = \left|\frac{dS}{dh}\right| \quad . \tag{4}$$

Ideally we would have a linear response curve

$$S = ax + b \quad . \tag{5}$$

Then

$$x = (S-b)/a \tag{6}$$

and

$$\left|\frac{dS}{dx}\right| = \left|a\right| \quad . \tag{7}$$

This simply says that to have high sensitivity we want a high magnitude slope of the response curve.

Suppose, however, S is nonlinear. We might have

$$S = A \cos^2 x \qquad (8)$$

as often happens in optics. If our task is to detect the condition x = 0, we have a problem because

$$\left| \frac{dS}{dx} \right| = 0 \qquad (9)$$

at x = 0. The maximum sensitivity is not where we want it but at $|x| = \pi/2$. If we want to find x = 0 to high accuracy we may wish to measure at two x's separated by π. When those two signals are equal we have a peak in S. In this case the sensitivity is $|A|$, not zero. This is a simple, mathematical illustration of intelligent use of the response curve.

Taking noise into account we can find a generalized matched filter (GMF) (1, 2, 3) with the property that

$$E\left[S(x-x')* \; GMF \; (x) \right] \simeq \delta(x-x') \qquad (10)$$

where * indicates convolution, $E[\;\cdot\;]$ indicates the expected value of the quantity in parenthesis, and $\delta(x-x')$ and zero elsewhere. That is we can determine x' from the measured noisy S(x-x') analytically. In this case we are using the whole response curve not just one or two points on it. Obviously this is preferable when it is possible and convenient.

Rather than do more mathematics, I think it would be more useful to give some examples of interest in machine tool metrology and use.

Consider the problem of locating a surface by bringing light to focus on it. We start with a point source of light, S; collimate it with a lens, L_1; and refocus it to a point with lens L_2 (see Figure 1). As we move the lens back and forth along the optical axis, the focussed point follows it. If the object intercepts a plane P, some of the light will be scattered backward through the system. If the P is precisely in the focal plane of P_2, the returned beam is focussed precisely on the source. The beamsplitter, BS, between S and L_1 allows us to separate S from its return image. If we place a S-sized pinhole in the returned image of S, and detect the returned signal R as a function of L_2 position, x; we will obtain a response curve somewhat like Figure 2. We have defined x=0 to be a point at which P_2 is focussed on P. Also we have called x positive when the focus of L_2 is beyond P. Clearly such a device provides a noncontacting probe of a surface (P) which we can use for many purposes. Our inquiry here is: How do we use the response curve R(x) with maximum sensitivity?

What we do not want to do is to adjust to x=0 by seeking the peak of the R(x) curve because the sensitivity is low. Several high sensitivity options are open to us, however.

First, we may restrict our attention to one side or the other of the x=0 region. Over a large range of x on either side we have $|dR/dx|$ near maximum. In that region, if we know we are in that region, it is easy to find the x which must correspond to R.

Second, we can drive L_2 (or P, for that matter) to achieve x=0 by going to the peak of R in the manner previously rejected. This time, however, we will monitor two out-of-focus return points as shown in Figure 3. The measurement points, x_1 and x_2, are chosen to be at x's such that $|dR/dx|$ is near maximum and $R(x_1) = R(x_2)$. Any deviation of δx from x=0 will produce signals $R(x_1 + \delta x)$ and $R(x_2 + \delta x)$. We will track

$$\Delta (\delta x) = R(x_1 + \delta x) - R(x_2 + \delta x)$$

$$\approx |2\ dR/dx|_{max}\ \delta x \qquad\qquad (11)$$

Note that the sign of Δ gives the sign of δx and the magnitude of Δ gives the magnitude of the correction. Thus we do not actually have to drive the system to x = 0 to find where x = 0 is. We simply get close and estimate accurately the correction.

Many other such schemes can be devised and applied to measurement of spatial, temporal, or spectral position. Our purpose here is not to show how but to show that effective use of the response curve is desirable and possible. The examples given above indicate the requirement common to all effective uses of the response curve: measurement at or near the peak sensitivity position.

Only one more remark on response curve use is needed to complete this introduction to their use in metrology. They can be used to measure at unsampled positions. Indeed the correction method previously described located x = 0 without having to be there. I have shown elsewhere (4) how to use angular alignment measurements at a few angles to obtain accurate angular measurements at intermediate angles. The response function R(x) may be available only with a sampling frequency Δx. If we want to measure x from such R samples to accuracy much better than Δx without adjusting x, then we must resort to some sort of curve fitting. A general scheme for doing this follows. We suppose that data can be gathered experimentally ahead of time on $\langle R(x) \rangle$, the expected value of R(x) and the covariance matrix $\langle R(x_1)(Rx_2) \rangle$ where x_1 and x_2 are samples of x taken on a grid of resolution smaller than the desired resolution. Assuming normal statistics we should be able to predict, on that small resolution scale, the probability of a given set of measurements $\{\mu\}$ on x's spaced by Δx for any value of x. Thus for x = x' the probability of obtaining a measurement set {m'} is

$$P(\{ m'\}|x') \qquad\qquad (12)$$

There are many ways of making an statistically meaningful choice of the best fit x'. One approach, probably the best if time permits, is to evaluate $P(\{m'\}|x')$. for all plausible x's and choose the maximum probability value. Another approach is to regard {m'} as a measurement vector. We can then use statistical pattern recognition to find a best fit (1, 2, 3, 4,).

III.OBTAINING A REFERENCE

Since all metrology is relative (comparing unknowns with what we hope or believe to be knowns), the obtaining of standards or references is vital. Standards of distance are subject to international agreement and "secondary standards" traceable to the true standards are widely available. This is not the case with standards of shape used in interferometry (optical shape comparison). With interferometry we can easily compare two spherical surfaces of roughly the same curvature to very great accuracy, but unless one of these surfaces is known by some other means we cannot extract absolute measurements from these relative measurements. Fortunately the laboratory grinding technique for spheres and flats is self correcting so good figure can be guaranteed.

On the other hand, how can we measure aspheres? If the reference sphere is close enough to the asphere, no problems arise. We turn now to the question of what to do when we want to measure an asphere which differs so much from the best reference sphere that classical interferometry breaks down. We need a perfect reference asphere, but how would we know if we had one? How could we measure it? For that we need a perfect reference asphere. To break out of this circularity problem we turn to computer holography.[5] That is we produce a two--dimensional hologram which creates the wavefront which would come from the perfect reference asphere. Because the hologram is two--dimensional, we can verify its accuracy using well established methods. Thus, the previously-noted cycle, its unmeasurable components, is broken. The breakthrough in recent years is that two companies have now produced hologram writers with sufficient absolute position accuracy $(1:10^5)$ to make such testing routine.[6]

A different but equally clever use of reference surfaces is the Ludman interferometer which references one side of a solid slab to the other side so that what is measured is not slab position or alignment but the parallelism of its two sides. This is a totally new use of interferometry. Recent developments have extended it to angle measurement as well.

IV. SPECTRAL AND TEMPORAL MEASUREMENT OF SPATIAL EFFECTS

Only in the last few years have there been systematic attempts to unify these spectral/temporal metrologies.   While the two types of measurement are intimately connected,   it is easiest to explain them separately.

There is a simple way to use time delay measurements to obtain the range to a given object. Clearly, if light must travel to and from a point a distance D away it takes a time

$$t = 2D/c. \tag{13}$$

With mode-locked lasers and streak cameras we can measure the range to hundreds of points in parallel with an accuracy in the millimeter range. A mode-locked laser of optical path length ℓ will produce repetitive pluses of period

$$\Delta t = 2\ell/c. \tag{14}$$

Using a Twyman-Green type of a structure, we can seek temporal coincidences between reference and object arms modulo Δt. With an ordinary laser rangefinder with resolution less than ℓ plus a picosecond rangefinder, we can locate features in absolute range to millimeter accuracy over large distances.

In the spectral domain, we can use the same Twyman-Green type of arrangement. Looking back toward the object from that region, we see the object with a flat mirror either behind it or before it. If the object itself were a flat mirror, this would look like a Fabry-Perot interferometer. For such a pair of mirrors separated by a range R, the spectral transmission curve has peaks separated by

$$\Delta\nu = c/2D. \tag{15}$$

Shining white light into such a system and spectrally analyzing it gives D as a function of position.

Note that if a source of spectral with $\Delta\nu$ is perfectly mode-locked it forms a pulse of duration

$$\Delta t \simeq 1/\Delta\nu. \tag{16}$$

With such a pulse we can just resolve a range

$$D = c\Delta t/2 = d/2\Delta\nu \tag{17}$$

Rearranging, we have

$$\Delta\nu = c/2D \tag{18}$$

as before. In this limit, these two totally different methods are equivalent. For prefectly understandable reasons [1-5] , these methods are really more complimentary than equivalent. Temporal methods are preferable for lower resolution, e.g. millimeter, resolution over long distances; while spectral methods are better for high resolution (wavelength domain) over small distances.

Thus spatial measurement need not be done in a spatial domain. There is a "natural encoding" with an "effective wavelength"

$$\lambda_e = c\Delta t = d/\Delta \nu. \tag{19}$$

The limiting resolution is, as usual,

$$\Delta D = \lambda_e/2 \tag{20}$$

V. ASSESSMENT

The trends noted and illustrated here are only three of many. Computer analysis, error cancellation, and optical preprocessing are among the trends I have not chosen to emphasize. All of these make metrology one of the most dynamic fields of optics.

To the optimism of this conclusion, I must add one discouraging note. Our educational institutes (at least in the USA) fail to teach optical metrology as a unified discipline. Why this is the case is hard to know but some obvious guesses include
- The "stigma" (at least for many academics) of immediate applicability,
- The lack of a suitable textbook,
- The failure of scientific societies to treat optical metrology as a unified discipline rather than as a "bag of tricks", and
- The historial fact that professors on metrology are not being produced in universities.
Rectifying this unfortunate situation it is a goal now being undertaken at University of Alabama in Huntsville.

REFERENCES
1. H.J. Caulfield and R. Haimes, Appl. Opt. 19, 181 (1980).
2. H.J. Caulfield, R. Haimes, and D. Casasent, Opt. Eng. 19, 152 (1980)
3. H.J. Caulfield and M.H. Weinberg, Appl. Opt. 21, 1699 (1982).
4. J.L. Horner and H.J. Caulfield, Appl. Opt. 21, 1599 (1982).
5. J.C. Wyant, "Holographic and Moiré Techniques" in D. Malcara, ed., Optical Shop Testing (John Wiley, NY, 1978) pp. 381-408.
6. H.J. Caulfield, Lasers and Applications, 1, 61 May (1983).

Fig. 1. Using retroreflection from a focussed point, we can determine when that point has been brought into coincidence with a plane normal to the optical axis.

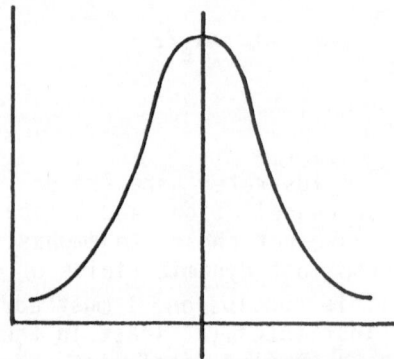

Fig. 2. The system operating curve, SOC, shows a peak at perfect coincidence and a decreased signal elsewhere.

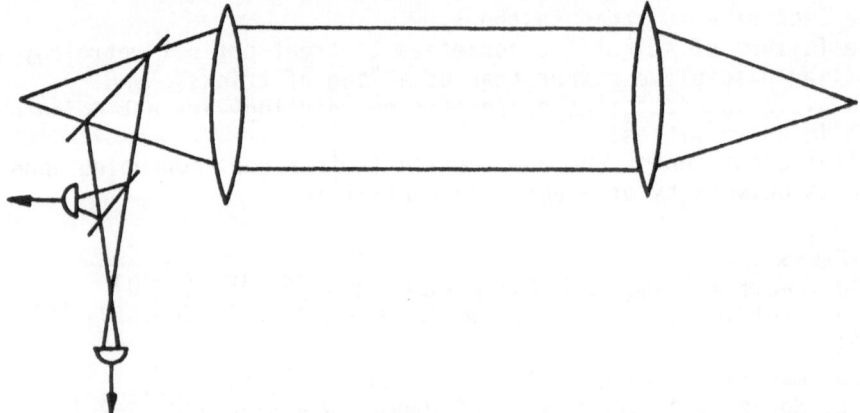

Fig. 3. By reading pre and post focus returns, we can operate away from the peak of the SOC and thus where the sensitivity is greatest.

BEAM DEFLECTION AND IMAGE FORMING SYSTEM

MEASUREMENT OF SURFACE FORM BY LASER PROFILOMETRY

A.E. Ennos

National Physical Laboratory,
Teddington, Middlesex, UK

It has up to now been almost accepted that precision measurement of the shape of a surface must be carried out by some form of
optical interferometry, which will give a pattern of contour fringes
that have to be interpreted in terms of height measurement from a
datum surface. Extraction of information from the pattern becomes
complicated and expensive when very high precision is required, as
for example when heterodyne methods are used. A more direct way of
obtaining the height information, which has proved to be extremely
sensitive, is to measure the local surface slope and to integrate
this along a line on the surface in order to yield the profile.
The method is an extension of the well known technique of autocollimation used to measure the out-of-flatness of engineers' surface
tables, but here the slope is obtained by reflecting a laser beam
from the surface itself, not from a mounted mirror resting on the
surface and stepped along it. Currently this non-contacting technique can achieve nanometre accuracy on flat or near-cylindrical
surfaces, and it has the advantage of providing absolute values of
profile height, related only to the direction of the beam from a
laser as datum.

1 PRINCIPLE OF THE METHOD AND ITS THEORETICAL ACCURACY

A laser beam is directed on to the surface at near normal
incidence and stepped along by incremental distances d, the diameter
of the beam. At each step the angle of reflection θ of the returned
beam is measured by autocollimation (Fig 1).

Relative to the origin, the profile height at a point distant X
along the line of measurement will be given by

14

$$h_X = \int_0^X \frac{\theta}{2} \cdot dx = \sum_{n=0}^{n=X/d} \frac{\theta_n}{2} \cdot d \qquad (1)$$

An essential of the method is that the means of stepping must not introduce any variations in the direction of the laser beam incident on the surface, nor must it drift during the period of measurement.

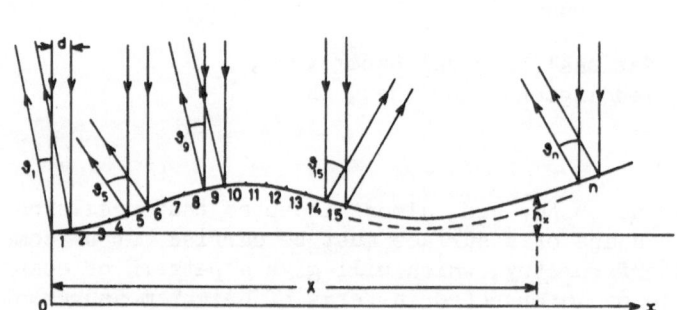

Fig 1.   Principle of surface profile measurement
         by laser autocollimation

The angle θ is measured by directing the laser beam on to a position-sensitive photodetector positioned at the focal point of a convex lens, and detecting the movement of the focused spot electronically. The size of the spot σ is dependent upon d, the beam diameter, and f, the focal length of the lens, as

$$\sigma \approx \frac{\lambda f}{d} \qquad (2)$$

It is characteristic of position-sensitive detectors, such as a bi-partite photocell, that they are in the limit capable of detecting some fractional displacement α of the diameter of the beam falling on them. So, the minimum detectable displacement is given by

$$\delta\sigma = \alpha\sigma \approx \frac{\alpha\lambda f}{d} \qquad (3)$$

The smallest detectable angle will then be

$$\delta\theta \;=\; \frac{\delta\sigma}{f} \;\simeq\; \frac{\alpha\lambda}{d} \tag{4}$$

At each stage of the integration performed to obtain the profile, the smallest detectable height change δh will thus be

$$\delta h \;=\; \frac{\delta\theta}{2} \cdot d \;\simeq\; \frac{\alpha\lambda}{2} \tag{5}$$

This shows that for every step of the integration the uncertainty in profile height is $\alpha\lambda/2$, independent of the laser beam diameter. The total uncertainty after n steps of integration will be $(\sqrt{n}).\alpha\lambda/2$. When a small beam diameter is used to give high spatial resolution a greater number of steps are needed to cover the same distance, so the uncertainty of measurement increases.

Instrumental Accuracy: With a bi-partite position sensitive detector and low noise amplifier, the value of α may typically be 1/2500, giving $\delta h \sim \lambda/5000 \sim 0.1$ nanometre. The potential accuracy of this method of profile measurement is thus inherently very high.

2 PRACTICAL INSTRUMENTS

2.1 Uniaxial System

The profilometer developed initially was designed for measuring the shape of X-ray mirrors (1). These are diamond-turned and lapped surfaces approximating to shallow cones. The system layout is shown in Fig 2. A low-power laser beam is directed down the axis of the mirror and reflected on to the surface by a pentagonal prism. This is fixed to an arm mounted on a precision slide so that it can be stepped along the axis. The reflected beam spreads out into a line of light which is directed on to the detection system via a beam-splitter. A quadrant photocell is arranged to detect only vertical movement of the horizontal line, and null setting of the signal is achieved by means of the optical micrometer (a tilting glass plate). The laser beam may alternatively be directed on to a fixed reflecting cylinder in order to monitor its direction.

Using a pentagonal prism for beam tracking ensures constancy of beam direction independent of angular movement due to imperfect slideways. The directional stability of the laser can be maintained to better than 1 microradian over the measurement period by housing it in a separate ventilated box, and the rest of the system is also insulated against drafts.

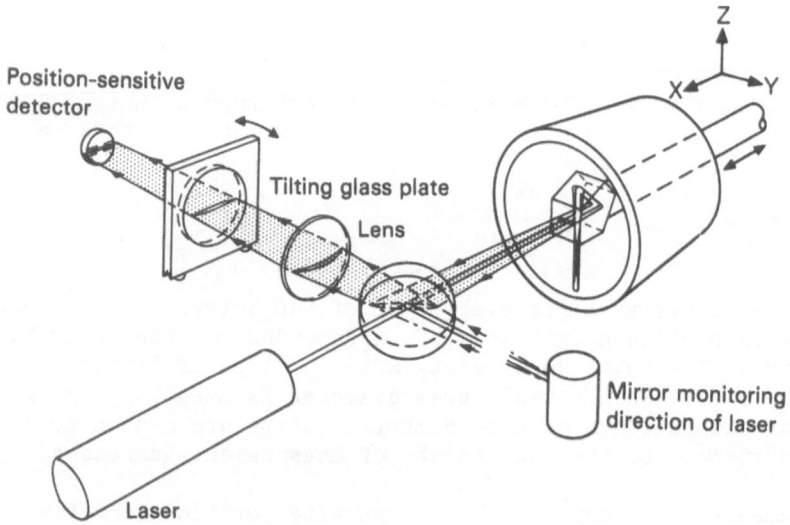

Fig 2.   System layout for measuring profile
          of X-ray mirrors

The profilometer system has been fully automated, with computer control of the stepping motorslide, servo-control of the tilt plate null setting, and direct read-out of its position by a moiré-grating linear transducer. The computer calculates the profile and norma-lises it to remove the effect of the initial slope value and to produce a symmetrical profile.

A typical profile measurement on an ellipsoidal surface is shown in Fig 3. The continuous and broken lines are separate measurements of the same profile recorded in reverse directions, the centre curves showing deviation from the design shape. Uncertainty of measurement is ± 2 nanometres over a 17 mm length.

With an unfocused laser beam, the lateral resolution is approximately 1 mm. The beam can be focused to a smaller probe size by using a split lens and slightly offsetting the reflected beam (Fig 4). The minimum spot size is limited by the defocusing effect due to scanning; typically, a 0.1 mm probe will scan over 15 mm without undue broadening. Fig 4 illustrates the modification to the system for measuring optical flats, where a 10 X magnification of the detected spot is necessary to match the size of the photo-cell.

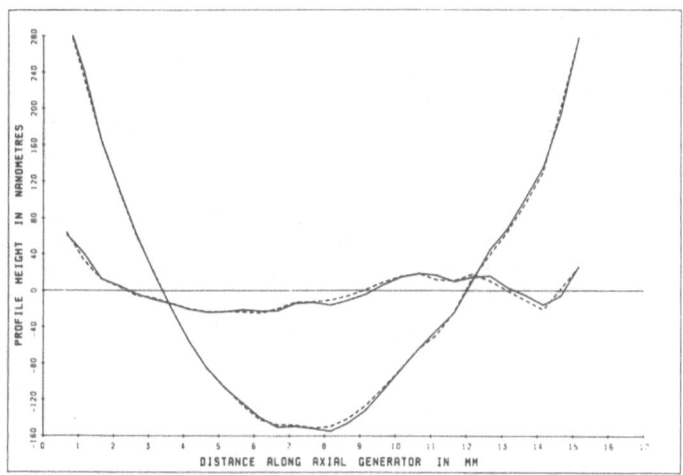

Fig 3.   Profile and error curves of ellipsoidal
          surface; _____, ..., two independent
          measurements

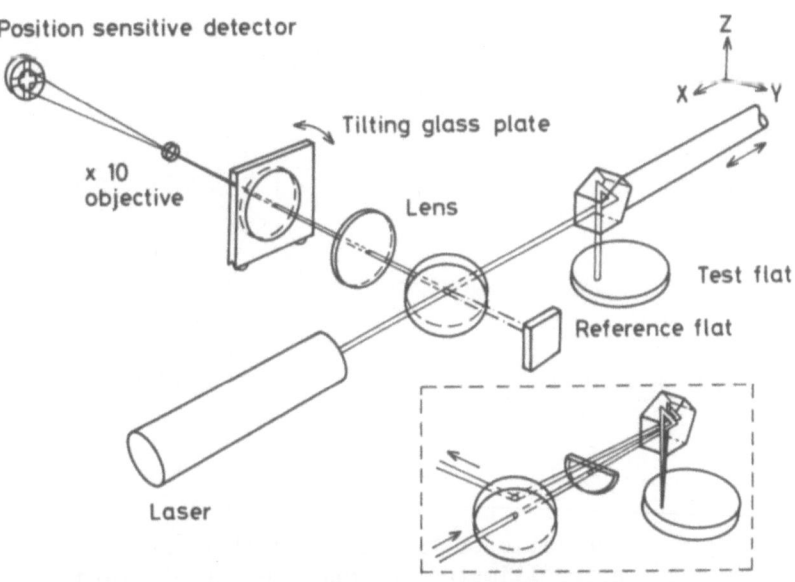

Fig 4.   System layout for measurement of optical
          flat; inset shows focusing arrangement for
          increased spatial resolution

2.2 Two-dimensional system

This system, under development, has been designed for high
precision profiling of optical flats in two dimensions. The laser
beam is stepped over the surface, raster fashion, by means of two
periscope prisms linked together by rotating joints which act like
an articulating arm (Fig 5). The periscopes do not alter the
direction of the scan beam for small angular motion imparted by
imperfect bearings. The null-setting optical micrometer plate can
be tilted in two orthogonal directions to measure the two components
of surface slope. The output of all four quadrants of the photodetector are used in this case.

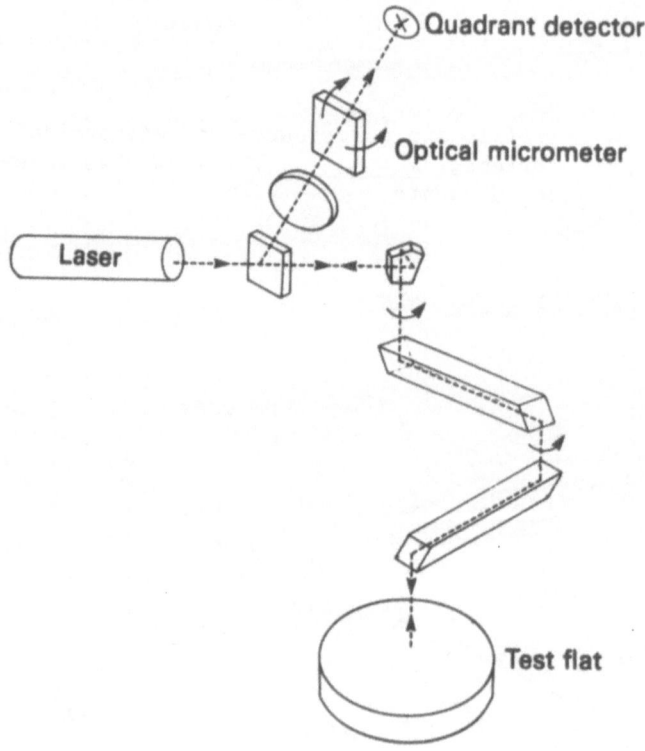

Fig 5.   Schematic of two-dimensional laser
          autocollimator for measurement of flats

For a nearly flat surface, high spatial resolution is not
generally required, so the scan beam is expanded to 3-4 mm diameter.
This also reduces the large amount of slope data that has to be
acquired to cover a two-dimensional surface.

3 FUTURE DEVELOPMENTS

The simplicity of the method and the direct digital read-out of surface height that it affords makes laser profilometry an attractive alternative to optical interferometry. Although present developments are restricted to the measurement of near flat or cylindrical surfaces, the possibility exists of extending the technique to more general shapes, such as aspherics, perhaps using more than one laser probe. In addition, rougher surfaces might also be measured to a lower accuracy by reflection of the scanning beam at high angles of incidence.

4 REFERENCE

1. Ennos, A.E. and Virdee, M.S. "Precision measurement of surface form by laser autocollimation", SPIE Vol. 398 - Industrial Applications of Laser Technology, pp 252-257, 1983.

THREE-DIMENSIONAL INSPECTION OF LARGE OBJECTS

G. Bickel, G. Häusler, M. Maul

Physikalisches Institut der Universität Erlangen-Nürnberg
Erwin-Rommel-Straße 1, 8520 Erlangen, Fed. Rep. of Germany

We describe an apparatus for shape inspection of large objects
(e.g., 2 x 2 x 2m^3). It is based on triangulation and overcomes
some limitations which are usually connected with such object
sizes.

For large objects and high resolution, the number of volume
elements which have to be addressed can be extremely large. Thus,
the problem of inventing 3-d measuring machines is essentially to
outwit the four-dimensional (x,y,z,t) space-bandwidth limitations
of optics and electronic hardware.

A systems analysis of 3-d optical inspection arrangements shows
that these devices can be considered as series of subsystems:
illumination - object - $z(x,y)$ - encoding-optics - transducer
and evaluation device. In such a series the weakest link determines the throughput.

Two main bottlenecks are:

- the limited depth of focus of optical imaging systems, which
  restricts either the depth of the object or the lateral
  resolution. This makes parallel addressing of the whole object
  difficult.

- The limitations of the transducer which converts optical
  signals into electronic signals for (e.g., digital) evaluation.
  Such transducers suffer from a low number of resolution elements. Here too, parallel addressing is difficult or impossible.

The usual way of overcoming both difficulties is by serial

addressing, e.g., by successive mechanical positioning (in x,y,z) of an optical or tactile sensor. Such mechanical point-by-point methods are accurate but slow.

We propose an apparatus, that overcomes to a large extent these drawbacks. It is based on the following principles:

1. The object is serially addressed (scanned) by a light pencil. The scattered light on the object forms a surface profile which can be observed from a suitable direction (triangulation). Using a light probe makes the system fast (as fast as the scanning mirror), in spite of the serial mode of addressing.

2. Both illumination and observation are performed via the same scanning mirror (see fig. 1), although the direction of illumination is different from the direction of observation. Thus, we can achieve the condition that, in spite of the scanning process, the imaged light spot stands still on the transducer if the object has a certain shape. This "reference shape" depends on the geometry of the apparatus and may conveniently be a plane or a sphere. The advantage of this scanning compensation is, that we can use a one-dimensional transducer for gathering the two-dimensional surface profile. Such transducers are commercially available up to 4000 pixels. All these pixels can be used for pick up of z-information, since x-y-information is independently encoded by the deflection angle of the scanning mirror.

3. The light pencil is generated by an axicon (fig. 2). With such a light pencil we can achieve high lateral resolution (small spot) on the object over a depth of several meters, which is much better than with conventional optics.

More details and different implementations of the principles described are explained in the german pat. appl. P33372519 of which preprints are available.

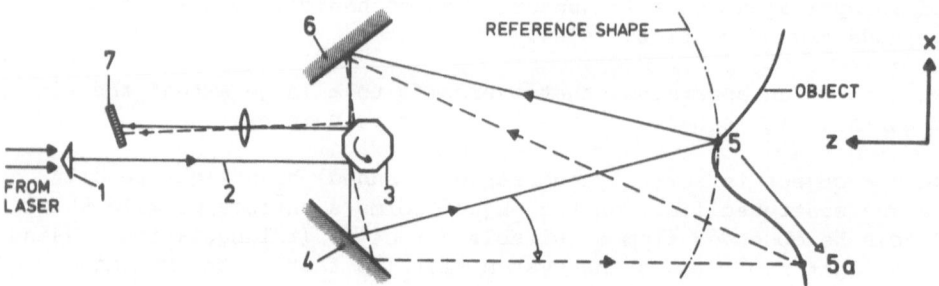

Fig. 1

Principle of triangulation with expanded depth and suppression of scan motion. Here a device is shown, where observation is performed in the plane of the scanning beam. The light pencil 2 generated by axicon 1 is directed onto the object via the rotating polygonal mirror 3 and the fixed mirror 4. The light spot on the object's surface (5 and 5a, respectively) is observed via the fixed mirror 6 and a different facet of the polygonal mirror 3. It is imaged onto a one-dimensional detector array 7 which is inclined to the optical axis to achieve a sharp image of each light spot irrespective of the shape of the object. The illuminating and the observing optical subsystem are coupled by use of the same polygonal mirror in such a way that the scan motion of the light spot (x-direction) is largely compensated in the detector plane. Thus, the (limited !) number of detector elements is reserved almost entirely for the measurement of the depth (z-coordinate) of the object.

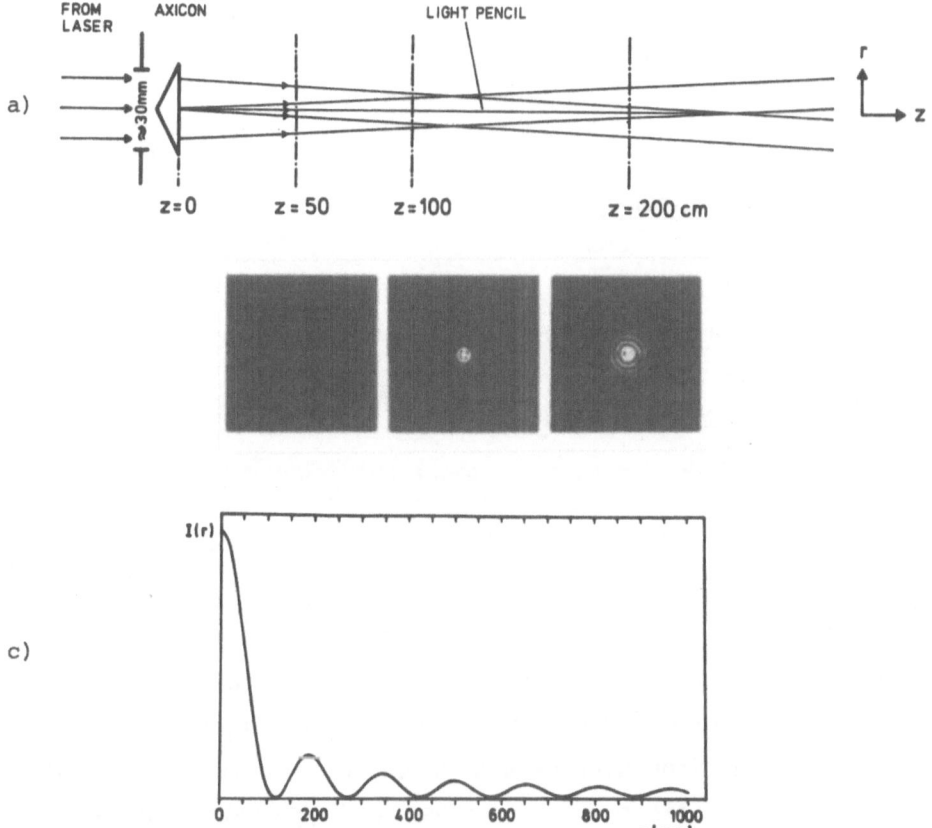

Fig. 2

Generation of a long, thin light pencil by an axicon (fig. 2a).
The axicon, illuminated by a plane wave generates an interference
pattern with a high central peak. The shape of this pattern does
not change significantly along the optical axis over a distance
of some meters (fig. 2b and 2c). The diameter of the central peak
may be as small as 100μm or less.

Laser Dimensional Metrology

O.D.D.Soares and A.O.S.Gomes

Centro de Física da Universidade do Porto
4000 PORTO, PORTUGAL

Abstract

Laser radiation characteristics and photodetectors performances can be combined to explore a range of dimensional metrologic applications covering a scale from micrometers to centimeters. Two complementary optical configurations are described and presented in combined arrangements to respond to an extended range of applications. Unidimensional and multidimensional metrologic applications are analysed. Introduction of galvanometric scanning mirrors and the use of the Laser beam polarization is considered. The analysis and discussion is mainly concentrated on design concept but directed to envisaged applications. Prototypes of the systems were tested with microprocessor incorporation for flexibility and convenient data processing.

Introduction

Recurrence to Laser radiation in metrology embodies the potential competitiveness of optical techniques: fiability; flexibility adequacy to the scale of measurement; high sensitivity; high precision; ease of use (no surface preparation) and real-time operation. Other major advantages also include:

    i) telemetric measuring capabilities without physical contact or contamination: mechanical, material or energetic.

    ii) measurement in inaccessible sites.

    iii) measurement in severe and aggressive environments.

    iv) measurement and monitoring with diverse and simultaneous magnification.

    v) simultaneous linear and angular measurement.

    vi) lightness and easiness of setting-up (including setting-up on production machines).

    vii) low levels of illumination without heating effects on the measuring devices.

    viii) direct recognition of the measurement pattern via image or its transform.

    ix) global field of measurement (parallel processing and real-time).

    x) long life and low cost and slow depreciation.

    xi) easy design concept with simpler training of operator.

    xii) no inertia and image based information that allows parallel display and in situ processing of data in short time on a range of parameters with high sensitivity and precision.

Furthermore, Laser radiation can present characteristics of unique inherent metrological potential that allow interesting schemes for measurement such as: geometric, radiometric, interferometric, diffractive and spectroscopic to quote but the most commonly employed. The Laser has progressively and steadily become the major component of a great variety of metrologic systems for inspection, testing, measurement and control in industry, science and bio-medicine.

Photodetectors have also reached such a development in performance that they offer great advantages in metrological instrumentation. It is then advisable to exploit combination of described attributes to conceive optical arrangements that being of simple design are adequate for envisaged metrological operations.

Dimensional metrology represents the most frequent range of measuring operations in industry, deserving then particular attention. Optical metrology techniques providing great flexibility of implementation offer the possibility of reducing complex metrologic operations to a combination of unidimensional metrologic operations. Methods and techniques based on unidimensional metrology are therefore of major concern.

Design concept are presented and discussed with a view to exploit the potential for the realization of metrologic instrumentation.

A compromise between scale versus accuracy of the measurement results, in general, in the consideration of different physical principles for metrologic operations that appear to be in essence of identical type of measurement. The complexity of the technique is kept however, within an acceptable level with inherent economic consequences.

For linear dimensions up to the order of $10^5 \times \lambda$ (λ the radiation wavelength) the analysis of the diffraction pattern provides measuring capabilities (1,2). For higher dimension values other methods are more appropriate. From those optical methods can perform in real-time, usually a must for series production at high rate.

Ideas and experiments were developed around two basic configurations which allow the coverage of a range of measurements from micrometers to centimeters:

i) a radiometric configuration designated as Unidimensional Metrologic

Image Dissector (3,4,5) uses a self-scanned photodiode-array with an A/D converter feeding a microprocessor with supporting peripherals. Appropriate optical system and informatic means for control, signal processing and display are added in accordance to the particular application: radiometry, diffractometry, surface analysis, interferometry, phase detection, speckle metrology, moirégrams, etc.

ii) a laser scanning configuration (5,6,7) of in plane translation movement type based on the directivity and low divergence of the laser beam designated as Micrometric Laser Scanner. Displacement of the beam is related to the linear dimension of the object to be gauged by monitoring the object-beam shadowing effect. Sensors and auto calibration features are introduced to realize corrections and overall reliability monitoring. Computer data processing allows flexibility both in static and dynamic measurements.

The complementarity of the two configurations is also explored by combination of the two principles in order to extend the range of applications. Furthermore, multidimensional operation is proposed by spatial multiplexing and laser beam polarization handling.

System implementation, operation principles and main characteristics are presented via examination of potential applications.

Unidimensional Metrologic Image Dissector (3,4,5)

Fig.1 represents, in block diagram, the arrangement of the linear image dissector where the optical system may vary in accordance to the particular application.

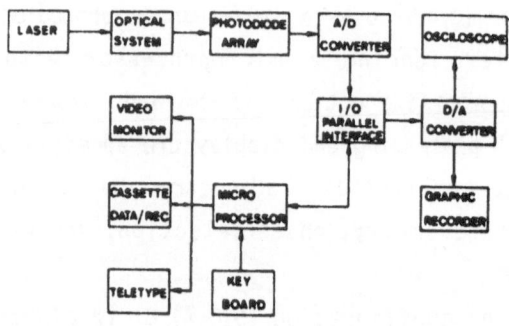

Fig.1: Block diagram representation of the image dissector

The photodiode array (256 photodiodes of 15 μm x 26 μm separated by 10 μm and linearly aligned) is the image receiver. The photosignal is sampled and A/D converted in 8 bit words stored in the microprocessor memory for subsequent processing, metrologic decoding and measurement.

The operator interacts via peripherals namely for input≡console of control, magnetic tape reader; and for output≡oscilloscope, recorder, TV monitor, printer and magnetic tape recorder. Fig.2 is an actual view of the experimental arrangement.

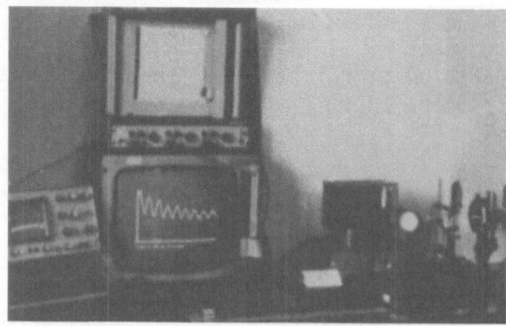

Fig.2: Image dissector experimental system

Some applications of the system include:

  i) Diffractometer
 ii) Interferometry and phase detection
iii) Surface analysis
 iv) Speckle metrology
  v) Moiré metrology
 vi) Spectrometry

that will be referred to in sequence.

i) Diffractometer

The Fraunhoffer diffraction pattern is well known to be related to the geometrical dimensions of the illuminated object. In particular, whenever the object has a regular structure and form, linear dimensions can be obtained from the analysis of the spatial spectrum of the diffraction pattern.   A diffractometer can then be assembled according to Fig.3 using the photodiode array to perform the radiometry of the diffraction pattern.

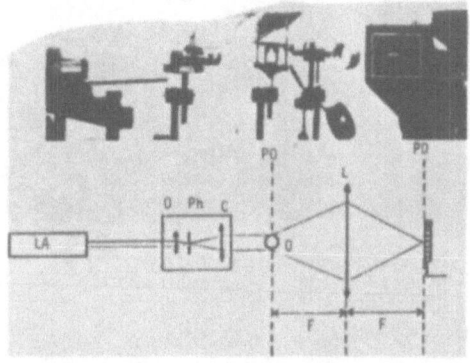

Fig.3: Fourier transform arrangement of the diffractometer.
         (LA - Laser; O - microscope objective; Ph- pin-hole;
         C - collimating lens; PO - object plane; PD - diffraction image
         plane; L - Fourier transform lens; F - focal length)

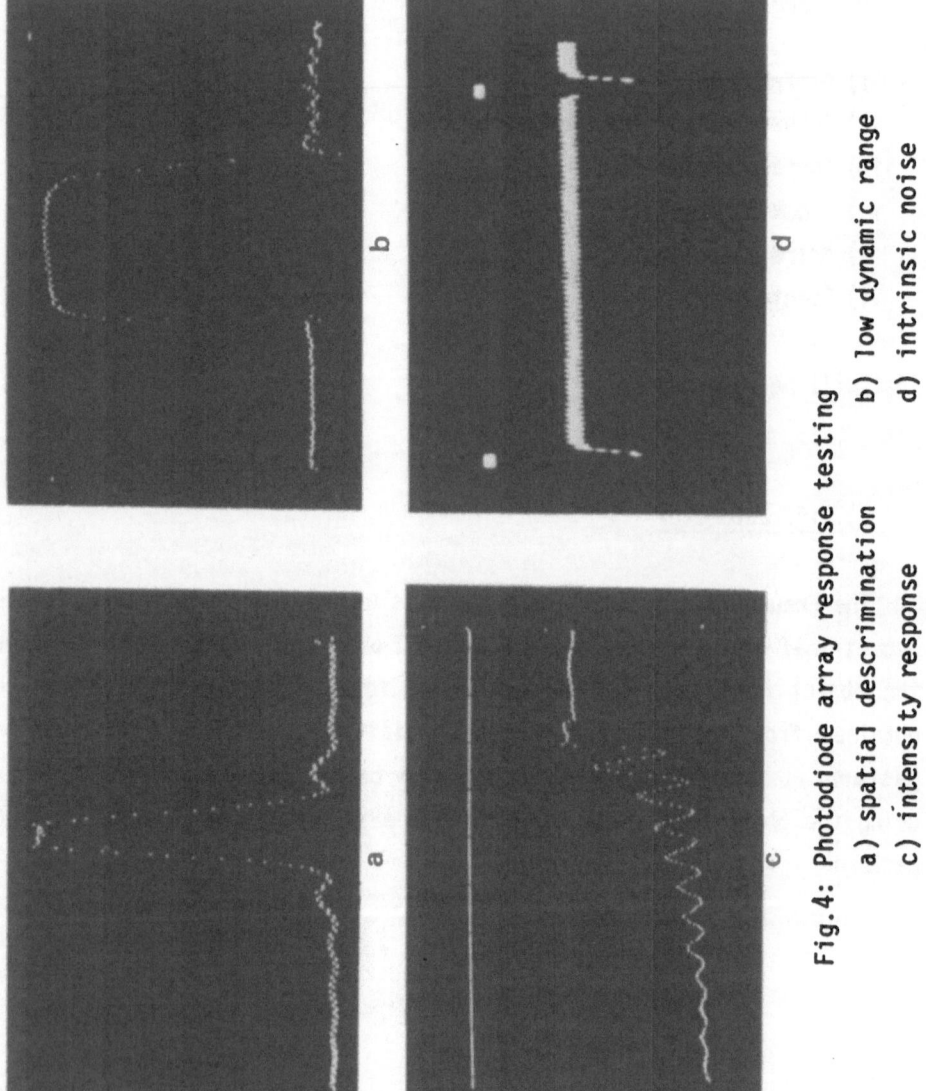

Fig.4: Photodiode array response testing

a) spatial descrimination   b) low dynamic range

c) intensity response   d) intrinsic noise

Decoding of metrologic information was achieved by two procedures: characteristic values location at the diffraction pattern such as the minima and subsequent calculus of the dimensional measurement, and the evaluation of the input aperture autocorrelation function with a F.F.T. algorithm over the diffraction intensity pattern.

The sampling of the photosignal contains perturbing noise as shown on Fig.4.

The S/N ratio is improved substantially (14 dB) by simple subtraction to the photo signal of the intrinsic noise combined with ambient light illumination, Fig.5.

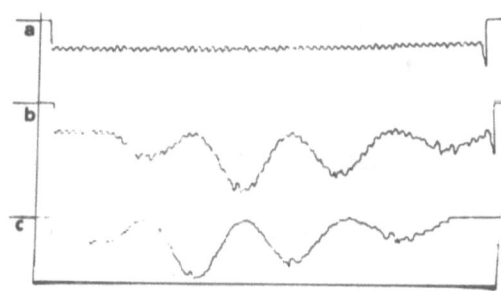

Fig.5: Arithmetic subtraction to improve S/N ratio in signal acquisition.
a) intrinsic noise combined with ambient light
b) photodiode array signal
c) S/N improvement by digital subtraction: $\left[b)-a)\right]$

The sampled photosignal is subjected to an averaging and other smoothing algorithms (e.g. minimum location by extrapolation of points of derivative signal inversion) over the discrete values so that extrapolated values from measurement match the theoretical calculations.

The system was calibrated to a precision level of the order of the μm and maximum errors within 1% for an average error of 0.6%.

Linearity response, Fig.6 and sensitivity curve Fig.7 were measured.

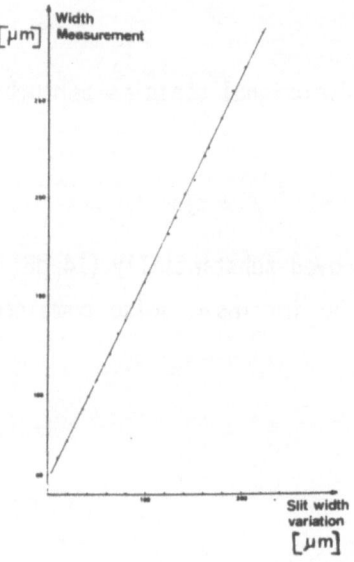

Fig.6: Linearity response by calibrated variable slit.

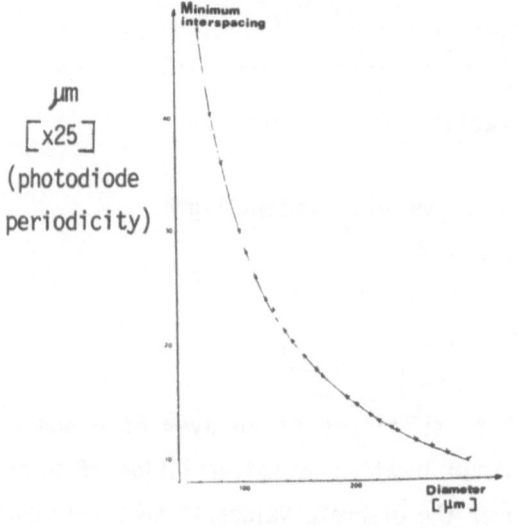

Fig.7: Sensitivity testing based on the algorithm of relative minima localization on the diffraction pattern.

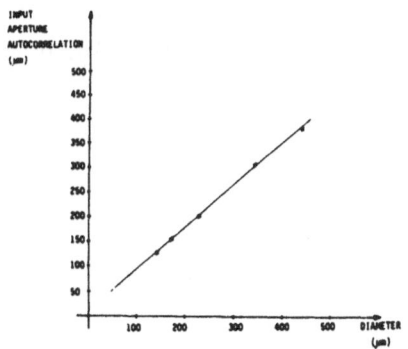

Fig.8: Sensitivity testing based on input aperture autocorrelation with FFT algorithm.

The system was tested to control the diameter of wires and width of slits using the arrangement of Fig.3 at 10KHz operating speed presenting a resolution of 1 μm, sensitivity of ± 0.5 μm, 1% linearity and repeatability better than 0.5 μm.

The optical arrangement shown on Fig.3 presents an invariant response for in plane translation of the object, transverse to the direction of the illuminating beam. Longitudinal displacements of the object may need to be monitored and eventually a correction would be introduced. The Fig.9 is self-explanatory of a possible design concept for this purpose.

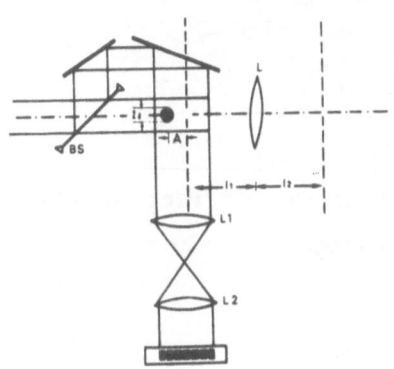

Fig.9: Optical arrangement for monitoring and correction of longitudinal displacement of the object being measured. (BS-beam splitter; A - object longitudinal displacement; L - lens; PA - photodiode array)

Dimensional control of piston rings was performed with the image dissector in accordance to norms (8). Fig.10 specifies some of the intervening parameters.

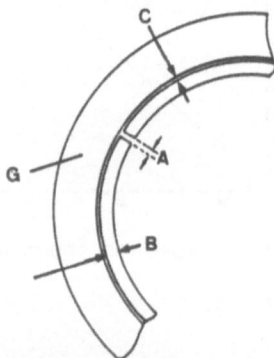

Fig.10: Dimensional control of piston rings
(G - ring gauge; A - dilatation gap; B - ring collar width;
C - adjustement gap)

The Fig.11 is the diffraction pattern for a dilation gap of 140 μm as displayed on the oscilloscope corresponding to a motor car piston ring.

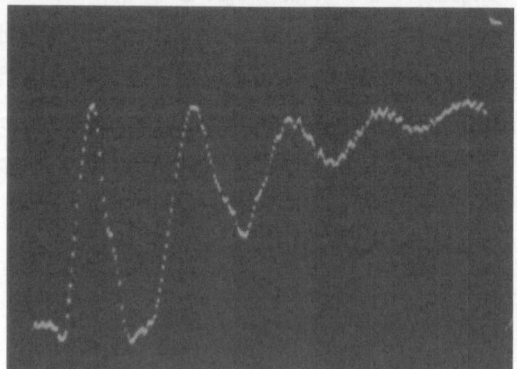

Fig.11: Intensity distribution of the diffraction pattern of the dilation gap of a piston ring.

ii) Interferometry and phase detection

Interferometric patterns in static and dynamic situations can be exploited with the image dissector. Examples will be drawn.

Thickness variations can be monitored and measured with simple arrangements as the one shown in Fig.12 for plane parallel glass plates.

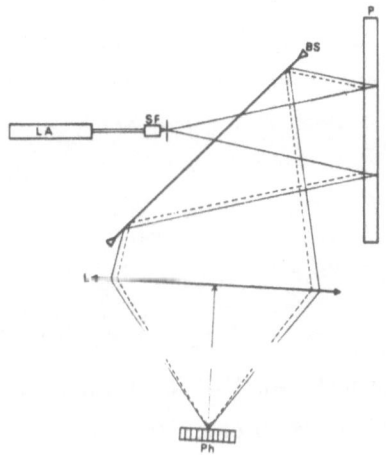

Fig.12: Experimental Arrangement for thickness variation on plane parallel glass plates.
(LA - laser; SF - spatial filter; BS - beam splitter; L-lens; Ph - photodiode array; P - plate under measurement)

Interference fringe locus is given by

$$(2K + 1) = \frac{4 d}{\lambda} \sqrt{n_i^2 - n^2 \sin \theta} \qquad (1)$$

$$k = 0,1,2,3,...$$

where variables refer to Fig.13. The Fig.14 is a visualization of fringe profile at the focal plane of collecting lens.

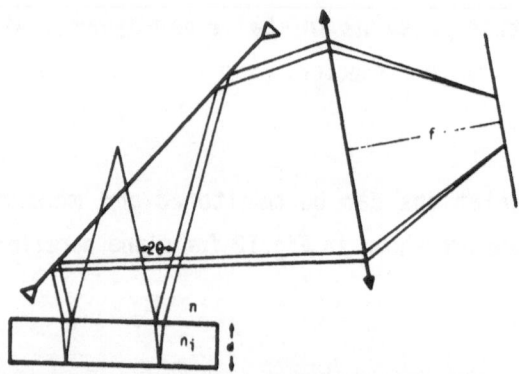

Fig.13: Geometric analysis of fringe formation for the shearing interferometer realized by plane parallel glass plate.

Fig.14: Fringe profile at the focal plane of imaging lens in Fig.12, in accordance with equation (1).

Tests were conducted with holographic glass plates (1.27mm thick, n_i = 1.503). Fig.15 is taken from a heating experiment from 18°C to 20°C where a 0.6 µm thickness variation occured.

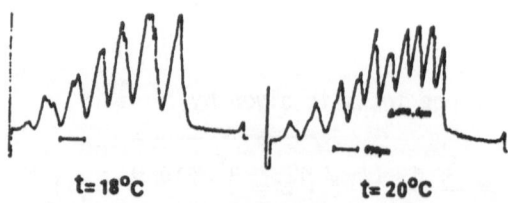

$t= 18°C$ $t= 20°C$

Fig.15: Graphic recording of interference fringe pattern evolution while heating a holographic glass plate.

Identically, thickness variation may be screened as in Fig.16 while Fig.17 accounts for a dynamical situation.

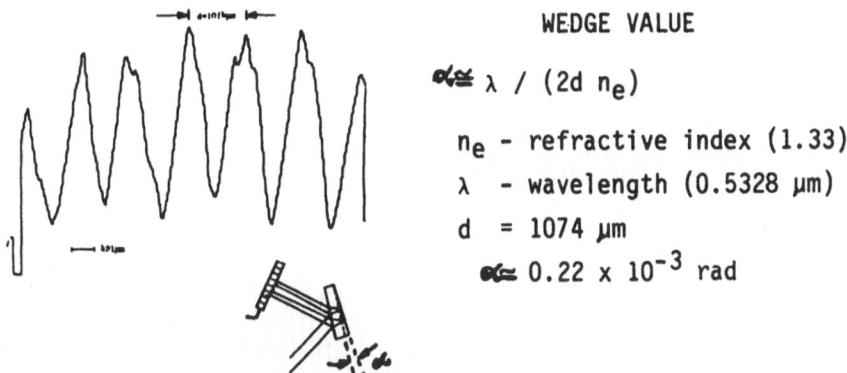

WEDGE VALUE

$\alpha \cong \lambda / (2d\ n_e)$

n_e - refractive index (1.33)
λ - wavelength (0.5328 µm)
d = 1074 µm
$\alpha \cong 0.22 \times 10^{-3}$ rad

Fig.16: Interferometric measurement of thickness variation on a glass sheet.

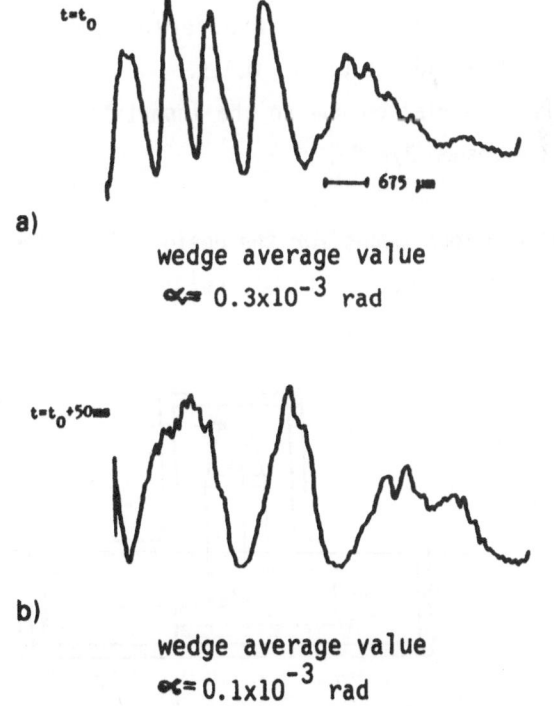

a)

wedge average value
$\alpha = 0.3 \times 10^{-3}$ rad

b)

wedge average value
$\alpha = 0.1 \times 10^{-3}$ rad

Fig.17: Interferometric monitoring of thickness evolution on a soap film.

38

Phase detection is also possible by monitoring the perturbation introduced on projected fringes (9). Fig.18 is a sample explanatory example which shows the effect of the insertion of a neutral density filter.

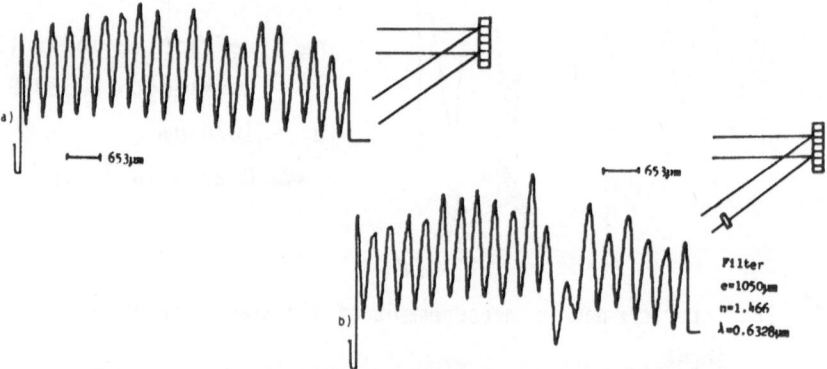

Fig.18: Projected interference fringes used for phase mapping.
a) Fringe pattern for interfering collimated beams at a relative incidence of θ = 0.12°.
b) Phase variation due to the insertion of the neutral filter (phase change 7/6 π).

This concept provides means for the design of sensors: thermal sensor - Fig.19; fluid flow sensor - Fig.20; reflective thin film thickness sensor - Fig.21; etc.

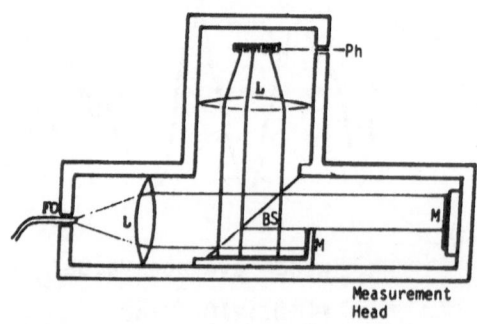

Fig.19: Thermal sensor
(FO - illuminating optical fiber; L - lens; BS - beam splitter; M - mirrors; Ph - photodiode array)

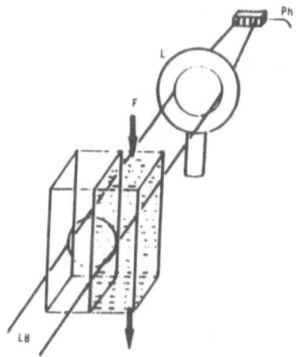

Fig.20: Fluid flow sensor

(F-fluid; Ph-photodiode array; LB-illuminating beam; L-lens)

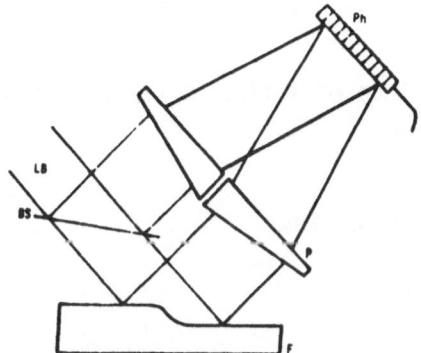

Fig.21: Reflective thin film thickness sensor

(F - reflective thin film; BS - beam splitter; P -prism;

Ph - photodiode array; LB - illuminating beam)

iii) <u>Surface analysis</u> (10,11,12)

Let us consider a surface rugosity given by a profile

$$z = h(x,y) \qquad\qquad (2)$$

with an average value z = 0. Assuming that h(x,y) is an aleatory distribution, stationary and ergodic with a normal distribution characterized by a standard deviation $\digamma$, autocorrelation $C(\mathfrak{G})$, then for a good conducting surface without shadowing or multireflection effects:

$$\frac{<\mid E^2\ (\vec{k}_{20})\mid>}{<\mid E_i^2\ (\vec{k}_{20})\mid>} = e^{-(\frac{4\ \pi}{\lambda}\ \digamma\cos\ \theta)^2} + f\left[\digamma, C(\mathfrak{G})\right] \quad (3)$$

where:

$E\ (\vec{k}_{20})$ is the amplitude of surface diffused field along $\vec{k}_{20}$ direction measured at the focal plane of lens, Fig.22

$E_i\ (\vec{k}_{20})$ is the amplitude of reflected field by a specular surface in same conditions

θ is the angle of incidence

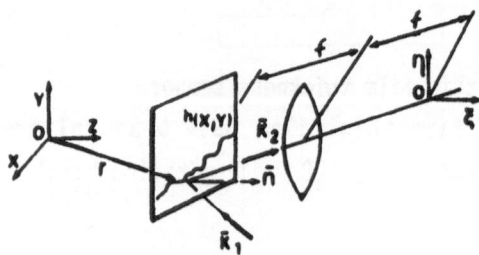

Fig.22: Parametric definition and optical arrangement for surface analysis.

For smooth surfaces and large values of θ (shalow incidence) $f\left[\digamma, C(\mathfrak{G})\right]$ is small, and $\digamma$ (r.m.s. surface rugosity) can be calculated from measurement of $<\mid E^2\ (\vec{k}_{20})\mid>$, $<\mid E_i^2(\vec{k}_{20})\mid>, \theta$ and λ . This model was used to compare rugosity of workshop surface standards - Etalons de Surface

Techniques LCA, Fig.23, utilizing the experimental arrangement of Fig.24.

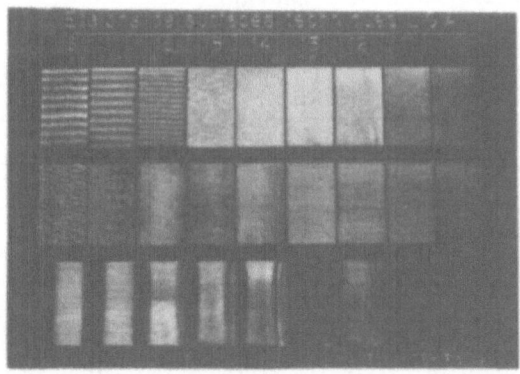

Fig.23: Workshop standards of surface rugosity

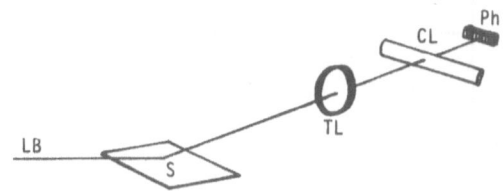

Fig.24: Experimental arrangement for surface rugosity measurement
(Ph - photodiode array; CL - cilindrical lens; TL - Fourier
transform lens; S - surface; LB - illuminating beam)

The rectified standard surfaces of the said 10,11,12 and 13, Fig.23
were intercompared for surface rugosity.

Typical results are condensed on Table I where surface 10 was taken as
comparison reference, and intensity distribution was measured with
photodiode array, Fig.25.

TABLE I

		SURFACES			
		10	11	12	13
INTENSITY (Relative value)		225	223	175	120
INTENSITY AFFER AMBIENT LIGHT CORRECTION $\langle|E^2(K_{20})|\rangle$		202	180	122	67
INCIDENCE ANGLE			82°	82°	82°
σ_{rms} [µm]	CALCULATED		0.136	0.263	0.385
	SPECIFIED	0.058	0.12	0.23	0.46
	DISAGREMENT		13%	14%	16%

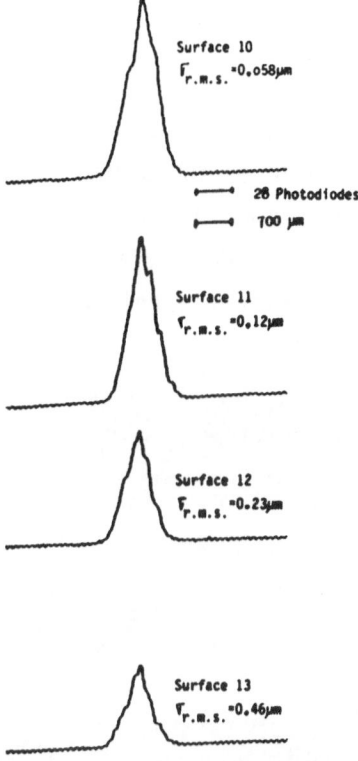

Fig.25: Photodiode array measurement of diffuse radiation distribution by standard rugosity surface. Distribution intensity shift is due to surface curvature.

While the previous discussion is but an introduction to the topic of optical measurement of engineering surfaces, the metrology and properties of surfaces are receiving considerable interest throughout the world. New international standards are being established and optical instrumentation is becoming available.

44

iv) Speckle metrology

Illuminating a rough surface with coherent radiation produces speckle. The speckle is characteristic of the surface structure and illuminating wavefront with dimensional scale related to the viewing optical system aperture. Speckle analysis provides surface characterization (13), Fig.26. Further, surface micro—movements relate to speckle pattern shifts, Fig.27.

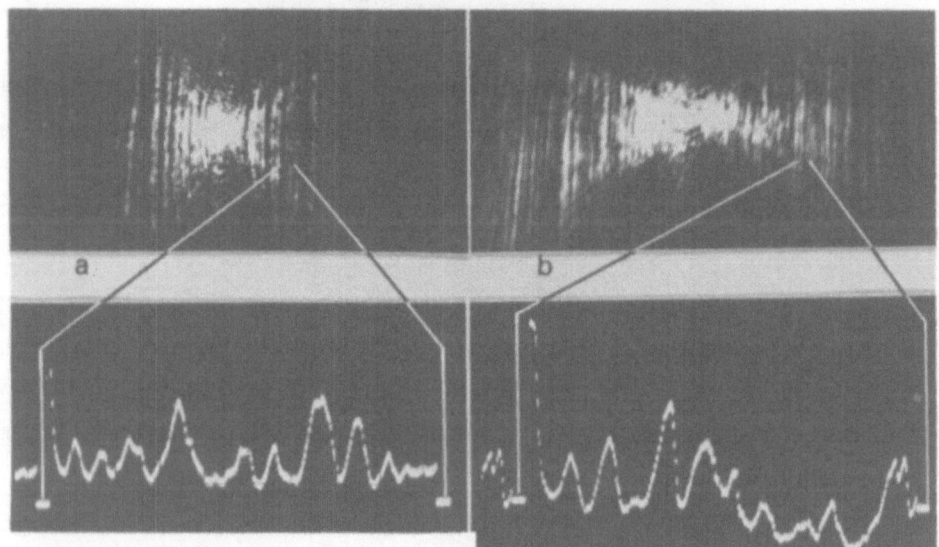

Fig.26: Example of direct comparison of surface finish by simple examination of surface speckle signature.
a) standard surface finish
b) sample control

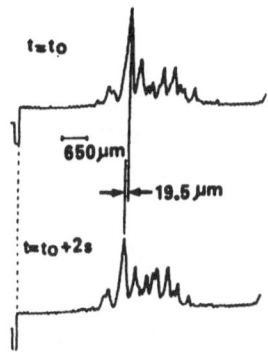

Fig.27: Speckle pattern shift after thermic loading of surface. Local dilation evaluated by speckle displacement is 19.5 μm corresponding to a heating of   T = 12ºC (optical magnification was 0.2)

v) Moiré metrology

Moirégrams exploitation can be accomplished in static and dynamic situations using the image dissector. Fig.28 is self-explanatory of the principle while it is obvious to recognize areas of application such as displacement, velocity and acceleration sensing by analysis of moirégram fringes evolution.

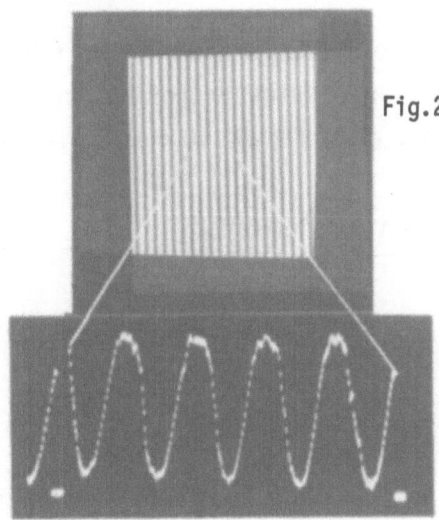

Fig.28: Moirégram analysis
    a)principle of the method
       with linear planar
       gratings

46

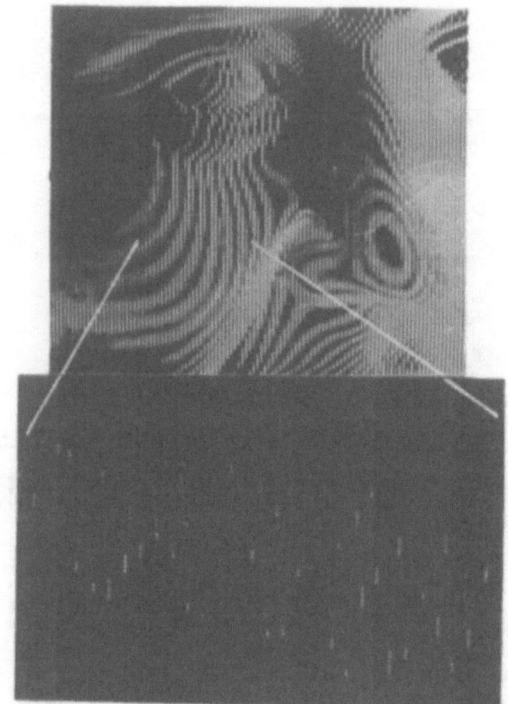

b)Moirégram exploitation

Fig.28: Moirégram analysis for further metrologic processing

vi) Spectrometry

The introduction of dispersive elements provides means for spectral radiation analysis whether in transmission or reflection. Spectral response of photodiode array is not flat, Fig.29, so that corrections may be necessary.

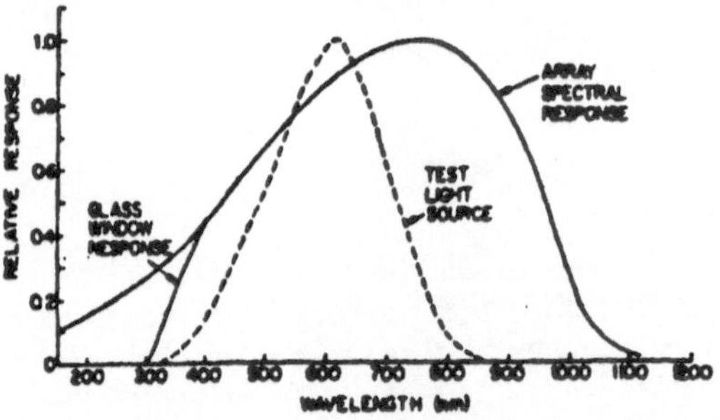

Fig.29: Spectral response of photodiode array as provided by manufacturer

Standard wavelengths can be used for calibration. Laser radiation lines are both convenient and simple to use, Fig.31. A dispersive set-up, is shown in Fig.30.

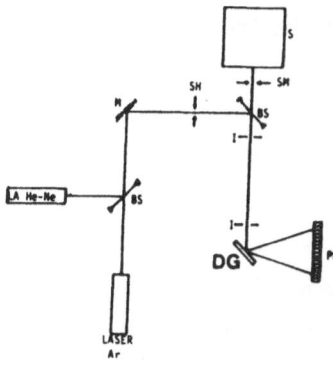

Fig.30: Spectrometric arrangement with a dispersive grating e.g. 20 lines/mm
(S - white light source; BS - beam splitter; I - Iris diaphragm; M - mirror; SH-SM -shutter;Ph-photodiode array;DG-dispersive grating)

The Fig.31 reports calibration of the set-up and wavelength bandwidth evaluation of a transmission yellow filter.

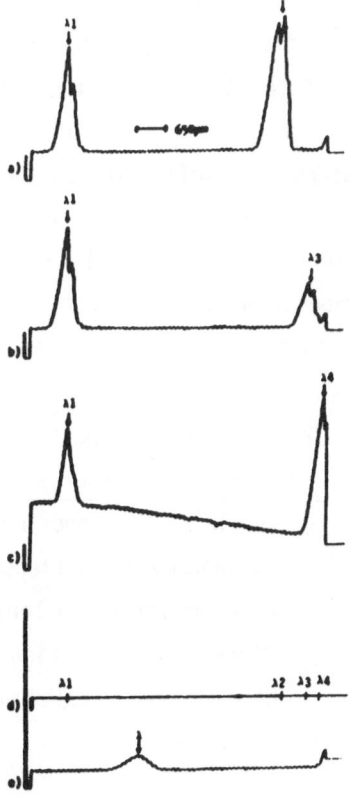

Fig.31: Spectrometric calibration of photodiode array:
a) to d;
e) yellow filter transmission spectrum
Colorimetric analysis is envisaged with the system by proper development of software.

λ_1 = 0.6328 μm λ_2 = 0.5145 μm
λ_3 = 0.4965 μm λ_4 = 0.488 μm
λ = 0.5896 μm estimated

Dynamical events and multidimensional measurements can also be dealt with the system. Stroboscopic illumination, synchronized with photodiode array controller and processor provides means for the application both in vibrational and transient analysis (14). Multidimensional metrology is performed by spatial multiplexing and will be treated further on.

Micrometric Laser Scanner (5,6,7)

The system operating principle is based on high directivity and low divergence of the laser beam. Once a regular displacement of the laser beam is known, in time and geometry, linear dimensions of the object to be gauged can be evaluated by the observation of the beam shadowing effect produced by the interposition of the object (time of flight scanner).

In the very simplified configuration where the laser beam moves at constant speed parallel to itself in a plane, the flying time corresponding to the shadow resulting from the interposed object measures directly the linear dimension of the object in the plane of the beam.

According to Fig.32 the laser beam scans by reflection on the moving mirror M of a galvanometric deflector driven by a sawtooth voltage. The lens L_3 corrects for the divergence otherwise introduced by collimating lens L_2. This guarantees invariance of measurement for object translation normal to the beam axis. The laser beam spot incidence on mirror M is in the focal plane of lens L_2.

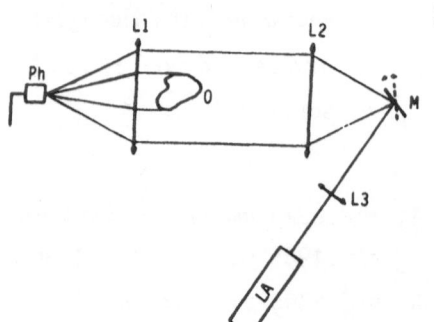

Fig.32: Principle of operation of the micrometric laser scanner (LA-laser;M-scanner mirror of galvanometric deflector;L_1 e L_2 - collimating lens ; L_3-divergence correcting lens;O-object for measurement;Ph-photodiode)

Practical implementation of the concept faces various problems such as: fluctuation of laser beam intensity and directivity; vibrating effects; noise inconveniences; mispositioning of the beam spot on the rotational axis of scanning mirror; non-constant velocity of the scanning mirror; the $\tan\theta$ error due to deflecting geometry and collimating lens performance, and thermal fluctuations.

To overcome these difficulties a more elaborate configuration was implemented (6), Fig.33, where an auto calibration branch was introduced to monitor and provide appropriate corrections.

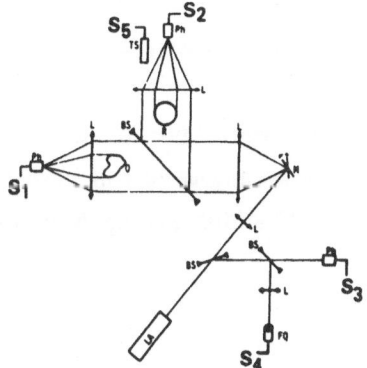

Fig.33: Implementation of micrometric laser scanner (6).
        (O-object under measurement; R-reference standard; LA-laser;
        TS-temperature sensor;L-lens;BS-beam splitter;Ph-photodiodes;
        FQ-photodiode of four sectors;M-scanning mirror;S-photosignals

In relation to Fig.34, the photosignal S_1 and S_2 are amplified and compared to a reference voltage adjusted in accordance to photosignal S_3. The comparators output drive the counters based on clock pulses. One counter reads the object dimension and the other counter provides measurement of reference standard in correspondence to the shadowing of the laser beam seen by photodiodes Ph_1 and Ph_2 respectively. Counters' values and photosignal S_4 and S_5 from laser beam monitoring for directivity and temperature fluctuation, respectively, are fed into microprocessor for final result output.

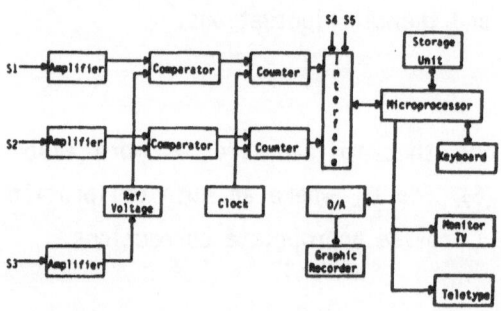

Fig.34: Detection and processing of signals for measurement with the micrometric laser scanner.

The system can be configured in various forms to exploit the flexibility of the optical arrangement and potentialities of the method. Fig.35 presents a rectro-reflection arrangement with folding over of the optical collimator.

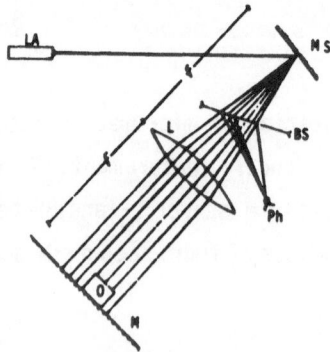

Fig.35: Rectro-reflection arrangement for the micrometric laser scanner
(M-flat mirror; O-object; LA-laser; L-collimating lens of f focal length; BS-beam splitter; Ph-photodetector; MS-scanning mirror)

Alternatively, the arrangement of Fig.36 may be used without folding over and having a collimated beam at object plane.

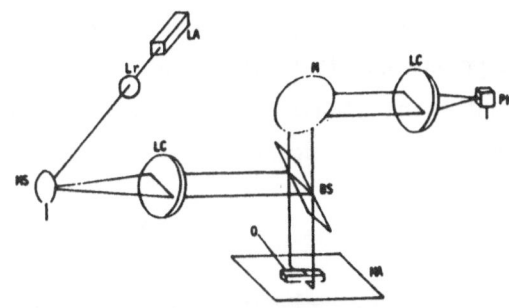

Fig.36: Rectro-reflection arrangement with double pass of laser beam. (M-flat mirror;O-object;LA-laser;LC-collimating lens;BS-beam splitter;Ph-photodetector;MS-scanning mirror;L$_r$-convergence correcting lens;MA-auxiliary mirror)

The arrangement of Fig.36 provides great flexibility as the object and the flat mirror can be placed, in principle, at any distance from the beam splitter. Typical results are shown in Fig.37 and 38.

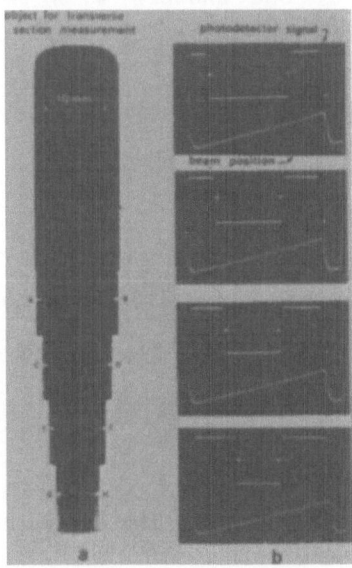

Fig.37: Typical results of measurement with the micrometric laser scanner.
a) calibrated sample
b) visualization of measurement: upper trace relates to the shadowing by the object; lower trace is the time scale, corresponds to the evolution of the scanning mirror deflection.

The system was operated up to 100Hz for the testing but speed must be increased to defeat vibration effects to be met in the industrial environment. Dimensional range from 1 to 50mm was exploited but lens diameter may be increased to accommodate larger dimensions. Resolution figures of 10 μm are easily achieved, and with stability control of laser and speed regulation improvement with a better scanning mirror of poligonal type (constant speed) better performances around 1 μm are envisaged.

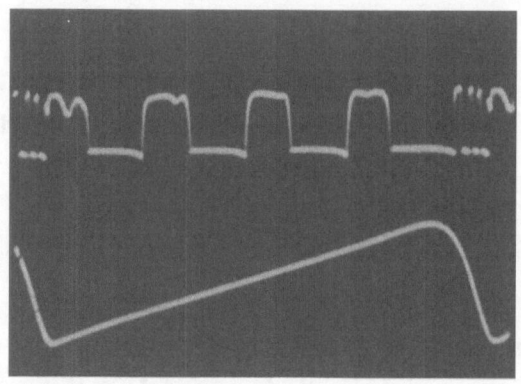

Fig.38: Simultaneous measurement of width (value B in Fig.10) of three piston rings placed on top of one another with same representation as in Fig.37.

Multidimensional Metrology (3,6)

The described systems can be configured for multidimensional metrology by multiplexing of the measuring branch with beam division or introducing means for XY scanning.

Surface metrology certainly requires XY scanning. The Fig.39 and Fig.40 refer to possible arrangements.

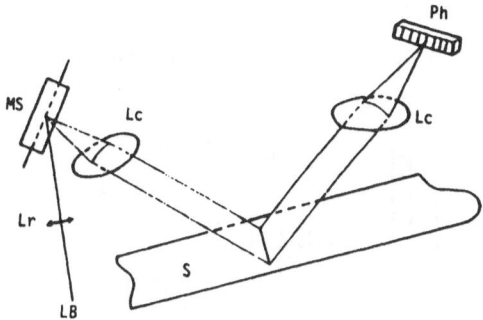

Fig.39: Surface scanning by rotation of the scanning mirror combined
with rotation axis bending within a plane normal to the plane
of beam incidence.
(LB-laser beam; L_r-convergence correcting lens; L_c-collimating
lens; Ph-photodiode array; S-surface; MS-scanning mirror)

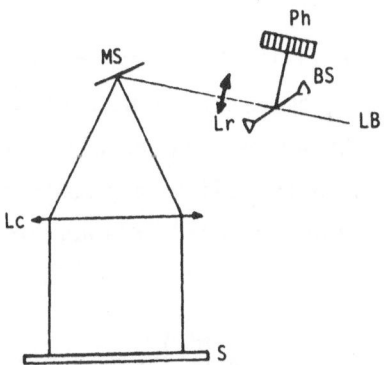

Fig.40: Surface scanning with rectro-reflection optical arrangement.
The surface realizes a translation movement.
(LB-laser beam; L_r-convergence correcting lens; L_c-collimating
lens; BS-beam splitter; Ph-photodiode array; S-surface;
MS-scanning mirror)

The photodiode array is essentially an unidirectional detector but by
rotation or translation it can be used as an orientable aperture providing
means for multidirectional analysis.

54

It is also obvious that several photodiode arrays could be used to explore different directions as shown in Fig.41.

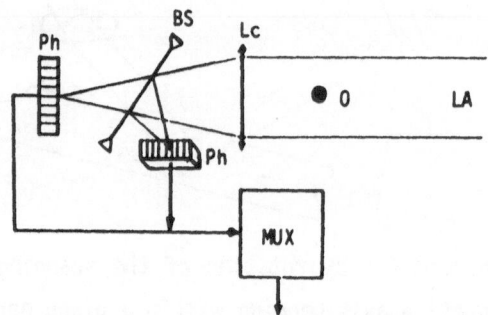

Fig.41: Spatial multiplexing for multidimensional analysis.
(LA-collimated laser beam; L_c-condensing lens; BS-beam splitter; O-object; Ph-photodiode array; Mux-multiplexer)

Polarization control of the laser beam can also be used to create two measuring branches. An example is given in Fig.42, where polarizing beam splitters separate the normal and parallel polarization component. Convenient switching of polarization components over the photodiode array provides the time multiplexing of multiple measurements.

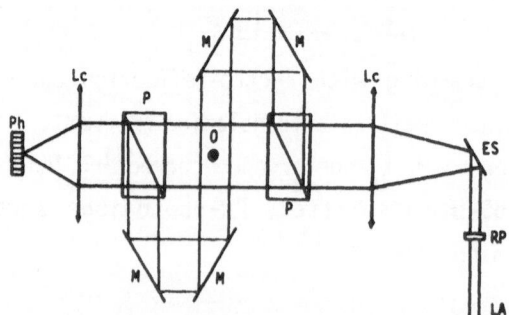

Fig.42: Polarization control of laser beam to generate two measuring branches.
(LA-linearly polarized laser beam; RP-retardation plate /4; ES-scanning mirror; L_c-collimating lens; P-polarizing beam splitter; M-mirror; O-object; Ph-photodiode array)

Polarization control combined with beam deflection may also be used to generate multiple branches for measurement, Fig.43.

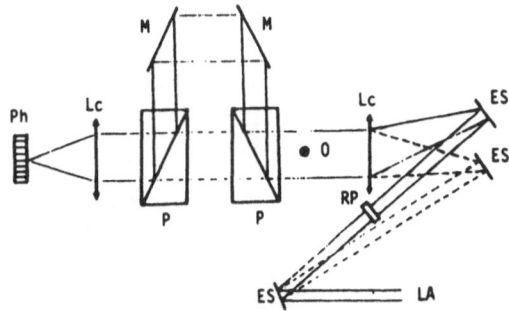

Fig.43: Combination of polarization control and beam deflection to generate two measuring branches.
(LA-linearly polarized laser beam, RP-retardation plate $\lambda/2$; ES-scanning mirror; L_c-collimating lens; P-polarizing beam splitter; Ph-photodiode array; O-object; M-mirror)

The use of multiple scanners is also capable of providing multidimensional measurement features as exemplified in Fig.44.

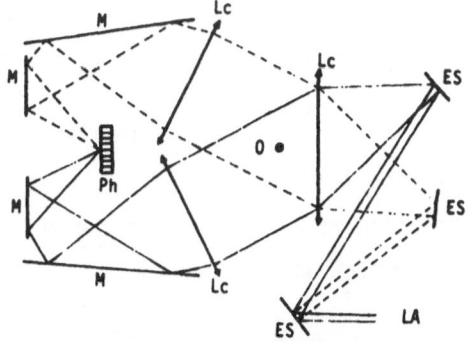

Fig.44: Multiple scanners arrangement for two-dimensional measurement.
(LA-laser beam; ES-scanning mirror; L_c-collimating lens; M-mirror; Ph-photodiode array; O-object)

The implementation of differential measurement is also of practical interest. The Fig.45 is an example showing schematically the principle for dimensional control of a wire extrusion.

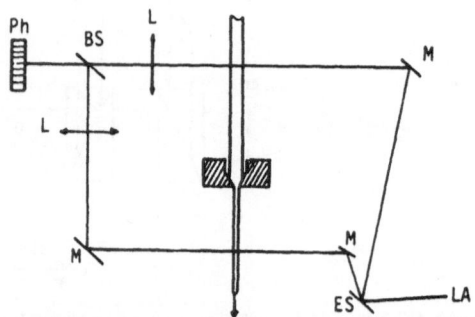

Fig.45: Differential measurement principle. Exemplification for a wire extrusion.
(LA-laser beam; M-mirror ; ES-deflecting mirror; L-Fourier transform lens; Ph-photodiode array; BS-beam splitter)

The number of examples and variety of applications could be extended but it would be only relevant in this context if considering specific applications.

Conclusion

Laser dimensional metrology offers appealing possibilities for industrial application in metrology, sensing and control with expeditious and flexible implementation.

The continuing lowering of price of lasers and of photodetectors with improved performances is likely to spur the spreading of a progressively increasing introduction of laser dimensional metrologic systems in multiple areas of science, industry and related fields.

Acknowledgements

Funding was provided by a research contract from JNICT 105.79.26 and INIC.

Valuable contributions by A.L.V.S.Lage and J.C.A.Fernandes are acknowledged as well as the technical assistance of J.S.Fernandes and L.Vilaça.

References

1. H.L.Kasdan, Opto-Electronic Dimensional Measurement
2. H.L.Kasdan, N.George, Linewidth Measurement by Diffraction Pattern Analysis, SPIE vol.80 (1976) 54-63
3. O.D.D.Soares, A.O.S.Gomes, Dissecador de Imagens Metrológico Unidimensional, Patent nº80074, I.N.P.I., Portugal (1985)
4. O.D.D.Soares,A.O.S.Gomes, Linear Photodiode Array Metrologic Laser System, TOOLPHOT 83, BEPA 1 (1983), 24-29
5. O.D.D.Soares, A.O.S.Gomes, Sistema Laser para Metrologia Dimensional, Internal Report (1984) 1-52
6. O.D.D.Soares, A.O.S.Gomes, Micrometro Laser de Varrimento, Patent nº 80335, I.N.P.I., Portugal (1985)
7. O.D.D.Soares, A.O.S.Gomes, Laser Scanner Micrometer, TOOLPHOT 83, BEPA 1 (1983), 29-34
8. DIN 70907 Kolbenringe Prüfung der Qualitätsmerkmale, Nov. (1974)
9. O.D.D.Soares,S.P.Almeida, Projection Interference Fringe Microscope, SPIE Vol. 427 (1983)

10. P.Beckmann, A.Spizzichino, The Scattering of Electromagnetic Waves from Rough Surfaces, Pergamon Press (1963)

11. F.Berny, C.Imbert, Determination Optique des Etads de Surface, Bulletim BNM (1976) 14-19

12. F.Milano, F.Rasello, An Optical Method for an On-line Evaluation of Machined Surface Finishing, Optica Acta 28 (1981), 111-123

13. Erf, R.K. (Edt), Speckle Metrology, Academic Press, NY (1978)

14. Soares, O.D.D.; Lage, A.L.V.S., Controllable Synchronized Multiple Pulse Illumination System for ESPI and Holography, Proc SPIE, 427 (1983)

DISTANCE SENSING VIA SPECTROGRAPHIC CODING

G. Molesini and F. Quercioli

Istituto Nazionale di Ottica
Largo E. Fermi 6, 50125 Firenze, Italy

A wholly optical distance sensor is presented for the monitoring of
surface position in ranges of interest for mechanical tooling
machines. The working principle is focus multiplexing by wavelength
encoding. Decoding and processing is performed by a spectrographic
unit equipped with a linear photodiode array. The device looks
suitable for engineering into a compact instrument for in-line
positioning of tools and workpieces, and for inspection and evaluation of surface profiles.

1. INTRODUCTION

Monitoring the position of mechanical surfaces with respect to
an inspecting head is of primary importance in robotics of computer
controlled tooling machines both for data acquisition on the initial
and final shape of the workpiece and for evaluation of microroughness
and profile of the surface after tooling.

Among the instrumental techniques for determination of surface
position and for reconstruction of surface profile the optical
approach is outstanding as it is non-contacting, versatile and fast.
Optical profilometers are usually designed as differential or absolute instruments, able to recover the surface profile after a series
of local determinations along a section of the sample under test.
The sample position with respect to the optical head is first
recorded, then the sample is laterally displaced by steps to let the

Soares, O.D.D. (ed), Optical Metrology
© *1987. Martinus Nijhoff Publishers, Dordrecht.*

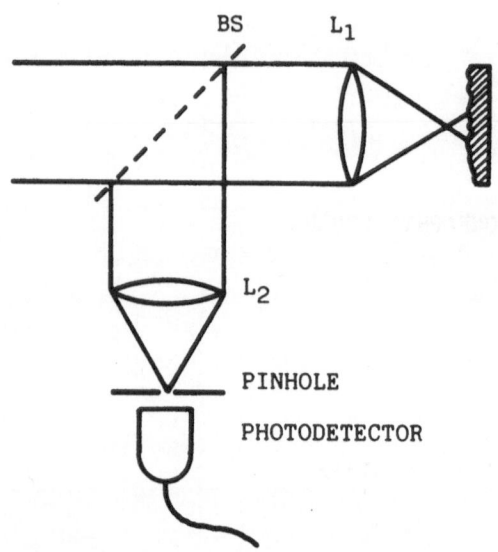

BS L$_1$

L$_2$

PINHOLE

PHOTODETECTOR

Fig. 1.
Schematic of
reflective
scanning
technique.

head to explore the surface along a step-line according to the
lateral translations. Samplings provide the surface profile reconstruction after proper data handling.

Optical profilometers usually focus light at the surface of the
sample under test and process the back-diffused light. The principle
of operation is based on a simple assumption which, although not
true in every case, gives in practice satisfactory results.
Referring to Figure 1, a collimated beam is focused by lens L1 at
the surface of the sample being tested. Part of the back-diffused
light is collected by L1 and partly separated by the beam splitter
BS. Lens L2 then focuses on a pinhole, selecting light from the
focus of L1. A photodetector behind the pinhole provides an electric
signal which is assumed to be maximum when the surface of the sample
is in focus [1]. Cases where such an assumption is invalid refer to
the possibility for the sample to have a local curvature whose
center exactly matches the focus of the probe beam. However such
cases are highly unlikely to occur in practice, as the tracking of
the surface is made in nearly a cat's eye configuration.

The above assumption suggests an operative procedure for
defining the position of a surface in depth. This may be accomplished by a relative displacement or oscillation of the focusing
optics with respect to the sample [2], or by further reasonable
assumptions on the intensity distribution of the back focused light
along the optical axis [3-9]. Thus the system is made to operate

as an optical probe, fingering the surface point by point to sense its distance from the measuring head.

The same assumption suggests also a different procedure to find out the surface position in front of the focusing lens. Instead of a single probe, serially scanning the optical axis in depth, a number of probe foci can be longitudinally operated in parallel, provided some coding is introduced to make the multiplexing operation suitably channeled. For this purpose the use of pseudocoloring techniques looks particularly appealing, as they allow for easy identification of wavelength encoded optical probes by means of spectrographic analysis of the power spectrum of the light back diffused [10]. In this work such an approach is described in some detail, and peculiarities shown by a laboratory prototype are reported.

2. PRINCIPLE OF OPERATION

The working principle of the wavelength encoded distance sensor is outlined in Figure 2. A white light source provides an illuminated

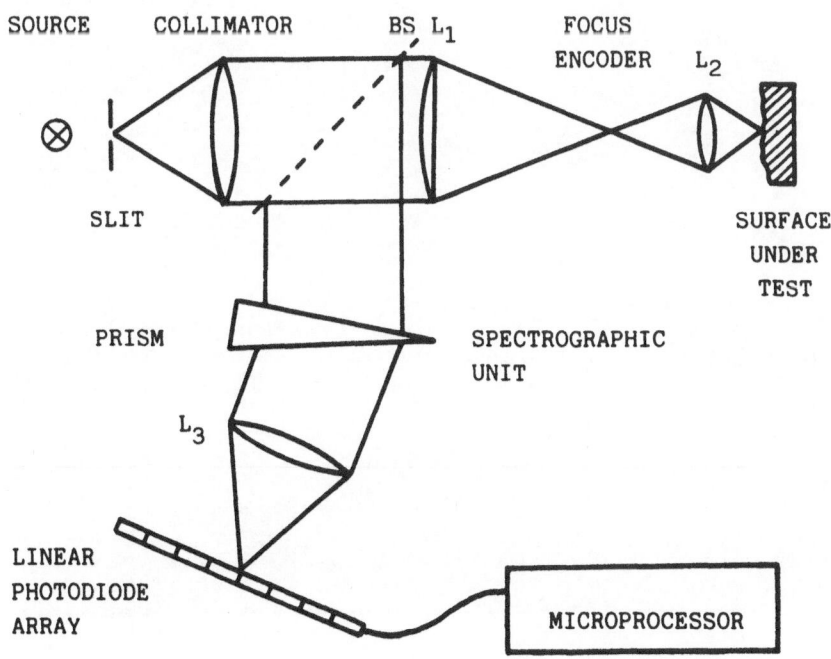

Fig. 2. Layout of the optical distance sensor.

object slit at the focus of a well corrected lens. The collimated
light is partly transmitted by a beam splitter and focused by means
of a plano-convex lens, thus introducing a certain amount of longitudinal chromatic aberration. In this way the focal distribution is
made of an axial series of slit images displaced in depth, coded in
different colors according to the lens' chromatic aberration. Then
a microscope objective compacts such a distribution about the surface
whose position has to be determined. The back diffused light is
partly collected by the same focusing optics and separated by the
beam splitter. Processing is carried out by means of a spectrographic
unit made of a dispersing prism and a lens. Finally the power
spectrum at the focal plane is displayed on a linear photodiode
array the output of which is read by a microprocessor which also
performs data analysis and evaluation.

As expected, maximum spectral power density and maximum signal
at the photodiodes are detected according to the probe wavelength
in focus at the surface being localized. Scanning the array provides

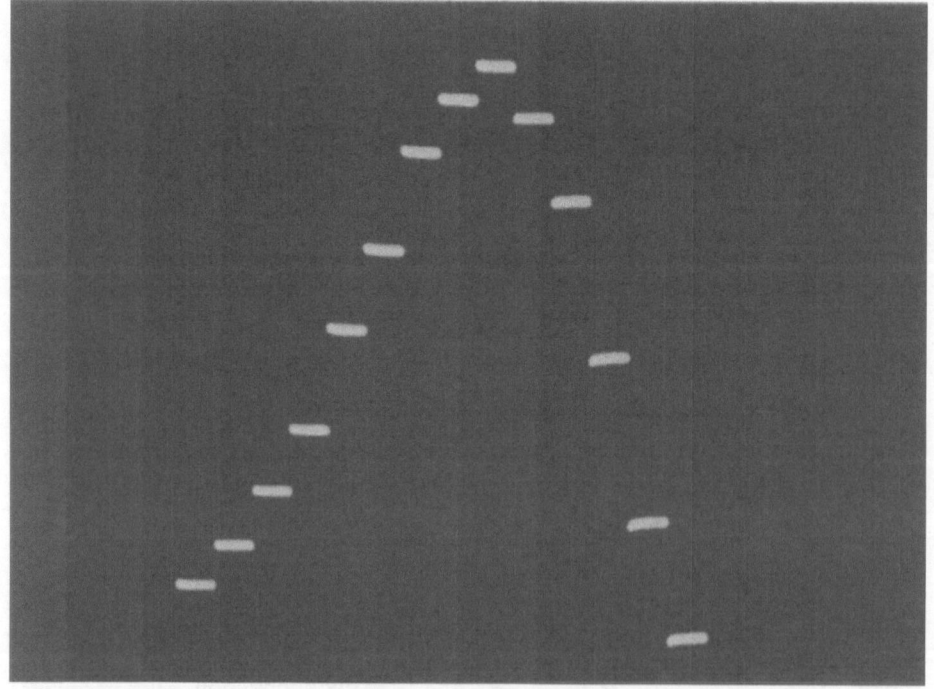

Fig. 3. Typical output at the oscilloscope after scanning the
photodiode linear array.

a bell-shaped histogram (Figure 3), to be manipulated with a proper
algorithm in order to achieve good sensitivity to height variations.
The peak position in the photodiode sequence indicates which wavelength is in focus at the surface, and thus reveals the distance of
the surface from the optical head.

3. SYSTEM PERFORMANCE

Referring to Figure 2, the parameters affecting the overall
performance are the lens focal lengths, the dispersion of the plano-
convex lens and of the prism, the source power spectrum, the photodiode sensitivity curve, the spectral absorptance of the sample and
the geometrical characteristics of the array. Though the expected
performance can be worked out on the basis of the above data, a more
realistic approach is to run a calibration test and record the system
response in the microprocessor for data reduction during operation
[10]. However, some specific features of the device can be considered
separately in order to evaluate the best configuration according to
the experimental conditions and to the accuracy requirements.

Lens L1 produces the focus encoding of the optical probes. To
reduce spherical aberration and improve dispersion a high refractive
index and low Abbe number for the glass should be chosen. If the
ultimate performance of the system is of concern, a specially
designed lens should be used. L1 defines the base feature of the
measuring head in terms of longitudinal dispersion and lateral resolution. This last characteristic clearly depends on the monochromatic
spot size and scales with f/N.

Lens L2 determines the actual dynamical range of measurement
and the actual lateral resolution as well. Namely, since the longitudinal magnification scales with the square of the lateral one, a
compacting operation of the dispersed images also improves lateral
resolution, but at the expense of a greatly reduced dynamical range.
In place of L2 it appears suitable to use a microscope objective,
as such lenses usually are well corrected for chromatic aberration
and can be easily interchanged profiting from the general stand-
ardization of the microscope configurations. This makes it simple
to change the range of operation of the device. Alternatively, if
the lens used is enough reliable, the working range could be varied
just moving L2 to a different axial position, thus changing the
magnification and the operating features.

The spectrographic unit affects the longitudinal sensitivity, as it controls the angular spread of the power spectrum on the linear photodiode array corresponding to the base longitudinal dispersion. This is mainly determined by the prism characteristics, as lens L3 only scales the lateral distribution without altering its ratio to the spot size. Lens L3 should then be chosen taking into account the actual geometrical characteristics of the photodiode array.

Data on a practical laboratory prototype and obtained performances are reported in Ref. 10. Of particular interest seem to be further possible improvements, aiming to obtain the surface profile along a line section in a single scan, without translating the sample by steps. This could be achieved using a matrix array in place of the linear one referred to here. Translation would provide information on 3-D reconstruction along a strip surface.

4. CONCLUSIONS

Spectrographic coding proves effective in focus multiplexing for distance sensing optical devices. The base configuration proposed here is reliable and versatile, as the discrete components separately affect the performance of the system in different ways.

Distance detection turns out to be fast enough for in-line real-time operation. Due also to the favourable characteristics of on-axis measuring devices, this approach looks suitable for a compact distance sensor, to be applied in simple robotics and in fine surface profiling.

REFERENCES

1. Caulfield, H.J., and D.L. Kryger, The Use of Microdensitometers as a Basis for Highly Accurate Metrology, Proc. Soc. Photo-Opt. Instrum. Eng. vol. 153 (1978) 23-26.
2. Arecchi, F.T., D. Bertani and S. Ciliberto, A Fast Versatile Optical Profilometer, Opt. Commun. 31 (1979) 263-266.
3. Simon, J., New Noncontact Devices for Measuring Small Micro-displacements, Appl. Opt. 9 (1970) 2337-2340.
4. Williams, T.L., A Scanning Gauge for Measuring the Form of Spherical and Aspherical Surfaces, Optica Acta 25 (1978) 1155-1166.

5. Sawatari, T., and R.B. Zipin, Optical Profile Transducer, Opt. Eng. 18 (1979) 222–225.
6. Fainman, Y., E. Lenz and J. Shamir, Optical Profilometer: A New Method for High Sensitivity and Wide Dynamic Range, Appl. Opt. 21 (1982) 3200–3208.
7. Gorecki, Ch., G. Tribillon and J. Mignot, Profilomètre Optique en Lumière Blanche, J. Optics (Paris) 14 (1983) 19–23.
8. Dobosz, M., Optical Profilometer: a Practical Approximate Method of Analysis, Appl. Opt. 22 (1983) 383–387.
9. Lou, D.Y., A. Martinez and D. Stanton, Surface Profile Measurement with a Dual-Beam Optical System, Appl. Opt. 23 (1984) 746–751.
10. Molesini, G., G. Pedrini, P. Poggi and F. Quercioli, Focus-Wavelength Encoded Optical Profilometer, Opt. Commun. 49 (1984) 229–233.

INTERFEROMETRY, VELOCIMETRY AND PHOTON CORRELATION

SOME RECENT DEVELOPMENTS IN LASER INTERFEROMETRY

Graham J. Siddall and Richard R. Baldwin

Hewlett-Packard Laboratories
1501 Page Mill Road
Palo Alto, CA   94304, U.S.A.

Abstract

An account is given of some new interferometric developments
which can be used to improve the performance of the laser interferometer as a position sensing device. The first of these developments, termed the double pass attachment, is an optical accessory
which doubles the resolution of the Hewlett-Packard linear and
plane mirror interferometers. Unlike previous attempts to extend
resolution optically, this device does not fold one of the two
interfering beams and hence does not introduce error due to its own
motion. The simple addition of a quarter wave plate to the attachment can be used to give a differential version of the plane mirror
interferometer. Various configurations of this differential interferometer, and their application to machine tool and integrated
circuit lithographic and inspection equipment, are discussed. A
new "wavelength tracking" device, based on differential interferometry, is described. This device directly and precisely monitors
changes in laser wavelength inside a highly stable mechanical
cavity. It can be easily incorporated, as an additional interferometer axis, into precision machines to give superior compensation
for changes in laser wavelength.

Introduction

The laser interferometer has been responsible for significant
advances in manufacturing technology. Its incorporation into equipment such as diamond turning machines and lithographic systems has
led to improvements in the accuracy of generation of aspheric optics
and in the micro-miniaturization of integrated circuits. The rate
of development of these emerging manufacturing technologies is such,

however, that increasing demands are being made on the accuracy of
the positional information supplied by the laser transducer.
Although the laser transducer output can be interpolated electronically to give a displacement resolution as high as 0.1 μin (2.5 nm),
the task of making this resolution meaningful necessitates a clear
understanding of the interferometer and its relationship to the
machine and operating environment. This paper describes new configurations of the interferometer which can lead to more meaningful
positional information, particularly in non-optimum operating
environments.

The polarizing interferometer, as typified by the Hewlett-
Packard (HP) laser interferometer system, is often referred to as a
Michelson interferometer. Although there are similarities between
the two, there is a subtle but important distinction. In the
Michelson interferometer, interference between the measurement and
reference beams occurs at the beamsplitter; in the polarizing interferometer the reference and measurement beams are orthogonally
polarized and interference does not take place until the two beams
are brought into the same plane of polarization. In the HP system
this occurs at the receiver. The significance of this difference
is that it is possible, in the case of the polarizing interferometer,
for beams to be transmitted more than once through one interferometer or serially through several interferometers with interference
between beams dictated solely by the location of the receiver(s).
This is clearly not possible in the case of the classical Michelson
interferometer.

The double pass attachment, described in the next section,
uses this principle to improve the resolution of the interferometer.
A simple modification to this attachment, described in the following
section, produces a differential version of the interferometer with
no increase in resolution but with several unique operational
advantages. Specific applications of the differential interferometer - to X-Y stage design and to laser wavelength compensation -
are described subsequently.

The double pass attachment

Previous attempts to improve resolution optically have involved folding the optical path of one[1] or both[2] of the interfering
beams to multiply the optical path difference introduced by translation of the moving reflector. The additional optics for folding
the light path or paths are not common to both measurement and
reference beams and hence unwanted motion of the optics introduces
measurement errors. In the double pass attachment, however, optical
path difference multiplication is achieved by folding both beams
simultaneously using the same optics; movement of these optics does
not introduce measurement error since it produces identical changes
in path length in each of the two beams.

Figure 1 shows the conventional polarizing linear interferometer and Figure 2 shows the interferometer with the addition of

the double pass optics. The laser beam entering the interferometer (Figure 1) is composed of two orthogonally polarized beams of differing optical frequencies F1 and F2, shown by continuous and hatched lines in the Figures. The polarizing beam-splitter reflects F2 (the reference beam) and transmits F1 (the measurement beam). After retroreflection from a movable cube-corner, the measurement beam returns to the interferometer where it joins the reference beam and travels along a common axis to the receiver. The two beams interfere to form fringes when they reach the demodulating polarizer in front of the photodetector in the receiver. Relative motion between the cube-corner and interferometer causes a difference in the Doppler shifts in the return frequencies and this is readily converted to a measurement of the change in optical path difference.

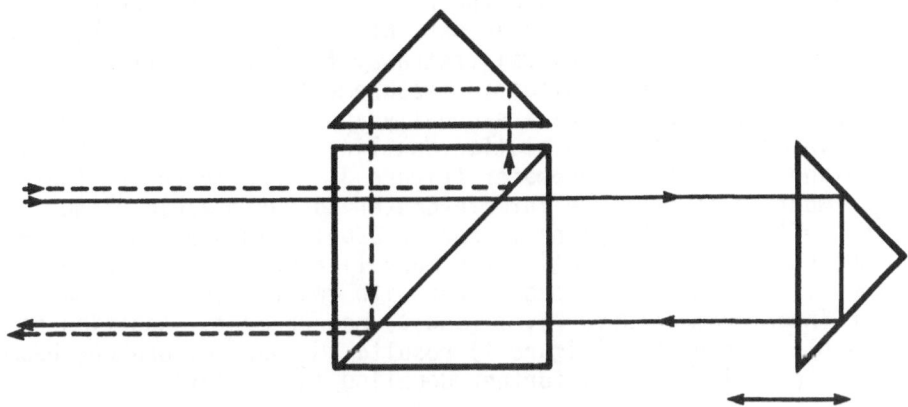

Figure 1. Conventional linear interferometer

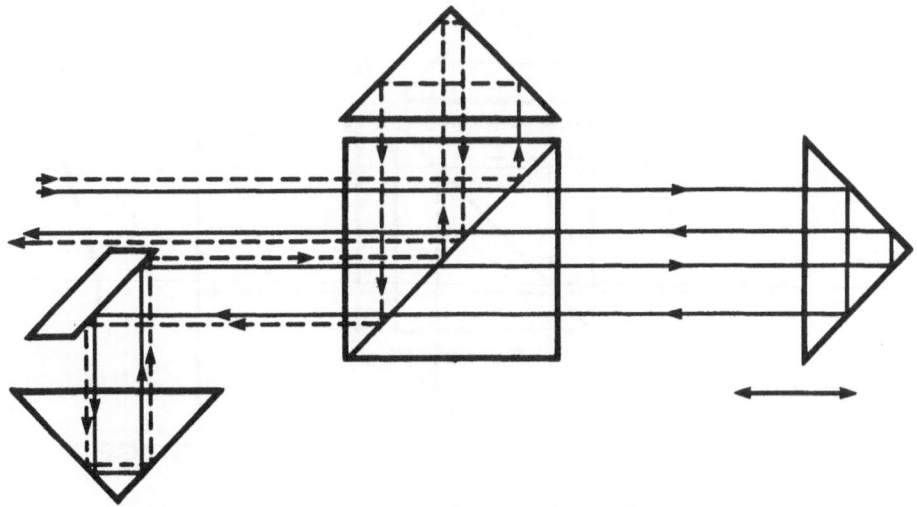

Figure 2. Double pass linear interferometer

In the double pass version of the linear interferometer (Figure 2), the measurement and reference beams are not intercepted by the receiver when they first exit the interferometer but are reflected back for a second pass through the interferometer by a beam bender and cube-corner. The size and location of the beam bender is chosen so that the beams, on their second exit from the interferometer, clear the beam bender on their way to the receiver. This arrangement doubles the optical path difference and hence improves the resolution by a factor of two. It is important to note, however, that the maximum slew rate or permissible velocity of the cube-corner is reduced by the same factor. This is something to bear in mind in schemes where the resolution is extended optically.

Since the additional double pass optics are common to both measurement and reference beams, the interferometer is insensitive to motions of these optics and their location and mounting are not critical. In a laboratory breadboard of the double pass scheme the optics could be mounted several feet away from the interferometer with no noticeable degradation in measurement stability.

The double pass optics are equally applicable to the plane mirror interferometer. Resolution doubling is inherent in the standard plane mirror interferometer (Figure 3) since there are two measurement beams from the interferometer to the movable plane mirror reflector. The plane mirror is usually mounted on a stage and the standard configuration for many lithographic and inspection systems is to use an L-shaped mirror with two (or three) plane mirror interferometers for X, Y (and θ) measurement. Addition of the double pass optics (Figure 4) results in four measurement beams to the stage mirror and a further doubling in resolution.

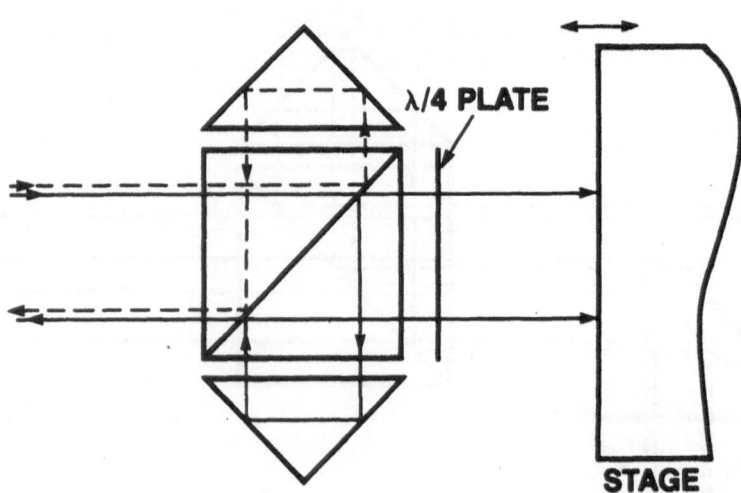

Figure 3. Conventional plane mirror interferometer

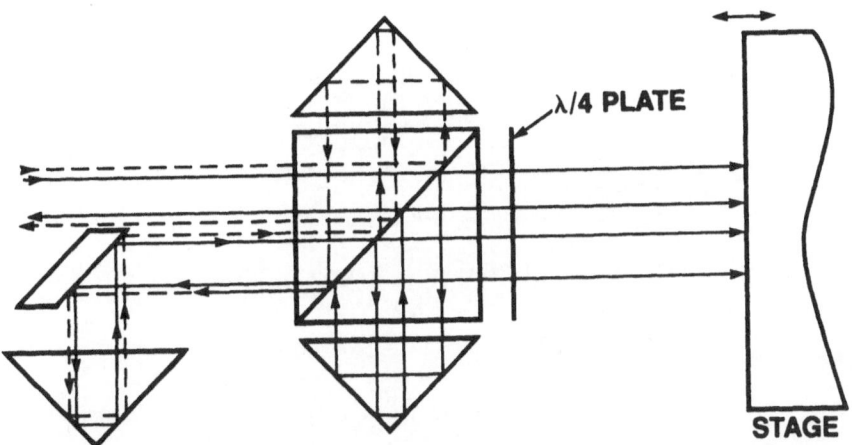

Figure 4.   Double pass plane mirror interferometer

In Figures 2 and 4, the optical paths for the double pass schemes are shown, for ray tracing convenience, in the plane of the paper.   In practice the aperture of the standard HP linear and plane mirror interferometers (20 mm) is not sufficient to accommodate four beam diameters, each of 6 mm, in a row, as shown in the Figures. To avoid vignetting of the beams it is better to use the beam configuration shown in Figure 5.   A photograph of the prototype double pass attachment, based on this configuration, is shown attached to a plane mirror interferometer in Figure 6.

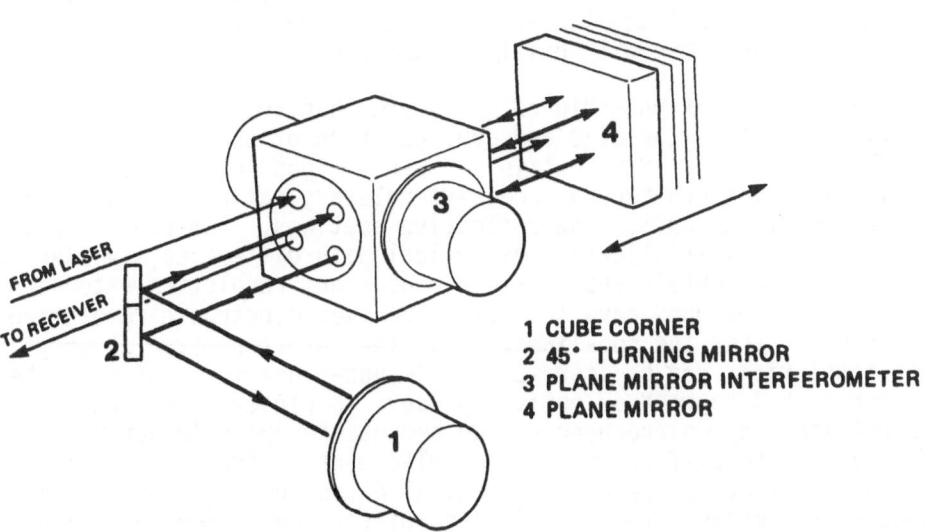

1 CUBE CORNER
2 45° TURNING MIRROR
3 PLANE MIRROR INTERFEROMETER
4 PLANE MIRROR

Figure 5.   Double pass plane mirror interferometer (isometric view)

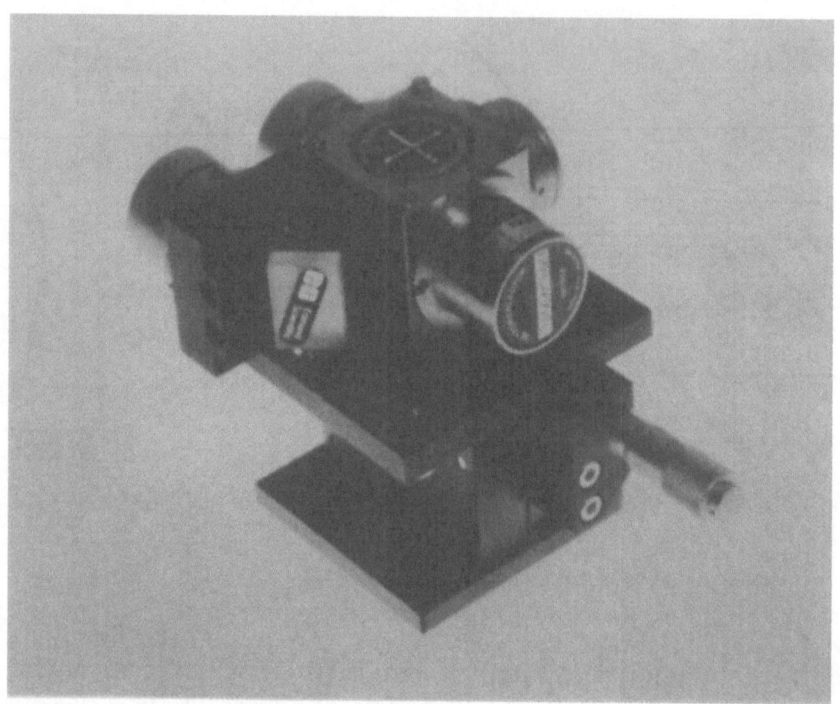

Figure 6.   Prototype double pass plane mirror interferometer

The attachment is an aluminum alloy bracket to which the beam
bender and cube-corner are attached, with two holes provided for the
input and output laser beams. It is a simple matter to mount the
attachment on either the linear or plane mirror interferometer using
two screws which mate with existing tapped holes in the interferometer housing.

Although it is possible to make more than two passes through
an interferometer, the size and configuration of the optics does
impose limitations. With multiple passes, light losses and wavefront distortion can become excessive unless great care is taken in
manufacturing the optics and reflective coatings. The reduction in
slew rate with multiplication of optical path difference may also be
restrictive in certain applications. However, resolution extension
by optical means does have the advantage over electronic resolution
extension in that the non-linearity of the interferometer at very
high resolutions can be improved. This non-linearity, which can be
as much as 5.4 degrees of optical phase[3], is caused by optical leakage between the two frequencies and amounts to about 10 nanometers
of optical path difference. The displacement equivalent of this
non-linearity can be reduced by optical extension of the resolution
since this increases the optical path difference caused by a given
displacement. In the case of the double pass plane mirror interferometer, with an optical path difference multiplication of eight,

the 10 nm non-linearity due to optical leakage corresponds to 1.2 nm of physical motion.

The differential double pass attachment

An interesting modification to the operation of the double pass plane mirror interferometer occurs when a suitably oriented quarter wave plate is inserted between the interferometer and the cube-corner of the double pass attachment. Instead of four measurement beams, two measurement beams and two reference beams are produced. This can be verified by ray tracing the schematic optical diagram in Figure 7. Although no longer giving resolution doubling, this arrangement has the advantage that it permits differential measurement between two plane mirrors, with the entire optical path outside the mirrors being common to both measurement and reference beams. Translation of the interferometer in the measurement direction and small displacements in the other five degrees of freedom have negligible effect on the measurement. This is particularly advantageous in many metrological applications where high stability is important.

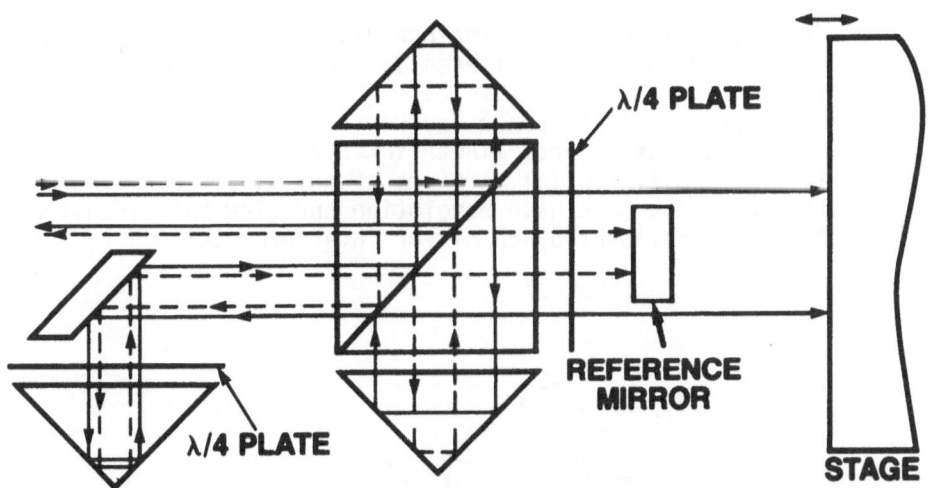

Figure 7. Differential plane mirror interferometer

An isometric view of the differential interferometer is shown in Figure 8. The pair of diagonally opposed reference beams are incident on the reference mirror, 5, and the pair of measurement beams pass through two holes in this mirror to the stage mirror, 6. The reference mirror can be made quite small and is much easier to mount close in to the stage mirror than the bulkier plane mirror interferometer. In many applications the deadpath (the optical path difference at the zero point of the stage) using the differential interferometer is limited only by the thickness of the reference mirror.

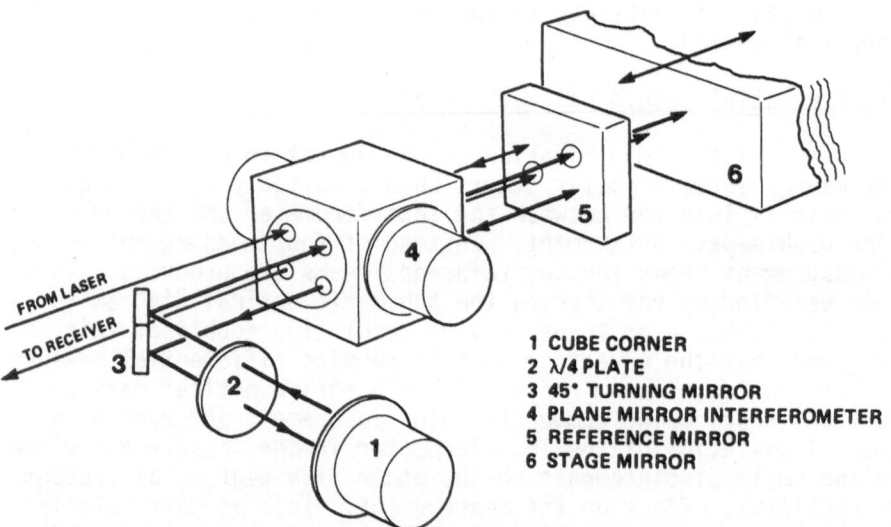

1 CUBE CORNER
2 λ/4 PLATE
3 45° TURNING MIRROR
4 PLANE MIRROR INTERFEROMETER
5 REFERENCE MIRROR
6 STAGE MIRROR

Figure 8.   Differential plane mirror interferometer (isometric view)

The differential interferometer can also be used with a trans-
latable cube-corner to give a differential version of the linear
interferometer (Figure 9).   In this case the measurement beam makes
two passes through the cube-corner and hence the resolution is
double that of the conventional linear interferometer in which only
one pass is made.   Differential versions of the linear and plane
mirror interferometers both have resolution and slew rate identical
to that of the conventional plane mirror interferometer.

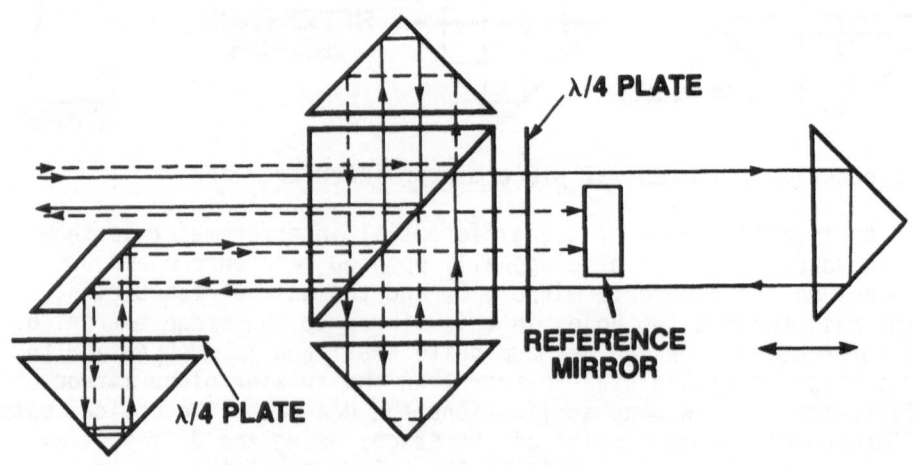

Figure 9.   Differential linear interferometer

Results

The stability of the differential plane mirror interferometer can be tested by simply removing the reference mirror. Both pairs of beams are now incident on the plane mirror and the optical path difference between them should be zero, irrespective of small motions of any component in the system. No elaborate fixturing is required. In fact, in the test to be described the plane mirror was attached to the end of a 6 inch plastic ruler mounted in a small vise!

To obtain high resolution in the measurement of path difference, the relative phase between the beams was measured using a Hewlett-Packard 3575A phase meter, which gave an effective resolution of 0.1 nm. An analog output was fed from the phase meter to a strip chart recorder for continuous monitoring of the phase difference.

Figure 10 shows the displacement equivalent of the phase difference variations over a period of five days. Each bar on the graph represents the maximum peak-to-valley variation, over a period of one hour, measured from the chart recorder. It is important to note that the data represents the stability of the total test system and includes contributions from the interferometer electronics, the phase meter, and the recorder, in addition to the interferometer. The test was carried out in a room with normal air conditioning, poor temperature control and no shielding of the interferometer from thermal effects. Despite the lack of precautions, the peak-to-valley drift was typically 1 nm during the night and weekend (with the air

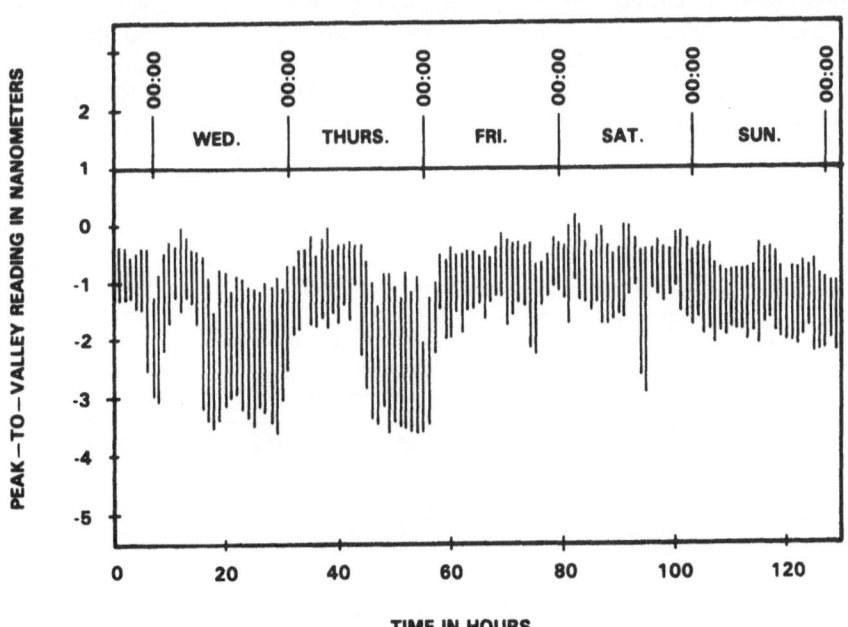

Figure 10. Stability of differential interferometer

conditioning off) and 2.5 nm during the day. No significant long
term drift was observed. Abrupt temperature changes (imposed, for
example, by spraying the interferometer with Freon refrigerant)
introduced temporary path differences of several nm. From these
observations it was concluded that the stability of the interferometer is limited mainly by temperature gradients, particularly
within optical components such as the cube-corners in which the
two beams are not coincident. Better environmental control should
reduce these figures even further.

Applications

1. X-Y Stage

The ability to measure relative motion between two surfaces to
very high precision has application, for example, in dilatometry,
material stability studies and high performance metrology stages.
An interesting example of the latter is given in Figure 11, which
shows one axis of a dual differential interferometer scheme for an
X-Y stage. The interferometer measures the differential position
of the stage with respect to two symmetrically disposed reference
mirrors. Since the optical path difference is zero when the stage
is centered between the two reference mirrors, there is no deadpath
at this point and laser wavelength changes have no effect. The
dual differential arrangement gives doubled resolution and the
system is unaffected by small motions of the folding optics and
interferometer since these are common to both measurement and reference beams. Thermal expansion effects are also reduced by the symmetry of the scheme. This arrangement can be used with an X-Y plane
mirror stage, unlike the differential cube-corner system proposed
by Tanimura[2], and should give impressive stability.

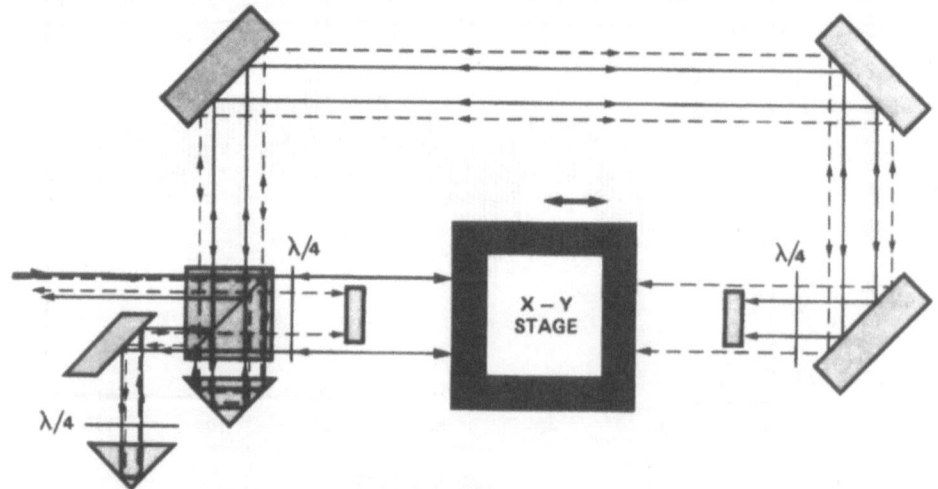

Figure 11. Differential interferometer for X-Y stage

2.   Wavelength tracking interferometer

Another application for the differential interferometer is to
track changes in wavelength of the laser source by monitoring
changes in the apparent length of a stable physical standard. A
convenient form for the latter is provided by an etalon in which the
two plane mirrors are optically contacted to the ends of a rigid,
stable tube (Figure 12). Such etalons, when constructed of an
ultralow expansion material such as Zerodur[TM], have been shown to
have long term stability of better than 1 part in 10^7 per year[4].
If the approximate length of the etalon is known, any change in
wavelength of the laser can be measured with high precision. It is
particularly important to correct for these changes when making displacement measurements in air since the index of refraction of air
is not constant and the laser wavelength is shifted by an amount
which varies non-linearly with air temperature, pressure and relative humidity. These changes can easily amount to several parts
per million and are usually compensated for by measuring the relevant air parameters with appropriate sensors and then calculating
the corresponding laser wavelength using empirically-derived
formulae[5,6]. With care, this method can give high accuracy[7], but
it is an indirect method of measurement and wavelength changes due
to other causes such as changes in the composition of air or of the
laser frequency itself, go undetected. The wavelength tracking
interferometer effectively overcomes these disadvantages and gives
greater ultimate accuracy in monitoring wavelength changes. A
photograph of the prototype wavelength tracking interferometer is
shown in Figure 13.

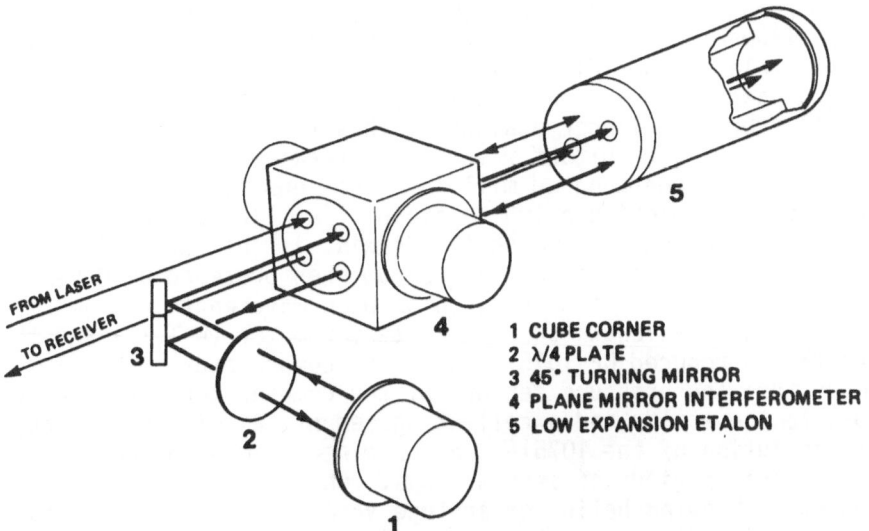

1 CUBE CORNER
2 λ/4 PLATE
3 45° TURNING MIRROR
4 PLANE MIRROR INTERFEROMETER
5 LOW EXPANSION ETALON

Figure 12. Wavelength tracking interferometer

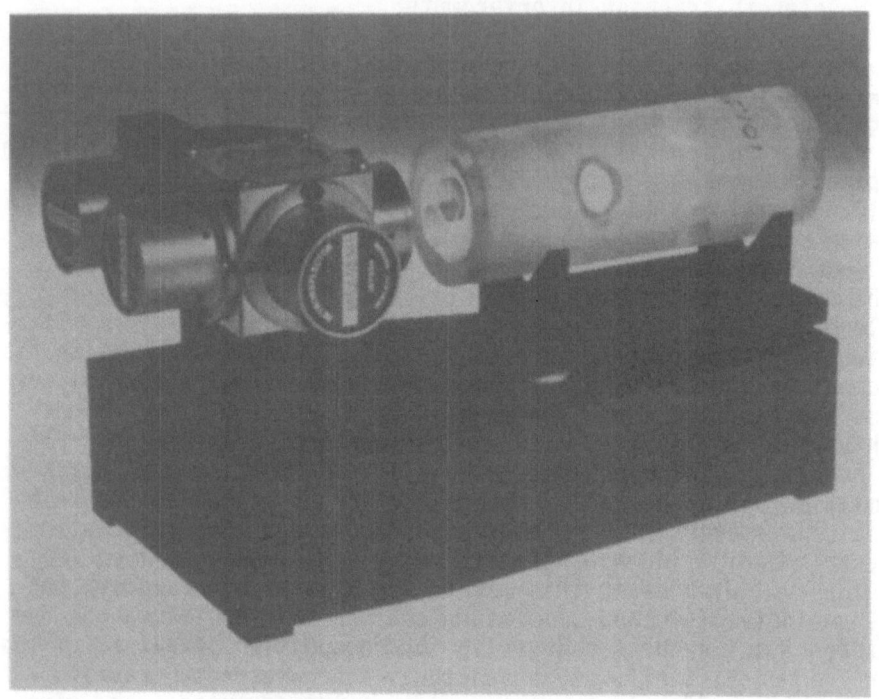

Figure 13.   Prototype wavelength tracking interferometer

The differential interferometer, on the left, is not sensitive to relative motion between it and the etalon since both interfering beams travel a common optical path outside the etalon. The etalon, to the right, is in this case 100 mm long and has holes drilled in it to allow air to circulate freely through its length. Fabrication of the etalon is not difficult since the parallelism of the end mirrors need be only a few arc minutes. It should be relatively easy to fabricate longer etalons of this type, up to a meter or so in length, for a resultant wavelength tracking precision of better than 1 part in one hundred million. The sensitivity of the proto-type system, employing a 100 mm etalon, is illustrated in Figure 14. The "V.O.L." number, shown on the Y-axis, is the ratio of the velocity of light in air to that in vacuum when preceded by 0.999.

Figure 15 shows a comparison between wavelength changes monitored by the wavelength tracking interferometer (W.T.I.) and the recently introduced HP 10751A automatic compensator. The air temperature and pressure sensors of the compensator were placed close to the etalon and excellent agreement, within the 1 part in 10^7 resolution of the 10751A, was obtained in monitoring wavelength changes over periods of several days. Changes in air composition, induced by blowing helium or acetone vapor across the laboratory, produced significant short-term wavelength changes, however, which went undetected, of course, by the 10751A.

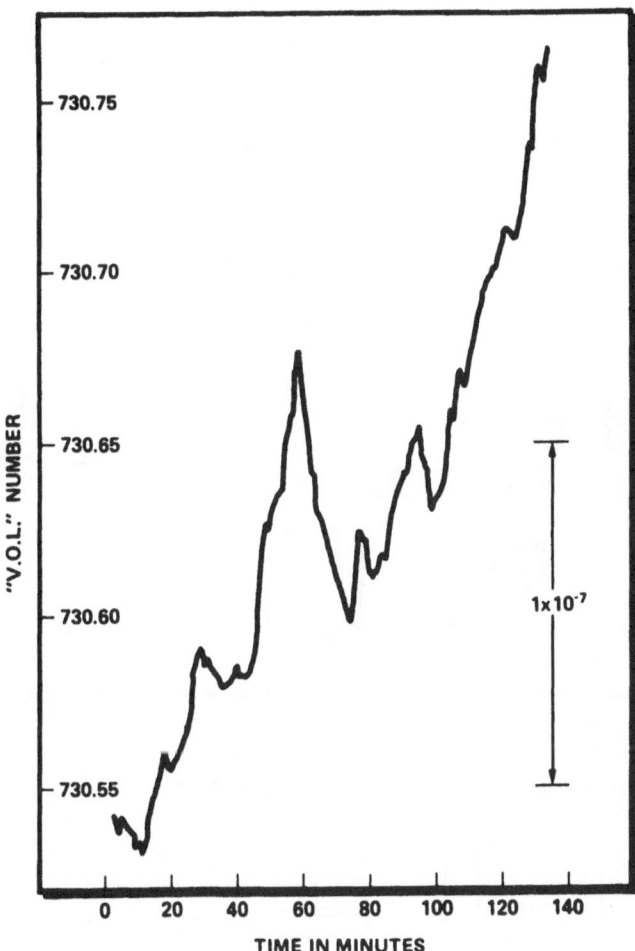

Figure 14.   Short term wavelength variations in air

A schematic layout of a lithographic stepper system, incorporating a wavelength tracking interferometer, is shown in Figure 16.   The wavelength tracking interferometer is mounted inside the stepper enclosure, close to the laser beam paths of the X-Y stage, and functions as an additional interferometer axis.   High resolution interpolator electronics are used to provide real-time information on wavelength changes to the control computer which then changes the stage scale factor accordingly.   Higher resolution interpolator electronics can be employed for wavelength tracking since a much lower slew rate bandwidth is acceptable.

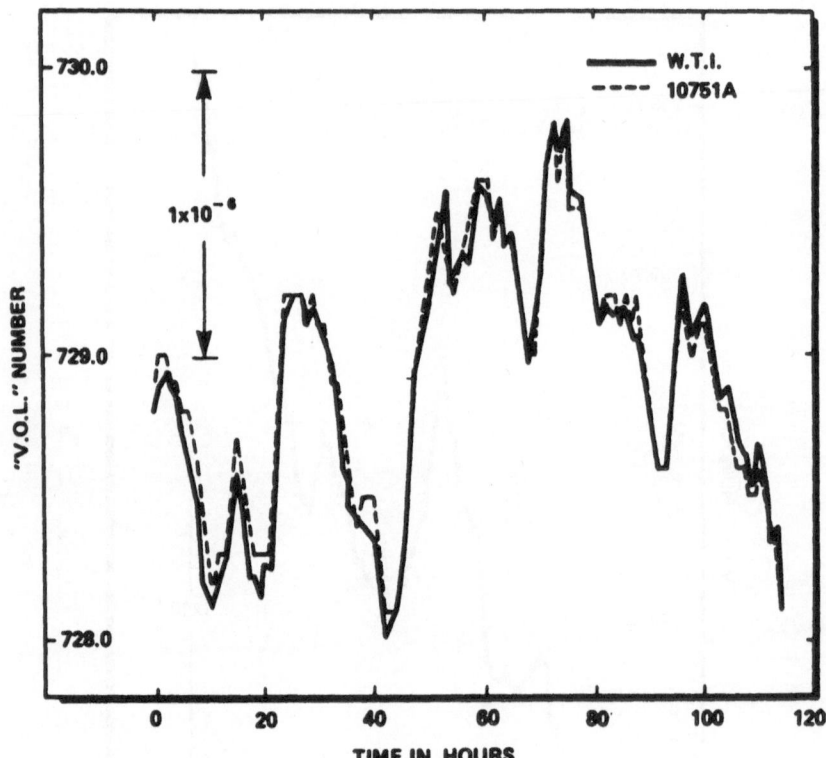

Figure 15.   Comparison of W.T.I. and 10751A

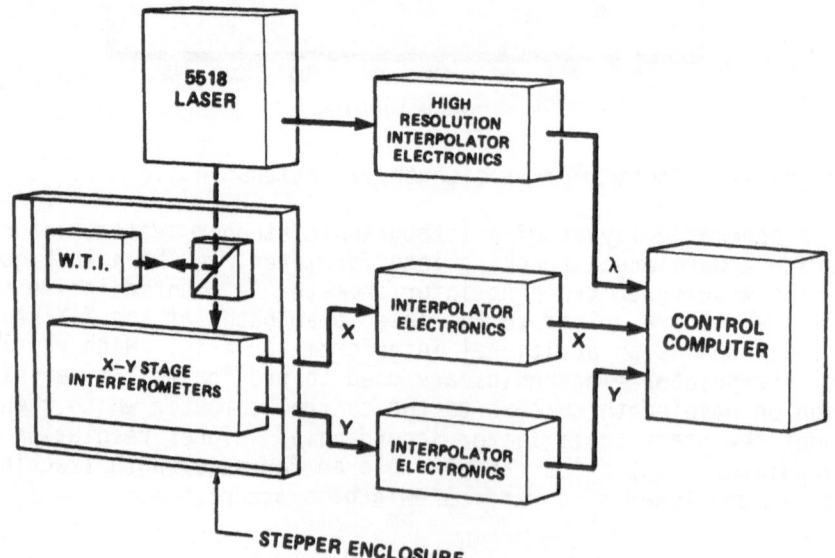

Figure 16.   Stepper system incorporating W.T.I.

Conclusions

Optical extension of interferometer resolution is possible using the double pass attachment in conjunction with the standard Hewlett-Packard linear or plane mirror interferometer. A differential version of the attachment results in improved interferometer stability and greater design freedom since location of the interferometer is no longer critical to performance. A wavelength tracking device, based on the differential interferometer, can be used to give superior compensation for laser wavelength changes.

Acknowledgements

The authors wish to acknowledge the technical support and encouragement of Len Cutler, Robin Giffard and Armand Neukermans of Hewlett-Packard Laboratories.

References

1. Murtz, M.V.R., Modification of Michelson Interferometer Using Only One Cube-Corner Prism, J. Opt. Soc. Am., Vol. 50, 83, 1960.

2. Tanimura, Y., A New Differential Interferometer with a Multiplied Optical Path Difference, Annals of the CIRP, Vo. 32, 449, 1983.

3. Quenelle, R.C., Nonlinearity in Interferometer Measurements, Hewlett-Packard Journal, P. 10, April 1983.

4. Berthold, J.W., Jacobs, S.F., Norton, M.A., Dimensional Stability of Fused Silica, Invar and Several Ultralow Thermal Expansion Materials, Applied Optics, Vol. 15, 1898, August 1976.

5. Edlen, B., The Refractive Index of Air, Metrologia, Vol. 2, No. 2, 71, 1966.

6. Jones, F., The Refractivity of Air, J. Res. NBS, Vol. 86, 27, 1981.

7. Estler, W.T., High-accuracy Displacement Interferometry in Air, NBS, Washington D.C. 20234, to be published.

LASER HETERODYNE PROBES

H.K. Wickramasinghe

IBM, T.J. Watson Research Center, P.O. Box 218, Yorktown Heights, New York, N.Y. 10598, U.S.A.

ABSTRACT

The technique Laser heterodyne probing goes to the late 1960's when it was first applied to visualise surface acoustic wave field distributions. Since then, the applications have broadened out into a number of other areal such as imaging, optical communications and metrology.

A heterodyne interferometer can be considered as a modified Michaelson interferometer in which the beam splitter element has been replaced by a Bragg cell. This has the effect of introducing an optical carrier frequency, thereby making it relatively simple to electronically filter noise sources such as microphonics and thermal fluctuations.

The talk will concentrate on introducing the basic principles of heterodyne probing in a simplified manner and derive equations which describe the sensibility in terms of systems parameters. Practical aspects of design will be discussed.

Examples of the application of the technique to measurements in the field of surface acoustic waves will be provided, demonstrating the very high sensibilities that can be achieved.

REFERENCES

1) E.G. Lean, "Interaction of Light and Acoustic Surface waves", Progr. Opt., XI, 123-166, (1973)

2) R.L. Whitman and A. Korpel, "Probing of acoustic surface waves by coherent light", Appl. Opt., 8, 1567-1576, (1969)

3) R.L. Whitman, L.J. Lamb and W.J. Bates, "Acoustic Surface Displacement Measurements on a Wedge-Shaped Transducer using an Optical Probe Technique", IEEE Trans., SU-15, 186-9, (1968)

4) R. de La Rue, R.F. Humphryes, I.M. Mason and E.A. Ash,"Acoustic Surface wave amplitude and phase measurements using laser probes", Proc. IEE, 119, nº2, 117-26 (Feb. 1972)

5) W. Puschert, "Optical detection of Amplitude and phase of mechanical displacements in the Angstrom range", Opt. Comm. 10 (4), 357-61, Apnd. 1974

6) H.K. Wickramasinghe and E.A. Ash, "Optical Probing of acoustic surface waves - Application to device diagnostics and to non--destructive testing", Proc. of MRI Symposium on Optical and Acoustic microelectronics, Polytechnic Institute of New York, Apr. 16-18, 1974

7) H.K. Wickramasinghe, Y. Martin, D.A.H. Spear and E.A. Ash, "Optical heterodyne techniques for Photoacoustic and Phototermal Detection", J. De Physique, 44 (10) P.P. C6 - 191-196, October 1983

8) S. Amen, E.A. Ash, U. Htoo, D. Murray and H.K. Wickramasinghe, "Laser Detection and Imaging Techniques for surface examination", Proc.of DARPA conference on Progress in quantitative NDE, San Diego, July 1979, P. 384-391

9) H.K.Wickramasinghe, Y. Martin, S. Ball and E.A. Ash, "Thermo-displacement imaging of current in thin film circuits", Electronics Letters, 5th August 1982, 18 (1&9, pp. 700-701)

SCANNING DIFFERENTIAL PHASE CONTRAST OPTICAL MICROSCOPE APPLICATION TO SURFACE STUDIES AND MICRO METROLOGY

H.K. Wickramasinghe *

IBM, T.J. Watson Research Center / P.O. Box 218,
Yorktown Heights, New York, N.Y. 10598   /   USA

ABSTRACT

The characteristics and theory of operation of a new scanning
differential phase contrast optical microscope are described, and
a number of results are presented.

High contrast micrographs of a polished steel sample are included,
showing clearly the grain boundaries, as well as some fine structure
within the grains. Micrographs are also presented of natural diamonds,
both in polished and unpolished forms. In the former, many polishing
lines are visible and in the latter, one clearly sees a large number
of stacking faults.

Results on the study of monolayers of Langmuir - Blodgett films
are also presented. The micrographs clearly show the boundaries, as
well as non-uniformities within the films. The ability of the System
to image objects showing refractive index variation is demonstrated
by producing micrographs of an exposed but undeveloped photoresist
film, and a partially doped silicon sample. In each case, a qualitative comparison is made with the differential interference (Nomarski)
micrograph of the same field of view.

Since the measured phase gradients can be related directely to the laser wavelength, the technique can be made quantitative. By integrating the detected signal with respect to time, it is possible to obtain a phase profile of the sample surface being scanned, to an accuracy of the order of 10^{-8} radians.

* work performed at University College London, England

References

1. G. Nomarski, Jour. Phys. Radium 16, 9 (1955).
2. H.K. Wickramasinghe, S. Ameri, and C.W. See, Electron Lett. 18, 22 (1982)
3. R,L. Whitman and A Korpel, Appl. Opt. 8, 1567 (1969).
4. R.M. De La Rue, R.F. Humpheryes, I. Mason, and E.A. Ash, Proc. IEE 119 (2), 117 (1972).
5. G.E. Sommargren and B.J. Thompson, Appl. Opt. 12 (9), 2138 (1973).
6. L.J. Laub, Jour. Opt. Soc. Amer. 62, 737 (1977).
7. K.B. Blodgett and I. Langmuir, Phys. Rev. 51, 964 (1973).
8. C.W. Pitt and L.M. Walpita, Electron. Lett. 12, 18 (1976).
9. G.E. Jellison Jr., F.A. Modine, C.W. White, R.F. Wood and R.T.Young Phys. Rev. Lett. 46, 1414 (1981).

LASER VELOCIMETRY

E. O. Schulz-DuBois

Institute of Applied Physics
University of Kiel, F. R. Germany

Abstract. In this lecture, several kinds of instrumentation for laser velocimetry are reviewed. As far as the laser optical marking of distances is concerned, the crossbeam setup and the dual focus setup are discussed. With respect to electronic circuity for the extraction of timing information, the frequency counter, the tracker with phase locked loop, the time to amplitude converter with multichannel analyzer, and the photon correlator are described. The presentation is tutorial and should enable the reader to work on velocimetry in the laboratory and to study original literature.

1. INTRODUCTION

In a stationary and homogenous flow field, the velocity v is given by a distance d and the time t that objects traveling with the fluid need to traverse it,

$$v = d/t . \tag{1}$$

In laser velocimetry, one takes advantage of small particles moving with the fluid; the distance d is marked in space by laser optical means, and the time t between photodetected light scattering signals is measured by electronic means. Several instruments for laser velocimetry differ in the laser optical marking of distance and in the electronic circuity for time measurement. In the following we survey the most popular approaches to both. The emphasis is on the principles involved, not so much on experimental details. Hopefully the student reader doesn't find it difficult

then to get acquainted with the details from the original literature and specialized textbooks [1, 2, 3, 4]. Finally the state of
the art in laser velocimetry is sketched.

2. LASER OPTICAL MARKING OF DISTANCE

2.1 Geometrical principles

Suppose for the moment that the laws of optical diffraction were
not valid so that one could form light beams of any shape. What
light beam geometries would be desirable then for optical velocimetry? Several useful geometries are sketched in Fig. 1. Consider
light beams in the form of two parallel planes A and B as shown in
Fig. 1a. A particle moving with velocity vector $\bar{v} = (v_x, v_y, v_z)$
crosses these planes at times t_A and t_B, giving rise to infinitely
sharp pulses of scattered light at both times, and obviously there
is

$$t_B - t_A = d/v_z . \tag{2}$$

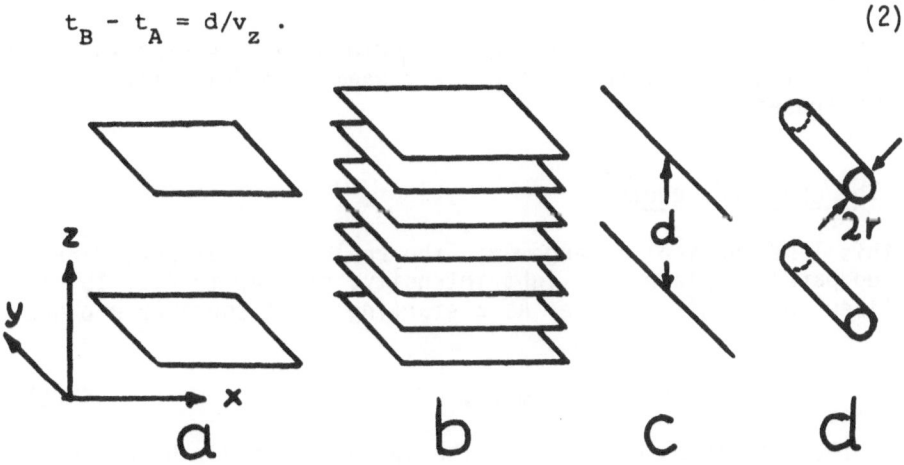

Fig. 1. Geometrical principles of laser velocimetry;
a) two parallel planes of light; b) periodic repetition of
parallel planes; c) two parallel straight lines; d) two
parallel cylinders of finite thickness and length

Note that the in-plane velocity components v_x and v_y do not show
up in the measured time difference. Hence velocimetry based on
this geometry strictly measured a Cartesian velocity component.
The same is true for a periodic repetition of parallel planes as
shown in Fig. 1b. The scattered light signal then would consist of

periodic pulses with repetition frequency f of radian frequency ω

$$\omega = 2\pi f = 2\pi \, v_z/d \; . \tag{3}$$

Next consider two infinitely thin parallel light beams as shown in Fig. 1c. In that case, a particle in principle could cross both beams only if its velocity component $v_x = o$, that is $v = (0, v_y, v_z)$, and the measured time difference is related only to the velocity component v_z as in equation (2). The probability of hitting an infinitely thin beam with a particle is zero. Hence, in order to obtain a finite rate of useful signals, one has to consider cylindrical beams with finite diameter $2r$ as sketched in Fig. 1d. Then, however, the velocity measured is not the Cartesian component v_z but rather $(v_z^2 + v_x^2)^{1/2}$ with the understanding that v_x is small,

$$| \, v_x \, | < 2r \, | \, v_z \, | \, / \, d \; , \tag{4}$$

otherwise a particle cannot pass both beams in succession.

These simplified considerations should make it easier to understand the function of realizable laser beam geometries for velocimetry purposes.

2.2 The crossbeam geometry

In this laser optical arrangement, the realization of periodically spaced parallel planes of light intensity, similar to Fig. 1b, is realized in a finite volume. As a starting point consider a plane wave

$$u_1 = u_o \exp (i \, k_1 \cdot r - i\omega t) \; , \tag{5}$$

where the quantities ω and t and the vectors k_1 and r have their usual meaning. It offers plane phase fronts defined by $k_1 \cdot r = $ const. As is well known, a plane wave is realizable only to an approximation. Laboratory waves usually diverge or converge, hence have spherical wavefronts. If one superposes on the wave u_1 of equation (5) another plane wave u_2 with wave vector k_2 of the same magnitude but different direction as k_1, the resulting field is

$$u_1 + u_2 = 2u_o \cos \left((k_1 - k_2) \cdot r/2 \right) \exp \left[i \, (k_1 + k_2) \cdot r/2 - i\omega t \right] \; . \tag{6}$$

Thus the time averaged intensity

$$| \, u_1 + u_2 \, |^2 = u_o^2 \left[1 + \cos \left((k_1 - k_2) \cdot r \right) \right] \tag{7}$$

follows a sinusoidal distribution of parallel planes with periodic spacing

$$d = 2\pi \, / \mid k_1 - k_2 \mid .$$ (8)

Laser output often is in the shape of a Gaussian beam [5]. Its wavefronts are spherical as long as the beam converges or diverges, but they are parallel planes at the smallest diameter, i. e. the beam waist. Therefore, in order to obtain a light intensity distribution on equally spaced parallel planes, one has to cross two Gaussian laser beams such that their waists overlap as sketched in Fig. 2a. The crossbeam pattern is produced from a single laser beam by a beam splitter prism and a focusing lens as shown in Fig. 2b. For Gaussian beams with 1/e Radius σ near the waist and

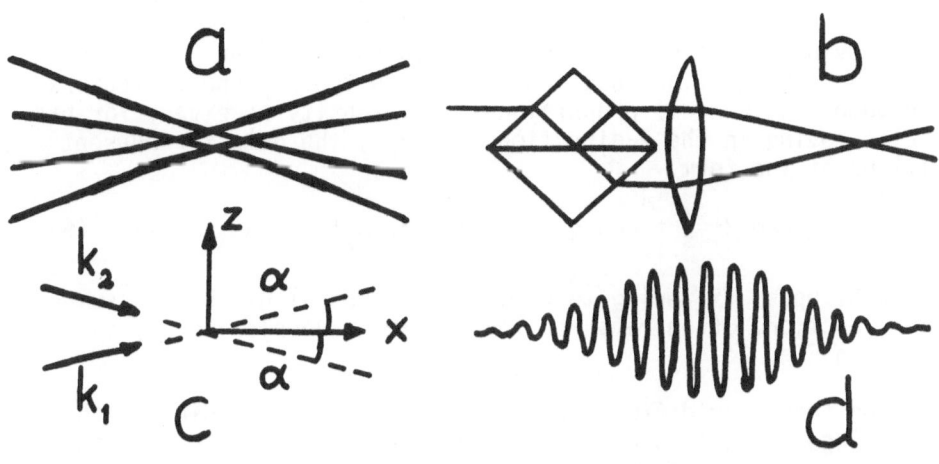

Fig. 2. The crossbeam optical setup; a) crossing of the beams at their waists; b) beam splitter and focussing lens; c) notation and coordinates for crossbeam geometry; d) laser Doppler burst, pedestal-free

the coordinates indicated in Fig. 2c, the intensity distribution is

$$| u_1 + u_2 |^2 = u_o^2 \exp\left[-\left(y^2 + (x \sin \alpha - z \cos \alpha)^2\right)/\sigma^2\right]$$

$$+ u_o^2 \exp\left[-\left(y^2 + (x \sin \alpha + z \cos \alpha)^2\right)/\sigma^2\right] \quad (8)$$

$$+ 2u_o^2 \exp\left(-(y^2+x^2\sin^2\alpha+z^2\cos^2\alpha)/\sigma^2\right) \cos\left(|\bar{k}_1-\bar{k}_2| \cdot z - (\omega_1-\omega_2)t\right).$$

In this formula, the first two terms describe the intensity due to both original beams. In velocimetry, light scattering associated with these terms has a slow time dependence that is usually referred to as a pedestal, and often in electronic processing this contribution is suppressed by highpass filtering. The third term is due to interference, it may add or subtract from the beam intensities depending on the sign of the cosine function; the magnitude of the interference term tapers off in a Gaussian ellipsoidal shape with semiaxes $\sigma/\sin \alpha$, σ, $\sigma/\cos \alpha$, respectively, in the x, y, and z-directions. Since the light intensity scattered by a moving particle is proportional to the incident intensity distribution equation (8), and since with typical photodetectors the output current is proportional to intensity received, for particles moving in the z-direction, $z = v_z t$, the detector current will be of the form

$$i = i_o \exp\left(-v_z^2 t^2 \cos^2\alpha/\sigma^2\right) \cos\left(|k_1-k_2| v_z t - (\omega_1-\omega_2)t\right) , \quad (9)$$

where highpass filtering to remove the pedestal has been assumed. This function, often referred to as a laser Doppler burst, is shown in Fig. 2d.

The frequency difference term in equation (8) indicates that the fringe pattern moves with the difference frequency between beams 1 and 2 if a frequency offset is introduced e. g. by Bragg shift. In this technique, the laser frequency ω_o of both beams is shifted, for example, to $\omega_1 = \omega_o + 2\pi \cdot 40.1$ MHz and $\omega_2 = \omega_o + 2\pi \cdot 40.0$ MHz by first order diffraction from acousto-optically induced phase gratings in two solids. In that case, the fringe pattern moves with 100 kHz, i. e. by one fringe spacing in a time of 10 μs. Bragg shifting offers two advantages: It shifts useful velocimetry signals to higher frequencies and thus permits a better discrimination against the pedestal. This is useful with low velocity measurements. It also permits a distinction of the sign of velocity, that is of direction. To see this, let $z = v_z t$ in the last cosine of equation (8). Then the detected signal will have the frequency $\omega_1-\omega_2 - (k_1-k_2) \cdot v_z$, that is, in our example it will be below 100 kHz for positive v_z and above 100 kHz for negative v_z. This is important in experiments where the velocity may change sign, for example in reciprocating or turbulent flows.

The crossbeam system described so far is also referred to as the real fringe system for the following reason. If one places a screen in the crossover region, a system of alternating bright and dark lines will be visible. However, since their distance usually is very small, for example 5 wavelengths λ for $\alpha = 6^{\circ}$, the fringe system is better observed by placing a microscope objective at the crossover region which projects an enlarged image of the fringe system onto a distant screen. The crossbeam laser velocimetry concept is originally due to H. J. Pfeifer at the Franco-German Institute St. Louis.

An alternative setup is the reference beam system, also referred to as the virtual fringe system. In that case one of the beams, for example k_1 in Fig. 2c, is made relatively weak and is permitted to fall directly onto the photodetector surface where it acts as the local oscillator in optical heterodyning. The other beam, k_2, illuminates moving particle and the resulting scattered light, as far as it originates in the sampling volume and has the direction of k_1, is superposed to the local oscillator on the photodetector. The resulting photodetector current is still of the form equation (9). The reference beam system was first implemented by H. Z. Cummins, then at Johns Hopkins University at Baltimore, in 1964 in the first demonstration of the laser Doppler principle. The most important differences are as follows. The real fringe system works best if there are very few particles in the sampling volume simultaneously, and it permits photodetection with very wide aperture in practically any direction, hence large received intensity. The reference beam system permits operation in cases when there are very many particles in the sampling volume; it requires a small receiving aperture for reasons of phase coherence in optical heterodyning, hence has to operate with small received intensity and permits little freedom in the choice of receiving directions.

2.3 The laser dual focus system

This system may be understood by reference to the idealization Fig. 1d. In reality the required laser light intensity distribution in space is that of two laterally displaced Gaussian waists or foci as shown in Fig. 3a. The solid lines indicate surfaces of constant intensity and the dashed lines delimit the contour of the beam. The essential feature is the high power concentration in two volumes of cigar shape. In principle, light scattering signals arise from particles crossing the Gaussian beams anywhere, but in practice, the light scattering signals from both cigar shaped volumes are so much stronger that those from other regions may be neglected. Typical values for the distance between

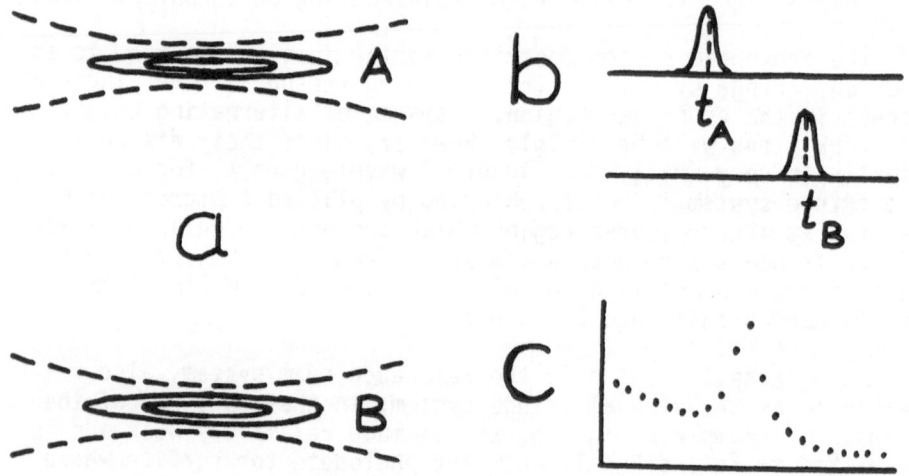

Fig. 3. The dual focus setup; a) lines of constant intensity (solid) and beam contours (broken); b) response of both detectors at the passage of a particle; c) display of contents of multichannel analyzer

both foci and the diameter of their waists are, respectively, 300 μm and 10 μm. Strong focussing of a Gaussian beam is associated with strong convergence before and strong divergence past the waist. This limits the length of the focal region to about 250 μm (this would be the distance between confocal points in a Gaussian beam [5]).

In the transmitting optics, two identical parallel and overlapping laser beams are generated whose axes are displaced laterally by e. g. 300 μm. This can be achieved by passing the original beam through a flat plate of a birefringeant material such as calcite. In this material, the ordinary and extraordinary beam are laterally displaced. Thus both beams have orthogonal polarization. The desired foci are then obtained by lenses which accomplish beam expansion first and then focussing. If it is undesirable to have two foci of orthogonal polarization, then a quarter wave plate may be used which will transform the two linear orthogonal polarizations both into circular polarization.

Several designs of receiving optics are possible and in practical use. Best performance in signal processing is obtained if separate receiving optics are provided for light scattered out of both foci. For a single particle passing both foci the electrical signals at the photodetectors then may look like Fig. 3b. Ideally they consist of Gaussian pulses centered at times t_A and t_B, and the object of signal processing electronics is to get a statistically significant mean and variance of the time difference. The optical realization of two separate receiving paths is simple, at least in principle. One makes use of reciprocity, that is, the same optics that in transmission lead to the separate foci,will in the reverse direction transmit scattered light out of the focal regions back into the transmitting optics. This return light has the same modal structure as the outgoing laser light so that it may be separated from the latter by semi-transparent mirrors and eventually is focussed onto two separate detectors.

A useful advantage of dual focus velocimetry is its insensitivity to flare, that is, to specular reflexion from metallic surfaces. It has been reported that velocimetry was possible at positions just one millimeter in front of flat metal blades in turbomachinery. This insensitivity is a consequence of the strong divergence of the laser beam beyond the focus. As a result the amount of reflected light entering the receiving path is quite small so that is doesn't make the measurement of particle transits impossible. The dual focus system, also called the laser transit system, was pioneered by R. Schodl at the DFVLR laboratories at Köln-Porz, F. R. Germany.

3. ELECTRONIC TIMING CIRCUITY

The velocimetry systems naturally fall into two categories: those that measure transit time (see equation (2)) in conjunction with the dual focus setup, and those that measure the so-called Doppler frequency (see equation (3)) in conjunction with the crossbeam geometry. Among frequency measuring systems, the spectrum analyzer is not discussed in detail; since there the local oscillator is swept, received signals are utilized only during the short time where both frequencies match. Thus this system is wasteful in terms of information, and it should be used only if nothing better is available.

3.1 Time-to-amplitude conversion and multichannel analyzer

In nuclear physics experiments, the time-to-amplitude converter (TAC) is a well-known device which is available in the form of an integrated circuit. At time t_A the charging of a capacitor is started and at time t_B it is stopped. The resulting voltage is proportional to time and it is sampled and converted into a digital signal by further electronic circuits. As time goes on, a series of digital numbers is generated in this way and fed to a multichannel analyzer.In this device,the frequency of occurrence of every digital number is noted;a so-called channel,that is a store, is available for every digital number and its content is increased by one unit for every occurrence of that number. In this way a probability distribution for transit times is obtained as in Fig. 3c. From the position of the peak the average velocity and from the width of the peak the variance of velocity may be obtained, the latter only with difficulty unless the peak is much broader than the so-called instrumental width.

As illustrated in Fig. 3c, the multichannel analyzer shows some background signals besides the desired peak. This background is due to events where one particle triggers the starting pulse at t_A, but another particle the stopping pulse at t_B. Since both particles are unrelated, the would-be transit times obtained in this way are statistically distributed, with short times being more likely than long ones. In practical measurements, this background may be significant. It then can be difficult to identify a peak on top of it.

3.2 Frequency measurement by the counter

A popular way of determining the frequency in a Doppler burst (see Fig. 2d) is by measuring the time between zero crossings. This works well only in favorable circumstances. The photodetector signal has to have a large signal-to-noise ratio, it should consist of isolated bursts coming from a low concentration of scattering particles, it should be free from pedestal by highpass filtering, and it should contain a good number of identifiable cycles. The counter is activated by a threshold device (a Schmitt trigger) as soon as the amplitude of the Doppler burst exceeds a given amplitude for the first time. It then counts the time between zero crossings for a number of cycles. A popular approach is to have two counters; both are started simultaneously, one measures the time needed for 8 cycles, the other for 5 cycles. Obviously these times should be in the ratio of 1.6. A logic circuit in the counter examines that ratio and admits only those measurements as valid where that ratio is found to within a given error margin of, for example, 10 per cent. The margin should be adjusted in accordance with the signal-to-noise ratio of the data and with gradients and turbulence of the flow. The output of counters usually consists of an analog voltage which is proportional to velocity. The voltage due to a measurement is kept constant at the output until the next valid measurement is made.

3.3 Frequency measurement by the tracker [6]

This device is similar to circuits used for demodulation of frequency modulation. The tracker contains a built-in oscillator which oscillates with constant amplitude but a frequency which is proportional to an applied voltage;it is called a voltage controlled oscillator (VOC).During operation it is the aim to have the oscillation of the VOC synchronized with an input Doppler burst. This refers not only to frequency,but also to a constant phase difference of $90°$.The synchronization is achieved by a feedback circuit known as second-order phase locked loop (PLL).In the PLL circuit,the voltage and the Doppler burst (after bandpass filtering and amplification) are multiplied.Assume that locking has been achieved; Then Doppler burst and VOC voltage are in phase quadrature.The product,averaged over one half cycle,is zero.If locking is not effective,a difference phase exists between both voltages.The product,again averaged over one half cycle,is a finite d.c. voltage. It is given to an integrator and then as control voltage to the VOC.The initial frequency and phase error is quickly reduced to zero and the feedback system maintains the input Doppler burst and the VOC voltage at the same frequency and phase quadrature.

The practical usefulness of a tracker depends on a number of facts. Obviously the signal-to-noise ratio of the input Doppler signal is important.If the input signal momentarily is too small for locking, the VOC output stays constant in frequency except that a small drift is unavoidable in practice.In that case an invalidity signal is derived by an additional circuit [6];it indicates that the output data at that time have no basis on a measured Doppler burst. If one works with turbulent flows,the measured velocity may change rather rapidly.To tune rapidly from one frequency to another,the PLL tracker needs a high slew rate.In practice this is not a significant limitation since slew rates of the order of 100 khz/ms can be realized.

Two outputs of the tracker may be used [6]. One is the control voltage of the VOC. To the extent that the VOC voltage-to-frequency characteristic is linear, it is directly proportional to the instantaneous velocity. In the case of a Bragg shift, it is proportional to velocity plus the fixed amount $d(\omega_1-\omega_2)/2\pi$. The other output is the VOC output. It is an alternating voltage of constant amplitude, but of frequency and phase equal to that of the input Doppler signals. This output is well suited for forming long-term averages or rate correlation signals [7].

3.4 Photon correlation for dual focus and crossbeam velocimetry

The subject of photon correlation is more fully treated in another lecture of this school and in the literature [3, 4]. It suffices here, therefore, to mention those aspects that are of most concern to applications in velocimetry.

In both optical setups, the photodetection by photomultipliers is run at such low light levels that the electrical signal consists of separate pulses, each one signifying the detection of a single photon. The pulses are standardized to a given amplitude and duration and then fed to a correlator. That device is a single purpose digital parallel processor which computes in real time the photon correlation function. Roughly speaking, this function is an approximation to the intensity correlation function of received light. In a crossbeam setup, the autocorrelation function of received light may look like Fig. 4a in a stationary laminar flow; the Gaussian roll-off of the curve is due to the Gaussian cross section of laser beams and sampling volume. The velocity

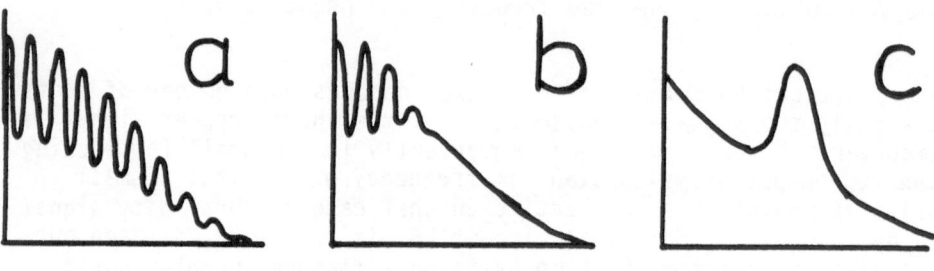

Fig. 4. Correlogram; a) crossbeam setup, laminar flow; b) crossbeam setup, turbulent flow; c) dual focus setup, cross correlation function

information is contained in the apparent modulation frequency of the correlogram. It may be obtained by Fourier transformation. For a turbulent flow, the modulation of the correlogram tapers off sooner as shown in Fig. 4b. In that case, the evaluation may yield an average value of the velocity and its variance, that is the degree of turbulence. With the dual focus system, the cross correlation function should be formed, it looks like Fig. 4c. From such a correlogram, a rough value of the transit time is read off easily, and a better estimate may be obtained by a fitting procedure which also may give the degree of turbulence.

In terms of practical usefulness, the correlator offers disadvantages and advantages. Among the disadvantages are the high price of the correlator itself and of photomultipliers, similarly the fact that the primary output after some measurement time

is a correlogram only which then must be evaluated before one
can have velocity and turbulence data. This is outweighed by several advantages. In many cases a smaller, cheaper laser suffices
as light cource. Photon correlation is able to work with a bare
minimum of received light. The utilization of detected photons
is optimal and "democratic", that is all photons enter the
final answer with the same statistical weight. Consequently there
is a well developed statistical theory of signal processing in
photon correlation so that one give results and error bounds with
a considerable degree of confidence. Finally there is a trade-off
between accuracy and measurement time: In one extreme, one may
get a velocity value after the passage of a single particle
which may take as little as one microsecond. The resulting velocity value has a large statistical uncertainty of maybe 10 per
cent. In the other extreme, one may measure over a period of
several seconds or minutes during which the correlator accumulates data. Then the remaining statistical errors are very small
and, given enough care in other details, velocity values may be
as accurate as 10^{-3} or 10^{-4}.

4. STATE OF THE ART

Laser velocimetry is a nonintrusive method of measuring fluid
velocities. It is thus suitable for hostile environments where
material probes cannot endure, for example in flames, explosions,
combustion,and acids.The distance between the optics and the sampling volume may vary between 1 km for telescopic systems used in
atmospheric research and 1 μm in a glass fiber system capable of
measuring the flow inside capillary blood vessels. Similarly the
range of velocities is enormous, from about 1 km/s in supersonic
flow as it occurs in wind tunnels designed for aircraft development or inside jet engines, to 100 nm/s for electrophoretic motion in emulsions or colloids and similar velocities for the
fluid inside living cells.

   This enormous range of possibilities has the consequence that
very few tasks in velocimetry may be solved by off-the-shelf
equipment. In most cases a careful consideration of the problem
and all pertinent boundary conditions is necessary. Not mentioned
in this lecture were the natural abundance or the necessity of
seeding light scattering particles; the problem of measurement
accuracy; or finally financial aspects in the choice of equipment. All these considerations make it likely that laser velocimetry will remain a field which offers interesting scientific
problems and employment opportunities for young scientists in
the future.

References

1. Durrani, T. S. and Greated, C. A., Laser systems in flow measurement, Plenum Press 1977
2. Durst, F., Melling, A., and Whitelaw, J. H., Principles and practice of laser Doppler anemometry, Academic Press 1976
3. Cummins, H. Z., and Pike, E. R., eds., Photon correlation spectroscopy and velocimetry, Plenum Press 1977
4. Schulz-DuBois, E. O., ed., Photon correlation techniques in fluid mechanics, Springer Verlag 1983
5. Boyd, G. D., and Gordon, J. P., Bell Syst. Tech. J. 40, 489 (1961)
6. Fedders, B. S., and Köneke, A., J. Phys. E: Sci. Instrum. 12, 766 (1979)
7. Schätzel, K., Optical Acta 27, 45 (1980)

VELOCITY MEASUREMENTS IN A ROTATING TANK USING AN LDV SYSTEM

M.S.Kılıçkaya and N.Ekem

Physics Department
Anatolian University
Eskişehir, Turkey

Abstract
Velocity measurements in a rotating tank which contains tap water have been accomplished by means of a Laser Doppler Velocimeter (LDV) system. The technique adopted the dual beam, or fringe method, observing the Doppler frequency shift caused by moving fluid particles in the probe volume.

The measurement system proved to be superior compared with conventional methods of velocity measurement in fluid dynamics, namely, the Pitot tube and the hot-wire anemometer.

1. Introduction
The conventional methods used to measure fluid velocities were mainly the Pitot tube and the hot-wire anemometer. However, there are some disadvantages when using these two devices. The Pitot tube measures the dynamic pressure of the fluid, which is proportional to the square of the velocity. Furthermore, the Pitot tube contains liquids, therefore it has a certain inertia, and cannot easily follow the fluctuations which occur in the velocity. If a Pitot tube were immersed in a rotating tank which contains water, the frictional and drag forces would be of the same order. This would lead to a drastic change in the flow. On the other hand, as soon as the Pitot tube is immersed in the tank, its body creates a wake behind it. The problem is that this wake may form a circular loop and the whole flow field may completely be altered.

The hot-wire anemometer system functions according to the rate of heat transfer between the hot-wire and the fluid medium. The hot-wire, heated by an electrical current, consists tungsten wire of 0.005 mm diameter and of 1-3 mm length. It is very sensitive to velocity fluctuations and is used essentially in turbulent flow analysis. The hot-wire anemometer has two common modes of operation, constant current and constant temperature. However, this method has many drawbacks which are listed below:

(i) it is necessary to calibrate the hot-wires before an experiment. This calibration procedure would be very time consuming and it can easily change during the experiment due to contamination,

(ii) the current-velocity characteristic of the hot-wire is non-linear and sharply increases at higher velocities (King's law),

(iii) the hot-wire would most probably be fragile during the experiment due to its fine structure.

Furthermore, the hot-wire anemometer system cannot give reliable results because it shows some ambiguity in current direction and it delivers heat to its holder by thermal conduction. For these reasons an absolute linear and non-contact fluid velocity measuring system is required. Such a system is the Laser Doppler Velocimeter (LDV). The first people to measure the velocity of a fluid by using the Doppler shift of laser light were Yeh and Cummins in 1964 [1].

2. Theory

The laser is a monochromatic and coherent source of light. The Laser Doppler Velocimeter (LDV) is based on the shift of the frequency of the light scattered from a moving particle to an observer. The fundamental mechanism of an LDV system is the "heterodyning" process. It consists of the mixing of a known reference light with the Doppler shifted radiation [2].

With an LDV system it is possible to measure velocities in media where scattering centers are present, in gases, a small addition of aerosols or fine solid particles will provide the required scattering.

Let us consider a particle moving with a velocity $\vec{v}$, illuminated by a laser beam. The laser light is of frequency f_o and wavelength λ_o (Fig.1). Let $\vec{\ell}_i$ be the unit vector corresponding to the direction of propagation of the illuminating beam.

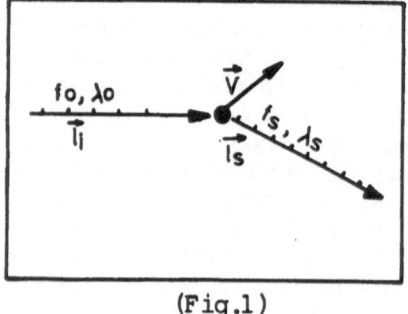

(Fig.1)

The light will be scattered in all directions. Let us consider the light scattered along a unit vector $\vec{\ell}_s$. It will be of frequency f_s and wavelength λ_s. The Doppler shift is:

$$f_D = f_o - f_s . \tag{1}$$

The relative speed between the wavefronts of the illuminating beam and the particle is: $c - \vec{v}.\vec{\ell}_i$ where $c = \lambda_o f_o$ is the speed of light in free space. The rate at which the wavefronts are intercepted by the moving particle is

$$f' = (c - \vec{v}.\vec{\ell}_i)/\lambda_o = f_o - (\vec{v}.\vec{\ell}_i)/\lambda_o . \tag{2}$$

The observer receiving only scattered light along $\vec{\ell}_s$ will intercept the wavefronts at a rate of:

$$f_s = (c + \vec{v}.\vec{\ell}_s)/\lambda_s = f' + (\vec{v}.\vec{\ell}_s)/\lambda_s . \tag{3}$$

Substituting the value of f' into Eq.(3) one finds:

$$f_s = f_o + \vec{v}\left(\frac{\vec{\ell}_s}{\lambda_s} - \frac{\vec{\ell}_i}{\lambda_o}\right) , \tag{4}$$

noting that $\lambda_s = \lambda_o$, Eq.(3) becomes,

$$f_s = f_o + \frac{\vec{v}}{\lambda_o} (\vec{\ell}_s - \vec{\ell}_i) . \tag{5}$$

Thus, the net change in frequency or Doppler shift, as viewed by a stationary observer is due to the relative speed of the particle and the orientation of the particle trajectory with respect to both the illuminating direction $\vec{\ell}_i$ and the scattered light direction $\vec{\ell}_s$. Eq.(5) reads,

$$f_D = f_s - f_o = \frac{\vec{v}}{\lambda_o} (\vec{\ell}_s - \vec{\ell}_i) . \tag{6}$$

There are three fundamental types of LDV systems, reference beam, dual beam and single beam. Of these we have chosen the dual beam, or "fringe" method, since it allows higher signal intensities, because the collection of light occurs in a wide solid angle.

Now let us consider the dual beam system in which two beams of equal intensity are illuminating the particles (Fig.2).

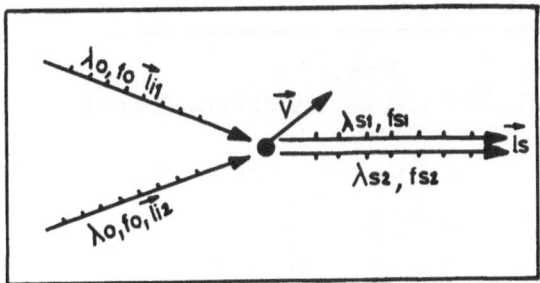

(Fig.2)

Each beam is scattered, resulting in two beams, (λ_{s_1}, f_{s_1}) and (λ_{s_2}, f_{s_2}), scattered in $\vec{\ell}_s$ direction. Using the same arguments as above we have,

$$f_{s_1} = f_o + \frac{\vec{v}}{\lambda_o} (\vec{\ell}_s - \vec{\ell}_{i_1}) , \tag{7}$$

$$f_{s_2} = f_o + \frac{\vec{v}}{\lambda_o} (\vec{\ell}_s - \vec{\ell}_{i_2}) , \tag{8}$$

$$f_D = f_{s_2} - f_{s_1} = \frac{\vec{v}}{\lambda_o} (\vec{\ell}_{i_1} - \vec{\ell}_{i_2}) . \tag{9}$$

Eq.(9) implies that the detected frequency is no longer dependent on the direction of scattering. The detected intensity at the surface of the photodetector will be much higher. On the other hand, for the same illuminating beam angle one will have lower frequencies to detect. This will greatly simplify the data acquisition system [3].

Let us now consider that the dual beam method can also be analyzed in another way, namely the "fringe" method. Assume that the two illuminating beams propagate in the absence of particles. When two laser beams of equal intensity are crossing, there is a formation of interference fringes in the intersection region (Fig.3).

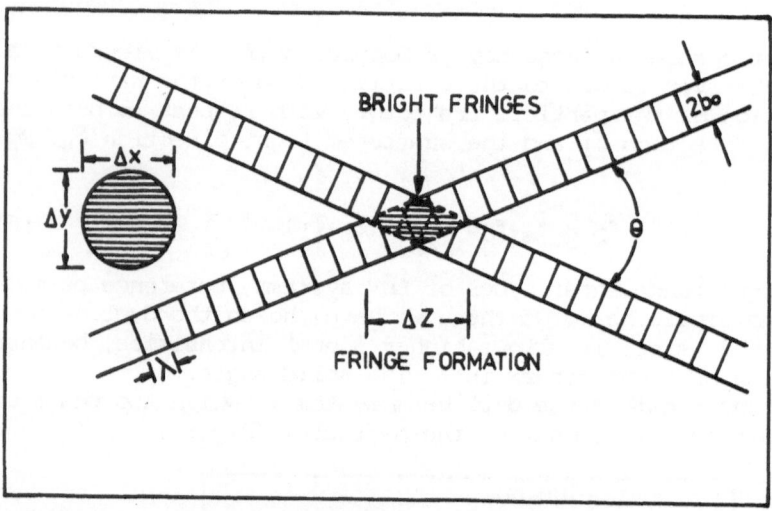

(Fig.3)

Referring to (Fig.3), one can easily obtain from the geometry of the system that the sampling volume:

$$\Delta x = 2b_o,$$

$$\left.\begin{array}{l} \Delta y = 2b_o/\cos\dfrac{\Theta}{2}, \\[2mm] \Delta z = 2b_o/\sin\dfrac{\Theta}{2}, \end{array}\right\} \tag{10}$$

The fringe spacing is given by,

$$d_F = \lambda_o/2n\sin\frac{\Theta}{2}, \tag{11}$$

where n is the refractive index of the medium under consideration.

The optical set-up may now be regarded as an interferometer. Because the frequency of the light is the same in the two beams, these fringes are stationary. If a solid particle passes through this region it will cross successively bright and dark fringes. Light will be scattered only when a bright fringe is crossed. A stationary observer will then "see" the particle only when it is illuminated, that is, when it is in a bright fringe. The rate at which the particle is crossing the fringes will thus be proportional to its velocity [4].

Actually, the laser beam has a Gaussian intensity distribution. This results in a fringe pattern which is also representing a Gaussian intensity distribution. The optical set-up required for the fringe mode is again an illuminating optic and a receiving optic. The illuminating optic produces two parallel beams which are converged to a common point by a lens (Fig.4).

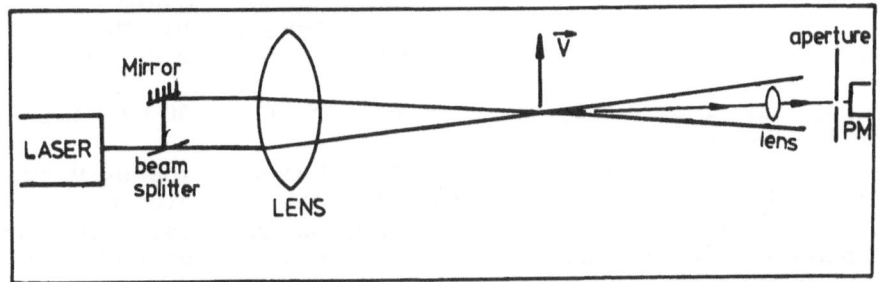

(Fig.4)

The receiving optic is essentially made of a lens which forms an image of the crossing point at the surface of a photosensitive detector. An aperture controls the spatial resolution of the system, eliminating all the light scattered from particles passing out of the crossing region. This is the region where interference fringes are formed and it is called the probe or sampling volume [5].

The signal, converted by a photodetector, can be displayed on a scope. Such a signal is made of an avarage D.C. signal and an A.C. modulation (Fig.5).

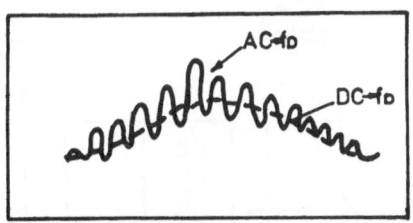

(Fig.5)

The measured velocity component is then given by,

$$v = f_D \cdot d_F = f_D \cdot \frac{\lambda_o}{2n\sin\frac{\Theta}{2}} \qquad (12)$$

3. Experiments

A block diagram of the experimental set-up is shown in (Fig.6). The laser source used was a Spectra Physics, Model-124 A, 15 mW He-Ne gas laser. The focus length of the first lens was 200 mm. The distance between the two lenses was 490 mm. The dimensions of the cylindrical glass tank were 103.5 mm. in outer diameter, a thickness of 3.55 mm and a height of 150 mm. The tank was completely transparent to light.

The experimental procedure was carried out in the following order:

(i) when the laser beam was on, the cylindrical tank was to rotated. The angular speed of the tank was adjusted with a variable resistor and its magnitude was determined by means of a stroboscope. The operating range was 190-203 rpm.

(ii) The scattered light beam was focused into the photomultiplier.

(iii) The signal from the photomultiplier was fed into a high pass filter which removed the pedestral frequency so that the signal that was fed into the counter was symmetrical. Then the signal was fed into a low pass filter which was used to supress part of the noise. A high pass filter of 1 MHz, and a low pass filter of 30 kHz, were used during the experiment.

(iv) The output of the low pass filter served as the input to the Doppler Data Processor (DDP) unit. This unit was a counter type frequency meter, with a variable gain amplifier as the first stage. The second stage of the unit was a fast comparator with a variable switch level which could be adjusted to separate the signal from the noise. Only the signals with an amplitude larger than the preset level were analyzed by the processor.

(v) The output of the processor was connected to an x-t recorder from which one can obtain the graph of the velocity with respect to time. This was actually the Doppler shifted frequency.

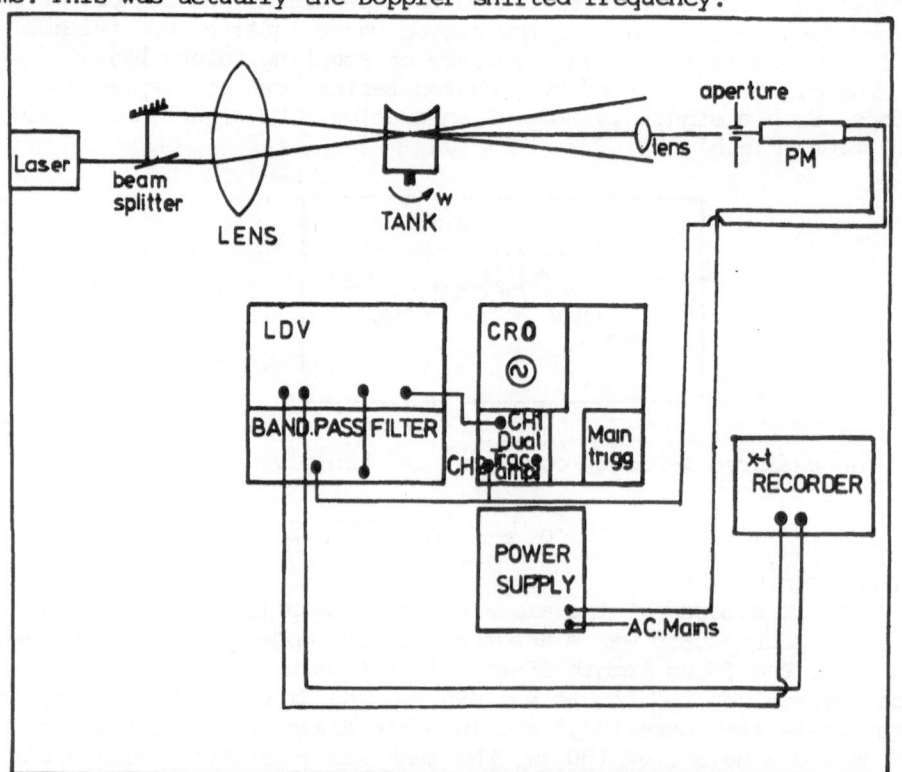

(Fig.6) Block diagram of the experimental set-up

A sample of data is shown in Table-1. The f_D values shown in the third column were obtained from the graph mentioned previously. The d_F values were calculated using a programmed calculator in conjunction with the related formula. The theoretical velocity in the last column was $v=\omega r$. The experimental velocities were obtained from Eq.(12). The results are represented in a linear graph (Fig.7).

4. Discussion

This experiment was accomplished using an LDV system, which is considered to be a new tool for velocity measurements in fluid dynamics, after the invention of the laser in 1960. The difference between experimental and theoretical results was mainly due to the special problems that an LDV system exhibits during the experiment. Since the system is a combination of optical, electronic and mechanical instruments, it has inherent sources of errors, which only an expert in this field could minimize. However, we have cited the main sources of errors, in an LDV system, which have been encountered many times in our experiments:

(i) Photomultiplier (PM); the signal to noise ratio (S/N) was maintained at the minimum value. (S/N) for the signal, at the output from the detector, determined by

(1) the incident signal power,

(2) inherent photodetector noise,

(3) pre-amplifier noise.

However, in our system it seemed that (S/N) increased with the current gain factor, at first very rapidly, and then slowly, until any further increase was limited by the saturation current.

(ii) Change in heterodyning efficiency or the coherence loss factor altered the results effectively.

(iii) Change in source coherent length.

(iv) Scattering from random scatterers; this effect should always be taken into account since it encompasses the fundamental optical scattering parameters of velocity distribution inside the scattering volume, the scattering angle, the amplitudes of the reference and the scattered beams, and the source linewidth or coherence length.

Actual detection volume, or probe volume, is not constant in time because of scatter center distribution. Determination of actual detection volume of, an LDV system, is more than a mere exercise in geometry. It involves a complex interplay between optics, electronics, and physical conditions [6].

(v) Transmission path coherence; the scattered and reference laser beams will potentially suffer from some degradation during their passages through the atmosphere and thus affecting their ability to produce coherent heterodyne signals at the photomultiplier mixer.

(vi) Alignment problems;

(1) angular misalignments between scattered and reference beams,

(2) angular misalignments between the light beam and the PM tube,

(3) transverse alignment between scattered and reference beams.

(4) Alignment in depth (focusing).

(5) Selection of beam spot sizes.

In order to reduce angular alignment problems we must use small beam spots at the photomultiplier tube, to reduce transverse alignment problems we must use large spots. Hence, the final choice must be based on a design judgement supplemented by operating experience.

(vii) Spectral broadening of scattered light;

(1) macroscopic processes: flow fluctuations and instabilities,

(2) microscopic processes: temperature broadening i.e. Doppler broadening due to relative motion of the particles and the observer. This effect was negligible because the massive size of the particles from which the Doppler shift measurements were made,

(3) instrumental broadening: finite aperture and finite scattering volume.

5. Results

According to the theoretical calculations that we have made, the velocity changes as a function of radial distance in the tank i.e., $v=f(r)$. The experimental data and a graph showing the results are provided in Table-1 and (Fig.7), respectively.

A good correlation was obtained between the theoretical and experimental results. However, it would be appropriate to state that many more points were not taken inside the tank due to technical difficulties. So, this meant that it was very difficult to reach the crossing point of laser beams very near the center. Also, it was not successful to make measurements very near the curved wall of the tank, because of the mechanical vibration present in the system. It was difficult to reduce these vibrational effects using the available means. It was decided that it was a challenging problem in such a rotating system. In the experiments that were conducted it seemed that a systematic error, due to the vibration of the tank, occured when it was rotating.

However, it is possible to say that, in spite of its complexity, an LDV system is a promising instrument for the measurement of velocities in various systems in industrial fluid dynamics.

Acknowledgements.
The support by the following organizations and individuals is gratefully acknowledged: Belgian Ministry of Education, General and Environmental Fluid Dynamics Department of the von Karman Institute, Prof.M.L.Reithmueller and present NATO/ASI "Optical Metrology" director Prof.Dr.O.D.D.Soares of the University of Porto, Portugal, whom whose kind help and encouragement motivated us to publish our research work.

References
1. Watrasiewicz, B.M. and Rudd, M.J., Laser Doppler Measurements (London: Butterworths, 1976) p.3.
2. Reithmueller, M.L., Laser Doppler Velocimetry (Brussels: von Karman Institute LS 73, 1975).
3. Trolinger, J.D., Laser Instrumentation for Flow Field Diagnostics (AGARD publ.no.186, 1974).
4. Durst,F., Melling,A., and Whitelaw,J.H., Principles and Practise of Laser Doppler Anemometry (New York: Academic Press, 1976).
5. Rolfe,E., et.al., LDV Instrument (NASA Report,No.CR-1199, 1968).

6. Lennert,A.E., et.al., Laser Technology in Aerodynamic Measurements
   (AGARD LS 49, 1972).

Table-1

Position of beam crossing point(mm)	Angular speed(ω) (rad/s)	Doppler frequency (f_D)(MHz)	Fringe spacing (d_F) (mm).10^{-3}	Velocity(m/s) Theoretical	Velocity(m/s) Experiment
43.79	20.42	0.62	1.367	0.894	0.847
43.02	21.25	0.62	1.377	0.914	0.854
42.24	19.89	0.61	1.387	0.840	0.846
41.45	20.73	0.61	1.397	0.859	0.852
40.64	20.73	0.59	1.408	0.842	0.830
39.83	20.10	0.58	1.418	0.800	0.822
39.00	19.79	0.56	1.428	0.771	0.800
38.15	20.62	0.55	1.439	0.786	0.792
37.30	20.73	0.55	1.450	0.773	0.797
36.43	19.89	0.50	1.460	0.724	0.730
35.55	21.04	0.48	1.471	0.747	0.706
34.65	20.83	0.45	1.483	0.721	0.667
33.74	20.62	0.45	1.494	0.695	0.672

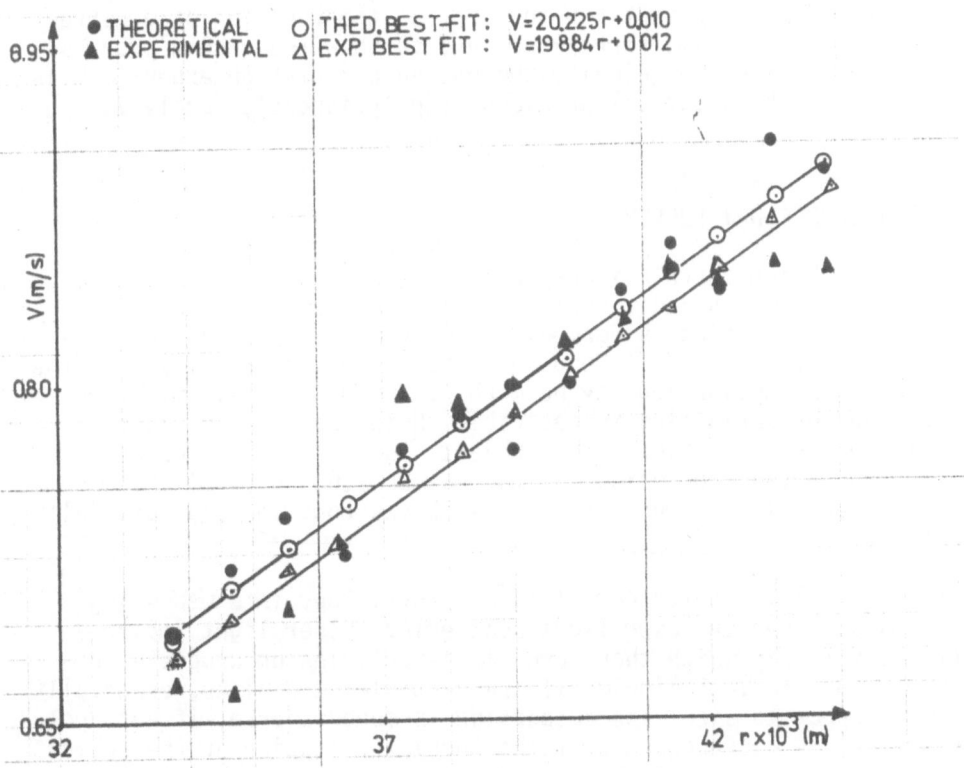

(Fig.7) The graph of the radial velocity.

SIZING OF MICROSCOPIC PARTICLES BY PHOTON CORRELATION

E. O. Schulz-DuBois

Institute of Applied Physics
University of Kiel, F. R. Germany

Abstract. In this lecture, the field of photon correlation spectroscopy is first surveyed in a tutorial fashion. Then the basic theorems by Wiener and Khintchine, by Siegert and by Mandel are reviewed. Some care is taken to describe the function of electronic digital real-time correlators and to show the recent progress in correlator design. Finally the photon correlation experiment is treated and the practical problems of polydispersity analysis are mentioned.

1. INTRODUCTORY SURVEY

The term photon correlation spectroscopy apparently refers to a technique in which individual photons are detected, in which a correlation function is formed, and where some spectroscopic information is obtained. To people without previous knowledge of the field, this explanation may be more confusing than illuminating, and a better definition is desirable. Note that no introductory text book is available on the subject. The interested reader is referred to the proceedings of two summer schools [1,2] and a recent topical conference [3]. These books contain some tutorial material.

Basically, photon correlation spectroscopy is a measuring technique based on laser light scattering. Laser light impinges on a cuvette in which there are small particles or droplets suspended in a solvent liquid. If the particles are alike, especially with respect to size, the suspension is monodisperse, if not, polydisperse. As a consequence of the thermal agitation of the solvent molecules with mean kinetic energy 1/2 kT per degree of freedom,

the particles undergo the so-called Brownian motion. As shown by
Einstein in his celebrated 1905 paper [4], Brownian motion is a
diffusion process; accordingly the mean square displacement of a
Cartesian coordinate x of a particle during the time τ is

$$< \left(x(t) - x(t-\tau) \right)^2 > = 2 \, D\tau \tag{1}$$

where the diffusion constant D is given by the Stokes-Einstein
formula

$$D = kT \, /6\pi \, \eta \, r \; ; \tag{2}$$

here k is Boltzmann's constant, T absolute temperature, η the solvent viscosity, r the particle radius, and the brackets < > indicate either ensemble or temporal averaging, they are equivalent
since the system is ergodic.

Light scattered from a sizeable number of particles is collected on the cathode surface of a photomultiplier tube. The aperture
of the receiving optics is relatively narrow so that the photodetection is coherent. This term may be explained as follows. Light
scattered from a single particle approaches the photocathode in
the form of a nearly plane wave. The same is true for the light
from another particle, but both wavefronts differ in their propagation direction. The detection is called coherent if the propagation directions from all scattering particles differ so little
that between every two incoming waves the optical phase difference
is constant over the detector surface with deviations small compared to $\pi/2$. If that is the case, the optical electric field is
the same everywhere on the photodetector surface. Then the probability of detecting a photon is proportional to the absolute
square of the sum of these field amplitudes. This is not so with
incoherent detection. In that case, the directions of incoming
waves differ so much that phase differences of 2π or more are observed over the area of the ohotodetector. The expression for intensity contains the squares of the individual field amplitudes
and cross terms. But, since the cross terms contain phase angles
which vary considerably over the area, they average out to zero.
Therefore, in incoherent detection, the probability of detecting
a photon is proportional to the sum of the squares of the individual field amplitudes, that is, to the sum of individual intensities.

Now, since the photo detection probability and hence the rate
of detected photoelectrons is controled by the sum of field amplitudes at the detector, that rate will change if the scattered
light contributions due to individual particles vary in their relative phases. The phase of scattered light depends on the location of the scattering particle. In fact, the phase is proportional to the particle coordinate in a coordinate system defined by

the incoming and the scattered light directions. Considerable
phase changes occur, if a particle moves by about a wavelength of
light. As a result, a formula similar to equation (1) can be given
for the meansquare phase shift. The relative phase shifts lead to
fluctuations in the detected intensity. In particular, the time
scale of the fluctuations is determined by the diffusion process
of the particles. The aim of the measuring technique is to find
the time scale of intensity fluctuations and to infer the diffusion constant and particle radius from it.

In spectroscopic terms, the fluctuating scattered light is
characterized by a linewidth increase compared to the incoming
laser light. The increase is very small, however, typically around
a kHz, so that it cannot be resolved by conventional spectroscopic
means such as a grating or interferometer. The small linewidth
corresponds to slow fluctuations of the intensity, typically on a
timescale of milliseconds. It should be possible, therefore, to
extract the linewidth information by an analysis of measured time
series of fluctuating intensity. The required theoretical tool for
the analysis is the correlation function. It is popular in the
description of stochastic signals such as turbulence or electrical
noise, provided these are stationary processes. It is assumed that
the reader is familiar with the mathematical properties of correlation functions so that it suffices, in the description of correlators below, to indicate how experimental estimators of correlation functions are formed on a discrete time basis by means of
digital electronics.

As it turns out, for a monodisperse suspension of particles
the scattered light photon correlation function is an exponential
function whose time constant is inversely proportional to the diffusion constant D. In the frequency domain, the scattered light
shows a Lorentzian line broadening with a linewidth proportional
to D. Thus a measurement of the time constant or of frequency
broadening will yield the diffusion constant, and from it, the
particle radius. For a polydisperse suspension, the correlation
function is a superposition of exponentials, each with the appropriate time constant and a weighting factor proportional to the
light scattering by the particular fraction. In principle a measured correlation function may then be split into a number of simple exponentials, if is assumed that the true distribution of particle sizes may be approximated by a histogram of concentrations
and sizes. In practice, the number of steps in the histogram is
quite limited. This is so because, roughly speaking, exponential
functions with time constants in the same range look very similar
so that it is difficult to tell them apart. In detail this depends
on the quality of the experimental data, that is on their signal-
to-noise ratio. The best ways of extracting polydispersity data
from correlation measurements are discussed controversially in the

literature [5] and more experience is needed to come to a definitive conclusion.

From a practical point of view, photon correlation spectroscopy is a non-intrusive method of measuring a particle size or, within limits, a size distribution. Thus it finds ready application in chemistry, pharmacy, biology, medicine, wherever the sizes of suspended particles or droplets are important. Note that more than half of the products of chemical industry, virtually all pharmaceuticals, all paints and beverages are colloidal suspensions or emulsions or at least have been so at some stage during fabrication. For quality control during production, continued checking of particle sizes is desirable, and in most cases, limited resolution may be sufficient for the purpose. Instruments of better resolution are required in research laboratories of the disciplines mentioned. An intriguing possibility is to study the growth of particles in vitro, for example when nutrients or chemicals are added. It may be estimated that at present there are nearly 100 laboratories or industrial plants in Western Europe which own photon correlation equipment, and the number is increasing. Also a diversification of the available instrumentation seems to be on the way to fit different requirements. By the same token, there is room for young scientists who are able to work with the method, who can adapt it to special circumstances, and who invent improvements of these optical correlation techniques.

In the remainder of this lecture, the physics of photon correlation spectroscopy is discussed in some detail. In a simplified fashion three important theorems are derived which are basic to this experimental technique. This seems necessary because they are not easily found in the textbook literature. Then the basic layout and function of digital correlators is explained. Some care is taken to outline the features, advantages and disadvantages of several designs. And finally the use of these techniques in photon correlation spectroscopy is discussed.

2.   BASIC RELATIONS OF PHOTON CORRELATION

The line of reasoning in presenting these three theorems is as follows. Suppose, firstly, one were able to measure instantaneous values of the optical field amplitude as a function of time. Then, to get the power spectrum out of measured amplitude fluctuations one needs the WIENER-KHINTCHINE theorem. But appreciate, secondly, that only intensity estimators may be measured as a function of time. Then one needs the SIEGERT relation to make the connection between correlations of amplitude and intensity fluctuations. Finally realize, thirdly, that intensity measurements in the optical range are carried out by the detection of single photons which is a quantum statistical process. One therefore needs the MANDEL formula to assess the reliability of measured photon correlation functions.

2.1 The Wiener-Khintchine theorem

It was derived in the 1930s independently by Wiener at M. I. T. in the USA and by Khintchine, member of the famous group of theorists at Moscow.

For a (possibly complex) stationary stochastic variable $A(t)$ the autocorrelation function is defined by

$$G^{(1)}(\tau) = \lim_{T \to \infty} \frac{1}{T} \int_0^T A(t) \, A^*(t-\tau) \, dt. \tag{3}$$

To form its Fourier transform one uses both definitions, introduces the gratuitous factor $1 = \exp(-i\omega t)\exp(i\omega t)$, replaces $t-\tau$ by s which causes some inconsequential changes in the limits of one of the integrals and finally, by interchanging the order of integrations and limits, the result is written as the product of Fourier transformed amplitudes. In this way one obtains

$$\lim_{T' \to \infty} \frac{1}{T'} \int_0^{T'} G^{(1)}(\tau) \exp(i\omega\tau) \, d\tau =$$

$$\lim_{T' \to \infty} \lim_{T \to \infty} \frac{1}{T'} \frac{1}{T} \iint A(t) \exp(i\omega t) A^*(t-\tau) \exp\left(-i\omega(t-\tau)\right) dt \, d\tau =$$

$$\left(\lim_{T \to \infty} \frac{1}{T} \int A(t) \exp(i\omega t) \, dt\right) \left(\lim_{T' \to \infty} \frac{1}{T'} \int A^*(s) \exp(-i\omega s) \, ds\right) =$$

$$A(\omega) \, A^*(\omega) = I^{(1)}(\omega) \, . \tag{4}$$

This theorem states that the Fourier transform of the amplitude autocorrelation function is the power spectrum. Its significance is obvious: The spectrum can be evaluated from time series of amplitude measurements.

2.2 The Siegert relation

It relates amplitude and intensity correlation functions for the case of Gaussian signal amplitudes

$$A(t) = \sum_k a_k \exp\left(i\varphi_k(t)\right) \, . \tag{5}$$

In a Gaussian signal the amplitude is composed of many, in the limit infinitely many terms, whose phases are distributed with equal
probability between $-\pi$ and $+\pi$, and any time dependence of the a_k
is unrelated to that of the phases. The amplitude $A(t)$ may present
the coherent superposition of light scattered by many particles in
a photon correlation experiment. By definition the amplitude autocorrelation function of this signal is

$$G^{(1)}(\tau) = \sum_{k,m} < a_k \, a_m \, \exp\left(i\varphi_k(t) - i\varphi_m(t-\tau)\right) > \tag{6}$$

where the averaging signs $< >$ indicate the limiting and integration processes as in equation (3). Now, for $k \neq m$ the phase
$\varphi_k - \varphi_m$ in equation (6) has any value between $-\pi$ and $+\pi$ with
equal probability. The complex exponential represents a point on
the unit circle in the complex plane, and since all these points
are equally likely, the expectation value of the exponential vanishes as long as $k \neq m$. Hence only terms with $k = m$ contribute so
that for a Gaussian signal

$$G^{(1)}(\tau) = \sum_k a_k^2 < \exp\left\{i\left(\varphi_k(t) - \varphi_k(t-\tau)\right)\right\} >. \tag{7}$$

With the intensity defined by

$$I(t) = A(t) \, A^*(t)$$

$$= \sum_{k,l} a_k a_l \, \exp\left\{i\left(\varphi_k(t) - \varphi_l(t)\right)\right\} \tag{8}$$

the intensity correlation function is

$$G^{(2)}(\tau) = < I(t) \, I(t-\tau) > = \tag{9}$$

$$< \sum_{klmn} a_k a_l a_m a_n \, \exp\left[i\left(\varphi_k(t) - \varphi_l(t) + \varphi_m(t-\tau) - \varphi_n(t-\tau)\right)\right] >$$

A similar argument as that leading from equation (6) to (7) shows
that the expectation of the exponential vanishes unless $k = l$ and
$m = n$ or $k = n$ and $l = m$. Summing these contributions in the
Gaussian limit leads to the Siegert relation

$$G^{(2)}(\tau) = \left(G^{(1)}(0)\right)^2 + \left| G^{(1)}(\tau) \right|^2 \tag{10}$$

where the first (second) term comes from the former (latter) choice of indices.

This formula was derived during World War II in connection with the statistical interpretation of radar return signals. Note that it is not universally valid, but only in the Gaussian limit. Thus there may be experimental situations where its uncritical use may lead to error. The absolute square in equation (10) shows that intensity correlation yields as much information as amplitude correlation only if the latter is a real function of time. Fortunately, in photon correlation spectroscopy this is the case. If, on the other hand, the amplitude correlation is a genuine complex function, the intensity correlation carries less information.

2.3 The Mandel formula

Photo detection is a probabilistic process which is described by Poisson statistics. This may be shown by invoking quantum electrodynamics and Glauber's P-representation of coherent states. Here a simpler discussion may suffice.

If I is a constant intensity, α the quantum efficiency and T the sampling time, then the number n of photoelectron pulses should have the expectation

$$< n > = \alpha I T . \tag{11}$$

Since n is Poisson distributed, the probability of finding n photoelectron pulses in the time T is

$$p(n;T) = \frac{(\alpha I T)^n}{n!} \exp (-\alpha I T) \tag{12}$$

Mandel's formula may be looked upon as a generalization of equation (12) for the case that the incoming intensity I itself is a fluctuating quantity described by a probability distribution $p(I)$. It reads

$$p(n;T) = \int \frac{(\alpha I T)^n}{n!} \exp (-\alpha I T) \, p(I) \, dI . \tag{13}$$

With this formula the so-called factorial moments can be calculated since summations and integrations can be carried out. One finds for the second factorial moment

$$< n(n-1) > = \sum_n n(n-1) \ p(n;T) = \alpha^2 T^2 < I^2 > . \tag{14}$$

Using these results for the product of uncorrelated photodetection events, that is with time differences larger than the dead-time of the photomultiplier, one finds from equation (11)

$$< n(t_1) \ n(t_2) > = \alpha^2 T^2 < I(t_1) \ I(t_2) > . \tag{15}$$

This shows that the expectation of the photon correlation function is, except for a scale factor, the same as the intensity correlation function. It is easy to show, however, that for the photon function the variance is larger. In practice this means that one has to measure for a longer time until the result has the desired accuracy. From equation (14) one finds for zero delay

$$< n(t_1)^2 > = \alpha^2 T^2 < I(t_1)^2 > + \alpha T < I(t_1) > . \tag{16}$$

The second term on the right hand side is called shot noise term because it also appears in noise theory. Here it leads to a discontinuous step in experimental correlograms between channels for zero and finite delay. This is one of the reasons that in some commercial correlators the channel zero is not implemented.

3. THE FUNCTION OF DIGITAL CORRELATORS

3.1 The digital correlation function

The mathematical definition of the intensity correlation function is, according to equations (3) and (9)

$$G^{(2)}(\tau) = \lim_{T \to \infty} \frac{1}{T} \int_0^T I(t) \ I(t-\tau) \ dt . \tag{17}$$

In this form, the function cannot be formed by digital electronics, hence some modifications are necessary.

1. The limit operation $T \to \infty$ is unrealistic. The use of a finite limit T means that one determines an estimator instead of the asymptotic correlation function.

2. The normalizing factor $1/T$ is omitted. Then the correlation function increases with measuring time by a continuous accumulation process.

3. Whereas in equation (17) running time t and delay time τ are continuous variables, in the correlator a discrete time basis is introduced. All time variables are integer multiples of a basic sample time Δt, for example

$$t = i\,\Delta t, \quad \tau = j\,\Delta t, \quad T = N\,\Delta t, \quad j_{max} = J \ .$$

4. Whereas the intensity $I(t)$ is a continuous variable, the number of photoelectrons $n'(i)$ counted in the interval between $i\Delta t$ and $(i+1)\,\Delta t$ obviously is an integer.

5. Often, depending on experimental conditions, on the choice of the sampling time Δt, and on the capabilities of the correlator, the true number $n'(t)$ may be too large for real time processing in the correlator. Several schemes are available for reducing $n'(i)$ to a manageable magnitude. All of them use counter circuitry. One may divide by M_1, take the integer part of the result as the reduced photon number

$$n(i) = n'(i)\ /\ M_1$$

and transfer the residue to the next interval $i+1$. One may subtract a constant M_2 if all $n'(i) \geq M_2$,

$$n(i) = N'(i) - M_2 \ .$$

An even more drastic reduction to one-bit format is required for the delayed photon count $n(i+j)$ in the so-called one-bit correlators as follows:

$$m(i+j) = \begin{cases} 1 & \text{if} \quad n(i+j) > n_o \\ 0 & \text{if} \quad n(i+j) \leq n_o \end{cases} .$$

If in this decision process, n_o is a fixed number, the process is called clipping, n_o is the clip level. If n_o varies statistically such that it assumes values between zero and the maximum of $n(i+j)$ with the same probability, the process is called random clipping. A similar process is scaling, in which photons are counted continuously on a counter with the upper limit equal to the maximum of $n(i+j)$. Then $m(i+j)$ is found as the number of overflows of the counter.

Among these methods, clipping is easy to implement but it leads to distorted correlograms unless the signal has well defined statistical properties. Random clipping does not introduce any distortions, but it requires fast random number generators which tend to be expensive. Sacaling is a more economic way of doing random clipping; it works without distortion only if the incoming signal pulses show strong shot noise.

In some correlators, similar clipping or scaling processes are used to reduce the incoming pulse sequence to digital four-bit signals. It should be appreciated that clipping or scaling imply a loss of information. One speaks of quantization noise to indicate the fact that resulting correlation functions are less accurate than they would be with the full photon signals. The loss of accuracy usually is small, however; in typical cases an increase in the number of samples, N, by 30 per cent is adequate to obtain the same accuracy with clipped or scaled signals.

With these modifications, the digital photon correlation function may be written

$$G^{(2)}(j) = \sum_{i=1}^{N} n(i)\, m(i-j) , \quad 1 \le j \le J. \tag{18}$$

The number of channels, J, in practical correlators is between 24 and 512 and may be increased beyond this number by "tricks". In experiments, the number of time samples, N, may be very large, for example 10^8 or 10^{10} .

3.2 Digital correlators

The space of progress is fast in electronics. The development of digital correlators has benefitted from that and no doubt will continue to do so. In this lecture it will suffice to discuss the historically earliest correlator, the Malvern one-bit device, and the most recent development, the ALV structurator/correlator. Other in-between developments will be mentioned only briefly.

The one-bit correlator [1,2] was developed around 1970 by Dr. E. R. Pike and associates of the Royal Signals and Radar Establishment in England. Since about that time it has been manufactured and marketed by Malvern Instruments Ltd., Malvern, UK. More than anything else it was the availability of this correlator which made photon correlation spectroscopy a practical laboratory measuring technique. To the best of our knowledge, this correlator is still being manufactured and sold despite its design date. The operating principle is illustrated in Fig. 1. Photoelectron pulses enter a clipping or scaling unit which issues a zero or one for every sampling time. These one-bit signals enter a shift register where, under control of the sample time clock, they are shifted from one location to the next. The content of every shift register cell controls the adjacent AND gate. Photoelectron pulses coming from the correlator input may enter those counter stores where a one is in the shift register while they may not if there is a zero. In this way the correlator builds up the so-called single-clipped correlation function, equation (18), in the stores. This correlator is fast, and hence usable for real-time operation, because the multiplication inherent in the definition of the correlation function is replaced

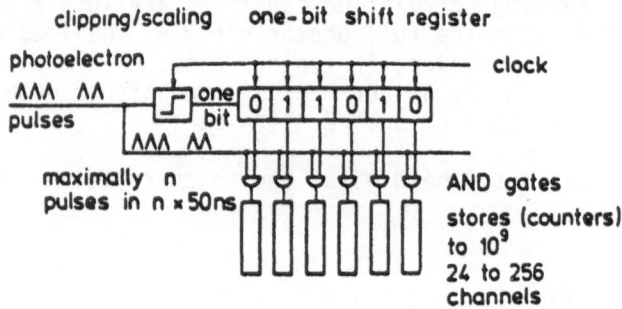

Fig. 1.   Scheme of the Malvern single-bit correlator

by the logical AND operation which is fast and economic compared to a genuine multidigit multiplication. More precisely this device is called an n x 1bit correlator because, referring to equation (18), the actual signal is represented by a sequence of $n(i)$ pulses while the delayed signal $m(i-j)$ is in one-bit format. The processing of one input pulse takes 50 ns, hence n input pulses require at least n times 50 ns. This minimum would be possible only with a derandomized sequence of pulses of 50 ns repetition time, otherwise the sampling time has to be longer. The sampling time Δt may be set longer than 50 ns.

A variation of this scheme is the n x 4bit correlator made by Langley-Ford of Amherst, Massachusetts, and by Malvern Instruments. Here the quantization of input signals is to the four-bit format and there is a four-bit shift register. The processing of one input pulse takes 100 ns, hence n pulses take n times 100 ns. If n can be as large as four-bit numbers go, namely up to n = 15, the fastest sampling time with derandomized pulses could be 1.5 µs. A similar hard-wired n x 4bit correlator is made by Brookhaven Instruments on Long Island, New York; its operation is controled by a microprocessor which offers added convenience with data input and output as well as with various operating modes and data evaluation.

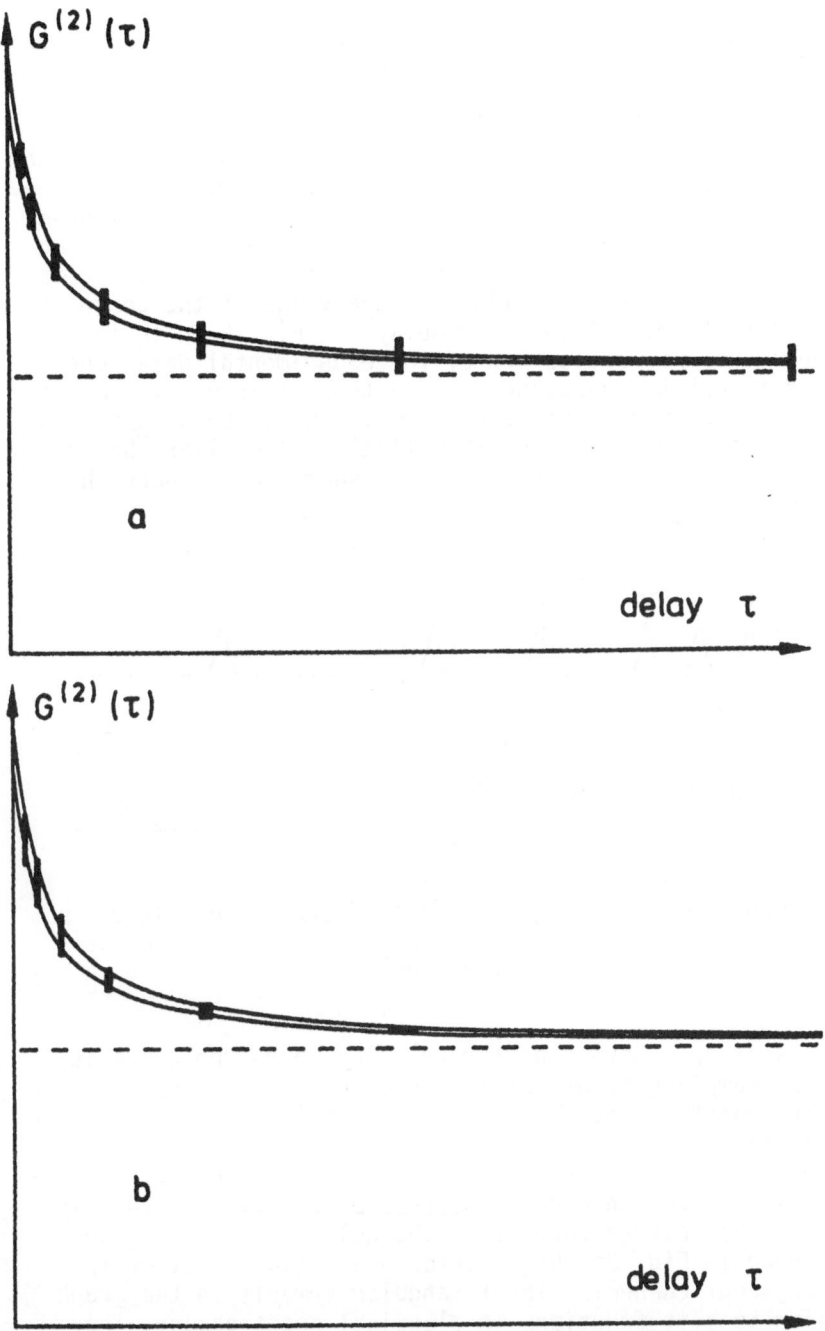

Fig. 2.   Correlogram sampled with monotonically increased delay time spacing; identification of experimental data with one of several theoretical curves; a) same statistical accuracy at all sampling points, b) better accuracy for larger delay

Another variation of the n x 1bit correlator is one with monotonically increasing delay time spacing, again developed by Dr.
Pike and associates and marketed by Malvern Instruments Ltd. It
differs from the scheme of Fig. 1 by sections of shift register
without AND gates and stores which are placed between those with
AND gates and stores. The purpose of this design may be explained
as follows. In photon correlation spectroscopy, the correlogram
typically looks like the curves of Fig. 2, it consists of a constant plus a superposition of decaying exponentials. The constant,
also referred to as the base line, is the value of the correlation
function in the limit of infinite delay, in Fig. 2 indicated by
the dashed line. For evaluation of the experimental data, one
would like to get the decaying curve with best accuracy, but at
the same time the number of measured data should be as small as
possible for economic reasons. Intuitively it is clear that one
needs closely spaced sampling points at short delay where the

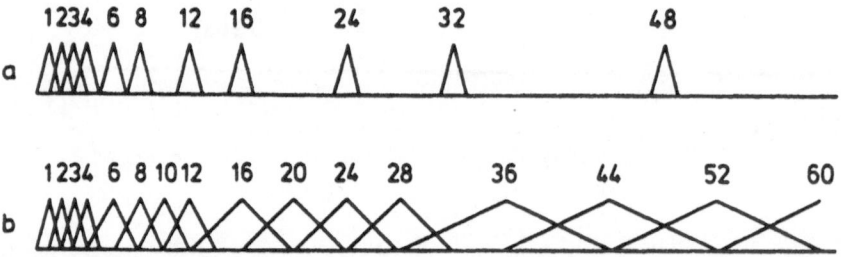

Fig. 3.   Monotonically increasing delay between correlator
channels and weighting of the data; a) in a commercial Malvern
correlator, b) in a recent Kiel design

function is steep, whereas at larger delay where the curve is
less steep, sampling points at larger spacing are adequate. Theoretical reasoning [5] suggests that the optimum spacing is given
by a geometric progression, i.e. consecutive delays increase by
a constant factor. For purposes of illustration, the delays between sampling points in Fig. 2 increase by a factor of two. In
reality, in the Malvern correlator the delay time spacing increases, as shown in Fig. 3a, by a factor near square root of two
between adjacent channels. The triangular symbols in the graph
are similar to slit functions in classical spectroscopy. They
indicate that the store contents for a nominal delay of τ contain
contributions from photons with relative delay between $\tau-\Delta t$ and
$\tau+\Delta t$ with triangular weighting. Thus all channels of that correlator collect data with the same statistical accuracy.

It turns out, however, that constant statistical accuracy is not really wanted. Suppose one wishes to fit a theoretical curve through experimental correlation data. Two possible curves each are shown in Fig. 2a and b. All possible curves converge towards each other and towards the base line. Hence the distinction of such curves requires better statistical accuracy for large delay. The error bars shown in the graphs, of constant size in Fig. 2a as for the Malvern correlator, and of decreasing size in Fig. 2b illustrate this point. The required smaller absolute error at larger delay may be achieved by sampling with a wider delay time window as illustrated in Fig. 3b. As a matter of fact, a doubling of the delay time window leads to fourfold statistical weight of the correlogram data and one over square root of two of the statistical error.

This feature is one of the options in a recent system, the Structurator/Correlator developed by Dr. K. Schätzel at Kiel and marketed by ALV of Langen, Germany. Since it operates under control of a built-in microprocessor, several different functions may be realized. The scheme of Fig. 4 shows functions which are partially realized in hardware, partially in software. Here both the undelayed and the delayed signals are present in four-bit format. Earlier experience in the Kiel laboratory had shown that the basic four-bit structurator/correlator may work at a sampling time Δt as short as 60 ns. In this time, the product of four-bit numbers $n(i)\,n(i-j)$ is added to each of the stores numbered by j. This is the correlator option. Alternatively, the square $\left(n(i) - n(i-j)\right)^2$ is added to the contents of store number j. This is the structurator option [6]. Both functions are similar although either one may be better suited in certain experimental situations [7]. The correlator is preferred if the photon shot noise limits the measurement accuracy. The structurator is preferred if drifts or slow fluctuations limit the accuracy. The arithmetic operations are fast because they do not involve true calculations but rather table look-up. The 256 possible outcomes of multiplibation or of difference squaring of four-bit numbers are stored in one read-only memory (ROM) each per hardware channel. The undelayed and the delayed four-bit signal form an eight-bit address to the ROM whose output appears rapidly and is added to the contents of a store for every channel. A ninth bit in the address is used to select either the correlator or the structurator option.

Now suppose one wants to work at a sampling time Δt of 6 µs compared to the 60 ns needed for the operation of the basic structurator/correlator. This sampling time is typical for particle sizing measurements by photon correlation. In that case the electronics could do their job in one per cent of the time and would sit idle for 99 per cent. To make the electronics work all the time, a temporary storage of the measured data is required. The

data come in the form of photoelectron pulses, they are converted to four-bit format every 6 μs (in our example) and written
into a random access memory (RAM). At the same time, previously
stored data are read every 60 ns from another section of the RAM.
They pass the four-bit shift register and the structure or correlation function is formed for a small number of channels, 1 to 4
in Fig. 4. Thereafter the same data may pass the shift register
again, and this time their delays correspond to channels 5 to 8,
whose stores are now connected to the ROMs. The next time, it
will be channels 9 to 12. In this way, the data can pass the hardware about 100 times in the same time span in which they were
collected originally. Hence a few hardware channels with the
necessary software controls suffice to obtain the same function
as a hardware correlator with several hundred channels. Another,

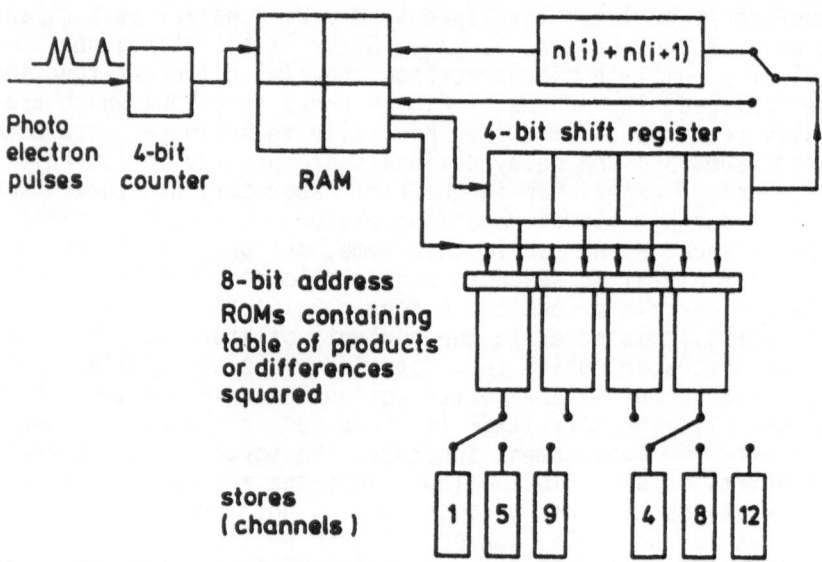

Fig. 4. Operating scheme of the recent Kiel four-bit struc-
turator/correlator design marketed by ALV

more interesting function may be obtained if, before their return
to the RAM, the data are added pairwise, $n(i) + n(i+1)$, $n(i+2)$
$+ n(i+3)$, ... Provisions are made that the sums do not exceed
the four-bit format. In this way a smaller number of input data
is generated, but the individual numbers have greater statistical
weight. Feeding these data again through the shift register gives
the structure or correlation function at delay times with doubled
spacing and fourfold statistical weight as suggested in Fig. 3b.
Continuing in this mode of operation, the length of the data

string decreases rapidly, every time by a factor two, so that the amount of required processing time decreases likewise. As a result, it is not difficult to obtain correlation functions on steadily increasing increments of delay time such that the longest practical delay is not just several thousand times Δt - this may be a practical limit if one wants to implement the delay by hardware shift registers as in the Malvern design - but even millions. Such a wide range of delay times is a necessity in research on composite suspension systems where the range of observable time constants goes from microseconds to seconds. A more complete description of this system is planned in a future publication by Dr. Schätzel.

For completeness, two more correlator schemes should be mentioned. A correlator for very short sampling times, $\Delta t = 5$ ns, was developed by Dr. W. T. Mayo Jr. of Spectron Laboratories, Costa Mesa, California and is marketed under the trade name Correlex. Its use is restricted to sparse signals, that is such small rates of detected photons that, in the average, there is a photo electron pulse in every tenth or hundredth sampling time only. Due to dead-time and afterpulsing effects, in a time of 5 ns after a photo detection event the nominal sensitivity of most photo multipliers is not fully restored. Hence, in order to get undistorted correlograms it is advisable to work with cross correlation, that is with two photo multipliers and two inputs to the correlator. This apparatus measures the elapsed time between single photon events and builds up the correlation function out of measured time differences.

For very slow processes, a modern minicomputer may perform the function of correlators. This requires special programming, and for best efficiency certain parts need be written in machine code. It should be mentioned that the structure function can be formed much more readily than the correlation function. For reasons of computing speed, the products $n(i) \, n(i-j)$ or the squares $\left(n(i) - n(i-j) \right)^2$ would not be calculated but obtained by table look-up from memory. With 8 bit numbers, there are 256 possible square numbers which may easily be stored whereas it is not practical to store the 64 000 possible products of 8 bit numbers. Some time ago, in the Kiel laboratory an 8 bit by 8 bit structurator was realized [8] on the basis of a commercial minicomputer which yields the structure function in real-time operation with 128 channels at $\Delta t = 1$ ms or with 1024 channels at $\Delta t = 10$ ms.

4. THE DYNAMIC LIGHT SCATTERING EXPERIMENT

Consider the experimental setup of Fig. 5. Laser light is scattered from particles undergoing Brownian motion and is received in the direction of the scattering angle by a photomultiplier whose output is fed into a correlator. The scattering geometry is also sketched in Fig. 4. A scattering wave vector

$$\bar{q} = \bar{k}_s - \bar{k}_i \tag{19a}$$

of magnitude given by the Bragg formula

$$q = (2\pi/\lambda)\ 2\ \sin\ \vartheta/2 \tag{19b}$$

is introduced and the coordinate x is measured in the direction of $\bar{q}$. The virtual fringe system shown in the sketch has fringe planes separated by

$$d = 2\pi/q\ . \tag{19c}$$

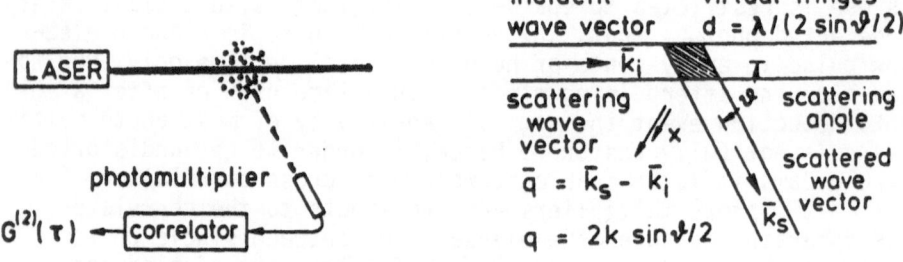

Fig. 5.   Experimental setup for photon correlation spectroscopy (left) and scattering geometry (right)

Particles aligned on these planes would scatter light in the direction of angle ϑ with the same phase. For a random distribution of particles, the consideration of optical phase delay shows that the light amplitude in the scattering direction is, except for an inconsequential common phase factor,

$$A(t) = \sum_k a_k\ \exp\left(iqx_k(t)\right)\ . \tag{20}$$

The summation extends only over particles k that are both illuminated by the laser and in view of the detector. The amplitude correlation function is

$$G^{(1)}(\tau) = \left\langle \sum_{k,m} a_k\ a_m\ \exp\left\{ iq\left(x_k(t) - x_m(t-\tau)\right)\right\} \right\rangle\ . \tag{21}$$

In forming the expectation indicated by < > , the cross terms with
k ≠ m do not contribute since the particles are distributed randomly in space, leading to all phases between $-\pi$ and $+\pi$ mod 2π with
equal probability, so that

$$G^{(1)}(\tau) = \sum_k a_k^2 < \exp\left\{ iq(x_k(t) - x_k(t-\tau)) \right\} >. \qquad (22)$$

The innermost bracket contains the displacement of particles
$x(t) - x(t-\tau)$ during time τ due to Brownian motion. It is known,
however, that this displacement is given by a normal distribution
with zero mean and, see equation (1), with mean square $2D\tau$. Formally one has to compute the expectation $< \exp(i\varphi) >$ for a normally distributed quantity φ with mean square σ^2. As one can see
by Fourier transforming the normal distribution and setting the
new variable equal to one, the result is $\exp(-\sigma^2/2)$, and hence
in the present case

$$G^{(1)}(\tau) = \sum_k a_k^2 \exp(-q^2 D\tau). \qquad (23)$$

From the Siegert relation, the intensity autocorrelation function
is

$$G^{(2)}(\tau) = \left(\sum_k a_k^2 \right)^2 \left\{ 1 + \exp(-2q^2 D\tau) \right\}. \qquad (24)$$

For a monodisperse suspension of particles, a measurement of $G^{(2)}$
yields the diffusion coefficient and from it, through the Stokes-
Einstein relation equation (2), the particle radius.

For a polydisperse suspension, $G^{(1)}$ becomes the superposition
of exponentials, each with a weighting factor proportional to the
light scattered by that fraction. Similarly, the photon correlation function $G^{(2)}$ is given by a constant plus square of this
superposition. From measured correlation functions, the underlying
polydispersity distribution may be obtained, at least in principle.
The success of such evaluations depends foremost on the degree of
noise in the correlograms, that is the scatter of data points
around an interpolating curve; on a relative scale, the noise
should be around 10^{-3} or less if one wants to get some significant
details on the distribution from the evaluation. Also important is
the available a-priori information; it may be known, for example,
that the distribution is unimodal with one peak only, or bimodal
with two humps, or particles with diameters above and below certain limits are not expected. Similarly the number and spacings of
the delay channels in the correlator are significant; as suggested
before, it is advantageous to have narrow spacings at short delay
which monotonically get wider for long delay. While these facts
are undisputed, the debate on the best mathematical procedure for

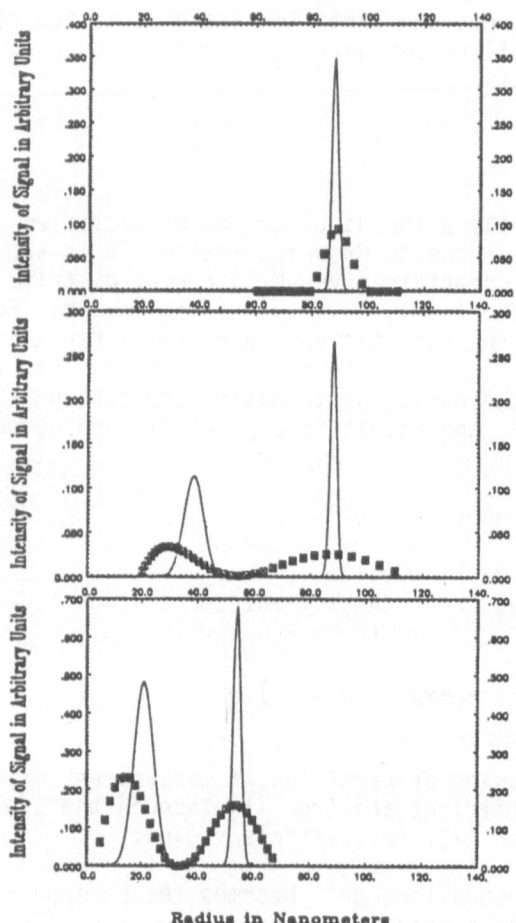

Fig. 6. Examples of size distributions measured by electron microscopy (thin solid lines) and evaluated by the CONTIN program from photon correlation data (heavy square symbols), taken from Stelzer et al [5]

polydispersity analysis is still going on. The six papers listed under reference [5] give a good feeling for the current state of that debate. Fig. 6, taken out of the paper by Stelzer et al [5], illustrates the state of the art. In the figure, thin solid lines indicate a unimodal and two bimodal size distributions which were determined by electron microscopy whereas the square dots show the size distribution as determined by photon correlation spectroscopy and subsequent computer analysis. The main flaw of photon correlation measurements as they were done here seems to be a broadening of the true distributions.

References

1. H. Z. Cummins and E. R. Pike, eds.,
   Photon correlation and light beating spectroscopy,
   Plenum Press, 1974
2. H. Z. Cummins and E. R. Pike, eds.,
   Photon correlation spectroscopy and velocimetry,
   Plenum Press, 1977
3. E. O. Schulz-DuBois, ed.,
   Photon correlation techniques in fluid mechanics,
   Springer Verlag, 1983
4. A. Einstein, Ann. Phys. 17, 549 (1905); 19, 371 (1906)
5. In reference 3, see chapters by N. Ostrowski and D. Sornette;
   M. Bertero and E. R. Pike; J. R. Ford and B. Chu;
   G. R. Danovich and I. N. Serdyuk; S. W. Provencher;
   E. Stelzer, H. Ruf, and E. Grell
6. E. O. Schulz-DuBois and I. Rehberg,
   Appl. Phys. 24, 323 (1981)
7. K. Schätzel, Optica Acta 30, 155 (1983)
8. See chapter by K. Schätzel in reference 3

IMAGE PROCESSING TECHNIQUES

OPTICAL FOURIER TRANSFORM CONSTRUCTION

Silverio P. Almeida and Srisuda Puang-ngern

Virginia Polytechnic Institute and State University
Department of Physics
Blacksburg, Virginia 24061

I. INTRODUCTION

In this paper we describe two methods of constructing optical
Fourier transforms for use in image processing. Both the coherent
parallel and converging beam Fourier transform methods are presented. The converging beam set-up was used to obtain results
presented in another paper (1) at this NATO school but its full
discussion at that time was postponed until now. The input objects
chosen for this study are those of snow crystals. They are most
appropriate since their Fourier transforms contain both some high
and low spatial frequencies, not to mention they are beautiful
patterns to observe.

Optical Fourier Transforms (OFT) of objects provide a powerful
method to analyze and manipulate the spatial frequency plane of an
object (2,3). One can alter the OFT plane by inserting high, low
or matched spatial filters in this plane. The inverse OFT can then
be performed and correlation experiments made using hybrid optical
systems. For example, filters can be used to remove unwanted
noise from a signal or a series of matched filters can be used to
perform pattern recognition studies (4). The OFT obtained from a
lens is not exact. Regardless of the quality of the lens, the
approximation is caused by finite extent of the lens. However, one
can get a good approximation, if one uses a sufficiently large lens
compared to the size of the transparency. Big lenses, even not a
special purpose Fourier lens which is corrected for aberrations, are
quite expensive. Joyeux and Lowenthal (5) have suggested that a
better Fourier transform can be obtained by using a converging beam
configuration instead of a parallel beam, when an ordinary convex
lens has to be used.

We shall perform some experiments to make a comparison between these two configurations. The Fourier transforms of a number of photographic transparencies of snow crystals with different spectral frequencies will be compared. The mathematical background of these two methods is also given.

II. A PARALLEL BEAM SET-UP

Let $t(x_0,y_0)$ represent the field distribution of the photographic transparency placed at the front focal plane of the Fourier transform (FT) lens with focal length f. When the transparency is illuminated by a parallel beam or plane wave of wavelength λ and unit amplitude, the field distribution in plane P_1 (Fig. 1) given by the Fresnel diffraction approximation (2) is

$$f(x_1,y_1) = (j\lambda f)^{-1}\exp(jkf)\int\int_{-\infty}^{\infty} t(x_0,y_0)\exp\{jk[(x_1-x_0)^2 +$$

$$(y_1-y_0)^2]/2f\}dx_0 dy_0 \tag{1}$$

where $k = 2\pi/\lambda$. After passing through the lens, the field distribution in plane P_2 becomes

$$f'(x_2,y_2) = f(x_2,y_2)P(x_2,y_2)\exp\{-jk(x_2^2 + y_2^2)/2f\} \tag{2}$$

where

$$P(x_2,y_2) = \begin{cases} 1, & x_2^2 + y_2^2 \leq a^2, \text{ a is the radius of the lens} \\ 0, & \text{otherwise.} \end{cases}$$

Again, under the Fresnel diffraction approximation, the field distribution in plane P_3 is given by

$$g(x_3,y_3) = (j\lambda f)^{-1}\exp(jkf)\int\int_{-\infty}^{\infty} f'(x_2,y_2)\exp\{jk[(x_3-x_2)^2$$

$$+ (y_3-y_2)^2]/2f\}dx_2 dy_2 \tag{3}$$

Using Eqs. (1) and (2), and assuming $P(x_2,y_2) \simeq 1$ for a sufficiently large lens compared to the size of the transparency, we get, after the integrations with respect to x_2 and y_2 are carried out,

$$g(x_3,y_3) = (j\lambda f)^{-1} \exp(jkf) \int\limits_{-\infty}^{\infty}\!\!\!\int t(x_0,y_0) \cdot$$

$$\exp\{-2\pi j(x_3x_0+y_3y_0)/\lambda f\}dx_0 dy_0$$

$$= (j\lambda f)^{-1} \exp(jkf)F(u,v) \qquad (4)$$

where $F(u,v) = \int\limits_{-\infty}^{\infty}\!\!\!\int t(x,y)\exp\{-2\pi j(ux+vy)dxdy$ represents the

FT of $t(x,y)$. Then $F(u,v)$ in Eq. (4) is the FT of $t(x_0y_0)$ evaluated at spatial frequencies $u = x_3/\lambda f$ and $v = y_3/\lambda f$. Hence, the intensity distribution in the back focal plane of the lens is proportional to the intensity distribution of the FT of the transparency as

$$I(x_3,y_3) \sim |g(x_3,y_3)|^2 = (\lambda f)^{-2}|F(u,v)|^2 \qquad (5)$$
$$\qquad\qquad\qquad\qquad\qquad u = x_3/\lambda f$$
$$\qquad\qquad\qquad\qquad\qquad v = y_3/\lambda f$$

III. A CONVERGING BEAM SET-UP

Let d_1 and d_2 be conjugate distances of the thin lens with focal length f. The relation between d_1 and d_2 is expressed as

$$1/d_1 + 1/d_2 = 1/f \qquad (6)$$

If a point source is placed on the central axis of the lens at the distance d_1 in front of the lens, after passing through the lens, a diverging beam will become a converging beam focusing on the axis of the lens at the distance d_2 behind the lens, provided $d_1 > f$. The field distribution immediately behind the photographic transparency placed in plane P_0 at the distance d from the focusing point of the converging beam (Fig. 2) is

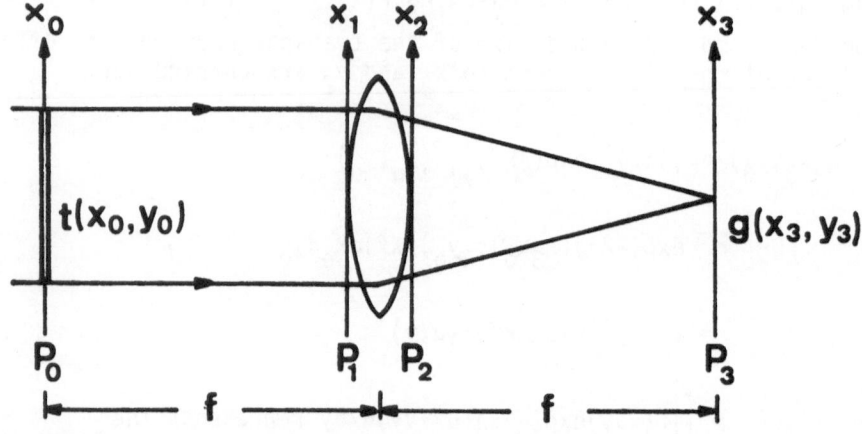

Fig. 1 Parallel beam configuration to perform a Fourier transform in the back focal plane of the lens.

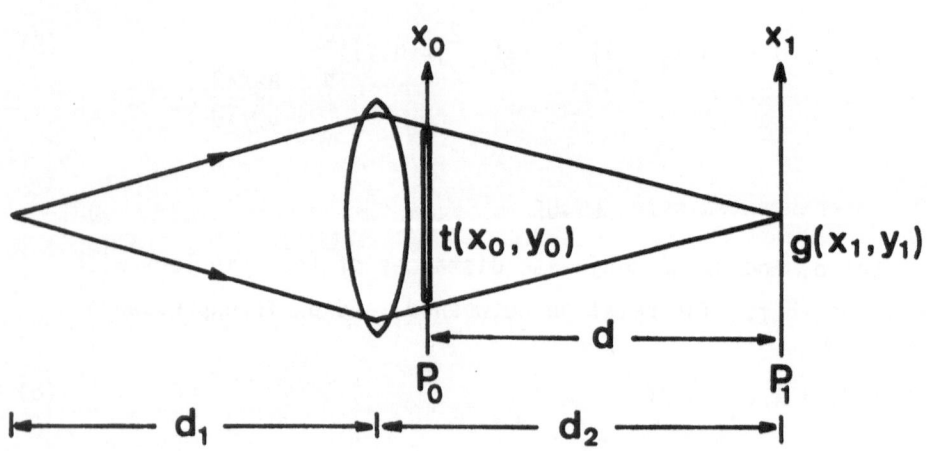

Fig. 2 Converging beam configuration to perform a Fourier transform in the conjugate plane of the point source.

$t(x_0,y_0)\exp\{-jk(x_0^2+y_0^2)/2d\}$. As in previous sections, the field distribution in plane P_1 is given by the Fresnel diffraction approximation. Then

$$g'(x_1,y_1) = (j\lambda d)^{-1}\exp(jkd)\int\int_{-\infty}^{\infty}t(x_0,y_0)\exp\{-jk(x_0^2+y_0^2)/2d\}\cdot$$

$$\exp\{jk[(x_1-x_0)^2+(y_1-y_0)^2]/2d\}dx_0dy_0$$

$$= (j\lambda d)^{-1}\exp(jkd)\exp\{jk(x_1^2+y_1^2)/2d\}\cdot$$

$$\int\int_{-\infty}^{\infty}t(x_0,y_0)\exp\{-2\pi j(x_1x_0+y_1y_0)/\lambda d\}dx_0dy_0$$

$$= (j\lambda d)^{-1}\exp(jkd)\exp\{jk(x_1^2+y_1^2)/2d\}F(u,v)\Big|_{\substack{u=x_1/\lambda d \\ v=y_1/\lambda d}} \qquad (7)$$

And the intensity distribution is given by

$$I'(x_1,y_1) = (\lambda d)^{-2}|F(u,v)|^2 \qquad\qquad (8)$$
$$\substack{u = x_1/\lambda f \\ v = y_1/\lambda f}$$

which is similar to the intensity distribution given by Eq. (5) for the parallel beam set-up. If $d = f$, $I'(x_1,y_1)$ in Eq. (4) and $I(x_3,y_3)$ in Eq. (7) are exactly the same. The difference between the parallel beam and converging beam set-up is that the FT obtained in the former system is on the back focal plane of the FT lens, whereas the FT obtained from the latter system is on the conjugate plane of the point source.

IV. SPATIAL FREQUENCIES AND SIZE OF THE LENS

In the FT plane, each point of light of the frequency spectrum represents a certain spatial frequency. It is related to an angle θ, by which the incident beam deviates from its original direction, by spatial frequency = $\sin\theta/\lambda$. Then $\theta = 0°$ or the central axis corresponds to the location of the zero frequency. The incident beam deviates by large angles, when the transparency possesses high spatial frequencies. If some of the deviated beam does not pass through the lens, the corresponding frequency will not be observed in the FT plane Fig. 3. Thus the frequency spectrum obtained is not the exact FT of the transparency. It is only approximate. If the lens is sufficiently large that all deviated beams caused by all spatial frequencies can pass through the lens, the exact FT can be obtained. Besides, parts of the transparency also have an effect upon the frequency spectrum. In Fig. 3, ray A may get through the lens, while ray A' may not.

Now let us consider the case in which the transparency is illuminated by a converging beams Fig. 4. The beams deviated from original directions by the same angle (same spatial frequency) from different parts of the transparency will focus at the same point in the FT. No deviated beam cannot reach the Fourier transform plane, even the transparency is nearly as large as the lens. The frequency spectrum will contain all spatial frequencies of the transparency, and the Fourier transform is exact.

V. EXPERIMENTAL PROCEDURES

Transparencies of snow crystals were prepared on 35 mm black and white film. The size of the snow crystal pattern is about 20 mm. Samples of four such patterns are shown in Fig. 5 with their respective Fourier transforms in Fig. 5. The optical set-up consisted of a convex lens L_1 (i.e. the FT lens) of focal length

$f = 390$ mm; and an aperture A (diameter = 25 mm). Both the same lens and aperture were used for the parallel and convergent set-ups. The illumination source was provided by a Helium-Neon (15 mW) laser. Shown in Fig. 6a, b are the parallel and convergent beam set-ups where the transparency T is located in front and just behind the lens L_1, respectively. In the parallel case the spatial frequencies are measured at the back focal plane (f) of lens L_1. While in the convergent case, they are measured at the distance d_2 (see Eq. 5).

VI. EXPERIMENTAL RESULTS AND DISCUSSION

Maximum spatial frequencies for each pattern of snow crystals are listed in Table I; where u_p and u_c represent the measured frequencies for the parallel and convergent beam set-ups respectively. The numbers for the snow crystals correspond to the numbers shown in Fig. 5. This figure shows both the snow flake patterns and their Fourier transforms taken with the convergent beam. The parallel beam FT patterns are not shown because they looked the same for patterns a, b and c. The snow flake d parallel FT lost some of its high spatial frequencies (i.e. 16-19 mm^{-1}) and was not the same as when performed in the convergent beam set-up.

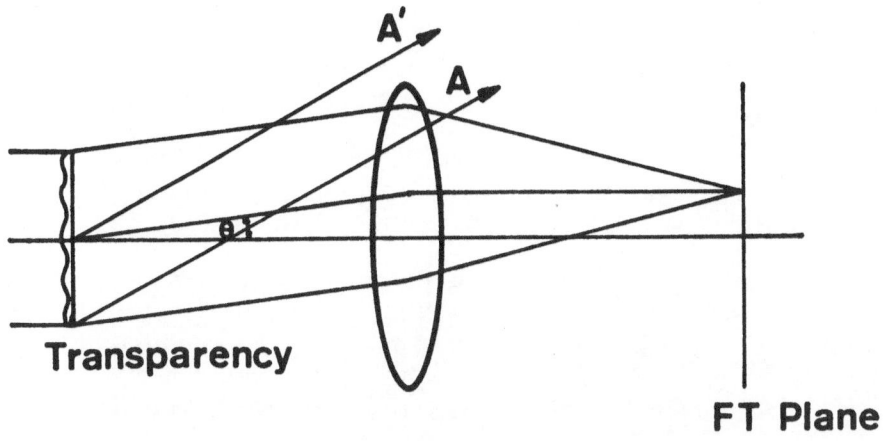

Fig. 3 Dependence of the frequency spectrum on the size of the lens in parallel beam configuration.

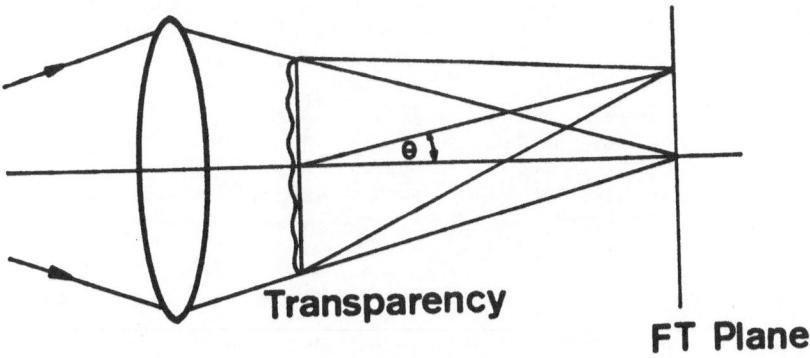

Fig. 4 Deviation of beams in converging beam configuration.

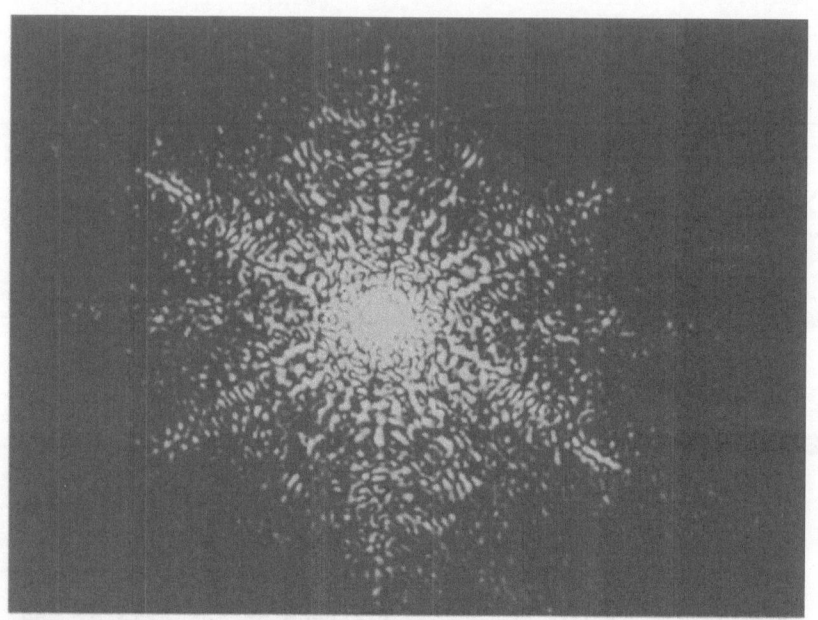

Fourier transform

Fig. 5a

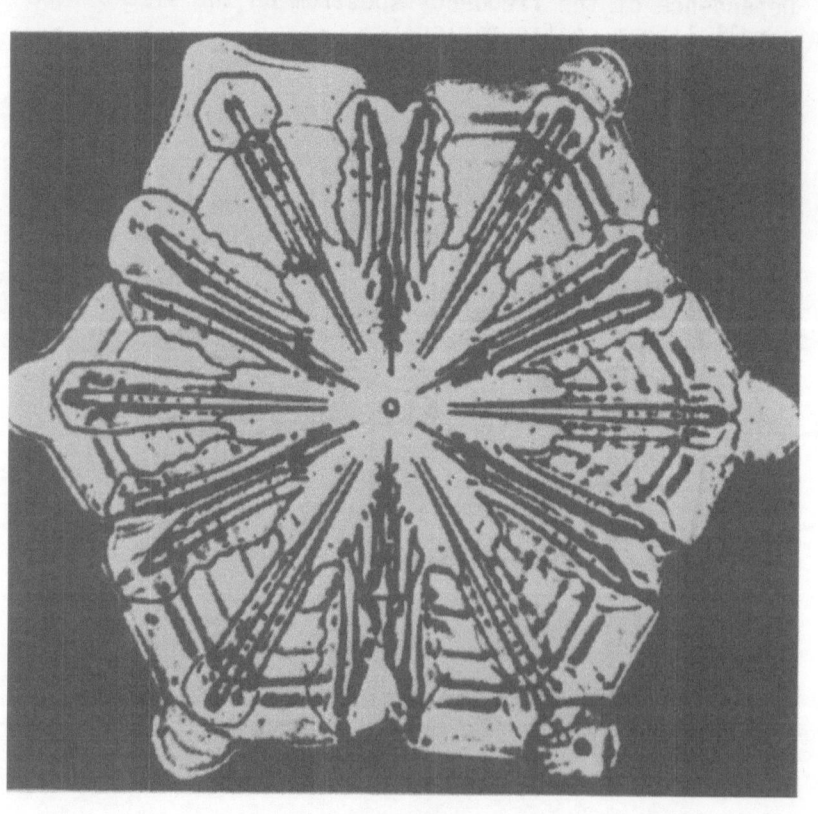

Snow crystal

Fourier transform

Fig. 5b

Snow crystal

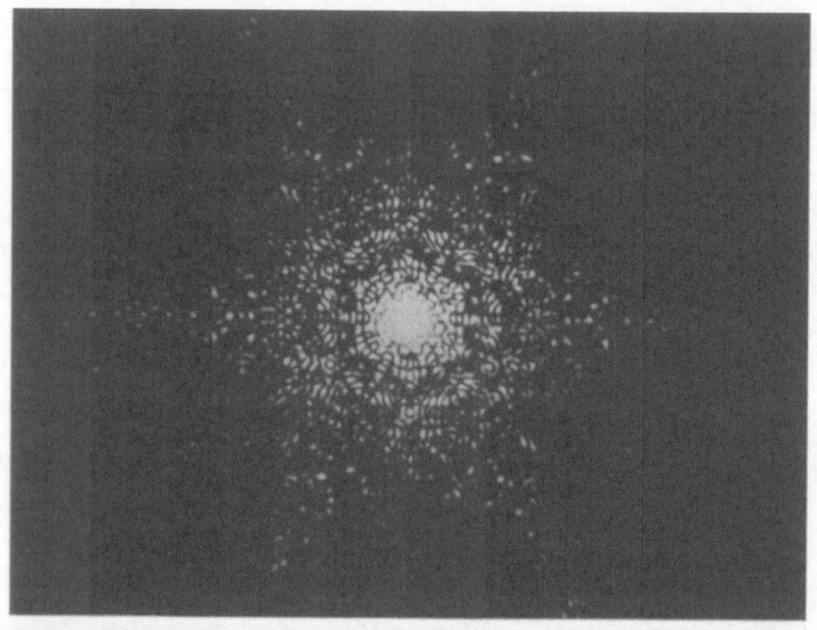

Snow crystal

Fig. 5c

Fourier transform

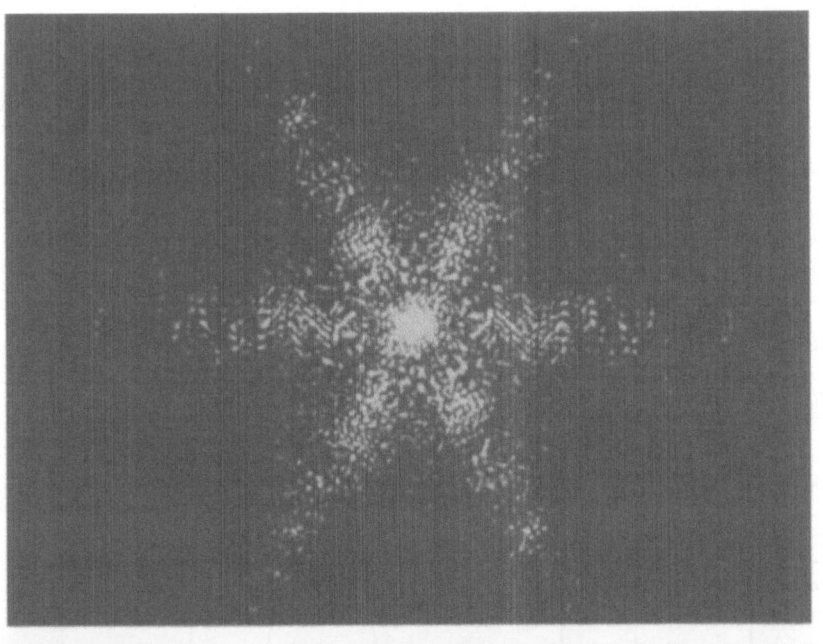

Fourier transform

Fig. 5d

Snow crystal

144

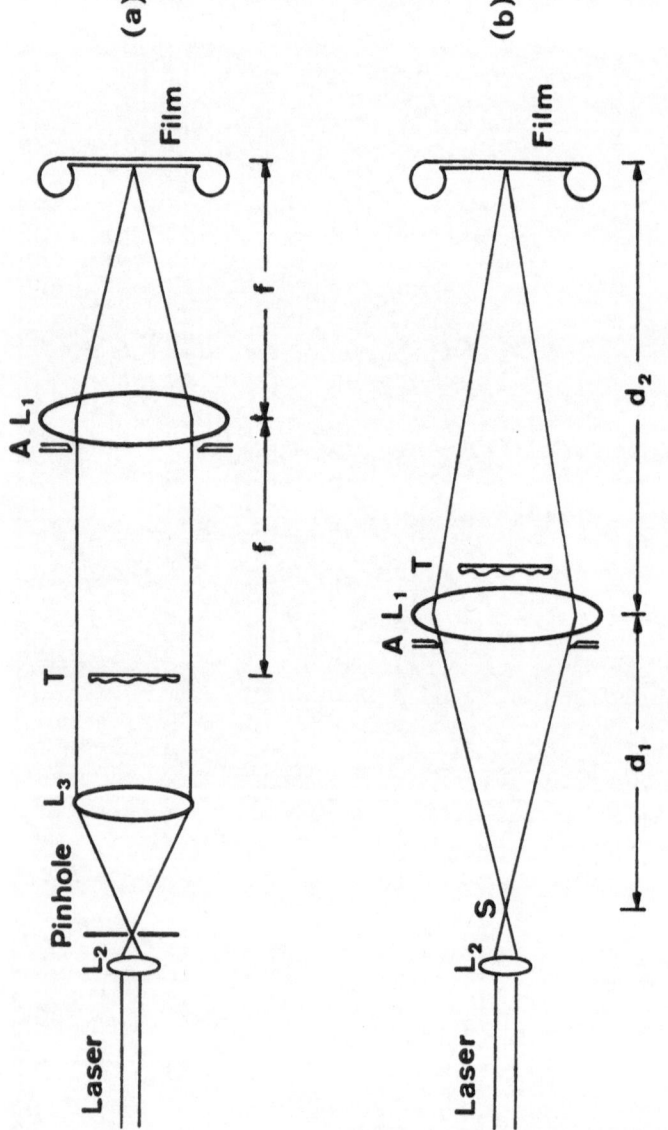

Fig. 6 (a) Parallel beam set-up and (b) converging beam set-up.

This loss results in a distortion of the inverse FT of the spatial frequency spectrum.

Although a given spatial frequency causes the light rays to deviate by a certain angle from their original direction, they reach the FT plane at different points in the two set-ups. According to Eq. (5), the beam corresponding to spatial frequency u will reach the FT plane in the parallel beam set-up at the point $x_f = u\lambda f$, and according to Eq. (7) for the converging beam set-up, $x_f = u\lambda d \simeq u\lambda d_2$. If d_1 in Eq. (6) satisfies $f < d_1 < 2f$, d_2 is greater than 2f. Then x_f will be greater than $2u\lambda f$ in the latter set-up. This makes it easier to observe the details of the FT and to block out the unwanted (noise) portion of the spatial frequency during the filtering process.

VII. CONCLUSION

In many cases the use of a converging beam set-up can produce as good or better optical Fourier transforms than, the classical, parallel beam set-up. The snow flake FT examples presented in this paper bear this out. The cost of a lens to do the converging beam FT will in many cases be lower than that necessary to achieve the same results by the parallel beam method. This is especially true at higher spatial frequencies. We noted that above 16 mm^{-1} frequencies the parallel method began to show distortions in the inverse FT of its frequencies. Therefore, a more expensive lens than the one used would be necessary to keep up with the converging beam image quality.

Table I Maximum spatial frequencies of each snow crystal.

No. of snow crystal	$u_p(mm^{-1})$	$u_c(mm^{-1})$
a	12	12
b	12	12
c	11	11
d	16	19

VIII. ACKNOWLEDGEMENTS

We would like to thank Dr. Luis M. Bernardo for his helpful discussions and technical assistance on this research. One of us

(S. P-n) is a Fulbright Fellow on leave from the Department of Physics, Mahidol University, Bangkok, Thailand.

REFERENCES

1. Almeida, Silverio P. Optical Metrology of Fish Scales, (presented at this NATO school).
2. Goodman, J. W. Introduction to Fourier Optics (McGraw-Hill Pub. Co., New York, 1968).
3. Stark, H. Applications of Optical Fourier Transforms. (Academic Press, New York, 1982).
4. Almeida, S. P. and H. Fujii. Applied Optics, Vol. 18, No. 10, 1663-1667 (1979).
5. Joyeux, D. and S. Lowenthal. Applied Optics, Vol. 21, No. 23, 4368-4372, (1982).

OPTICAL METROLOGY OF FISH SCALES

Silverio P. Almeida

Virginia Polytechnic Institute and State University
Department of Physics
Blacksburg, Virginia 24061

I. INTRODUCTION

The study of fish scales presents an interesting problem in
optical processing. These scales are composed of a series of
annuli differing in number, spacing and contour shape. It turns
out that different species of fish as well as their age can show
differences in their scale structure. By performing the Fourier
transform of the scale, small structural differences in the scale
can be better analyzed in the transform plane due to the properties
of the Fourier transform. We have obtained preliminary results
on some scales and compared their optical Fourier transforms in
order to see if, indeed, their structural differences can be
observed and analyzed in this transform plane.

II. OPTICAL FOURIER TRANSFORMS

Shown in fig. 1 is a block diagram of the procedure used for
the study of the fish scale structure using the Fourier transform
method. The scale input is a positive photographic transparency
made from a scale impression. The optical Fourier transform of
this scale is produced by the method shown in fig. 2 and will be
treated in a separate paper (1). The mathematical details of the
optical Fourier transform are not the immediate subject of this
discussion. They have been well documented and are given in the
references (2,3). We shall concentrate our study on the input
scales and their transform differences. Once the Fourier transform of the scale is made optically, it is imaged onto the vidicon
where it is then digitized by an 8-bit A/D (analogue to digital)
converter. The digitized picture is composed of a pixel array
of 256 by 256 dots; each pixel in this matrix is, itself, sliced

SCALE AGE – VIA OPTICAL FOURIER TRANSFORM ANALYSIS

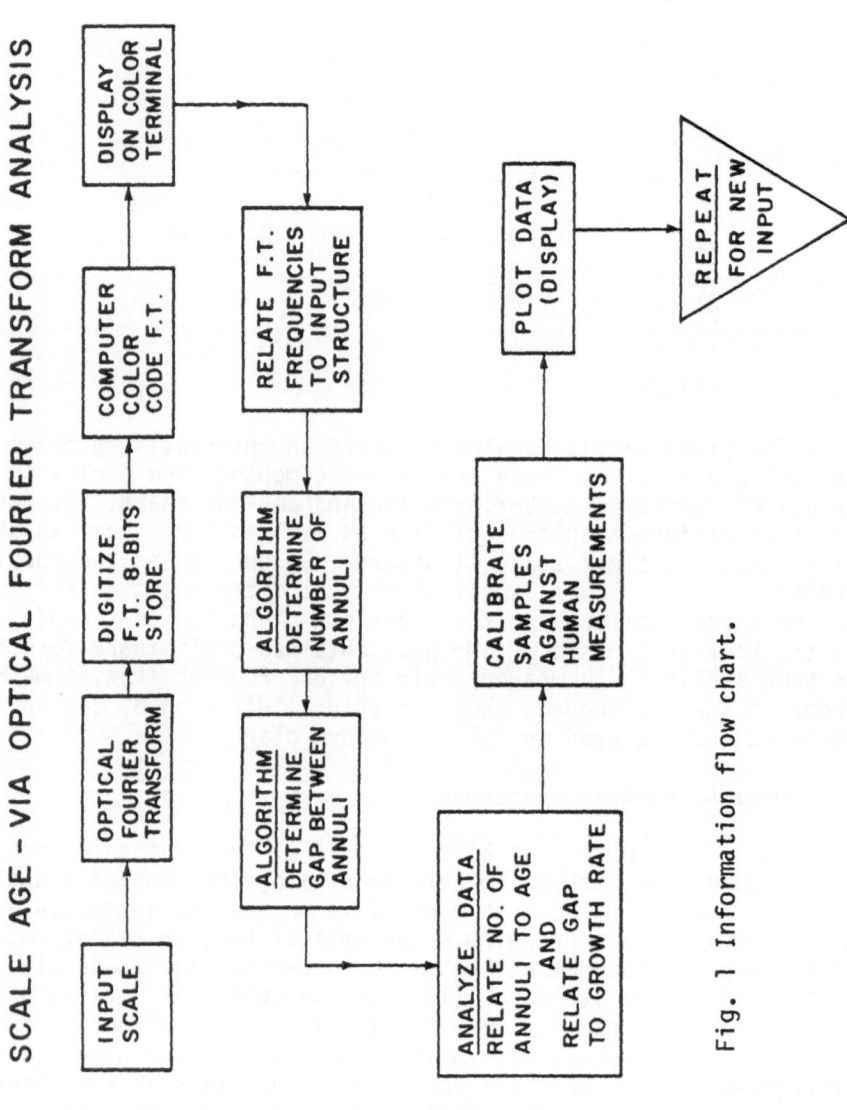

Fig. 1 Information flow chart.

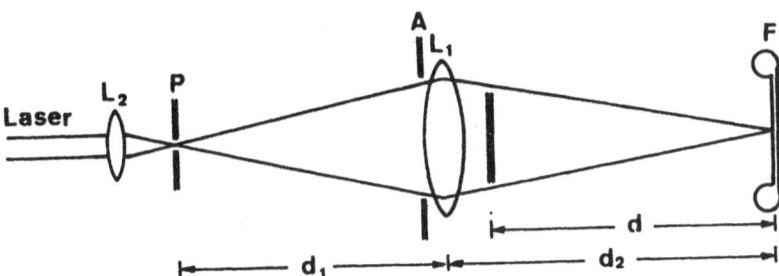

Fig. 2 Converging beam Fourier transform setup.

into 256 gray levels of light intensity. We have, therefore, mapped out the scale onto the digitized Fourier transform plane and stored this information into the micro-computer for analysis. At this point we can either display the results of the transform in black and white on the TV-monitors or in color. In this paper we shall consider only the black and white display.

Figures 3, 4 show the scale pictures of two Haddock fish used as input to the optical processor. Each is about 5 years old but they were caught in different waters. Their corresponding Fourier transforms are presented in fig. 3a, 4a. The pictures have not been normalized to output intensity and therefore should not be compared on this basis. They illustrate the total Fourier transform of the scales. We found, however, that it may not be necessary to use the total scale but simply a slice out of it. This enabled us to reduce the input data considerably and concentrate on the salient features of the scale. Figures 3b, 4b show the sampled scale inputs with their corresponding Fourier transforms given in fig. 3c, 4c. These sampled transforms were also digitized (8-bits) and stored in the micro-computer for comparison.

The computer was used to plot the sampled Fourier transform data shown in fig. 3c, 4c. Three dimensional plots of the data is presented in fig. 3d, 4d. Position is designated by the x, y coordinates; the z-coordinate represents the intensity of the pixel (where a range of 0-255 gray levels is possible).

III. RESULTS AND CONCLUSION

The optical Fourier transform photographs and their three

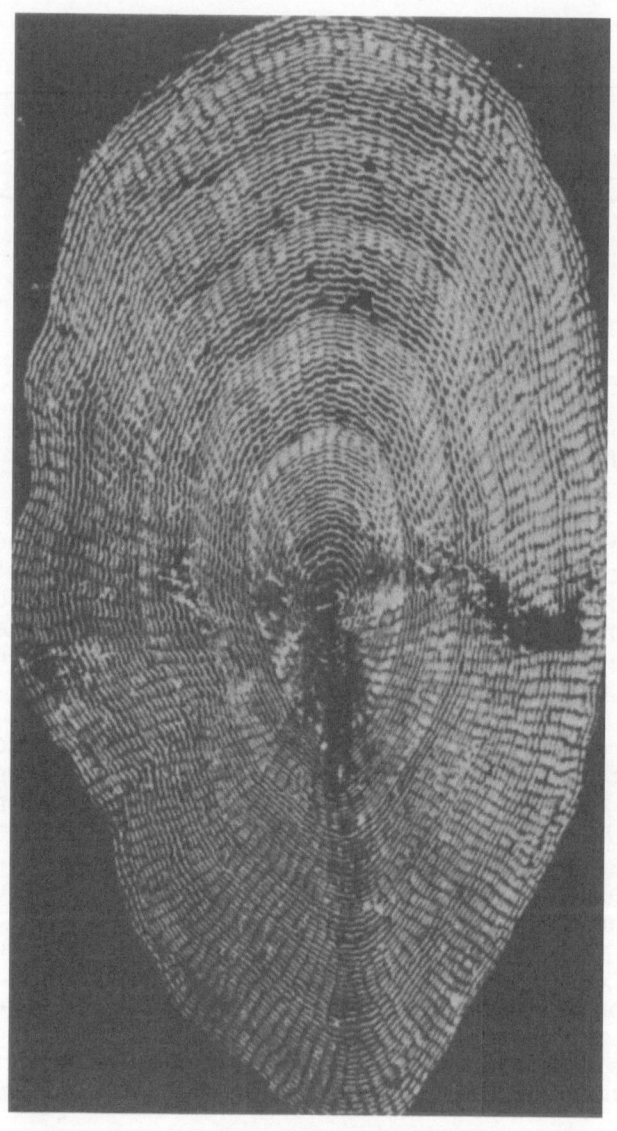

Fig. 3 Fish scale (Haddock) 5 years old.

Fig. 4 Fish scale (Haddock) 5 years old.
Different water than fig. 3 scale.

152

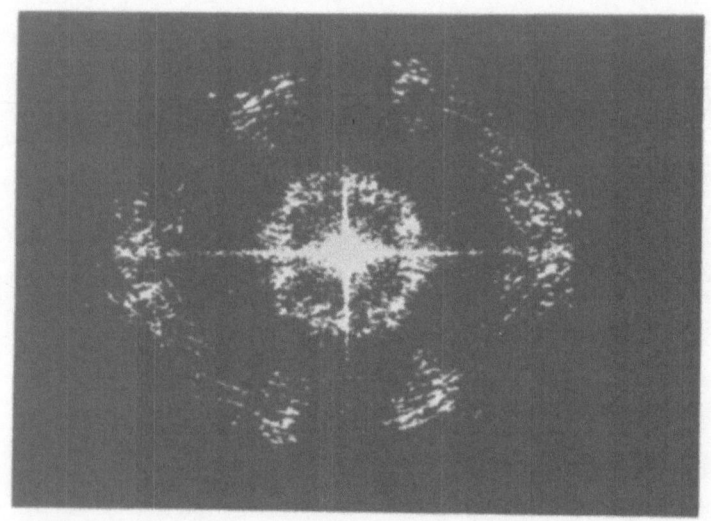

Fig. 4a Optical Fourier transform of fig. 4 scale.

Fig. 3a Optical Fourier transform of fig. 3 scale.

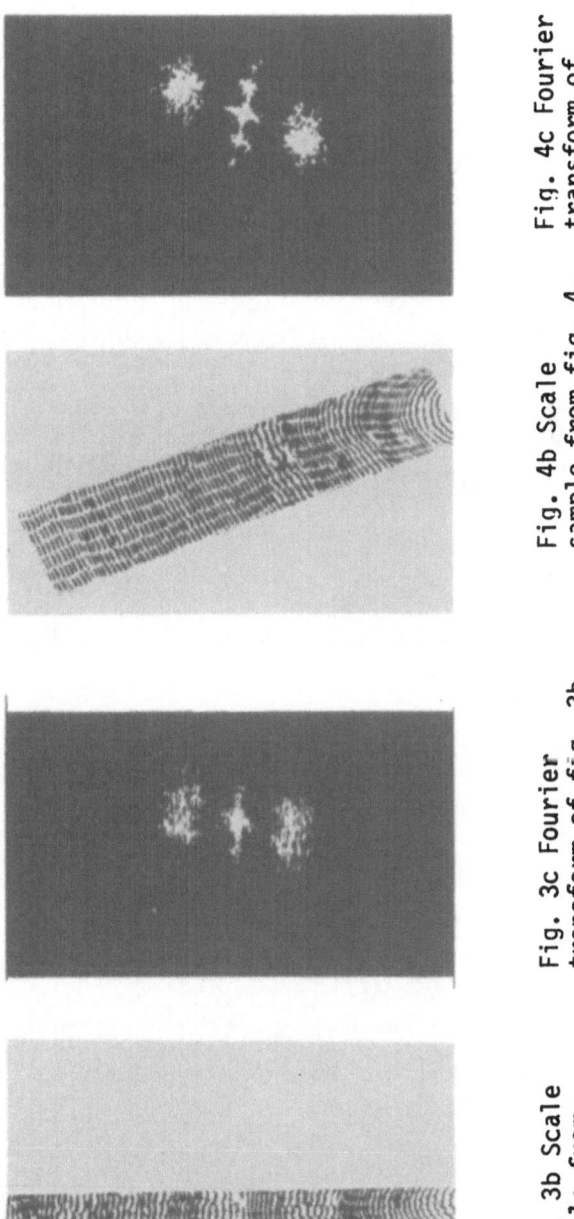

Fig. 4c Fourier transform of 4b.

Fig. 4b Scale sample from fig. 4.

Fig. 3c Fourier transform of fig. 3b.

Fig. 3b Scale sample from fig. 3.

154

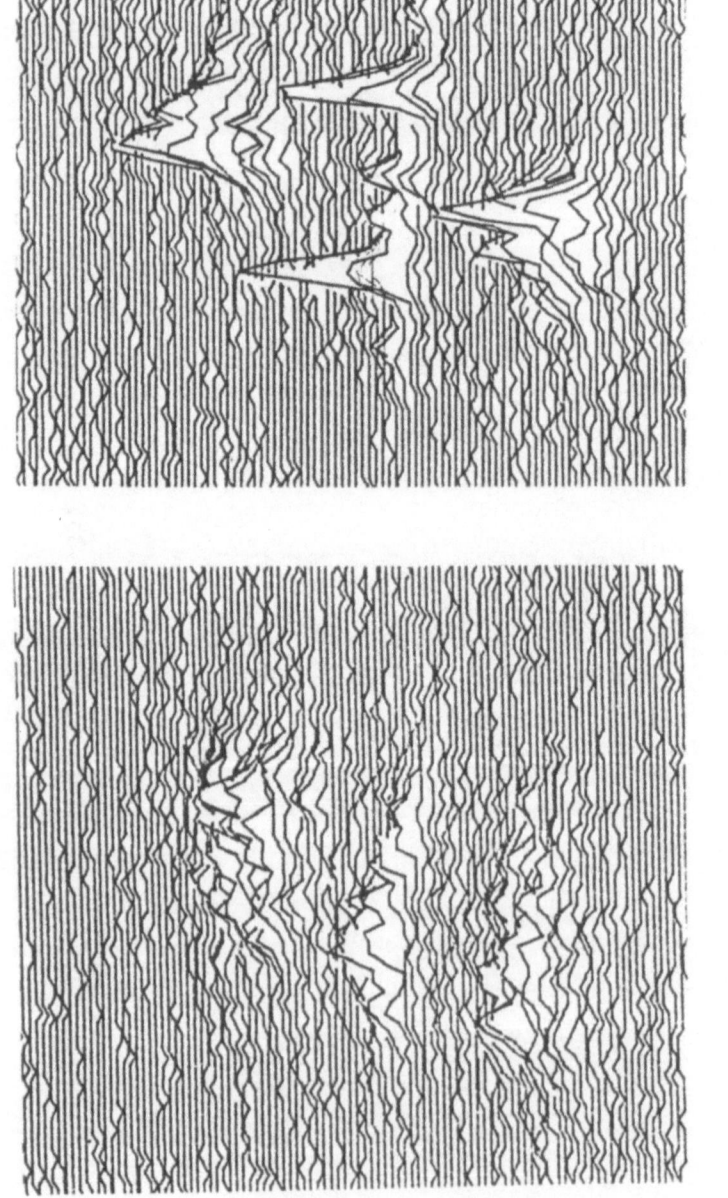

Fig. 4d 3-Dimensional plot of fig. 4c.
(Fourier transform).

Fig. 3d 3-Dimensional plot of fig. 3c.
(Fourier transform).

dimensional plots show scale stock differences (i.e. same species and age but caught in different waters). The exact correlation between the transform plane and details on the scales are currently underway. The new samples of scales being used represent the same fish species caught in the same waters but differing only in age. We shall try to quantitatively correlate the Fourier transforms of the scales for age signatures. The scale intensity will also be calibrated to some appropriate normalization factor so that an accurate Fourier transform comparison can be made.

IV. ACKNOWLEDGEMENTS

It is a pleasure to thank Dr. Luis M. Bernardo and Srisuda Puang-ngern for help and discussions on this research. We are also grateful to Dr. Ambrose Jearld, Jr. for providing us with the samples and for fruitful discussions on the fish scale problems. This research was partly supported by the contract: DOC NA-83-FA-C-0029.

REFERENCES

1. Puang-ngern, Srisuda and Silverio P. Almeida. Converging Beam Optical Fourier Transforms (Presented at this NATO School as a separate paper).
2. Goodman, J. W. Introduction to Fourier Optics (McGraw-Hill Publishing Co., New York, 1968).
3. Stark, H. Applications of Optical Fourier Transforms (Academic Press, New York, 1982).

HYBRID COHERENT OPTICAL AND ELECTRONIC FILTERING

Silverio P. Almeida

Virginia Polytechnic Institute and State University
Department of Physics
Blacksburg, Virginia   24061

I. INTRODUCTION

The term "hybrid" in optical processing refers here to the coupling of a computer and electronics to an optical system. One can combine in a single system the attractive features of the fast parallel processing that optics provides with the flexibility and analytical power afforded by the computer. We shall describe such a system that was designed to produce and analyze optical Fourier transforms and holographic filters for the purposes of pattern recognition and image studies of objects. Due to the nature of such a hybrid system we shall cover a variety of topics in this paper. Namely, discussions on optical Fourier transform setups, digitization of images, graphical representation of the data by on-line computer plotting, hardware involved in such a system, software necessary to process the optical information and finally some applications. This paper will be, therefore, a description of the system; the details of the mathematics involved in the Fourier process (1,2) and electronic hardware are given in the references.

II. OPTICAL IMAGE PROCESSING VIA DIGITIZATION

A. Fourier transform digitization

Perhaps the simplist operation one can use the hybrid processor for, is to image the object onto a vidicon and display it on a television monitor (TVM). The next step would be to connect the vidicon to a digitizer (say 6 or 8 bits) and transmit the data to a mini-computer. Shown in fig. 1 is an overall view of

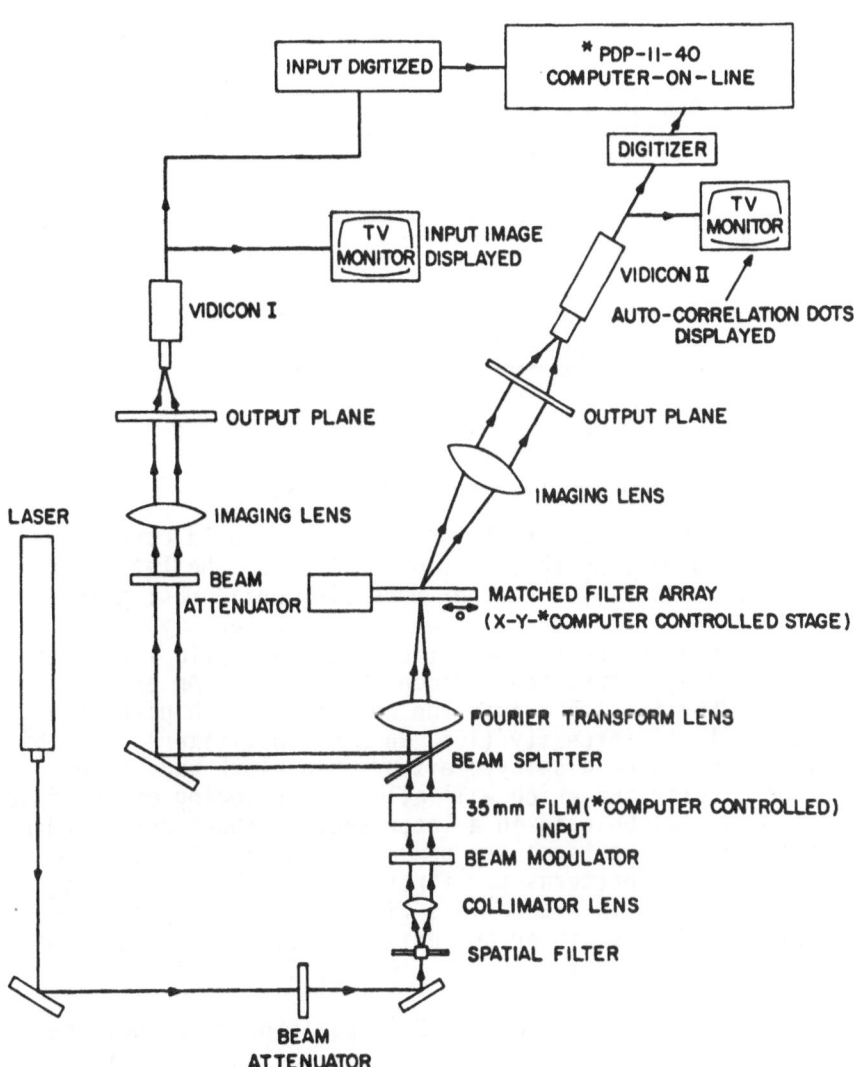

Fig. 1 Hybrid optical processor layout.

158

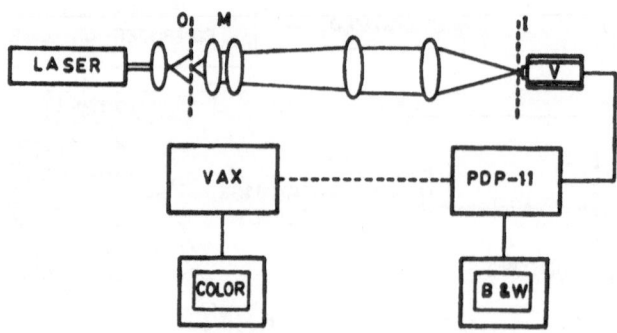

Fig. 2 Optical Fourier transform (OFT) system.

the hybrid system. The method just discussed would correspond to
the information passing through vidicon I (i.e. the left-hand
portion of the diagram). Another possible configuration is
shown in fig. 2. Here we see an object (transparency) O
illuminated by a coherent source (laser). The optical Fourier
transform is imaged onto the vidicon camera (V). An array of
pixels (256x256) is stored into the computer. Each pixel has
256 gray levels of intensity (i.e. an 8-bit digitizer). At this
point the data can be displayed either on a black and white
monitor or a color one which will allow color coding of the intensities. More will be said in a later section about color coding.
Photographs of two patterns are presented in fig. 3. In this
and most cases the patterns are first photographed onto a high
contrast film 35 mm in size. This film (i.e. the input) can
either be on a film strip or an individual slide. It is then
inserted into the object plane (O) in fig. 2.

The vidicon camera V in fig. 2 is positioned so that one
can either move it to record the image of the input, or the
Fourier transform of the input. Figure 4a shows twenty one
patterns all with the same orientation. They have the same shape
except for size differences. Their averaged Fourier transform
is presented in fig. 4b. The averaged transforms are useful for
group pattern studies. Once the digitized Fourier transform is
stored in the computer it is possible to make comparison studies
of them in order to look for similarities and differences among
the input patterns. Shown in fig. 5a and 5b are the plotted
differences between two transforms. The computer is programmed so
that each of the two transforms is exactly superimposed then the
subtraction of the two transforms is performed. The spatial

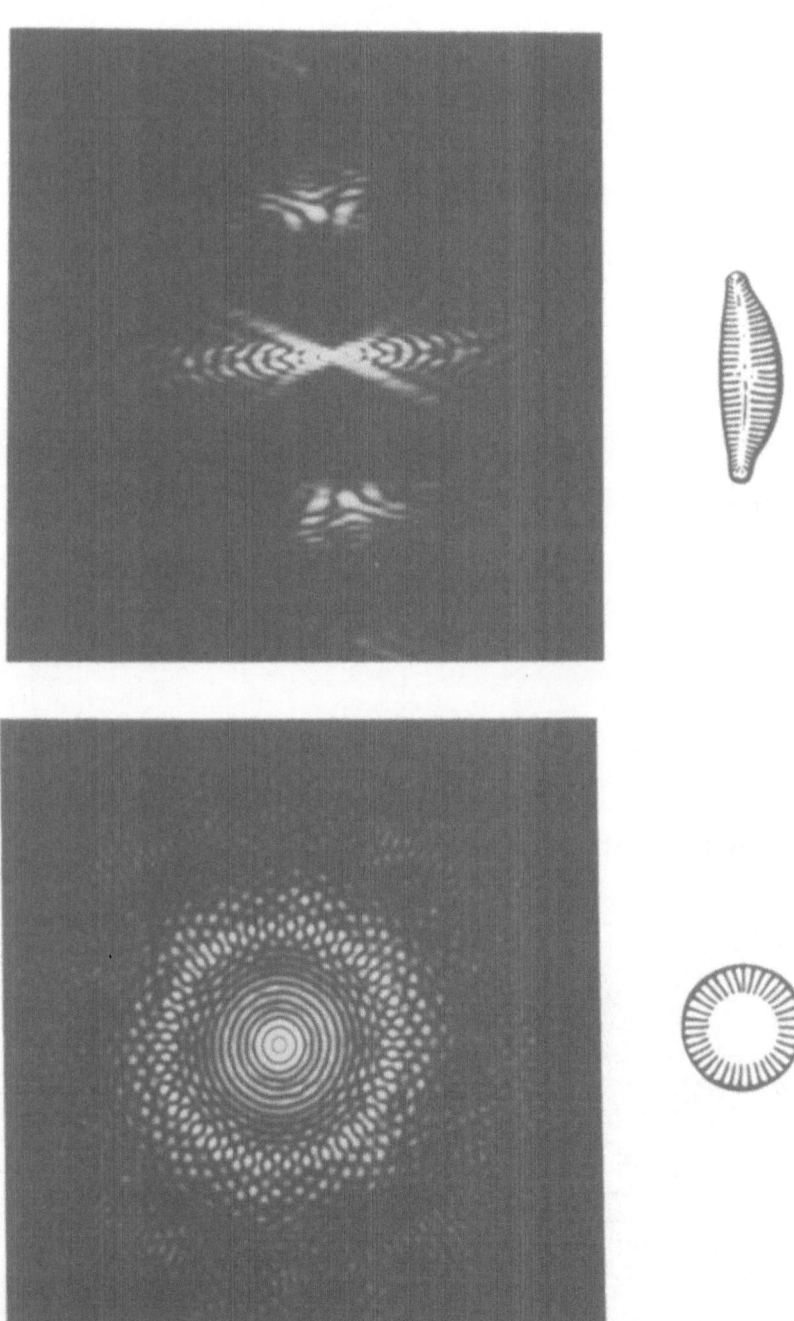

Fig. 3 Input patterns and their OFT.

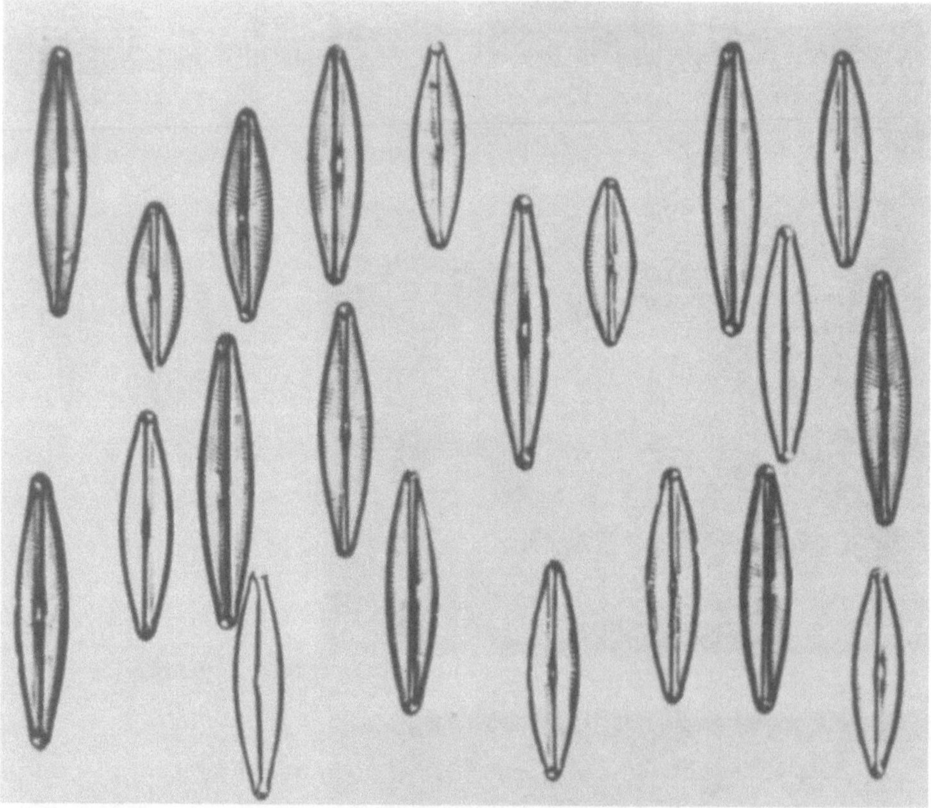

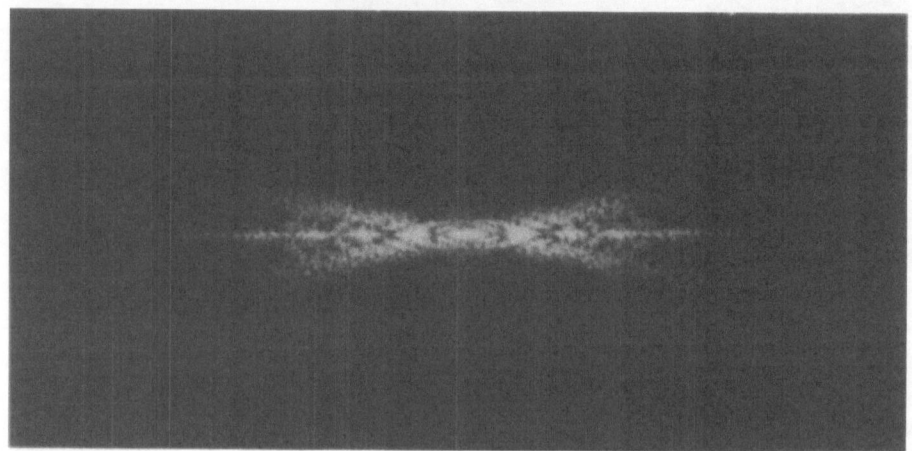

Fig. 4a   Similar input patterns.

Fig. 4b   Averaged Fourier transform of (a).

Fig. 5a Differences between digitized OFT patterns.

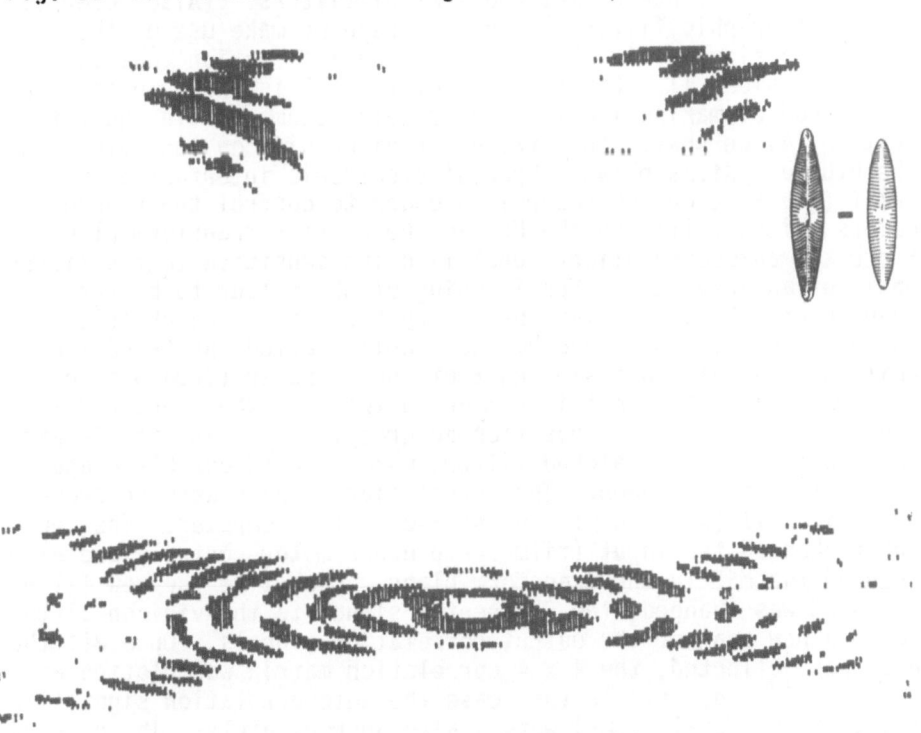

Fig. 5b Same as (a); different patterns.

frequency results shown in fig. 5, therefore, allow one to measure the coordinates of the various remaining pixel groups and calculate where on the original inputs the patterns are similar or different in structure. It is also possible to plot fig. 5 as a function of intensities by simply thresholding the transform pixels before plotting them (3).

Once the Fourier transform optical setup is made one can position a second lens to perform a transform of the transform. It is almost an inverse transform except that the coordinates are reversed. The double transform setup has the advantage that one can then modify the Fourier transform plane with some type of high or low spatial frequency filter before the second transform is performed. The filtered picture can be plotted the same way that is shown in fig. 5. This is a powerful method of studying the various spatial frequency effects and their contributions to a given pattern.

B. Matched spatial filters (MSF)

The hybrid optical processor plays an important part in the construction and use of matched spatial filters. (Also referred to as holographic filters.) In this case we make use of the right hand side of fig. 1. That is, the information that is being sent to vidicon II. Construction of the MSF is shown in fig. 6b. The method of making the filters is well documented in the lit- erature (4) we shall concentrate our discussion on the role that the hybrid systems plays. Special electronic interfacing was built (see section on hardware) in order to control the stepper motors which positioned the MSF in the Fourier transform plane. Since the autocorrelation signal is quite sensitive to the filter position an accurate x,y positioning of the filter to better than about 2.5 microns was necessary for this research (5). Through a special designed hardware unit, called the MPX-Buffer unit the central processing unit of the computer (CPU) was able to communicate through tailor made software to the stepper motors and achieve the desired position accuracy. Shown in fig. 6a are four patterns whose matched filters were used to correlate against the four patterns shown. The correlation signals were recorded by vidicon II (see fig. 1) and stored in the computer. The com- puter changed the input (film strip under motor control) and also changed the MSF in the transform plane. Each time the new filter or input was changed, the CPU sent a signal to the vidicon II to begin a new scan of the output correlation signals. Once all the data was collected, the 4 x 4 correlation matrix was plotted as is shown in fig. 6a. In this case the autocorrelation signals (i.e. the diagonal terms) were scaled appropriately. The hybrid system is a tremendous value in such studies since it eliminates the very tedious and time consuming process of accurate position-

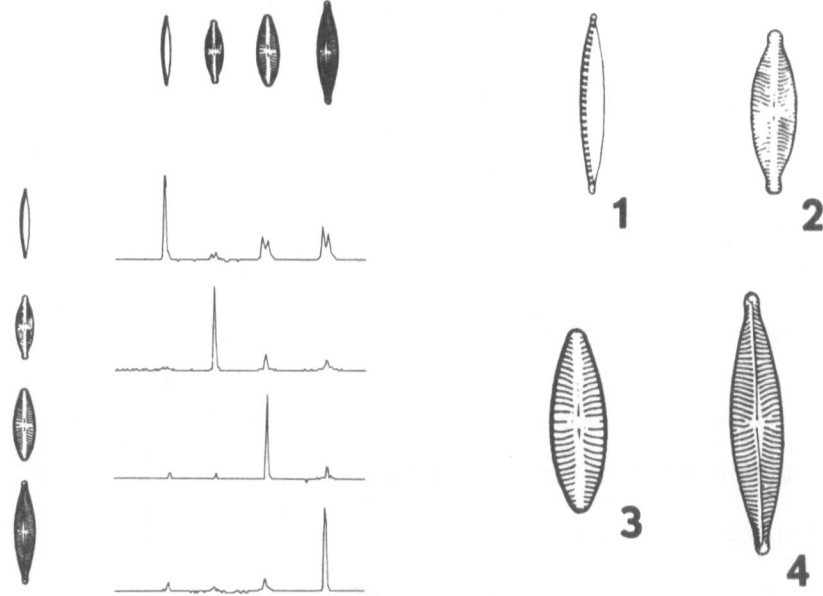

Fig. 6a Correlation signals (normalized) for matched spatial filtering of patterns shown.

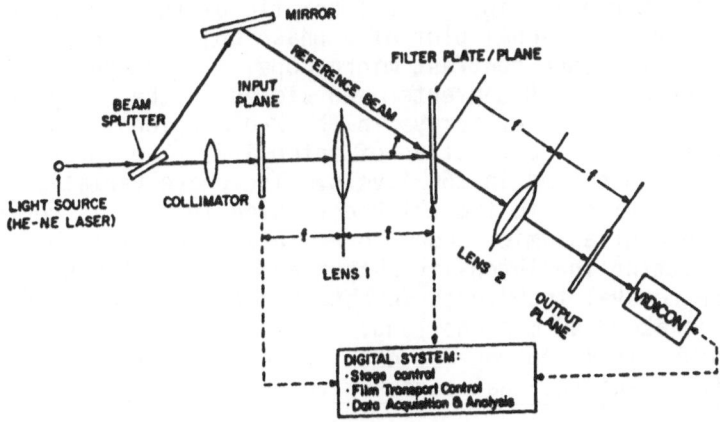

Fig. 6b Matched spatial filter setup.

ing by hand micrometers. Of course, the construction of the hybrid system itself is a considerable amount of work and cost. One must, therefore, balance off the various trade-offs in such research. Other examples of the MSF correlation signals as seen by the vidicon II are shown in fig. 7. One way to reduce the MSF frequency data is shown in fig. 8. Here the horseshoe shape figure was substituted for the actual pattern above it. The simulated MSF used primarily low spatial frequencies and was able to remove some of the filter sensitivity to scale variations and also rotations (6). Again, the hybrid system had rotating stages under computer control through the MPX-Buffer unit.

III. Graphical representation of digital data

On-line optical processing has another advantage in that the digitized data stored in the computer can be easily accessed to be plotted. The plots can be presented in such a way as to best extract the information from the data. That is, one can use 2-dimensional hidden line removal techniques and also rotate the plot in space to get the best perspective. More will be said about the software programs in the following section. We shall now give an example of 3-dimensional plotting.

A. Phase objects

Phase objects while more difficult to work with than most other objects play an important role in biology and medicine. By using the Zernike phase contrast method we are able to study such objects.(1). Shown in fig. 9a is schematic of the procedure. Fig. 9b is a 2-dimensional plot of a phase object which was imaged out of an interference contrast microscope. The image was digitized via an 8-bit A/D converter and stored in the computer. Each of the 256x256 pixel arrays in the 2-dimensional plot was plotted in fig. 9b as a function of intensity. The intensity of each pixel was recorded in 256 levels. The phase variation of the object can be related to the thickness which in turn is converted by the phase contrast microscope to an intensity variable. Therefore, by plotting the intensity output and using 3-dimensional hidden line removal techniques we obtained the specimen's size in real life is about 40 microns long. A microscopist would normally have to focus his microscope at each different depth in order to view the full 3-dimensional perspective shown in fig. 9c.

B. Color coding for intensity

Once the digital intensity information is available one can move a step beyond just plotting the data as was shown in fig. 5 and fig. 9b, 9c. It is possible (we have done this but the plots are not shown due to the lack of color in this publication) to color code the 2- and 3-dimensional plots for intensity. This

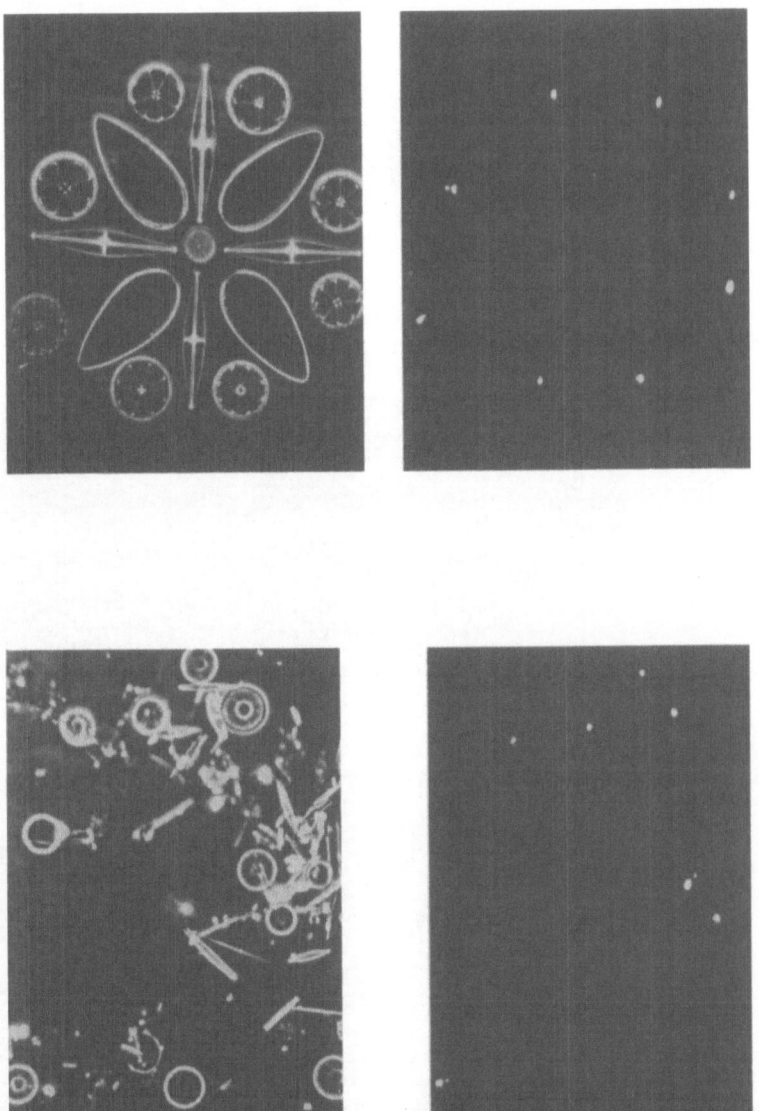

Fig. 7 Bright dots correspond to the autocorrelation signals obtained for the matched spatial filter. Location of the input patterns is in the same relative position as for the dots.

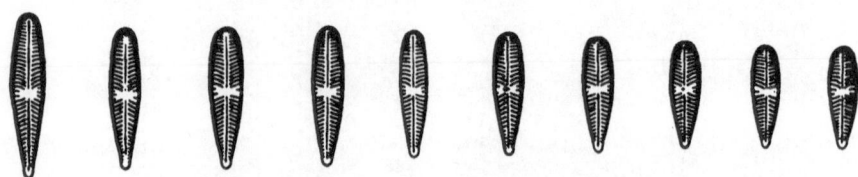

Fig. 8a   Same input patterns except for size.

Fig. 8b   Autocorrelation signals using filter for object shown.

Fig. 8c   Correlation signals using a spatial filter for the horseshoe.

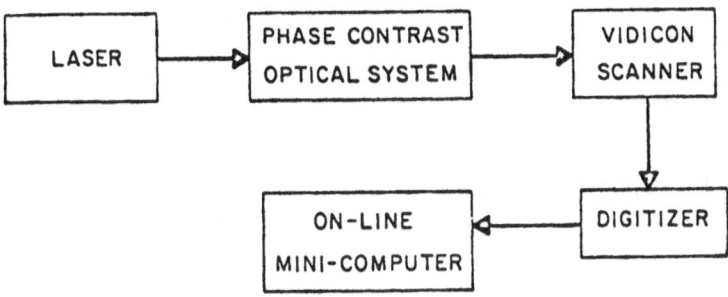

Fig. 9a Phase contrast setup.

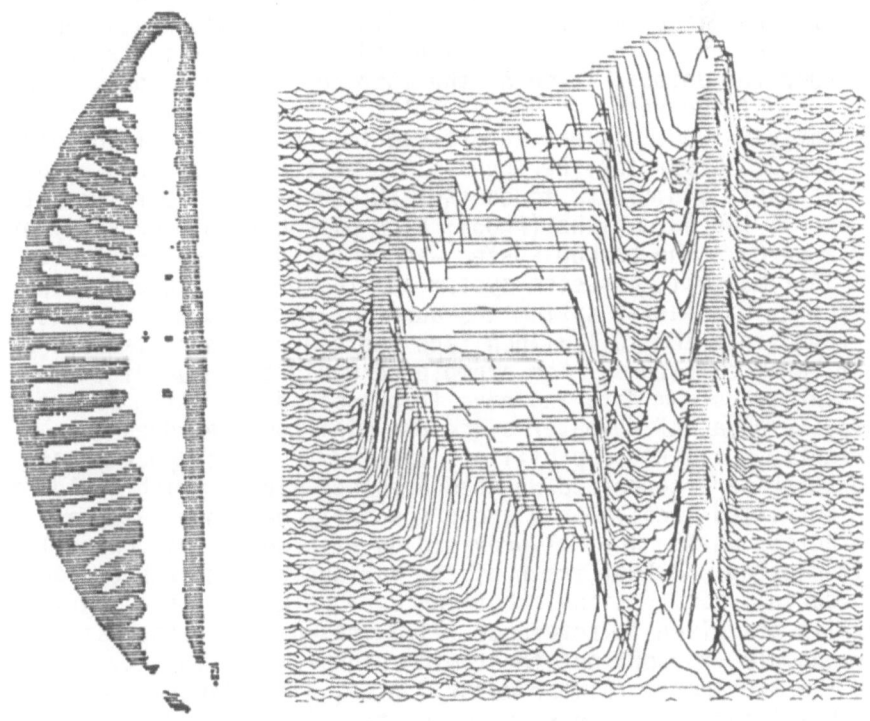

Fig. 9b Two-dimensional plot of
the phase object intensity.

Fig. 9c Three-dimensional
plot of the same object.
Intensity is the 3rd
dimension.

technique greatly enhances the data for visual observation. Important intensity differences are highlighted in different colors. It allows one to reprocess the data in such a way as to extract the maximum amount of information from the image. This method greatly facilitates the enhancement of edge and contour plots.

IV. HYBRID HARDWARE

The particular hardware configuration necessary to carry out optical processing will vary depending on the research objectives. We shall describe those which pertain to our research on the study of optical Fourier transforms and color coding. However, many of the components used here are in fact common to other areas of optical processing. Shown in fig. 10a is layout for the information flow pertaining to the matched spatial filter research we discussed previously. Fig. 10b is block diagram showing the hardware configuration and the various interfaces to the computer unibus. Some of the components used in our, later version, of the research are as follows. The TV camera was the Hamamatsu C-1000. It was interfaced to the DEC-PDP-11/40 computer via the DEC-Laboratory Peripheral System (LPS). The LPS used one of its 8-bit digitizers to do the A/D conversion. Data was stored on one of two hard disks (RK05); each disk having a capacity of 1.2 M words. The backup system was composed of two media; the DEC-TA11 magnetic cassette system and the DEC-TM magnetic tape system. This latter system was our link to the other computer systems since the 10 1/2-inch tape reels are rather standard on most large computers. For example we would take our digitized Fourier transform data and have it processed for color coding analysis on a VAX computer since our mini-computer does not currently have this capability. Insofar as the stepper motors and stages are concerned we used Aerotech stages which could either be controlled locally or via our specially designed interface MPX-Buffer unit. The MPX provided us with additional channel multiplexing and kept track of the various timing problems occurring during sharing signals between various motors, etc. peripherals. Details are given in reference (7). The unit was responsible, for example, to sort out the timing signals from the x,y motors and the 360 degree rotator motor. Also, via software commands the MPX triggered the next scan for the vidicon, and advanced the film strip to the next input. When all this was done, a command signal was given to readout the correlation signals and store the results onto disk. Shown in fig. 11a is a photograph of the MSF stage with its filters (up to a 10 x 10 array of filters could be placed on this 2 x 2 inch plate). Computer and stepper motors positioned the filters to better than 2.5 micron accuracy in the x,y coordinates. Fig. 11b shows a photograph of the MPX-Buffer unit. It contained both D/A (digital to analogue) and BCD (binary coded channels) to control the various motors on the optics table.

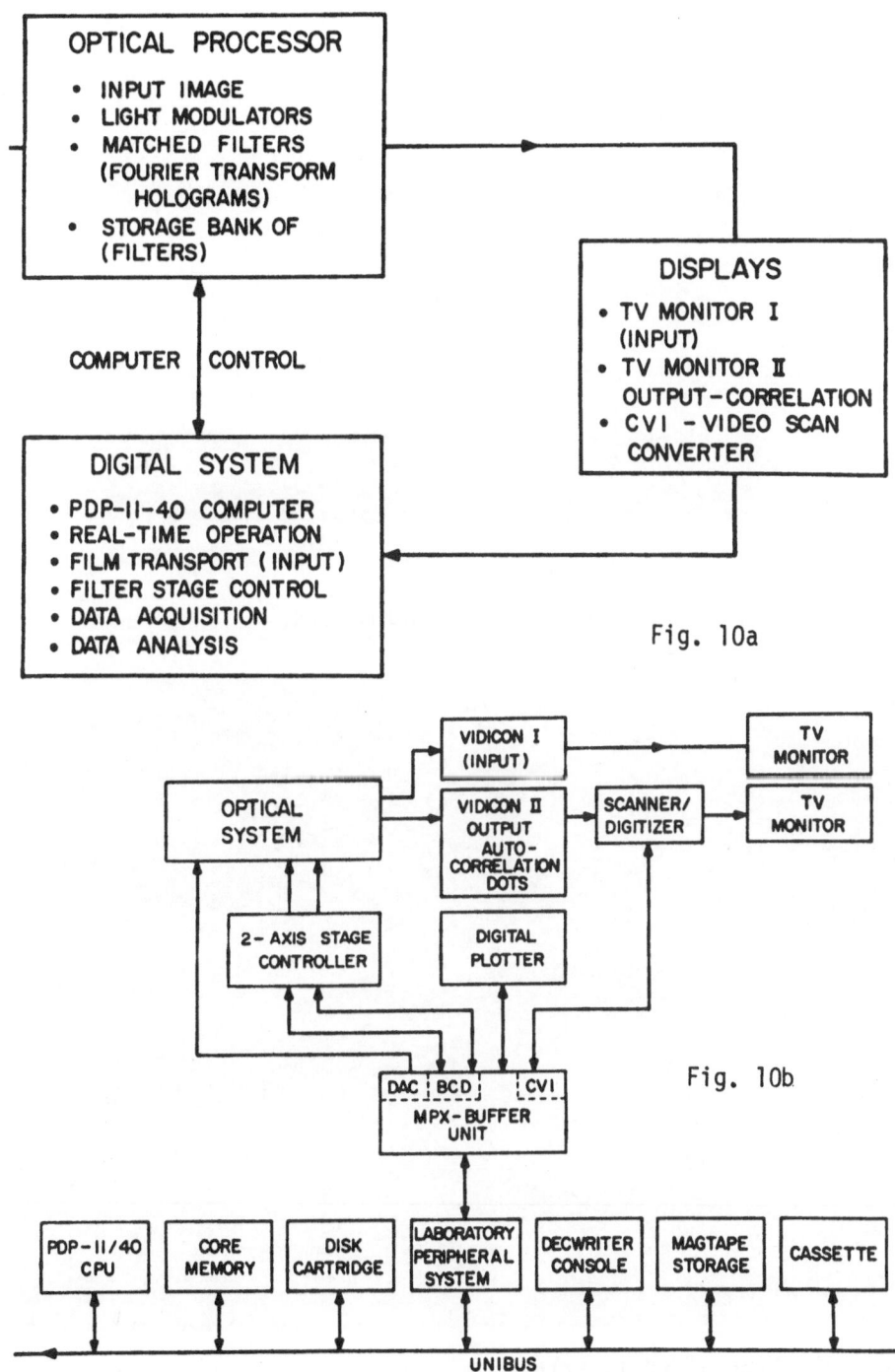

Fig. 10a

Fig. 10b

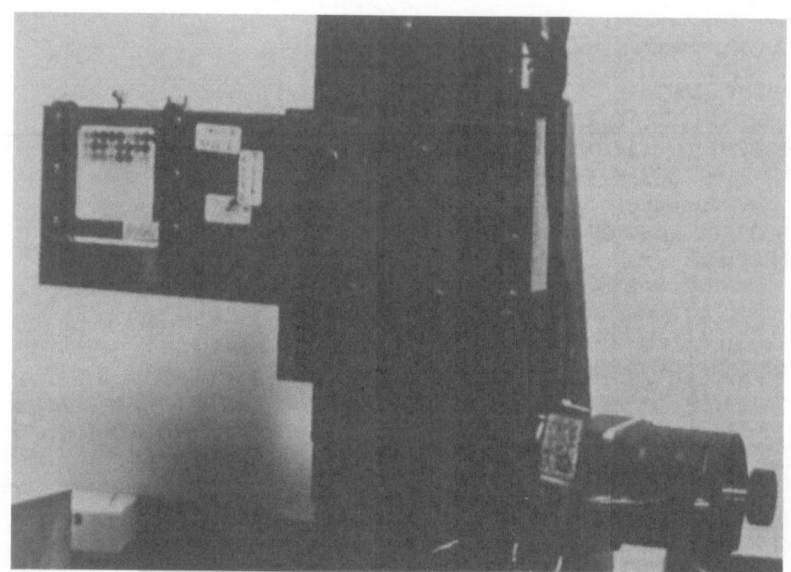

Fig. 11a Filter stage and positioning motors.

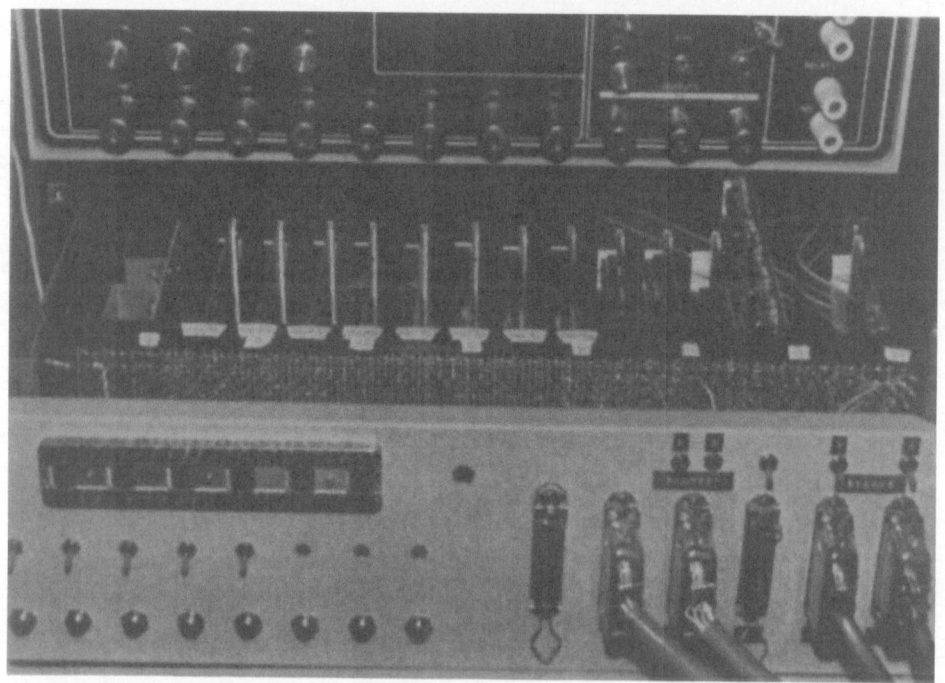

Fig. 11b MPX-Buffer interface.

V. HYBRID SOFTWARE

The software necessary to control the various stepper motors and deal with the digitization data were all written in PDP-11 MACRO-11 assembly language. The main operating system for the PDP-11 system computer is the DEC-RT-11 program. Once the lower level machine language programs were written they were used as subroutines in Fortran language to simplify operators usage of the processor. The next level up from Fortran was to develope a program package which utilized all the lower level subroutines but which was itself a "MENU" driven package. This software is called Image Enhancement Device System "IEDS2" and is capable upon a single letter command of reading a file, writing a file, displaying a file and getting a video input. The commands would be: <R>, <W>, <D>, <V> etc. Under <D> display for example there would be another menu listed. This menu would enable the operator to do picture processing and manipulation. That is, one could use picture processing algorithms such as the operators: <D>ITHER, <C>ONSTRAINED AVERAGE, <R>OBERTS CROSS GRADIENT, <T>HRESHOLD, <M>EASURER, <J>OYSTICK, <Z>OOM, <S>CREEN ERASE, <P>LOT HISTOGRAM, and <E>XIT. These are some of the possible software processing operations one could perform on the digitized data stored in the computer. Various programs to remove noise, inverse contrast, rotate a plotted figure and others were also available. The final results were printed on a digital HP-7210A plotter which was under the MPX-Buffer control. In the case of color displays we either took the data over to the VAX computer for picture processing, since those programs were more involved, or we transmitted the data via computer terminals. Once the data was stored in the VAX computer we could communicate with the computer via a fiber optics data link and bring up the color displays in our own laboratory. The programs used in this case were formulated in the Spatial Data Analysis Laboratory and run under the program called "GIPSY".

VI. SUMMARY AND CONCLUSION

We have described a hybrid optical processor and how it can be used to study optical Fourier transforms, spatial filter correlations, phase contrast objects and color coding. The advantages of such a system are that one has great flexibility in analyzing the data that the optical system provides it. Alone, using optics, one could not carry out so much analysis. However, the speed with which one can do optical parallel processing comes to a great slow down at the computer interface. In the end the slowest component in the system dictates the over all speed of the processing. In our case it was the stepper motors and, line scanners and the serial data channels of the computer. This is a trade-off for having the possibility of performing additional picture processing by the software programs. A hybrid system

can add, as we have seen, new dimensions to the data processing.
The optical components processing the data can greatly reduce the
burden on the computer processing. Hence, the proper marriage
between the two systems can very well solve problems that neither
one alone is capable of handling.

References

1. Goodman, J. W. Introduction to Fourier Optics (McGraw-Hill
    Pub. Co., New York, 1968).
2. Stark, H. Applications of Optical Fourier Transforms
    (Academic Press, New York, 1982).
3. Almeida, Silverio P. and Hitoshi Fujii, Applied Optics,
    Vol. 18, No. 10, 1663-1667, (1979).
4. Vander Lugt, A. B., Trans. IEEE, Vol. IT-10, 139-145, (1964).
5. Almeida, Silverio P. and Kim-Tzong Eu, Applied Optics,
    Vol. 15, No. 2, 510-515, (1976).
6. Fujii, Hitoshi and Silverio P. Almeida, Applied Optics,
    Vol. 18, No. 10, 1659-1662, (1979).
7. Almeida, Silverio P., James Kim-Tzong Eu and Peichung F.
    Lai, Trans. IEEE, Vol. IM-26, No. 4, 312-316 (1977).

HYBRID COHERENT OPTICAL AND ELECTRONICAL OBJECT RECOGNITION

R. Dändliker and K. Hess

Institut de Microtechnique de l'Université,
CH-2000 Neuchâtel, Switzerland.

The described hybrid system combines the high speed of parallel
Fourier processing (coherent optics) and the flexibility of digital
programming. This is realized by replacing the matched optical
filter (hardware) in the Fourier plane by electronic processing
(software) of amplitude and phase detected in that same plane. The
theoretical and experimental results give evidence of the feasibility of this concept for on-line object tracking and recognition,
such as robot vision. Learning and programming are easily done by
presenting the reference objects at their respective reference
positions and electronically memorizing the corresponding signatures
and phase values.

1. INTRODUCTION

Coherent optical image processing uses essentially the 2-D
Fourier transform properties of coherent optical imaging and
appropriate filters in the Fourier plane to modify the image by
convolution or to extract correlation. It offers high-speed,
on-line, real-time operation at high spatial resolution due to its
inherent parallel processing of 2-D information. However, its practical
applications are limited mainly by the lack of flexibility of the
filter function, which has to be realized in the hardware.

A new class of hybrid system combines the high speed Fourier
transform properties of coherent optics with the flexibility of
digital programming for filtering. This can be done by placing the
interface between optics and electronics in the Fourier plane and
by carefully selecting the revelant information before optoelec-

Soares, O.D.D. (ed), Optical Metrology
© *1987. Martinus Nijhoff Publishers, Dordrecht.*

tronic detection in order to reduce the amount of data to be
processed electronically. Very promising system determine the 2-D
irradiance moments of a pattern, or an object [1], from which the
invariant moments can be deduced for efficient pattern recognition
independent of orientation and size [2].

The use of this concept in on-line object tracking and
recognition for robot vision will be described in the following
sections. Heterodyne detection in the Fourier plane is used to
determine the phase of the spectrum which is necessary to calculate
the object position.

2. REAL-TIME OBJECT TRACKING AND RECOGNITION

The basic task in robot vision is to identify an object and to
determine its position from the 2-D image seen by the eye of the
robot. As shown in Fig.1 the position of an object is given by the
two coordinates x_s and y_s of its center of gravity and its angular
orientation α_o. The object has to be identified within the ensemble
of a limited number of known objects O_n. Real time means that the
object must be recognized and located within less than 1 sec.

Solutions to the problem using coherent optical correlation
have been described [3,4]. An incoherent image of the scene,
containing the diffusely scattering object to be recognized, is
formed by a first lens on the input face of an incoherent-to-
coherent image transducer, in this case a liquid crystal light
valve (LCLV). The coherent image is then optically Fourier
transformed by a second lens. The object recognition is performed
with the help of a holographically recorded matched filter in the
Fourier plane which produces a correlation spot after transformation by a lens.

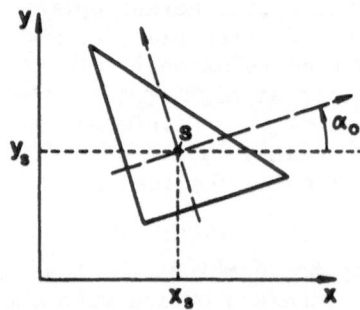

Fig.1. Angular orientation α_o and lateral position x_s, y_s of the 2-D
image of an object.

Since the Fourier transform is not invariant to rotation, it is necessary to rotate the light field with respect to the filter. This can be done by an optical image rotator between the image transducer and the Fourier plane [4]. To identify several objects, individual holographic filters have to be employed for each object.

The main drawbacks of coherent optical correlation systems for object recognition and tracking may be summarized as follows:

-- Holographic recording of the matched filters
   (hardware programming);
-- Sequential searching for the angular orientation by
   producing image rotation;
-- Individually matched filters for each object to be recognized
   (hardware addressing).

The hybrid system presented in Fig. 2 eliminates these drawbacks by replacing the matched holographic filter (hardware) with electronic processing (software) of amplitude and phase detected in the Fourier transform [5,6].

Recognition and tracking is achieved as follows: In a first Fourier plane the power spectrum (intensity), which is known to be shift invariant, is observed [Fig.3(a)]. A circular scan with a diode array detector yields a 1-D signature $P(\phi)$ [Fig.3(b)] of the object. The angular orientation α_0 and the identification of the object are then easily obtained by calculating the correlation with the stored signatures $P_n(\phi)$ of the reference objects. In a second Fourier plane the phase of the spectrum is measured by heterodyning with a reference wave shifted in frequency by about 40 kHz. With

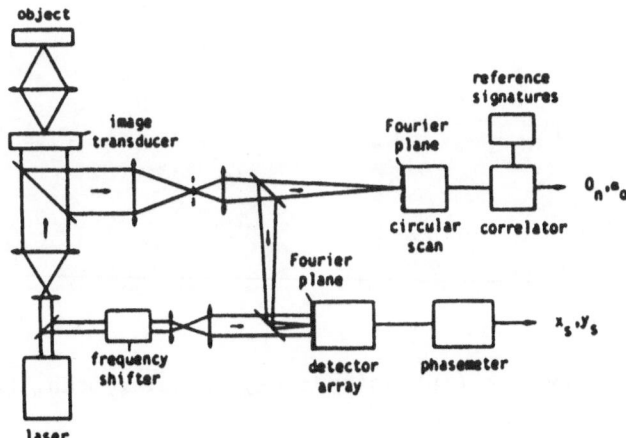

Fig.2. Hybrid optical-electronic system for robot vision.

two pairs of photodetectors [Fig.5(a)] the linear phase shifts due to the translation x_s, y_s of the object can be determined in two orthogonal directions. The optical phase shifts are transformed by the heterodyne detection into phase differences of sinusoidal signals [Fig.5(b)] which are easily measured electronically.

3. POWER SPECTRUM: IDENTIFICATION AND ANGULAR ORIENTATION

As shown in Fig.2, the power spectrum is observed in a first Fourier plane by looking at the intensity of the diffraction pattern [Fig.3(a)]. This is known to be shift invariant and centered at the origin of the Fourier plane. The complex amplitude of the light field in that plane is proportional to the 2-D Fourier transform,

$$\hat{O}(\vec{p}) = \int d^2x \; O(\vec{x}) \; \exp(-i2\pi\vec{p}\cdot\vec{x}), \tag{1}$$

of the coherent object function $O(\vec{x})$ in the input plane. The power spectrum $P(\rho,\phi)$, expressed in polar coordinates ρ and ϕ, is then given by

$$P(\rho,\phi) = |\hat{O}(\vec{p})|^2, \quad \rho = |\vec{p}|, \tag{2}$$

independently of the lateral position $\vec{x}_s$ of the object $O(\vec{x} - \vec{x}_s)$. Rotating the object, however, rotates the power spectrum by the same angle.

 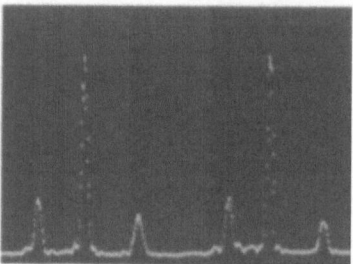

Fig.3. Power spectrum (intensity): (a) circular scan, (b) signature $P_0(\phi)$.

The angular orientation α_o (Fig.1) and the identification of the object are obtained from a circular scan of the power spectrum [Fig.3(a)] at radius ρ_o. This yields a 1-D signature [Fig.3(b)],

$$P_o(\phi) = P(\rho_o,\phi), \tag{3}$$

which can be compared with the previously stored signatures $P_n(\phi)$ of the reference objects O_n. For this purpose the 1-D correlations are calculated:

$$C_n(\alpha) = \int d\phi P_o(\phi) P_n(\phi-\alpha)/p_o p_n, \tag{4}$$

$$\text{with } p_o^2 = \int d\phi P_o^2(\phi) \text{ and } p_n^2 = \int d\phi P_n^2(\phi).$$

The height of the peaks $C_n(\alpha_{max})$ is used to identify the object, whereas the position α_{max} gives the angular position α_o. Positive identification of an object can be verified through the threshold (C_r) condition

$$C_r \leq C_n(\alpha_{max}) \leq 1, \tag{5}$$

which must be fulfilled for one and only one reference object O_n.

Fig.3 shows the power spectrum $P(\rho,\phi)$ and the signature $P_o(\phi)$ of a triangular object. The scan was made ba a Reticon RO-720 circular photodiode array [6]. It has 720 diodes of 250 μm radial length, 20 μm width, and 30 μm spacing (center to center) on a circle of 3.44 mm radius. The scan radius ρ_o was chosen to be about six times the average radius ρ_c of the zero-order diffraction spot $(\rho_o \simeq 6\rho_c)$, which is inversely proportional to the overall size of the object. Since the fine structures of the spectrum have about half the size of the zero-order spot, the detector ring covers about one-third of such an element and the peaks in the signature $P_o(\phi)$ [Fig.3(b)] are about 20 diodes wide.

For experimental tests of the concept, a set of four binary objects $O(\vec{x})$ (structured openings in metal plates), representing different classes of shape, was chosen. Two examples of detected signatures $(\rho_o = 10\,\rho_c)$ are shown in Figs.4(a) and (b), together with their corresponding objects in the upper right corner. Since the object functions $O(\vec{x})$ are real valued, the power spectra are symmetric (Fig.3), and thus the signatures consist of two equal halves. The two halves were superposed to get a shorter signature of better quality. The result was digitized and stored as an array of 360 nine-bit words.Remember,that this symmetry causes a 180° ambiguity in the determination of the object orientation,which can be removed by looking at the asymmetry of the phase distribution in the Fourier plane [1].

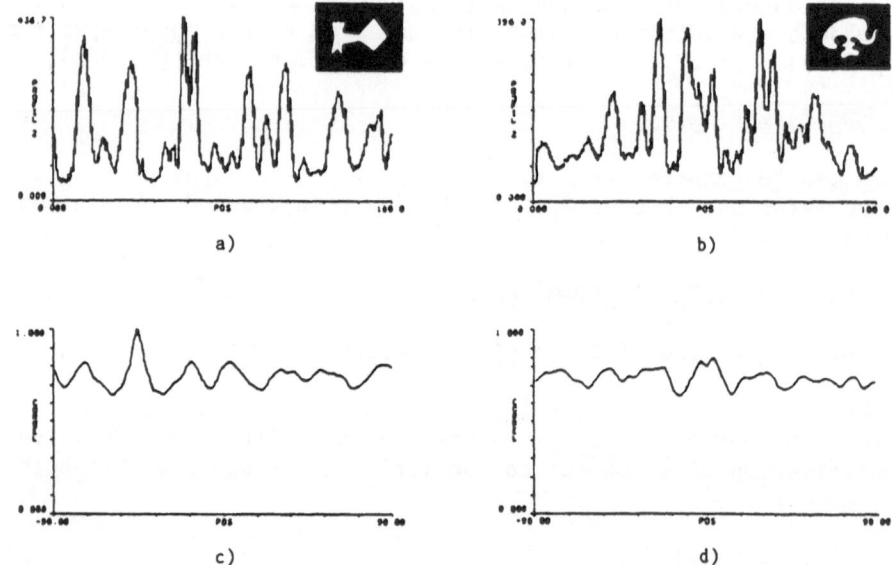

Fig.4. (a) and (b) digitized signatures of the two objects shown in the upper right corners, (c) correlation of signatures for the same object [Fig.4(a)] in two different positions(The correlation peak indicates the angular position with an accuracy of $\Delta\alpha = \pm 0.5°$),(d) correlation for two different objects [Figs.4(a) and (b)].

The mutual correlations of the profiles were calculated according to Eq.(4) to test the reliability of the object recognition concept. The accuracy of the determination of the object orientation and other important data can also be deduced from these correlation functions. Fig.4(c) shows a correlation of two signatures belonging to the same object. The object was rotated by 45° and somewhat shifted between the two records of the profiles used for the correlations. The profiles are equal to the one given in Fig.4(a), except that they correspond to different object positions. The peak value of that correlation is 0.99 and indicates the angular position with a precision of $\pm 0.5°$.The correlation of the signatures of two different objects, namely the ones shown in Figs.4(a) and (b), is presented in Fig.4(d). It has no pronounced peak and the values stay always below 0.86.

A statistical analysis of the experimental results proves that the described concept offers both reliable object recognition and accurate detection of the object orientation. Scan radii ranging from 1.5 to 15 times the average extension of the zero order diffraction spot are well suited for signature detection. Reliable object discrimination requires a minimum angular resolution to detect the fine structures in the power spectrum. About 60 points per half circle are necessary for a mean scan radius. The object orientation can be determined with an accuracy of about 0.5° if the detection and the correlation are performed with the corresponding resolution. Neither the object identification nor the determination of the orientation is considerably impaired by the use of binary reference signatures.

4. PHASE OF THE SPECTRUM: OBJECT POSITION

As seen from Fig.2, the phase of the spectrum is measured in a second Fourier plane by heterodyning with a reference wave, the frequency of which is shifted by acoustooptic modulators [7]. With the two pairs of photodetectors shown in Fig.5(a) the linear phase shifts due to the translation $\vec{x}_s$ of the object can be determined in two orthogonal directions. The parallel fringes in the zero-order diffraction spot [Fig.5(a)] result from the translation of the object. In heterodyne detection (frequency shifted reference wave) this fringe pattern moves at constant speed across the photodetectors, yielding the sinusoidal signals presented in Fig.5(b) for the electronic phase measurement. The heterodyne frequency shift was 40 kHz in this example.

The shift theorem gives, for the Fourier transformation $\hat{Q}(\vec{p})$ of the displaced object $Q(\vec{x}) = O(\vec{x} - \vec{x}_s)$, the relation

$$\hat{Q}(\vec{p}) = \hat{O}(\vec{p}) \exp(-i2\pi\vec{x}_s\cdot\vec{p}), \tag{6}$$

where $\hat{O}(\vec{p})$ is the Fourier transformation of the object $O(\vec{x})$ at its reference position. The corresponding phase difference Ψ [Fig.5(b)]

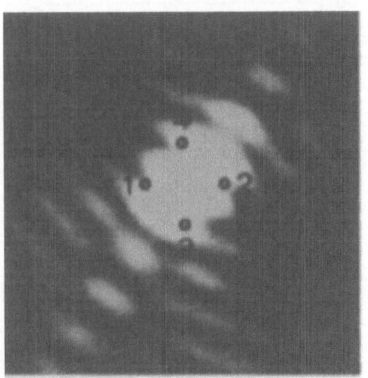 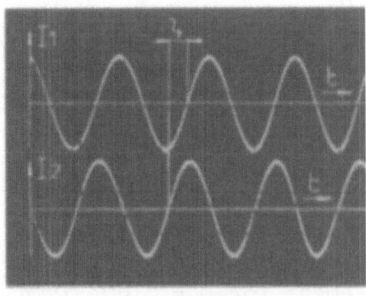

Fig.5. Displacement fringes and phase: (a) detectors in the zero-order spot, (b) heterodyne signals for phase measurement.

measured with two detectors separated by $\Delta\vec{p}$ in the Fourier plane is therefore proportional to the displacement $\vec{x}_s$ of the object, namely, $\psi = 2\pi\Delta\vec{p}\cdot\vec{x}_s$, or in orthogonal components

$$\psi_x = 2\pi\Delta p x_s, \qquad \psi_y = 2\pi\Delta q y_s. \tag{7}$$

The experimental verification of the relation is presented in Fig.6 for a circular object of radius R. The separation of the detectors was chosen as $|\Delta\vec{p}| = a = 1/(2R)$. The geometrical separation of the detectors in the Fourier plane was 2.5 mm and the diameter of their active area was 75 μm. The analog output ψ of a commercially available phasemeter, which has an integration time of 0.1 sec, is plotted vs the normalized shift d/R of the object. The estimated error is $\Delta\psi = 1^0$, which corresponds to $\Delta d/R = 0.56\%$ for the object position. For unambiguous determination of the object position, the separation $\Delta\vec{p}$ of the detector has to be chosen so that the measured phase for the maximum displacement $(\Delta x_s)_{max}$ is smaller than 180^0, which yields $d_{max} = \pm R$ for the example in Fig.6.

So far, however, the object-specific wave-front curvature in the zero-order diffration spot has been neglected. The corresponding phase terms have been investigated by means of the Taylor expansion of the Fourier transform, which is related to the moments of the object function [6]. The results prove that the wave-front curvature does not cause serious errors in the centroid location measurement and indicate how the 180^0 ambiguity in the determination of the angular object position may be resolved by checking the third-order phase terms.

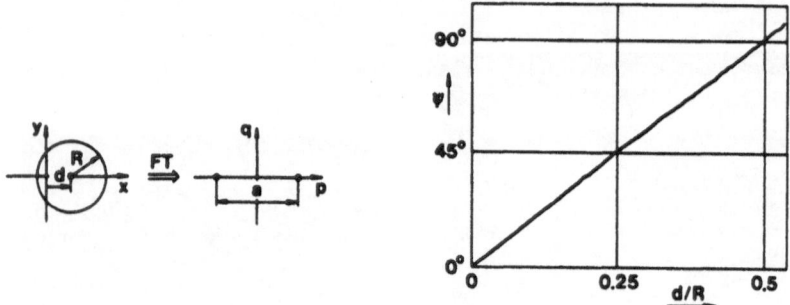

Fig.6. Experimental verification of the shift theorem by heterodyne measurment of the phase ψ. Detector separation is a = 1/2R. Maximum displacement for $\psi = \pm 180^0$ is $d_{max} = \pm R$. The accuracy of the phase measurement is $\Delta\psi = 1^0$. The corresponding resolution of the displacement is $\Delta d/R = 0.6\%$.

Quasi—heterodyne techniques allow CCD diode arrays and digital processing to be used instead of individual photodetectors and phasemeters for the amplitude and phase measurement in the Fourier plane. Extracting up to the sixth-order moments, i.e. 28 complex-value chractersitic parameters of the object function, seems to be possible with a 7 x 7 element diode array. This would allow the recently proposed methods of invariant moments [2,7] for pattern recognition to be implemented efficiently.

REFERENCES

1. Teague, M.R. Optical Calculation of Irradiance Moments, Appl.Opt. 19 (1980) 1353-1356.
2. Casasent, D. and D. Psaltis, Optical Pattern Recognition Using Normalized Invariant Moments, Proc.Soc.Photo—Opt.Instrum.Eng. vol. 201 (1979) 107-114.
3. Gara, A.D. Real—Time Tracking of Moving Objects by Optical Correlation, Appl.Opt. 18 (1979) 172-174.
4. Gara, A.D. Optical Computing for Image Processing, in Computer Vision and Sensor-Based Robots (Plenum, New York, 1979), pp.207-234.
5. Dändliker, R., K. Hess, Th. Sidler, Hybrid Optoelectronic Object Recognition, CLEO'82 (Opt.Soc.Am., Washington D.C., 1982), pp.68-70.
6. Dändliker, R., K. Hess, Th. Sidler, Hybrid Coherent Optical and Electronic Object Recognition, Appl.Opt. 22 (1983) 2081-2086.
7. Teague, M.R., Image Analysis via the General Theory of Moments, J.Opt.Soc.Am. 70 (1980) 920-930.

Dynamical Digital Memory for Holography, Moiré and E.S.P.I.

O.D.D. Soares, A.L.V.S. Lage, A.O.S. Gomes
Centro de Física
J.C.D.M. Santos
Departamento de Engenharia Electrotécnica
Universidade do Porto, Portugal

Abstract

A concept of design and construction of a real time dynamical digital TV-frame-storage-memory is presented. Features of the system include - real time dynamical arithmetic and logic operations based on a prototype configuration of a single frame store capacity, and microprocessor control; random and individual access to every pixel; image scaling operations based on time multiplexing and partition; synchronization with TV-camera operation and synchronism with microprocessor control. Design and implementation is oriented to transient studies by recourse to memory partition. External frame triggering and stroboscopic illumination permits selection of a time interval for analysis. Relevant principles of operation and construction are detailed. Potencial applications are outlined.

Introduction

Moiré, multiple-beam interferometry, holographic interferometry and laser speckle methods are well established techniques for surface 3D-contouring and vibration or deformation measurement.

A recognized disadvantage is the use of photographic recording both from the inconveniences of the need for photosensitive material processing and the subsequent numerical data extraction.

Whenever spatial resolution requirements are within range, TV camera image acquisition presents known advantages, particularly if a digital image processor (1) is associated.

It is described a dynamical digital processor (2,3,4) developed for fringe pattern processing in data handling from interferograms, moirégrams, holograms and specklegrams.

A presented prototype configuration of the system, Fig.1 permits image acquisition with simultaneous monitor display at the maximum TV camera rate using a frame-store-memory (5) of 256x256 pixel and 6 bit/pixel level resolution. Subtraction, addition and logic operations can be performed in real time by recourse to specific hardware implementation at the arithmetic and logic unit (A.L.U.).

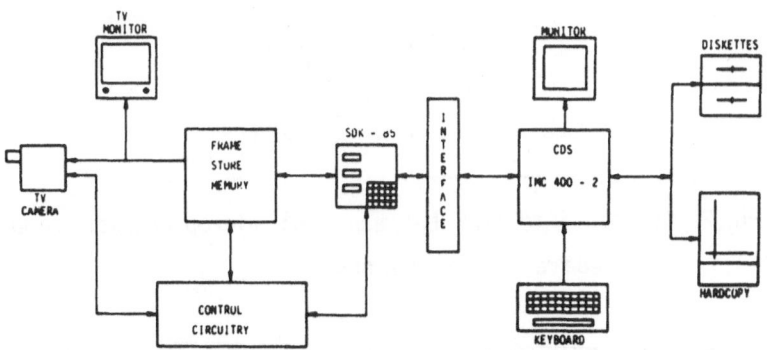

Fig.1: Prototype system configuration

Partition, in real time, of the frame-store-memory and subdivision of the display area of the video monitor is possible.

This allows several frames to cover sequentially the TV monitor sub-areas with some loss of resolution. It corresponds to a multiplexing in time of information via spatial demultiplexing which results in one TV frame time compression by sequencial multiframe display, a feature particularly intended for transient studies.

The frame-store-memory, in the prototype configuration, is directly controlled by one 8085 INTEL microprocessor. The designed 8085 microprocessor (6) addressing capacity is 64 Kbyte. However, the introduction of an extra-line between the microprocessor and the frame-store-memory extending the addressing capacity of the microprocessor permits one to use effectively the frame-store-memory as an extended part of the 64 Kbyte of the microprocessor memory whenever convenient for data handling. Fig. 2 explains the practical implementation of this design philosophy.

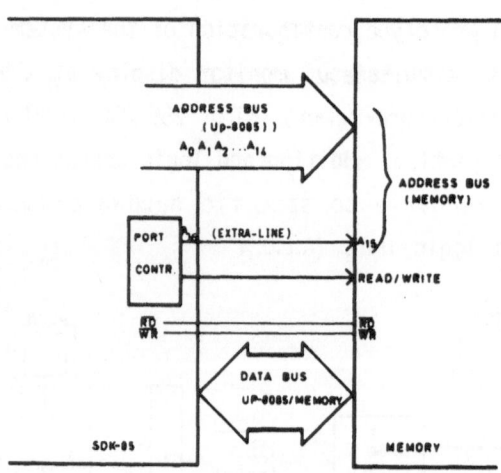

Fig.2: Control interlink between 8085 microprocessor memory and designed frame-store-memory

Individual and random access to any of the 64 Kbyte positions of the frame-store-memory is available at any operation time. This implies that a previous machine coded instruction indicates the selected half of the frame-store-memory area where the information resides by activating the extra-line interlink.

It is proposed to introduce video camera image acquisition control by the microprocessor that will be combined with implemented hardware of the look-up-table (LUT) type so that the pixel memorized intensity can be modified at will, in real time.

The system is appropriately interfaced to higher capacity processors for further image processing operations.

Fig.1 represents one configuration being used with a CDS IMC 400-2 microcomputer (7) and several peripherals. The parallel interface enables a fast transfering of data and programmes between the 8085 microprocessor and the IMC 400-2 microcomputer, namely:

i) machine coded programmes (non-volatile storage) combined with programme development at higher level language.

ii) to load machine coded programmes into the 8085 microprocessor memory by selection from the library at the microcomputer memory to accomplish faster image processing.

iii) to store an image from the frame-store-memory into a diskette and its inverse operation.

iv) to output data to a hardcopy peripheral (plotter used in this case).

v) to control the entire system by entering a sequence of commands in the keyboard of the IMC 400-2. These commands may realize image processing operations (1) such as:

a) histogram calculation and its manipulation.

b) filtering.

c) edge enhancement.

d) skeletoning.

e) arithmetic operations between images.

vi) to introduce image processing algorithms (build up of macrocommands)

Operation Principle

The basic prototype configuration of the frame-store-memory is shown in Fig.3. A common video camera not necessarily with external synchronism capacity may be used for image acquisition. The video signal is sampled and digitized by a video A/D converter (TRW - TDC 1014 J) at a chosen maximum rate of 10 MHz. The RAM memory was implemented, in the prototype configuration, with 2114A-4 static RAM chips (1 024 x 4 bit) with access time of 200 ns.

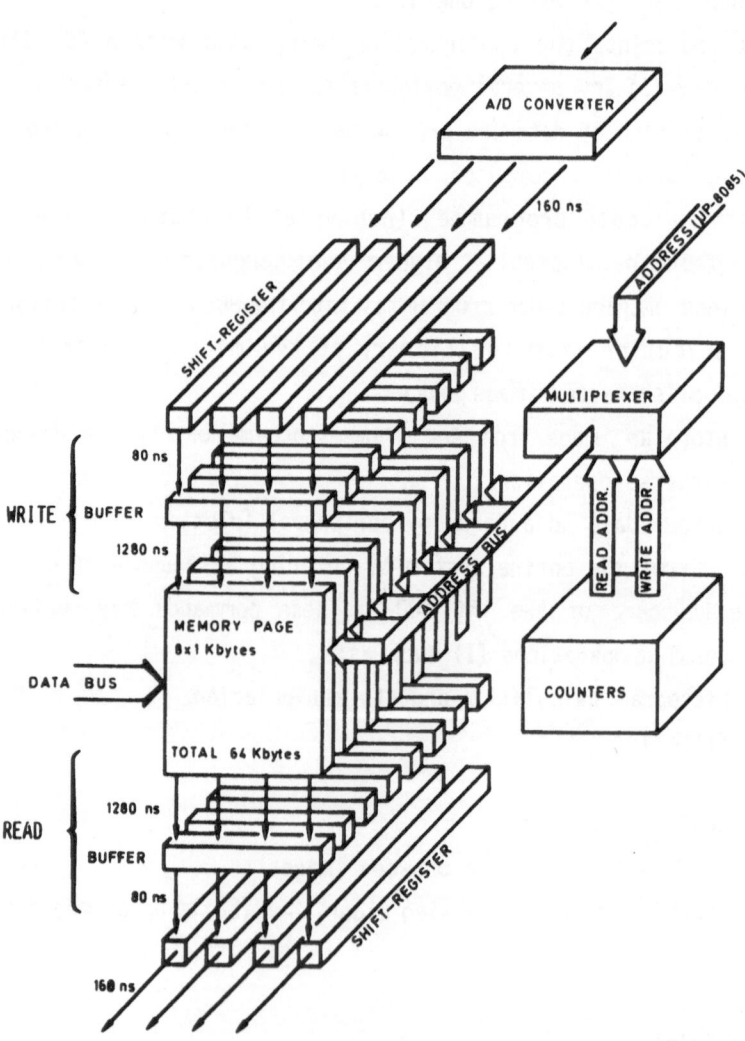

Fig.3: Frame-store-memory prototype configuration with 4 bit/pixel representation (2).

To perform, in real time, the writing time at the frame-storage-memory has to be expanded by some hardware arrangement.

According to Fig. 3 data from the A/D converter is stacked, in parallel, during a time corresponding to 8 clock pulses of the A/D converter, on four shift-registers of 8 bit. When filled up, the data stored is transfered to 8 buffers of 8 bit per buffer where it remains over the next 8 clock pulses. This results in a writing time in the memory 8 times greater than the clock pulse interval of the A/D converter. Similarly, an inverse sequence is used to read from the memory as also shown in Fig.3.

The memory then becomes organized in 8 pages of 8 Kbyte where each byte makes 2 bits available for the coding of the image data (e.g. color) in view of its use for ulterior processing.

Fig.4 describes why it is necessary to allocate four A/D converter clock pulses for data transfer from memory to READ buffer (Fig.3) and four other clock pulses for WRITE buffer to memory transfer in an obvious alternate sequence. It should be noticed that this clock time management would not be required were it made a duplication of the frame store-memory.

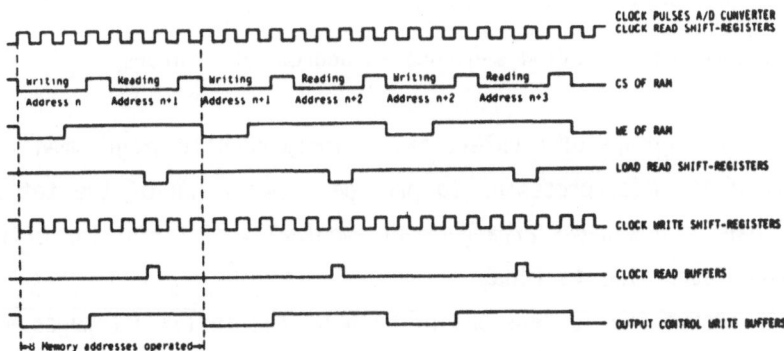

Fig.4: Timely interlaced write-reading operations for real time single frame storage configuration.

Furthermore, the memory address has to be read 12 clock pulses before its writing in order to perform, in real time, operations between an image stored on the frame-storage and the actual image seen by the video camera, combined with the storing of the result back into the frame-storage without loss of data. For this purpose dephased read and write addresses are multiplexed and generated on two separated group of counters, in a time sequence shown in Fig.5.

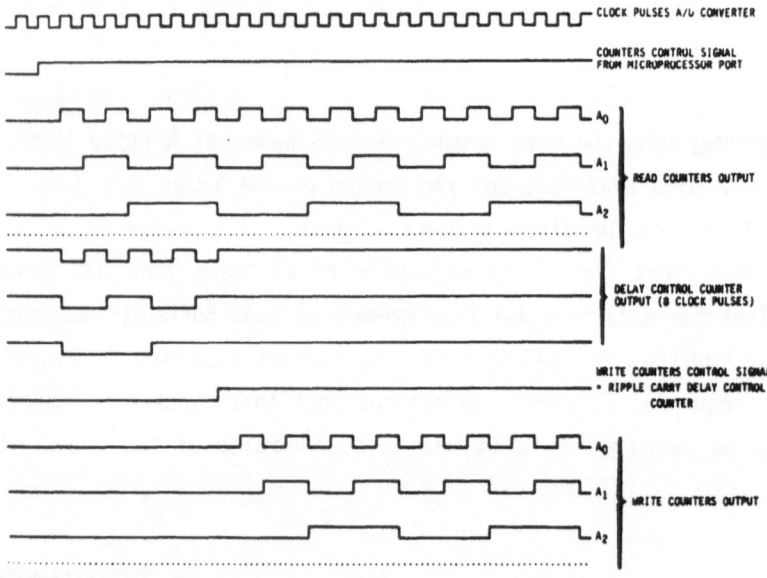

Fig.5: Counters operation sequence as address generators

The two groups of counters can be independently programmed via output ports of the microprocessor to provide a partition of the total memory positions in new pages arrangemment eventually different for writting and reading, according to Fig.6.

To ensure the dimensions of each of the individual pages match the actual dimensions of the image, horizontal and vertical dimensions of each original image pixel can be independently programmed by division of the TV line position clock pulse, frequency f_r, and line numbering clock pulse, frequency f_l, under control of 8 microprocessors ports, Fig.7. In this way effective pixel areas with dimensions 16x16 greater than the usual dimension in the 256x256 pixel image can be accommodated.

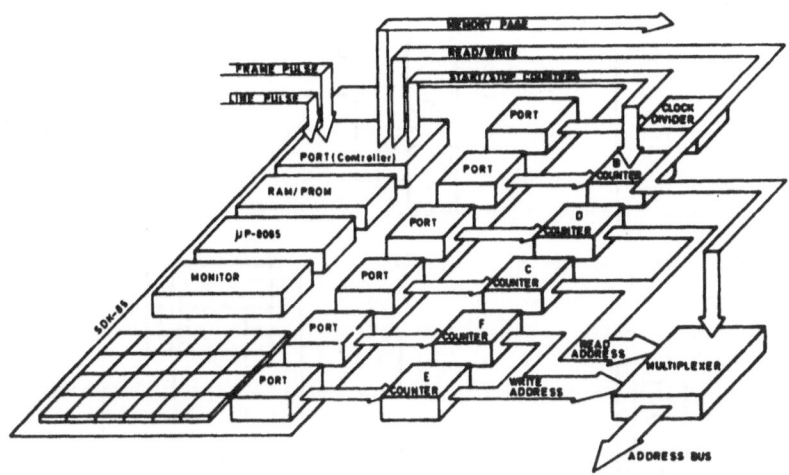

Fig.6: Windowing effect generation with combined action of counters and microprocessor control.

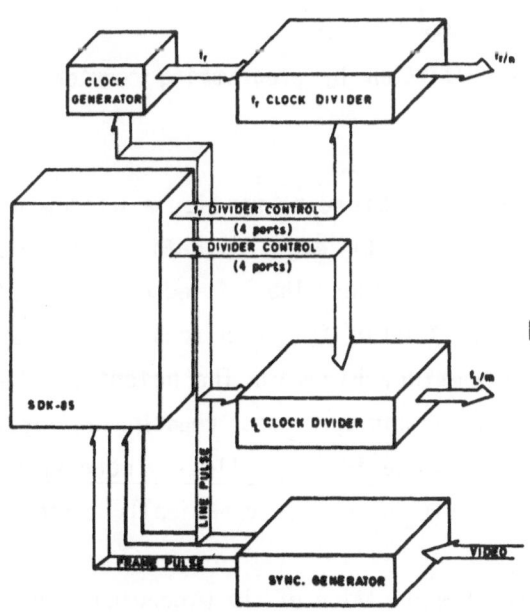

Fig.7: Effective pixel dimension control by programmed clock division under the control of microprocessor ports.

190

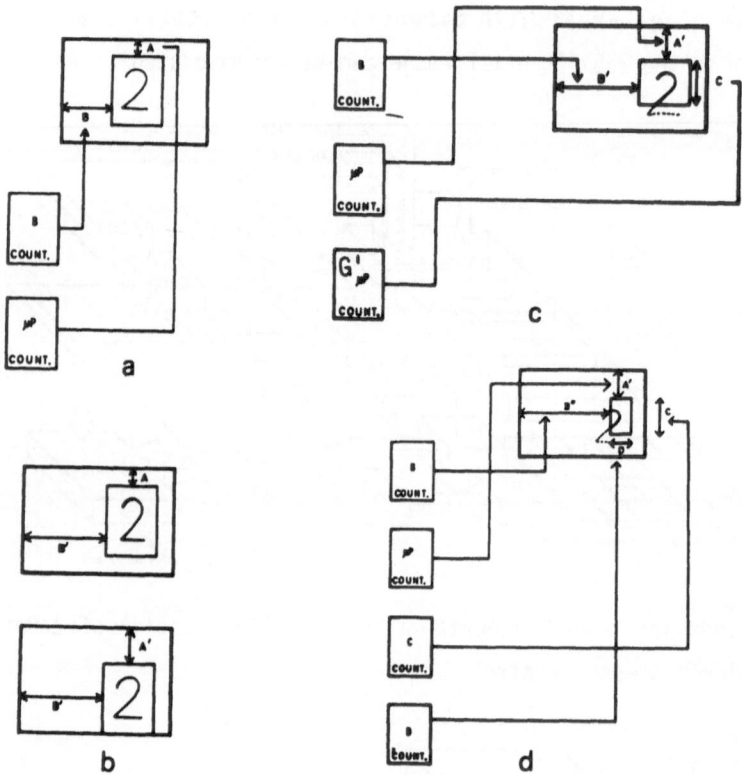

Fig.8: Window size and position definition by programming counters and
microprocessor action

Counters action, Fig.6 and 8, is combined to realize a windowing
effect. In reference to Fig.8, the upper edge is defined by the
microprocessor counting of the frame TV-lines. The left edge is defined
by counter B, Fig.6 and 8 a). The position of image on monitor display
area is chosen by counters programming, Fig.8 b). The height of the
displayed image is determined either by microprocessor counting G', Fig.8
c) or by counter C, Fig.6. Width of image is controlled by counter D,
Fig.6. Therefore, window area and position becomes dynamically defined,
Fig.8 d).

Fig.9 represents schematically an application of the windowing effect.
A four frame sequence is displayed on the monitor with a composition
previously choosen. The windowing effect is pertinent to the study of

transient phenomena, Fig.10.

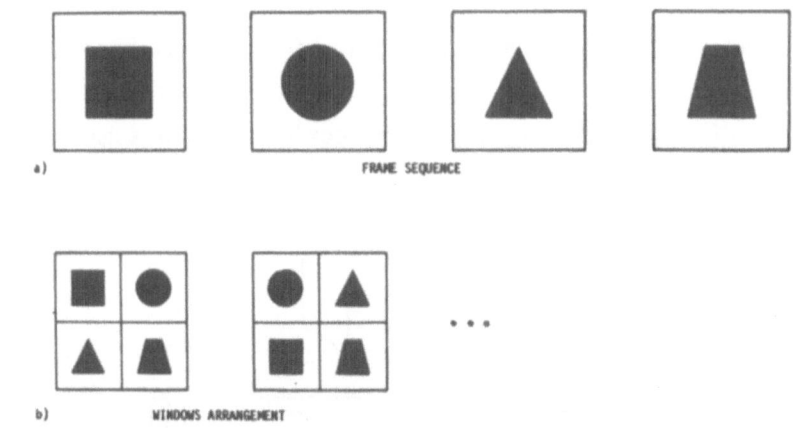

Fig.9: Windowing effect on the Tv-monitor

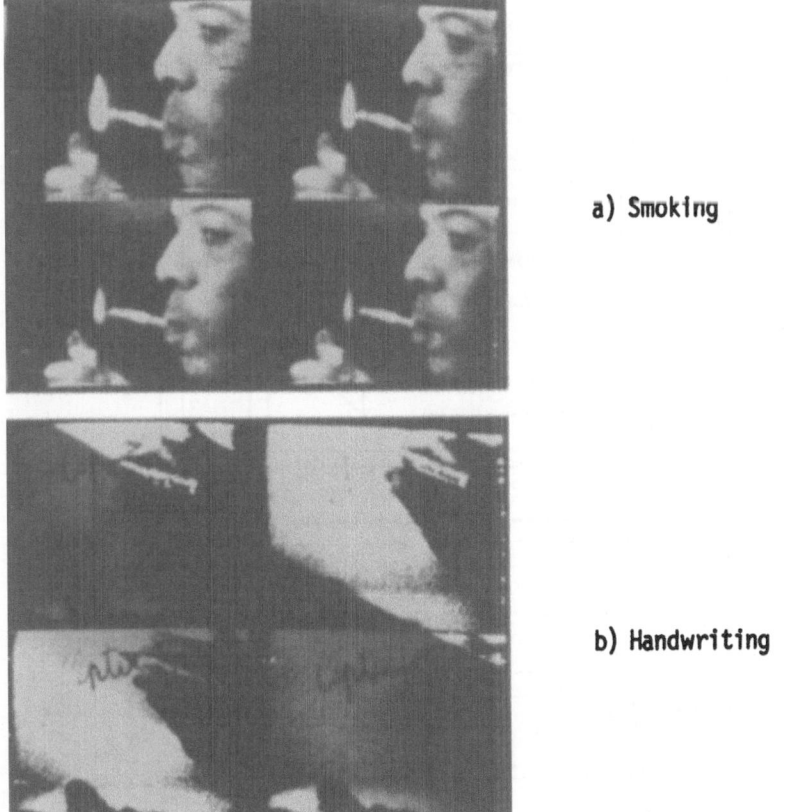

a) Smoking

b) Handwriting

Fig.10: Windowing effect to record a sequence of daily transient phenomena

Real-time Operation

Real time operation of the system relies on time management achieved under microprocessor control and adequate data flow between frame-store-memory, A.L.U., and control circuitry, combined with a multiplexer operation shown in Fig.11.

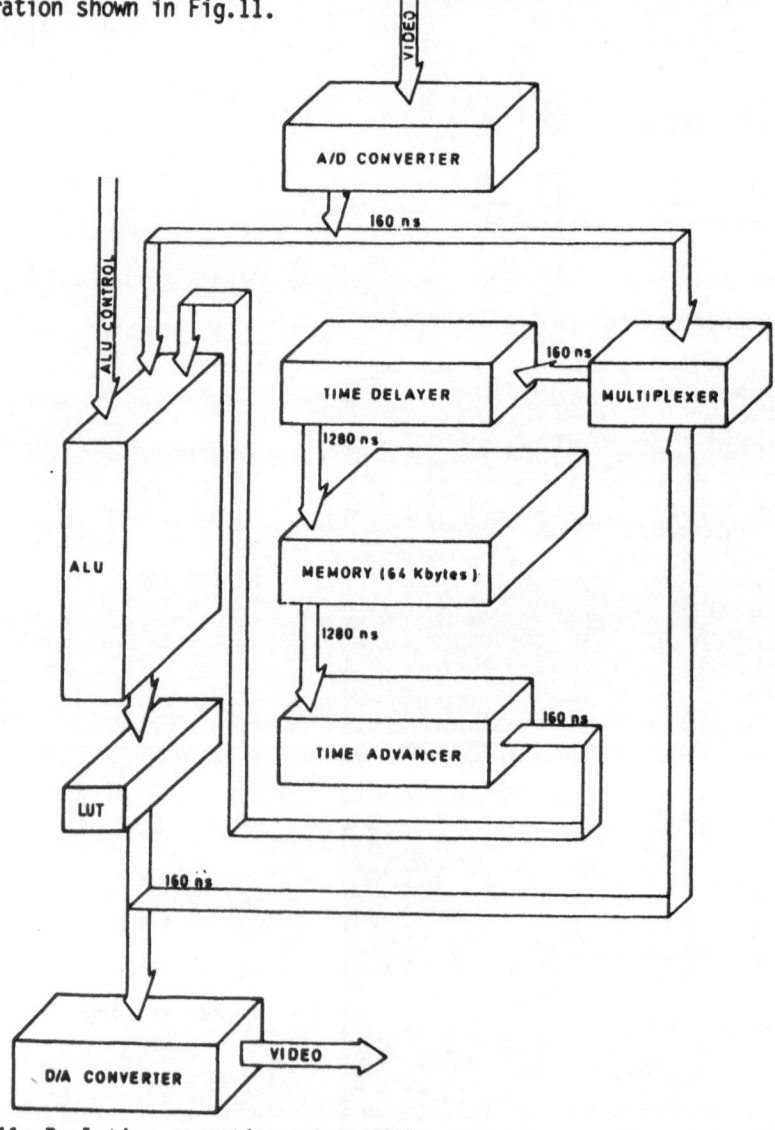

Fig.11: Real time operations data flow

Programming the data flow sequence within each TV frame results in the desired image processing operation. Fig.12 presents the particular case for dynamical sequential image subtraction and storage of the corresponding result.

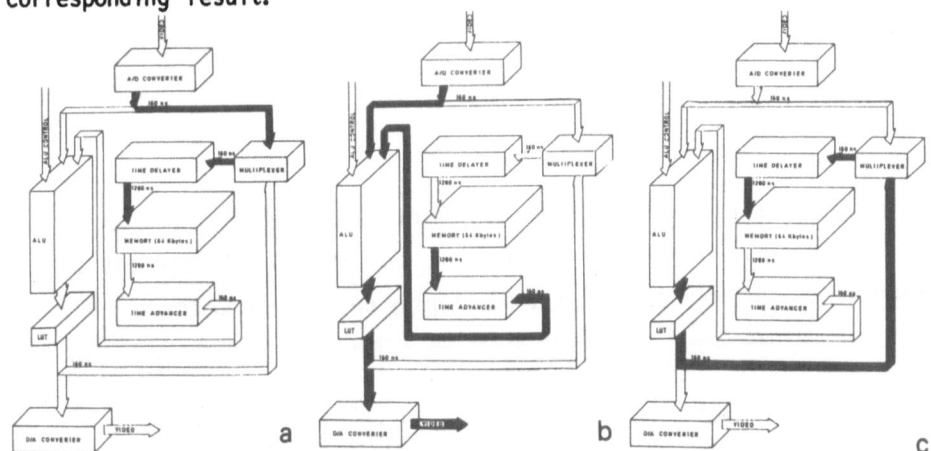

Fig.12: Dynamical sequential image subtraction
a) Storage of 1st. image frame.
b) Subtraction of image frame seen by the camera from stored image frame.
c) Storage of the subtraction result (see Fig.11 for details).

Convenient simultaneous programming of the two group of counters and the partition of the frame-storage-memory provides flexibility to ensure real time display of processed images with time evolution of the phenomenon being studied.

Real time operations which are possible include:

i) simple storage in the frame-storage-memory of the image seen by the video camera.

ii) dynamical display of the image processing result from an operation between stored image and actual image seen by the video camera.

iii) sequence ii) with storage of the result.

iv) sequence ii) with storage of the actual image seen by video camera providing real time continuous operation involving consecutive images.

v) storage of the image seen by the video camera in anyone of the pages

194

resulting from partition of the frame-storage-memory.

vi) display of the result of an operation involving one of the images stored on one of the pages resulting from frame-storage-memory partition, and the image seen by the camera.

vii) sequence vi) with storage of the result on the involved page.

viii) sequence vi) with storage of the image seen by the video camera on the involved page.

Fig.13 is a pictorial representation of the above described operations taking a sequence of four image frames, and showing the content on the frame-storage-memory and the actual display on the TV monitor.

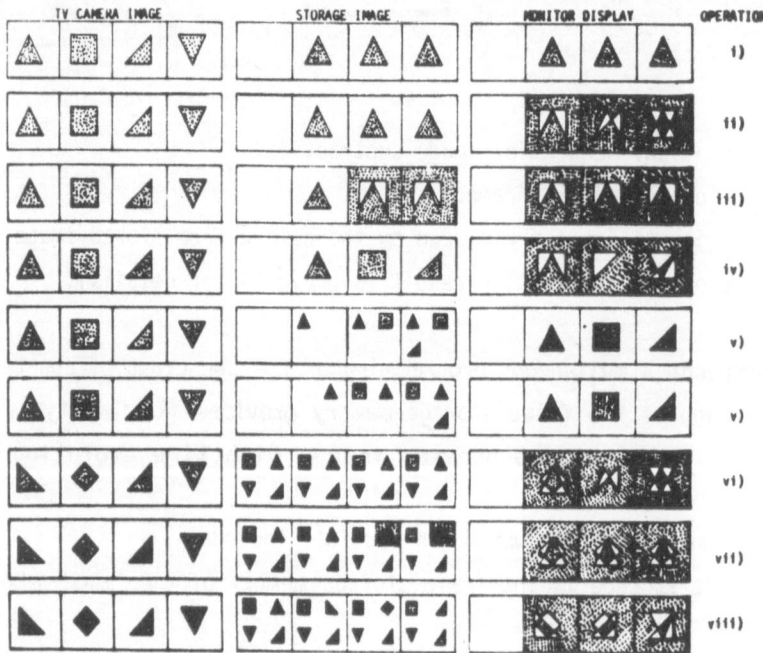

Fig.13: Pictorial representation of real-time operation with a sequence of four image frames as described in the text

Examples of Application

The dynamical digital memory has been in use in research activities reported elsewhere. A scattered sample of examples of its application in specific fields will be briefly reported.

Fig.14 is a sample of image subtraction. An example of transient studies in biomechanics is reported in Fig.15. Moiré evaluation techniques (8) are presented in Fig.16 for contouring with a change of reference. Moiré derivation is illustrated in Fig. 17. Transient analysis by E.S.P.I. with memory action and synchronized pulsed illumination (10) of forced vibrations on a loudspeaker is photographed in Fig.18 from video recording tape.

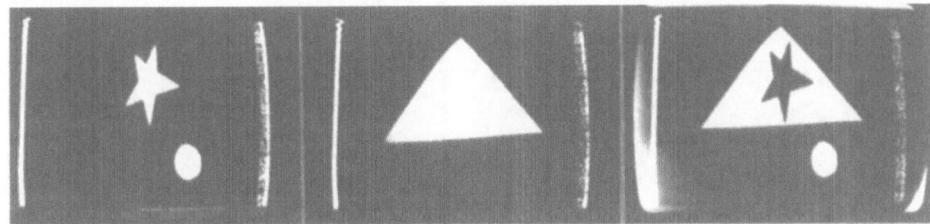

Fig.14: Image subtraction
a) additive
b) subtractive
c) result

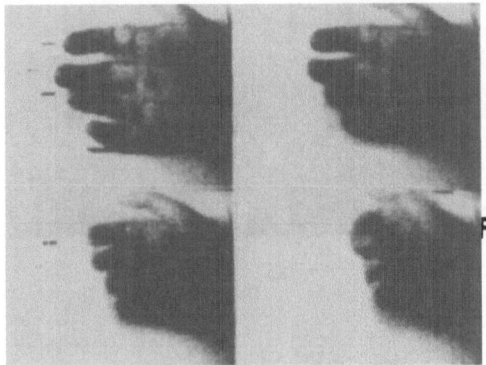

Fig.15: Transients in biomechanics with windowing effect

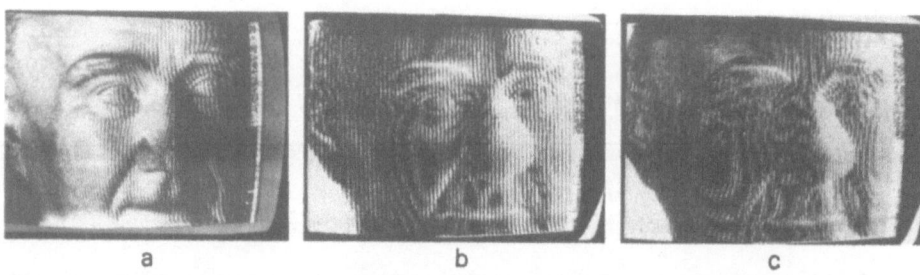

Fig.16: Contouring by Moiré with change of reference (8)

    a) plane grating projected on the surface.

    b) moiré contouring referred to a plane.

    c) change of reference from a plane to a curved surface.

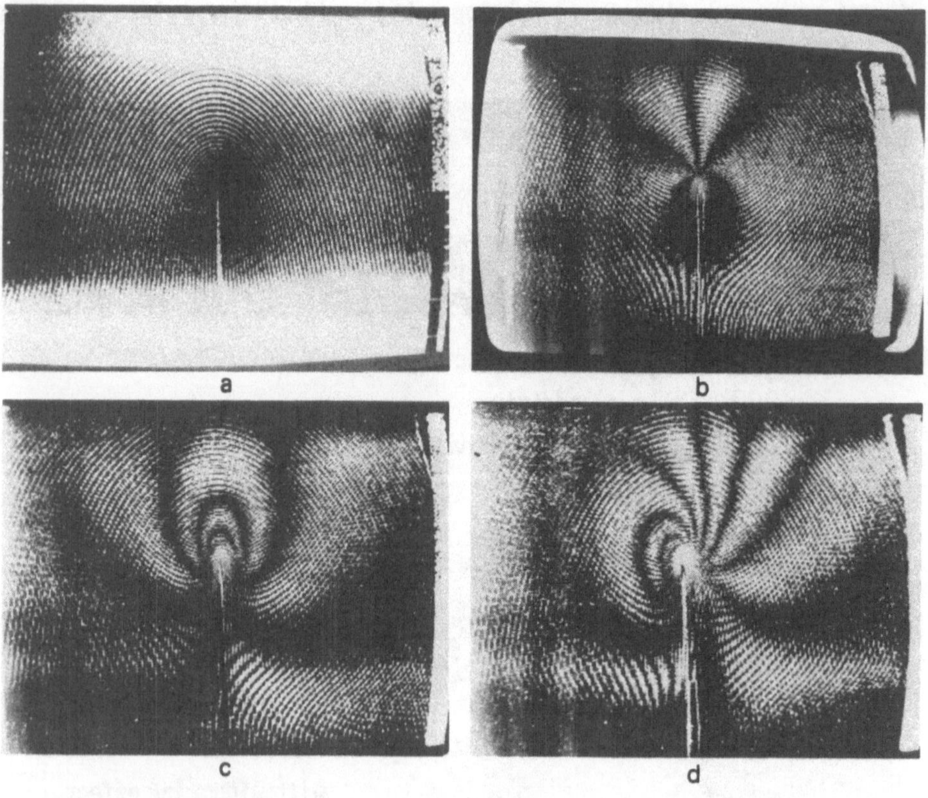

Fig.17: Moiré derivation

    a) Fringe pattern   b) $\frac{\partial}{\partial x}$ horizontal direction

    c) $\frac{\partial}{\partial y}$ vertical direction   d) $\frac{\partial 2}{\partial x \partial y}$

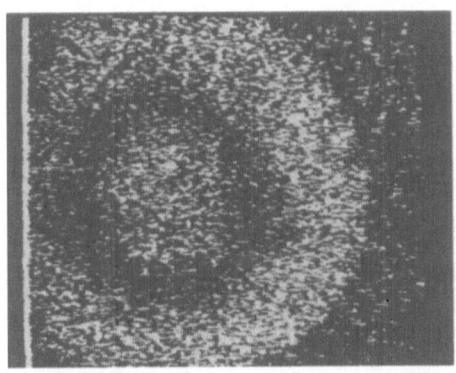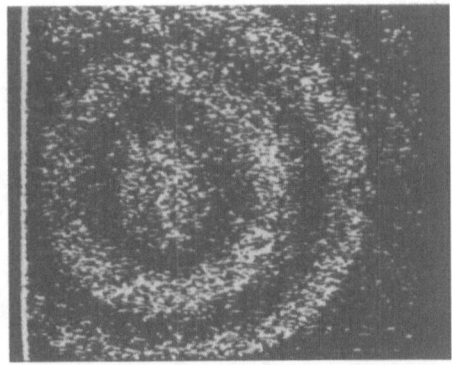

Fig.18: Hand held loudspeaker with forced vibration when visualized by E.S.P.I. with synchronized pulsed illumination (11)

Conclusion

A real time dynamical digital memory concept of design for frame-storage-memory was introduced. A prototype configuration has been described with detailed analysis of implementation and operation capabilities. Examples of application have been given to show its potential for transient studies and dynamical analysis of images, in general. Metrologic applications envisaged for the system relate to research with holographic, moiré and E.S.P.I. techniques.

Acknowledgements

Research was supported by a grant from I.N.I.C.

J. S. Fernandes and L. Vilaça provided skilled technical assistance.

References

1 - Rosenfeld A., Picture Processing by Control, Academic Press (1969)

2 - Soares, O.D.D.; Lage, A.L.V.S.; Gomes, A.O.S., Memória Digital Dinamica Operável em Tempo Real, Patent pending

3 - Soares, O.D.D.; Lage, A.L.V.S.; Gomes, A.O.S., Digital TV - Frame Dynamical Memory for Transient Studies, COMPUTIM 84, Strasbourg (1984)

4 - Soares, O.D.D.; Lage, A.L.V.S., Use of TV - Frame - Memory on Electronic Speckle Pattern Interferometry Applied to Orthopedics, SPIE vol.348 (1982) 838 - 844

5 - Klingman, E., Microprocessor System Design, Prentice-Hall (1977)

6 - INTEL MCS - 80/85 Family User's Manual (1979)

7 - IMC 400/2 Computer System Manual, CDS Electronics B.V., Vlaardingen, The Netherlands (1979)

8 - Soares, O.D.D.; Lage, A.L.V.S.; Bernardo, L.M., Moiré Evaluation of Pulse Illuminated Interferograms by Synchronized Video Recording, SPIE vol.491 (1984)

9 - Soares, O.D.D.; Lage, A.L.V.S.; Bernardo, L.M., Moiré-Evaluation with Fringe Patterns of Interferograms, Holograms,Moirégrams and Specklegrams, Optical Metrology, Martinus Nijhoff (1985)

10 - Soares, O.D.D.; Lage, A.L.V.S.,Método de Análise Diferencial e Integrativa em Holografia, Patent Nº 80 333, INPI, Portugal (1985)

11 - Soares, O.D.D.; Lage, A.L.V,S.; Controllable Synchronized Multipulse Illumination System for ESPI and Holography, SPIE vol. 427 (1983)

12 - Soares, O.D.D.; Lage, A.L.V.S.; Improvements on Electronic Speckle Pattern Interferometry, Optical Metrology, Martinus Nijhoff (1985)

AUTOMATION IN MANUFACTURING INTELLIGENT SENSING

Professor Bruce Batchelor

UWIST, Cardiff

P W Heywood

British Robotic Systems Ltd
London

1. INTRODUCTION

Intelligent sensing is the means whereby information regarding an object or process is automatically analysed and decisions made in order either to accept or reject the part or process in a manufacturing environment.

The most important intelligent sensing field at this point in time is vision sensing. That is, the replacing of the function of human vision in the manufacturing process. Tactile sensing is still in the laboratory/development stage and is not yet seriously being utilised in real manufacturing processes. However this will be the case in the not too distant future.

Other forms of sensing input to intelligent machines in the future could be:

Acoustic
X-Ray
Ultrasonic
Radar
Microwave, etc

This paper concerns itself mainly with image processing although the subjects of Artificial Intelligence and Expert Systems will be covered, both of which will make significant impact on the intelligent sensing scene.

Practical vision sensing for manufacturing automation has only been commercially available since around 1978. The majority of suppliers in the field tend to be small to medium sized companies (20-70 employees) when compared with the larger manufacturing organisations who are their main customers. The majority of companies have only been operating for about 4 years and their numbers are increasing rapidly, particularly in the US.

Market survey evidence indicates that the growth rate for vision systems supply will increase by about 100% per year. In the USA alone the market has increased from about $8M in 1981 to an anticipated $80M in 1984. This growth rate is a good indication of the importance now being attached to vision sensing by manufacturing industry.

It is generally understood that the rest of the world market is the same as the US market and that the European market is 50% that of the US market.

2. CONSTITUENTS OF IMAGE SENSING

There are three main stages in image sensing:

(a) Image acquisition
(b) Image processing
(c) Decision-making

(a) Image Acquisition

This is normally the most difficult of the three stages. It consists of using a combination of cameras, illumination, optics and part presentation in order to ensure that a robust image is acquired for further processing. Robustness means that the image must not be too susceptible to ambient lighting variations and other relevant variations as well as not requiring specialised and delicate illumination or optics. A significant amount of knowledgeable application engineering expertise is required to design a suitable image acquisition system. This immediately creates a problem in the business because such engineers are not readily available and this is expected to be one of the factors that could influence growth in the future.

(b) Image Processing

Once a suitable image has been acquired then the image requires processing mathematically by algorithmic means in order to determine whether the dimension, feature, position, orientation, texture etc are correct or not. Early vision systems worked on a 2-level "binary" system, that is images acquired were immediately thresholded to provide a black and white picture or the binary picture was produced by silhouette illumination. Such binary

picture is easy to work with and can provide a rapid computational
response. However in the real world very few problems can be
solved by such binary analysis and therefore "multi grey level"
analysis has to be applied. Typically multi grey level means
that a picture is analysed from between 64 and 256 levels of
light intensity. 64 grey levels are often adequate for current
manufacturing applications. However 256 grey levels are required
for more refined use such as crack detection.

In order to determine the particular image processing
algorithm required to solve a specific problem, a sophisticated
tool is often employed. One such tool is described in Section 3.

(c) Decision-Making

This is usually the easiest part of the process once the image
analysis stage has produced specific information which can then be
used to effect control. Control can be a GO/NO GO gate mechanism,
robot guidance coordinates generated, indexing table, polar
coordinates provided, or simply the control of conveyors etc to
effect the automatic process in the manufacturing cell.

3. IMAGE PROCESSING ALGORITHM DEVELOPMENT - AUTOVIEW VIKING

3.1 Laboratory and Lighting

It is important to stress that algorithm development is
hindered by poor illumination and image capture techniques. Any
laboratory work must have as a minimum:

Newvicon camera for low light level images.
Frame for mounting samples and a fairly solid work bench.
Moveable lights that can be dimmed.
Light box for silhouette work.
Selection of lenses which have aperture control, close up
lenses and extension tubes.
The capability to exclude all unwanted light, even if this is
done by heavy black curtains.

This minimum set will allow a wide range of problems to be
investigated. The lighting and image capture techniques are
important to solving the particular image processing problem
at hand. It is pointless trying to process a poor image if five
minutes extra work with the lighting will produce a simpler image
to analyse. AUTOVIEW Viking can also be useful when setting up
the initial illumination as a simple combination of commands will
check for uniform lighting.

The various techniques available for lighting are numerous. Many are the application of everyday experience, experimentation and common sense. For example, to examine a highly polished surface for scratches glancing illumination is used. The scratch changes the reflectivity and hence looks different. An example would be when looking at a record for scratches. It is held close to the eye with the light at a glancing angle. An attempt to duplicate this lighting arrangement is made in the laboratory when we are inspecting plastic components for scratches.

A good example of the different lighting techniques available is shown in the inspection of glassware and liquids in glassware.

o   Cracks and defects in glass can be discovered very easily by back lighting with a broad light source like a light box. Any cracks then show up as very dark lines which can be easily processed out.

o   Particles and bubbles can be seen if the sample is lit orthogonally to the viewing angle. The results of this lighting technique are shown in Photograph 1.

o   The edges of glassware can be detected using dark field illumination, where an opaque mask is placed between the glassware and a broad light source. This effect can be seen clearly in Photograph 2 which shows a milk bottle illuminated in this way.

3.2 AUTOVIEW Viking - Detailed Description

3.2.1 Hardware. The main hardware components of a Viking system are shown in Figure 1. The standard Viking framestore is a Matrox which digitises and displays an image that is 512 x 512 pixels (picture elements) with 8 bits per pixel (this gives 256 grey levels). Pseudo colour output (that is each of the 256 levels possible are assigned a different colour for display on a colour monitor) and the ability to select from four camera inputs under software control.

The disks for backing store can be either twin floppy disks or a floppy/Winchester combination.

The display output and keyboard input device is a VT220.

3.2.2 Software. AUTOVIEW Viking software can be regarded as nothing more complex than a large collection of highly modular units that can be called in turn, either at the direction of keyboard input or through some stored sequence of commands. This structure is how the software is actually written with each vision command being coded as a single entity that can be

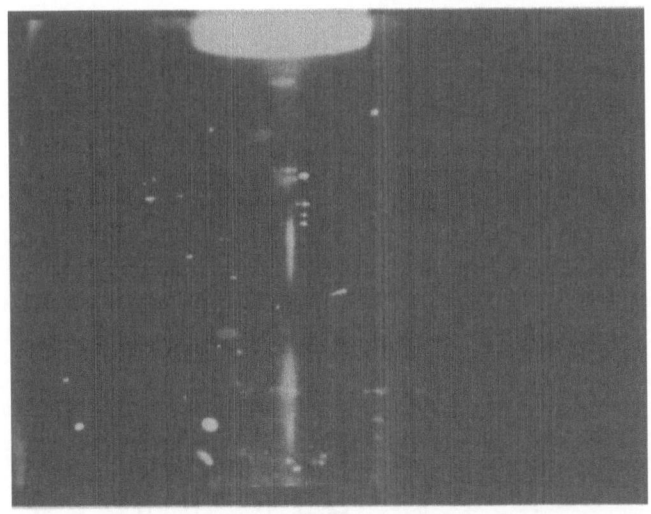

Photo 1   Particle inspection in fluids. The light is from
           below and the container is viewed from the side.

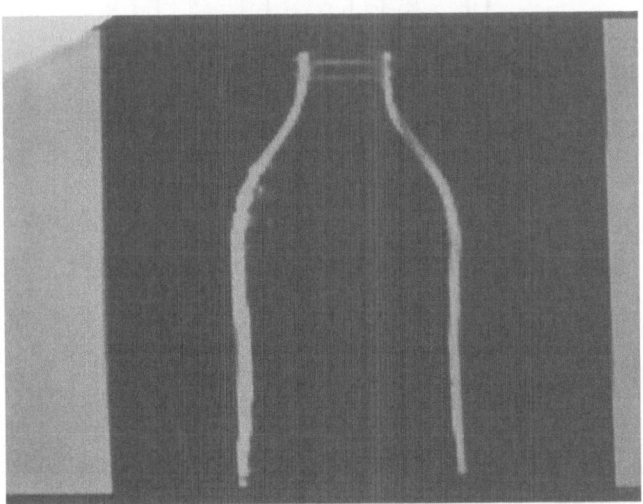

Photo 2   Milk Bottle illuminated from behind showing the
           edges clearly.

FIGURE 1

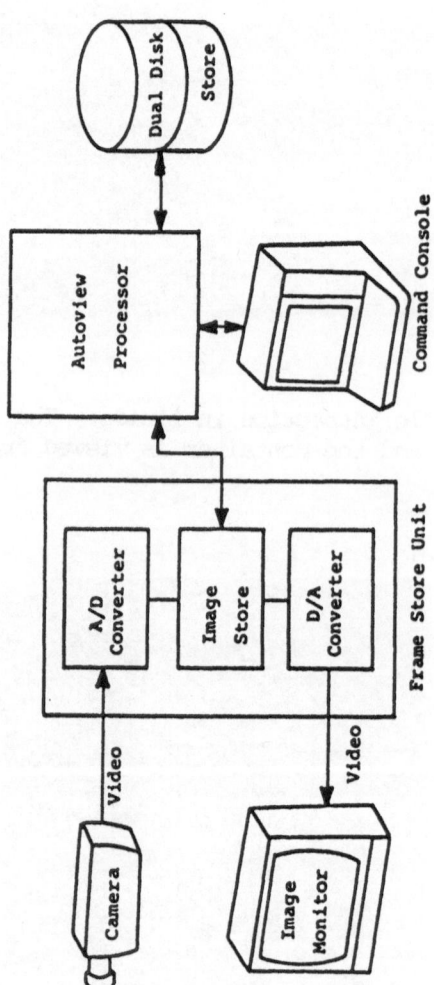

MAIN HARDWARE COMPONENTS OF THE AUTOVIEW VIKING SYSTEM

compiled separately. The vision modules are loaded and executed
by one of two command interpreters. The simpler of these consists
of a set of mnemonics with associated parameters. The command
interpreter decodes the parameters and calls the appropriate vision
processing function. The higher level is a standard PROLOG
interpreter that has extra functions built in to allow it to
call the lower level command interpreter to execute vision
functions. The PROLOG interpreter extends the power and flexibility of the vision system. For example, it allows more complex
scenes to be analysed as will be described in Section 5.
It also allows for sophisticated help functions to be built into
the Viking system to guide the inexperienced user in his attempts
to solve a vision processing problem.

3.2.3 Basic Commands. There are about 200 basic commands which
will not be described in detail. However, the main classificat-
ions are as follows:

Image Input/Output	This provides commands to write and read from disc and framestore, etc.
Macros	This section provides commands to define, edit and print macros.
Register Arithmetic	This section provides simple arithmetic operation on the registers.
Basic Picture Arithmetic	Simple picture processing operation like adding two images, finding the maximum of two images.
Picture Transformation	A wide range of operations, including changing from cartesian to radial images contrast enhancement, picture integrate across the image, aspect adjustment, expand and contract grey areas into black etc.
Filters	High, low pass and general purpose 3 x 3 filters.
Gradients	A selection of gradient operators over a 3 x 3 window.
Test Images	Wedge of intensity, cone of intensity, sine variation of intensity etc.
Histogram Manipulate	A collection of routines for generating and manipulating the histogram buffer, including cumulative totals, smoothing

and trough finding.

Template Matching   A sparse filter operator that is not limited to the 3 x 3 window of the other filters, that can be used for template Matching.

Decision-Making   A set of routines for matching the contents of the registers with stored values, these can be used for recognition tasks.

Program Control   Labels and jumps to alter command sequences.

Debugging and System Control   Help information, cursor control, debugging information etc.

Thresholding   Simple and Dynamic Thresholds

Feature Extraction   Angles of lines, region analysis, medial line, connectivity etc.

Transfer Function   Table Look-Up for transformations for example a log transform.

The basic command sequence is very wide and a large range of vision processing tasks can be performed by just using these basic commands.

3.3 Typical Examples of Use of Image Processing Laboratory

The examples given below are not necessarily complicated and many can be performed with a few basic commands. These are included to show some of the real problems that have been solved using vision sensing.

1. Photos 3-5 show a sequence checking for cracks in a metal con-rod. The con-rod is magnetised and coated with a fluid that is magnetic and fluorescent when illuminated with UV light. The image processing part attempts to distinguish cracks which are long and thin from blobs which are caused by reflections.

2. Subtracting an image from a master image is a very powerful inspection technique if the two images can be aligned correctly. Photos 6-8 show such an inspection. Photo 8 shows there is a one line difference that is due to movement. In an industrial system these one line differences would have to be allowed for before rejecting the part.

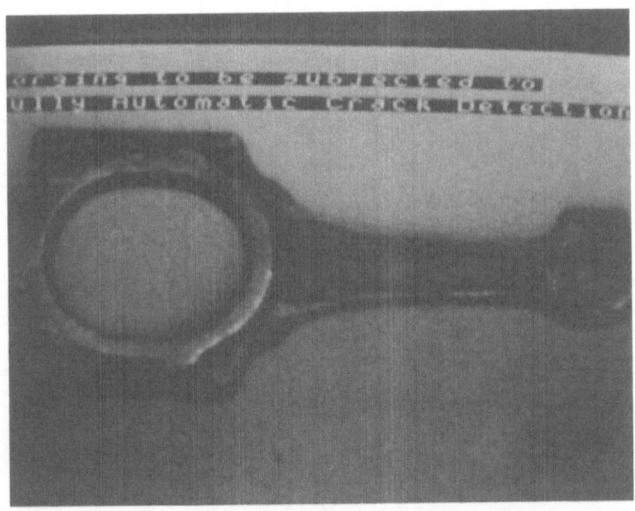

<u>Photo 3</u> Normal view of casting

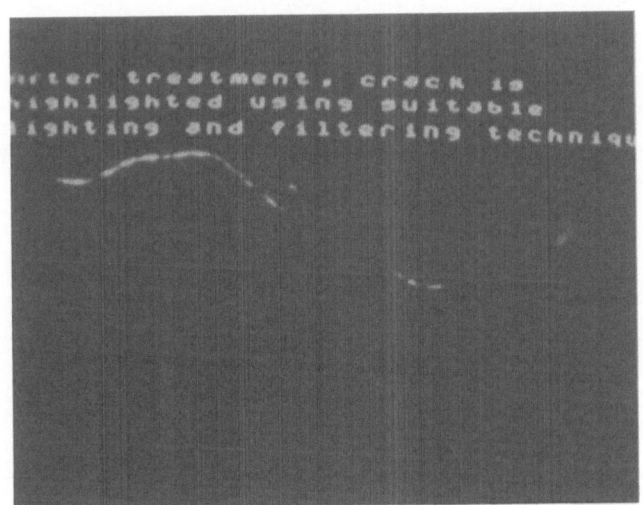

<u>Photo 4</u> Casting after treatment and viewed under UV light.

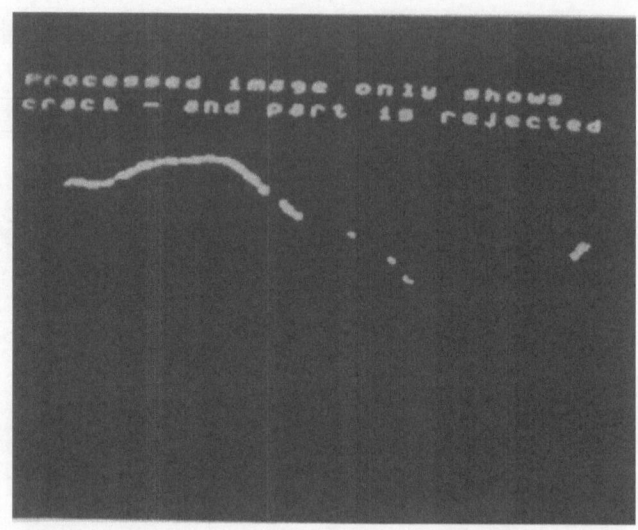

Photo 5   Processed Image from Photo 4

Photo 6 Master LCD Display for later comparison

Photo 7 Image to be compared with defects

210

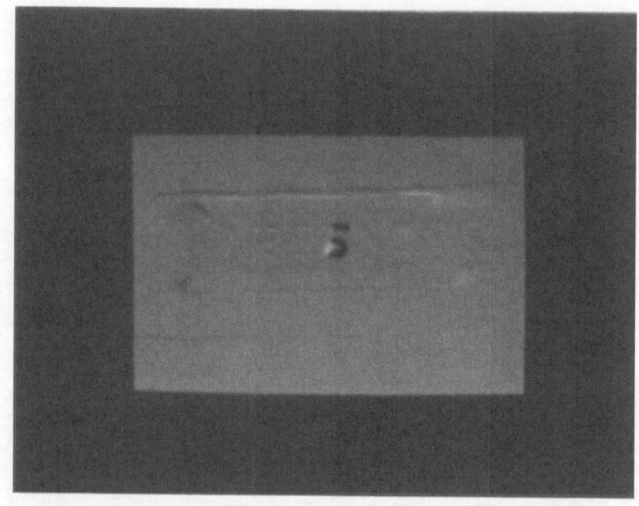

Photo 8   Defects between Photos 6 and 7
highlighted

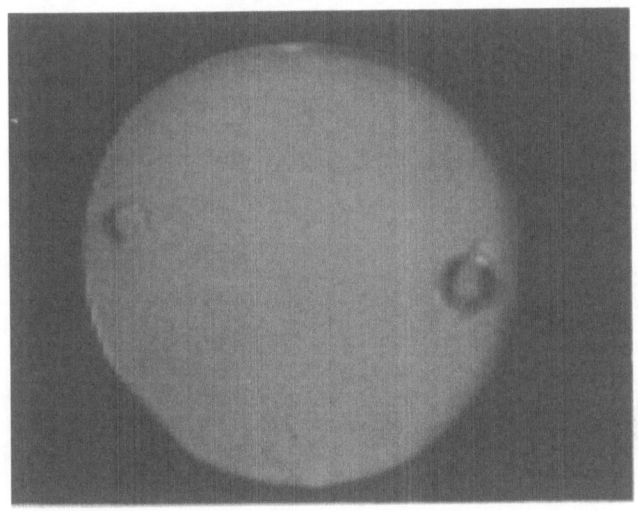

Photo 9 View of top of barrel

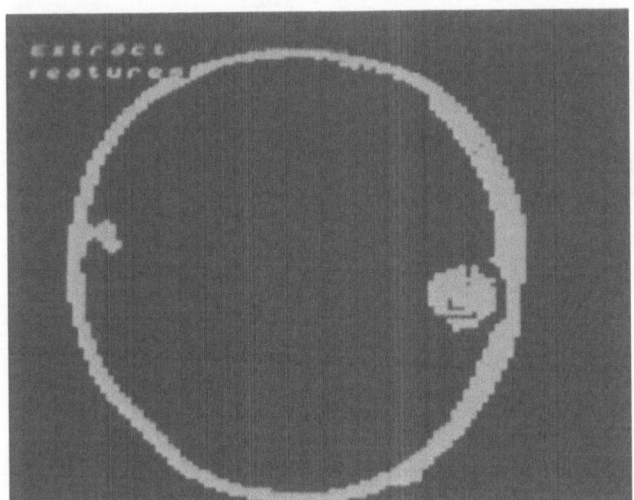

Photo 10 Edges detected and thresholded from Photo 9

Photo 11 Filling hole detected from Photo 9

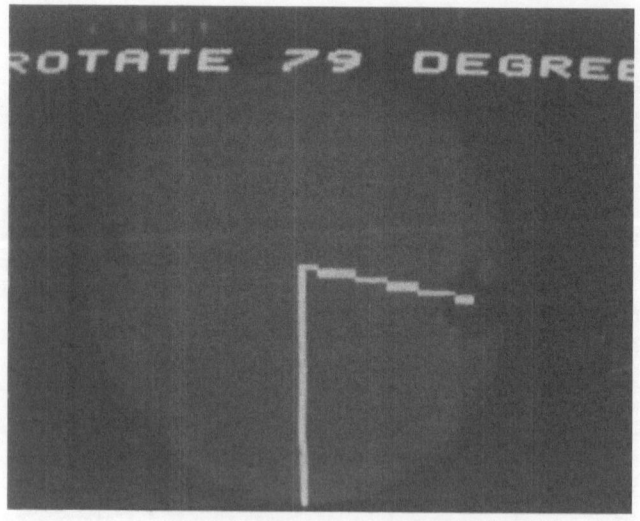

Photo 12 Rotation required shown on original from Photo 9.

3. Photos 9-12 show a simple orientation problem. The fill
    holes had to be found and the barrel rotated to bring the
    larger hole in line with a filling nozzle. Photo 9 shows
    the original image. The edges are found and thresholded,
    then the filling cap is located as the largest region inside
    the circumference.

    It is then simple to calculate the angle to turn as shown
in Photo 12.

    With all the above examples, vision solutions are produced
quickly and easily. The algorithms for all of the above
sequences were developed in less than two hours and could be
demonstrated to customers who then have increased confidence
that the final system will work.

4. APPLICATIONS

    The following applications are provided as an example of
real life situations that have been resolved by vision sensing.

4.1 Vision Sensing for FMS - The SCAMP Project

    The SCAMP FMS consists of a mixed range of CNC and convent-
ional machine tools which completely process a wide range of
turned components. Robots are used for automatic loading of all
machines which are linked by a palletised conveyor system, the
whole being fully integrated into a flexible transfer line
under computer control. The vision sensing system is used to
identify and orientate castings received from a robot in order
that they can be loaded automatically into a machine tool. The
use of visual sensing eliminates the need to employ mechanical
locating devices to orientate parts and enables new parts to be
introduced rapidly to a production line. Traditional mechanical
or opto-electronic locating devices to perform the orientation
process would not be practical or cost-effective because of the
range of components that may be presented to the system. A
common feature of visual sensing applications, SCAMP included,
is the need to provide facilities for teaching the system new
recognition tasks. Teaching has to be able to be carried out
by production personnel without any detailed knowledge of the
vision sensing system.

4.2 Razor Blade Inspection

    A manufacturer has recently launched a new product, a
disposable razor. In order to be able to compete successfully
with imported razors which have been assembled using cheap labour,
they decided to totally automate the assembly process. A vision

system has been installed to provide one hundred percent inspection of all production within the assembly machine.

The assembly machine consists of an indexing conveyor which has several work stations for the final assembly. After assembly the complete razors are dropped onto a conveyor where they are transported to an automated packing machine. The vision system is used to check each razor after final assembly to check that all components are present and assembled correctly. In particular it is needed to detect any faults which would make the razor dangerous to use, such as an incorrectly positioned blade or broken guard bar.

The vision sensor performs the manual inspection normally carried out as the components are packed, and also inspects each razor assembly in the same detail as can only be currently achieved on a sample basis.

The particular product in question was also assembled with the blade in the closed position and so, even if packed manually, the operator would be unable to detect any faults associated with the blade.

A common feature of vision sensing systems is the need to be able to teach the system to inspect new or modified components, and to be able to adjust the sensitivity of the inspection process. Teaching has to be able to be carried out by production engineers without any detailed knowledge of the vision sensing system.

4.3 FMS - Stack Picking

The FMS cell is designed to produce a range of machined parts from the appropriate rough castings. All the parts are circular and made of ferro-magnetic material. Two robots move the parts from machine to machine within the cell. There is also a supervisory computer that directs all machining operations. The vision system is treated as one of the sensor peripherals of the central supervisory computer.

4.3.1 The Vision System. The vision system has three tasks to perform in the cell. These are: unloading the pallets, centering the part, and orientating the part.

When a new pallet arrives at the FMS cell the part identity is keyed into the vision system. This allows pre-taught values to be loaded into the vision system from the database. This database contains information such as the part radius and thickness, together with data extracted at a teaching phase.

Communication with the central cell controller is via

parallel I30 lines. At the highest level the cell controller can request one of the three vision functions.

4.3.2. Stack Picking. The function of the stack picking module is to direct the robot to pick the most accessible part from a bin of parts with all parts arranged in stacks.

The first time that the cell controller requests the coordinates of a part from the bin, the vision system obtains an image of the whole bin and determines the location of all parts that are visible. It then generates a removal strategy by assigning the parts to be extracted first as those closest to the centre of the bin. On subsequent calls to the bin picking function only, the area around the part of interest is looked at and not the whole bin. The coordinates are communicated to the robot via a serial link. This is achieved by the robot sending a set of coordinates which the vision system modifies to show the correct location of the part. The height information is not given by the vision system. Height information is obtained by a proximity sensor on the gripper.

4.3.3. Part Centering. Having extracted the part from the pallet, the part is placed on a table for the second vision function. Here the centre of the part is estimated to $\pm$ 2mm. This represents an accuracy of better than 1% of the radius for the largest parts. The robot is then instructed to pick the part up by sensing the x,y coordinates to the robot over the serial link. The part is then loaded into a numerically controlled lathe.

4.3.4. Part Orientation. The robot places the part on an indexing table and the part orientation function selected. The component is then rotated through 360° in regular angular steps under the depth map camera. The turntable is controlled by the vision system computer. The system takes initial images over 3° intervals. Data from the depth map image of the component so formed is then compared with taught data held in the database. The orientation is then known to $\pm 3^{\circ}$.

In order to achieve greater angular accuracy the system scans a particular section of the component with finer angular resolution. The section of the component chosen includes one or more features that uniquely identify the orientation of that component. Data from the depth map image created will again be compared to taught data held in the database, and the orientation found to within $\pm \frac{1}{2}^{\circ}$. When the orientation of the part has been established the part will be turned through an appropriate angle, such that when the robot picks up the part it will be in the correct attitude for loading into the next machine in the cycle.

The light for the depth map system is set up to project a
line onto a radius of the turntable surface under the camera.
When a part is placed on the table this line appears displaced
by an amount related to the thickness of the part.

4.3.5. Calibration Procedures. Pallet picking and regripping
require that the vision system sends coordinate data which is in
robot 'world' coordinates. To achieve this coordinate transforms
are derived in order to relate the camera field of view to the
robot 'world' coordinates. The calibration procedure provides
the means of deriving these transforms. This procedure
should only have to be performed when the relationship between
the robot and the relevant camera is changed.

To perform a calibration cycle the operator places the
calibration target under the camera. Upon instruction from
the operator the vision system finds the centre of the target
and stores the information. The operator then moves the
robot such that a pointer attached to the gripper is
coincident with the centre of the target.

The robot coordinates are transmitted to the vision system
over the serial link. This process is repeated four times,
the target being moved round the corners of the field of view.
With the four sets of coordinates the vision system calculates
the transform required to relate the field of view to the robot
'world' coordinates. The transform is tested by placing the
target in the field of view and directing the robot to the
target centre. If the transform is acceptable it is stored on
disc for subsequent use.

5. FUTURE TRENDS

5.1 Artificial Intelligence/Expert Systems

Unfortunately Artificial Intelligence means all things to
all men. It is of vital importance that practical use is made
of AI and Expert Systems. Serious work is proceeding on AI with
respect to vision sensing. In this type of application it is
known as Image Understanding.

Image analysis can best be explained by looking at an
elementary task, illustrated in Figure 2. Suppose the task is
to recognise an object shaped like an 'L' and distinguish
it from other objects. Using conventional image processing
a first step might be to check the area of the object by counting
the number of black pixels. This would reject any incomplete
objects, and if this were the only possible variation it would
be an adequate test.

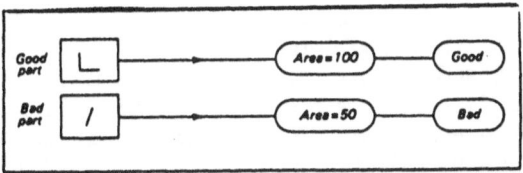

FIGURE 2

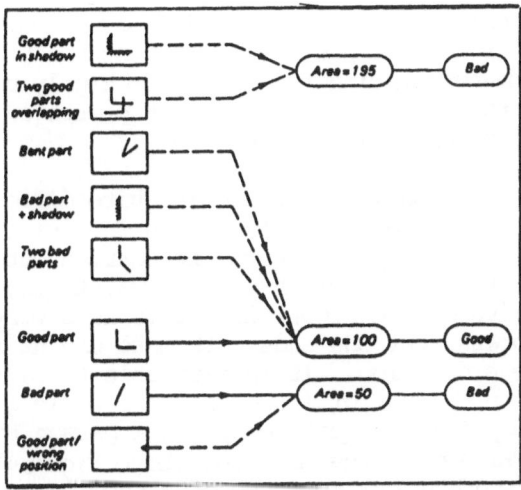

FIGURE 3

THE CAT SAT ON THE MAT

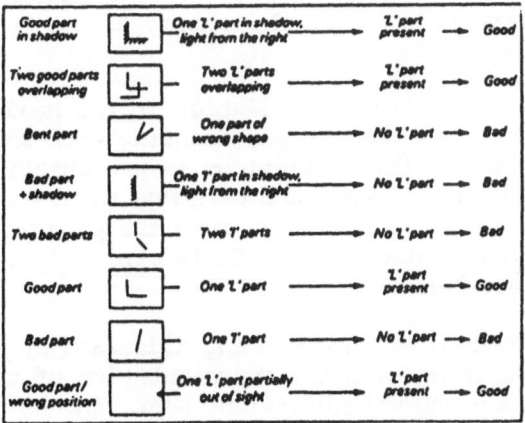

FIGURE 4

However, if the 'manufacturing' conditions were more variable, one might encounter parts casting a shadow which could affect the apparent area, they could be bent, there might be two good parts overlapping and so on. The main possibilities are shown in Figure 3 which indicates how some good parts will be rejected and some bad parts accepted. Some of these errors could be removed by further tests, others would require constraints on the mechanical presentation, the number of parts presented and the lighting.

These constraints are inflexible and may be expensive or even impossible if for example one is reading handwritten characters. In manufacturing too it may be highly uneconomical to constrain parts so that they are always presented in a particular position and orientation.

The image understanding approach is quite different. Instead of applying a series of quantitative rules involving computing numbers of pixels and comparing them, it carries out a qualitative description of the image in a similar way to the description a person would give, as in Figure 13. This is a very much more complex process, but one ends with more useful information on which to base a decision. It is also able to deal with situations where no amount of image analysis could give an answer. For example it is quite easy for a person to read the sentence: the cat sat on the mat (see above Figure 4), even though the As and Hs are identical, because of a knowledge of what words are possible.

So the image understanding approach will be able to cope with unconstrained situations by the generation of complex 'rich' descriptions of complex images and using knowledge of how the objects look - how the camera and lighting affect the image, the effect of gravity and so on.

The knowledge is already available, in the head of the system designer, who knows how cameras and lighting work and has to anticipate all the possible variations in presentation. The aim of artificial intelligence and of image understanding in particular is to put that knowledge into the program which controls the vision system so that it is available for use while the program is running.

The main problems which are being addressed by artificial intelligence have to do with the representation of this large amount of knowledge and with efficient searching so that one can get at it when it is needed.

One very important result of the introduction of knowledge-based software will be that the user will be able to teach the system very much more easily. It will be like the difference in giving instructions to a person who is an expert and to somebody who knows nothing about the subject. An expert will be able to act on very brief 'shorthand' instructions, whereas the non-specialist will have to be given full, detailed descriptions of what must be done.

The effects of such developments will be felt first within the organisation supplying systems, in helping to produce software for new applications very much faster than at present. In the longer term, one can envisage buying a vision system off-the-shelf which, on the basis of relatively non-technical instruction, will be able to carry out any of a large variety of inspection or control tasks.

More powerful or comprehensive solutions will be possible to problems which can already be tackled in a relatively limited way. In crack detection, for example, it should be possible to analyse each crack that is detected and decide whether it is sufficiently important, taking into account its size, position and so on, for the part to be rejected - in effect replicating and judgement of the human inspector.

Longer-Term Possibilities

Beyond this, and thinking only in terms of machine vision, one can envisage the possibility of finding practical solutions to generalised problems which are well outside the scope of present-day systems - like reading of handwriting - which would greatly enlarge the range of applications for machine vision. The generalised bin-picking problem has already been mentioned, and artificial intelligence software will make an important contribution to its practical solution.

Another important development might be the creation of an interface to computer-aided design so that, for example, a vision system would be able to use dimensional and tolerance data input from a CAD database to carry out inspection of workpieces. The consequences for flexible automation would be enormous in companies producing short-batch complex work in great variety - for example in the aerospace industry.

Artificial Intelligence could also conceivably link CAD with assembly, so that a vision-assisted robot assembly station could carry out an assembly operation from a computer-generated assembly "drawing" and then inspect its own handiwork.

5.2 Optical Processing

There is an increasing interest in the use of optics to

process image information prior to computer analysis. This is a
new field and development is mainly for military purposes.
However it is envisaged that within the next few years component
costs will be reduced to a level that enables engineers to consider
using optical processing as a front-end operation prior to
computer analysis.

The obvious benefits of optical processing are infinite
resolution and "instant" processing.

5.3 Higher Speed Hardware

There will always be requirements to improve computer
processing speeds. The introduction of genuine parallel program-
mable processors is already producing some improvements (up to
150 times processing capability in real manufacturing situations).
However this is only the beginning. Already 512 x 1 and 96 by 96
processor configurations are available. However in the latter
case the problem of fully utilising the capabilities of such
a processor are enormous. There has to be a complete "rethink"
on computer languages and operating systems. This major
conceptual change, when researchers have produced it, will
probably be difficult to absorb by existing software people in
the main.

6. CONCLUSION

Intelligent processing, particularly vision processing,
is now well established technology on the factory floor. However
it is still in its infancy. The scope of vision sensing is enormous and the potential for the future knows no bounds, particularly when one considers the speed at which developments are
proceeding.

Because intelligent sensing requires a mixture of diverse
skills - electronic engineering, computer engineering, software,
mechanical engineering, industrial psychology, optics, camera
technology, other sensor technology etc - it will require an
increasing supply of highly qualified engineers who are able to
approach the problem from a system concept point of view. Urgent
attention needs to be given to the educational processes that
will be required in order to produce such people.

Acknowledgements

The Authors wish to acknowledge the contributions made by
Dr David Mott, Dr Graham Page, David Upcott of BRSL and
Simon Cotter of UWIST.

MOIRE METHODS

MOIRE' INTERFEROMETRY AND ITS DEVELOPMENTS IN STRAIN ANALYSIS

L. Pirodda

Dipartimento di Ingegneria Meccanica
Università di Cagliari, Italy

ABSTRACT

An optical technique is presented for the direct and real-
time determination of unit surface strains. The technique is based
on the following main operations: (a) recording a high density reflection grating on the undeformed surface, (b) obtaining a shearing interferogram of one of the diffracted beams from the deformed surface, (c) compensating for out-of-plane displacement and for
surface roughness by means of real-time phase conjugation.

1. INTRODUCTION

Much effort has been expended in an interdisciplinary collaboration between applied optists and mechanists in order to develop non-destructive testing methods which afford a whole field
quantitative picture of the state of stress or strain in machine
or structural components.

The classical techniques, like photoelasticity and moiré,
have been supplemented, over the past twenty years, by modern approaches which exploit the properties of coherent laser light.
Among these holographic and speckle interferometry are probably
the most popular.

Generally speaking the above techniques, at least in their

simplest implementation, are not adequate to directly provide the stress analyst with the data he ultimately needs, i.e. the derivatives of the in-plane displacement components. The user of three dimensional holographic interferometry, for instance, has to perform a rather complex series of calculations in order to draw from the fringe map first the general set of displacement components and subsequently the in-plane components and their derivatives.

It is not therefore surprising that a remarkable amount of ingenuity has been devoted by a host of distinguished authors to developing new optical techniques, which in general are more or less complex modifications of the basic ones, providing directly the in-plane displacement components or ultimately their derivatives.

The technique becoming known as "moiré interferometry" or "high sensitivity moire" follows this trend (see [1]-[4]). Its basic features are: a) recording on the surface of the undeformed specimen a grating with density ranging from 500 to 2000 line pairs/mm and supposed to undergo the same deformation. This requires applying to the specimen some sort of photo-sensitive coating and exposing it to an interferometrically modulated light field; b) performing a second exposure when the specimen is in the deformed state so that a second grating is superimposed on the first (double exposure technique), or c) observing the deformed specimen (with one single deformed grating) under illumination by an appropriately modulated light field (real-time technique).

Since in both cases b) and c) two gratings are superposed, moiré fringes are to be expected, mapping one in-plane displacement component according to the classical geometrical moiré theory. However, a rigorous analysis of the phenomena involved, with particular concern to questions of visibility of the fringes, can only be made in terms of wave optics, as illustrated in the following section 3.

The technique provides separately the in-plane displacement components in a comparatively simple way, although recording the grating on the specimen cannot be considered so far a straightforward operation (see [5] and [6] for this important aspect of the technique which cannot find space here). However, the derivatives of the displacement must still be calculated. It should also be pointed out that the method is highly sensitive to rigid body motions, particularly to in-plane rotations.

A further development which provides the derivatives directly, possibly in real time and with little or no sensitivity to rigid motions is presented in this contribution. Its basic points are briefly outlined below.

The surface of the specimen and the grating, referred to an x,y,z coordinate system as in Fig.1a (z normal to the surface, y parallel to the lines of the grating), is assumed to be initially plane. If a plane light wave-front impinges on the undeformed grating, one reflected and two first order diffracted plane wave-fronts emerge (higher diffraction orders are neglected). If the surface (and hence the grating) deforms, the reflected and diffracted wave-fronts are no longer plane. The "warping" of the re-

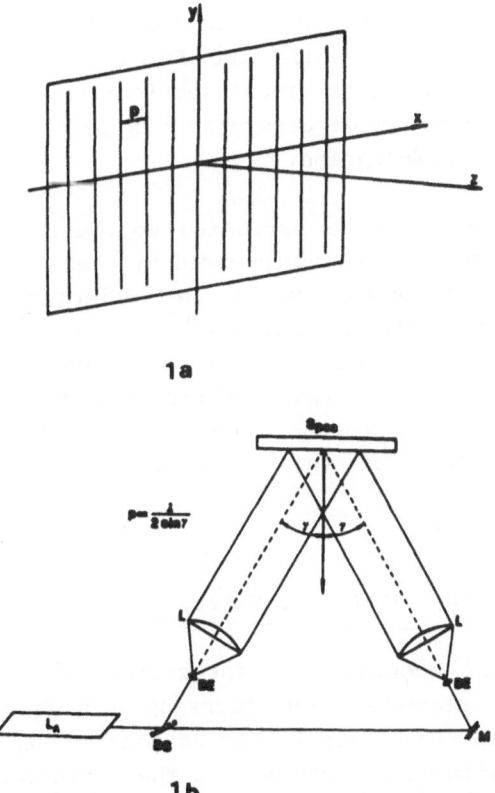

1a

1b

Fig.1 - Plane surface with recorded grating referred to x,y,z coordinates.

flected wave is due to the out-of-plane displacement component w, while the warping of the diffracted ones is due both to w and to the in-plane component (according to the normal to the lines of the grating) u. The warped waves can be considered as phase objects and thus one of the methods that afford an intensity image of such objects can be used in order to extract from them the required information about the u(x,y) and w(x,y) unknown functions. This is in fact the case of the moiré-interferometric techniques, as the discussion in Section 3 will point out.

In the alternative presented here, the intensity image of the warped waves is obtained by shearing interferometry. This is a real-time technique, practically insensitive to rigid-body motions.

The shearing interferogram of the reflected wave presents a series of fringes mapping the partial derivative δw/δs according to the direction of shearing (s is the linear coordinate in the direction of shearing), while the fringes of the interferogram of a diffracted wave map a linear combination of δw/δs and δu/δs. The two derivatives can be separated via point by point calculation upon both interferograms.

The separation can be performed optically and in real-time if the above procedure is supplemented by real-time phase conjugation. The phase conjugate wave of the reflected (0th order) wave is used to reilluminate the surface. The new 0th order (retroreflected) wave is restored as a plane wave, while one of the corresponding diffracted waves (under certain conditions to be defined later) is only warped by the in-plane displacement component. A shearing interferogram of this wave is therefore a map of the δu/δs derivative.

2. THE REFLECTED AND DIFFRACTED WAVES

Let us first consider the theoretical case of an optically flat reflecting plane surface carrying a high density grating. With reference to Figs.1 and 2, let us consider an impinging monochromatic wave with wavelength λ, and the reflected and diffracted waves in the particular case where all propagation directions are on the xz plane.

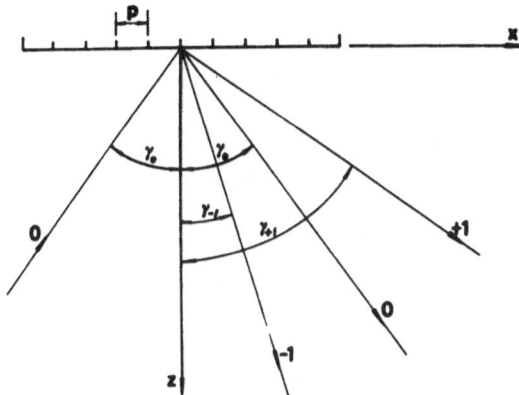

Fig.2 - Impinging, reflected and first order diffracted waves.

The diffraction law for the two 1st order diffracted waves is [7]:

$$\sin \gamma_1 = \sin \gamma_0 \pm \frac{\lambda}{p} \tag{1}$$

The following two specific cases will be dealt with later on:

(a) Retrodiffraction

$$\sin \gamma_0 = \frac{\lambda}{2p} \; ; \qquad \sin \gamma_{-1} = -\sin \gamma_0 \tag{2}$$

The -1 wave is diffracted in the same direction and opposite verse as the impinging wave. This condition is satisfied by either of the two beams used to record the grating, in the case of Fig.1(b).

(b) Normal diffraction

$$\sin \gamma_0 = \frac{\lambda}{p} \; ; \qquad \sin \gamma_{-1} = 0 \tag{3}$$

The -1 wave is diffracted perpendicularly to the surface. Let us now suppose that the surface and the grating undergo the same deformation, defined by the u(x,y), v(x,y), w(x,y) displacement components. The reflected and diffracted waves are "warped" by an amount which can be measured by the phase differences produced, in a plane normal to the propagation direction, by the w and u components.

The optical path difference produced by w(x,y) upon two rays 1 and 1' of the same diffracted wave can be evaluated from Fig.3.

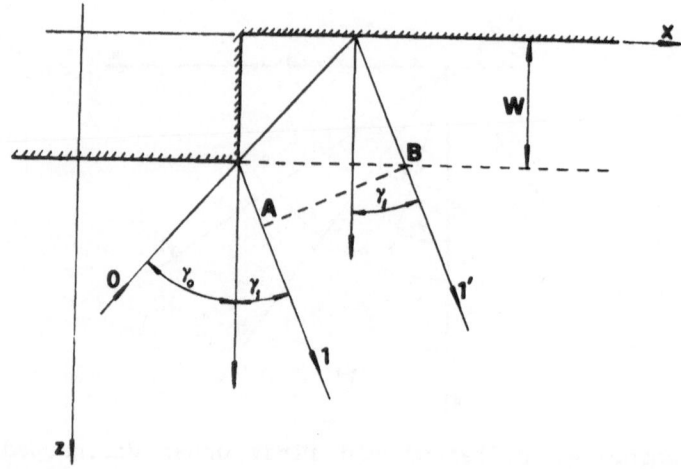

Fig.3 - Evaluation of the optical path difference due to w(x,y).

This difference is expressed by:

$$\delta_{1w} = w\left[\frac{1}{\cos\gamma_0} + \frac{1}{\cos\gamma_1} - (\tan\gamma_0 + \tan\gamma_1)\sin\gamma_1\right] = w\delta_1\left(\gamma_0, \frac{\lambda}{p}\right) \quad (4)$$

For the reflected wave $(\gamma_1 = \gamma_0)$ we have:

$$\delta_{0w} = w\,2\cos\gamma_0 = w\delta_0(\gamma_0) \quad (5)$$

One particular case, which is relevant for later applications, is where

$$\delta_{1w} = \delta_{0w}, \quad \text{or} \quad \delta_1\left(\gamma_0, \frac{\lambda}{p}\right) = \delta_0(\gamma_0)$$

It can be shown, by making use of Eqs.(1),(4) and (5) that this condition is met for one of the diffracted waves when

$$\frac{\lambda}{p} = 4\sin\gamma_0\cos^2\gamma_0 \quad ; \quad 0 \leqslant \gamma_0 \leqslant 35° \quad (6)$$

Setting for instance $\frac{\lambda}{p} = 1$, we have for Eqs.(6) and (1): $\gamma_0 = 17°$; $\gamma_1 = -45°$.

The effect of u(x,y) can be evaluated referring to Fig.4, where 0 is the direction of an impinging plane wave, 1 the direction of one diffracted 1st order wave, x'y' the plane normal to the

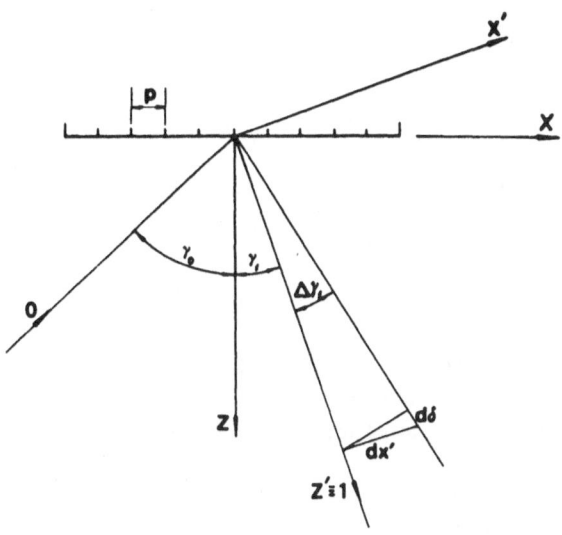

Fig.4 - Evaluation of the optical path difference due to u(x,y)

latter direction. By differentiating Eq.(1) with respect to p we have:

$$\Delta\gamma_1 = \mp \frac{\lambda}{p^2} \frac{\Delta p}{\cos \gamma_1} = \mp \frac{\lambda}{p} \frac{\varepsilon_x}{\cos \gamma_1} \left(\frac{\Delta p}{p} = \varepsilon_x \right)$$

Hence

$$d\delta = -\Delta\gamma_1 dx' = -\Delta\gamma_1 dx \cos \gamma_1 = \pm \frac{\lambda}{p} \varepsilon_x dx$$

and

$$\delta = \delta_{1u} = \int_0^x d\delta = \pm \frac{\lambda}{p} \int_0^x \varepsilon_x dx = \pm \frac{\lambda}{p} u(x,y) \tag{7}$$

is the optical path difference produced in the points of the x'y' plane by the u(x,y) displacement component. In the computation of the signs above, provision has been made for the fact that for $\Delta\gamma_1$ positive $d\delta/dx'$ is negative, i.e. $d\delta$ corresponds to a phase lag.

The complex amplitudes for the reflected and diffracted waves are:

$$a_0 = A_0 \exp\left(i \frac{2\pi}{\lambda} \delta_{0w} \right) = A_0 \exp\left[i \frac{2\pi}{\lambda} \delta_0 (\gamma_0) w \right]$$

$$a_{+1} = A_{+1} \exp\left[i \frac{2\pi}{\lambda} (\delta_{1w} + \delta_{1u}) \right] = A_{+1} \exp\left[i \frac{2\pi}{\lambda} \delta_1 \left(\gamma_0, \frac{\lambda}{p} \right) w + i \frac{2\pi}{p} u \right]$$

$$(8)$$

$$a_{-1} = A_{-1} \exp\left[i \frac{2\pi}{\lambda} (\delta_{1w} + \delta_{1u}) \right] = A_{-1} \exp\left[i \frac{2\pi}{\lambda} \delta_1 \left(\gamma_0, \frac{\lambda}{p} \right) w - i \frac{2\pi}{p} u \right]$$

where A is the real amplitude factor.

It must be pointed out that, if the surface is not optically flat, the w factor is the sum of the out-of-plane displacement component and the local surface irregularity.

3. MOIRE' INTERFEROMETRY

Moiré interferometric techniques will be presented by means of the three cases illustrated in Figs.5, 6, 7 respectively. The first case (Fig.5) is that of double exposure. The photosensitive coat of the specimen is first doubly exposed to the interferometric set-up in the loaded and unloaded conditions consecutively, so that two gratings are recorded. For one impinging wave the two recorded gratings diffract two waves along the same diffraction direction, say -1. The two waves interfere giving rise to interference (moiré) fringes. The amplitude of the resultant diffracted -1 wave is:

$$a_{-1} = A_{-1} \exp\left(i \frac{2\pi}{\lambda} \delta_{1w} \right) + A_{-1} \exp\left[i \frac{2\pi}{\lambda} \delta_{1w} - i \frac{2\pi}{p} u(x,y) \right] \qquad (9)$$

where w represents the local surface irregularity if the observation is made in the unloaded condition and the surface irregularity plus out-of-plane displacement if the observation is made in the loaded condition. In both cases the intensity is:

$$I = a \cdot a^* = 2A_{-1}^2 \left(1 + \cos \frac{2\pi}{p} u \right) \qquad (10)$$

The intensity maxima and minima occur along fringes of (theoretical) visibility 1, which are loci of equal value of u , with sensitivity factor p. The surface irregularities and the out-of-plane displacement have no influence on the fringes and their visibility.

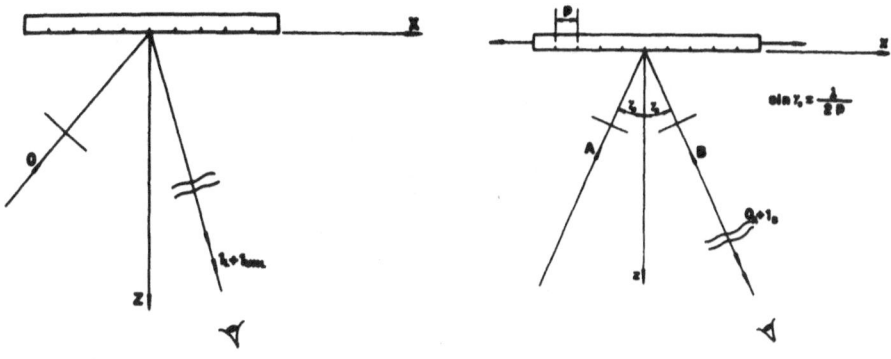

Fig. 5 - Moiré interferometry; double exposure.

Fig. 6 - Moiré interferometry in real time; reflected plus retro-
      diffracted waves.

Figure 6 illustrates one of the proposed [3] real-time observation techniques. A high density grating is previously recorded
over the unloaded specimen. The deformed grating on the loaded
specimen is illuminated in the same interferometric set-up as that
used for recording and observed according to one of the directions
of the recording beams. One retrodiffracted and one reflected wave
propagate and interfere in this direction. The resultant amplitude
is:

$$a = a_0 + a_{-1} = A_0 \exp\left(i \frac{2\pi}{\lambda} \delta_{0w}\right) + A_{-1} \exp\left[i \frac{2\pi}{\lambda} (\delta_{1w} + \delta_{1u})\right] \tag{11}$$

The expression for the corresponding intensity can be shown to be:

$$I = (A_0 - A_{-1})^2 + 2A_{-1}A_0 \left\{ 1 + \cos \frac{2\pi}{\lambda} \left[2w\left(\frac{1}{\cos \gamma_0} - \cos \gamma_0\right) - \frac{\lambda}{p} u\right]\right\} \tag{12}$$

since for the retrodiffracted beam

$$(\gamma_1 = - \gamma_0) \delta_{1w} = \frac{2w}{\cos \gamma_0}$$

The intensity is modulated according to fringes, the visibility
of which is:

$$V = \frac{I_{max} - I_{min}}{I_{max} + I_{min}} = \frac{2A_0 A_{-1}}{A_0^2 + A_{-1}^2}$$

Furthermore, the fringes are not independent of the out-of-plane component w or of surface irregularities. Only for $1/\cos \gamma_0 = \cos \gamma_0$, i.e. for small angles γ_0, is such dependence negligible. However, this is not the case of interferometric moiré, demanding dense gratings and large γ_0.

If the coated surface departs from optical flatness, the corresponding term in w introduces, according to Eq.(12), localized disturbances which can easily impair the quality of the fringes.

For the above reasons the real-time set-up depicted in Fig.6 is hardly to be commended. Much better results can be obtained by the real-time set-up [4] represented in Fig.7. Here the loaded specimen, carrying the deformed grating, is illuminated by two symmetrical beams at two equal angles, γ_0 satisfying the normal diffraction condition (Eq.(3)), and is observed according to the normal, where the two -1 diffracted beams interfere. Their resultant amplitude is:

$$a = A_{-1} \exp \left[i \frac{2\pi}{\lambda} (\delta_{1w} + \delta_{1u}) \right] + A_{-1} \exp \left[i \frac{2\pi}{\lambda} (\delta_{1w} - \delta_{1u}) \right] \qquad (13)$$

To explain the opposite signs for the δ_{1u} term one must consider that a given in-plane deformation of the grating causes opposite

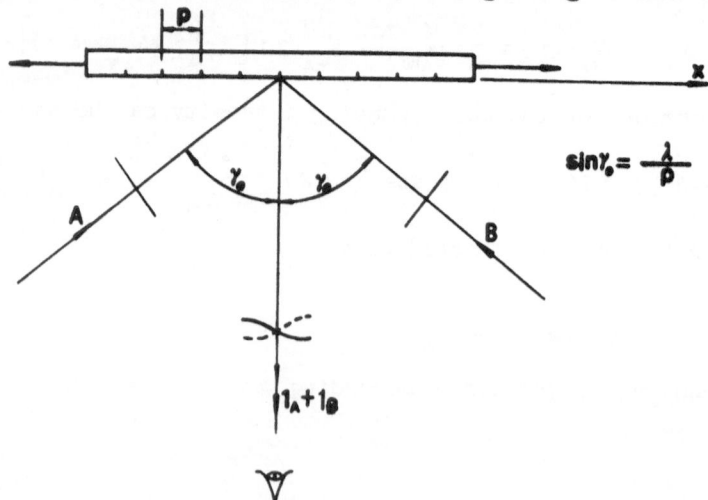

Fig.7 - Moiré interferometry in real time; normal diffraction

rotations of the diffracted beams, hence opposite phase-lag gradients according to x.

From (13) we have for the intensity:

$$I = A_{-1}^2 2\left[1 + \cos\left(\frac{2\pi}{\lambda} 2\delta_{1u}\right)\right] = A_{-1}^2 2\left[1 + \cos\left(2\pi\frac{2u}{p}\right)\right] \tag{14}$$

The intensity is modulated according to fringes of visibility one, which are loci of equal value of the in-plane displacement component u, with sensitivity factor p/2. The set-up just examined is therefore insensitive to out-of-plane deformation and to surface roughness and provides doubled sensitivity for a given p. However, since the interferometric set-ups for recording and observing are different, the recorded and "live" gratings must be micrometrically aligned.

One common drawback of all the techniques examined in this section is their sensitivity to rigid-body motion of the specimen, between the exposures (in the case of the double exposure technique) or during observation. In fact, some components of the motion of the grating with respect to the illuminating beams cause a relative motion of the interfering diffracted or reflected beams. For example, a rotation of the specimen around the z axis during observation in the set-up in Fig.7 causes the appearance of fringes normal to the lines of the grating, physically explainable in terms of wave optics or also, according to the classical moiré theory, due to the relative rotation of two superimposed grids, having pitch p and p/2 respectively.

Another adverse aspect, when dealing with strain analysis problems is the need to calculate the derivatives of the displacement components in order to get the unit strains, which means that the optical technique is only the preliminary to a more complex and cumbersome process.

4. DIFFRACTION PLUS SHEARING INTERFEROMETRY

An intensity image of a warped reflected or diffracted wave can be provided alternatively by shearing interferometry, i.e. by an optical device capable of splitting the wave into two identical ones having a common path, and a mutual spatial shift. This operation is schematically depicted in Fig.8 for a -1 diffracted wave.

234

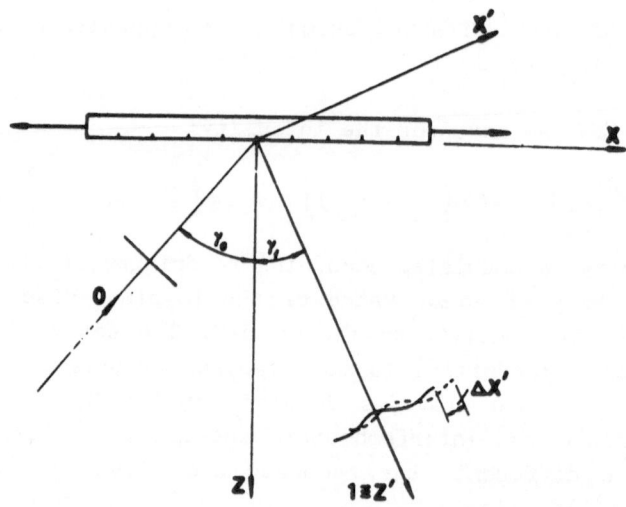

Fig.8 – Application of shearing interferometry to a diffracted wave

Here the mutual shift is a linear one $\Delta x'$ in the direction x'. A practical implementation of the operation is described later. The resultant amplitude of the two mutually shifted waves propagating in the direction -1 is:

$$a = A_{-1} \exp\left[i\, \frac{2\pi}{\lambda}\, (\delta_{1w} + \delta_{1u}) \right]$$

$$+ A_{-1} \exp\left[i\, \frac{2\pi}{\lambda}\, (\delta_{1w} + \Delta\delta_{1w} + \delta_{1u} + \Delta\delta_{1u}) \right]$$

$$= A_{-1} \exp\left[i\, \frac{2\pi}{\lambda}\, (\delta_{1w} + \delta_{1u}) \right] \left\{ 1 + \exp\left[i\, \frac{2\pi}{\lambda}\, (\Delta\delta_{1w} + \Delta\delta_{1u}) \right] \right\}$$

Setting:

$$\Delta\delta_{1w} \simeq \frac{\partial\delta_{1w}}{\partial x'}\, \Delta x' = \frac{\partial\delta_{1w}}{\partial x}\, \Delta x$$

$$\Delta\delta_{1u} \simeq \frac{\partial\delta_{1u}}{\partial x'}\, \Delta x' = \frac{\partial\delta_{1u}}{\partial x}\, \Delta x$$

and evaluating the intensity, we find:

$$I = A_{-1}^{2}\, 2 \left[1 + \cos \frac{2\pi}{\lambda}\, \Delta x \left(\frac{\partial\delta_{1w}}{\partial x} + \frac{\partial\delta_{1u}}{\partial x} \right) \right] \tag{15}$$

The intensity is modulated according to fringes, along which:

$$\frac{\partial \delta_{1w}}{\partial x} + \frac{\partial \delta_{1u}}{\partial x} = \frac{\partial w}{\partial x} \delta_1 - \frac{\partial u}{\partial x} \frac{\lambda}{p} = \text{constant} = n \frac{\lambda}{\Delta x} \qquad (16)$$

where n is an integer.

By similar argument we can find that, if the shearing operation is performed upon a reflected wave, the intensity fringes are
loci of equal value for:

$$\frac{\partial w}{\partial x} = n \frac{\lambda}{\delta_0 \Delta x} \qquad (17)$$

The problem of determining the derivatives of the displacement components is solved, in general, by simultaneous application of the
shearing technique to one reflected and one diffracted wave and
subsequent separation of the two unknowns by means of Eqs.(16) and
(17).

In addition to the advantage of direct optical determination
of the derivatives, the technique outlined above is insensitive to
rigid-body motion. In fact, the only effect of any component of the
motion of the grating is a change in the direction of the reflected
or diffracted beam, which has no consequence on the interferogram,
unless the motion is so significant as to require a readjustment
of the interferometer, unlikely in the case of strain analysis problems.

Conversely, the need to resort to numerical calculation in order to separate the derivatives appears a real disadvantage. However, the computation involved, i.e. solving simple linear sets of
equations, can be performed to any degree of accuracy, which is not
the case with the computation of derivatives.

Another drawback of the technique, in the form presented in
this paragraph, is its sensitivity to the roughness of the surface
of the grating, which in turn depends on the original roughness of
the specimen and on the process of deposing the photosensitive
coating and of recording the grating. If the roughness component
in w, Eq.(16), is not sufficiently small, the quality of the fringes deteriorates proportionately.

A practical implementation of a shearing interferometer is

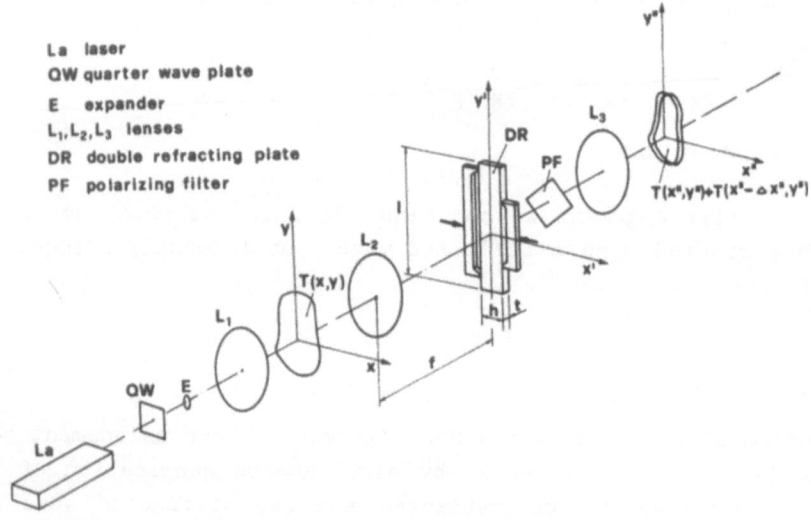

La   laser
QW quarter wave plate
E   expander
L_1, L_2, L_3 lenses
DR   double refracting plate
PF   polarizing filter

Fig.9 - Implementation of a shearing interferometer.

presented in Fig.9, where for representation purposes the transmission mode of operation is considered [8]. A plane coherent wave is warped after crossing the transparency T(x,y) (or equivalently after reflection or diffraction upon a reflective grating). The splitting of the wave is accomplished by double refraction in the focal (Fourier) plane of the lens L_2. The double refracting element is a stress birefringent polycarbonate plate subjected to uniform flexure. The two waves emerging from the plate are orthogonally polarized and possess a mutual phase shift linearly varying along x'. Since a linear phase shift in the Fourier plane corresponds to a spatial shift in the plane x y or in the (image) plane x"y" (by virtue of the shift theorem of Fourier transform theory), the device of Fig.9 is suitable for performing the shearing operation.

The set-up is completed by a quarter-wave plate which circularly polarizes the linearly polarized laser beam, a polaroid filter used as analyzer and a second imaging lens L_3.

Figure 10 illustrates an example of shearing interferometry applied to a diffracted wave by means of the set-up in Fig.9. The specimen was a thin transmitting strip axially stretched and pierced with a hole. Due to the transmission mode of operation, to the small thickness of the specimen and to its smoothness, the in-

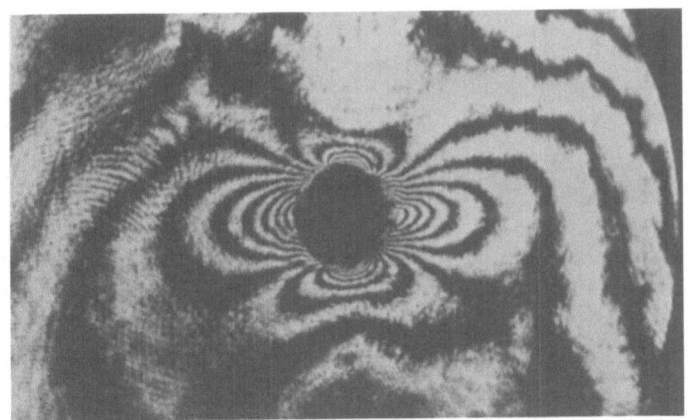

Fig.10 - Unit strain fringes in a transparent specimen.

fluence of w was negligible, hence the fringes map $\partial u/\partial x$.

5. DIFFRACTION PLUS SHEARING INTERFEROMETRY PLUS PHASE CONJUGATION

The principle of a direct and real-time technique for in-plane strain determination is schematically represented in Fig.11. A plane wave 0 impinges upon a loaded specimen carrying a (deformed) grating. The angle of incidence γ_0 satisfies the condition (6), so that $\delta_{1w} = \delta_{0w}$. The reflected wave 0' is made to impinge on a phase conjugator: a device which 'reflects' the conjugate of any impinging wave. The conjugate 0'⁻ of 0' impinges back on the specimen and its diffracted -1 wave is subjected to the shearing interferometric process.

The amplitudes of 0' and 0'⁻ are respectively:

$$a = A_0 \exp\left(i \frac{2\pi}{\lambda} \delta_{0w}\right) = A_0 \exp\left(i \frac{2\pi}{\lambda} \delta_{1w}\right)$$

$$a = A_0 \exp\left(-i \frac{2\pi}{\lambda} \delta_{1w}\right)$$

Hence the amplitude of -1⁻ is:

$$a = A_{-1} \exp\left[i \frac{2\pi}{\lambda} (\delta_{1w} - \delta_{1w} + \delta_{1u})\right] = A_{-1} \exp\ i \frac{2\pi}{\lambda} \delta_{1u}$$

while the resultant amplitude at the image plane of the shearing interferometer is:

$$a = A_{-1} \exp\left(i \frac{2\pi}{\lambda} \delta_{1u}\right) + A_{-1} \exp\left[i \frac{2\pi}{\lambda} \left(\delta_{1u} + \frac{\partial\delta_{1u}}{\partial x} \Delta x\right)\right]$$

with intensity

$$I = A_{-1}^2 2\left(1 + \cos \frac{2\pi}{\lambda} \frac{\partial\delta_{1u}}{\partial x} \Delta x\right) \tag{18}$$

Taking Eq.(7) into account we can see that the intensity fringes are loci of equal value for $\partial u/\partial x$, with sensitivity factor $p/\Delta x$

The practical implementation of the principle outlined in Fig. 11 is at hand, since real-time phase conjugation is now an established optical technique [9]. The type of phase conjugator suitable for the application could basically be a platelet of a photorefractive medium where the equivalent of a hologram is recorded in real time by O' as the object wave plus an additional reference wave r. The conjugate r⁻ of r reconstructs the conjugate O'⁻ of O'. If r is a plane wave, r⁻ can be simply obtained by means of a plane mirror.

Figure 12 presents the proposed optical set-up which implements the principle, and which needs little further elucidation. The photorefractive component is a small slab (say 15x15x3 mm) cut from a monocrystal of BSO ($Bi_{12}SiO_{20}$), which is sufficiently sensitive at the wavelengths and intensities obtainable from an argon ion laser. A d.c. electric field of about 5kV/cm must be applied to the crystal, as shown in the figure. Since the active surface of the BSO crystals, currently available at not too prohibitive

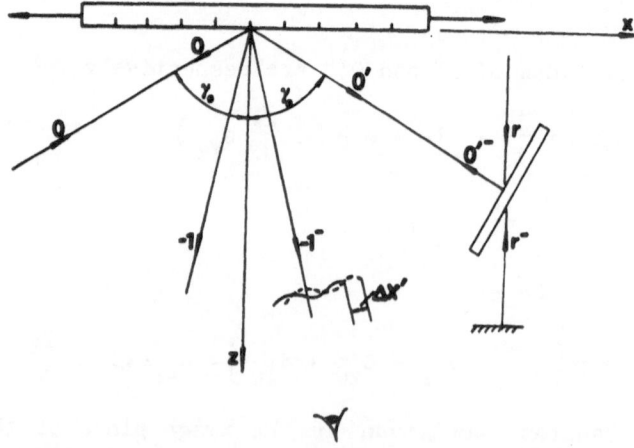

Fig.11   Principle of shearing interferometry plus phase conjugation

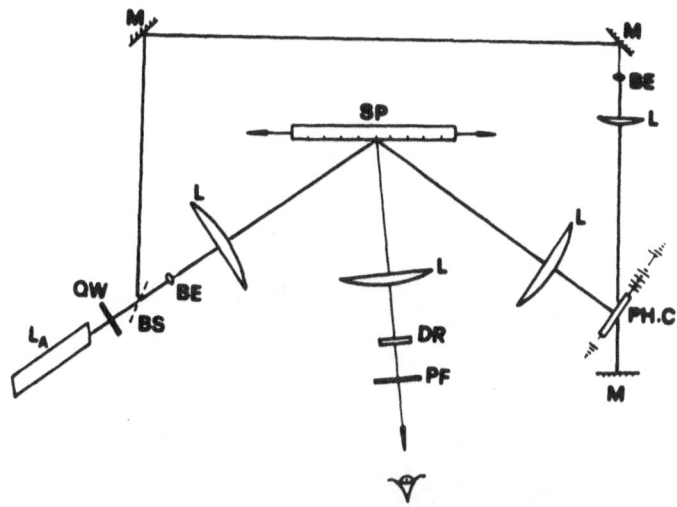

Fig.12 - Complete set-up for direct strain analysis.

costs, is of the order of 2 cm², a lens, L_4, is generally required in order to match the said surface to the field of the specimen. Focussing the light by this lens also ensures the necessary intensity on the crystal.

It should be pointed out that the real-time phase conjugator automatically and instantly compensates for the abberations introduced by the lens, L_4, and for any rigid motion of the specimen.

The apparatus in Fig.12, although relatively complex, seems nevertheless to be the only one capable of giving the unit strains components, in real time, with sufficient sensitivity and without the errors due to rigid motion. The practical implementation of such apparatus is currently being pursued. Meanwhile, the principle of Fig.11 has been positively tested by using an hologram as phase conjugator. A conventional hologram has first been recorded by the 0' and r waves. Subsequently, upon illumination by r^- the 0^- wave has been reconstructed and used to illuminate the specimen.

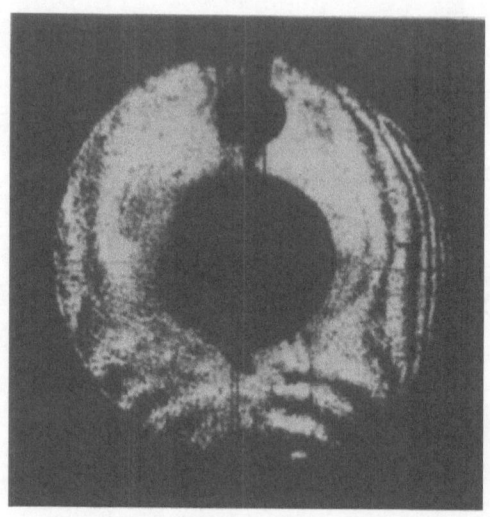

Fig. 13 - Shearing interferogram after phase conjugation

A shearing interferogram of the diffracted wave satisfying Eq. (6) was then taken and it is shown in Fig.13. The object was a ring cut radially and loaded in the same cut by two equal opposite tangential forces. The lines of the grating were normal to the forces, i.e. parallel to the cut. The curves of equal value of ϵ_x are clearly visible in the lower part of the figure and they match the theoretical results according to the theory of large curvature solids. Although the quality of the interferogram could be better, since when the experiment was performed there was no means available to exactly re-position the hologram and fulfil Eq.(6), the result seems nevertheless convincing.

REFERENCES

1. Boone, P.M.. Surface Deformation Measurements using Deformation following Holograms. Nouv. Rev. d'Opt. Appl., 1 (1970) 10.
2. Cook, R.W.E. Recording Distortion of Irregularly Shaped Objects, Optics and Laser Tech., 3 (1971) 71-3.
3. Wadsworth, N., Marchant, M. and Billing, B. Real Time Observation of In-Plane Displacements of Opaque Surfaces. Optics and Laser Techn., 5 (1973) 119.

4. Post, D. Developments in Moiré Interferometry. Optical Engng, 21 (1982) 458–67.
5. Bartolini, R.A. Characteristics of Relief Phase Holograms recorded in Photoresists. Appl. Opt., 13 (1974) 129–39.
6. De Caluwé, M. Brush Coating of Resists for Normal and High Frequencies Moiré Techniques. Strain, 11 (1975) 26–30.
7. Rossi, B. Optics (Addison Wesley Publishing Co., Reading, Mass. U.S.A., 1957).
8. Pirodda, L. Optical Methods of Non-Destructive Testing in Italy: A Short Selection, Industrial Application of Holographic Non-Destructive Testing, J. Ebbeni (Ed)., Proc. SPIE 349 (1982) 167–85.
9. Pepper, D.M. Nonlinear Optical Phase Conjugation. Optical Engng, 21 (1982) 156–81.

MEASUREMENTS OF FRACTURE TOUGHNESS AND INTEGRITY BY MOIRE INTERFEROMETRY

J. McKelvie, C.A. Walker, P.MacKenzie, T.G.F. Gray

Dept. of Mechanics of Materials, University of Strathclyde

1.   FRACTURE TOUGHNESS.

Fracture Toughness is a concept representing the resistance of a material, having a pre-existing crack, to the propagation of the crack through the material. It is a material property essential in Fracture Mechanics.

For the case of materials having a completely linear-elastic stress-strain relationship, there is a relatively straightforward treatment, the Linear Elastic Fracture Mechanics (LEFM). The Fracture Toughness of such materials is specified by the "critical Stress Intensity Factor". However, such materials have a much lower fracture resistance than those exhibiting substantial ductility prior to fracture. The toughness of these ductile materials is measured by two alternative parameters called the "Critical Crack Opening Displacement, δ_c" and the "Critical J-value, J_c" respectively.

Critical COD is something of an empirical parameter based upon the knowledge of the behaviour of a crack in a tough material whenever a stress field is applied such as to open the crack:- The crack initially opens, without propagating, in a manner illustrated schematically in Fig.1, which shows the length referred to as the COD. The COD test consists of loading a specimen, of standard geometry and having a fatigue crack pre-induced in it, in a standard monotonic manner, and determining the value of COD at which crack propagation commences (e.g. Fig.2). This value is the "critical" COD. or δ_c, and is a measure of the fracture toughness.

The J Concept is more fundamental. It was defined by Rice [1], and it represents the energy available, within the body, for the propagation of the crack.

J is defined as:-

$$J = \oint_\Gamma (W dy - T_i . \partial u_i / \partial x \, ds)$$

where Γ is a contour enclosing the crack tip,
      W   is the strain-energy density,
      T   is the traction vector normal to Γ
      u   is the displacement vector,
      x   is the coordinate direction parallel to the crack,
      s   is an element of the contour path.

As with COD, there is a critical value, Jc, at which, under monotonic loading, the crack starts to propagate.

2. FRACTURE INTEGRITY.

If a structure has a crack, (or a flaw), then an important question surrounds its utility. If it can be demonstrated that, under its service-stress condition, the crack will not propagate, then there is every reason to leave the piece in service, since replacement may be an extremely expensive business.

It is a routine purpose of Fracture Mechanics to examine this question. Having a knowledge of the material's Fracture Toughness and of the structure's geometry and loads, judgements on Integrity can be provided. Thus, for example, values of maximum allowable defect size can be stipulated for material inspection purposes.

3. MEASUREMENTS OF FRACTURE TOUGHNESS AND INTEGRITY.

Methods of Measurements of δ_c and J_c are described in Brit.Standard 5762 : 1979 "Methods for crack opening displacement (COD) testing" and ASTM E.24.01.09 respectively. COD is usually measured by attaching a clip gauge which records the "crack mouth" opening, and then, by considerations of geometry, (including crack length), the COD is calculated.

J measurement is more complicated (see for example, [2]).

The more usual methods use the relation

$$J = \left|\frac{\partial U}{\partial a}\right|_q \qquad \text{where} \quad \begin{aligned} U &= \text{strain energy} \\ a &= \text{crack length} \\ q &= \text{load-point deflection} \end{aligned}$$

It involves plotting load against load-point deflection and measuring crack length differences, either as between a number of different specimens or on the one specimen as a crack is sequentially grown to different lengths. Other methods using strains measured round a contour have been reported, [3], [4], but these are related to specific specimen geometries, due to the inability to measure $\partial v/\partial x$. One solution to the latter difficulty is the use of a photo-elastic coating technique [5], but the form of the raw data necessitates circuitous derivation of the required quantities, leaving the method open to considerable potential errors.

It has been unfortunate that whereas J has a more fundamental basis in theory than δ as a fracture criterion parameter, its measurement is much more complex.

The critical values of δ and J are usually arrived at by plotting a so-called "R-curve", wherein the parameter is ordinate and the crack-length is abscissa, and then extrapolating back to "zero crack-length" (i.e. crack growth initiation). Fig. [3] shows a typical R-curve.

Fracture Integrity measurements are in principle possible. One would measure the fracture parameter (δ or J) on the structure under actual or simulated service loading, and compare the value obtained with the critical value. In practice it has not been possible to measure J in this situation, and measurement of COD presents problems, since it is essentially an empirical parameter and strictly therefore relates to a specific specimen geometry and, further, requires a measurement of the position of the crack tip.

4. MOIRE INTERFEROMETRY.

Whole-field Moire Interferometric methods discussed in this paper are capable not only of measuring COD [6], but are also particularly suitable for strain-based J determination, as flexibility of choice is offered with respect to integral paths and measurement locations, the only geometrical restriction being that the integration contour includes the crack tip. Moreover, the troublesome rotation term $\partial v/\partial x$ can be determined directly, as the measurements are displacement-based.

In the work reported here, a 3-beam laser Moire´ method was used to measure all the desired displacement/strain quantities directly and the energy and traction terms in equation 1 were

then computed through a linear elastic constitutive relation as
in the cited work. The apparatus illuminates a 50 mm diameter
field around the crack tip, within which any valid and
convenient analysis path can be chosen.

Experimental procedure.

The details of the technique for recording displacement fields
are described in [6]. In brief, a 1000 lines/mm epoxy cross-
grating is bonded to the specimen surface in the cracked region
with one of the grating directions aligned parallel to the crack
plane. A reasonably flat specimen surface is required over the
50 mm diameter analysis zone. The specimen is loaded and the
grating is then interrogated at each load step to produce
displacement fringe patterns for the x, y and 45° directions, by
using, in turn, each of the pairs of 50 mm beams produced by the
3-beam interferometer. Interference of each beam pair produces
a stable "reference" grid for that direction. The three fringe
patterns at a given load are photographed as shown in figures
4a, 4b, 4c. The sensitivity of the system is one micron per
fringe for x and y displacements and $1/\sqrt{2}$ micron for the 45°
displacements.

Analysis of fringe photographs.

Identical integral contour paths are drawn on each of the three
displacement field photographs corresponding to a given load
level, and identical points are marked out around each contour
path. The strains ε_x, ε_y and ε_{45}, together with the
rotations $\partial v/\partial x$ are determined at each point from the general
formula

$$\frac{\partial u_i}{\partial x_j} = \frac{ns}{\ell}$$

where n is the number of u_i fringes measured over distance ℓ
in the x_j direction and the sensitivity is s. The analysis is
made much easier if a digitising tablet is used, linked to a
microcomputer.

Principal stresses and tractions relative to the contour path
can then be derived via the elastic stress/strain relations,
leading in turn to evaluation of the strain energy term W and
the traction terms $T_j.(\partial u_j/\partial x)$ at each point round the path.
Spline curves are fitted to the discrete values of W and $T.(\partial u_j/\partial x)$
and integrated round the complete contour to determine J.
The individual terms of the formulation are given in the
Appendix. In practice, it is convenient to choose a rectangular
path, as the number of terms which have to be evaluated on

horizontal and vertical portions of the contour is greatly
reduced.

Effect of whole body rotation.

It is impossible in practice to prevent whole body rotation of
the specimen from its original position and such rotations are
of course included in the local values of $\partial v / \partial x$ measured.
However, analysis for small rotations shows that since the whole
body rotation is constant throughout the field, the term
integrates to zero round the contour.

Evaluation of technique.

A test to evaluate the accuracy of the interferometric method
was conducted on pin-loaded single-edge-notch steel specimens
within the plane strain, small-scale yielding regime. The
dimensions of the rectangular contour used are shown in figure
1a and, typically, 25 evaluation points were marked off at
approximately 3 mm intervals. (This excludes the part of the
path coincident with the free edge where $T_i = 0$ and W is very
small for the contour chosen). As an additional precaution the
computational process was carried out for the contour shown
dashed which does not enclose the crack. This yielded a
residual value of 0.3 kN/m at a load corresponding to a J
contour evaluation of 10.4 kN/m.

The values of J measured by the interferometric method were then
compared with values calculated from the measured load through a
strip yielding model for the geometry (7). (For the range of
loading considered, this is little different from the standard
small-scale-yielding plastic-zone size-correction). It can be
seen from Table 1 that the measured value of J is always within
15% of the theoretical value. (It should be said that J_c for
the material concerned is an order of magnitude greater than the
maximum J measured in this work).

CONCLUSION.

Moire´ Interferometry has been applied to the measurement of
Rice´s J-Integral. The method is time-consuming but in
principle may be automated through video recording of the fringe
patterns and on-line processing of the signals. The method is
not geometry dependant and is therefore adaptable to the
measurement of both Fracture Toughness and Fracture Integrity.

Since the technique has previously been shown to provide a
simple method of measuring COD, a direct comparison of the two
parameters is possible in arbitrary geometries.

REFERENCES

1. Rice, J.R. "A Path Independent Integral and the Approximate Analysis of Strain Concentration by Notches and Cracks". Trans ASME Ser.E., Jnl. App.Mech, 35 (1968) 379-386.

2. Landes, J.D. Begley, J.A. "Experimental Methods for Elastic-Plastic and Post-yield Fracture Toughness Measurements", in "Post-Yield Fracture Mechanics, LATZKO, D.G.H. editor, App. Sc. Publishers, London, 1979.

3. Kawahara, W.A., Brandon, S.L., "J-Integral Evaluation by Resistance Strain Gauges". SESA/JSME, Spring Meeting, (May 1982).

4. Read, D.T. "Experimental Method for Direct Evaluation of the J-Contour Integral", in ASTM, STP.791, II 199-213 (1983).

5. Muller Th., Gross, D., "Experimental Investigations of the Path Independence of the J-Integral for Large Plastic Zones" in "Fracture and Fatigue". J.C. Radon, Editor, Pergamon Press, (1980).

6. McDonach, A., McKelvie, J., MacKenzie, P.M., Walker, C.A., "Improved Moire Interferometry and Applications in Fracture Mechanics, Residual Stress, and Damaged Composites", Exp.Tech., 7, 20-24 (1983).

7. Gray, T.G.F., "A Closed-Form Approach to the Assessment of Practical Crack Propagation Problems", in "Fracture Mechanics in Engineering Practice", P. Stanley, Editor, App. Sc. Pub. Ltd., (1977).

Load (kn)	K_I (LEFM) (MN m$^{-3/2}$)	K_I^* (elastic/plastic, ref 7) (MN m$^{-3/2}$)	K_I^{*2}/E kN m^{-1}	$J_{experimental}$ kN m^{-1}	Error %
141	34.8	35.1	6.07	6.02	-1%
177	43.7	44.4	9.7	8.2	-15%
187–192	46.8	47.6	11.2	10.4	-7%

Comparison of theoretical J value with fringe-determined J

TABLE 1

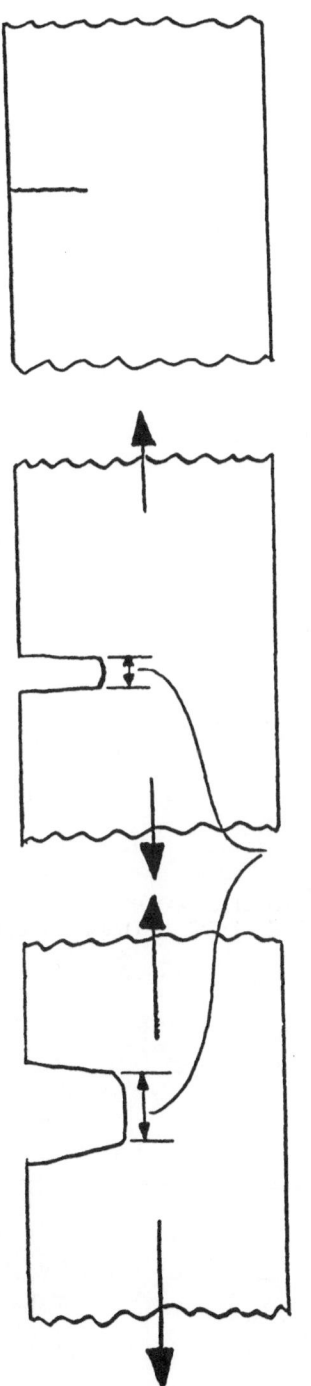

CRACK OPENING
DISPLACEMENT

FIG.1.

SCHEMATIC ILLUSTRATING CRACK OPENING
MECHANISM IN A DUCTILE MATERIAL BELOW
"CRITICAL" CONDITION.

250

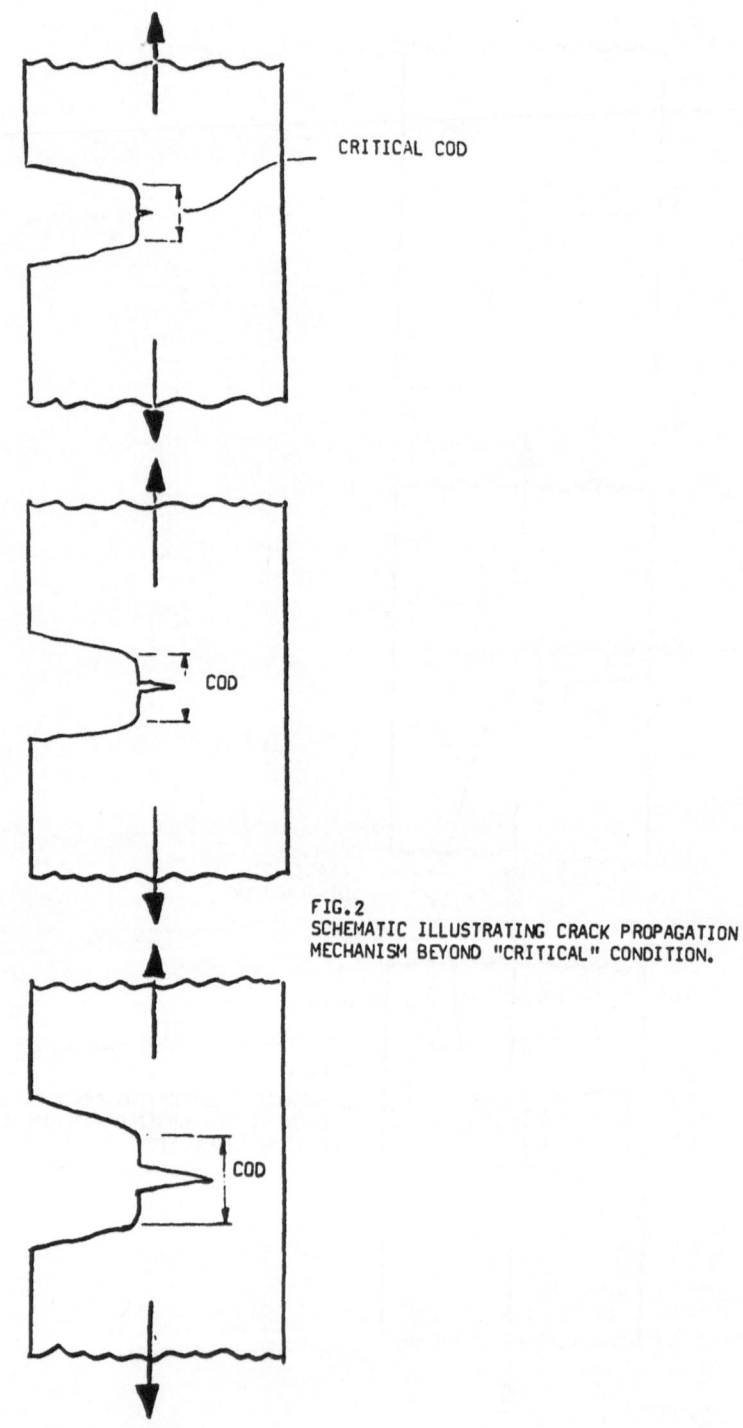

CRITICAL COD

COD

COD

FIG.2
SCHEMATIC ILLUSTRATING CRACK PROPAGATION
MECHANISM BEYOND "CRITICAL" CONDITION.

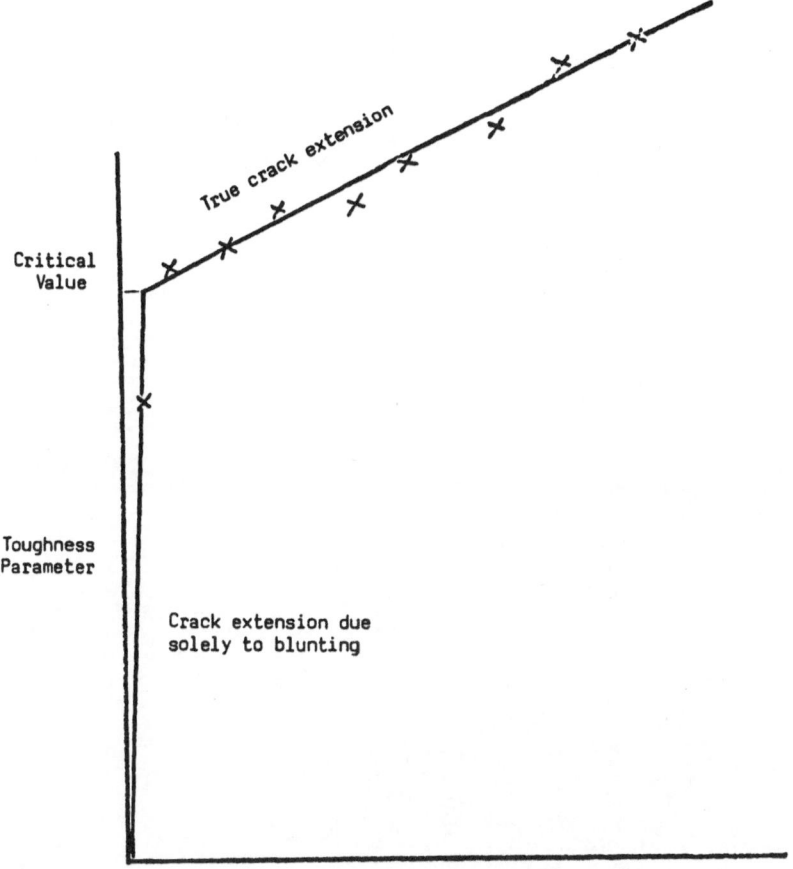

FIG.3   TYPICAL   R - CURVE.

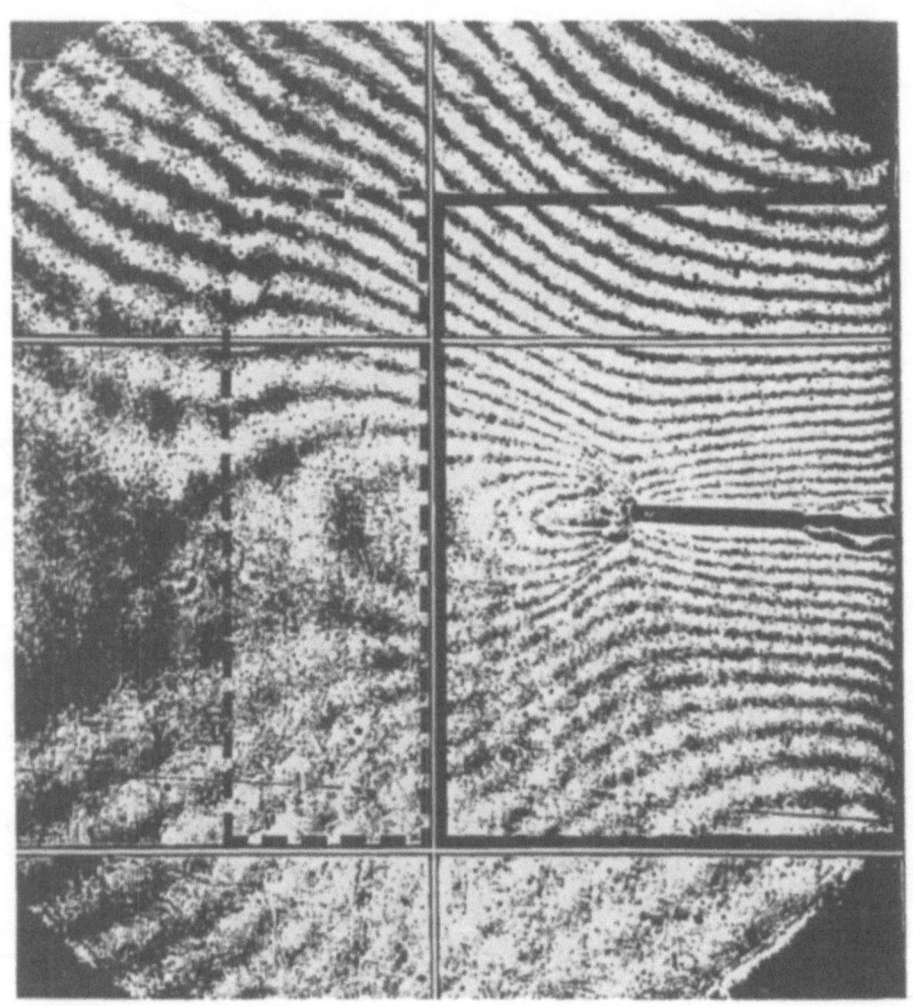

≡≡≡	FIDUCIAL LINES (2 CM. APART)
▬▬▬	J-CONTOUR
▪ ▬▪▪▬▪	"RESIDUAL" CONTOUR

x - component fringe pattern at 190 kN load

Figure 4a.

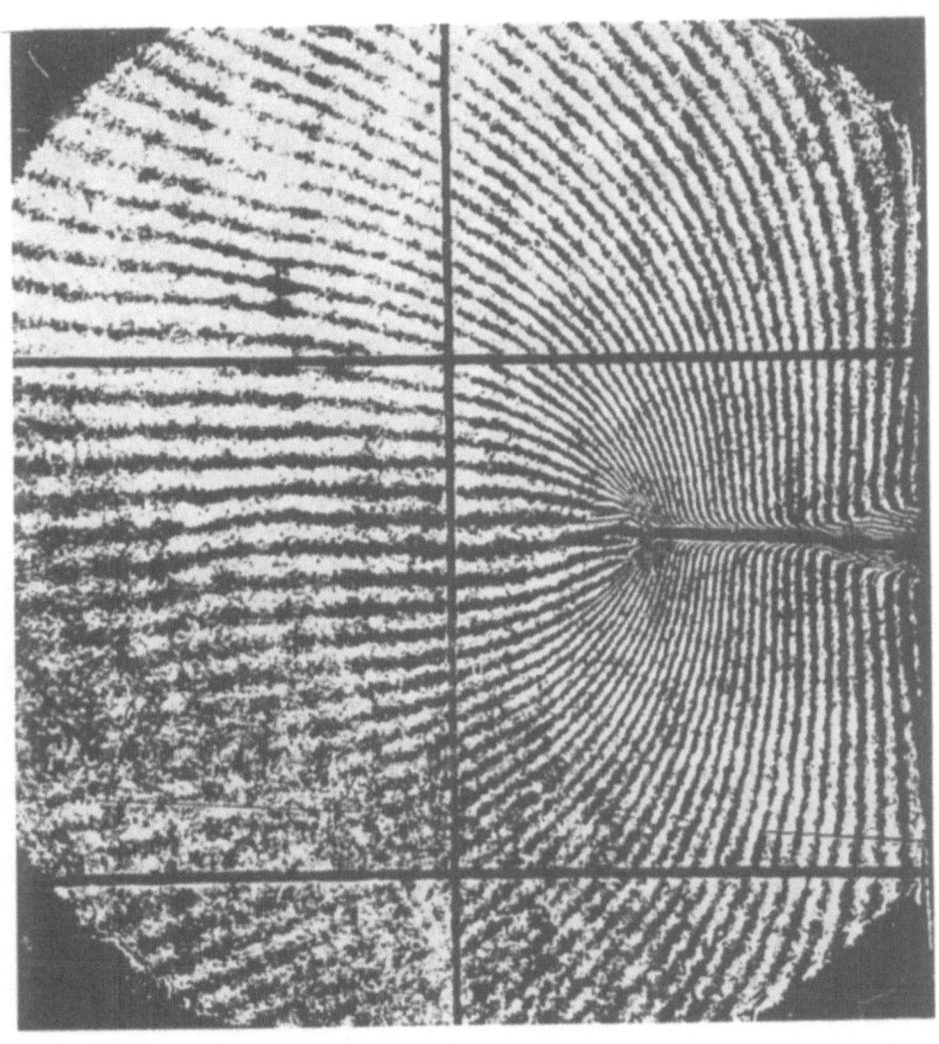

y - component fringe pattern

__Figure 4b.__

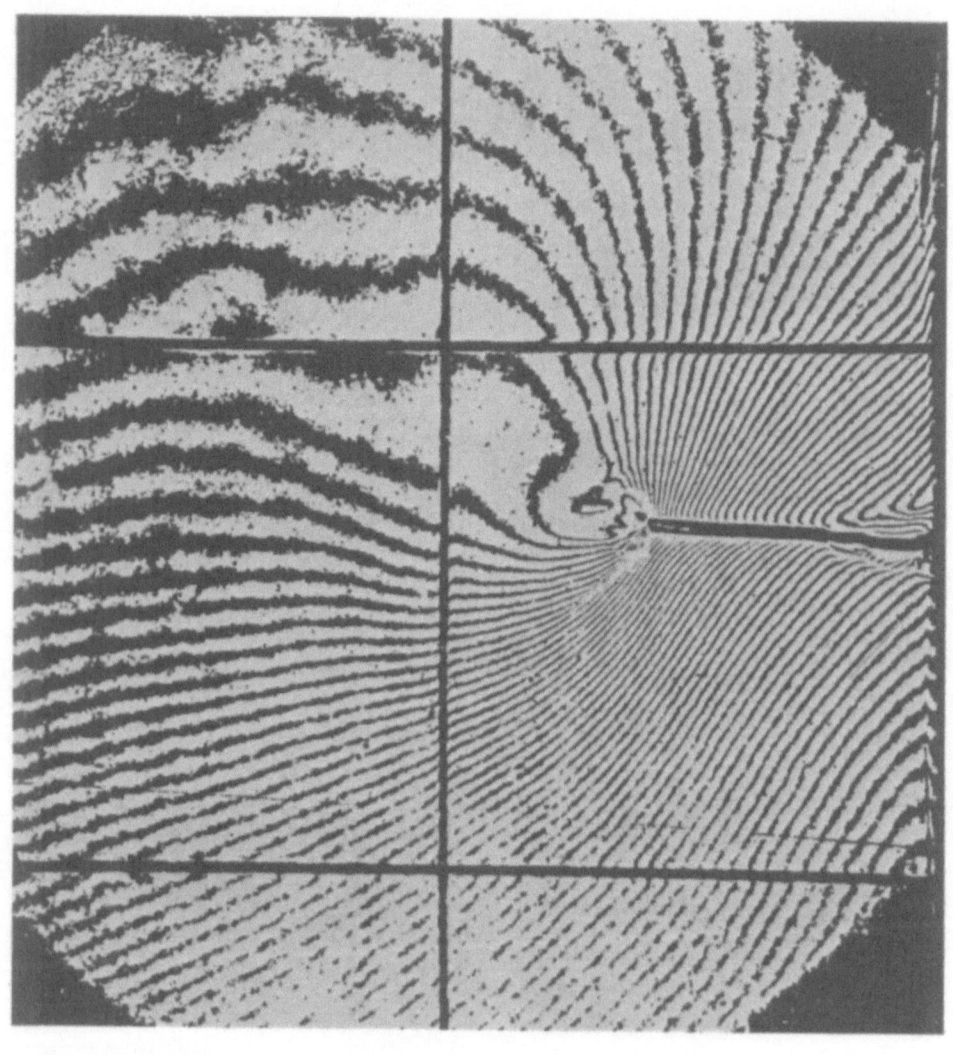

45° component fringe pattern

Figure 4c.

APPENDIX   **J – computation**

Given 3 values of direct strain at a point, the principal strains, and so the principal stresses, can be evaluated and hence the strain energy, from

$$W = (\sigma_1^2 + \sigma_2^2 - 2\nu\sigma_1\sigma_2)/E$$

In the case of the traction term, the stress values σ_x, σ_y and τ_{xy} are evaluated from the 3 strains, whereupon

$$T_i \cdot \left(\frac{\partial u_i}{\partial x}\right) = \left(\sigma_x \varepsilon_x + \tau_{xy} \frac{\partial v}{\partial x}\right)\cos\theta + \left(\sigma_y \frac{\partial v}{\partial x} + \tau_{xy}\varepsilon_x\right)\sin\theta$$

where θ is the angle between the normal to the contour and the x axis.

PHASE-MEASURING MOIRÉ TOPOGRAPHY

G T Reid

National Engineering Laboratory
East Kilbride, Glasgow

1   INTRODUCTION

A number of full-field optical methods which exploit interferometric, holographic, speckle or moiré fringe phenomena have been
developed for the measurement of surface form and deformation.
Between them, these methods offer a wide range of sensitivities
together with the ability to visualize the quantity of interest
through a fringe pattern which describes the height or deformation
contours of the surface.

The idea of linking a full-field measuring system to a digital
image analyser is being pursued in a number of laboratories[1-5]
and measuring machines which use this technology are commercially
available[6-8].

The performance of any automatic full-field measuring system is
partially dependent upon the quality of the image which is presented to the computer. Problems in obtaining high quality fringe
patterns from 'unco-operative' objects (ie those with optical
properties which are, to some extent, unsuited to this type of
measuring machine) create a major obstacle to the widespread adop-
tion of this technology. In a laboratory environment, or in
isolated industrial applications, a matt white coating can sometimes be applied to the object to give it a uniformly diffuse
surface. This is an unattractive procedure when rapid measurements of mass-produced components are required.

This paper describes a method for the measurement of three-
dimensional surface form which relies on a phase-measuring version
of projection moiré contouring, operating in conjunction with a

digital image analyser. Fringe analysis using phase, rather than intensity, measurement significantly reduces the sensitivity of the method to variations in the optical properties of the object and allows fine subdivision of the contour interval. The use of a charge coupled device (CCD) video camera which is linked to a micro-computer allows direct capture of dimensional information from the object without the intermediate photographic stage found in many alternative contouring techniques.

2   PHASE MEASURING MOIRÉ TOPOGRAPHY: THE Auto-MATE SYSTEM

A number of projection moiré methods have been proposed for surface contouring and a variety of mathematical treatments of the methods are available in the literature.

Most authors have described apparatus which is similar to that shown schematically in Fig. 1. A grating is projected onto the object while an image of the object is formed in the plane of a reference grating. The reference grating is a photograph of the projection of a grating onto a flat surface. Interaction of the image with the reference grating causes a number of moiré fringes to appear superimposed upon the surface of the object. If a number of geometric conditions are met, the fringes describe contours of equal height as shown in Fig. 2.

Rather than analyse the fringe pattern by measuring the locations of the fringe centres, we adopt an approach which is closely analogous to phase-measuring interferometry[9,10].

The Auto-MATE (Automatic Moiré Analysis for Topographic Evaluation) system is shown schematically in Fig. 3. A CCD video camera feeds an image into a framestore. The image, now in digital form, is read into the computer memory. The projection grating, of pitch P_p, is moved through a distance P_p/k (that is the phase of the projection grating is moved through $2\pi/k$) and a second image is read into the computer. This process is repeated until the projection grating has moved through $(k - 1)$ equal steps and a total of k images has been stored in the computer memory so that k intensity levels have been stored at each pixel location.

To derive the phase, $\phi(x, y)$, of the moiré fringe pattern at pixel coordinates (x, y), we evaluate the Fourier series coefficients $\alpha_1(x, y)$ and $\beta_1(x, y)$ thus:

$$\alpha_1(x, y) = \sum_{N=0}^{k} I_N(x, y)\cos(2\pi N/k)$$

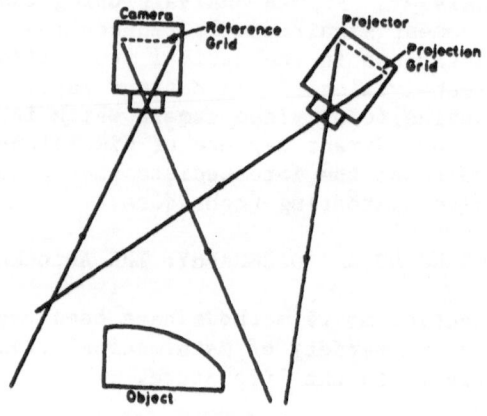

Fig 1   Projection Moire Topography-Schematic

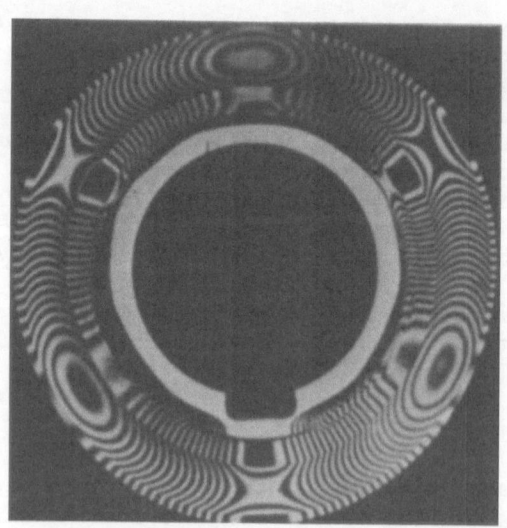

Fig 2   Projection Moire Contour Map of a Three Lobed Cam

$$\beta_1(x, y) = \sum_{N=0}^{k} I_N(x, y)\sin(2\pi N/k)$$

where $I_N(x, y)$ is the intensity level at pixel (x, y) at image number N. The phase is then given by

$$\phi(x, y) = \arctan\{\beta_1(x, y)/\alpha_1(x, y)\}. \tag{1}$$

By evaluating equation (1) at each pixel location, we transform the contour pattern into a phase distribution which is insensitive to the optical properties of the object and which unambiguously describes the concavity or convexity of the surface. Furthermore, since phase values are available at every pixel location, measurements can be made over the entire surface and not just at the fringe centres.

3   GEOMETRIC DESIGN OF THE PROJECTION MOIRÉ SYSTEM

Strong similarities between projection moiré contouring and close range photogrammetry have been noted elsewhere[11 12]. The analysis given below broadly follows that of Indesawa, Yatagai and Soma[13] but in this paper we aim to derive equations which describe projection moiré topography as a variant of close range photogrammetry in which one of the photogrammetric cameras has been replaced by a grating projector. This approach is attractive for two reasons: firstly it provides a concise, yet readily applicable, analysis of the method, and secondly it allows existing software[14], originally developed for photogrammetric work, to be modified for use in the analysis of projection moiré fringe patterns.

Referring to Fig. 4, we define a coordinate system (X_p, Y_p, Z_p) with its origin at the perspective centre of the projector and with the Z_p axis lying along the optical axis of the projector. The axes (x_p, y_p) of the projected grating are parallel to axes (X_p, Y_p) and this grating occupies the plane $Z_p = -a_p$. The lines on the projected grating are parallel to y_p.

Consider an analogous arrangement for the camera. The perspective centre of the camera serves as the origin of the coordinates (X, Y, Z) which describe the surface of the object. The Z-axis lies along the optical axis of the camera and the axes (x_c, y_c) of the reference grating are parallel to (X, Y). The reference grating occupies the image plane of the camera at $Z = -a_c$ and the grating lines are parallel to y_c.

The camera coordinates are rotated through angles u, v and w (about the X, Y and Z axes respectively) with respect to the

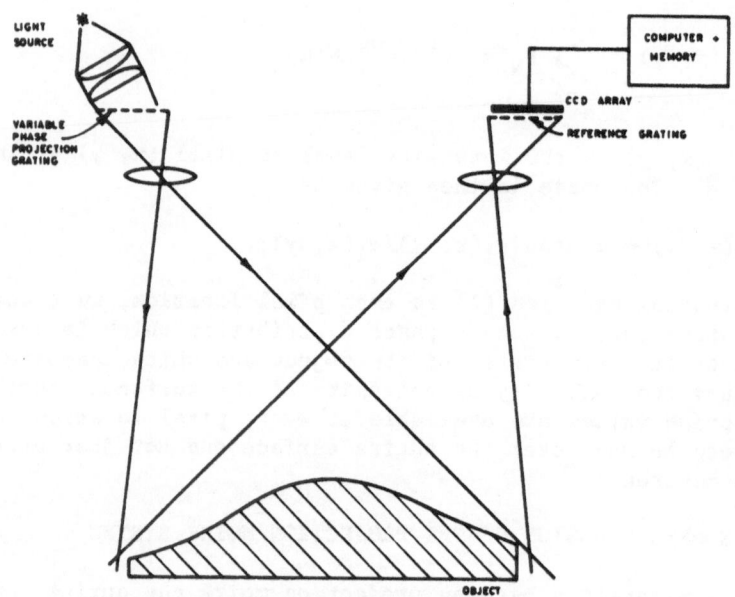

Fig 3   Phase-Measuring Moire

Topography : The Auto-MATE System

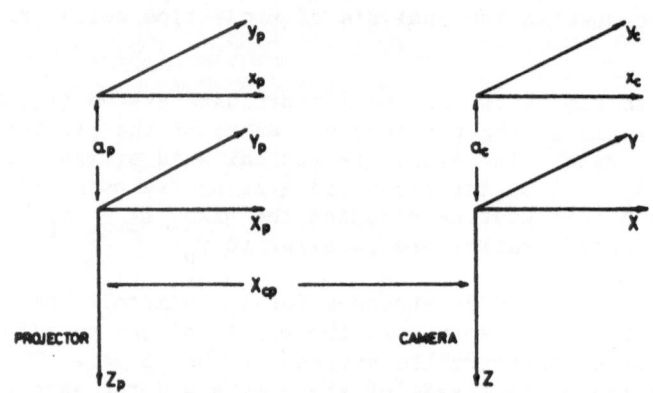

Fig 4   Projection Moire Topography

with Constrained Geometry

projector coordinates and the origins of (X, Y, Z) and
(X_p, Y_p, Z_p) are linked by the vector (X_{cp}, Y_{cp}, Z_{cp}).

The relationship between a point (x_c, y_c) in the image plane of
the camera and a point (X, Y, Z) on the object is given by

$$X = \frac{Z x_c}{a_c}$$

$$Y = \frac{Z y_c}{a_c}$$

(2)

and the point (X, Y, Z) is related to a coincident point
(X_p, Y_p, Z_p), in the projector coordinates, through the transformation equation

$$\begin{bmatrix} X_p \\ Y_p \\ Z_p \end{bmatrix} = \begin{bmatrix} M_{11} & M_{12} & M_{13} \\ M_{21} & M_{22} & M_{23} \\ M_{31} & M_{32} & M_{33} \end{bmatrix} \begin{bmatrix} X - X_{cp} \\ Y - Y_{cp} \\ Z - Z_{cp} \end{bmatrix}.$$

(3)

Where M_{ij} are the elements of the three-dimensional transformation
matrix.

The relationship between the point (x_p, y_p) on the projection
grating and the point (X_p, Y_p, Z_p) on the surface of the object is

$$x_p = \frac{Z_p x_p}{a_p}$$

$$Y_p = \frac{Z_p y_p}{a_p}.$$

(4)

If the projection and reference gratings have pitches P_p and P_c
respectively and the grating lines are indexed by $(n_p + \delta_p)$ and
$(n_c + \delta_c)$, where n_p and n_c represent line numbers and δ_p and δ_c
represent fractional distance within a grating pitch then

$$x_p = P_p(n_p + \delta_p)$$

(5)

$$x_c = P_c(n_c + \delta_c).$$

(6)

We define the $(n + \delta)$th moiré fringe as occurring when the
$(n_p + \delta_p)$th line on the projected grating is imaged onto the
$(n_c + \delta_c)$th line on the reference grating so that

$$n = (n_c - n_p)$$

(7)

$$\delta = (\delta_c - \delta_p).$$

(8)

Substituting equations (5), (7) and (8) into equation (6) we deduce that

$$x_p = \frac{P_p}{P_c} \{x_c - P_c(n + \delta)\}. \tag{9}$$

To derive a relationship between a point (x_c, y_c) on the reference grating, where the $(n + \delta)$th moiré fringe appears, and the corresponding point (X, Y, Z) on the surface of the object, we substitute for X_p, Z_p and x_p in equations (5) from equations (3) and (9). Choosing $(n + \delta)$ as the subject of the resulting expression we get

$$(n + \delta)$$

$$= \frac{x_c}{P_c} - \frac{a_p}{P_p} \left\{ \frac{M_{11}(X - X_{cp}) + M_{12}(Y - Y_{cp}) + M_{13}(Z - Z_{cp})}{M_{31}(X - X_{cp}) + M_{32}(Y - Y_{cp}) + M_{33}(Z - Z_{cp})} \right\}. \tag{10}$$

Equations (2) and (10) describe the behaviour of a projection moiré contouring system of arbitrary geometry. We note the strong structural resemblance of equation (10) to the corresponding equations (11) for photogrammetry[14].

$$x = -f \left\{ \frac{M_{11}(X - X_o) + M_{12}(Y - Y_o) + M_{13}(Z - Z_o)}{M_{31} X - X_o) + M_{32}(Y - Y_o) + M_{33}(Z - Z_o)} \right\}$$

$$y = -f \left\{ \frac{M_{21}(X - X_o) + M_{22}(Y - Y_o) + M_{23}(Z - Z_o)}{M_{31}(X - X_o) + M_{32}(Y - Y_o) + M_{33}(Z - Z_o)} \right\}. \tag{11}$$

Equations (11) express the image plane coordinates (x, y) in terms of the focussing distance f (which is equivalent to a_p in equation (10), the vector $(X_o, Y_o, Z_o)^T$ which links the perspective centres of two photogrammetric cameras (equivalent to $(X_{cp}, Y_{cp}, Z_{cp})^T$ in equation (3)) and the elements M_{ij} of the transformation matrix.

From Fig. 4 we see that the geometry of the apparatus can be constrained by choosing $Z_{cp} = Y_{cp} = 0$ and by ensuring that the optical axes of the camera and projector are parallel. In addition, we have ensured that $a_p/P_p = a_c/P_c$. Under these conditions, equation (10) reduces to

$$(n + \delta) = \frac{a_c X_{cp}}{P_c Z} \tag{12}$$

or

$$Z = \frac{a_c X_{cp}}{P_c(n + \delta)}. \tag{13}$$

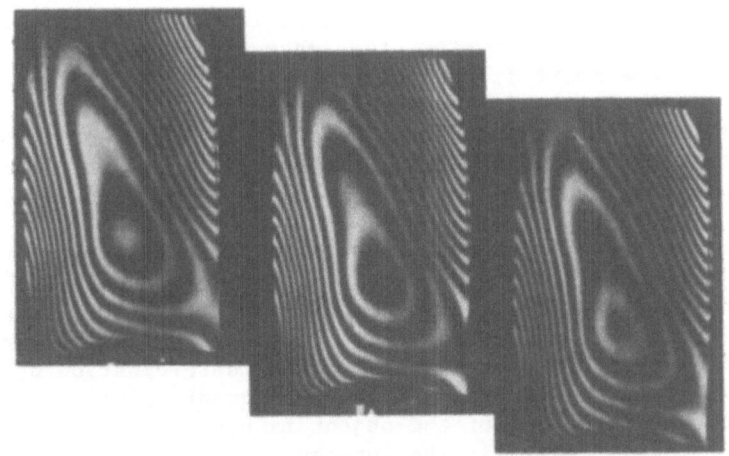

Fig 5   Moire Fringe Contour Maps of a Fan Blade.
The Photographs were taken at Three Different
Positions of the Projection Grating

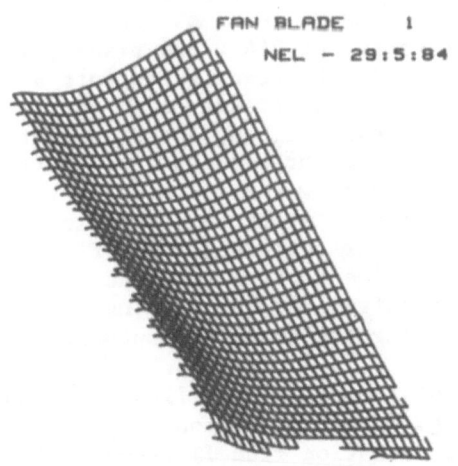

Fig 6   Isometric View of the Fan Blade
Shown in Fig 5

Equations (2) and (13) explicitly express the three-dimensional
coordinates of the surface of the object in terms of the geometry
of the projection moiré system, the image plane coordinates
(x_c, y_c) and the moiré fringe number $(n + \delta)$. Assuming that the
geometry of the system is known and that the image plane
coordinates can be obtained from the pixel locations of the CCD
camera then, by reading the fringe number at each pixel, the
three-dimensional form of the object can be calculated.

All objects are mounted at an approximately known perpendicular
distance from the line joining the perspective centres of the
camera and projector. The absolute value of the fringe integer,
n, is determined from equation (12) for one point on the surface.
Errors of several per cent in this initial determination of n have
little effect on the accuracy of the subsequent fringe analysis so
that precise positioning of the object is unnecessary. The
relative fringe integer (ie the integer value of the fringe count
referred to an arbitrary zero fringe on the contour map) and the
fringe fraction are added to, or subtracted from, the absolute
fringe integer to provide values of $(n + \delta)$ at each pixel and
thereby allow evaluation of equations (2) and (13).

4   RESULTS

Fig. 5 shows a set of projection moiré contour maps of a fan blade
which was formed with projection grating phases of 0, $2\pi/3$ and
$4\pi/3$. The contour interval is 2 mm. The contour patterns were
detected by a CCD camera of 244 × 320 pixels, fed into a 512 × 512
× 8 bit framestore and subsequently fed into a micro-computer.
Equation (1) was evaluated at each pixel and, using the previously
known geometric constants X_{cp}, a_c and P_p, equations (2) and (13)
were used to provide the three-dimensional surface coordinates.
The height of the surface was calculated to within 1/180th of the
contour interval. Fig. 6 shows an isometric view of the object,
derived from the contour patterns, while Fig. 7 shows a number of
cross-sectional profiles derived from the same images. No manual
intervention was required between placing the object in front of
the system and viewing the graphical output.

The capabilities of Auto-MATE are illustrated more explicitly in
Figs 8 and 9. Fig. 8 shows a contour map of an aluminium plate.
Since the maximum deviation from flatness is less than the contour
interval, less than one moiré fringe covers the vertical range of
the surface. Since uneven illumination of the surface will also
produce variations in the image intensity, we cannot be sure that
the information contained in Fig. 8 is entirely due to the surface
topography. Furthermore, we see that it is impossible to dis-
criminate between elevations and depressions from Fig. 8 alone.
Fig. 9 shows a graphical representation of the aluminium plate
which was obtained by phase-measuring moiré topography. By

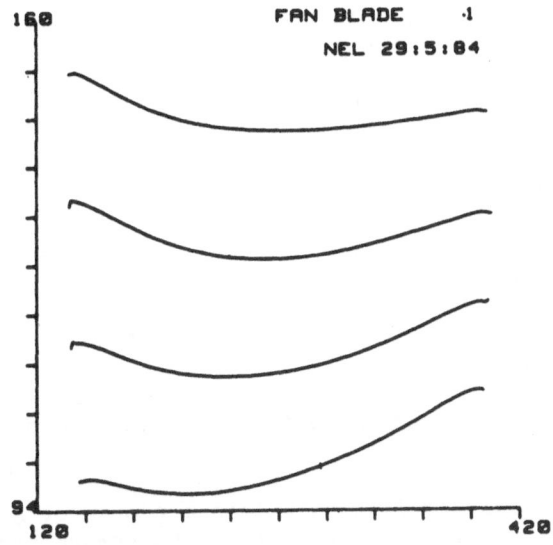

Fig 7   Cross-Sectional Profiles of the
Fan Blade Shown in Fig 5

Fig 8   Moire Fringe
Contour Map of an
Aluminium Plate

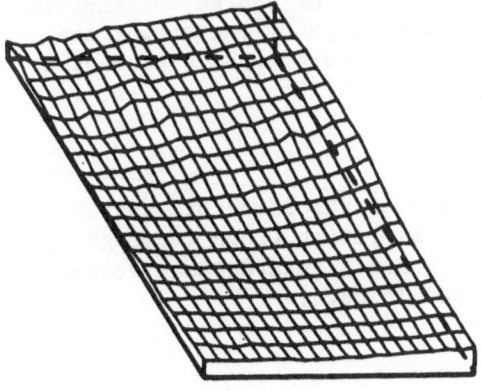

Fig 9   Isometric View
of the Aluminium Plate
Shown in Fig 8

transforming the moiré fringe pattern into a phase distribution, we have removed the effects of uneven illumination and have unambiguously described the surface topography. The sensitivity of the phase calculation has also revealed a level of detail which was not available from the conventional moiré fringe pattern.

5   DISCUSSION

We have demonstrated the feasibility of phase-measuring moiré topography with on-line image analysis. The optical geometry can be regarded as a variant of close range photogrammetry while the contour analysis uses techniques derived from phase-measuring interferometry.

Two factors predominate in defining the performance of the system:

a   Equation (13) is invalid unless a number of geometric conditions are met. Careful alignment of the camera and projector is therefore required if the contours are to accurately represent planar sections through the object. In practice, the extent to which the contours can be usefully subdivided is limited by uncertainties in our knowledge of the system geometry.

b   The resolving power of the CCD camera is limited by the number (320 × 244) of pixels in the detector array. This comparatively low resolution has two effects on the performance of the system. Firstly, although the geometric stability of the camera ensures repeatability in measurements of the lateral coordinates of surface points, the positions of the edges of the object cannot be accurately determined. Secondly, since the camera cannot resolve the closely packed contours which appear on high surface gradients, the system is limited to the measurement of objects with slopes of a few tens of degrees or less. The resolution of closely packed contours can, of course, be improved by increasing the magnification of the image or increasing the contour interval, or both, so that fewer contours are present in the field of view.

ACKNOWLEDGEMENTS

This paper is published by permission of the Director, National Engineering Laboratory. The paper is British Crown copyright. The work was funded by the Metrology and Standards Requirements Board of the Department of Trade and Industry.

REFERENCES

1   GASVIK, K. J. Moiré Technique by Means of Digital Image
     Processing. Appl. Opt., 22, (1983) 3543-3548.

2   MERTZ, L. Real-time Fringe-pattern Analysis. Appl. Opt.,
     22, (1983) 1535-1539.

3   BECKER, F., MEIER, G. E. A. and H. WEGNER. Automatic
     Evaluation of Interferograms. Proc. SPIE, 359, (1982) 386-
     393.

4   TAKEDA, M., HIDEKI, I. and S. KOBAYASHI. Fourier-transform
     Method of Fringe-pattern Analysis for Computer-based
     Topography and Interferometry. J. Opt. Soc. Am., 72, (1982)
     156-160.

5   YATAGAI, T., INDESAWA, M., YAMAASHI, Y. and M. SUZUKI.
     Interactive Fringe Analysis System: Applications to Moiré
     Contourgram and Interferogram.   Opt. Engng, 21, (1982) 901-
     906.

6   HURDEN, A. P. M. Vibration Mode Analysis Using Electronic
     Speckle Pattern Interferometry. Opt. Laser Technol., 14,
     (1982) 21-25.

7   Interferogram Interpretation and Evaluation Handbook. (Zygo
     Corporation, Middlefield, Ct, USA, n.d.)

8   INOUE, S., TSUJI, H., OTSUKA, Y., SUZUKI, H. and A. SHINOTO.
     Moiré Topography for Early Detection of Scoliosis. (Fujinon
     Technical Information, Fuji Photo Optical Co., Omiya City,
     Japan, n.d.)

9   BRUNING, J. H. Fringe Scanning Interferometers, in: D.
     Malacara, ed., Optical Shop Testing, (New York, Wiley, 1978.)

10   WYANT, J. C. Interferometric Optical Metrology: Basic
     Principles and New Systems. Laser Focus, 18, (1982) 65-71.

11   PERRIN, J. C. and A. THOMAS. Electronic Processing of Moiré
     Fringes: Application to Moiré Topography and Comparison with
     Photogrammetry. Appl. Opt., 18, (1979) 563-574.

12   FROBIN, W. and E. HIERHOLZER. Calibration and Model
     Reconstruction in Analytical Close-range Photogrammetry.
     Part II: Special Evaluation Procedures for Rasterstereography
     and Moiré Topography. Photogrammetric Engineering and Remote
     Sensing, 48, (1982) 215-220.

13   INDESAWA, M., YATAGAI, T. and T. SOMA.   Scanning Moiré Method
      and Automatic Measurement of 3-D Shapes.   Appl. Opt., 16,
      (1977) 2152-2162.

14   MARZAN, G. T. and H. M. KARARA.   A Computer Program for
      Direct Linear Transformation Solution of the Colinearity
      Condition and Some Applications of it.   (Proc. Close Range
      Photogrammetric Systems Conf., 1975, Falls Church, Va, USA)
      pp 420-476.

HOLOGRAPHIC METROLOGY

THE HOLO-DIAGRAM, SANDWICH-HOLOGRAPHY AND LIGHT-IN-FLIGHT RECORDING

Nils Abramson

Industrial Metrology, Dept. of Production Engineering, The Royal
Institute of Technology, S-100 44 Stockholm, Sweden

1. The "holo-diagram" a practical device for the making and the evaluation of holograms.

Abstract

The engineer is interested in the measurement of small deformations of large machine parts, for instance, the deformation caused
by force and temperature of large slide-ways in machine-tools.
For this purpose it was thought that hologram interferometry would
be particularly suitable. However, no general method appeared to
be available either for the making or for the evaluation of holograms recording large objects, therefore the method described in
this paper was worked out. It is based on the use of a special
diagram which the author has named the "holo-diagram". This can
be used to make both ordinary holograms and Lippman holograms. We
have found that it simplifies the evaluation of interference holograms for measuring dimension, deformation and vibration. This
work also inspired new ideas for the design of interferometers.

1.1. Introduction

In a hologram, the information about the object is recorded in
the form of interference fringes on a photographic plate. These
fringes are formed when diffuse light from the object and an easily reproduceable light beam (the reference beam) fall upon the
same place at the same time.

For fringes to be produced on the photographic plate they must be
in a fixed position in relation to its surface during most of the
exposure time. In other words, the phase relation between the

two light beams has to be constant during this time. One of the conditions for the phase relation to be constant is that the difference in path-length for the two light components must be shorter than the coherence length of the light in use. The other condition is that this distance is constant during most of the exposure.

If any part of the object moves so that the corresponding lines on the photographic plate are wiped out, no light will be deflected to this part when the finished hologram is illuminated with laser light (reconstructed). Therefore this part looks dark in the reconstruction.

The hologram can be exposed twice, with half of the exposure time used for the recording of the undeformed object and the other half for the recording of the deformed object. When this is done the reconstruction of the hologram will give a picture of the object with dark sections on all those parts that have made such a movement that corresponding interference fringes have moved half the fringe separation or a multiple of this. In this way we get an interference hologram with dark fringes which, like the contour lines on a map, show the amplitude of displacement of any part of the object.

If the object is both illuminated and looked upon in a direction that is parallel to the direction of the deformation, the distance between two dark fringes will correspond to a displacement of half the wavelength of the light in use. In all other cases the corresponding displacement is greater than this value [6]. The following diagram (fig. 1) has been constructed in order to find a simple and practical way to evaluate the correlation between the position of the fringes and the amplitude of displacement and also to simplify the making of the hologram.

1.2. Construction of the diagram

A is the point (fig. 1) from which the divergent laser beam originates (e.g. a spatial filter). B is the point on a photographic plate that is to be examined after the exposure of the hologram (e.g. the centre of the plate). Locate the object C in the diagram so that its position in respect to A and B corresponds to the real position. Light radiates from A and some of it illuminates C and is reflected from there to B. Let the object C be just a mathematical point. The path-length for the light rays from A to B via C is constant if C is moved along an ellipse with the focal points A and B. This distance changes, however, as soon as we move C in a direction perpendicular to the periphery of the ellipse. In reality the ellipses of the diagram are intersections between ellipsoids and the plane on which we work.

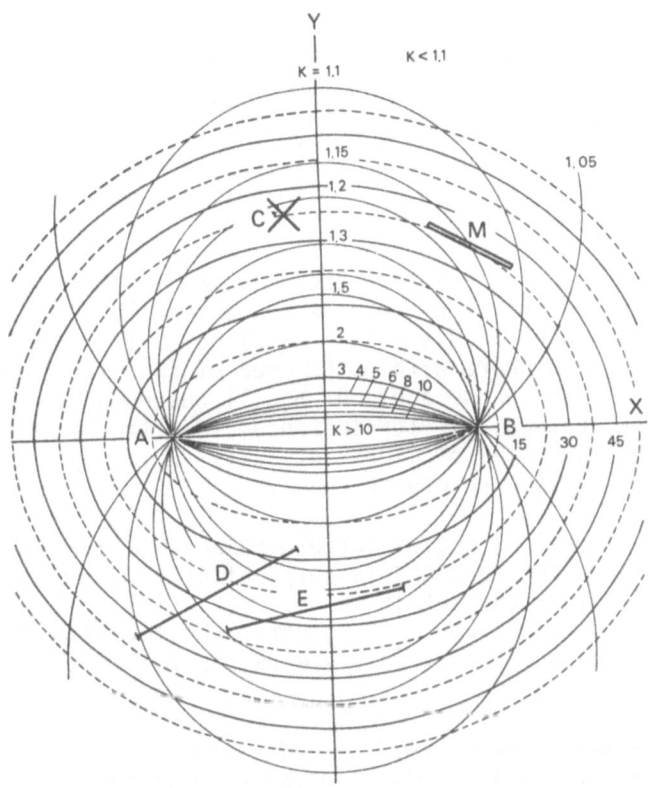

Fig. 1. The holo-diagram. A is the point from which the divergent laser beam originates, B is the centre of the holographic plate.

Let us now place a mirror M somewhere along the ellipse on which C is situated. If the surface of the mirror is parallel to the periphery of this ellipse it will reflect light from A to B. This reflected light forms the reference beam. The path-length of the light via object C is the same as via mirror M. Therefore, the light rays are in phase and there will be a maximum of light at point B on the photographic plate. If the object C is moved in a direction perpendicular to the periphery of the ellipses, the intensity of light at B will alternate between a maximum and a minimum value.

Let us assume for a moment that the ellipses in the diagram are so closely spaced that they intersect the x-axis in points separated by a distance of only half a wavelength. If a point object is moved around anywhere within the area of the diagram and illuminated from A, the phase angle at B will change 360º each time an ellipse is crossed. If, for instance, 10 ellipses are crossed there will be 10 cycles of light intensity variation (if B is also illuminated by a reference beam).

The greatest number of cycles corresponding to a given movement is produced when the movement is perpendicular to the ellipse with the maximum at the x-axis to the right side of B. The nearer to the centre of the diagram the greater the movement that is required to give one cycle. This yields the following result in the hologram.

Along the x-axis on the right side of B the distance between two dark fringes represents a displacement of half the wavelength (λ). In all other places the distance represents a greater displacement. Let us name the displacement k $\frac{1}{2}\lambda$. The nearner to the centre of the diagram the more the value of k increases and in the centre is infinite. The position line of constant k-value is formed by arcs of circles, which are drawn in the diagram with the k-value printed near to the y-axis.

If the object is large, the light that is reflected from its different parts must travel different distances. The larger the object the greater the length of coherence is required. When the difference in distance for the light via the object and via the mirror is greater than the coherence length there will no longer be any interference fringes at B.

The diagram in fig. 1 is constructed in such a way that if a reference mirror is placed anywhere along one ellipse the coherence length is just enough that an object placed on a second adjacent ellipse can be recorded in the hologram.

1.3. How to use the holo-diagram

Optimal utilization of the coherence length

The light travels from A to B via the object C. Locate C in the diagram (fig. 1) so that its position in respect to A and B corresponds to the real position. The distance between two ellipses corresponds to a difference in light-path from A to B of 15 cm (when the distance between A and B is 100 cm). Usually the coherence length from a long HeNe gas laser is at most 30 cm. Therefore the object may at most fill the space within four elliptic areas, or in other words it may cross four but not five ellipses. The object D is too large to be recorded in the hologram but if it is changed to position E it can be recorded without any trouble.

If the real distance between spatial filter and photographic plate is, for example, 50 cm there is a scale-factor of two. The object may then at the most cut eight but not nine ellipses. By use of the diagram it is possible to learn how to record in a hologram an object many meters long.

Selection of fringe separation

If the coherence length of the light in use is sufficient, the object can be positioned with regard only to the wanted fringe separation. If the k-value is, for example, 3, the distance between two fringes will correspond to a displacement of about 1 μm if a HeNe gas laser is used. If extreme accuracy is needed it must be determined exactly, which part of the photographic plate corresponds to B.

The simplest way to utilize the diagram is to use a permanent foundation with the diagram outlined directly on the surface where the holograms are made. Then one can see at once where to position the object to optimize the conditions.

Positioning of the reference mirror

To get full use of the coherence length it is necessary to position the reference mirror at the optimal place. The best position is anywhere along the ellipse that intersects the object in its middle, or if the coherence length is adequate, at the place where highest resolution is wanted. This is only strictly true if the object and reference mirror use the same divergent laser beam.

The surface of the mirror should be parallel to the ellipse on which it is placed. If the resolution of the photographic plate is low the angle separating mirror and object as seen from B is limited.

Evaluating the dark fringes

Count the number of dark fringes between a fixed (not displaced) point on the object and the point whose displacement is to be evaluated. Multiply this number by the k-value of the object. This gives the displacement expressed as the number of half wavelengths.

The evaluated displacement is the projection of the real displacement onto the normal of the ellipse that passes through the object. One way to find this normal is to look for the intersection of the k-circle through the object and the negative y-axis. The normal passes through this intersection.

Minimizing the sensitivity to unwanted movements

If the object moves or vibrates the hologram might be destroyed. Therefore, one should try to position the object in such a way in the diagram that the sensitivity to movement is as low as possible. This is the case if the object is positioned so that the direction of displacement is parallel to the ellipse that intersects the object. If the unwanted displacement has not preferred direction the object should be positioned where the k-value is as high as possible.

2. Sandwich holography: an analogue method for the evaluation of holographic fringes

2.1. Introduction

Holographic interferometry is becoming more and more used by mechanical engineers, e.g. for nondestructive testing. When the conventional double exposure method is used the following practical evaluation problems have to be solved.

Though it exists a great number of methods (e.g. the holo-diagram) to evaluate the direction of object displacement in all three coordinates, the sign of the displacement cannot be found because the hologram contains no information on which exposure was made first.

If many different situations are to be compared, e.g. during a step-by-step loading of the object, a new hologram has to be exposed before and after each load change that is to be studied. New combinations cannot be made during reconstruction.

When the load is applied, unwanted motion of the whole object and its fixture very often complicate the fringe pattern in such a way that the wanted information is hidden.

If the deformation of only one part of the object is to be studied the evaluation is complicated by fringes caused by the total object deformation.

The stress and strain caused by bending is represented by the change of fringe separation over the total object surface, thus the derivative of the fringe separation has to be calculated, which might cause difficulties.

The ordinary fringes represent digital information, fractions of a fringe are very difficult to measure which results in low accuracy especially if only a few fringes are seen.

If object shape is to be measured by holographic contouring the direction of the intersecting interference surfaces becomes fixed during exposure and cannot be changed during reconstruction. A rotation of the intersecting planes during reconstruction would be useful, e.g. for checking the parallelism of different object surfaces.

It appears as if the sandwich hologram represents the solution to all these different problems. So far, however, we have only studied sandwich holography for some special cases (object displacement parallel to the direction of illumination and observation) but it still appears to be an extremely useful method. It embodies all the possibilities of fringe manipulation during reconstruction. In the following we will present some of our recent results.

2.2. Making the sandwich hologram

One hologram plate was placed in a special plate holder equipped with three contact points that locate the plate surface while three pins locate two sides of the plate. A second hologram plate was then placed in contact with the first, covering its surface. Both plates had their emulsion towards the object (fig. 2). The emulsions were therefore separated by the glass base of the front plate. When the plates were exposed simultaneously both the object beam and the reference beam reaching the back plate passed through the front plate which was not covered by any anti-halo layer. The two exposed plates were taken away and two new plates were placed in the plate holder.

The object consisted of three vertical steel bars (fig. 3) that were fixed by screws at their lower ends to a rigid frame which surrounded the bars and was supposed to function as an undeformable fixed reference surface. The middle bar was also supported at the top end. A force was applied to the middle of the bars. After the deformation of the three objects the second pair of plates

278

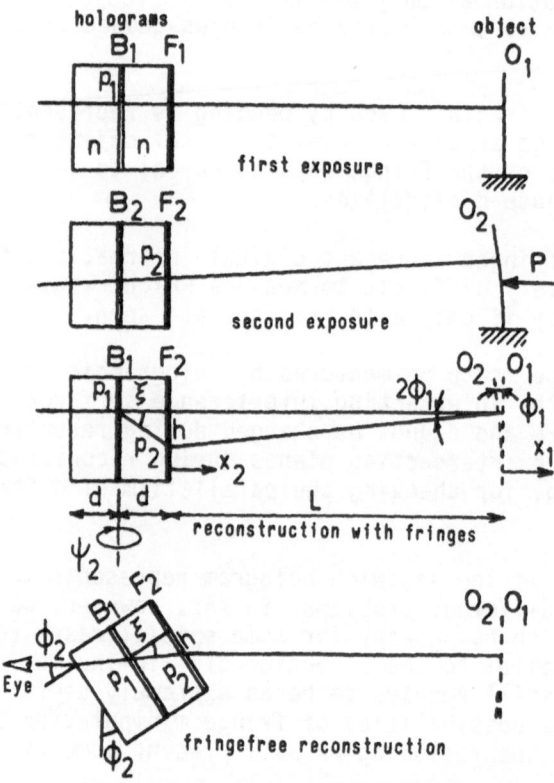

Fig. 2. The top of the object O is tilted through an angle Φ_1 by the force P. Hence, a speckle ray from one object point is moved the vertical distance h from p_1 to p_2. B and F are the emulsions of back and front hologram plates respectively. Glass plate thickness is d and refractive index in n. Φ_2 and Ψ_2 represent sandwich rotation around a horizontal and a vertical axis respectively. The identical reference and reconstruction beams are excluded from the figure.

Fig. 3. Rigid motion of the total objects has hidden the information concerning deformations. If this hologram had been an ordinary double exposure it would have been judged a failure.

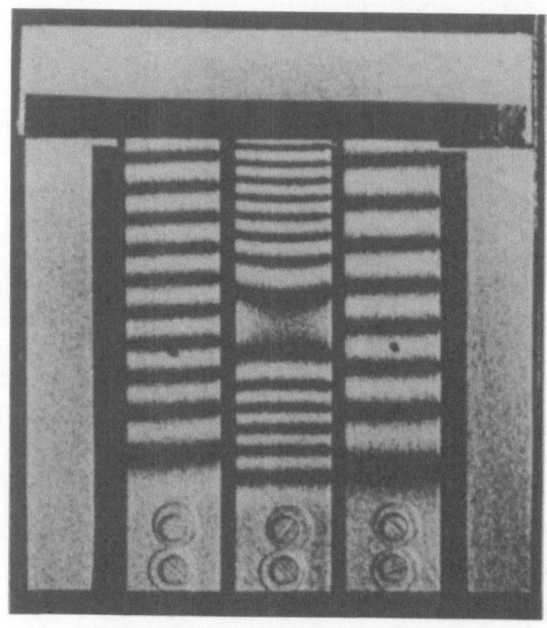

Fig. 4. The same reconstruction as that of fig. 3 but a small
tilt of the sandwich hologram has eliminated the fringes on the
rigid frame. The fringe pattern now correlates well with the expected deformations. The object consists of three bars, the lower
ends of which are screwed to a rigid frame. The middle bar is
also supported at the top end. Between the two exposures forces
were applied at the middle of the bars. The right bar is deflected
away from the observer. The other two bars are deflected towards
the observer.

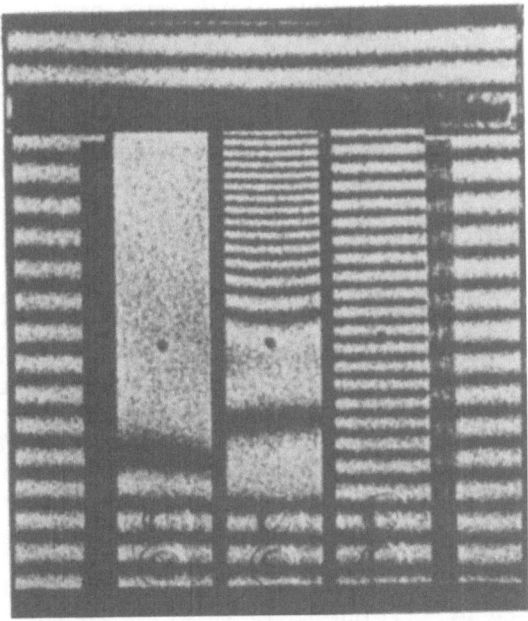

Fig. 5. The same reconstruction as that of fig. 4 but the sandwich
hologram has been tilted so that its top moved towards the observer.
At a certain tilt angle (Φ_2) the top of the left bar (which was
deflected with an angle Φ_1 towards the hologram) is fringe free.

was exposed. Thereafter all four plates were processed and the back plate of the first exposure (B_1 of fig. 2) was repositioned in the plate holder behind the front plate of the second exposure (F_2). The two plates were bonded together to form a sandwich hologram and reconstructed by a laser beam with direction and divergence identical to those of the reference beam during exposure.

2.3. Elimination of spurious fringes

Strange looking fringes were seen on all three bars (fig. 3). It was very difficult to find any correlation between this fringe pattern and the expected deformations. Straight, inclined fringes that covered the supposedly fixed frame disclosed the situation. The total object, including the frame, had made unwanted rigid motion. Even after this fact had been found it was difficult to evaluate the fringes on the steel bars. If the hologram had been an ordinary double exposure it would simply have been considered a failure.

The sandwich hologram, however, has the property that its fringes can be manipulated by tilting it in relation to the reconstruction beam. By holding the sandwich by hand and tilting it in different directions it was easy to find the reconstruction angle that made the rigid frame appear fringe free (fig. 4). The fringe pattern on the three bars now represented the deformation of the bars in relation to the frame and correlated well to what could be expected from their preknown load situation. Thus we had now changed our failure into something that would usually be considered a successful hologram which any holographer would be content with.

2.4. Sign and magnitude of object tilt

But a lot of information is still missing. From fig. 4 we cannot find out which bar was bent forwards or backwards. What has happened to the bar in the middle? Should we add all fringes from the top to the bottom to find the deflection? Or should we let the fringe number change sign halfway? By tilting the sandwich hologram so that its top moved towards the observer the number of fringes on the left bar decreased while the number on the right bar increased (fig. 5). At a certain sandwich tilt angle the top of the left bar became totally fringe free. The rigid frame now was covered by fringes having the same spacing and direction as those that had been eliminated from the left bar. From this rather large sandwich tilt the much smaller tilt angle of the bar could be calculated.

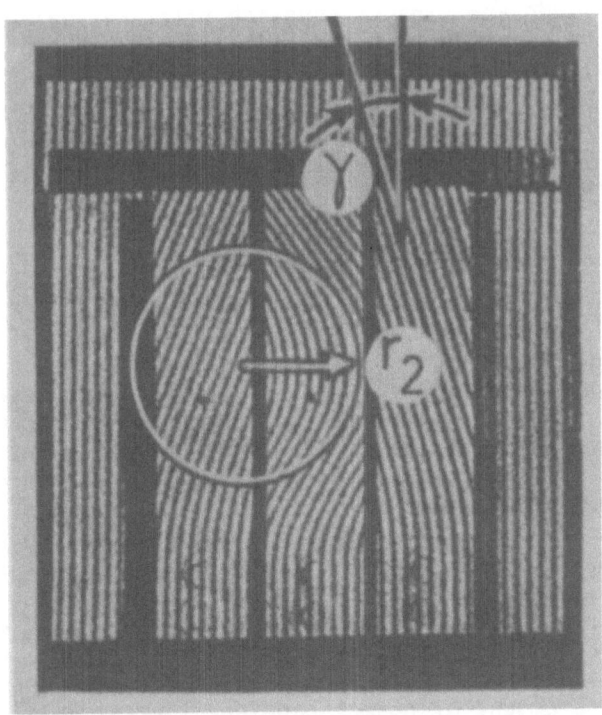

Fig. 6. The same reconstruction as that of fig. 4 but the sandwich
has been rotated through the angle Ψ_2 around a vertical axis. The
angle (γ) between the fringes and a vertical line is a measure of
object tilt angle (Φ_1). The radius of curvature of the fringes
(r_2) is a mesure of object bending radius (r_1), from which the
bending moment and maximal object strain can be calculated. Observe the large information content of fig. 6 compared to fig. 3.

As both magnitude and sign of sandwich tilt are analogous to
object tilt we conclude that the left bar had tilted towards the
hologram while the right bar had tilted away from it. The number
of fringes on the lower part of the middle bar decreased while
those on the top increased thus proving that its lower part had
tilted towards the hologram while its top had tilted in the opposite direction.

2.5. Maximal strain by fringe rotation

The sandwich hologram was once more tilted so that the rigid
frame appear fringe free. Then it was slightly rotated around a
vertical axis (Ψ_2 of fig. 2). The rigid frame became covered by
vertical fringes while the fringes of the left bar became inclined
to the right, the fringes of the bar on the right became inclined
to the left and the fringes of the middle bar bent into a "spoon
like" shape. This reconstruction (fig. 6) presents in one single
view much more information that can be found from fig. 3 which represents the information content of a conventional double exposed
hologram. The inclination of the fringes at each part of the
object is a measure of both sign and magnitude of object tilt.
The fringes are curved over those parts of the object where the
tilt angle varies. Thus the curvature of the fringes represents
the derivative of the conventional fringe frequency and is a
measure of the amount of object bending. There exists a simple
relation between fringe curvature and object bending moment from
which object stress and strain can be evaluated.

The fringe motion caused by sandwich tilt can be explained by the
study   of one single speckle ray which reacts to object movements as if it was reflected by a small mirror fixed to its surface
(fig. 2). Object bending will cause the speckle to be recorded
at different positions on the two exposed hologram plates. Any
shearing or tilt of the sandwich that during reconstruction repo-
sitions the two recordings of the speckle along the line of sight
towards the object, eliminates the effect of the original speckle
displacement and thus also eliminates the fringes indicating
object motion. Thus, fringes on the reconstructed object are dependent on tilt direction, either subtracted from or added to
the original fringe pattern. The spacing and direction of the
resulting fringes are analogous to a moiré effect of the original fringe pattern and the fringe pattern caused by sheared observation.

The following equations are limited to the study of displacements caused by object tilt that is parallel to the direction of
illumination and observation. It has also been assumed that the
sandwich tilt angle Φ_2 is so small that sin Φ_2 is practically equal
to Φ_2 .

$$\phi_1 = \frac{d}{2Ln} \cdot \phi_2 \tag{1}$$

$$\phi_1 = \frac{d}{2Ln} \cdot \Psi_2 \cdot tg\gamma \tag{2}$$

$$r_1 = \frac{2Ln}{d} \cdot \frac{1}{\Psi_2} \cdot r_2 \tag{3}$$

$$\sigma = \frac{d}{2Ln} \cdot \frac{\Psi_2 Et}{2} \cdot \frac{1}{r_2} \tag{4}$$

The symbols are defined in fig. 2 and fig. 6

ϕ_1 = object tilt.
ϕ_2 = sandwich tilt.
d = thickness of hologram glass plate
L = distance from object to hologram plate.
n = refractive index of hologram plate.
Ψ_2 = sandwich rotation around vertical axis
γ = fringe inclination to vertical.
r_1 = object bending radius.
r_2 = fringe radius of curvature.
σ = object stress.
E = modulus of elasticity.
t = object thickness.

3. Holography using picosecond lightpulses

3.1. Introduction

"Light-in-flight recording by holography" can be used for measurement of the three-dimensional shape of objects. If the wavefront (pulsefront) of a short pulse is flat it will produce a thin flat sheet of light. The intersection of this sheet by a three-dimensional body will produce on its surface a thin line representing a cross-section. It is advantageous if this cross-section represents a flat surface which will be the case if illumination and observation are collimated or made from large distances.

Thus "light-in-flight recording by holography" can be used as a contouring method that differs from other similar methods (e.g. shadow-moiré or two-wavelength holography) in that it produces

Fig. 7. The propeller of an ordinary fan was used for our contouring experiment. It was made of pressed steel sheet and had become rather deformed and unsymmetric by hard handling. To simplify the recording the propeller was covered by retroreflective paint. During exposure it was, of course, not hand-held, but rigidly fixed.

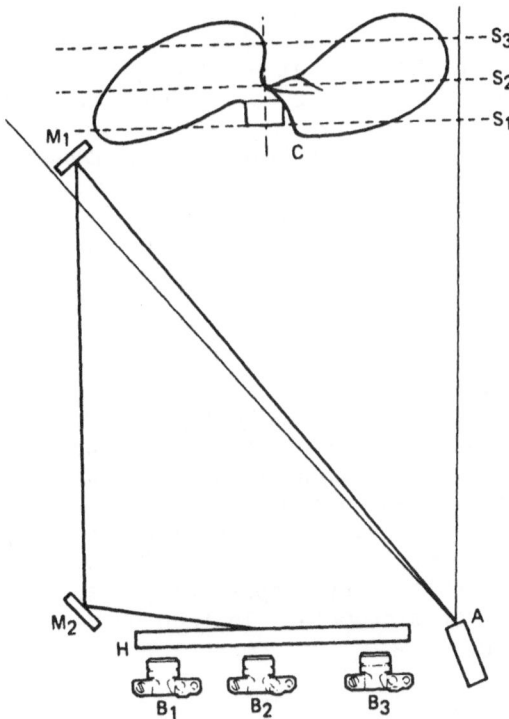

Fig. 8. A schematic view of the holographic set up. The propeller (C) of Fig. 7 is illuminated by the divergent beam from the pico-second laser (A). The reference is reflected by the two mirrors (M_1 and M_2) towards the hologram plate (H). In the actual experiment the distance between H and C was much longer compared to the size of H and C, than it is shown in this drawing.

not a set of fringes but just one single fringe. This fact has
the great disadvantage that one single image represents just one
single cross-section. However, it has the great advantage that
this single cross-section can be moved about in depth. By moving
the point of observation behind the hologram plate from left to
right the intersecting sheet of light can be moved forward or
backward. The determination of fringe order number by finding
the zero-order fringe does no longer present any problem as there
exists only the one single zero-order fringe. Therefore this new
way of contouring has great advantages especially when the observation and calculation should be made by computers.

It exists already lasers producing pulses shorter than one
pico-second and soon there will be lasers in the region of 0.3
picoseconds representing pulselengths of 0.1 mm. Using heterodyne-
like methods it should be no problem to find the centre of the
contouring line with a precision of 10% which corresponds to
about 10 μm. Using such an equipment the three-dimensional shape
of e.g. a propeller, a turbine blade or a car body could be measured with high accuracy. This way one could say that "light-in-
flight recording by holography" represents a radar method which
works in the region of micrometers instead of meters or kilometers
like ordinary radar.

3.2. Practical results

The propeller of an ordinary fan (fig. 7) was used for our contouring experiment. It was made of pressed steel sheet and had become
rather deformed and unsymmetric by hard handling. To simplify
the recording the propeller was covered by retroreflective paint.
During exposure it was, of course, not hand-held, but rigidly
fixed. The illumination and the observation directions were approximately parallel to the shaft of the propeller. The light-source
was a Spectra Physics argon laser (Model 171) which pumped a dye
laser (Model 375) placed in a cavity of the same length as that
of the argon laser. The whole set up was referred to as a "Synchronously pumped mode-locked dye laser". When used to expose
our holograms it produced pulses of some 10 picoseconds (3 mm of
length) separated by twice the cavity length or some two meters.
The output power was 20 milliWatt and the exposure time was around
five seconds on Agfa-Gevaert 10 E 75 plates.
A schematic view of the holographic set up is seen in Fig. 8.
The propeller (C) is illuminated by the divergent beam from the
picosecond laser (A). The reference is reflected by the two mirrors
(M$_1$ and M$_2$) towards the hologram plate (H). In the actual experiment the distance between H and C was much longer when compared to
the size of H and C, than it is shown in the drawing.

The reconstruction of the hologram disclosed the propeller inter-

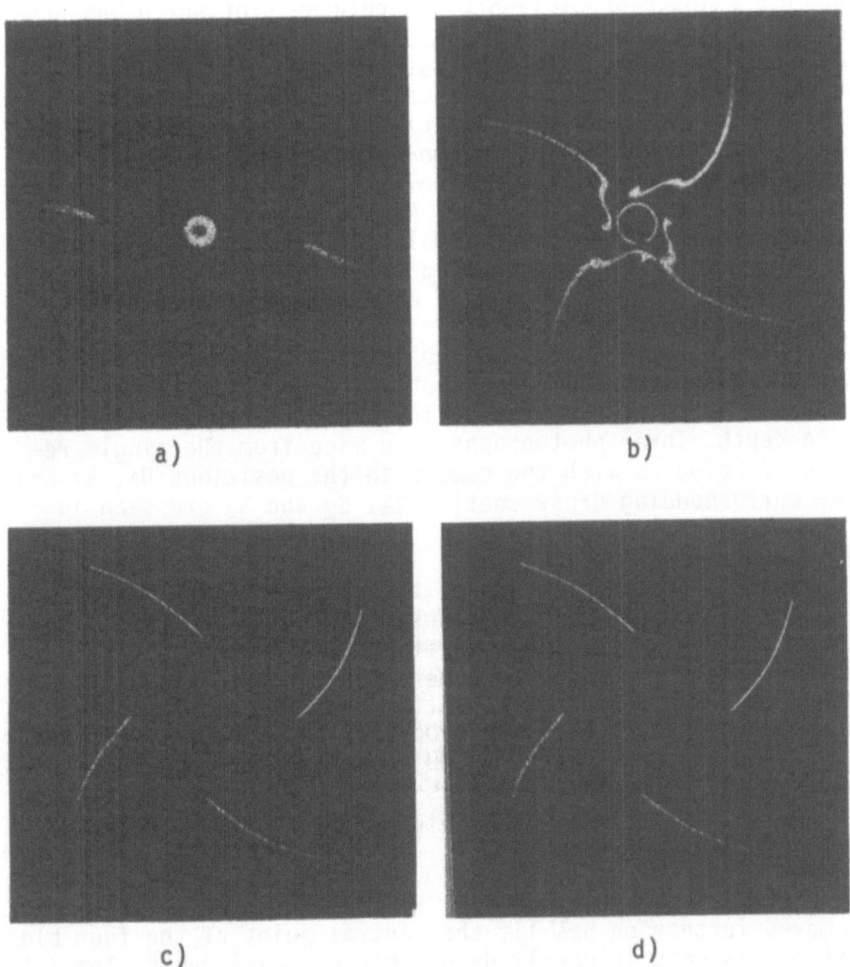

a)　　　　　　　　b)

c)　　　　　　　　d)

Fig. 9. (a) This photo was taken during reconstruction with exposure of the camera positioned at B_1 which corresponds to the cross-section S_1 (Fig. 8). The sheet of light has just reached the hub of the propeller and two of its blades. (b) The camera at B_2 of Fig. 8 produced the crossection S_2. The hub has no light and remains into darkness while the light pulse has moved further so that the four blades and their central joint are intersected. (c) The light sheet has just passed the central joint and cross-section of the blades reveal their three-dimensional shape. (d) The intersection S_3 photographed with the camera at B_3 shows how the light sheet has passed further on and soon will leave the propeller altogether.

sected by a thin sheet of light, the thickness of which was equal
to half the pulse length. This would be the ideal case if this intersecting sheet of light had been flat, which would be the case if
illumination and observation were collimated or made from infinite distances. In the actual experiment the intersecting surfaces were parts of ellipsoids of the holodiagram, one focal point
(A) being the point of illumination, the other (B) the point of
observation at the hologram plate. As the object (C) and the
separation between A and B were small compared to the distance
from C to A and B, the intersecting sheet of light could be approximated into a section of sphere with almost zero curvature.

For every certain distance from the left of the hologram plate there exists a corresponding cross-section in depth so that by moving
the observation point sideways behind the plate the crossection
moves in depth. Three photographs were made from the single reconstructed hologram with the camera in the positions B_1, B_2 and
B_3. The corresponding cross-section S_1, S_2 and S_3 are seen in
Fig. 9 a, b and d.

Fig. 9 a) shows the circular hub of the propeller with its hole
for the shaft. Two of the rather unsymmetrical blades are also
seen just touched by the light sheet S_1. The photo was taken during
reconstruction by exposing the camera positioned at B_1 which corresponds to the cross-section S_1 (fig. 8). The sheet of light
has just reached the hub of the propeller and two of its blades.
The following photography shows the cross-section some 50 picoseconds later while the last photo is exposed when the light pulse is
further down to the right of the plate representing a delay of still
another 100 picoseconds.

Fig. 9 b) reveals how the hub goes into darkness while the light
sheet moved further on has lit the central joint of the four blades.
Its out-of-flatness is easily detectable and measurable. The intersection of the four blades reveals their inclinations and their
assymetries. When the intersection is continuously moving past the
propeller altogether (fig. 9 c).

In order to produce a thin intersection of the object it is important that the observation aperture is small (at least as small
as the pulse length along the plate). The reason is that different parts of the lens observe the objects as it was recorded
at different points of time, when the sheet of light had reached
different dephs.

The best way to make the photographs of the reconstructed image
is either by photographing the virtual image using a vertical
slit aperture at the plate surface, or by using the real image,
illuminating the plate by a laser-beam which is preferably elon-

gated along a vertical axis.

When the virtual image is photographed it is of great importance
that the camera lens is close to the hologram plate intersecting
plane it would appear inclined otherwise. The reason for this tilt
is that different parts of the object are observed through different
parts of the plate. Thus they are recorded at different points of
time when the light sheet had reached different dephts. The possibility to tilt the intersecting plane by changing the separation
between camera and hologram plate can, when carefully planned, be
advantageous during the evaluation.

3.3. Light focused by a lens

The first experiments had been made in Sweden using an argon
laser with short coherence length. Even this primitive equipment
produced, as already described, results that were superior to
everything published before in the field of recording light in
flight.

In spite of the similarities between light of short coherence
length and that of short pulse length I did, however, wish to use
light that without any doubt consisted of just one single picosecond
pulse. To produce such a pulse with energy enough to expose my holographic scenery I needed a very powerful ruby laser. Such a laser
exists, however, it is not yet commercially available and none
was available for my experiment. Thus I decided instead to use a mode-locked dye-laser that produced a train of picosecond pulses.

Spectra Physics Inc. at Mountain View California outside San Fran-
cisco, had kindly arranged so that I should be able to use their
laboratory and their equipment.

The light source was a mode-locked argon laser (Model 171) which
pumped a dye laser (Model 375) placed in a cavity of the same
length as that of the argon laser. As referred this "Synchronously
pumped mode-locked dye laser" produced pulses of some 10 picoseconds (3 millimeters of length) separated by twice the cavity length or by some two meters.

From earlier experiments I knew that for light in flight recording the ratio between reference beam and object beam
should be much lower than that of ordinary holograms (some ten to
one), but that it should be as low as about one to ten was a surpri-
se. One rather obvious explanation could be that as only a thin line
of light is recorded, only that light should be taken into account
when the intensity of the object beam is measured. That light
was about one hundredth of the total object light so that could

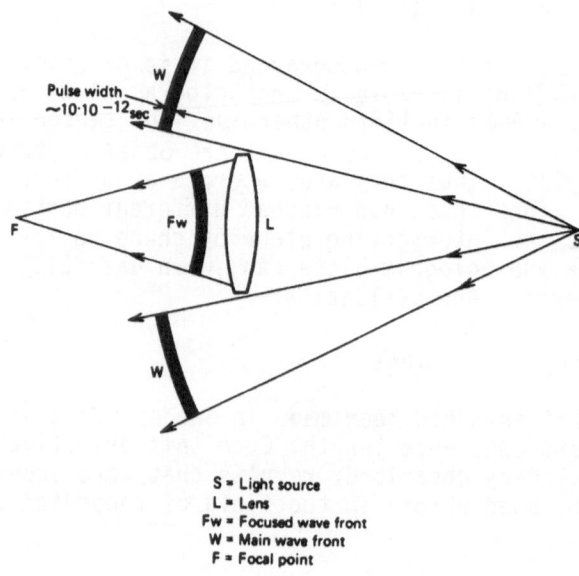

Pulse width
~10·10⁻¹²sec

W

Fw   L

F

S

S = Light source
L = Lens
Fw = Focused wave front
W = Main wave front
F = Focal point

Fig. 10. A diagram showing what is to be expected when a spherical wavefront (W) emanating from the point source (S) is focused by the lens (L) towards the focal point (F). Light moves slower through glass than air, therefore the wavefront that has passed through the lens (Fw) is delayed. As the lens is thickest in its middle, the originally convex wavefront is transmitted into a concave one.

explain the result. The object light that arrived to the plate at those moments when it was not simultaneously illuminated by the reference beam had no influence on the recording of the hologram, it simply worked as an added incoherent illumination which darkened the plate.

Fig. 10 is a diagram showing what is to be expected when a spherical wavefront (W) emanating from the point source (S) is focused by the lens (L) towards the focal point (F). Light moves slower through glass than air, therefore the wavefront that has passed through the lens (Fw) is delayed. As the lens is thickest in the middle, the originally convex wavefront is transformed into a concave one.

Some examples of the recording of a short light-pulse that passes through a positive lens are seen in accordance to fig. 10, and in the photographs of fig. 11 a-d. The diagram of fig. 10, shows the expected result of the experiment. The spherical wave expands from the point source (S) and moves to left. The curved line (W) represents the spherical wavefront intersected by the flat observation screen, a whitepainted plate of aluminium with a size of about 20 cm x 30 cm. A cylindrical lens (L) is fixed with its axis normal to the screen so that it focuses the light which arrives almost parallel to the screen surface. As light travels slower through glass than through air the light that has passed through the lens (Fw) will be delayed as compared to the original wavefront (W). The lens is thickest in its centre which results in the light that passes this part is delayed most. Thus the originally convex wavefront is transformed into a concave shape. As the wavefront is travelling perpendicularly to the lens plane the concave shape means that the light is travelling towards the focal point (F) which is in the center of the curved line.

Figure 11 is an experimental result performed as schematically shown by fig. 10.

a) A three millimeter (10 picoseconds light pulse) long pulse-front representing the spherical wavefront traveling from right to left. It sweeps almost parallel to the observation screen and has not yet reached the cylindrical lens.

b) The wavefront has just partly passed through the lens (which is not seen in the photo). Only the top and bottom parts of Fw (fig. 10) have passed through the glass.

c) The light has left the lens and is focused towards the focal point. Observe the resemblance to the diagram of fig. 10.

294

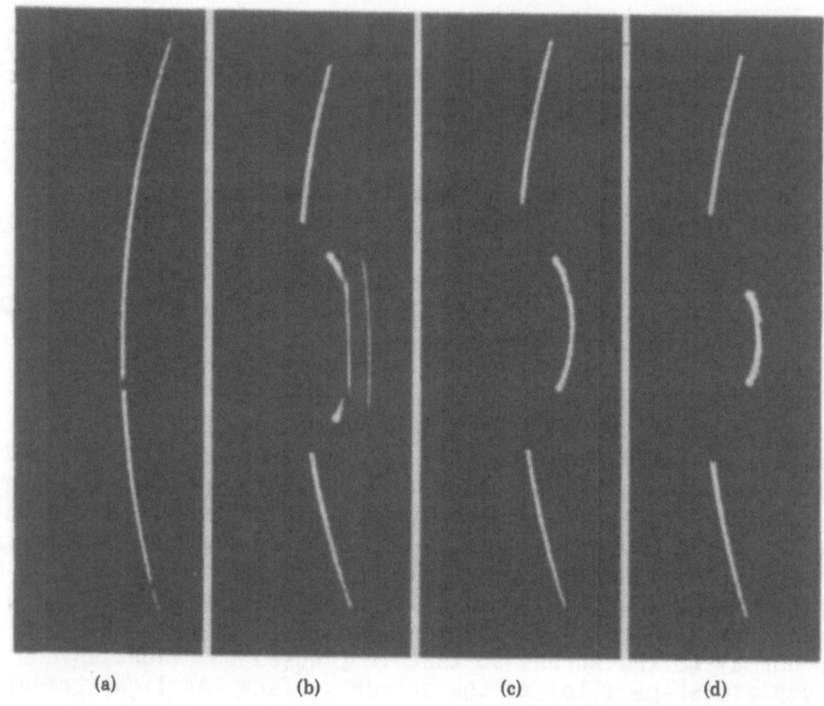

(a)   (b)   (c)   (d)

Fig. 11. (a) The practical result of the experiment described in
Fig. 10. A three millimeter (10 picosecond light pulse) long
pulsefront representing the spherical wavefront travels from right
to left. It sweeps almost parallel to the observation screen and
has not yet reached the cylindrical lens. (b) The wavefront has
just partly passed through the lens (which is not seen in the
photo). Only the top and bottom parts of Fw (Fig. 10) have passed
through the glass. (c) The light has left the lens and is focused
towards the focal point. Observe the resemblance to the diagram of
Fig. 10. (d) The length and the radius of curvature of the wavefront are decreasing as it travels away from the lens, towards the
focal point.

d) This exposure is made just before the wavefront has reached
   the focal point.

The photographs of fig. 11 show how extremely well the experiment
does verify the expected results. Pictures taken at earlier moments
(photographed further to the right through the hologram plate)
and at later moments (photographed further to the left) show
clearly the focusing effect of the lens.

<u>Reference</u>:

Nils Abramson, <u>The Making and Evaluation of Holograms</u>
(Academic Press, London, 1981).

QUANTITATIVE INTERPRETATION OF TIME-AVERAGE HOLOGRAMS IN VIBRATION ANALYSIS

by

Ryszard J. Pryputniewicz
Center for Holographic Studies and Laser Technology
Department of Mechanical Engineering
Worcester Polytechnic Institute
Worcester, MA 01609

ABSTRACT

This paper deals with quantitative interpretation of time-average holograms of vibrating objects. It is shown that the images obtained during reconstruction of such holograms are modulated by a system of fringes described by the square of the zero-order Bessel function of the first kind. Procedure for quantitative interpretation of these fringes is described and is illustrated with examples. The experimental results, obtained directly from the time-average holograms, show very good agreement with an exact solution of the differential equation of a vibrating beam.

1. INTRODUCTION

Transverse vibrations of beams are of primary interest in many engineering applications. However, the conventional methods, both analytical
and experimental, are rather limited when it comes to vibration analysis.
In general, the analytical methods, including the finite element techniques,
require accurate knowledge of boundary conditions and material properties.
The experimental techniques, on the other hand, are invasive and frequently take
into consideration only a few points, on the studied structure. The results
obtained from analysis of these limited data are then "extrapolated" to
predict the dynamic behavior of the entire structure. Also, as the vibration frequency increases, the corresponding displacement amplitude decreases making it difficult to measure with conventional instrumentation.
In 1965, however, the technique of hologram interferometry was developed
[1] which provided means of measuring vibrations of very small amplitudes.

In hologram interferometry, two (or more) waves emitted, e. g., from the
laser, interact with each other because of the differences in their path
lengths. The most popular technique for recording of holograms is the
off-axis method [2]. In this arrangement, Fig. 1, the highly coherent and
monochromatic laser light is divided into two beams by means of a beamsplitter. One of the beams, in the case shown that transmitted through the
beamsplitter, is expanded and filtered (by means of a microscope objective
and a pin-hole assembly) and is directed (by mirrors) to illuminate the
object to be recorded. This beam, modulated by reflection from the scene
being recorded, is called the object beam. It caries instantaneous information about the object's configuration. The other beam, i. e., that
reflected from the beamsplitter, is also expanded and spatially filtered,
but it is directed by mirror(s) in such a way that it does not interfere
with the object. This is the reference beam. The reference beam is also
directed to overlap with the object beam. The two beams interfere in the
regions of space where they overlap. A photosensitive medium placed in
this region records the resulting interference pattern. The developed
medium, when reilluminated with the original reference beam, faithfully
reconstructs the image of the recorded object.

In practice, multiple object configurations are recorded in the same photosensitive medium. Therefore, during reconstruction of such holograms multiple images are produced, i. e., one image for each recorded object configuration. Since each of these images is reconstructed in coherent light,
they interfere with each other in any region of space where they superpose.
As a result of this, the reconstructed image of the object is seen covered
by a set of bright and dark interference fringes. These fringes are a direct
measure of changes in the object's position and/or shape that occured while
the hologram was being recorded.

Depending on the particular application, the interference fringes may be
recorded using one of the following techniques [3]: (i) double-exposure hologram interferometry, (ii) real-time hologram interferometry, (iii) time average hologram interferometry; the time average technique can be further subdivided into (i) stroboscopic time-average hologram interferometry and (ii)
continuous time-average hologram interferometry. Each of the above techniques

has certain advantages over the others in specific vibration measurement applications.

The double-exposure technique is uniquely suited for studies of transient vibrations; it can be used equally well in studies of resonating objects. However, it relays on the use of a pulsed laser which delivers two sub-microsecond pulses synchronized with object's motion. During reconstruction of a double-exposure hologram, the object's image is modulated by cosinusoidal fringes. Although interpretation of such fringes is straightforward [4, 5] cost of the pulsed laser system is rather high.

The real-time method requires the use of a liquid gate plate holder, or some mechanical micropositioner, to precisely locate developed hologram. The image reconstructed from this hologram, when superposed onto the original object, interferes with the light field modulated by vibrating object, at the instant the object moves. This method is particularly useful in identification of object resonances.

The stroboscopic time-average hologram interferometry is really an extension of the double-exposure method, except that now a continuous wave (CW) laser can be used to record a hologram. In this application, the CW laser beam is "chopped" into short pulses synchronized with the object frequency; the pulse length depends on the nature of the vibration studied. To effectively use this method, object vibration must be monitored continuously to assure proper characteristics of the illuminating beam. This synchronization must be maintained over many vibration cycles, to provide for sufficient exposure of the photosensitive medium. Although interference fringes produced during reconstruction of a stroboscopic hologram are cosinusoidal and are straight-forward to analyze, the electronic apparatus, needed to produce good quality images, may be complex and expensive.

In continuous-time average hologram interferometry, a single holographic recording of an object, undergoing a cyclic vibration, is made. With the (continuous) exposure time long in comparison to one period of the vibration cycle, the hologram effectively records an ensemble of images corresponding to the time average of all positions of the object while it is vibrating. During reconstruction of such a hologram, the interference occurs between the entire ensemble of images, with the images recorded near zero velocity (i. e., maximum displacement) contributing most strongly to the reconstruction. The interference fringes observed are of unequal brightness. In fact, they vary according to the square of the zero order Bessel function of the first kind, J_0^2, as will be shown in subsequent sections.

The continuous time-average hologram interferometry is the most popular of the holographic methods discussed herein, when it comes to vibration analysis. The existing holographic laboratories are, in general, well equipped to perform continuous time-average studies. The apparatus is the same as that used in recording of conventional holograms, except for the mechanism to "drive" the object. The driving mechanism can be a piezoelectric shaker, a loud speaker, a magnetic oscillator, a flowing fluid, etc. However, regardless of the method used to excite the object, the (continuously) time-averaged interference fringes are the same in nature. In the following sections, the discussion will center on quantitative interpretation of the (continuous) time-average holograms.

2. INTERPRETATION OF TIME-AVERAGE HOLOGRAMS

The time-average holograms can be recorded using a setup similar to that shown in Fig. 1. Let the complex light field, $F_i(x,y,)^*$ illuminating the object be represented as

$$F_i(x,y,z) = A_i(x,y,z) \exp\left[i\phi_i(x,y,z)\right] , \qquad (1)$$

where A_i and ϕ_i are the amplitude and phase, respectively, of the illuminating light field propagating in the Cartesian space defined by the x,y, and z coordinates. The complex light field, $F_o(x,y,z)$, produced by reflection from the stationary object is

$$F_o(x,y,z) = A_o(x,y,z) \exp\left[i\phi_o(x,y,z)\right] , \qquad (2)$$

where A_o and ϕ_o are the amplitude and phase, respectively, of the light field modulated by interaction with the object.

The displacement, $L_t(x,y,z,t)$, of the vibrating object is a function of time t, and for cosinusoidal excitation can be expressed as

$$L_t(x,y,z,t) = L_o(x,y,z) \cos(\omega t) . \qquad (3)$$

In Eq. 3, L_0 is the magnitude of the displacement of the object which is vibrating at a circular frequency ω. This object motion causes temporal changes, $\Omega_t(x,y,z,t)$, in phase of the light field reflected from the object. These phase changes can be represented by a scalar product of $L_t(x,y,z,t)$ with the sensitivity vector $K(x,y,z)$ [3-8]:

$$\Omega_t(x,y,z,t) = K(x,y,z) \cdot L_t(x,y,z,t) , \qquad (4)$$

where

$$K(x,y,z) = K_2(x,y,z) - K_1(x,y,z) \qquad (5)$$

with $K_1(x,y,z)$ and $K_2(x,y,z)$ being illumination and observation propagation vectors. These vectors are easily defined in terms of position vectors R_1, R_2 and R_p locating point source of illumination, point of observation, and point on the object, respectively, that is,

$$K_1(x,y,z) = K_1 = k\frac{R_p - R_1}{|R_p - R_1|} = k\frac{(x_p - x_1)\hat{i} + (y_p - y_1)\hat{j} + (z_p - z_1)\hat{k}}{\left[(x_p - x_1)^2 + (y_p - y_1)^2 + (z_p - z_1)^2\right]^{1/2}} = k\hat{K}_1 , \qquad (6)$$

$$K_2(x,y,z) = K_2 = k\frac{R_2 - R_p}{|R_2 - R_p|} = k\frac{(x_2 - x_p)\hat{i} + (y_2 - y_p)\hat{j} + (z_2 - z_p)\hat{k}}{\left[(x_2 - x_p)^2 + (y_2 - y_p)^2 + (z_2 - z_p)^2\right]^{1/2}} = k\hat{K}_2 , \qquad (7)$$

---

* Double underlined quantities indicate vectors which are shown by boldface symbols in equations.

In Eqs 6 and 7, $\hat{R}_1$ and $\hat{R}_2$ are the unit illumination and observation vectors, respectively, k is the magnitude of these vectors, defined as

$$\left|\vec{K_i}\right| = \left|\vec{K_2}\right| = k = \frac{2\pi}{\lambda} \quad . \tag{8}$$

with λ being the wavelength of the laser light, while the position vectors are given by

$$\mathbf{R} = x\hat{i} + y\hat{j} + z\hat{k} \quad , \tag{9}$$

where $\hat{i}$, $\hat{j}$, $\hat{k}$ are unit vectors in the Cartesian coordinate system.

From the foregoing discussion it is clear that the complex light field, $\underline{E}_v(x,y,z,t)$, propagating away from the vibrating object can be obtained by combining Eqs 2 and 4, that is,

$$\mathbf{F}_v(x, y, z, t) = A_o(x, y, z) \exp\left[i\phi_o(x, y, z) + i\Omega_t(x, y, z, t)\right] \quad . \tag{10}$$

Let the complex reference field, $\underline{F}_r(x,y,z)$, be defined as

$$\mathbf{F}_r(x, y, z) = A_r(x, y, z) \exp\left[i\phi_r(x, y, z)\right] \quad , \tag{11}$$

where A_r and ϕ_r are amplitude and phase of the reference beam, respectively.

Thus, at any instant of time, when the light fields given by Eqs 10 and 11 are brought to interfere with each other in the hologram plane, the resulting complex field $F_h(x,y,z,t)$ is

$$\mathbf{F}_h(x, y, z, t) = A_o(x, y, z) \exp\left[i\phi_o(x, y, z) + i\Omega_t(x, y, z, t)\right] +$$
$$+ A_r(x, y, z) \exp\left[i\phi_r(x, y, z)\right] \quad . \tag{12}$$

The photosensitive medium, on which hologram is recorded, responds to intensity, $I_h(x,y,z,t)$, of this field. The value of this time-varying intensity is expressed as the product of $F_h(x,y,z,t)$ and its conjugate $F_h^*(x,y,z,t)$, that is,

$$I_h = \mathbf{F}_h \mathbf{F}_h^* = \left[A_o\exp(i\phi_o+i\Omega_t) + A_r\exp(i\phi_r)\right]\left[A_o\exp(-i\phi_o-i\Omega_t) + A_r\exp(-i\phi_r)\right]$$
$$= A_o^2 + A_r^2 + \mathbf{F}_v\mathbf{F}_r^* + \mathbf{F}_r\mathbf{F}_v^* \quad , \tag{13}$$

where the arguments (x,y,z) and (x,y,z,t) were omitted for simplification.

The image recorded within the photosensitive medium is the time-average of I_h over the exposure time T, thus

$$\frac{1}{T}\int_0^T I_h\, dt = A_o^2 + A_r^2 + \frac{1}{T}\mathbf{F}_r\int_0^T \mathbf{F}_v^*\, dt + \frac{1}{T}\mathbf{F}_r^*\int_0^T \mathbf{F}_v\, dt \quad . \tag{14}$$

When the time-average hologram is developed and illuminated by the original reference field F_r, one obtains:

$$F_r \frac{1}{T}\int_0^T I_h dt = A_o^2 F_r + A_r^2 F_r + A_r^2 \exp(2i\phi_r)\frac{1}{T}\int_0^T F_v^* dt + A_r^2 \frac{1}{T}\int_0^T F_v dt \quad . \tag{15}$$

The first two terms on the right hand side of Eq. 15 represent an attenuated undiffracted reconstruction field, the third term gives rise to the conjugate image, while the fourth term is proportional to the time-average of the original object field and describes formation of the virtual image. Therefore, the developed hologram, when illuminated by E_r, produces on object wave which has a complex amplitude proportional to the time-average of E_v over the time interval T, that is,

$$\frac{1}{T}\int_0^T F_v dt = \frac{1}{T}\int_0^T A_o(x,y,z)\exp\left[i\phi_o(x,y,z) + \Omega_t(x,y,z,t)\right]dt$$

$$= F_o(x,y,z)\frac{1}{T}\int_0^T \exp\left[i\Omega_t(x,y,z,t)\right] \quad , \tag{16}$$

where Eq. 2 was used to simplify the relationship.

The time-average integral appearing in Eq. 16 is called the characteristic function and is denoted by M_T [9], that is,

$$\frac{1}{T}\int_0^T \exp\left[i\Omega_t(x,y,z,t)\right]dt = M_T(x,y,z) \quad . \tag{17}$$

The integral of Eq. 17 may be evaluated as $J_0(|\Omega_t|)$, the zero-order Bessel function of the first kind of the magnitude of the argument Ω_t. The magnitude of Ω_t can be obtained by expanding Eq. 4 and using Eqs 3 and 5, that is,

$$|\Omega_t| = |K \cdot L_t| = |(K_2 - K_1) \cdot L_t|$$

$$= k\left[(\hat{K}_{2_x} - \hat{K}_{1_x})L_x + (\hat{K}_{2_y} - \hat{K}_{1_y})L_y + (\hat{K}_{2_z} - \hat{K}_{1_z})L_z\right] \quad , \tag{18}$$

where $\hat{K}_{1x}$, $\hat{K}_{2x}$, $\hat{K}_{1y}$, ... are the components (direction cosines) of the unit illumination and propagation vectors, defined in Eqs 6 and 7, while L_x, L_y, and L_z represent the components of L_t.

Thus, according to Eqs 16 and 17, the reconstructed complex amplitude responsible for formation of the virtual image (see Eq. 15), is proportional to $E_o \cdot M_T$, and the corresponding intensity, $I_{im}(x,y,z)$, of the reconstructed image is

$$I_{im}(x,y,z) = \left[F_o(x,y,z)\right]^2 \left[M_T(x,y,z)\right]^2 = A_o^2 J_0^2(|\Omega_t|) \quad . \tag{19}$$

Equation 19 shows that the virtual image obtained during reconstruction of a time-average hologram is modulated by a system of fringes described by the square of the zero-order Bessel function of the first kind. Thus, for non-trivial value of A_0, centers of the dark fringes will be located at those points on the object's surface where $J_0(|\Omega_t|)$ equals zero, as shown in Fig. 2. The first 30 zeros of the J_0 are given in Table 1.

The images formed during reconstruction of time-average holograms of cosinusoidal motions are modulated according to the variations of J_0^2, shown in Fig. 2. Figure 3 indicates that the J_0 fringes differ from the cosinusoidal fringes obtained in double-exposure hologram interferometry. One of these differences is that the zero-order fringe is much brighter than the other J_0 fringes. This fringe represents the stationary points on the vibrating object and thus allows easy identification of nodes. The brightness of other fringes, as well as their spacing, decrease with increasing fringe order and can be directly related to the mode shapes.

Application of the foregoing analysis, in quantitative interpretation of transverse vibrations of a cantilever beam, is discussed in Sect. 3.

3. QUANTITATIVE INTERPRETATION OF TIME-AVERAGE HOLOGRAMS OF A CANTILEVER BEAM

The following is a representative example of a cantilever beam that was fixed at the lower end while its upper end was free. The cosinusoidal excitation was applied at the free end, by means of a loud speaker. The excitation was always set in such a way that the beam's motion was in the direction parallel to the z-axis, see Fig. 1. For this case, Eq. 3 became:

$$L_1(0, 0, z, t) = L_0(0, 0, z) \cos(\omega t) \quad , \tag{20}$$

while Eq. 18 yielded,

$$|\Omega_1| = k(\hat{K}_{2z} - \hat{K}_{1z})L_z \quad , \tag{21}$$

where k is given by Eq. 8, $\hat{K}_{1z}$ and $\hat{K}_{2z}$ can be computed from Eqs 6 and 7 as

$$\hat{K}_{1z} = \frac{z_p - z_1}{\left[(x_p - x_1)^2 + (y_p - y_1)^2 + (z_p - z_1)^2\right]^{1/2}} \quad , \tag{22}$$

and

$$\hat{K}_{2z} = \frac{z_2 - z_p}{\left[(x_2 - x_p)^2 + (y_2 - y_p)^2 + (z_2 - z_p)^2\right]^{1/2}} \quad , \tag{23}$$

respectively, and L_z is the vibration amplitude.

The goal of the analysis is to determine L_z which gives the mode shape. However, before this can be done, Eqs 22 and 23 have to be evaluated, at every point of interest on the vibrating object.

For the case of retro-reflective illumination and observation, parallel to the z-axis, the quantity $(\hat{R}_{2z}-\hat{R}_{1z})$ has the maximum value of 2, as determined from Eqs 22 and 23. Thus, combination of Eqs 8 and 21, yields

$$L_z = \frac{\lambda}{4\pi}\left|\Omega_i\right| \quad . \tag{24}$$

The values of L_z, for various orders of J_0 finges and $\lambda=0.6328$ µm, as computed from Eq. 24, are shown in Table 1.

For any other geometry, where the directions of illumination and observation are not parallel to the z-axis, the quantity $(\hat{R}_{2z}-\hat{R}_{1z})$ will always be less than 2. Its actual magnitude will depend on the magnitudes of angles γ_1 and γ_2 that the directions of $\underline{K}_1$ and $\underline{K}_2$, respectively, make with the direction of $\underline{L}$, as shown in Fig. 4. That is, for every case when directions of illumination and observation deviate from being parallel to the direction of motion, the L_z computed from Eq. 21 will always be greater than that given by Eq. 24, for the same order of the J_0 fringe. The percentage error, PE, between the value of L_z, for the non retro-reflective case, and the value of $L_{z_{rr}}$, for the retro-reflective case, can be defined as

$$PE = \left(\frac{L_z - L_{z_{rr}}}{L_{z_{rr}}}\right) \times 100\% \quad . \tag{25}$$

Equation 25 was evaluated for various values of $(\hat{R}_{2z}-\hat{R}_{1z})$ and the results are shown in Fig. 5. This figure indicates that Eq. 24 will yield L_z to within 5% of the value actually experienced by the object as long as $(\hat{R}_{2z}-\hat{R}_{1z})>1.9$. This result, in turn, allows to determine, from Fig. 4, that when $|\gamma_2|=0^0$, $|\gamma_1|$ should be less than or equal to 26^0, while for $|\gamma_2|=15^0$, $|\gamma_1|<21^0$. By reciprocity, the following pairs of angles will be true: $|\gamma_1|=0^0$, $|\gamma_2|<26^0$ and $|\gamma_1|=15^0$, $|\gamma_2|<21^0$. Figure 4 also indicates that experimental setups with $|\gamma_1|>26^0$ (or, by reciprocity, with $|\gamma_2|>26^0$) will always yield $(\hat{R}_{2z}-\hat{R}_{1z})<1.9$, which will result in errors (as defined by Eq. 25) in excess of 5%.

One way to compensate for the errors resulting from setup geometry is to use the following formula

$$L_z = \left(1 + \frac{PE}{100}\right)L_{z_{rr}} \quad , \tag{26}$$

where $L_{z_{rr}}$ is the displacement corresponding to the order of the J_0 fringe crossing the point of interest on the particular object, as given in Table 1, while PE is determined from Figs 4 and 5. Therefore, L_z obtained from Eq. 26 is the displacement at a given point on the object, compensated for system geometry.

Another way to achieve the same result is to use Eqs 21 to 23 directly, as discussed in Sect. 5.

4. EXPERIMENTAL SETUP

The cantilever beam used in this study was rigidly fixed at the bottom and free at the top. Its free length was 160 mm (6.3in), it was 28.58 mm (1.125 in) wide and 3.18 mm (0.125 in) thick. The cantilever beam, made of 6061-T6 aluminum, was excited acoustically at its free end. The origin of the right-handed rectangular Cartesian coordinate system was located at the fixed end of the cantilever beam with the positive z-axis pointing toward the hologram. In this coordinate system, the position vectors $\underline{R}_1$ and $\underline{R}_2$ defining point source of illumination and point of observation, respectively, were determined to be

$$\mathbf{R}_1 = 216\hat{i} + 23\hat{j} + 1321\hat{k} \tag{27}$$

and

$$\mathbf{R}_2 = -13\hat{i} + 38\hat{j} + 806\hat{k} \quad , \tag{28}$$

where coefficients are distances in millimeters. The experimental setup, shown schematically in Fig. 1, was used to record time-average holograms of the vibrating cantilever beam. These holograms were analyzed and the results are discussed in Sect. 5.

5. EXPERIMENTAL RESULTS

The excitation frequency was varied until the cantilever beam resonance was achieved. These experimentally obtained resonance frequencies compared very well with the frequencies computed using the beam theory, Fig. 6

Representative time-average holograms were recorded at the first three resonance frequencies with the corresponding results shown in Figs 7 to 9, respectively. For example, for the first flexure mode, Fig. 7, the center of the fifth dark fringe was located 67 mm above the fixed end. That is, for this point on the object,

$$\mathbf{R}_P = 67\hat{j} \quad , \tag{29}$$

Substitution of Eqs 27 to 29 into Eqs 22 and 23 yielded

$$\hat{K}_{1_z} = -0.9864 \tag{30}$$

and

$$\hat{K}_{2_z} = 0.9992 \quad , \tag{31}$$

respectively.

Subsequent substitution of Eqs 8, 30, and 31 into Eq. 21, together with the values of $\lambda=0.6328$ μm (wavelength of the He-Ne laser used to record and re-

construct the time-average holograms, in this study) and $|\Omega_t|=14.93092$
(corresponding the fifth J_0 fringe, Table 1) gave the following result:

$$L_z = \frac{\lambda}{2\pi(\hat{K}_{2_z} - \hat{K}_{1_z})} |\Omega_t| = 0.757 \; \mu m \; . \tag{32}$$

Equation 32 indicates that the L_z corrected for setup geometry is only 0.7%
greater than that for retro-reflective illumination and observation as shown
in Table 1. For the center of the 10th dark fringe, located 102 mm above the
fixed end of the cantilever beam, one obtains $L_z=1.557 \; \mu m$ which is 0.9% above
the value given in Table 1. Finally, the center of the 19th dark fringe,
at 157 mm, corresponds to a displacement of 3.010 μm which is 1.5% larger
than the displacement for retro-reflective illumination and observation for
n=19, as given in Table 1.

In a similar manner, displacements at other points on the cantilever beam
were determined and are shown in Fig. 7. For comparison, the mode shape of
the cantilever beam vibrating at its first flexure frequency was determined
analytically using the beam theory. The agreement between the theory and
experiments is very good.

Following the procedure described above, mode shapes for the second and the
third flexure frequencies were determined, as shown in Figs 8 and 9, respectively. Again, the agreement between theory and the experimental results,
obtained from time-average holograms, is very good.

6. CONCLUSIONS

The results presented herein indicate that images obtained during reconstruction of time-average holograms are modulated by a system of fringes described
by the square of the zero-order Bessel function of the first kind. These
images can be readily interpreted to obtain quantitative information on resonance frequencies and corresponding mode shapes of vibrating objects.
The representative examples show very good agreement between the experimental
results obtained from time-average holograms and the analytical results based
on the beam theory. It should be remembered, that these results were determined
for a cantilever beam of constant cross-section for which an exact solution of
the differential equation can be obtained. However, for beams of varying cross-
section, as well as for objects of complex geometry or with complicated boundary conditions, exact analytical solutions may be unobtainable. The experimental procedures of time-average holography, on the other hand, are independent of object geometry or its boundary conditions. As such, once a hologram
of a vibrating object has been recorded, it can always be interpreted to obtain
quantitative results following the procedure described herein. Furthermore,
the experimental results obtained from time-average holograms can be used to
define boundary conditions needed to "run" the analytical methods, e.g.,
finite element computer codes. This combination of time-average holography
(or other optical methods) with the finite element method will lead to the
development, of new design systems. Such hybrid systems, possessing both

experimental and analytical capabilities, will be the next generation of tools that engineers will use to create new designs for the future.

7. REFERENCES

1. Powell, R. L., and Stetson, K. A., "Interferometric vibration analysis by wave-front reconstruction," J. Opt. Soc. Am., 55:1593 (1965).

2. Leith, E. N., and Upatnieks, J., "Wavefront reconstruction with continuous-tone objects," J. Opt. Soc. Am., 53:1377 (1963).

3. Pryputniewicz, R. J., Laser Holography, Worcester Polytechnic Institute, Department of Mechanical Engineering, Worcester, MA (1979).

4. Stetson, K. A., "A rigorous treatment of the fringes of hologram interferometry," Optik, 29:386 (1969).

5. Pryputniewicz, R. J., and Stetson, K. A., "Fundamentals and applications of laser speckle and hologram interferometry," Worcester Polytechnic Institute, Department of Mechanical Engineering, Worcester, MA (1980).

6. Stetson, K. A., "Fringe interpretation for hologram interferometry of rigid-body motions and homogeneous deformations," J. Opt. Soc. Am., 64:1 (1974).

7. Stetson, K. A., "Homogeneous deformations: determination by fringe vectors in hologram interferometry," Appl. Opt., 14:2256 (1975).

8. Pryputniewicz, R. J., and Stetson, K. A., "Holographic strain analysis: extension of fringe vector method to include perspective," Appl. Opt., 15:725 (1976).

9. Stetson, K. A., "Effects of beam modulation on fringe loci and localization in time-average hologram interferometry," J. Opt. Soc. Am., 60:1378 (1970).

Table 1. Zeros of J_0 and values of corresponding L_{zrr}. (The values of L_{zrr} were determined for the case of retroreflective illumination and observation and for $\lambda = 0.6328$ μm.)

n	$|\Omega_t|$	L_{zrr} (μm)
1	2.40483	0.121
2	5.52008	0.278
3	8.65373	0.436
4	11.79153	0.594
5	14.93092	0.752
6	18.07106	0.910
7	21.21264	1.068
8	24.35247	1.226
9	27.49348	1.385
10	30.63461	1.543
11	33.77582	1.701
12	36.91790	1.859
13	40.05843	2.017
14	43.19979	2.175
15	46.34119	2.334
16	49.48261	2.492
17	52.62405	2.650
18	55.76551	2.808
19	58.90698	2.966
20	62.04847	3.125
21	65.18996	3.283
22	68.33147	3.441
23	71.47298	3.599
24	74.61450	3.757
25	77.75603	3.916
26	80.89756	4.074
27	84.03909	4.232
28	87.18063	4.390
29	90.32217	4.548
30	93.46372	4.707

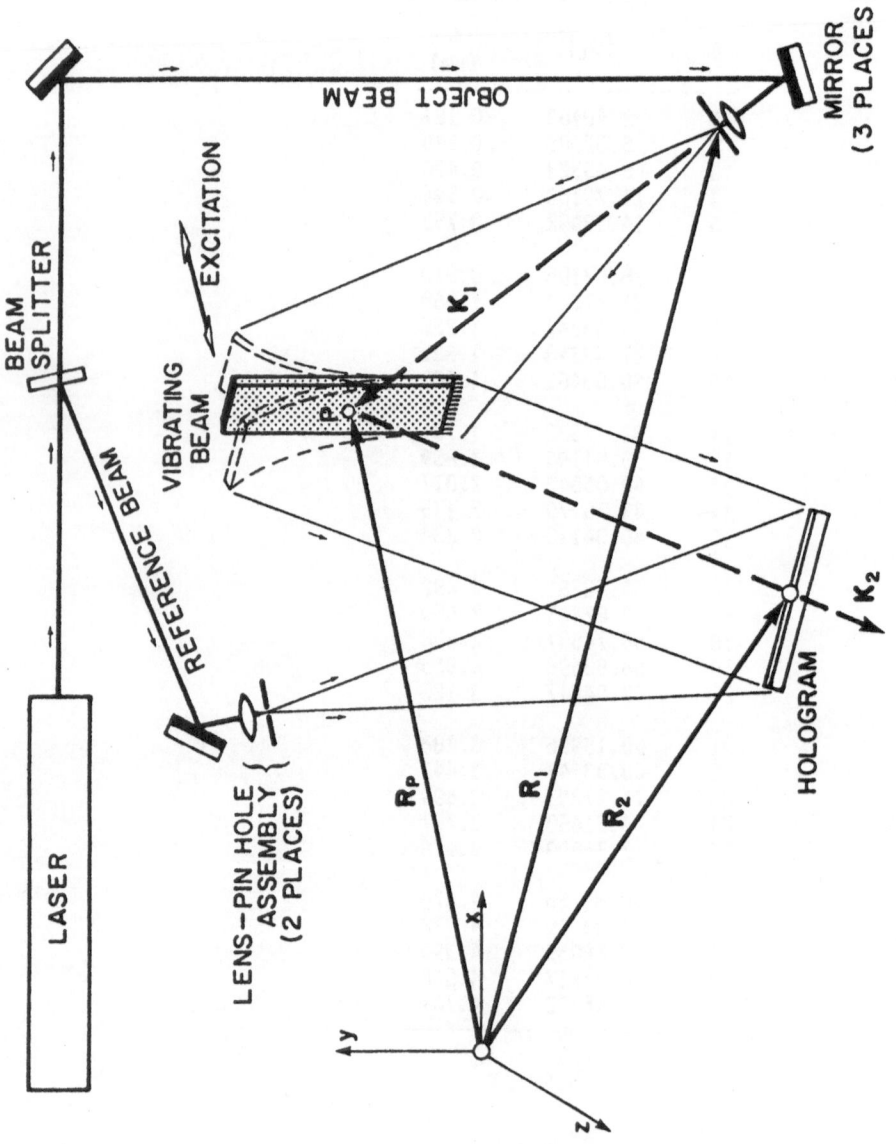

Fig. 1. Schematic representation of a setup for recording and reconstruction of continuous time-average holograms. $\underline{K}_1$ is the vector giving direction of illumination (during recording of a hologram), from point source defined by position vector $\underline{R}_1$ to a point P on the object specified by $\underline{R}_P$. The observation vector $\underline{K}_2$ is propagating from P to a point on hologram at $\underline{R}_2$, through which the reconstructed image is viewed.

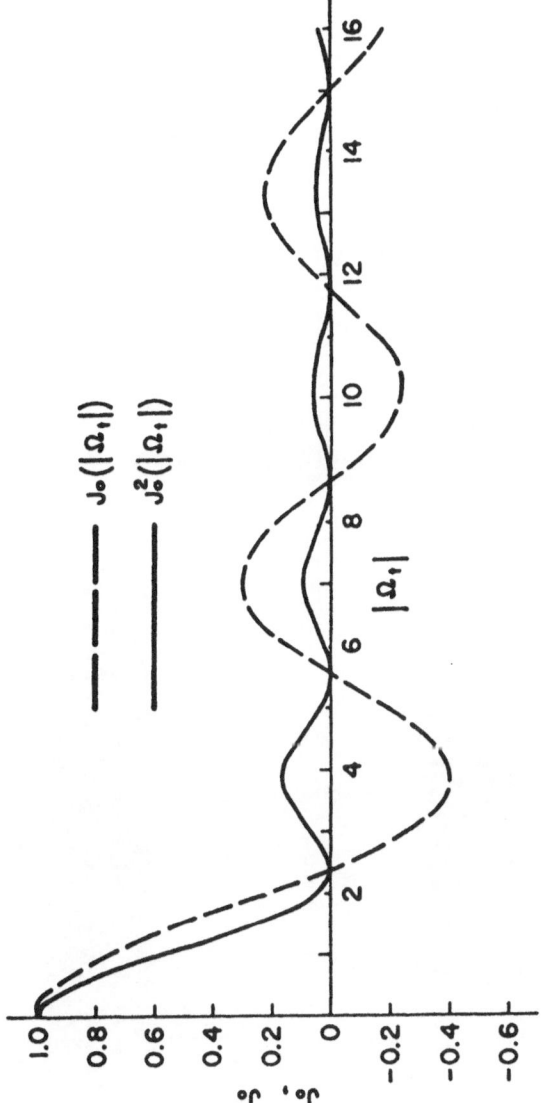

Fig. 2. The zero order Bessel function of the first kind and its square, defining location of centers of dark fringes seen during reconstruction of continuous time-average holograms of vibrating objects.

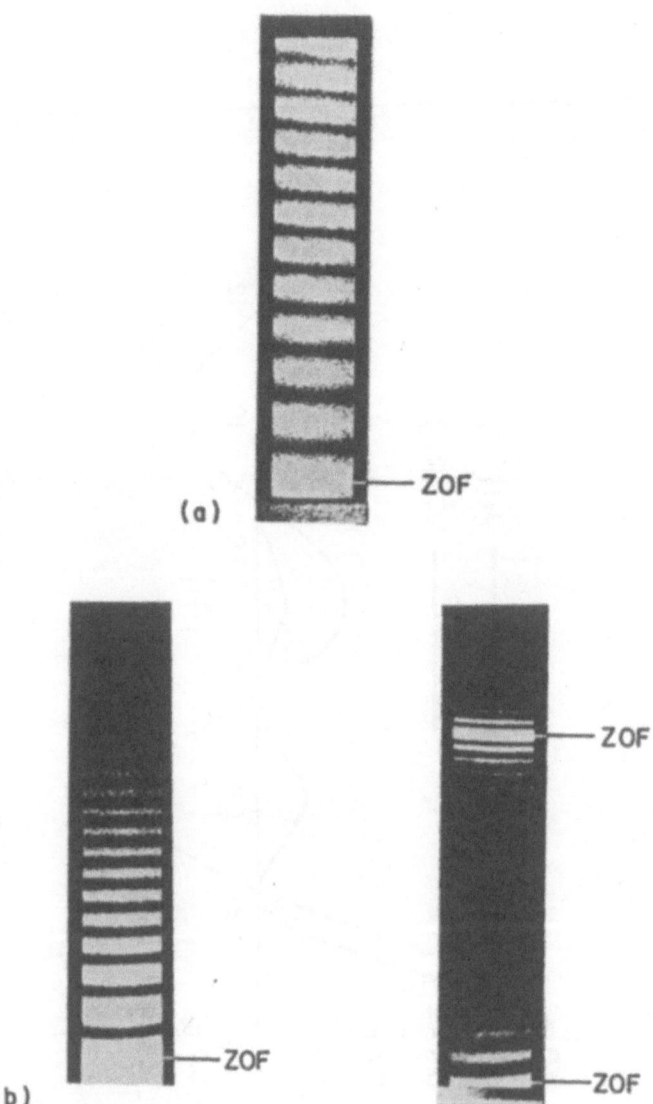

Fig. 3. Fringes obtained during reconstruction of:
(a) conventional double-exposure hologram, showing
cosinusoidal intensity variation; note, the zero-
order-fringe (ZOF) is as bright as other fringes.
(b) continuous time-average holograms, showing J_o^2
intensity variation; note, that the ZOF's are
much brighter than the higher order J_o fringes.

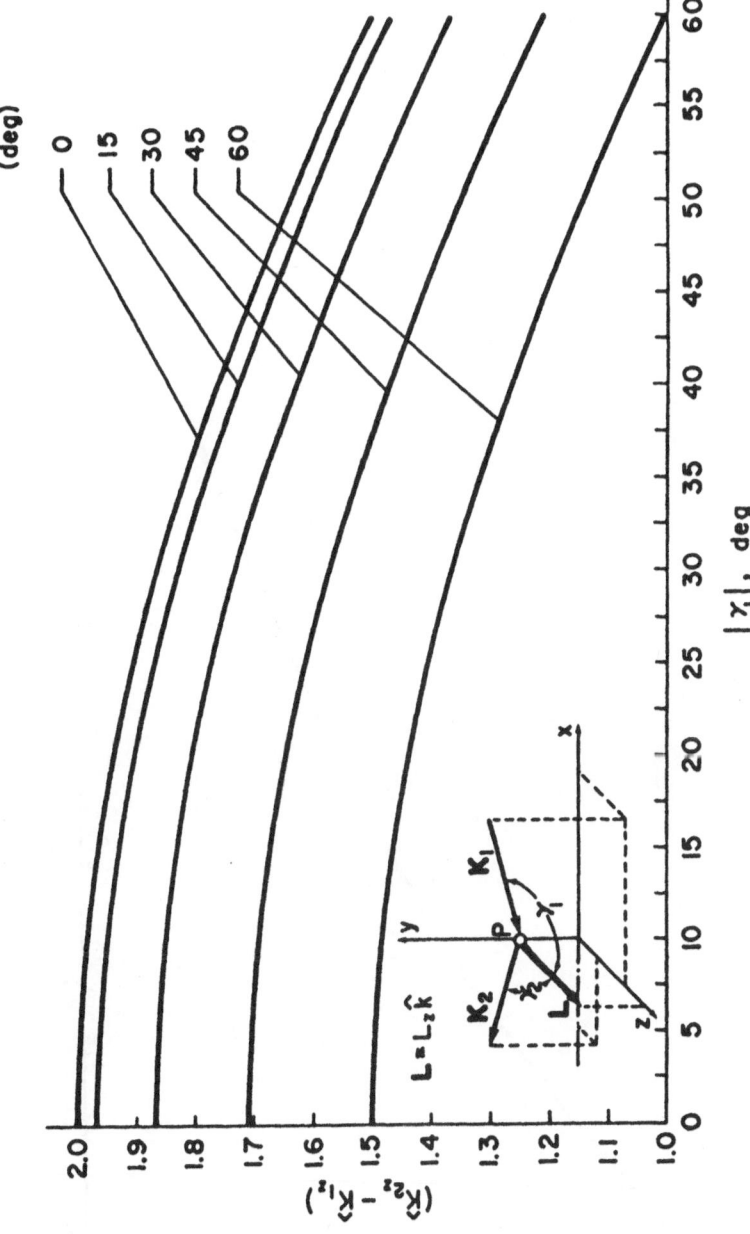

Fig. 4. Values of $(\hat{K}_{2_Z} - \hat{K}_{1_Z})$ as a function of γ_1 and γ_2.

312

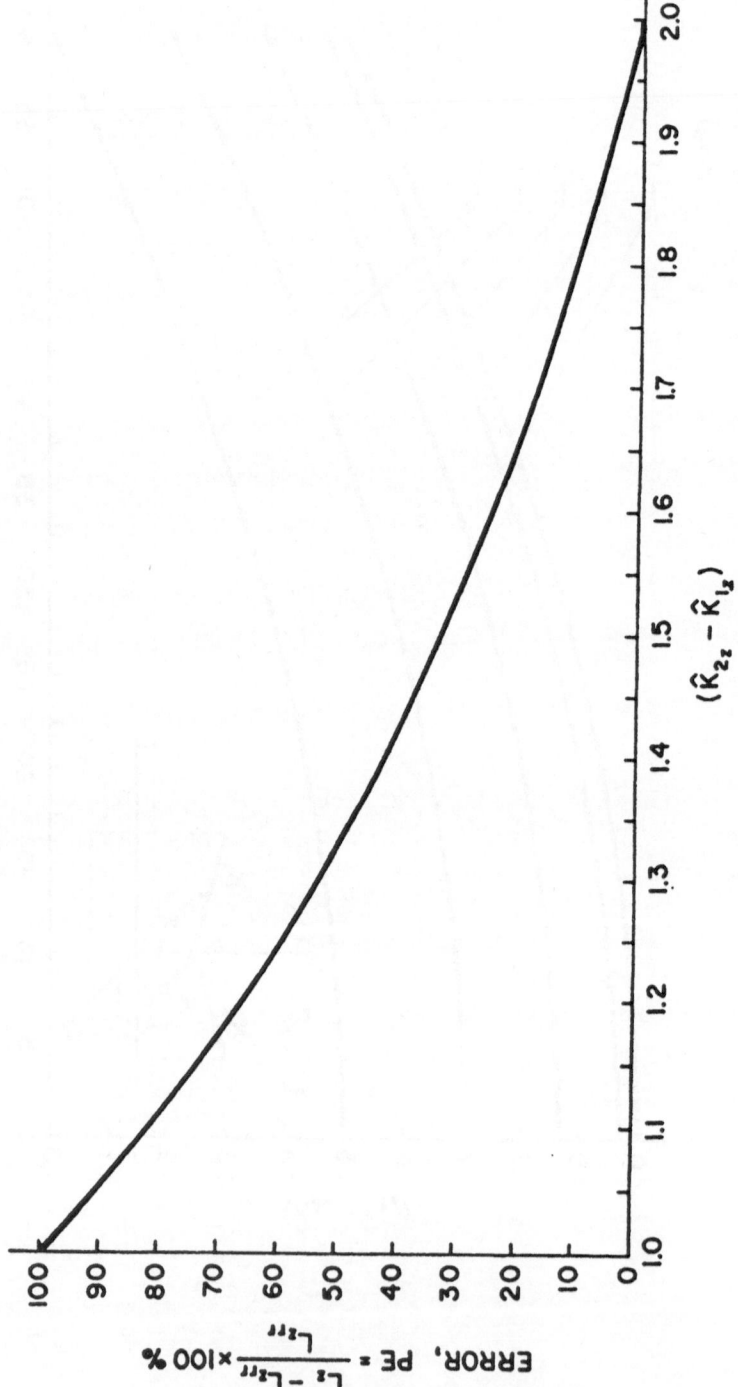

Fig. 5. Percentage error in L_z, with respect to $L_{z_{rr}}$, as a function of $(\hat{K}_{2_z} - \hat{K}_{1_z})$.

313

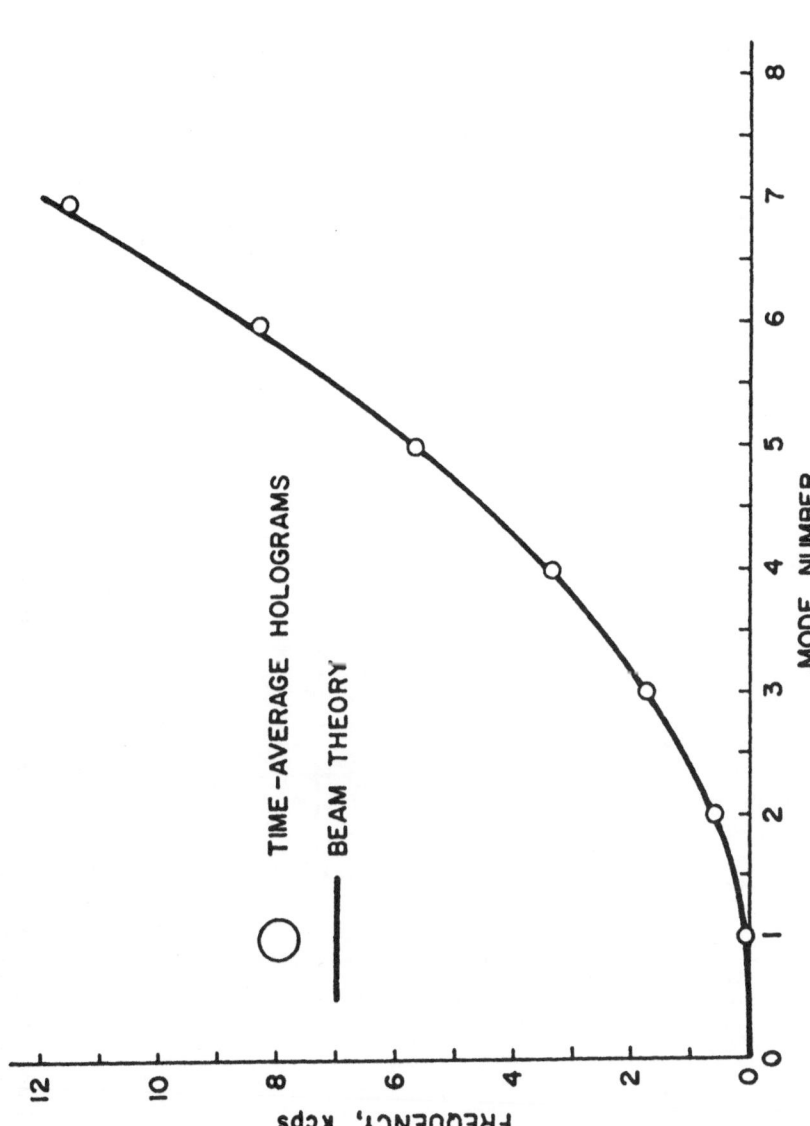

Fig. 6. Resonance frequency of a vibrating cantilever beam as a function of mode number. Comparison between the experimental results obtained from time-average holograms with the beam theory.

314

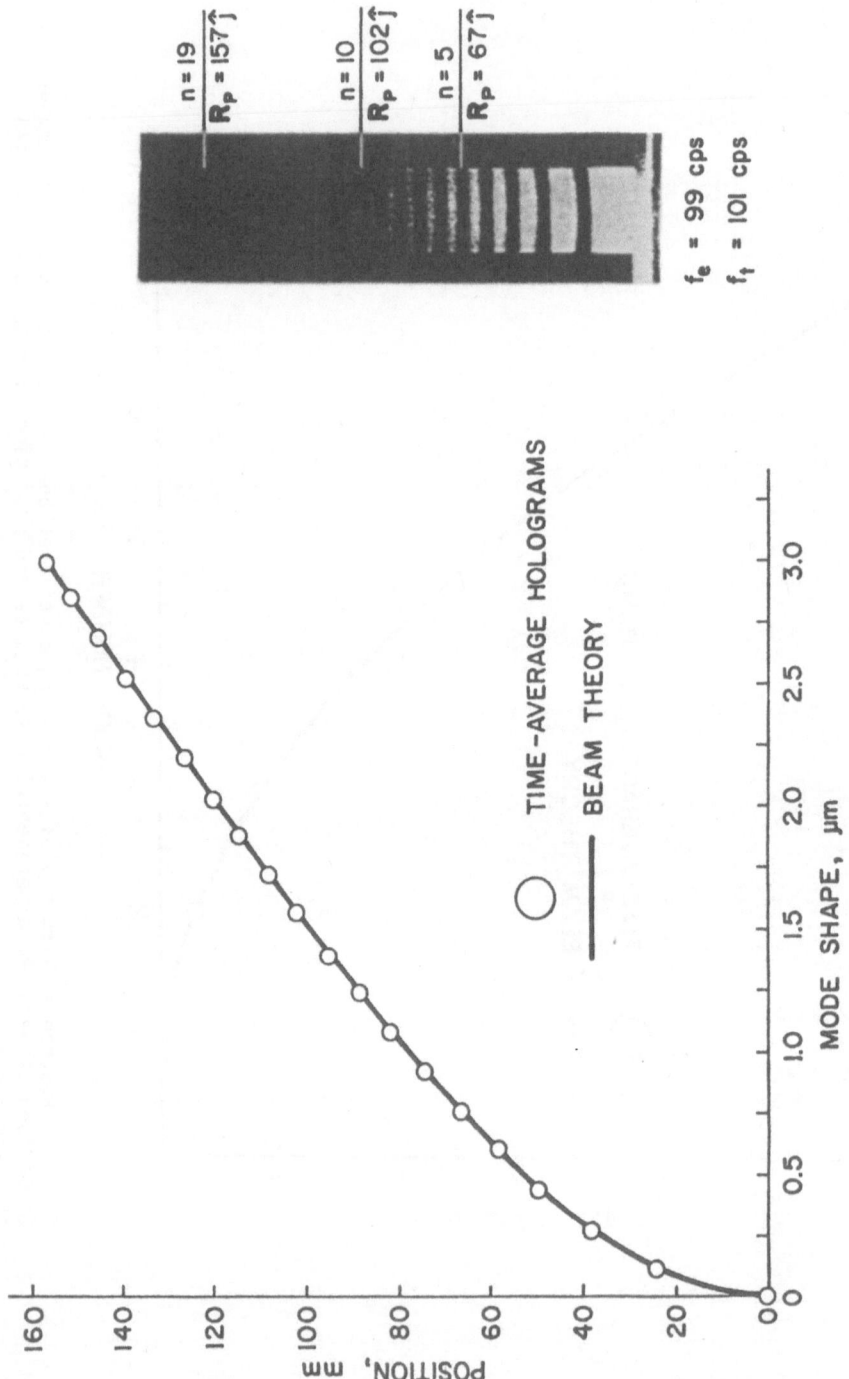

Fig. 7. Shape of the first flexure mode of a vibrating cantilever beam. Comparison between the experimental results obtained from time-average hologram with the beam theory.

315

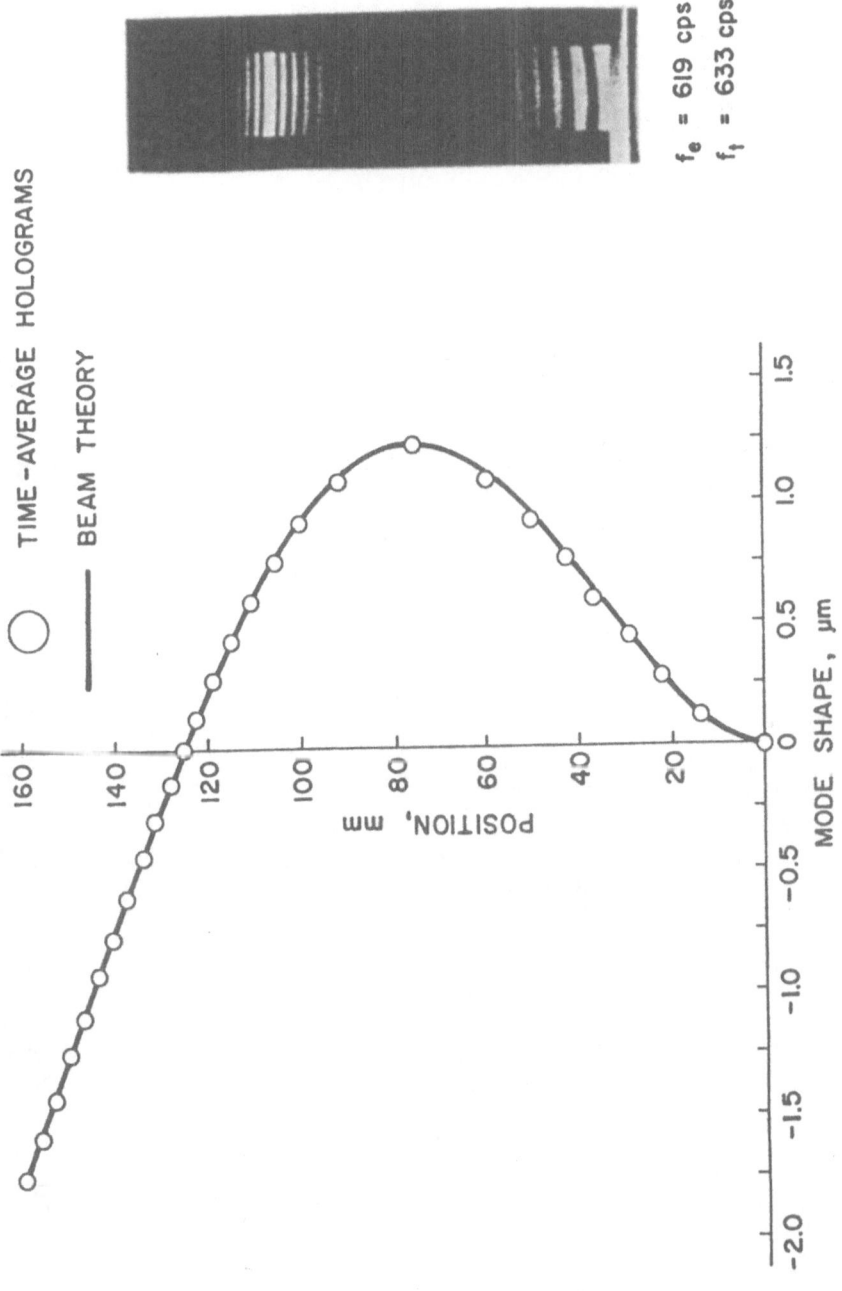

TIME-AVERAGE HOLOGRAMS

BEAM THEORY

f_e = 619 cps

f_t = 633 cps

Fig. 8. Shape of the second flexure mode of a vibrating cantilever beam. Comparison between the experimental results obtained from time-average hologram with the beam theory.

316

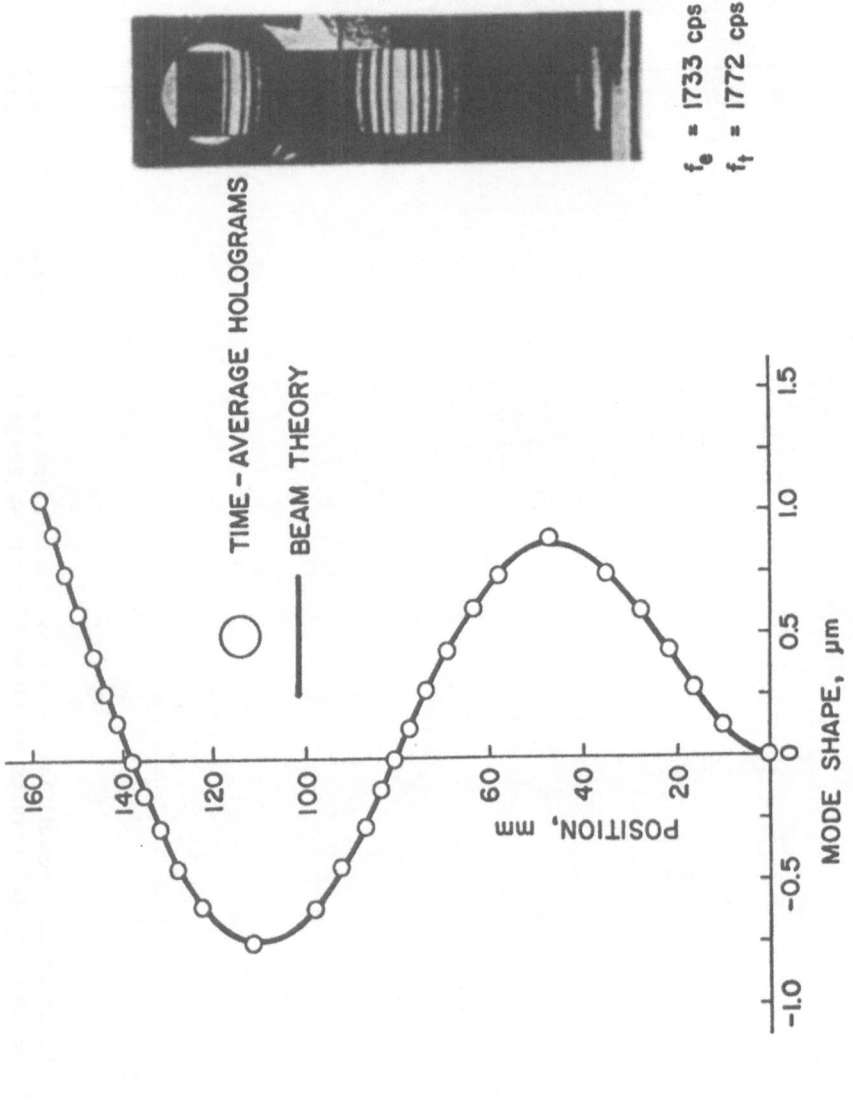

Fig. 9. Shape of the third flexure mode of a vibrating cantilever beam. Comparison
between the experimental results obtained from time-average hologram with the beam theory.

MATRIX METHODS IN HOLOGRAM INTERFEROMETRY AND SPECKLE METROLOGY

Karl A. Stetson

United Technologies Research Center
East Hartford, Connecticut, U.S.A. 06108

1 INTRODUCTION

The technologies of hologram interferometry and speckle metrology are useful tools for structural analysis in the field of mechanical engineering. In many situations, these methods provide simple and direct analyses of the deformations of mechanical systems with only a minimum of mathematical computation. With more complex mechanical systems, however, the mathematics required to relate forces and deflections increases in complexity. Systems of inter-related components give rise to systems of coupled equations which lead, logically, to matrix formulations. Furthermore, fracture mechanics, which is one of today's most important and practical technologies, deals largely with tensor quantities such as stress and strain. When strain measurement and the deformations of three-dimensional structures are considered, it is not surprising that the analysis of hologram interferometry and speckle metrology should involve the same matrix formalism as the mechanical systems themselves.

The purpose of these lectures is to clarify those matrix methods that are practical and helpful in the analysis of hologram interferometry and speckle metrology. Despite the complexities that have appeared in the literature, there are basically only two or three types of matrix operations that occur in these analyses. As mentioned already, strain is intrinsically a tensor and described by a matrix. Rotation can be described alternatively as a vector or as a matrix, and it is often combined with strain in its matrix form to characterize homogeneous deformations. The next most commonly encountered matrix is that which describes the shadow cast by a vector on a surface, i.e. the projection matrix. Finally, the technique of least-square-error analysis generates a family of matrices, all of which possess a common form. Let us begin, therefore, with a review of the basic definitions and operations behind these matrix methods and proceed to examples of how to employ them in optical metrology.

2 BASIC VECTOR AND MATRIX OPERATIONS

2.1 Vectors and Scalar Products

Vectors are quantities such as displacement or velocity that possess both magnitude and direction. They will be designated by underlined characters, for example, $\underline{A}$. A unit vector is defined as having a magnitude of 1.0 and will be designated by a hat, for example, $\hat{a}$. The unit vectors, $\hat{i}$, $\hat{j}$ and $\hat{k}$, pointing along the x, y, and z axes of a coordinate system may be used to express a vector in terms of its components in those directions. For example,

$$(1) \qquad \underline{A} = \hat{i}A_x + \hat{j}A_y + \hat{k}A_z,$$

where A_x, A_y, and A_z are the x, y, and z components of $\underline{A}$.

Three commonly encountered vectors are: $\underline{L}$ which is the displacement of an object point, $\underline{K}_1$ which is the propagation vector of the illumination of an object, and $\underline{K}_2$ which is the propagation vector of the light traveling from the object to the observer. The vector difference of the second and third appears so often that it is given a separate identity and called the sensitivity vector $\underline{K}$.

$$(2) \qquad \underline{K} = \underline{K}_2 - \underline{K}_1.$$

The scalar product of the sensitivity vector $\underline{K}$ and displacement $\underline{L}$ yields the fringe-locus function, Ω, constant values of which define the contours of fringes in hologram interferometry by defining zeros of the characteristic fringe function, $M(\Omega)$. Ω also equals the decrease in optical path length from the illumination source to the object point to the observer that is caused by the object displacement.

(3) $\Omega = \underline{K} \cdot \underline{L} = K_x L_x + K_y L_y + K_z L_z$

$M(\Omega)$ is the function which appears to multiply the field that would otherwise be reconstructed by a hologram of the stationary object. For double-exposure holography, with $\underline{L}$ equal to the total displacement between exposures,

(4) $M(\Omega) = \cos(\Omega/2).$

For time-average holography of a vibrating object, with $\underline{L}$ equal to the peak-to-peak displacement,

(5) $M(\Omega) = J_0(\Omega/2),$

where J_0 is the zero-order Bessel function of the first kind. In problems dealing with vibration, it is often convenient to redefine $\underline{L}$ as the vibration amplitude and replace $\Omega/2$ by Ω in Eq.(5).

2.2 Vector Products and Matrix Operators

An example of a vector product of two vectors occurs in the computation of object displacements due to a small vectorial rotation $\underline{\Theta}$. (The direction of $\underline{\Theta}$ lies along the rotation axis, in the righthand convention, and its magnitude equals the rotation angle in radians.) Let $\underline{R}$ be a position vector from an origin, located on the rotation axis, to some point on the object. To first-order approximation

(6) $\underline{L} = \underline{\Theta} \times \underline{R},$

where $\underline{\Theta} = \hat{\imath}\Theta_x + \hat{\jmath}\Theta_y + \hat{k}\Theta_z.$

The righthand side of Eq.(6) may be thought of as an operator, $\underline{\Theta}\times$, that transforms the vector $\underline{R}$ into the vector $\underline{L}$. Operators that transform vectors in this way may be represented by matrix transforms, and thus the rotation may be represented by the matrix

(7) $[\underline{\Theta}] = \begin{bmatrix} 0 & -\Theta_z & \Theta_y \\ \Theta_z & 0 & -\Theta_x \\ -\Theta_y & \Theta_x & 0 \end{bmatrix},$

and Eq.(6) may be rewritten as

320

(8) $\underline{L} = [\underline{\Theta}] \, \underline{R}$.

When a martix operates upon a vector, the vector may be regarded as either a row or a column matrix depending upon whether the vector premultiplies or postmultiplies the matrix respectively. Unless the matrix is symmetrical, these two operations result in different transformations, which, may be in some way related. Reversing the order of $\underline{L}$ and $[\underline{\Theta}]$ in Eq.(8) results in a negative rotation and therefore negative displacements. This is equivalent, of course, to reversing the order of the vector product in Eq.(6). This leads to the conclusion that either $\underline{\Theta}x$ or $x\underline{R}$ in Eq.(6) can be represented by a matrix transform. Thus object displacements can be thought of as being generated by a rotational transformation of space or as a spatial transformation of rotation. This duality can prove useful in numerous situations.

The rotation matrix represents a singular transformation, and this is one of its important properties. This means that once a vector has been transformed by this operation, it cannot be restored to its original form by an inverse transformation. Physically this follows from the fact that the operation of a vector product destroys the parallel components of the vectors involved. Mathematically it follows from the fact that the reciprocal of the determinant of a matrix is a scale factor for its inverse, and the determinant of the rotation matrix is zero. This will be discussed later in more detail.

2.3 Projection Matrices and Matric Products

There are a number of situations in hologram interferometry and speckle metrology that involve the apparent displacement of points on the object surface. Apparent displacement is defined as the projection of the object displacement vector onto a plane normal to the observation direction. Speckle photogrammetry, and photogrammetry in general, provide a simple illustration. Displacement of an object point toward the camera lens does not generate image displacement; therefore, in a photograph, it is only possible to measure the projection of the object displacements on the image surface. In photogrammetric analysis, it is necessary to describe these projections mathematically, especially if two or more photographs are used to determine the true object displacements. In holography, these projections play an important role in fringe localization and in the apparent orientation and spacing of photographed fringes.

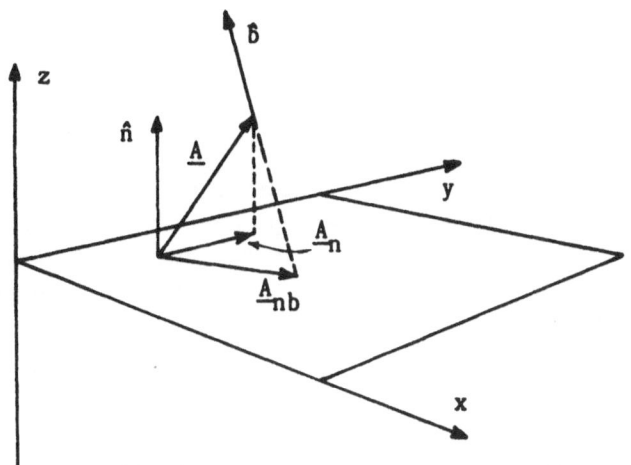

Fig. 1. The projection of a vector onto a surface.

The operation of projecting a vector onto a plane consists of subtracting from it another vector that lies along the line of projection and has sufficient magnitude that the resultant lies in the plane of the surface (see Fig. 1). This is represented by Eq.(9).

$$(9) \qquad \underline{A}_{nb} = \underline{A} - \hat{B}X,$$

where $\underline{A}_{nb}$ is the projection of vector $\underline{A}$ on the surface normal to $\hat{n}$ from a direction parallel to unit vector $\hat{B}$, and X is an unknown magnitude. The value of X may be found by taking the scalar product of both sides of Eq.(9) with the surface normal vector, $\hat{n}$. The result equals zero because $\underline{A}_{nb}$ is perpendicular to $\hat{n}$ by definition.

$$(10) \qquad \hat{n} \cdot \underline{A}_{nb} = 0 = \hat{n} \cdot \underline{A} - (\hat{n} \cdot \hat{B})X.$$

This may be solved for X to give

$$(11) \qquad X = (\hat{n} \cdot \underline{A})/(\hat{n} \cdot \hat{B}),$$

which may be substituted into Eq.(9) to give

$$(12) \qquad \underline{A}_{nb} = \underline{A} - \hat{B}(\hat{n} \cdot \underline{A})/(\hat{n} \cdot \hat{B}).$$

Note that for the normal projection onto the surface, $\hat{B} = \hat{n}$, so that the last term is simply the out-of-plane component of $\underline{A}$ in which case Eq.(12) becomes

(13) $\quad \underline{A}_{nb} = \underline{A} - \hat{n}(\hat{n} \cdot \underline{A})$.

The operation of projecting a matrix onto a surface, like the vector product operation, can be represented advantageously by a matrix transformation. The matrix can be derived two ways. First, consider the triple vector product, $\hat{n}x(\hat{b}x\underline{A})$. In expanded form this is

(14) $\quad \hat{n}x(\hat{b}x\underline{A}) = -\underline{A}(\hat{b} \cdot \hat{n}) + \hat{b}(\hat{n} \cdot \underline{A})$.

If Eq.(14) is divided by the scalar quantity $-(\hat{b} \cdot \hat{n})$, the result is identical to Eq.(12). Thus

(15) $\quad \underline{A}_{nb} = -\hat{n}x(\hat{b}x\underline{A})/(\hat{b} \cdot \hat{n})$.

The two successive vector products in the righthand side of Eq.(15), $-\hat{n}x(\hat{b}x$, may be represented as two successive matrix operations in the form described previously for object rotations. This gives

(16) $\quad -\hat{n}x(\hat{b}x = \begin{bmatrix} 0 & -n_z & n_y \\ n_z & 0 & -n_x \\ -n_y & n_x & 0 \end{bmatrix} \begin{bmatrix} 0 & -b_z & b_y \\ b_z & 0 & -b_x \\ -b_y & b_x & 0 \end{bmatrix}$,

or, in compact notation

(17) $\quad -\hat{n}x(\hat{b}x = -[n][b]$.

The elements of the matrix represented by Eqs. (16) or (17) may be evaluated and simplified by use of the fact that the sum of the squares of the components of $\hat{n}$ or $\hat{b}$ is unity by definition. The result is

(18) $\quad -[n][b] = \begin{bmatrix} 1-b_x n_x & -b_x n_y & -b_x n_z \\ -b_y n_x & 1-b_y n_y & -b_y n_z \\ -b_z n_x & -b_z n_y & 1-b_z n_z \end{bmatrix}$.

The matrix in brackets equals the identity matrix (a matrix with major diagonal elements = 1 and all others zero) minus a matrix composed of all nine possible products of $\hat{b}$ and $\hat{n}$. This last matrix defines a third product between two vectors in addition to the more familiar scalar and vector products. It may be called an outer product, a dyadic product, or a matric product, and it may be represented by the symbol $\blacksquare$. Thus Eq.(17) may be written as

(19) $-\hat{n}x(\hat{b}x = -[n][b] = ([I] - \hat{b}\blacksquare\hat{n})$.

This leads to the expression for the projection matrix

(20) $[P_{bn}] = -[n][b]/(\hat{b}\cdot\hat{n}) = [I] - \hat{b}\blacksquare\hat{n}/(\hat{b}\cdot\hat{n})$,

where $[P_{bn}]$ may be defined as the oblique projection matrix. When $\hat{b} = \hat{n}$, this operator becomes the normal projection, $[P_n]$, where a single subscript indicates the vector common to both the surface and the axis of projection.

(21) $[P_{nn}] = -[n][n] = [I] - \hat{n}\blacksquare\hat{n} = [P_n]$.

This matrix may also be identified essentially by inspection from Eq.(12). Consider the product $\hat{b}(\hat{n}\cdot\underline{A})$. It is possible, in general, to rewrite vector products as matrix products. This can be made easier to visualize by writing the vector components as rows or columns in square matrices, with the rest of the elements set to zero. For the product $\hat{b}(\hat{n}\cdot\underline{A})$ this becomes

(22) $\hat{b}(\hat{n}\cdot\underline{A}) = \begin{bmatrix} b_x & 0 & 0 \\ b_y & 0 & 0 \\ b_z & 0 & 0 \end{bmatrix} \begin{bmatrix} n_x & n_y & n_z \\ 0 & 0 & 0 \\ 0 & 0 & 0 \end{bmatrix} \begin{bmatrix} A_x & 0 & 0 \\ A_y & 0 & 0 \\ A_z & 0 & 0 \end{bmatrix}$.

It is clear by inspection of Eq.(22) that the product of the second and third matrices yields a matrix with all elements equal to zero except the upper lefthand element, and its value equals the scalar product of $\hat{n}$ and $\underline{A}$. Matrix products are associative, however, and the product of the first two matrices may be evaluated before multiplication with the third. The product of the first two matrices generates a 3x3 matrix of all nine possible products of the elements of $\hat{b}$ and $\hat{n}$, i.e. the matrix represented by $\hat{b}\blacksquare\hat{n}$. It is worth pointing out that this operator is also written as $\hat{b}\hat{n}$ in some texts.

Projection matrices have interesting properties. First of all they are singular, which is not surprising because it is possible for more than one vector to cast the same shadow on a surface, and, therefore, the inverse transformation cannot be defined.

Next consider symmetry. The normal projection matrix, e.g. $[P_n]$, is symmetric, and, therefore,

$$(23) \qquad \underline{A}[P_n] = [P_n]\underline{A},$$

and

$$(24) \qquad [P_n] = [P_n]^T,$$

where $[P_n]^T$ is the transpose of $[P_n]$.

The oblique projection matrix, however, is not symmetric, and its transpose may be obtained by exchanging the order of the unit vectors in the matric product. Therefore,

$$(25) \qquad [P_{bn}]^T = [P_{nb}].$$

Exchanging these two vectors exchanges the direction along which the projection is made and the surface onto which the projection falls. By examination of the results of $[P_{bn}]\underline{A}$ and $\underline{A}[P_{bn}]$, it can be seen that again the roles of $\hat{b}$ and $\hat{n}$ have been reversed. This means that when a vector postmultiplies $[P_{bn}]$, it becomes projected along a direction parallel to $\hat{b}$ onto a plane perpendicular to $\hat{n}$, and when it premultiplies $[P_{bn}]$ it is projected along $\hat{n}$ onto a plane perpendicular to $\hat{b}$. When $[P_{bn}]$ is transposed, everything is reversed. This duality makes it somewhat complicated to keep track of how the projection matrices operate. The order of the two subscripts, which is taken from the order of the two vectors in the matric product, is important. The subscript furthest from the vector denotes the axis of the projection while the one nearest the vector denotes the normal to the plane on which it falls. The phrase, "along the furthest onto the nearest" may help the memory. Note that in Eq.(20), where the the projection matrix is expressed also in terms of two vector-product operators, the order of the b and n are reversed.

Finally, repeated projections of a vector along the same axis may be simplified by eliminating the surfaces associated with the intermediate projections. This leads to the following useful identities:

(26) $[P_{nb}][P_{nc}] = [P_{nb}]$, and $[P_{bn}][P_{cn}] = [P_{cn}]$.

Note that the repeated index is either to the left on both transforms or to the right on both transforms, otherwise the axes of the two projections are not aligned. These relationships can be derived mathematically by writing the transforms in the form of Eq(20) and expanding terms.

(27) $[P_{bn}][P_{cn}] = ([I] - \mathbf{B}\hat{n}/(\mathbf{B}\cdot\hat{n}))([I] - \mathbf{C}\hat{n}/(\mathbf{C}\cdot\hat{n}))$

$= [I] - \mathbf{B}\hat{n}/(\mathbf{B}\cdot\hat{n}) - \mathbf{C}\hat{n}/(\mathbf{C}\cdot\hat{n}) + (\mathbf{B}\hat{n})(\mathbf{C}\hat{n})/(\mathbf{B}\cdot\hat{n})(\mathbf{C}\cdot\hat{n})$

If the product $(\mathbf{B}\hat{n})(\mathbf{C}\hat{n})$ is expanded in terms of elementary matrices, it can be shown that

(28) $(\mathbf{B}\hat{n})(\mathbf{C}\hat{n}) = (\mathbf{B}\hat{n})(\mathbf{C}\cdot\hat{n})$.

Thus the fourth term cancels the second in Eq.(27) leaving only the projection matrix $[P_{cn}]$. When sequential normal and oblique projections share a common vector, the following identities result:

(29) $[P_n][P_{nb}] = [P_{bn}][P_n] = [P_n]$,

$[P_n][P_{bn}] = [P_{bn}]$, and

$[P_{nb}][P_n] = [P_{nb}]$.

Also, because of its symmetry,

(30) $[P_n][P_n]^T = [P_n]^T[P_n] = [P_n] = [P_n]^T$.

2.4 Least-Square-Error Analysis

In hologram interferometry and speckle metrology it is often necessary to solve sets of equations for a set of unknown parameters. In a typical example, a hologram may be recorded of an object whose points may be displaced various amounts in various directions. The object may be observed from various directions through the hologram, each of which defines a sensitivity vector as in Eq.(2). Measurement of the fringe order at a particular point on the object, as observed from a particular direction, allows determination of that component of the object displacement parallel to corresponding sensitivity vector. This can be done by means of Eq.(3); however, it is first necessary to define fringe order in terms of the fringe-locus function. The irradiance of the recon-structed image will be proportional to the square of the charac-teristic fringe function, i.e. $M^2(\Omega)$. The bright and dark fringes are defined by the zeros of the derivative of this function with respect to Ω. Let a prime denote the derivative of a function with respect to its argument. The derivative may be written as

$$(31) \qquad 2M(\Omega)M'(\Omega) = 0.$$

There are two sets of roots to Eq.(31) associated with dark and bright fringes respectively.

dark fringes: $\quad M(\Omega_n)=0$, and

bright fringes: $\quad M'(\Omega_n)=0$.

For cosinusoidal fringes, defined by Eq(4), it is convenient to write

$$(32) \qquad \Omega_n = 2\pi n,$$

from which it is possible to make the following identifications of fringe order.

For bright fringes,
$\qquad n = -2, -1, 0, +1, +2$, etc., i.e. integers.

For dark fringes,
$\qquad n = -3/2, -1/2, +1/2, +3/2$. etc., i.e. half integers.

For Bessel function fringes due to vibration, the fringe orders are associated with the zeros, maxima, and minima of the zero-order Bessel function. Numerical fringe order may usually be determined visually and then related to the corresponding value of the fringe-locus function by Eq.(32). Fractional values of fringe order may be interpolated to within values of about 0.2.

Data from different directions of observations can be combined to yield the vectorial object displacement by solution of a set of simultaneous equations, each having the form of Eq.(3). Because the components of $\underline{L}$ appear as common factors in these equations, they can be combined in matrix form as

(33) $[K]\underline{L} = \{\Omega\}.$

Each row of the matrix $[K]$ contains the components of a sensitivity vector and each element of the vector $\{\Omega\}$ is the fringe-locus function corresponding to the observed fringe order. If three observation directions are used which define three noncoplanar sensitivity vectors, it is possible to compute the inverse of $[K]$, i.e. the matrix $[K]^{-1}$ defined such that $[K][K]^{-1} = [I]$, the identity matrix. When Eq.(33) is multiplied by the inverse K-matrix, the result is the solution for the vector displacement $\underline{L}$,

(34) $\underline{L} = [K]^{-1} \{\Omega\}.$

Although Eq.(34) presents a solution to the problem of determining vectorial displacements, the solution is strongly dependent upon the accuracy of the data, i.e., the observed fringe orders and therefore $\{\Omega\}$. This dependency can be reduced dramatically by use of data from more than three directions of observation to generate an overdetermined set of equations, i.e. a set where there are more equations than unknowns. These may be solved for that vector $\underline{L}$ which minimizes the sum of the squares of the errors generated when it is substituted back into all of the equations. This solution has a form that is very straightforward, easy to remember, and which can be applied to many situations. The following is probably the most compact derivation of it.

When Eq.(33) has more than three rows (i.e when there are more than three observations) it will not generally be satisfied by any single value of $\underline{L}$. We seek, however, that value of $\underline{L}$ which will generate the minimum of the total squared errors when substituted into Eq.(33). This substitution generates error vector, $\{E\}$, which may be written as

(35) $\{E\} = [K]\underline{L} - \{\Omega\}.$

$\{E\}$ will have as many elements as there are observations. Denote the sum of the squares of the elements of the error vector as ΣE^2. This equals the scalar product of {E} with itself.

(36) $\Sigma E^2 = \{E\} \cdot \{E\} = ([K]\underline{L} - \{\Omega\}) \cdot ([K]\underline{L} - \{\Omega\})$

ΣE^2 will be minimum when its partial derivatives with respect to the components of $\underline{L}$ are simultaneously zero.

It will be helpful at this point to digress briefly on vectorial derivatives. Most students of vector calculus are familiar with the gradient operator $\nabla = \hat{\imath}\partial/\partial x + \hat{\jmath}\partial/\partial y + \hat{k}\partial/\partial z$. Common operations with it include: 1) the gradient of a scalar ∇a which yields a vector equal to, and pointing in the direction of, the maximum rate of change of the scalar funtion, 2) divergence of a vector $\nabla \cdot \underline{A}$ which yields a scalar, and 3) curl of a vector $\nabla \times \underline{A}$ which yields a vector. In addition, it may operate in the manner of a matric product, i.e. $\nabla \blacksquare \underline{A}$, in which case it yields a tensor. This operation will be referred to here as a matric derivative.

It is also possible to construct other gradient operators that take derivatives with respect to other variables. In the present situation, it is useful to define $\nabla_L = \hat{\imath}\partial/\partial L_x + \hat{\jmath}\partial/\partial L_y + \hat{k}\partial/\partial L_z$, which is a gradient operator with respect to the components of $\underline{L}$. With this operator it is possible to define the minimum of the square error expressed in Eq.(36) as

(37) $\nabla_L(\{E\} \cdot \{E\}) = 0$.

The resulting vector may be written as

(38) $2[\nabla_L \blacksquare \{E\}]\{E\} = 0$.

Substitution from Eq.(35) gives, for the matix of Eq.(38),

(39) $[\nabla_L \blacksquare \{E\}] = [\nabla_L \blacksquare ([K]\underline{L} - \{\Omega\})]$.

Neither [K] nor {Ω} are variables with respect to the components of $\underline{L}$, so the result of the matric derivative is

(40) $[\nabla_L \blacksquare ([K]\underline{L} - \{\Omega\})] = [\nabla_L \blacksquare \underline{L}][K]^T$.

It is relatively simple to show that the matric derivative of a vector with respect to its components yields the identity matrix, i.e. $[\nabla_L \bullet \underline{L}] = [I]$. Therefore,

$$(41) \qquad [\nabla_L \bullet ([K]\underline{L} - \{\Omega\})] = [K]^T,$$

and Eq.(38) becomes

$$(42) \qquad [K]^T([K]\underline{L} - \{\Omega\}) = 0.$$

This may be solved for $\underline{L}$ to give

$$(43) \qquad \underline{L} = [[K]^T[K]]^{-1}[K]^T\{\Omega\}.$$

Although this solution is derived here in terms of a three element vector that must satisfy more than three equations for minimum squared error, the form of the solution is extremely general and may be applied to many similar problems. The product of the matrix [K] by its transpose reduces it to a square matrix which is usually well conditioned and can be inverted with good accuracy. The product of the transposed $[K]^T$ matrix and the data vector $\{\Omega\}$ reduces the vector's length to match the rank of the reduced matrix. Thus, the computations are reduced in complexity.

3 APPLICATIONS

3.1 Coordinate Transformations

Before entering into the various applications of matrix methods to hologram interferometry and speckle metrology, it will be helpful to discuss the issue of coordinate systems. Vector and tensor relationships are generally independent of coordinate systems; however, when numerical computations are performed they must be done in a consistent set of coordinates or remarkable nonsense will result. Often, however, there are preferred coordinates for one part of a physical system that differ from those of another part, and in these cases it is necessary to make transformations between them. The vectors and tensors themselves remain the same, of course, and only the coordinate systems change. By analogy, a quantity of liquid is the same whether measured in pints or liters, but the number required to represent it changes depending upon the measuring system. Thus, when we speak of transforming vectors and

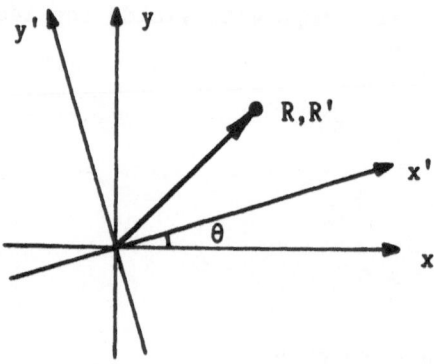

Fig. 2 An example of rotation of coordinate systems.

tensors from one coordinate system to another, we actually mean transforming their numerical values.

Translations of coordinate systems, without rotation, are essentially trivial. The values of vectors in the new system equal those in the old minus the vector translation. Matrices are invariant with coordinate translations. The most common transformation, therefore, is rotation of rectangular coordinates. Let us begin with a simple rotation about the z axis as shown in Fig.2.

Let $\underline{R}$ be the value of a vector in the x,y,z coordinate system and $\underline{R}'$ be its value in the x',y',z' system. If the second system has been rotated by an angle θ relative to the first, simple trigonometry shows that

(44) $\quad \underline{R} = \begin{bmatrix} \cos\theta & -\sin\theta & 0 \\ \sin\theta & \cos\theta & 0 \\ 0 & 0 & 1 \end{bmatrix} \underline{R}' = [T] \, \underline{R}'.$

The matrix [T] relating $\underline{R}$ and $\underline{R}'$ is a coordinate rotation matrix. A very important property of the coordinate rotation matrix is that its inverse equals its transpose, i.e. $[T]^{-1} = [T]^{T}$. As a consequence of this, if a vector postmultiplies [T], the coordinate system rotates clockwise by the angle θ, and if it premultiplies it rotates counterclockwise. It is easy to remember the form of [T] in Eq.(44) but often hard to remember where to put the minus sign. If it is nearest the vector, it rotates the coordinates clockwise.

Coordinate rotations transform the values of matrix operators as well as vectors and in a manner that is more complex. It can be derived from the vector transformation as follows. Assume that some object displacement $\underline{L}$ results from some homogeneous transformation $[f]$ of the spatial coordinates $\underline{R}$ of an object, i.e.

(45) $\underline{L} = [f]\underline{R}.$

Substitution for $\underline{R}$ via Eq.(44) gives

(46) $\underline{L} = [f][T]\underline{R}',$

where $\underline{R}'$, is the vector describing the object coordinates in the primed system. The object displacement, $\underline{L}'$ in the primed system is related to its corresponding values, $\underline{L}$, by

(47) $\underline{L}' = [T]^{-1}\underline{L} = [T]^{-1}[f][T]\underline{R}'.$

The matrix $[T]^{-1}[f][T]$ operates upon $\underline{R}'$ to yield $\underline{L}'$ in the same way that $[T]$ operated upon $\underline{R}$ to yield $\underline{L}$. It represents the values of the matrix $[f']$ in the primed coordinate system, and therefore,

(48) $[f'] = [T]^{-1}[f][T].$

As shown here, the coordinates rotate counterclockwise from the unprimed to the primed coordinates. Exchange the matrix and its inverse and the coordinates rotate clockwise. For the simple rotation of Eq.(44) we see again that if the minus signs are nearest the matrix, the coordinates rotate clockwise.

For coordinate systems that are arbitrarily oriented there are two ways to formulate the appropriate transformation matrix. First, the matrix can be written in terms of the unit vectors of the two coordinate systems, $\hat{\imath}$, $\hat{\jmath}$, $\hat{k}$ and $\hat{\imath}'$, $\hat{\jmath}'$, $\hat{k}'$ as

(49) $[T] = \begin{bmatrix} \hat{\imath}\cdot\hat{\imath}' & \hat{\imath}\cdot\hat{\jmath}' & \hat{\imath}\cdot\hat{k}' \\ \hat{\jmath}\cdot\hat{\imath}' & \hat{\jmath}\cdot\hat{\jmath}' & \hat{\jmath}\cdot\hat{k}' \\ \hat{k}\cdot\hat{\imath}' & \hat{k}\cdot\hat{\jmath}' & \hat{k}\cdot\hat{k}' \end{bmatrix}.$

The matrix [T] is an array of the components of the unit vectors $\hat{\imath}$, $\hat{\jmath}$, and $\hat{k}$ expressed in terms of the unit vectors of the primed coordinate system. This matrix has a similar form to the matric product, and it may be helpful to propose a notation for it. Let $\hat{u}$ represent the three unit vectors $\hat{\imath}$, $\hat{\jmath}$, $\hat{k}$ and let $\hat{u}'$ represent the unit vectors $\hat{\imath}'$, $\hat{\jmath}'$, $\hat{k}'$. The array of scalar products may be indicated by

(50) $[T] = [\hat{u}:\hat{u}']$.

The advantage of this notation is that it indicates which way the rotation goes. If the vector is on the left, the rotation is from unprimed to primed coordinates; if it is on the right, the rotation is from primed to unprimed. Reversing the order of $\hat{u}$ and $\hat{u}'$ transposes, and therefore inverts, the matrix. In this notation we may rewrite Eq.(48) as

(51) $[f'] = [\hat{u}':\hat{u}][f][\hat{u}:\hat{u}']$.

Because the unprimed unit vectors are to the inside, the operation transforms from unprimed to primed coordinates.

The alternative method of formulating coordinate rotation matrices is by the use of Euler rotation angles. These involve, for example, rotation about the z axis, followed by rotation about the new x axis, and finally rotation about the new z axis. There are considerable variations with respect to the definitions of the angles involved and the reader is advised to consult the literature for further information. It will suffice here to point out that such an operation is represented by the product of three coordinate rotation matrices, each with a form similar to Eq.(44).

3.2 Perspective in Photogrammetry

Like the human eye, all camera systems possess spherical perspective except for telecentric systems that have been designed to eliminate it. This means that object points are observed along rays that radiate from a common point at the center of the entrance pupil. Unlike the eye, the camera usually forms an image on a flat film plane that is normal to the lens axis. This geometry provides a simple introduction to the use of oblique projection matrices, and illustrates the use of a preferred coordinate system.

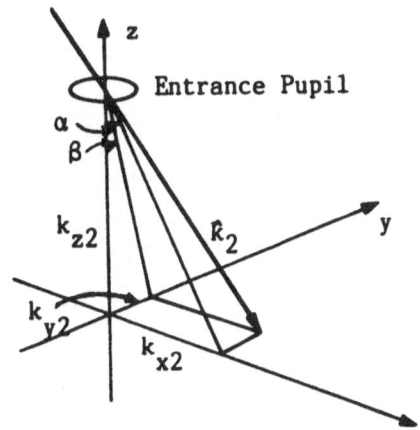

Fig 3. Photographic perspective through a lens pupil.

Consider Fig. 3 where an object point is being observed through an entrance pupil along the direction of unit vector $\hat{k}_2$. The coordinate system is drawn so that the z direction is aligned with the lens axis and the x,y plane is the image plane. This may be called the surface-normal coordinate system. If the object point undergoes a small but measurable displacement $\underline{L}$, the corresponding image displacement, $\underline{L}_{im}$, will be

$$(52) \quad \underline{L}_{im} = [P_{kn}]\underline{L},$$

where $\hat{n}$ is the surface normal to the x,y plane, i.e. $\hat{n}=\hat{k}$. This results in a considerable simplification of the oblique projection matrix.

$$(53) \quad [P_{kn}] = \begin{bmatrix} 1 & 0 & -k_{x2}/k_{z2} \\ 0 & 1 & -k_{y2}/k_{z2} \\ 0 & 0 & 0 \end{bmatrix}.$$

The two terms in the third column of this matrix can be identified as the tangents of the angles to the z axis shown in Fig. 3 as α and β. Therefore,

$$(54) \quad [P_{kn}] = \begin{bmatrix} 1 & 0 & -\tan\alpha \\ 0 & 1 & -\tan\beta \\ 0 & 0 & 0 \end{bmatrix}.$$

When this is used in Eq.(52), the result is

(55) $\underline{L}_{im} = \hat{\imath}(L_x - L_z\tan\alpha) + \hat{\jmath}(L_y - L_z\tan\beta).$

This equation shows the extent to which object displacements in the z direction admix with those in the x and y directions as the points of observation move away from the center of the field. If coordinates were chosen that were oblique to the lens axis, this physical insight would be lost, although the computations would still be the correct. Note also that for normal projection, the two angles become zero and

(56) $[P_{kn}] = \begin{bmatrix} 1 & 0 & 0 \\ 0 & 1 & 0 \\ 0 & 0 & 0 \end{bmatrix}.$

In the surface-normal coordinate system, the normal projection is an identity matrix with its third element equal to zero. This is exactly what should be expected because the normal projection matrix should merely reproduce the original vector minus its z component.

3.3 Photogrammetric Analysis

If image displacements of points on an object are photographically measured from two or more directions, then it is possible to compute the vectorial object displacements from the resulting information. This information exists in the form of a set of equations of the form of Eq.(52), and they may be stacked vertically to give

(57) $\{\underline{L}_{im}^r\} = [P_{kn}^r]\underline{L}.$

The superscript r is used here to denote individual sets of three equations associated with each photographic observation. Even for two observations there are more equations than the three unknowns of $\underline{L}$, so the least-square-error technique of the previous section should be employed. Multiplying both sides by the transpose of the matrix of projection matrices yields

(58) $[P_{kn}^r]^T\{\underline{L}_{im}^r\} = [P_{kn}^r]^T[P_{kn}^r]\underline{L}.$

This may be simplified by noting that when $[P_{kn}^r]$ is transposed, it goes from a column of 3x3 matrices to a row of the same matrices transposed. The products in Eq.(57), therefore, involve only vectors and matrices of the same superscript r. Thus $[P_{kn}^r]^T(\underline{L}_{im}^r)$ = $\Sigma_r[P_{nk}^r]\underline{L}_{im}^r$, and $[P_{kn}^r]^T[P_{kn}^r] = \Sigma_r[P_{nk}^r][P_{kn}^r]$. From the results of the previous section, the solution for $\underline{L}$ for least square error is

$$(59) \qquad \underline{L} = [\Sigma_r[P_{nk}^r][P_{kn}^r]]^{-1} \{\Sigma_r[P_{nk}^r]\underline{L}_{im}^r\}.$$

If the photographic observations are telecentric, or if they are sufficiently near the center of field to neglect obliquity, Eq.(59) simplifies still further. The fact that $[P_{nk}^r]$ becomes $[P_n^r]$ allows use of Eq.(30) to give

$$(60) \qquad \underline{L} = [\Sigma_r[P_n^r]]^{-1} \{\Sigma_r\underline{L}_{im}^r\}.$$

Perhaps the most common use for Eqs.(59) and (60) is in photogrammetry of speckle photographs. In this technique, the object is lit by laser light and the image speckles are used as markers of object points in double-exposure photographic recordings. The photographs may be analyzed for image motion which is related to vectorial object motion by projection. There are techniques in holography that measure object displacements transverse to a line of sight, and these equations may be applied there as well.

3.4 Fringe Vectors and Strain Analysis

A fringe vector can be used to describe the fringes that appear on the surface of an object when it undergoes a homogeneous deformation. A deformation is homogeneous when it consists of a rotation plus a uniform strain. Even when strains and rotations are not uniform, they may be approximately so over a local region, and a fringe vector can be used to describe the fringes on a local region of an object. Historically, the concept grew out of observations of fringes on objects that had been given rotations. These appeared to lie along the intersection of the object surface and sets of equally spaced planes. Such planes may be described as lying normal to a vector, which may be derived as follows.

Let us return to Eq.(3) and expand the fringe locus function in a Taylor series about some point on the object surface, which shall be the center of a coordinate system. For simplicity, let us use a superscript variable to indicate partial differentiation.

(61) $\quad \Omega = \underline{K} \cdot \underline{L} = \Omega_o + \Omega_x^x \Delta x + \Omega_y^y \Delta y + \Omega_z^z \Delta z.$

Ω_o is the value of the fringe locus function generated by the displacement of the origin of the coordinates. The three last terms in Eq.(61) are of the form of a scalar product of the space vector, $\underline{R}$, with the gradient of the fringe locus function Ω. The gradient of the fringe locus function, however, may be identified as the fringe vector, $\underline{K}_f$. Therefore,

(62) $\quad \Omega = \underline{K} \cdot \underline{L} = \Omega_o + \nabla\Omega \cdot \underline{R} = \Omega_o + \underline{K}_f \cdot \underline{R}.$

Equation (62) expresses the spatial variation of the fringe locus function as the scalar product of the space vector and a fringe vector, and a number of consequences follow. Any particular fringe must lie at points in space where this scalar product has a constant value. This occurs where the component of the vector $\underline{R}$ parallel to $\underline{K}_f$ is constant, and this defines a plane in space. If the fringe function is periodic, its roots define equally spaced planes, parallel to the fringe vector, that appear to intersect the object surface. The spacing of these fringe laminae in three-dimensional space, d, is the distance by which $\underline{R}$ must be increased in the $\underline{K}_f$ direction to increase the argument of the fringe function by one cycle. This is expressed by

(63) $\quad 2\pi = |\underline{K}_f| d,$

which may be solved for d to give

(64) $\quad d = 2\pi / |\underline{K}_f|.$

The spacing between fringe laminae is inversely proportional to the magnitude of the fringe vector, and the quantity, $|\underline{K}_f|/2\pi$ may be interpreted as the spatial frequency of the fringes in three-dimensional space.

Fringe vectors defined by Eq.(62) may be used in the analysis of holographic fringes to determine strain. It is assumed that the deformation of the object, over a local region, is homogeneous and can be represented by a transformation as defined in Eq.(45). The matrix [f] is the sum of a symmetric strain matrix, [ε], and an antisymmetric rotation matrix, [Θ]. It can be related to the derivatives of the displacement vector as follows. Transpose Eq.(45) and operate upon both sides with the matric derivative, $\nabla_{\!o}$.

(65) $\nabla_{\!o}\underline{L} = \nabla_{\!o}\underline{R}[f]^T$.

$\nabla_{\!o}\underline{R}$ equals the identity matrix, $[I]$, which leaves the result,

(66) $[f] = [\nabla_{\!o}\underline{L}]^T$, or

(67) $[f] = \begin{bmatrix} L_x^x & L_x^y & L_x^z \\ L_y^x & L_y^y & L_y^z \\ L_z^x & L_z^y & L_z^z \end{bmatrix}$.

Once these derivatives have been determined, strain and rotation may be determined by

(68) $[\varepsilon] = (1/2)([f]+[f]^T)$, and $[\Theta] = (1/2)([f]-[f]^T)$.

A relationship involving the strain-rotation matrix $[f]$ and the fringe vectors may be derived by taking the gradient of the fringe locus function as expressed in Eq.(62). Ω_o is a constant and its derivatives are zero. Therefore,

(69) $\nabla(\underline{K}_f \cdot \underline{R}) = \nabla(\underline{K} \cdot \underline{L}) = [\nabla_{\!o}\underline{L}]\underline{K} + [\nabla_{\!o}\underline{K}]\underline{L}$.

The righthand side of Eq.(69) follows directly from the product rule for differentiation. When this expansion is applied to the left side, the result is merely $\underline{K}_f$ because $\underline{K}_f$ is assumed to be constant over the region of consideration and $\nabla_{\!o}\underline{R}$ equals the identity matrix, $[I]$.

The first term of the righthand side of Eq.(69) is the transpose of the strain-rotation matrix times the sensitivity vector. The second term results from illumination of the object by spherical wavefronts and observation with spherical perspective, both of which generate linear variations of the sensitivity vector. The resulting matrix has an interesting form that may be obtained by substituting the definition of the sensitivity vector.

(70) $[\nabla_{\!o}\underline{K}] = [\nabla_{\!o}\underline{K}_2] - [\nabla_{\!o}\underline{K}_1]$.

The propagation vectors for illumination and observation have by definition a constant magnitude $2\pi/\lambda = k$ which may be factored out to leave unit vectors. Of the nine possible derivatives of the components of a unit vector, those three taken in the direction in which it points must be zero. These unit vectors are associated with points in space, namely the illumination source and the center of perspective for observation. As a consequence, the derivatives in directions transverse to the pointing direction are inversely proportional to the distance origins of the unit vectors. This leads heuristically to the relationship derived by Schumann and Dubas[1] that the matric derivative of a unit vector is a scaled normal projection matrix, i.e.

(71) $[\nabla \mathbf{\hat{k}}_2] = [P_{k2}]/R_{ob}$, and $[\nabla \mathbf{\hat{k}}_1] = [P_{k1}]/R_{i11}$,

where R_{ob} and R_{i11} are the distances from the object surface to the observer's entrance pupil and the illumination source respectively. It is possible, with these relationships, to rewrite the transpose of Eq.(70) as

(72) $\underline{K}_f = \underline{K}[f] + k(\underline{L}_{ob}/R_{ob} - \underline{L}_{i11}/R_{i11})$,

where $\underline{L}_{i11}$ and $\underline{L}_{i11}$ are the projections of the object displacement onto planes normal to the observer and normal to the illumination.

If Eq.(72) is normalized by the factor k it can be simplified by redefinition of the terms.

$\underline{k}_f = \underline{K}_f/k$ is the normalized fringe vector,

$\underline{k}_\sigma = \underline{K}/k$ is the normalized sensitivity vector,

$\underline{\sigma}_{ob} = \underline{L}_{ob}/R_{ob}$ is the vectorial angle subtended by the object displacement as seen by the observer, and

$\underline{\sigma}_{i11} = \underline{L}_{i11}/R_{i11}$ is the vectorial angle subtended by the object displacement as seen from the illumination direction. With these definitions, Eq.(72) may be rewritten as

(73) $\underline{k}_\sigma[f] = \underline{k}_f - \underline{\sigma}_{ob} + \underline{\sigma}_{i11}$.

It was shown in section 2.4 that if an object point could be observed with three or more independent sensitivity vectors,

the resulting equations (of Eq.(3) in form) could be combined into
a matrix equation and solved for the object displacement vector.
The result was expressed as Eq.(43). Although Eq.(73) is already
a matrix equation, the same approach can be taken. Assume that
observations are made with three or more sensitivity vectors and
that the fringe vector and the object displacement are determined
for each. The resulting equations can be combined into the following
matrix form.

(74) $[k_\sigma][f] = [k_f] - [\sigma_{ob}] + [\sigma_{ill}].$

The solution for $[f]$ that gives the least square error is, by the
arguments presented in section 2.4,

(75) $[f] = [[k_\sigma^T][k_\sigma]]^{-1}[k_\sigma^T]([k_f] - [\sigma_{ob}] + [\sigma_{ill}]).$

All that is required to implement Eq.(75) is a method for
determining the fringe vectors associated with different sensitivity
vectors. The most practical method is to observe the differential
fringe orders between various geometrical locations, $\Delta \underline{R}$, on the
object. These may be converted to differential increments of the
fringe locus function, $\Delta \Omega$ and the following matrix equation can be
set up.

(76) $\{\Delta \Omega\} = [\Delta R]\ \underline{K}_f,$

where the vectors $\Delta \underline{R}$ form rows of the matrix $[\Delta R]$. With three or
more independent values for $\Delta \underline{R}$, the solution for least square error is

(77) $\underline{K}_f = [[\Delta R]^T][\Delta R]]^{-1}[\Delta R]^T]\{\Delta \Omega\}.$

3.5 Observed Fringe Vectors and Surface Strain

Fringes on an object that can be characterized by a fringe
vector appear similar to those projected onto it by means of two
interfering beams (except for the lack or correspondence with the
shadows). In either case, however, the fringes in the field actually
being observed exist as laminae that radiate from the center of the
lens pupil. In hologram interferometry in fact, these are the only
fringes that have a real physical meaning and the concept of fringe
laminae perpendicular to the fringe vector is merely a method for

describing their appearance. The observed fringes can be described by an observed fringe vector, $\underline{K}_{fob}$, perpendicular to the viewing direction, which can be easily determined from fringes on a photograph. It is helpful, therefore, to establish the connection between observed fringe vectors and their corresponding true fringe vectors.

It is possible to define two different fringe locus functions in three-dimensional space, Ω and Ω_{ob}, using the true and observed fringe vectors respectively.

(78) $\Omega(\underline{R}) = \Omega_o + \underline{K}_f \cdot \underline{R}$, and $\Omega(\underline{R})_{ob} = \Omega_o + \underline{K}_{fob} \cdot \underline{R}$.

The key to the connection between true and observed fringe vectors is that they must both create the same fringes on the object surface. The object surface is defined by the vector $\underline{R}_s = [P_n]\underline{R}$, which may be substituted for $\underline{R}$ in Eq.(78). The result is

(79) $\Omega(\underline{R}_s) = \Omega_o + \underline{K}_f[P_n]\underline{R}$, and $\Omega(\underline{R}_s)_{ob} = \Omega_o + \underline{K}_{fob}[P_n]\underline{R}$.

The two fringe locus functions in Eqs.(79) can be equal only if

(80) $\underline{K}_f[P_n] = \underline{K}_{fob}[P_n]$.

If the vectors in Eq.(80) are projected along the object surface normal, $\hat{n}$, onto a plane perpendicular to the viewing direction, $\hat{k}_2$, the result is

(81) $\underline{K}_f[P_n][P_{kn}] = \underline{K}_{fob}[P_n][P_{kn}]$.

The identity $[P_n][P_{kn}] = [P_{kn}]$ may now be used to simplify Eq.(81). Note, however, that the observed fringe vector already lies in a plane normal to observation direction and it is unchanged by any further projection into that plane. This leaves

(82) $\underline{K}_f[P_{kn}] = \underline{K}_{fob}$.

This relationship between observed and true fringe vectors in hologram interferometry is also useful in describing the fringes observed in moiré techniques using projected fringes. A further insight into this relationship can be gained by substituting for $\underline{K}_{fob}$ in the relationship for Ω_{ob}.

(83) $\qquad \Omega(\underline{R})_{ob} = \Omega_o + \underline{K}_f[P_{kn}]\underline{R}.$

The observed fringe locus function may be thought of as the scalar product of the true fringe vector and the oblique projection of the space vector along the viewing direction onto the object surface. Alternatively, it may be thought of as the scalar product of the true space vector and the projection of the true fringe vector along the surface normal onto a plane perpendicular to the viewing direction. The bidirectionality of the oblique projection matrix becomes particularly interesting in Eq.(83).

Strain gages, in general, only measure strain over a small segment of an object surface and provide no information about out-of-plane deformation. When such a segment is considered in hologram interferometry, it may be impossible to determine a true fringe vector due to the flatness of the surface segment. Only the normal projection of the fringe vector onto the object surface can be defined. Furthermore, it is generally impossible to distinguish between out-of-plane rotation and shear so that only the normal projection of the strain rotation matrix onto the object surface can be defined as well. This normal projection is sufficient, however, to define surface strain. Consider, then, the result of operating upon Eq.(75) with a normal projection onto the object surface.

(84) $\quad [f][P_n] = [[k_\sigma^T][k_\sigma]]^{-1}[k_\sigma^T]([k_f] - [\sigma_{ob}] + [\sigma_{i11}])[P_n].$

$[f][P_n]$ may be identified as a surface strain-rotation matrix, $[f_s]$. The identity of Eq.(80) makes it possible to substitute the observed fringe vector in place of the true fringe vector in Eq.(84) to give

(85) $\quad [f_s] = [[k_\sigma^T][k_\sigma]]^{-1}[k_\sigma^T]([k_{fob}] - [\sigma_{ob}] + [\sigma_{i11}])[P_n].$

Equation (85) provides a direct means of evaluating the surface strain-rotation matrix from observed fringe vectors that may be measured by means of photographs.

4 CONCLUSION

This material has provided an introduction to the use of matrix methods in hologram interferometry. Unfortunately, the space is not available to permit a completely exhaustive treatment of this subject. It may be hoped, however, that the reader may have gained sufficient familiarity with these operations to explore their use in other applications.

Reference

1. Schumann, W. and M. Dubas. Holographic Interferometry (Berlin Heidelberg New York, Springer-Verlag, 1979).

OPTICAL METROLOGY AND COMPUTER TOMOGRAPHY FOR MEASUREMENT OF TEMPERATURE AND DENSITY

Charles M. Vest

Department of Mechanical Engineering and Applied Mechanics
The University of Michigan
Ann Arbor, Michigan 48109, USA

1. INTRODUCTION

This paper is a brief tutorial introduction to the application of holographic interferometry to measurements in transparent media. Emphasis is on quantitative evaluation of interferograms. Interferometric measurements of transparent media are relevant to industrial and research work in stress analysis, flow visualization, aerodynamics, heat transfer and plasma diagnostics. Of particular interest is the use of computer tomography for analysis of interferometric data to produce measurements of two- and three-dimensional fields. Although the discussion of the field of holographic interferometry is general, most illustrative applications are drawn from work in the author's laboratory at The University of Michigan.

2. HOLOGRAPHIC INTERFEROMETRY [1,2]

Holographic interferometry is the interferometric comparison of two coherent waves, at least one of which is a holographic reconstruction. The most common technique is that of two-exposure holographic interferometry. This technique can be applied to flow visualization or measurement using the system shown in Figure 1. Two holographic exposures are recorded on the same film plate. During the first exposure the object wave passes through the test section while it contains a quiescent fluid with homogeneous refractive index n_O. For example, if this is an aerodynamic experiment, the windtunnel would be either turned off, or else the air would be flowing undisturbed. During the second exposure the air would be flowing past a test model, thereby inducing changes in the density and refractive index distributions.

When this two-exposure hologram is developed and illuminated with a wave of laser light (reconstruction wave), identical to that used as the reference wave during recording, both the first (undisturbed) and second (disturbed) object waves will be reconstructed simultaneously. Because these two waves are mutually coherent, they will interfere, giving rise to a fringe pattern which is indicative of the refractive index, and therefore density, distribution in the test section.

Having this qualitative description of the technique of holographic interferometry as background, we now proceed to discuss the underlying principle. Holography is a linear process in the sense that two or more optical waves can be recorded sequentially in time and later can be reconstructed simultaneously. Therefore, the sum, difference, or even time average of a sequence of waves

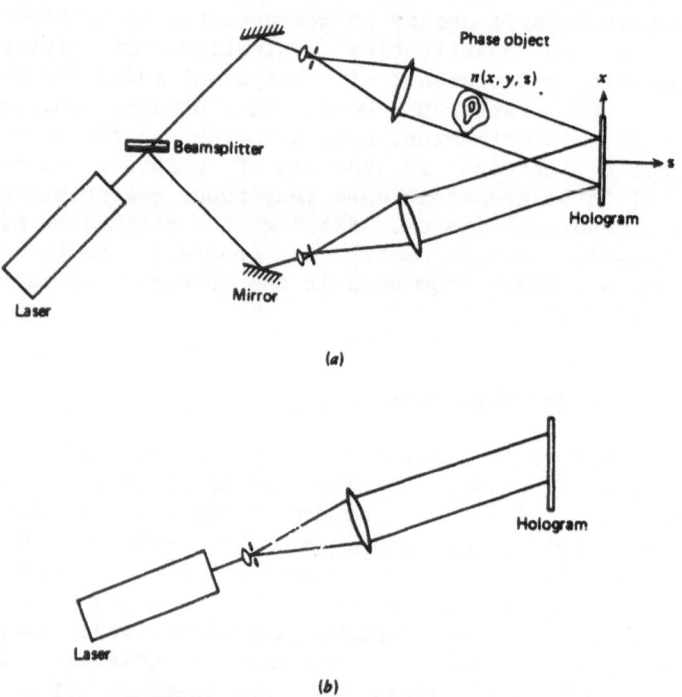

(a)

(b)

Figure 1. Typical off-axis configuration for holographic interferometry. (a) System for recording. (b) System for reconstruction.

can be formed. For example, at some time t, the off-axis holographic system shown in Figure 1(a) can be used to record an optical wave whose complex amplitude in the hologram plane is $\underline{U}_1$ (x,y). At time t₂ a second wave, $\underline{U}_2$(x,y), can be recorded on the same photographic plate. The process is simply holographic double exposure; the plate is first exposed to $\underline{U}_1$(x,y) together with a reference wave $\underline{U}_R$(x,y), and then is exposed to $\underline{U}_2$(x,y) together with $\underline{U}_R$(x,y). When the hologram formed by developing this plate is illuminated by $\underline{U}_R$(x,y), the complex amplitude of the reconstructed wave will be proportional to $\underline{U}_1$(x,y) + $\underline{U}_2$(x,y), and the irradiance will be proportional to

$$I(x,y) = \left| U_1(x,y) + U_2(x,y) \right|^2 \qquad (1)$$

In applications to interferometry, $\underline{U}_1 = \underline{U}_0$(x,y) represents the light transmitted through an undisturbed test section and $\underline{U}_2 = \underline{U}_0'$(x,y) represents light which has passed through the same test section while the flow event is occurring. Slight changes of the object wave primarily affect the phase of $\underline{U}_0$, so we write

$$\underline{U}_0(x,y) = a(x,y) \exp\left[-i\Phi(x,y)\right] \qquad (2)$$
$$\underline{U}'(x,y) = a(x,y) \exp\left\{-i[\Phi(x,y) + \Delta\Phi(x,y)]\right\}$$

The irradiance of the reconstructed wave, equation (1) then becomes

$$I(x,y) = \left| a(x,y) \exp\left[-i\Phi(x,y)\right] \right.$$
$$\left. + a(x,y) \exp\left\{-i[\Phi(x,y) + \Delta\Phi(x,y)]\right\} \right|^2 \qquad (3)$$
$$I(x,y) = 2\,a^2(x,y)\left\{1 + \cos\left[\Delta\Phi(x,y)\right]\right\}$$

Equation (3) represents the irradiance of the object, a^2(x,y), modulated by a fringe pattern 2{1 + cos[Δφ(x,y)]}. Dark fringes are contours of constant values of Δφ which are odd integer multiples of π. Bright fringes are contours of constant values of Δφ which are even integer multiples of π. In applications to transparent media Δφ is the optical pathlength change due to the change of fluid density distribution between the two exposures:

$$\Delta\Phi(x,y) = \int[n(x,y,z) - n_0]\,dz \qquad (4)$$

Here we consider briefly the formation of interference fringe patterns and, more importantly, the manner in which $n-n_0$ is determined from them. There are three types of distributions of refractive index which arise in practice:

1. Two-dimensional distributions with no variation in the direction of the optical (z) axis.

2. Radially symmetric distributions.

3. Asymmetric distributions.

An example of the first case is the measurement of a boundary layer in which the density distribution is described approximately by

$$\rho = \rho_0 - \rho_1 e^{-ay} \tag{5}$$

Correspondingly the refractive index distribution would be of the form

$$n(y) = n_0 - n_1 e^{-ay} \tag{6}$$

where n_0 is the refractive index far from the boundary. This is also the value of the homogeneous refractive index distribution at the time of the first exposure of a two-exposure holographic interferogram. Figure 2 is a graphical representation of this distribution and of the corresponding fringe pattern.

To analyze this interferogram, we assign the order number $N = 0$ to the bright fringe outside the boundary layer because this is a region whose density remained unchanged between exposures. The subsequent bright fringes are assigned consecutive integer values as shown in the figure. Strictly speaking, these integer order numbers are assigned at the location of the center of each bright fringe. Half-integer orders, 0.5, 1.5, 2.5, --- are assigned to the center of each dark fringe. At any location the optical pathlength difference is

$$\Delta\Phi = N\lambda. \tag{7}$$

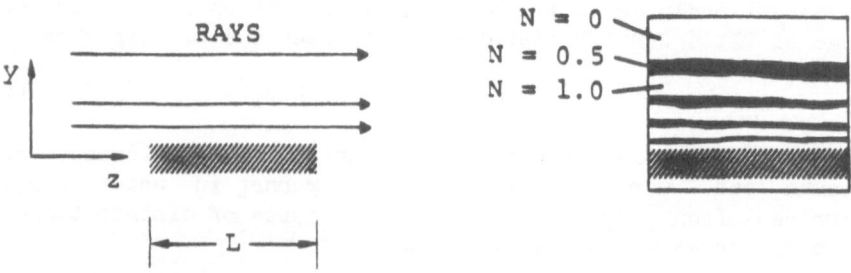

Figure 2. Boundary layer type refractive index distribution and schematic representation of the corresponding fringe pattern.

There is a sign ambiguity in fringe order numbers which arises because $\pm\ \Delta\Phi$ yield identical fringe patterns. Often the experimenter has sufficient knowledge of the field being examined to infer the appropriate sign.

Once $\Delta\Phi$ has been evaluated for an interferogram, Equation (4) must be inverted to determine $n(x,y,z)-n_0$. In the present case the unknown change of refractive index is independent of z and the length of the test region is L, so

$$\lambda N(x,y) = \int_0^L [n(x,y) - n_0] \, dz$$

$$= [n(x,y) - n_0] \cdot L.$$

Therefore

$$n(x,y) - n_0 = N(x,y) \cdot \lambda/L. \tag{8}$$

This simple relation is often useful because flow around airfoils, temperature distributions near long heated cylinders, and many other fields encountered in practice are approximately two-dimensional.

One of the most important practical advantages of using holographic interferometry is its <u>phase error cancellation property</u>. In a classical interferometer, for example a Mach-Zehnder interferometer, only one wave passes through the phase object. It then interferes with a plane comparison wave which has travelled a different path through the interferometer. In a holographic interferometer, like that shown in Figure 1, both the object and comparison waves travel across the same object space.

Holographic interferometers are therefore <u>single path interferometers</u>. This important feature permits the use of test sections with windows of rather poor optical quality. If a test section window is not optically flat and homogeneous, it will introduce a <u>pathlength error</u>, or noise, $\Delta\Phi_n$ (x,y) into an optical wave passing through it. Let $\Delta\Phi(x,y)$ represent the pathlength difference due to the phase object inside the test section. It is contours of $\Delta\Phi(x,y)$ which we desire the interferometer to display as a fringe pattern. In a Mach-Zehnder interferometer, the fringe pattern for this test section and phase object would be

$$I(x,y) = 2 \left\{1 + \cos \frac{2\pi}{\lambda} [\Delta\Phi(x,y) + \Delta\Phi_n (x,y)]\right\}. \tag{9}$$

Thus pathlength variations in test section windows lead to errors in the fringe pattern which can be eliminated only by using optically flat, homogeneous windows so that $\Delta\Phi_n(x,y)$ = constant. In a holographic interferometer only <u>changes</u> in pathlength between exposures are displayed. (See equation 3.) Since the test section windows are present during both exposures the effect of

$\Delta\Phi_n(x,y)$ is cancelled, and the interference pattern is given by the intensity distribution in equation (9) with $\Delta\Phi_n = 0$.

3. RELATION OF REFRACTIVE INDEX TO OTHER PHYSICAL PROPERTIES

Once an interferogram has been analyzed to determine a spatial distribution of refractive index, this must in turn be related to the physical property of interest. Here we briefly review basic relations between refractive index and physical properties of transparent media.

In flow visualization and aerodynamics, the property of interest usually is mass density ρ, which is related to refractive index n by the Gladstone-Dale relation [3],

$$n-1 = K\rho. \tag{10}$$

In Equation (10) K is the Gladstone-Dale constant of the gas. K is usually only a very weak function of λ, the wavelength of light.

In plasma diagnostics one must deal with mixtures of atoms, positive ions and free electrons. The property to be determined is the electron number density N_e, which is the number of free electrons per cm^3. A modified form of Equation (10) can be used:

$$n = K_A\rho_A + K_I\rho_I + n_e \tag{11}$$

where the subscript A denotes atoms and I denotes positive ions. K_A and K_I are nearly independent of λ, but the refractive index contribution of the electrons, n_e, is very dispersive, i.e. dependent on λ^4 :

$$n_e = 1-4.46 \times 10^{-14} \lambda^2 Ne. \tag{12}$$

where λ is measured in cm. Because of dispersion, a common technique is to make measurements of n using two different wavelengths, λ_1 and λ_2, then

$$n(\lambda_1) - n(\lambda_2) \simeq n_e(\lambda_1) - n_e(\lambda_2) \tag{13}$$
$$= -4.46 \times 10^{-14}(\lambda_1^2 - \lambda_2^2)N_e.$$

Holographic interferometry can be used to produce an interferometric fringe pattern indicative of the difference $n(\lambda_1) - n(\lambda_2)$ in a single exposure. This is done by producing light of wavelength and, by frequency doubling, $\lambda_2 = 2\lambda_1$ from a single laser pulse. The resulting single-exposure hologram is processed to have a nonlinear amplitude transmittance response to exposure. The second-order diffracted wave from the hologram recorded with λ_1 coincides and interferes with the first-order wave from the hologram recorded with λ_2 to form the desired interference pattern [5].

To study heat transfer in gases at constant pressure, refractive index can be related to temperature by substituting the ideal gas equation of state $P = \rho RT/M$ into the Gladstone-Dale relation, Equation (10) to obtain

$$n - 1 = KMP/RT. \tag{14}$$

In Equation (14) M is the molecular weight, P is the pressure, R is the universal gas constant and T is the absolute temperature.

The refractive index of liquids generally obeys the Lorentz-Lorenz relation

$$\frac{n^2 - 1}{\rho(n^2 + 2)} = \bar{r}(\lambda) \tag{15}$$

where $\bar{r}(\lambda)$ is the specific refractivity of the liquid. Because is not a simple function of T, empirical relations between n and T must be developed. These can be quite complicated, for example the Tilton-Taylor data for water is expressed as an algebraic relation with 13 empirical constants. An accurate relation for water using $\lambda = 632.8$ nm is

$$n - 1.3331733 = -(1.936T + 0.1699T^2) \times 10^{-5} \tag{16}$$

where T is the temperature in degrees Celsius [6].

Optical stress analysis of birefringent solids is based on the Maxwell-Neumann stress-optic law [7]:

$$n_1 - n_0 = A\sigma_1 + B(\sigma_2 + \sigma_3)$$
$$n_2 - n_0 = A\sigma_2 + B(\sigma_1 + \sigma_3) \tag{17}$$
$$n_3 - n_0 = A\sigma_3 + B(\sigma_1 + \sigma_2)$$

In Equation (17), n_i is the refractive index for light polarized in the direction of σ_i and n_0 is the uniform refractive index prior to stressing. Holographic techniques can be used to advantage for photoelasticity in order to separate the isochromatic fringes from the isopachic fringes [7-9] however, in this paper attention is restricted to isotropic materials for which A = B:

$$n - n_0 = A(\sigma_1 + \sigma_2 + \sigma_3) + A\sigma_{kk}. \tag{18}$$

4. APPLICATIONS

In this section four applications of holographic interferometry to flow visualization and measurement in transparent media are considered. They are selected to include one-dimensional, radially symmetric, and asymmetric distributions and to illustrate measurement of several different physical properties.

4.1 FLOW VISUALIZATION IN STRATIFIED MEDIA

Figure 3 is a two-exposure holographic interferogram of a steady, low-speed, gravity-dominated two-dimensional flow of a stratified fluid. A tank (1.22 x 0.15 x 0.31m) with sidewalls made of commercial plate glass was filled with salt water such that its density varied linearly in the vertical direction. The stratification was stable, that is heavy on the bottom and light on the top. A horizontal cylinder of rectangular cross section was towed slowly from left to right. The first holographic exposure was recorded as the cylinder passed through the illuminated region. Next the salt water was removed and the tank was refilled with pure water.

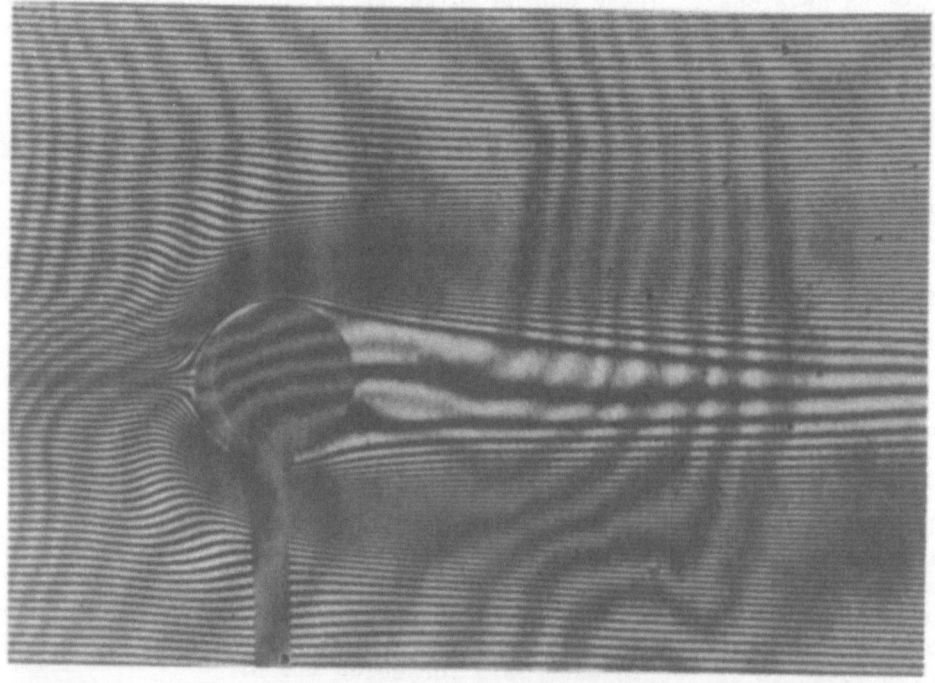

Figure 3.   Interferogram displaying streamlines of a low-speed stratified flow as a rectangular cylinder is towed from left to right.

Because there is no appreciable density change in the direction normal to the plane of Figure 3, the fringes in this figure are curves of constant density (isopychnics). Because the flow is steady and low speed as well, the fringes are also streamlines. Quantitative analysis of interferograms from this experiment enabled velocity profiles to be determined [10].

4.2 INTERNAL PRESSURE MEASUREMENT IN TRANSPARENT SOLIDS

Figures 4 and 5 are interferograms of a block of acrylic, 10 x 10 x 5 cm, which was submerged in a glass tank containing a mixture of mineral oil and immersion oil whose refractive index nearly matched that of the unstressed acrylic [11].

We studied two different loading configurations. In the first, a pressure p was applied uniformly over a small circular area of diameter 6.35 mm in the center of th top surface of the acrylic. In the second configuration, a pressure was applied uniformly over an annular region centered on the top surface of the acrylic. The annular region had an 11.1 mm inside diameter and 12.7 mm outside diameter.

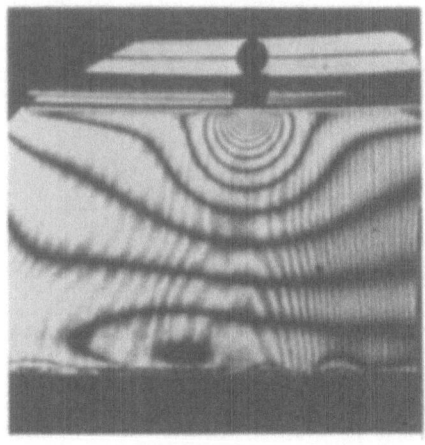

Figure 4. Holographic interferogram of an acrylic block subjected to a pressure of 41.4 MPa over a circular area with 6.35 mm diameter.

Figure 5. Holographic interferogram of an acrylic block subjected to a pressure of 44.2 Mpa over an annular region with 11.1 mm ID and 12.7mm OD.

Acrylic is a nearly isotropic material, so the stress-optic relation reduces to Equation (18). Furthermore, both loading configurations result in radially-symmetric distributions of the sum of principal stresses σ_{kk} . Hence the refractive index $n(r) - n_0$ in any horizontal plane z = constant depends only on r (See Figure 6.) In this case the stress-induced optical pathlength change of a light ray passing through the specimen in the x direction is

$$N(y;\, z)\lambda = \int [n(r) - n_0]dx = 2 \int_y^R \frac{[n(r) - n_0]r\,dr}{\sqrt{r^2 - y^2}} \qquad (19)$$

Equation (19) is a form of Abel's integral equation and has the inversion

$$n(r) - n_o = -\frac{\lambda}{\pi}\int_r^R \frac{(\partial N/\partial y)\,dy}{\sqrt{y^2 - r^2}} + \frac{n(R) - n_o}{\pi\sqrt{R^2 - r^2}} \qquad (20)$$

The second term of Equation (20) accounts for a step change in refractive index at the boundary r = R; such a step change cannot be evaluated by interferometry but may be known from physical boundary conditions. If the stresses vanish at the boundary, the first term alone gives the distribution. Assuming this to be the case, Equations (18) and (20) can be combined to give

$$\sigma_{kk}(r;\, z) = -\frac{\lambda}{\pi A}\int_r^R \frac{(\partial N/\partial y)\,dy}{\sqrt{y^2 - r^2}} \qquad (21)$$

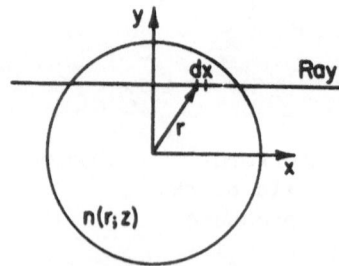

Figure 6.   Nomenclature for analysis of axisymmetric stress distribution.

Figure 4 is a typical interferogram of the object loaded uniformly over a circular region. The distribution of σ_{kk} was determined in three horizontal planes at distances of 4.76 mm, 6.35 mm, and 9.53 mm below the top surface. (It should be noted that some optical problems are apparent near the top surface. Imperfect refractive index matching and slight object wave misalignment caused total internal reflection which led to the thin dark band at the top of the fringe pattern. Vertical distances should be measured from the center of this band.) For each horizontal plane the interferogram was scanned using a microscope translated by a micrometer screw. Data consisting of the location of the centers of each dark and light fringe and the corresponding fringe orders N were recorded.

Direct application of Equation (21) to invert these data is not feasible because the values of N are discrete and not very numerous and are therefore likely to produce inaccurate values of the derivative dN/dy. Several approaches to approximate Abel inversion of discrete data have been reported in the literature. In this study we fitted cubic splines to the measured values of fringe order. $N(y; z)$ was then represented between each data point by a cubic polynomial which was analytically inverted using Equation (21). A computer code was developed to accept the measured data, compute the coefficients for the spline functions, effect the Abel inversion, and convert the resulting distribution of $n - n_0$ to that of σ_{kk} using Equation (21) with $A = 3.17 \times 10^{-8}$ m^2 /kN.

Figure 7 displays values of σ_{kk} /p determined by analysis of the interferogram of Figure 4. The solid curves in this figure represent theoretical distributions of σ_{kk} /p based on the theory of elasticity. The agreement of theory and experiment is very good at $z/R = 1.0$ and 1.5, but an error of 9 percent occurs at $r = 0$, $z/R = 0.75$. Factors which may contribute to this error include nonuniformity of applied pressure, effects of small optical misalignment, and the increased effect of small measurement errors due to the fine fringe spacing in this region.

Figure 5 is an interferogram of the specimen loaded uniformly over an annular region. The experimental results for this case are presented in Figure 8 for $z/R = 0.15$. The experimental results are in rather good agreement with the theoretical results.

4.3 MEASUREMENT OF THREE-DIMENSIONAL TEMPERATURE AND DENSITY FIELDS

An important feature of holographic techniques is the ability to easily record interferograms which can be viewed from many different directions. The technique used to analyze such multipleview interferograms is known as computer tomography. A

354

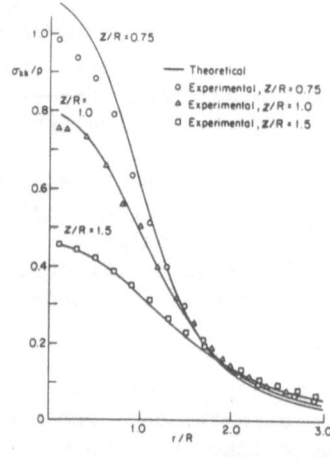
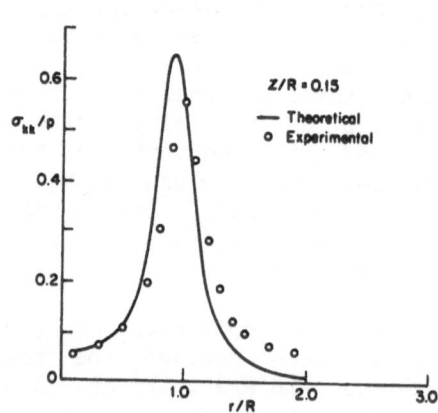

Figure 7.   Distribution of σ_{kk}/P
corresponding to
Figure 4.

Figure 8.   Distribution of
σ_{kk}/P corresponding to Figure 5.

Mach-Zehnder interferometer can be used to record a sequence of interferograms, each from a different viewing direction, but holographic techniques generally are more convenient.   If the flow field is not too large, a single holographic interferogram may contain all the required data.

Figure 9 shows schematically four different ways in which multiple-view interferograms can be recorded. In the first (Figure 9(a)), the test section is illuminated by a ground glass, or opalized glass, which diffuses the object wave of a holographic interferometer.   Two-exposure holographic interferograms recorded with such illumination can be viewed from any direction compatible with the size of the diffusing screen and the aperture of the hologram. Such interferograms are dramatic displays of the density field which can be viewed with the unaided eye.   One can observe the fringe patterns shifting and changing dynamically as one's viewing direction is varied. However, such interferograms have the disadvantage that laser speckle is present in the image.   Because the interference fringes may be localized in very convoluted surfaces in space it may be necessary to use a camera or viewing system with a high f-number to photograph or observe the fringes. Such high f-number systems cause large laser speckle.   If the speckle size approaches the width of the interference fringes, they cannot be observed and counted accurately.   For this reason, the second system (Figure 9(b)) can be used.   In this system phase gratings are used to break a single plane object wave into several plane waves, each travelling in a different direction.

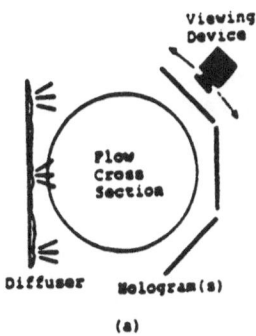
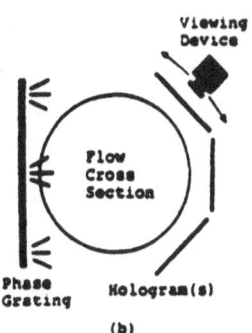

(a)

(b)

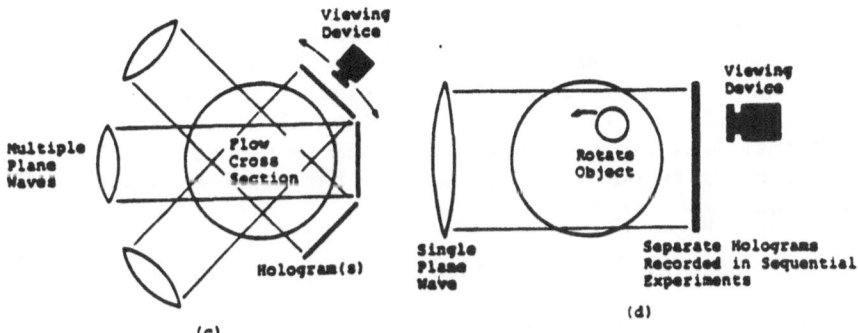

(c)

(d)

Figure 9. Systems for forming interferograms which can be viewed
from several directions. (a) Diffuse-illumination holography. (b) Phase-grating illumination holography.
(c) Multiple-illumination holography. (d) Single
object wave holography with a rotating object.

A third system is shown schematically in Figure 9(c). Several
individual plane waves are used to illuminate the test sections.
With each of the three systems described so far the required data
can be recorded in a single pulsed-laser holographic interferogram;
hence, transient events can be studied. If the flow is steady or
repeatable, the fourth system (Figure 9(d)) can be used. Here the
object field rotates relative to a fixed object wave. This
approach can also be used with a Mach-Zehnder interferometer.
Later in this section an application will be described in which

this scheme is applied to the flow field near an object (a helicopter rotor blade) which naturally rotates.

We use the technique of computer tomography to analyze a sequence of interferograms, each corresponding to a different viewing direction. Here we discuss the problem of reconstucting the density distribution in a plane. In fact, this can be one planar slice of a three-dimensional flow field. By reconstructing the density in each of a sequence of parallel planes, the three-dimensional structure can be determined [12-15].

Computer tomography is an important technique for reconstructing two-dimensional or three-dimensional fields from measurements of line integrals (such as optical pathlength) through the field. It arises in several types of remote sensing and has been under development since the late 1960's. However, its origins are in the mathematical analysis of Radon published in 1917, work in radio astronomy published by Bracewell in 1956, and applications to medical imaging in the mid 1960's by Cormack and by Hounsfield, who received the Nobel prize for their works.

In the analysis that follows it is assumed that the density distribution of interest is a two-dimensional function $\rho(x,y)$, which can represent either a cylindrical distribution or, more generally, the distribution in a horizontal plane within a three-dimensional flow field. With reference to Figure 10, the optical pathlength difference along a line through the object can be written as

$$\Delta\Phi(p,\theta) = K\int\int f(r,\phi)\delta[p - r\sin(\phi - \theta)]dxdy$$
$$= \lambda N(p,\theta)$$

(22)

where θ defines the ray direction and p its location. When $\Delta\Phi(p,\theta)$ is known for all lines through the object, it defines the two-dimensional Radon transform of $f(r, \phi) = \rho(r, \phi) - \rho_0$. Reconstruction of $f(r,\phi)$ therefore is accomplished by computing the inverse Radon transform:

$$\rho(r,\phi)-\rho_0 = \frac{K}{2\pi^2}\int_{-\frac{\pi}{2}}^{\frac{\pi}{2}}d\theta\int_{-\infty}^{\infty}\frac{(\partial N/\partial p)dp}{r\sin(\phi-\theta)-P}$$

(23)

In real applications the function $N(p, \theta)$ is replaced by a set of discrete measured values N which contain measurement errors. As in the case of the Abel transform, several techniques have been devised for computing approximate reconstructions from such data. These fall into three categories: implicit methods, explicit methods and Fourier transform methods. In implicit methods $\rho(r,\phi)$

357

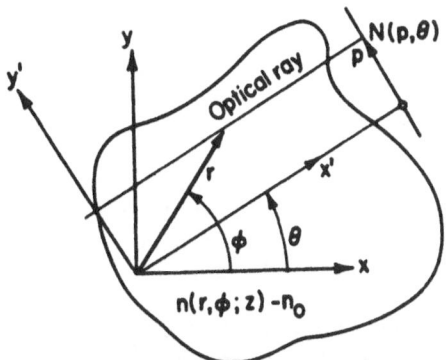

Figure 10. Nomenclature for analysis of asymmetric distributions.

is represented by a series expansion or as a grid of rectangular
elements within each of which $(\rho - \rho_0)$ has a constant value f. The
right-hand side of Equation (22) then is approximated as a
summation:

$$\Delta\Phi_i = \sum_k A_{ki} f_k \qquad (24)$$

where f is either a series coefficient or the value of
$(\rho - \rho_0)$ in the k th rectangular element. The coefficients
A_{ki} are derived from Equation (22) by integration. Reconstruction
is then accomplished by solving this set of linear algebraic
equations for f.

In explicit methods, the fringe order $N(p, \theta)$ or its
derivative, is represented by a series expansion using the measured
values N_i , and the right-hand side of Equation (23) is
approximated as a summation. The most common methods of this type
are the convolution or filtered-back-projection methods in which
the reconstruction is computed as a weighted sum of values of N:

$$f_i = \sum_k W_{ki} N_k \qquad (25)$$

Fourier transform methods are based on the Central Section
Theorem, due to Bracewell, which states that the one-dimensional
Fourier transform with respect to p of $N(p, \theta)$ with θ held constant
equals the two-dimensional Fourier transform of $f(r, \phi)$ along a line
through the origin of the transform plane in a direction

perpendicular to the projection direction θ = constant. Hence the reconstruction can be accomplished by computing a one-dimensional fast Fourier transform (FFT) of the values of ΔΦ for each available direction, interpolating them onto a rectangular grid in the transform plane and then computing a two-dimensional inverse FFT.

To illustrate the use of holographic interferometry and computer tomography we cite first an experiment conducted by Radulovic and Vest [12] in which interacting free convective plumes above two electrically heated disks were studied. The optical system was of the type shown in Figure 9(b). The glass test section was filled with deionized distilled water. Two horizontal circular disks, 10.5 mm in diameter were submerged in the water with their centers separated by 16 mm.

The objective of the experiment was to measure the temperature distribution in a horizontal plane 3 cm above the plane of the heated disks. Ten small thermocouple probes were inserted into the water to provide a test of the accuracy of the interferometric measurement.

A holographic interferogram was recorded by exposing the photographic plate to the reference and object waves simultaneously before the heaters were energized. After the heaters were energized and a steady state was attained a second exposure was made on the same plate. When the resulting hologram was developed, the object waves were reconstucted by illuminating it with the reference wave. Along each viewing direction a fringe pattern could be observed with the telecentric viewing system. The position of each fringe was measured by scanning each view with the vernier microscope. 17 views were used, and approximately 55 fringe locations were recorded in each view. Due to refraction, the angular range of the views was reduced to 46° in the test section.

Two views of the holographic interferogram are shown in Figure 11. These views are separated by 23°. Three pins and a sphere which were used as fiducial marks for determining positions and viewing directions are visible at the bottoms of these photographs. The thermocouple probes can be seen extending downward into the plume.

The computer code used to analyze the interferometric data was based on the grid method described in Ref. 13. A total of 933 fringe positions were recorded for reconstruction of the temperature field in one plane. From these data the refractive index was determined at 27 points in the plane of interest by solving 933 algebraic equations in 27 unknowns by a least mean square algorithm. This high redundancy was required for good accuracy because the effective viewing angle was only 92° . Once

the refractive index change f(x,y) was determined the temperature
was calculated using Equation (16).

Figure 12 is a computer plot of the isothermal contours in a
horizontal plane 3.0 cm above the surface of the heated disks. The
location and size of several thermocouple beads are also noted in
this figure. The average difference between the optical and
thermocouple measurements is about 0.8 C. The four small isotherms
at the edges of this figure are artifacts due to "ringing". These
represent errors of about 0.5 C.

The combination of holographic interferometry and computer
tomography has been illustrated by a small scale experiment. As an
example of a large-scale experiment, we cite work currently being
done by J. Kittleson and Yung Yu [16] at the U. S. Army
Aeromechanics Laboratory at Ames Research Center in California.
The objective of their work is to measure the three-dimensional
aerodynamic density distribution in the transonic flow field near a
scaled-down helicopter rotor. The object wave is approximately 1 m
in diameter and is derived from a pulsed ruby laser.
Interferograms are recorded as the blade rotates through this
single, fixed object wave. Figure 13 is a schematic diagram of the
system. A sequence of interferograms, each recorded at a different
rotational position of the blade, provides the data necessary for
tomographic reconstruction of the flow.

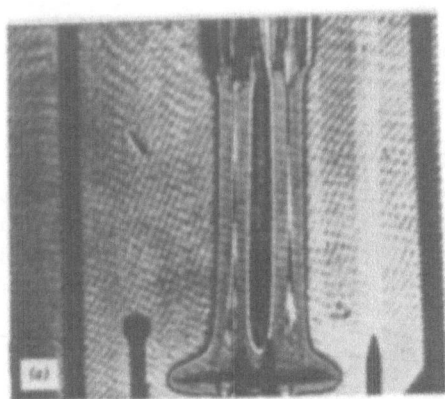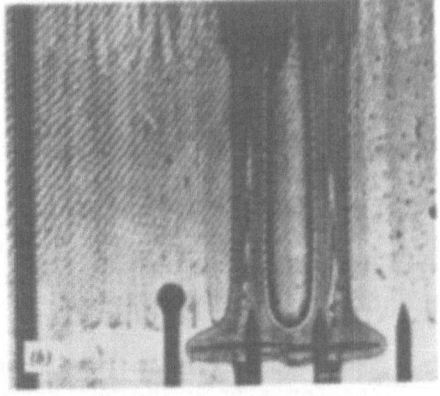

Figure 11. Multidirectional holographic interferometry of an
            asymmetric thermal plume rising in water above two
            heated disks. (a) Interferogram viewed from one
            direction. (b) Interferogram recorded when the same
            hologram is viewed from a different direction. The two
            directions differ by 23.

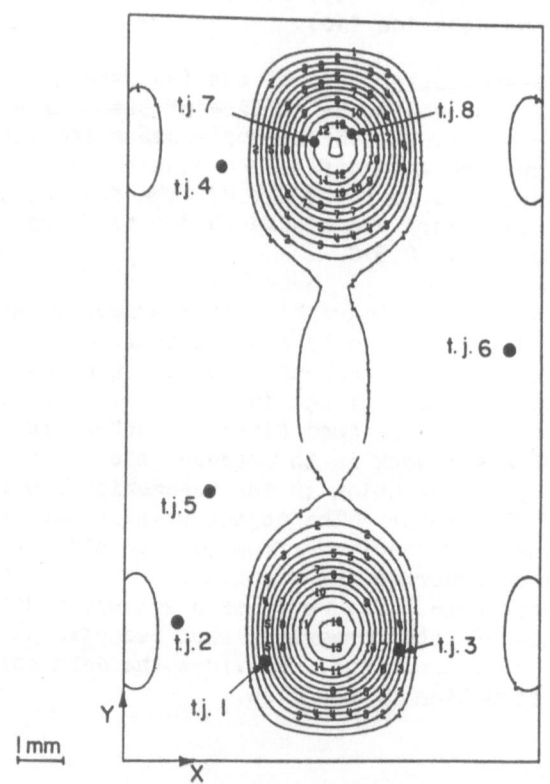

Figure 12.   Tomographic reconstruction from a collection of interferograms including those in Figure 11. The curves are
isotherms, and the dots marked t.j. are thermocouple
locations.

Figure 14(a) is an interferogram recorded with the blade at
180 , that is perpendicular to the plane of this figure.   Figure
14(b) is an interferogram recorded at 186 .   As this experiment
progresses, data will be recorded over an entire 180   range of
viewing directions and computer tomography will be used   to
reconstruct the density field.

Recent advances in the use of holographic interferometry and
computer tomography include attempts to correct for the effects of
strong refraction. When steep gradients of refractive index occur
in the flow field, the probing rays are bent. When the rays are
not straight lines, normal tomography procedures are not valid.
Cha and Vest [17] have used a combination of imaging and iterative
computational techniques to improve the accuracy of reconstruction
when strong refraction occurs. Their technique was applied to
measurement of very thin electrolytic boundary layers. Snyder and

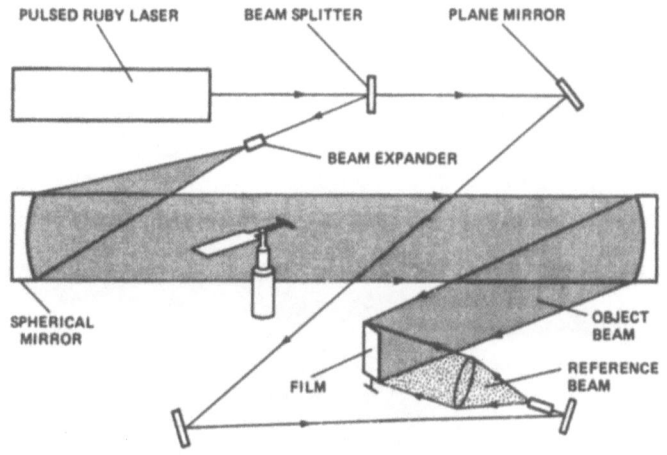

Figure 13. Schematic diagram of holographic system used to study
transonic flow near a helicopter rotor. (Courtesy of
J. Kittleson and Yung Yu.)

Hesselink [18] have also developed an iterative correction
technique which can be used when refraction effects are small but
troublesome because of large physical distances from the object to
the interferogram.   Their technique was applied to the helicopter
rotor experiments described above.

362

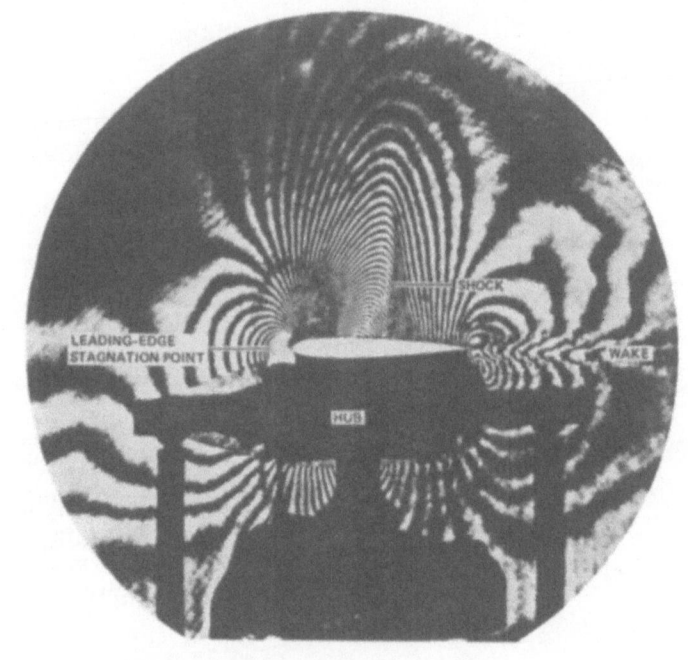

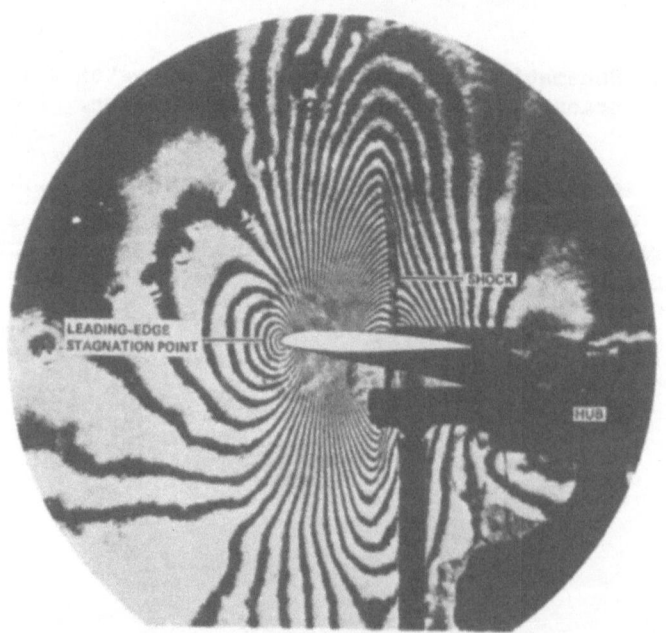

Figure 14. Interferograms of flow near the helicopter rotor
blade. (a) Viewing direction 180° . (b) Viewing
direction 186° .

REFERENCES

1. C. M. Vest, <u>Holographic Interferometry</u>, Wiley, New York, 1979.

2. Yu. I. Ostrovsky, G. V. Ostrovskaya and M. M. Butusov, <u>Interferometry by Holography</u>, Springer Verlag, Berlin, 1980.

3. W. Merzkirch, <u>Flow Visualization</u>, Academic Press, New York, 1974.

4. V. Ascoli-Bartoli, Plasma diagnostics based on refractivity, in <u>Physics of Hot Plasmas</u>, B. J. Rye and J. C. Taylor (Eds), Plenum Press, New York, p. 405.

5. R. J. Radley, Jr., Two-wavelength holography for measuring plasma electron density, <u>Phys. Fluids</u>, <u>18</u>, 175-179 (1975).

6. P. T. Radulovic, <u>Holographic Interferometry of Asymmetric Temperature or Density Fields</u>, doctoral dissertation, The University of Michigan, Ann Arbor, 1977.

7. P. S. Theocaris and E. E. Gdontos, Matrix Methods of Photoelasticity, Springer-Verlag, Berlin, 1979.

8. J. Ebbeni, J. Coenen and A. Hermanne, New Analysis of holophotoelastic patterns and their application, <u>J. Strain Anal.</u>, <u>11</u>, 11-17 (1976).

9. J. D. Hovansian, Variable isochomatic/isopachic-fringe visibility in photoelasticity, <u>Exp. Mech.</u>, <u>14</u>, 233-236 (1974).

10. W. R. Debler and C. M. Vest, Visualization of a stratified flow by holographic interferometry, <u>Proc. Roy. Soc. (London)</u>, A358, 1-16 (1977).

11. C. M. Vest and E. A. Ural, The role of interferometry and tomography in stress analysis of transparent media, in <u>Proc. of the 1981 Spring Meeting, Society of Experimental Stress Analysis</u>, SESA, Brookfield Center, CT, 1981, pp. 242-247.

12. P. T. Radulovic and C. M. Vest, Measurement of three-dimensional temperature fields by holographic interferometry, in <u>Applications of Holography and Optical Data Processing</u>, E. Marom and A. A. Friesem (Eds), Pergamon Press, Oxford, 1977, pp. 241-249.

13. D. W. Sweeney and C. M. Vest, Reconstruction of three-dimensional refractive index fields from multi-directional interferometric data, _Appl. Opt_. 12, 2649 (1973).

14. R. D. Matulka and D. J. Collins, Determination of three-dimensional density fields from holographic interferometry, _J. Appl. Phys_., 42, 1109-1119 (1971).

15. H.-G. Junginger and W. VanHaeringer, Calculation of three-dimensional refractive-index field using phase integrals, _Opt. Commun_., _5_, 1-4 (1972).

16. J. Kittleson, A holographic interferometry technique for measuring transonic flow near a rotor blade, NASA TN84405 (1983).

17. S. Cha and C. M. Vest, Tomographic reconstruction of strongly refracting fields and its application to interferometric measurement of boundary layers, _Appl. Opt_., _20_, 2787-2794 (1981).

18. R. Snyder and L. Hesselink, Optical tomography for flow visualization of the density field around a revolving helicopter rotor blade (submitted to _Appl. Opt_., 1984).

UNIFICATION OF FEM WITH LASER EXPERIMENTATION

Hayrettin Kardestuncer
University of Connecticut
Storrs, Connecticut 06268

Ryszard J. Pryputniewicz
Worcester Polytechnic Institute
Worcester, Massachusetts 01609

Unification of finite element methods with laser
experimentation is presented. It is pointed out that
most engineering problems contain regions in which
finite element modeling encounters difficulties due
to nonlinearities, irregular boundaries, ambiguous
energy functionals, etc. Measurements obtained by
laser experimentation, particularly in these
regions, can be digitized and automatically
incorporated into the finite element modeling to
improve results. Unification is possible in solid
mechanics, fluid mechanics, gas dynamics, heat
transfer, and in an everincreasing number of other
fields.

INTRODUCTION

Solution methodologies for engineering problems can, in
general, be categorized as experimental, analytical, and
numerical. In the recent past, the emphasis appears to have
shifted from the first to the last. Certainly, each
methodology has considerable advantage over the others for a
given class of problems and each makes use of the others for
verification of the results. In many cases, even the data
furnished by one methodology has been utilized by the others.
In spite of recent advances in number crunching equipment
which have drawn considerable attention to numerical

methodologies, in particular to the finite element methods,
the importance of experimental, analytical, or semi-analytical
methods has not diminished.

Today's demands for optimum and reliable design are, to a
great extent, satisfied by application of finite element
methods. In these applications, the finite element methods
are used to solve problems for which exact solutions are
nonexisting, or are very difficult to obtain. Also, the
finite element methods are the only way to analyze complex
three-dimensional structures, response of which to applied
load system cannot be predicted in any other way. However,
results obtained by the finite element methods are subject to
the boundary conditions used, rely greatly on the accurate
knowledge of material properties, depend on accurate
representation of the object's geometry, and are sensitive to
the shape and size of elements employed in modeling. All of
the information necessary to "run" the finite element models
is obtained either from published data (for example, material
properties), from design specifications (object's geometry),
or from experimental studies (boundary conditions, shape and
size of elements).

As is often the case with new and powerful methods, the finite
element method has been over-used, perhaps even misused. Only
recently have we begun to realize that virtually all versions
of FEM contain some shortcomings. As a result, the need for
unifying (merging, coupling) FEM in the physical and time
domain with other methods has begun to manifest itself (see,
for instance, Kardestuncer (1975, 1978, 1979, 1980, 1982), and
Zienkiewicz et al. (1977, 1980). Here we are interested in
exploring the unification of laser experimentation with FEM in
space and time simultaneously.

Other experimental techniques which can be used in conjunction
with the finite element methods employ strain gauges,
photoelastic procedures, etc. These experimental procedures,

although conventionally used, do not provide all the information necessary to reliably model an object's response to the applied load system by the finite element methods. For example, strain gauges give only pointwise information for the surface of the object directly under them; to obtain a complete strain mapping a large number of strain gauges must be bonded to the surface of interest. The procedure, moreover, is invasive and interferes with the object's performance. In photoelastic modeling, on the other hand, an oversimplification is made because the object is formed in plastic which has properties totally different from those of the actual material the object is made of. Although such a model, when observed under polarized light, is very useful in identification of stress fields, it does not represent the true response of the object to the applied force system.

Ideally, what is needed is an experimental method that would provide necessary displacements and/or deformations at any point on the investigated object. Also, the results should be provided in three dimensions with high accuracy and precision in such a way as not to interfere with the object's performance.

Recent advances in the field of optoelectronics have led to development of methods satisfying the above requirements. These methods utilize lasers as a source of light and as such can be called the laser methods. Although there are several laser methods available today, of particular interest to finite element analysis are: (i) hologram interferometry, (ii) heterodyne hologram interferometry, (iii) laser speckle metrology, (iv) fiber optic metrology, and (v) directed light beam metrology. Each of these methods has certain characteristics which make it particularly useful in specific experiments. In general, however, all of these methods allow highly accurate, precise, noninvasive, rapid determination of the object's response to the applied load system.

In this paper, laser methods are described, including their representative applications, with particular emphasis on their unification with the finite element methods to improve the results.

UNIFICATION IN ERROR ANALYSIS

The most important issue in any approximate procedure in engineering is the accuracy of the results. How good are they? What are the upper bounds of errors? Such questions have always been asked though answers have not always been found. Nevertheless, problems were solved and systems were put into service. The easiest response to these questions has always been the use of a factor of safety (FS) big enough to accommodate all uncertainties. How big should it be has, of course, been another question. If it was big enough, the engineer was successful; if not, he was doomed.

An alternative approach to these questions has been to experiment (full scale, half scale, whatever) before putting the system into service. Recognizing that things designed and built yesterday are not as complex as those designed and built today, experimentation and the choice of FS were relatively easier tasks than they are today. In recent years, however, the availability of numerical tools (both in respect to methodology and equipment) has enabled engineers to design very complicated systems by successfully solving very difficult problems. Nevertheless, one question raised above still remains: how good are the results?

When we examine the finite element methods, for instance, we find that error sources are quite numerous. Basically, these sources can be categorized as mathematical modeling, discretization, and manipulation (Melosh and Utku (in print)). In addition, each of these has many subdivisions of error sources. To address all and come up with a generally acceptable methodology for error bounds might very well be the most difficult task in numerical methods today.

Some of the error sources are rather general--tool-dependent (i. e., they include errors due to equipment, methodology, solution procedures, etc.); others are very specific--problem-dependent (i. e., they include physical and geometric characteristics of the domain). The latter are more difficult to deal with.

Many have addressed the question of error bounds for problems of the first kind in finite element methods; in particular, are works by Babuska and Rheinboldt (1977, 1978, 1980), Szabo and Mehta (1978), Peano et al. (1979), Utku and Melosh (1984), and a very fine work on a posteriori error analysis by Kelly et al. (1983). Error bounds and controls for problems of the second kind, in particular for those which are time and path dependent, have yet to be established.

When it comes to the finite element methods, certainly h and p (mesh size and order of polynomials, respectively) are the more important (or, the easier) parameters to play with in estimating or even minimizing error bounds which are due to discretization only. The work by Kelly et al. (1983), cited earlier, estimates and minimizes error bounds based on information obtained during the solution process itself. Using two independent error measures consisting of an error indication and an error estimation, they establish certain criteria for where to refine a given mesh and when to stop adaptive processes. The programs developed using either or both (the latter, often referred to as the pony express policy, is claimed to be the better) are called self-adaptive processes because they require no interaction on the part of the user. Supposedly, it is also more practical and less expensive than theoretical a priori error estimates and classical approaches requiring multiple analyses. Self-adaptive processes, however, are tested for linear and

self-adjoint boundary value problems only. One of the main
features of <u>a posteriori</u> error analysis presented by Kelly et
al. (1983) is that it involves local rather than global
computations. It also necessitates establishing an energy
functional beforehand for a given problem.

In spite of many useful properties of energy functionals,
there are a good many problems for which one can not come up
with a functional which is valid for all stages of the
problem. Furthermore, in self-adaptive processes, the
coefficients characterizing the physical domain must be
well-defined and their variations in respect to time and path
must be sufficiently smooth. Otherwise, codes developed based
on h- and p-processes will be insufficient. Nevertheless,
they cover at least one aspect of error analysis and
minimization. The fact remains, however, that development of
fully automatic self-adaptive processes is one of the most
crucial needs of finite element computations today. To
achieve this, one must not develop algorithms based on the
computed information alone. Instead, information based on
actual measurements made during the processing must also be
incorporated into the algorithm. These measurements
(observations) should be employed not for veryfying the
computed values of the unknown function (as is often done) but
for estimating and even controlling errors.

When discretizing the domain, engineers generally pay
attention to certain regions of the domain which are critical
or very sensitive to changes in h- and p-refinement
parameters. Localized error norms in these regions may
fluctuate drastically or even diverge as in the case of
ill-conditioned systems of equations. If u_c and u_m represent
the computed and measured values of the unknown function,
respectively, then the error can be defined as either
$e_c = u - u_c$ or $e_m = u - u_m$, where u is the correct answer.
Moreover, if the measurements are of very high precision, then
we suggest that the latter be employed for error estimates and

for adaptive processes in the critical regions of the domain.

In structural mechanics problems, the energy of the error corresponding to a particular solution is given as

$$\eta^2 = -\int_{\Omega} e \, r \, d\Omega \,, \tag{1}$$

where r represents the residual forces (Kelly et al. (1983)). Following the same procedure as in Kelly et al. (1983) one can obtain the possible refinement on u by using a hierarchic mode N_{p+1} (the finite element basis function for the polynomial of degree p+1). Since the energy absorbed by this additional mode is assumed to be directly proportional to the corresponding force and inversely proportional to the stiffness, the above equation for the ith element becomes

$$\eta_i^2 = \frac{\left[\int N_{p+1} \, r_i \, d\Omega_i \right]^2}{K_{ii}^{p+1}} \,. \tag{2}$$

Hence, this procedure suggests that among all the available N_{p+1}, the one that gives the greatest error decrease should be chosen as the new refinement. One should, however, make sure that N_{p+1} is not orthogonal to r_i otherwise $\eta_i = 0$ indicating that the estimate may be deceptive.

UNIFICATION IN EVALUATING ELEMENT MATRICES

Experimental techniques can also be used for direct evaluation of the element stiffness and/or flexibility coefficients. In particular, when an element's shape is irregular (i. e., possesses curved lines and surfaces, which is often the case at the free boundaries) or when the material is anisotropic, composite, nonlinear, or stratified, the computed stiffness matrix, even using higher order isoparametric elements (implying p-refinements), may not yield the accuracy desired.

The h-refinements for those elements would, on the other hand, increase the number of equations to be solved, thereby decreasing the accuracy of the results. In the case of solid elements, for instance, (whether one-, two-, or three-dimensional), the stiffness matrices can be obtained experimentally. For this we shall refer to Castigliano's second theorem in tensor notation (Kardestuncer (1977)). Thus,

$$p^{iq} = \frac{\partial W}{\partial u_{iq}} , \tag{3}$$

where

$$W = \frac{1}{2} p^{iq} u_{iq} , \quad i = 1, ..., n , \quad q = 1, 2, 3 \tag{4}$$

represents strain energy stored in the element. Substituting Eq. 4 into Eq. 3 and keeping in mind that

$$\frac{\partial u_{jr}}{\partial u_{iq}} = 1 \quad \text{for } i, q = j, r, \text{ respectively; } \quad \text{zero otherwise,}$$

the result is

$$p^{iq} = \frac{\partial p^{iq}}{\partial u_{jr}} u_{jr} , \tag{5}$$

which can be written as

$$p^{iq} = K^{ijqr} u_{jr} . \tag{6}$$

Kardestuncer (1969) has investigated the tensorial properties of Eq. 6 and the similarities with the following well-known tensorial equation in solid mechanics

$$\sigma^{iq} = E^{ijqr} \epsilon_{jr} . \tag{7}$$

Note that Eq. 7 is a physical equation (Hooke's Law) without geometry (i. e., direction but no distance) whereas Eq. 6 contains geometry as well as physics.

The bivalent version of the quadrivalent tensor on the right hand side of Eq. 6 is the stiffness matrix of the element. This equation indicates that the stiffness (or flexibility) matrix coefficients can be determined by observing (measuring) the variation of p^{iq} in respect to u_{jr} or vice versa. Note that in this equation i and j represent the nodes (the integration points in the standard FEM) of the element and q and r are the directions of coordinate axes (local and global). Today, there are many high precision instruments that can evaluate the stiffness (or flexibility) matrix coefficients of an element of any shape and material. Here, we emphasize the use of laser technology for such evaluation. Since the measurements are continuous (independent of time and path), the stiffness matrix coefficients for those elements (highly nonlinear both in respect to time and path) can be determined at any increment of time and/or load. These coefficients can then be incorporated into the global K prior to the solution procedure.

Let us assume that the overall final stiffness matrix is partitioned as follows

$$
\begin{bmatrix} p_1 \\ \hline p_{II} \\ \hline p_{III} \end{bmatrix} = \begin{bmatrix} \left[K_{I,I}\right]_{c,m} & \left[K_{I,II}\right]_{c,m} & \left[K_{I,III}\right]_c \\ \hline & \left[K_{II,II}\right]_{c,m} & \left[K_{II,III}\right]_c \\ \hline \text{SYMMETRIC} & & \left[K_{III,III}\right]_c \end{bmatrix} \begin{bmatrix} u_{I,c} \\ u_{II,c} \\ u_{III,m} \end{bmatrix} \quad , \tag{8}
$$

where the subscripts c and m identify the computed and measured entities, respectively. Note that in certain portions of K, the measured and computed elements are coupled and identified with subscripts c,m. Equation 8 can be further reduced to

$$
\begin{bmatrix} p_i - \begin{bmatrix} K_{i,m} \end{bmatrix}_c u_{m,m} \\[2ex] p_u - \begin{bmatrix} K_{u,m} \end{bmatrix}_c u_{m,m} \end{bmatrix} = \begin{bmatrix} \begin{bmatrix} K_{i,i} \end{bmatrix}_{c,m} & \begin{bmatrix} K_{i,u} \end{bmatrix}_{c,m} \\[2ex] \begin{bmatrix} K_{u,i} \end{bmatrix}_{c,m} & \begin{bmatrix} K_{u,u} \end{bmatrix}_{c,m} \end{bmatrix} \begin{bmatrix} u_{i,c} \\[2ex] u_{u,c} \end{bmatrix} \quad . \tag{9}
$$

Since the left hand side of this equation contains all the known entities (whether given, computed, or measured), its solution is possible and will yield the unknown values of the function at the nodes where no measurements have been taken.

Once we have determined $u_{i,c}$ and $u_{u,c}$, we can compute the residual force vector as

$$
r = p_m - \begin{bmatrix} K_{m,i} \end{bmatrix}_c u_{i,c} - \begin{bmatrix} K_{m,u} \end{bmatrix}_c u_{u,c} - \begin{bmatrix} K_{m,m} \end{bmatrix}_c u_{m,m} \neq 0 \quad . \tag{10}
$$

The components of this vector corresponding to element i can then be utilized in Eq. 2 to determine the next refinement for the adaptive processes presented by Kelly et al. (1983). We shall now present various laser methods to obtain the measurements mentioned above.

HOLOGRAM INTERFEROMETRY

The most useful of all methods of hologram interferometry, for finite element applications, is the two-beam, off-axis method (Pryputniewicz (1979, 1982a)). In this method, the laser beam is divided into two beams, as shown in Fig. 1. One of the beams is made to interact with the object, or scene being recorded (the so-called object beam) while the other beam does not interact with the object at all. In fact, the second beam is a reference beam against which the object beam is recorded. A setup for recording holograms is made so that the object beam and the reference beam overlap in a given region of space. As a result of this, the two beams interfere with each other and the resulting interference pattern is recorded in a suitable medium (Smith (1977)). The exposed medium, upon processing, becomes a hologram. The hologram is reconstructed with the same setup that was used in recording, except that now it is illuminated with the reference beam alone. Of the images produced during reconstruction, the most applicable to finite element analysis is the virtual image. To observe the virtual image, the reconstruction should be viewed through the hologram as if it were a window. The image is seen in the space which was occupied by the object while the hologram was recorded, even though the original object had since been removed. The image observed has all the visual characteristics of the original object. In fact, there is no visual test that can differentiate between the two.

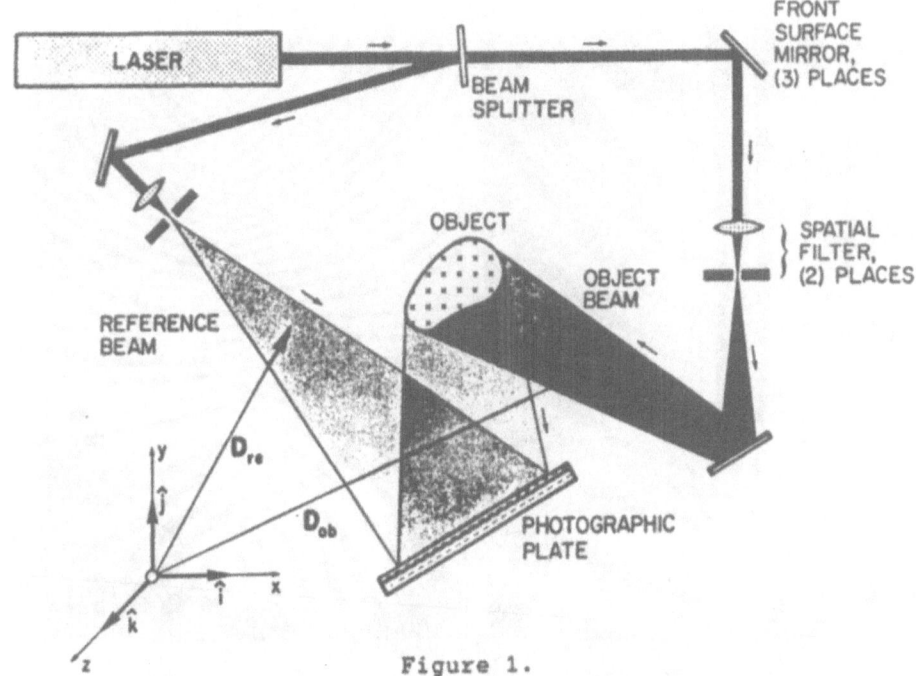

Figure 1.

Setup for recording and reconstruction of holograms. Directions of propagation of the object beam and the reference beam are defined by position vectors **D** specified in respect to the x-y-z coordinate system.

There are three basic variations of hologram interferometry:
(i) real-time, (ii) time-average, and (iii) double-exposure.

Real-time hologram interferometry involves recording a single
exposure hologram as shown in Fig. 1, processing it, and
reconstructing it by illumination with the original reference
beam. The reconstructed image is superimposed onto the
original object which is also illuminated with the same beam
as used in recording the hologram. If the object is now even
slightly displaced and/or deformed, interferometric comparison
between the holographically reconstructed image and the new
state of the object occurs instantaneously (Fig. 2). The
particular advantage of the real-time method is that different
types of motion, dynamic as well as static, can be studied
with a single holographic exposure.

Figure 2.
Images obtained using real-time hologram interferometry:
studies of microcracks in porous, ceramic components.

In <u>time-average interferometry</u> a single holographic recording of an object undergoing a cyclic vibration is made. With the exposure time long compared to one period of the vibration cycle, the hologram effectively records an ensemble of images corresponding to the time-average of all positions of the object during its vibration. In the reconstruction of such a hologram, interference occurs between the entire ensemble of the recorded images, with the images recorded near zero velocity contributing most strongly. As such, images reconstructed from the time-average hologram have intensity distribution given by the zero-order Bessel function (Fig. 3). In the case of stroboscopic illumination of a vibrating object, however, cosinusoidal intensity distributions are obtained.

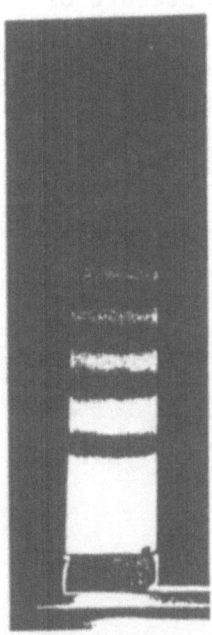

Figure 3.
Time-average hologram
of a vibrating
cantilever beam:
the first flexure
mode.

The <u>double-exposure hologram interferometry</u>, which can be considered to be a special case of the time-average method (where only two exposures of the object are made in the same medium), is the most widely used of all holographic methods. In this method, the object is displaced and/or deformed between the two exposures. Therefore, the object beam during the second exposure is different from that used in making the first exposure. During reconstruction of the double-exposure hologram, both object beams are faithfully reconstructed, forming images of the object's initial and final positions. Since these images are formed in coherent laser light, they interfere with each other forming a pattern of bright and dark fringes resulting in cosinusoidal intensity variation of the image (Fig. 4). These fringes are a direct measure of changes in the object's position and/or shape which occured between the two exposures.

Figure 4.
Double-exposure hologram
of a hydraulic cylinder:
pressure between the
exposures was increased
from 5,100 psi to 5,800 psi.

QUANTITATIVE INTERPRETATION OF HOLOGRAMS

There are a number of methods dealing with interpretation of the fringes observed within the holographically reconstructed image (Stetson (1979), Schuman and Dubas (1979), Vest (1978), Pryputniewicz (1980a), Pryputniewicz and Stetson (1976, 1980)). The most general of these methods employs multiple observations of the holographic image. It results in displacement vector u expressed as a product of the inverse of the matrix formed by the sum of the projection matrices B with the matrix representing the sum of the observed vectors u_{ob} (Stetson (1979), Pryputniewicz and Stetson (1980), Pryputniewicz (1980a))

$$u = \left[\sum_{i=1}^{n} B^i \right]^{-1} \left(\sum_{i=1}^{n} u_{ob}^i \right) \quad . \tag{11}$$

In Eq. 11, i denotes the observation number with n being the total number of observations while u_{ob} is measured in the plane normal to the direction of observation and is defined by the corresponding B. The projection matrix B^i, for the ith direction of observation, can be either one of the following two types. If the projection is made in a direction parallel to the direction of observation, then the projection is normal; if it is not, the projection is oblique. The normal projection is defined as a difference between the identity matrix and the matrix resulting from the dyadic product of the ith unit observation vector with itself. In the case of the oblique projection, the matrix is formed by the dyadic product of the object's surface unit normal vector with the unit vector defining the particular direction of observation.

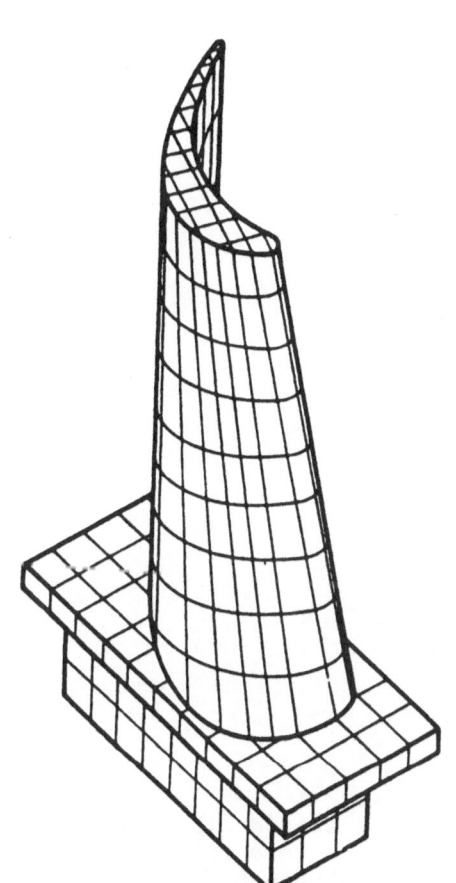

Figure 5.
Typical finite element
breakup of an airfoil.

Of particular interest in FEM modeling (Fig. 5) is the application of double-exposure hologram interferometry in determination of strains and rotations (Pryputniewicz and Stetson (1976)). In this case, the strain-rotation matrix f is determined directly from the parameters S (defining illumination and observation geometry) and S_f (defining shape and distribution of fringes seen during reconstruction of a hologram)

$$f = \left[S^T S \right]^{-1} \left[S^T S_f \right] \qquad (12)$$

Decomposition of the matrix f into the symmetric part e and the antisymmetric part θ gives strains and rotations, respectively.

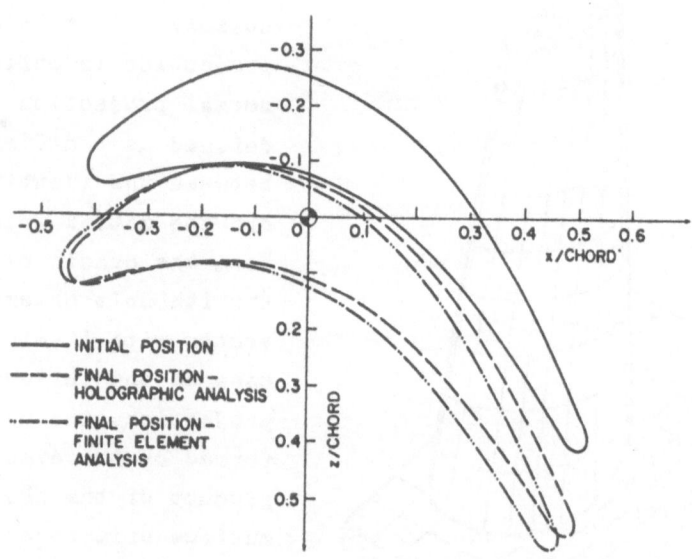

Figure 6.
Displacements of a radially loaded airfoil:
the finite element computations were subject to
the boundary conditions obtained from
the double-exposure holograms.

The matrix S_f appearing in Eq. 12 consists of fringe vectors, one for each direction of observation, which can be computed from the fringe patterns produced during reconstruction of the holograms, that is,

$$\omega = S_f \cdot D \quad , \qquad (13)$$

where ω is the fringe-locus function, constant values of which
define fringe loci on the object's surface, and D specifies
coordinates at which the specific values of ω were determined.
Knowledge of the fringe vector is essential in quantitative
interpretation of holograms (Fig. 6). The fringe vector, as
expressed in Eq. 13 and used in Eq. 12, is also helpful is
determining the system's optimum geometry for recording of
holograms.

It should be noted that analysis of holographically produced
fringes does not depend on material properties at all. In
fact, the holographic procedures are particularly suited for
quantitative determination of a material's constitutive
behavior.

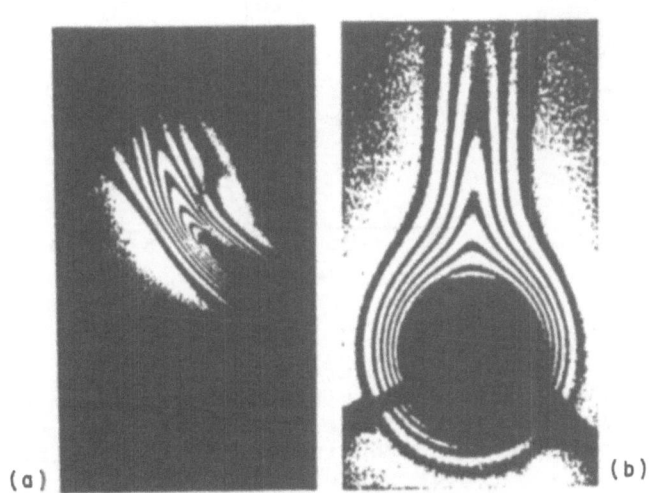

(a)   (b)

Figure 7.
Double-exposure holograms of: (a) heated inclined plate,
(b) heated horizontal rod.

Hologram interferometry is also very useful in heat transfer
studies. For example, Fig. 7 shows typical images recorded
during studies of heat transfer characteristics of flat and

curved surfaces. From reconstructed images, temperature distributions can be determined to within a fraction of one degree, anywhere within the image, without any interference whatsoever with the studied "space".

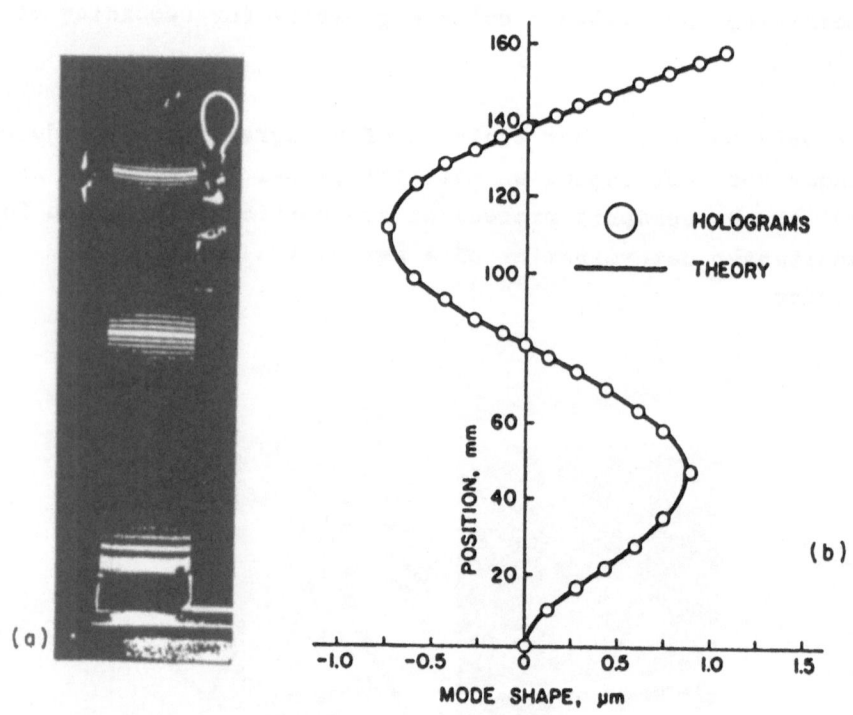

(a)

(b)

Figure 8.
Quantitative study of the vibrating beams: (a) the time average hologram of the cantilever beam, (b) displacements corresponding to the third flexure mode shown in (a).

In the case of time-average hologram interferometry, displacements are found from (Pryputniewicz (in print))

$$u_z = \frac{\lambda}{2\pi(\hat{S}_{2_z} - \hat{S}_{1_z})} |\omega_1| \quad , \tag{14}$$

where λ is the laser wavelength, $|\omega_t|$ is the argument of the zero order Bessel function, while $\hat{S}_{1Z}$ and $\hat{S}_{2Z}$ are components of the unit vectors defining directions of illumination and observation, respectively. Typical results for a vibrating cantilever beam are given in Fig. 8 showing good agreement with the theoretical predictions. In the case shown, the beam is vibrating in the third flexure mode as vividly depicted by the hologram (Fig. 8-a) where nodes are demarcated by the brightest fringes and antinodes by the darkest fringes; for this mode, the theoretically predicted frequency was 1772 cps, while that determined experimentally was 1733 cps. Also, Fig. 8-b shows very good agreement between the mode shape determined from the hologram and the mode shape predicted by a theoretical model, which was developed to simulate beam vibrations. However, it should be noted that the theoretical computation was subject to the boundary conditions which were provided from the results obtained directly from the holograms. As shown in Fig. 8-b, the maximum displacement of the beam, vibrating in the third flexure mode, is 1.05 microns.

In the manner similar to that described above, mode shapes at other frequencies can be studied using methods of hologram interferometry.

HETERODYNE HOLOGRAM INTERFEROMETRY

Heterodyne hologram interferometry is similar to the double-exposure hologram interferometry in that, there also, two images of an object, at different states of stress, are recorded in the same medium. However, each of these images is recorded with a different reference beam, in such a way that the reference beams can later be reconstructed independently (Dandliker et al. (1976), Pryputniewicz (1982b)). This allows introduction, during the reconstruction process, of a small frequency shift between the two reconstructed and interfering light fields, resulting in an intensity modulation at a beat

frequency of these light fields, for any point within the
interference pattern.

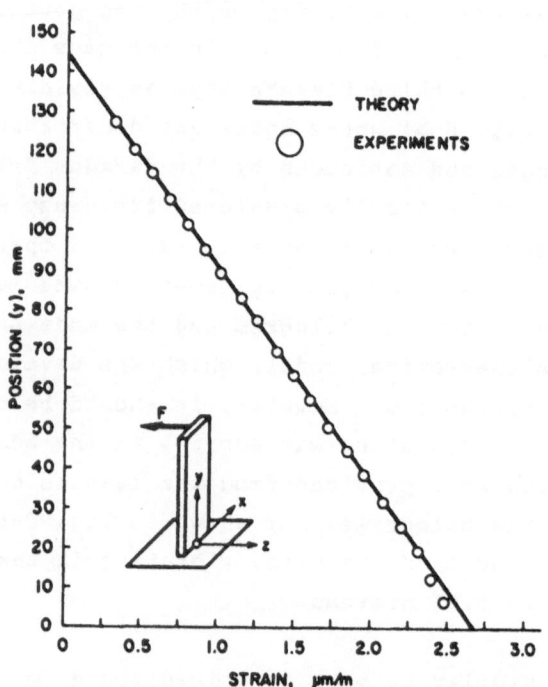

Figure 9.
Strains determined from a
heterodyne hologram of a loaded
cantilever beam.

The optical phase difference, corresponding to the
displacement and/or deformation recorded within the hologram
being reconstructed, is converted into the phase of the beat
frequency of the two interfering light fields. This phase, in
turn, is interpolated optoelectronically, resulting in
determination of fringe orders to within 1/1000 of one fringe.
This high accuracy in determining fringe orders leads to
determination of displacements to within 0.3 nm, and strains
to within 0.000,02 % (Pryputniewicz (1982a, 1982b)).

Representative results obtained using heterodyne hologram
interferometry are shown in Fig. 9. In this case, a prismatic
cantilever beam was loaded in the direction normal to its
neutral plane, between the exposures of the heterodyne
hologram. Resulting interferograms were scanned by placing a
fiber-optic detector probe in the image plane formed by a lens
placed between the hologram and the detector (Pryputniewicz
(1982b). The resulting phase measurements were then processed
using the equations relating them to parameters characterizing
the system used to record, reconstruct, and scan the
heterodyne hologram.

Figure 9 shows that the results obtained from the heterodyne
holograms correlate very well with the theory. It should be
noted that the measured strains ranged from 0.3 microns/m to
2.5 microns/m and were well below the values that can be
reliably detected by conventionally used strain measuring
devices. Also, the results presented in Fig. 9 were obtained
without contacting the object at all, and without interfering
with it in any other way. All measurements were made remotely
by scanning the object's image, thus producing the results in
a truly noninvasive manner.

SPECKLE METROLOGY

Any object illuminated with laser light will seem to have a
granular appearance. That is, its surface will appear to be
covered with fine randomly distributed light and dark
irregular spots. If the observer moves, these spots appear to
twinkle and move relative to the object. This phenomenon is
caused by each point on the object scattering some light
toward the observer. In fact, the laser light scattered by
one point on the object's surface interferes with the light
scattered by other points. In any region of space where these
light fields overlap, a random pattern of interference spots
is observed. These interference spots are known as

"speckles". The size of the speckles depends on optical properties of the imaging system and directly influences the accuracy of measurements: the finer the speckles the higher the accuracy.

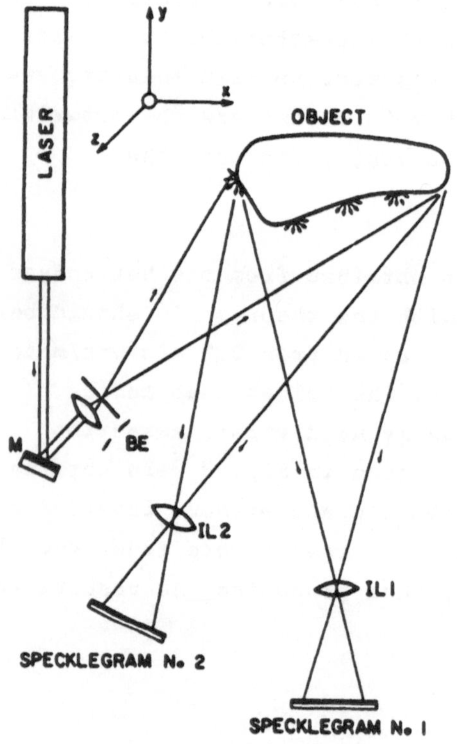

Specklegrams are recorded by illuminating an object with a single laser beam; no reference beam is used (Fig. 10). The light scattered by the object (or transmitting medium in the case of fluid flow or gas dynamics applications) is imaged from one or more directions onto a high resolution recording medium. For interferometric purposes, two exposures are made in the same medium to record the object's initial and final configurations, unless tandem specklegrams are used where each configuration is recorded in separate media which are later "sandwiched" together.

Figure 10.
Setup for simultaneous recording of two specklegrams from two different directions.

Developed specklegrams are analyzed by sending a narrow laser beam directly through the specklegram (Fig. 11). Upon passing through the specklegram, the illuminating beam diffracts and forms a halo which is modulated by Young's fringes (Fig. 12). The frequency of Young's fringes is directly proportional to the magnitude of the displacement recorded by the specklegram, while their direction is normal to the direction of this displacement.

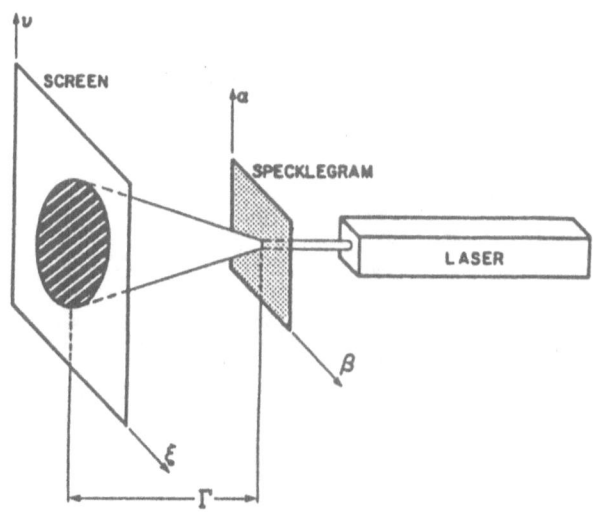

Figure 11.
Setup for reconstruction of specklegrams.

Recent studies (Stetson (1978), Pryputniewicz and Stetson (1980), Pryputniewicz (1980b)) show that the equations governing determination of displacements from specklegrams are exactly the same as those used for quantitative interpretations of holograms. That is, Eq. 11 applies directly in quantitative speckle metrology. This equation indicates that two specklegrams recorded from different directions are sufficient to compute three-dimensional displacements of loaded objects.

Figure 12.
Typical Young's fringe pattern observed during reconstruction of a double-exposure specklegram.

The parameters necessary to interpret specklegrams are
obtained directly from the geometry of the recording and
reconstructing systems (Figs 10 and 11, respectively) and from
the observed Young's fringes (Fig. 12).

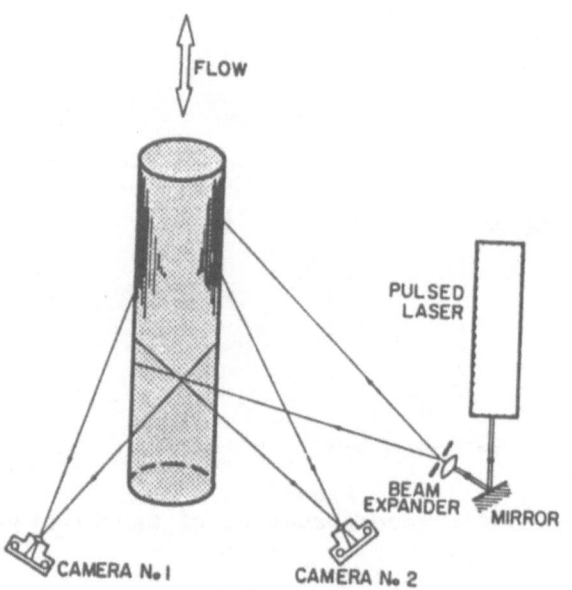

Figure 13.
Setup for simultaneous recording of two
specklegrams in fluid flow analysis.

The speckle metrology finds particular applications in studies
of three-dimensional displacements of solid objects, in
studies of fluid flow (Fig. 13), and in gas dynamics. In
these applications, the speckle methods allow recording of the
displacement and/or deformation pattern over the entire
surface of the object, permit recording of the entire velocity
profiles or the thermal profiles and are particularly suited
to studies of dynamic as well as transient behaviors.

COMPUTER AIDED INTERPRETATION OF LASER IMAGES

In FEM modeling, coordinates of nodal points are known. To specify boundary conditions at these nodes, their position has to be established and reproduced while using the experimental methods. One of such methods involves scanning the holographically reconstructed image (or a diffraction halo obtained during reconstruction of a specklegram) with a computer compatible video digitizer, as shown in Fig. 14. The digitizer, in addition to converting the scene being observed into a composite analog video signal which is viewed on a monitor, produces a digital signal that is transmitted directly to a computer. The computer, in turn, rapidly reads the electronic signal corresponding to the video image being digitized. It processes the digitized data, producing plots of intensity distribution within the image plane. Data characterizing these intensity distributions, together with other pertinent parameters, are used in quantitative interpretation of laser images. These results can be obtained

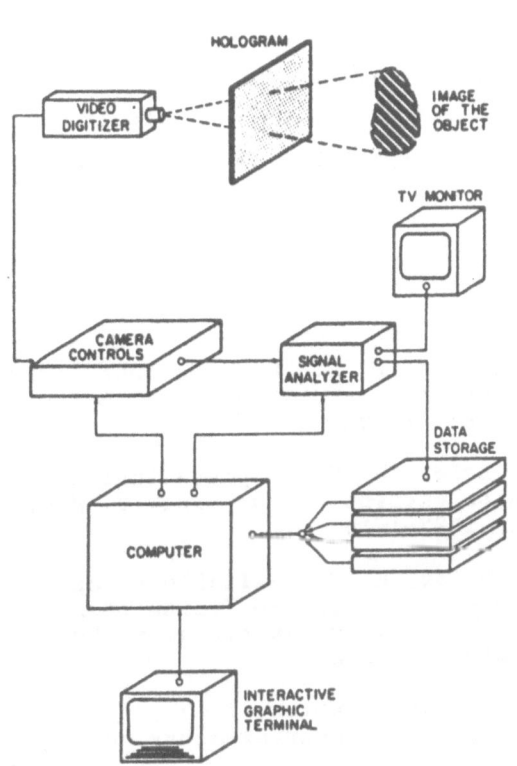

Figure 14.
Schematic of a computer controlled system for automated interpretation of holograms.

for any point within the reconstructed image by simply
instructing the computer to perform calculations for a point,
or a number of points, at specified coordinates.

A system such as that shown in Fig. 14 will provide a unique
capability for unification of finite element methods with
laser experimentation. As such, it will lead to the
development of a fully automated system for quantitative
analysis of structural deformations, which will provide highly
accurate and precise results at any point on the surface of
the studied objects.

REFERENCES

1. Babuska, I., and Rheinboldt, W. C., Computational aspects
    of the finite element method, in: Mathematical Software,
    Vol. III (Academic Press, New York, 1977).

2. Babuska, I., and Rheinboldt, W. C., A posteriori error
    estimates for the finite element method, Int. J. Num.
    Meth. Engr., 12 (1978) 1597-1615.

3. Babuska, I., and Rheinboldt, W. C., Reliable error
    estimation and mesh adaptation for the finite element
    method, in: Oden, J. T. (ed.), Computational Methods in
    Nonlinear Mechanics (1980) 67-108.

4. Dandliker, R., Marom, E., and Mottier, F. M.,
    Two-reference beam holographic interferometry, J. Opt.
    Soc. Am., 66 (1976) 23-30.

5. Kardestuncer, H., Tensors in discrete mechanics, Tensor
    Quarterly - TSGB, 20 (1969) 1-9.

6. Kardestuncer, H., Descrete Mechanics: A Unified Approach
    Springer-Verlag, Vienna, 1975).

7. Kardestuncer, H., Proceedings of the UFEM Symposium
    Series (University of Connecticut, Storrs, CT, 1978,
    1979, 1980, 1982).

8. Kardestuncer, H., Tensors versus matrices in discrete
    mechanics, in: Branin, F. H., Jr., and Huseyin, K.
    (eds.), Problem Analysis in Science and Engineering
    (Academic Press, New York, 1977).

9. Kelly, D. W., de Gago, J. P., Zienkiewicz, O. C., and

Babuska, I., A posteriori error analysis and adaptive processes in the finite element method: Part I -- Error analysis, Part II -- Adaptive mesh refinement, Int. J. Num. Meth. Engr., 19 (1983) 1593-1619.

10. Melosh, R. J., and Utku, S., Efficient finite element analysis, to appear in: Kardestuncer, H. (ed.), Finite Element Handbook (McGraw-Hill, New York).

11. Peano, A. G., Pasini, A., Riccioni, R., and Sardella, L., Adaptive approximation in finite element structural analysis, Comp. & Struct., 10 (1979) 332-342.

12. Pryputniewicz, R. J., Laser Holography (Worcester Polytechnic Institute, Worcester, MA, 1979).

13. Pryputniewicz, R. J., State-of-the-art in hologrammetry and related fields, Internat. Arch. Photogram., 23 (1980a) 620-629.

14. Pryputniewicz, R. J., Projection matrices in specklegraphic analysis, SPIE, 243 (1980b) 158-164.

15. Pryputniewicz, R. J., Unification of FEM modeling with laser experimentation, in: Kardestuncer, H. (ed.), Finite Elements - Finite Differences and Calculus of Variations, (University of Connecticut, Storrs, CT, 1982a).

16. Pryputniewicz, R. J., High precision hologrammetry, Internat. Arch. Photogram., 24 (1982b) 377-386.

17. Pryputniewicz, R. J., Quantitative interpretation of time-average holograms in vibration analysis, in print.

18. Pryputniewicz, R. J., and Stetson, K. A., Holographic strain analysis: extension of fringe-vector method to include perspective, Appl. Opt., 15 (1976) 725-728.

19. Pryputniewicz, R. J., and Stetson, K. A., Fundamentals and Applications of Laser Speckle and Hologram Interferometry (Worcester Polytechnic Institute, Worcester, MA, 1980).

20. Schuman, W., and Dubas, M., Holographic Interferometry (Springer-Verlag, Berlin, 1979).

21. Smith, H. M., Holographic Recording Materials (Springer-Verlag, Berlin, 1977).

22. Stetson, K. A., Miscellaneous topics in speckle metrology, in: Erf, R. K. (ed.), Speckle Metrology (Academic Press, New York, 1978).

23. Stetson, K. A., The use of projection matrices in hologram interferometry, J. Opt. Soc. Am., 69 (1979)

1705-1710.

24.   Szabo, B. A., and Mehta, A. U., P-convergence finite
      element approximations in fracture mechanics, Int. J.
      Num. Meth. Engr., 12 (1978) 551-560.

25.   Utku, S., and Melosh, R. J., Solution errors in finite
      element analysis, Comp. & Struct., 18 (1984) 379-393.

26.   Vest, C. M., Holographic Interferometry (Wiley, New York,
      1978).

27.   Zienkiewicz, O. C., Kelly, D. W., and Bettess, P., The
      coupling of the finite element method and boundary
      solution procedures, Int. J. Num. Meth. Engr., 11 (1977)
      355-373.

28.   Zienkiewicz, O. C., Kelly, D. W., and Bettess, P.,
      Marriage a la mode -- the best of both worlds (Finite
      elements and boundary integrals) in: Glowinski, R.,
      Rodin, E. Y., and Zienkiewicz, O. C. (eds.), Energy
      Methods in Finite Element Methods, Ch. 5 (John Wiley, New
      York, 1980).

Moiré Evaluation with Fringe Patterns of Interferograms, Holograms, Moirégrams and Specklegrams

O.D.D. Soares, A.L.V.S. Lage and L.M. Bernardo
Centro de Física, Universidade do Porto, 4000 Porto, Portugal

Abstract

Interferometric based techniques are known to present practical difficulties of sensitivity adjustment, range of measurement and setting--up practicalities. Moiré evaluation techniques have been progressively introduced to solve some of the difficulties and to make available some of the benefits of moiré studied methods. A discussion of moiré evaluation techniques is presented under different perspectives: moiré reference transform, differential moiré, integrative moiré, and moiré derivation. Examples and fields of application of the technique are analised in view of the novel concepts and methods described. The advantages and limitations of the metrologic moiré evaluation techniques are investigated showing profitable aspects for practical application.

Introduction

Moiré evaluation techniques are receiving growing attention in optical metrology (1-12). This is possibly a consequence of two central aspects:

i) The moiré effect is of a fundamental nature, i.e. a direct consequence of spatial frequency spectra comparison.

ii) moiré techniques, in principle, are not subject to severe correlation range limitations and coherent conditions but only to information correlation criterion, Shanon' sampling theorem requesites and visibility threshold.

Consequently, moiré tecniques lead, in general, to a more versatile setting-up and a more flexible adjustment of the desired sensitivity.

Interferometric techniques, however, carry in principle, phase information relative to the light source, and readily produce high contrast fringes inducing great attractiveness in practical applications. Furthermore, the spatial frequency carrier is high, in the order of the inverse of the radiation wavelength so that it results in methods of high and short range sensitivity measurement, and imposes usually very high performances of the optical system and recording medium.

A compromise between benefits and limitations among moiré and interferometric techniques has been actively sought by several authors (4,5,8,10). The dynamical moiré evaluation technique (10) presented here for holography (and in principle for speckle metrology) is also a contribution to solving some of the practical difficulties of the interferometric methods while exploring the advantages of moiré analysis.

Dynamical moiré evaluation (10,12)

The principles of metrology by holography and speckle methods can, in theory, be reduced to the comparison of interferograms (regular and random pattern) or in more general terms to a certain kind of intermodulation of spatial frequencies resulting in a lower frequency spatial spectrum in a direct analogy to moiré methods.

However, both holographic and speckle developed techniques present severe inconveniences of practical importance:

i) the global movement, rigid body motion or the integrative displacement from localized body deformation, mask the effective local deformation or displacement as the actual wavefronts being correlated correspond to the object surface before and after global transformation. Such effects have to be controlled and eventually eliminated. Moiré evaluation techniques are of suitable use towards that aim. Sandwich holography (4), in particular, is one of the solutions already developed.

ii) in repositioning the hologram as for real-time hologrametric interferometry, fringes appear due to minute errors from replacement and fluctuations of the experiment environment parameters that induce wavefront disturbances of the order of the quantities being measured.

iii) in time extended analysis all interferometric techniques require extreme mechanical stability of the entire set-up, thus reducing drastically the possibilities outside laboratory application.

iv) in the time average method the fringe visibility decreases with amplitude of vibration. The alternative method of stroboscopic holography with double pulse requires a large set of holograms to explore the vibration cycle as well as in the case of transient phenomena to perform an extended analysis.

v) in a process where small changes in the microscopic surface structure may occur they lead to decorrelation and rule out the application of interferometric techniques.

vi) speckle techniques present, in principle, a measurement range smaller than the speckle grain size. A large scale phenomenon requires a submultiple division of the interval of measurement and a multiple step by step analysis.

A combination of moiré evaluation techniques with hologrametric interferometry is proposed to overcome these limitations. It is thought that the method bears capabilities of aplication to speckle metrology if fringe quality can be conveniently improved.

These techniques comprise:

i) moiré reference transform

ii) differential moiré

iii) integrative moiré

iv) moiré derivation

and will be presented by way of examples.

The moiré effect could be generated both at the image and Fourier spectrum domain.

The moiré techniques produce, in general, a broadening of the spatial frequency spectrum so that filtering operations are necessary for contrast improvement. This corresponds, in fact, to the spatial separation of real and virtual image in holography.

In view of the applications dealing with dynamical phenomena, details of the tecniques to be described were specially conceived for situations where synchronized illumination pulses (13) would be used.Transients can then be properly analysed by recourse to video recording with synchronized pulse illumination. The introduction of a fast dynamic memory (14) for the image digitizing and moiré evaluation permits a following of transient patterns within speed limits. Various combinations of pulses (15,16) may be used and are being currently studied (13).

Experimental results

i) Moiré reference transform (12)

Interference fringe patterns represent a very odd kind of multiplicative moiré effect. The space within the correlation volume is metrologically coded by the interfering fields.

Let us consider first for the sake of simplicity the case of two plane waves as generalization may be derived by Fourier analysis.

The pitch of the propagation volume gratings are the corresponding lengths along the studied directions that correspond to the wavelength of the radiation when projected along the propagation direction.

Inside the correlation volume of interference a resulting moiré volume grating is generated which is stationary. A plane intersection is, in general, said to be the interference pattern or interferogram.

It is possible to give a different reading of the interferogram. A reference wavefront (or grating) interferes (results in moiré effect of multiplicative kind plus phase interference effect) with an object wavefront. In other words the object wavefront is represented or mapped with respect to the reference or carrier wavefront through the interference pattern.

The question arises, can one change the reference without repeating the hypothetical experiment? Moiré interference is one answer. Fig. 1 describes the well known fact that an interference pattern of two plane waves can be generated in an infinite variety of well defined ways.

While generalization is valid for each individual plane wave of object and reference wavefront spectrum as it is formally recognized by Fourier analysis, the effect of intermodulation does show how complex the situation becomes for the analysis of the general case. However, it is also known that if the reference is changed and the interference pattern is used as diffraction grating with orders of diffraction separable (filtering possibility) there is then a diffracted wave that represents the new object wave which if interfered with the modified reference wave, would generate the same interference pattern (invariance of boundary conditions and principle of holographic reconstruction).

Conversely, if the object wavefront is to be maintained and the reference is to be changed the new corresponding interference pattern can be evaluated by moiré effect between original interference pattern and the interference pattern resulting from the interference between the two reference wavefronts in discussion.

The designation of which of them is the reference wavefront is, in principle, arbitrary so that one may reason in the complementary way. If one maintains the reference wavefront and modifies the object wavefront the result is the change of the interferogram. The moiré effect between the two interferograms represents the interference of the two object wavefronts. In other words the final interference pattern could have been obtained from the first by moiré effect had the change expressed in the moiré pattern been known. In the case of light being diffracted by an object this could refer to a modification of the object surface which could be subsequently taken as the reference to describe further modifications of the illuminated surface. These considerations do apply within the correlation interference volume of the wavefronts, as it is the case of holographic interferometry.

Sandwich holography (4) is a case where there is a correlated change of reference for the two correlated holograms of the same object but corresponding to different configurations. The reference change is produced by the tilting of the holograms. This tilting of the sandwich is seen differently by each hologram due to the glass substrate of holograms and consequently a relative change of reference between the two holograms is observed simulating an effective tilting between reconstructed images of the two holograms.

The change of reference and object wavefront is seen in reconstruction by interferometric fringes produced between reconstructed wavefront and object diffracting wavefront (real-time set-up) so that it may be compensated by appropriate movement of the holographic plate. This has been earlier suggested (17) as a potential means to the establishment of metrologic applications.

Fig.2 is an interferometric hologram representing the global deformation of a radius bone subject to non-axial load, Fig.2 a). The Fig.2 c) shows the result when flexural bending is subtracted by moiré evaluation so that localized deformations are revealed. A digital memory (14) was used to performed subtraction in identical condition in Fig.2 d), e) and f). It should be noticed that adequate patterns may be generated electronically, eventually by recourse to computer so that special operations can be performed in view of the analysis of a particular problem.

Fig.3 shows a human hand in movement illuminated by a stroboscopic light source synchronized with TV frame sequence. Shadow fringes from a linear grating are projected on the hand, Fig.3 a). The hand topography is mapped in relation to a reference plane fringe coded and electronically generated. This is accomplished with moiré subtraction at the dynamical digital memory (14) of the hand projected fringes and reference plane fringes. Fig.3 c) provides the result when the reference plane is changed.

The concepts developed permit one to envisage methods which bring about:

  i) elimination of the influence on information display created by unwanted effects,

 ii) discrimination of composed effects (in plane against out of plane components, rotations and translations rigid body motions versus deformations);

iii) change of reference for adequacy of scale and direction of sensitivity.

Applications of the method include subtraction of the effect of rigid body movement, discriminated visualization of the movements of parts of composed structures, and comparison of form of surfaces.

ii) Differential moiré (10,11,12)

In a simplified form one may assume the recording of an object at three stages of evolution maintaining coherence or information correlation between the three images through a common reference image or carrier. Let us consider the first image as representing a spatial spectrum that is used as the carrier. In holography the corresponding reconstructed image from first recording will produce two interferograms (correlated specklegrams or moirégrams are also viable) when interfered with diffracted light from the object at second and third stage of evolution. Comparison (additive or multiplicative) between interferograms will generate moiré fringes equivalent to the interferogram that results directly from interference between the said second and third object image. In principle, the result obtained is independent of the choice of the first image so that in real-time technique it will occur desensitizing to: repositioning, mechanical instabilities and other disturbing effects, while global movements do not interfere. This assumes that perturbations of the kind are irrelevant between second and third image.

The example cited on previous section of radius deformation under non-axial load can be interpreted also as differential moiré.

Pulsing the illumination beam (12) allows the study of vibrations over the complete cycle with a sole hologram and lower levels of illumination. The illumination beam is pulsed according to selected instants of the vibrating cycle.

According to Fig.4 the first pulse of light can create a hologram or, in general, an image, I_1. To make the description simpler let us continue to consider the case of holographic interferometry. A second pulse of light produces a holographic interferogram I_{12} say using real-time holographic interferometry. A third pulse of light provides a second holographic interferogram, I_{13}. Additive moiré (or multiplicative) originates a moirégram that represents the deformation between second and third light pulse, I_{23}.

Fig.5 shows the comparison between a set of two membranes under deformation. The Fig.5 d) represents the result of differential moiré and Fig.5 c) the corresponding double exposure interferometric hologram at same exposures of the second and third light pulse. Upper and lower membrane diverge in number of fringes, i. e. spatial frequency carrier so that contrast is different and implications of the Shanon'sampling theorem are also reflected in the moiré fringe visibility.

Fig.6 reports the application of differential moiré to the holographic study of vibrating membrane (12).

Fig.6 b) is the holographic restitution as seen on the TV-monitor of the vibrating membrane at rest. Fig.6 f) is the holographic image at pulsed illumination after the hologram was tilted (θ_z = 0.5 mrad) and translated (Δx = 100 μm) to increase the spatial frequency carrier for sensitivity, and moiré fringe contrast improvement, combined with reference transform for adequate filtering of background fringes. Fig.6 e) results from differential moiré between Fig. 6 f) and Fig. 6 g) interferograms of two deformation states of the vibration mode. Fig.6 k) and 1) correspond to the use of the same initial hologram to study other resonant modes equally.

Invariance on the first image (the carrier) largely attenuates repositioning errors (if processing in situ is not used e.g. thermoplastic film), and desensitizes to disturbances, defects and aberrations, contributing to an increase of the measurement range.

The recording of a sole initial hologram permits the lengthy analysis of: deformation, vibration cycle and the study of the resonant mode spectrum. Ambiguity in the direction of the movement (hill or valley) is eliminated in relative terms by convenient coding of the object space through the known displacement and filtering of the hologram.

The width of the opaque and transmission bands of the interferograms may be different resulting in drastic reduction of moiré fringes intensity. In applying the method care should be taken to obtain sharp fringes particularly when additive moiré fringes are to be generated. Image digitizing and enhancing techniques can also be used but at the expense of processing speed.

Applications include vibration analysis, loading-unloading cycles, hysteresis, measurement of local deformation amplitude by recognition of the tilting of the object tangent plane, and fringe interpolation by increasing the spatial frequency carrier.

The inconveniences of the use of a traditional non-erasable recording medium can advantageously (latent memory and spatial frequency resolution improvement) be overcome by recourse to phase wave conjugation techniques (18). Furthermore, it is advisable to use a dynamical digital memory (14) for quasi-real-time differential moiré analysis.

Fig.7 shows schematically a real-time experimental arrangement in transmission configuration. Two holograms from a surface (front surface of a plane mirror) before and after deformation are stored in the crystal. The interferogram resulting from reconstruction of stored holograms is shown in Fig.8 a).If a further deformation of the surface occurs and real-time reconstruction of the hologram interferes with object wavefront, it results in the pattern of Fig.8 b) which represents the moiré effect as it would be obtained combining the interferogram of Fig.8 a) with the interferogram representing the surface evolution between two configurations of the said deformation, Fig.8 c).

Fig.9 refers to the quasi-real-time implementation of the technique using the digital dynamic memory. Only the first hologram is then stored in the crystal. Fig.9 a) is the interferogram in real-time reconstruction of the surface interfered with deformed configuration. This interferogram is stored in the digital memory.

Another deformed configuration generates a second interferogram, Fig.9 b) received on the vidicon target and digitally subtracted from the stored interferogram in the memory. This incoherent operation produces a moiré pattern, Fig.9 c) seen on the TV-monitor.

This implementation of the technique presents the advantage that a first hologram can be memorized in the crystal for instance at the rest position of a vibrating structure. Illumination intensity does not represent then a major problem. Second and third exposure can be done with low levels of light intensity so that Laser power requirements are less stringent, or alternatively, larger surfaces can be examined. Therefore, the method is appropriate for the study of cyclic phenomena where availability of illumination intensity is precarious.

Analogous results can be sought for other means of information representation such as correlated moirégrams.

Generalization of the principle is, in theory, possible as shown by the diagram of Fig.10.

Contouring is also a differential method so that it deserves consideration but because of its special nature will not be dealt herein.

iii) Integrative moiré (10,12)

Fig.11 exemplifies the concept of integrative moiré. The first image, I_1, is compared to a second exposure I_2, while an assumed deformation or another stage of evolution is taking place. This results in an interferogram or moirégram, I_{12}. The second image itself, I_2, is compared with a third one, I_3, corresponding to a further stage of deformation, resulting in a new fringe pattern, I_{23}. The said second common image may be used to transpose the correlation to the extremes of the interval. The product of the two interferograms or moirégrams, $I_{12} \times I_{23}$, results in a new interferogram corresponding to the comparison of the first and third image, I_{13}. The process corresponds then to the extension of correlation at the expense of an intermediate common reference state.

The concept can be generalized, Fig.12, so that a larger interval may be covered if physical limitations of the method are not violated (fringe resolution, contrast threshold, Shanon 'sampling theorem, combined filtering requirements).

Fig.13 illustrates an example of integrative moiré evaluation for deformation on circular membranes.

It is possible to imagine various ways of achieving the required result under dynamical situation. Fig.14 represents schematically a conceived principle of implementation using phase wave conjugation with control of the fading time of the recording in a combination of two crystals and adequate switching of illumination pulses. A proper combination of synchronized pulse illumination and phase wave conjugation techniques would represent a rather flexible tool for vibration studies.

This tecnique allows one to extend the range of coverage of a phenomenon making it pausible for instance to correlate two stages of deformation that combined would exceed the holographic interferometric correlation range. This is achieved by multiplication of double exposure holographic interferograms that have in common a single exposure.

An example of envisaged application is the study of large amplitude vibrations. The first exposure, I_2, would correspond to an intermediate amplitude value as referred to in Fig.15.

Limitations of the method result from the maximum number of fringes that it is possible to handle in practice (in holography around 100);

availability of spatial frequency spectrum gap for proper filtering (eventually by diffraction) and Shanon' sampling theorem conditions. Consequently, it is advisable that spatial frequencies involved are of the same order of magnitude for easiness in filtering. Visibility of fringes is also essential for good results implying the need for fringe enhancement at some intermediate stages.

Applications of the technique refer to the cases where decomposition of the interval of the study is a necessity deriving from method correlation range or it is convenient to bring intermediate stages into evidence. Recurrence to phase wave conjugation techniques would bring the best of benefits to practical applications of the method whenever coherent optical means are selected.

iv) Moiré derivation (10,11,12)

Fringes are produced, in general, to represent the locus of a particular value relative to a specific physical property, e.g. points of equal deformation on a surface.

Furthermore, a constant step increase is observed on the characterizing locus value attributable to the successive fringe orders. Consequently, derivatives in a direction can be obtained by moiré interference with a fringe pattern at two positions corresponding to a translation along the direction to be considered for derivation. This kind of spatial derivation is of current use (19,20), e.g. evaluation of stress from strain mapped by holography (21,22), moiré (23) or speckle interferometry (24).

Fig.16 shows spatial derivatives obtained by this method using a dynamical digital memory (14,15). In plane and out of plane displacement of the pattern can be, in principle, generated through adequate software (radial gradients). In a sense the method is analogous to a coarse spatial heterodyning but presents extreme simplicity, global field evaluation, and almost a real-time technique.

Applications are found in literature corresponding to shearing interferometers, and on the evaluation of gradients along axis coordinates (19,20) or radial direction.

It is also possible to conceive a moiré derivation in time (14,15) as described by Fig.17. Derivation in time permits the mapping of velocities and accelerations.

The derivation can, in principle, be extended to higher orders in accordance with visibility of fringes among other physical limitations, Fig.17

Specklegrams

Speckle correlation fringes present poor quality for further processing operations. There are however methods for improving the coarse fringes up to comparable holographic interferometry fringe quality without significant reduction of resolution (25,26).

These methods make it possible to apply moiré evaluation concepts to specklegrams. Furthermore, speckle correlation fringes can be considered a result of moiré effect between random distributed gratings so that, in principle, one may look for some more basic operation of the moiré evaluation type with speckle patterns. Attention is being given to this difficult problem.

Conclusion

Moiré effect being intrinsically a basic operation resulting from spatial frequency spectra intermodulation serves both to explain physical phenomena whenever the spatial frequency concept is part of the model, and to search for new features or improve performance of experimental methods.

Interferometric based techniques were analysed in the perspective of their moiré analogy. It has brought to evidence some of the advantages of considering moiré evaluation, in the forms of: moiré reference transform, differential moiré, integrative moiré and moiré derivation.

Features directed to practical applications were derived from the analysis which led to the exploring of the evaluation techniques and open up for discussion interesting aspects, deserving further study.

Acknowledgements

The authors acknowledge the support of the Deutsche Gesselschaft fur Techniche Zusammenarbeit (GTZ) and donation of equipment that made it possible to pursue the research while complemented by a research grant from INIC.

Contribution to the experiments by A.O.S. Gomes were valuable as was the technical assistance of J.S. Fernandes and L.M. Vilaça.

Some of the work was carried out at the University of Münster.

The authors thank Prof. S.P. Almeida of Virginia Polytechnic Institute and State University for the BSO Crystal used for the experiments.

Part of the results have already been published elsewhere (12).

References

1. Theocaris, P.S., Moiré fringes: A Powerful Measuring Device, in Applied Mechanics Surveys, Sportan Books Inc., Washington DC (1966) pp 613-626

2. Theocaris, P.S., Moiré Fringes in Strain Analysis, Pergamon Press Oxford, 1969

3. Takasaki, H., Moiré Topography, Appl. Opt. 9 (1970) 1467-1472

4. Abramson, N., The Making and Evaluation of Holograms, Academic Press, N.Y. (1981)

5. Sciammarella, C.A., Holographic-moiré in Optical Methods in Mechanics of Solids, Ed. A. Lagarde, Sijthoff & Noordhoff (1981)

6. Ikeda, T and Terada, H., Development of the Moiré Method with Special Reference to its Application in Biosterometrics Opt. and Laser Technol. 13 (1981), 302-306

7. Pirodda, L., Optical Methods of Non-destructive Testing in Italy - a Short Selection,Proc SPIE 349 (1982), 167-185

8. Post, D., Developments in Moiré Interferometry, Opt. Eng. 21 (1982),458-467

9. Reid, G. T., Moiré Fringes in Metrology, Opt. and Laser Technol. 5 (1984) 63-93

10. Soares, O.D.D.; Lage, A.L.V.S.; Método de Análise Diferencial e Integrativa em Holografia Interferométrica por Técnica Moiré, Patent Nr. 80333, I.N.P.I., Portugal (1985)

11. Soares, O.D.D.; Lage, A.L.V.S., Moiré Hologrametry, Optics in Modern Science and Technology, ICO - 13, Sapporo (1984)

12. Soares, O.D.D.; Lage, A.L.V.S. and Bernardo, L.M., Moiré Evaluation of Pulse Illuminated Interferograms by Synchronized Video Recording, Proc SPIE, 491 (1984)

13. Soares, O.D.D.; Lage, A.L.V.S., Controllable Synchronized Multiple Pulse Illumination System for ESPI and Holography, Proc SPIE, 427 (1983)

14. Soares, O.D.D.; Lage, A.L.V.S., Use of TV-Frame Memory on Electronic Speckle Pattern Interferometry Applied to Orthopedics,Proc SPIE, 348 (1982), 838-844

    Soares, O.D.D.; Lage, A.L.V.S.; Gomes, A.O.S.; Santos, J.M., Dynamical Digital Memory for Holography, Moiré and ESPI, Optical Metrology, Martinus Nijhoff (1985)

    Soares, O.D.D.; Lage, A.L.V.S.; Gomes, A.O.S., Memória Dinâmica Operável em Tempo Real, Patent Pending

15. Chopra, K.N. and Bhatnagar, Quadruple-Exposure Technique in Stroboscopic Holographic Interferometry, Appl. Opt., 13 (1974), 2467-2470

16. Vickram, C.S, A Triple-Exposure Technique to Reduce Recording Time in Stroboscopic Holographic Interferometry, Opt. Commun., 6 (1972), 296--299

17. Soares, O.D.D., Hologram Repositioning by an Interferometric Technique, Appl. Opt., 18 (1979), 3838-3840

18. Gunter, P., Holography, Coherent Light Amplification and Optical Phase Conjugation with Photorefractive Materials, Physics Reports, 93 (1982), 199-299

19. Durelli, A. and Parks, V, Moiré Patterns of Partial Derivation of Displacement Components, J. Appl. Mech XII (1966)

20. Chirico, C.; Ginesu, F. and Pirodda, L., Optical Differentiation with White Light Diffracted Wavefronts, Optics in Modern Science and Technology, ICO - 13, Sapporo (1984)

21. Boone, P. and Verbiest, R., Application of Hologram Interferometry to Plate Deformation and Translation Measurements, Optica Acta, 16 (1969) 555-567

22. Rastogi, P.K., A Real-Time Holographic Moiré Technique for the Measurement of Slabs Change, Optica Acta, 31 (1984), 159-167

23. Dantu, P., Moiré du Deuxiéme Ordre Methode Permettant d'Obtenir Directement les Lignes d'Egale Dilatation, Revue fr. Mec., 17 (1966),1

24. Rastogi, P.K., Speckle Metrology Techniques: a Parametric Examination of the Observed Fringes, Optical Eng., 21 (1982), 411-426

25. Martienssen, W.; Spiller, S., Holographic Reconstruction without Granulation, Phys. Lett., 24A (1967), 126

26. Lokberg, O.J.; Slettemoen, G.A., Improved Fringe Definition by Speckle Averaging in ESPI, Optics in Modern Science and Technology, Conference Digest, ICO - 13, Sapporo (1984)

410

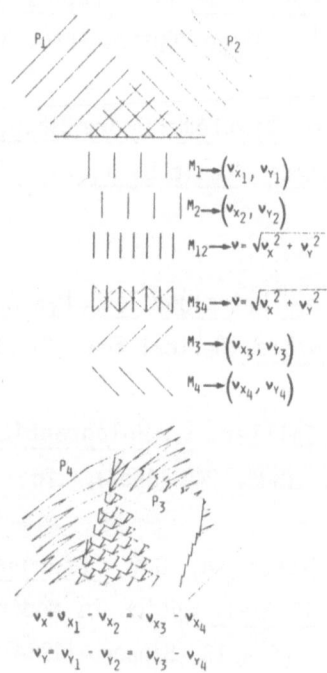

$$\nu_x = \nu_{x_1} - \nu_{x_2} = \nu_{x_3} - \nu_{x_4}$$
$$\nu_y = \nu_{y_1} - \nu_{y_2} = \nu_{y_3} - \nu_{y_4}$$

Fig.1: Plane interference pattern generated by two plane waves.
Description by moiré analogy of reference transformation
  i) the two plane waves P_1 and P_2 produce the interference
     M_{12} (spatial frequency $\nu \equiv (\nu_x, \nu_y)$) which is the moiré
     effect of M_1 (spatial frequency (ν_{x_1}, ν_{y_1})) and M_2 (spa-
     tial frequency (ν_{x_2}, ν_{y_2})), the corresponding plane gra-
     tings of wavefronts P_1 and P_2
 ii) the same interference pattern can be generated by moiré
     effect of M_{34} (spatial frequency $\nu \equiv (\nu_x, \nu_y)$) relative to
     grating M_3 (spatial frequency (ν_{x_3}, ν_{y_3})) and M_4 (spatial
     frequency (ν_{x_4}, ν_{y_4})) resulting from plane wave P_3 and P_4.

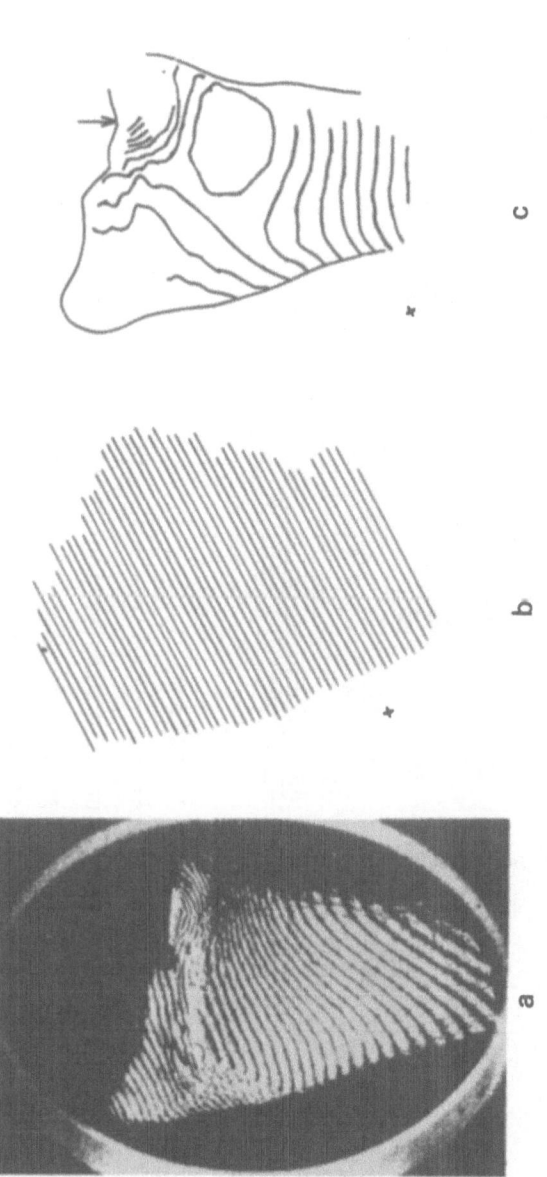

Fig.2: Moiré reference transform
a) radius bone under non-axial load—global effect (holographic interferogram)
b) flexural bending contribution evaluated on free moving part of the bone
c) localized deformations of bone surfaces

412

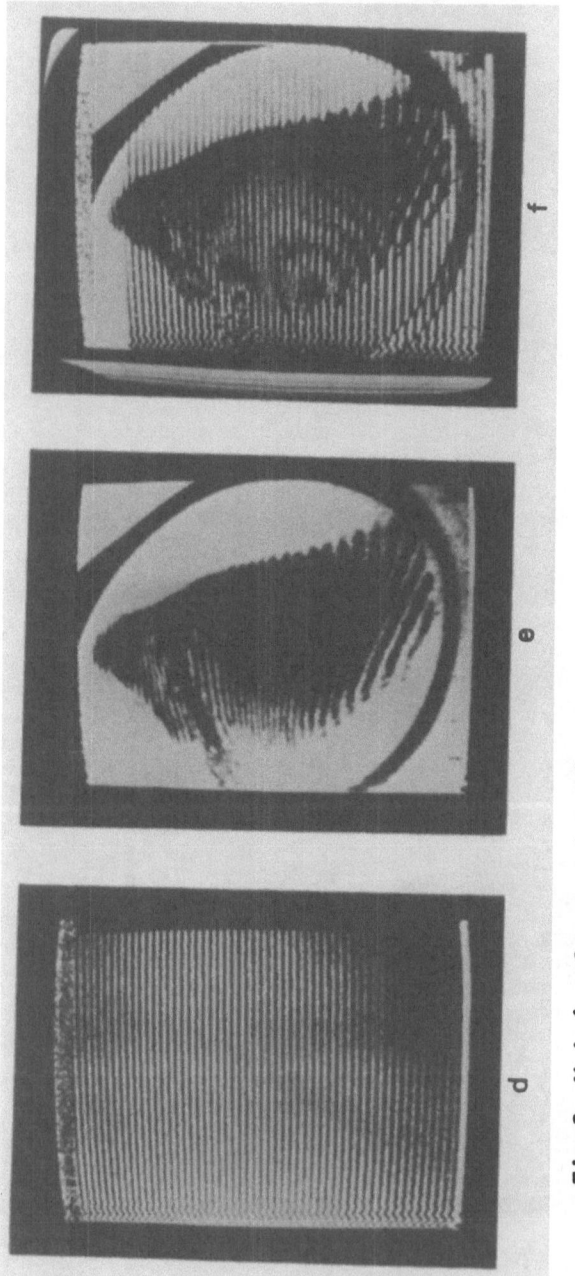

Fig.2: Moiré reference transform

d) electronic generation of rigid body motion fringes

e) interference pattern acquisition of global effect of loading

f) differential effect corresponding to pure deformation as
   evaluated by digital memory (14)

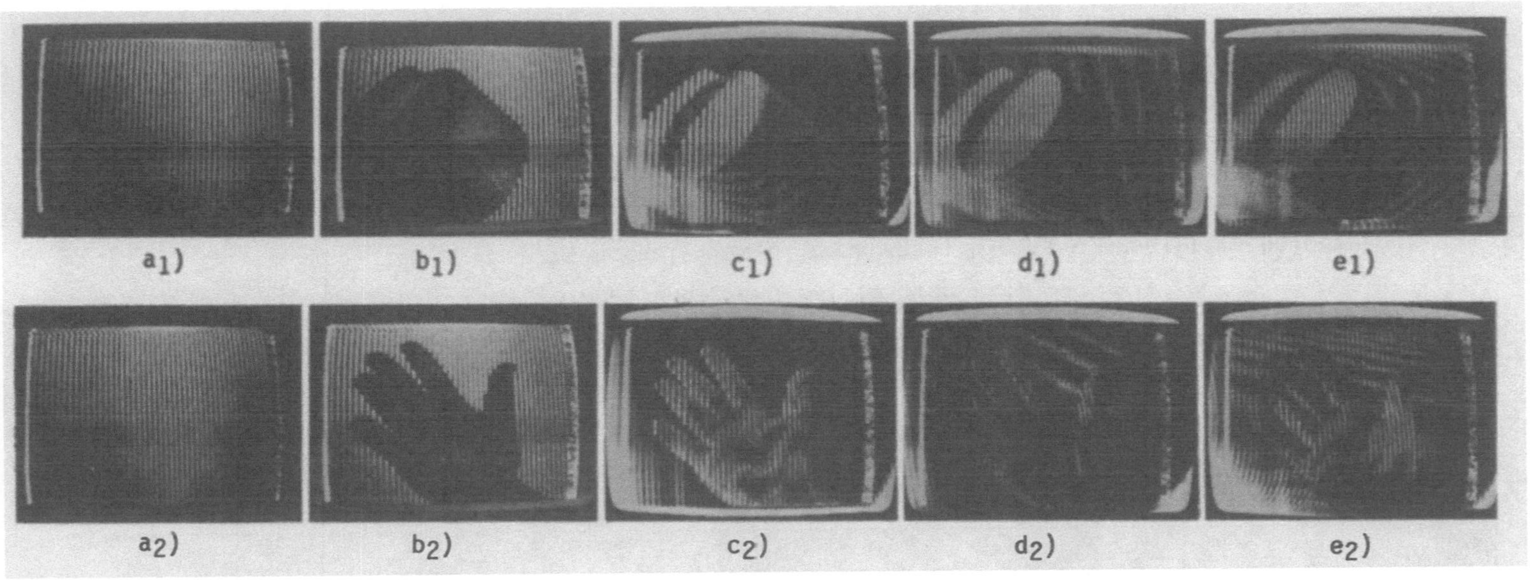

413

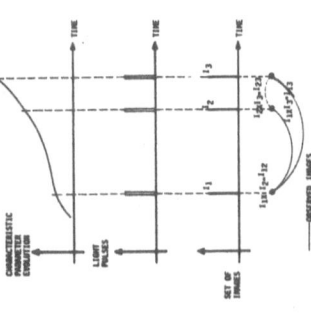

Fig.3: Moiré reference transform on a testing cylinder (upper row) and
with stroboscopic illumination (lower row) on a human hand.

a) reference plane for surface topography

b) shadow fringes projected with grating as in a)

c) moiré topography from a) and b)

d) change of reference plane⊟moiré evaluated

e) reference curved surface transform as in d)

Fig.4: Differential Moiré principle (10)

414

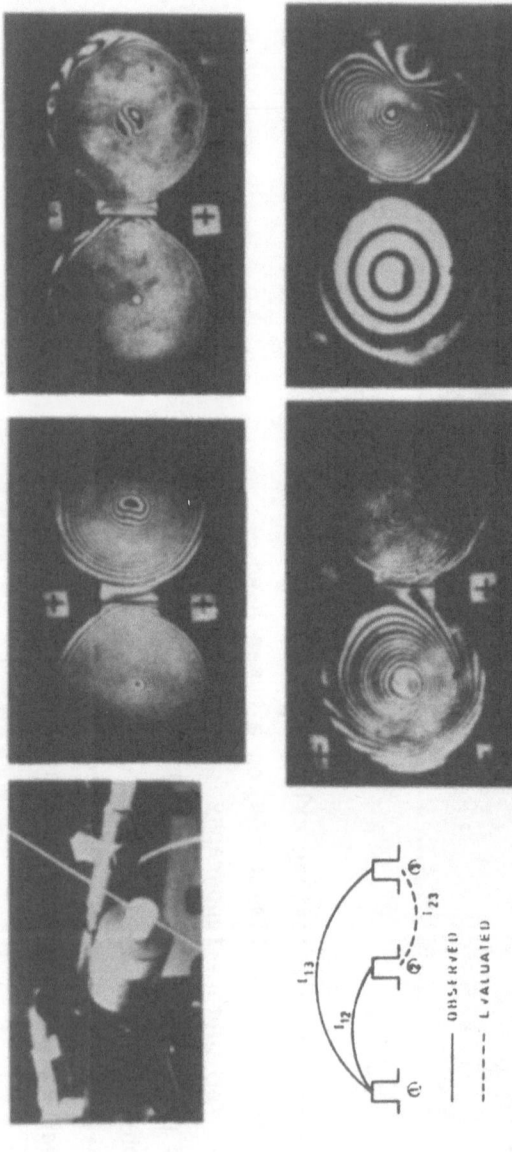

Fig.5: Differential moiré evaluation for a couple of circular membranes under central loading (12)
a) actual view of membranes; b) and c) real-time holographic interferograms I_{12}, and I_{13}, respectively; d) differential moiré evaluation of I_{23}; e) equivalent double exposure hologram H_{23}

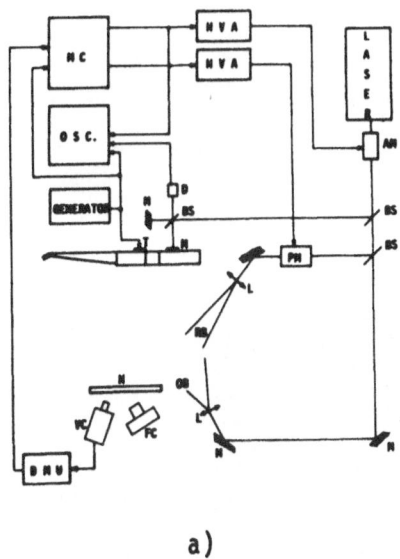

a)

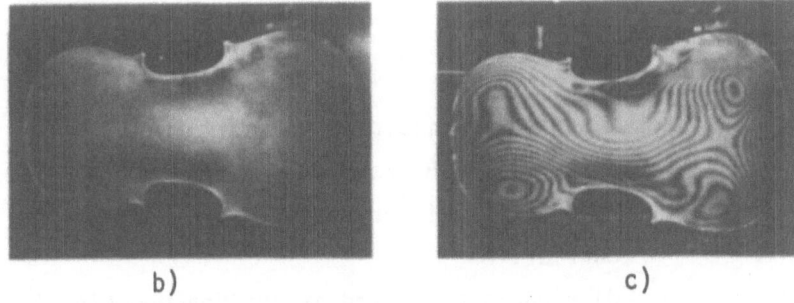

b)   c)

Fig. 6: Differential moiré evaluation in holographic study
         of a vibrating membrane[12] - violin.
         a) experimental set-up
         b) holographic restitution (first exposure with mem-
            brane static)
         c) real-time interferogram of the membrane
            (object beam + holographic restitution)

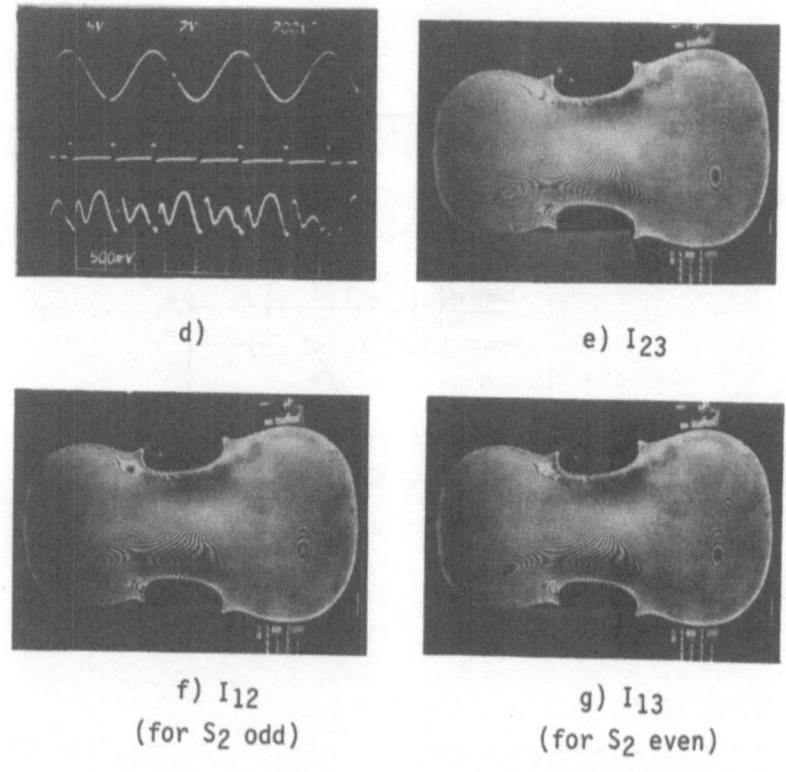

d)

e) I_{23}

f) I_{12}
(for S_2 odd)

g) I_{13}
(for S_2 even)

Fig. 6 cont.: Hologram is tilted around vertical
axis (θ_z = 0.5 mrad) and translated (Δx = 100 µm)
to create a higher spatial frequency of the
carrier.
d) signals presented at oscilloscope Fig. 6 a)
for experiment control:
S_1 - vibration generator signal
S_2 - amplitude modulator driving pulses
S_3 - Michelson interferometer signal at photodiode D, Fig. 6 a) for amplitude and phase control of effective vibration
e) moiré evaluated resonant mode at 1.7 KHz.
f) and g) combined patterns for moiré evaluation
Fig. 4

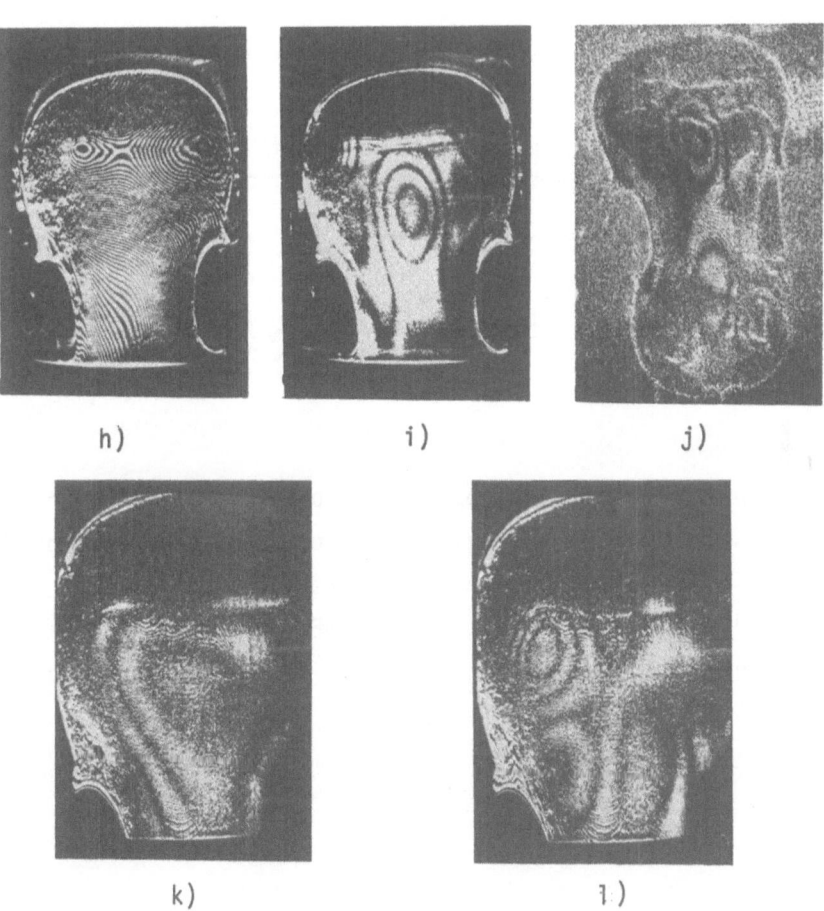

h)   i)   j)

k)   l)

Fig. 6 cont.: Filtering of moiré patterns

        h) filtering by electronic means of f)

        i) filtering by electronic means of e) from
           real-time observation. Moiré fringes are
           clearly visible at TV monitor and carrier
           is filtered.

        j) photograph e) processed by an optical filter

        k) resonant mode 0.8 KHz as in e)

        l) resonant mode 2.4 KHz as in e)

Note: resonant modes in k) and l) were examined one week later
      to test the method.

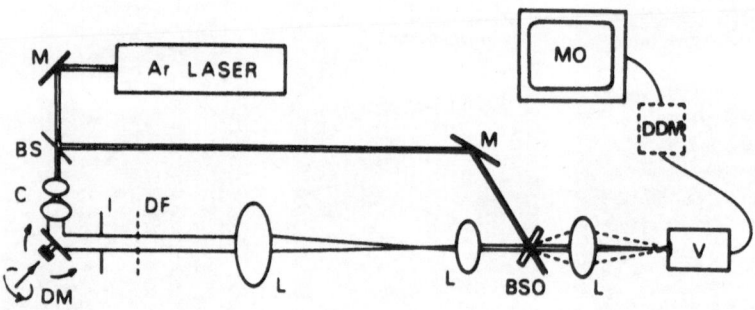

Fig.7: Set-up lay-out for the observation of differential incoherent
       (storage of the digital dynamic memory) and coherent moiré-
       grams (M - mirror, BS - beam-splitter, C - collimator,
       DM - deformable mirror, I - iris, DF - neutral density filter,
       L - lens,   BSO - $Bi_{12}SiO_{20}$ crystal, V - vidicon, DDM - dynamic
       digital memory, MO - monitor)

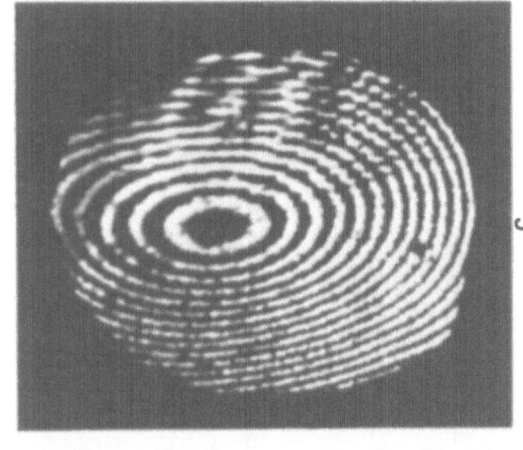
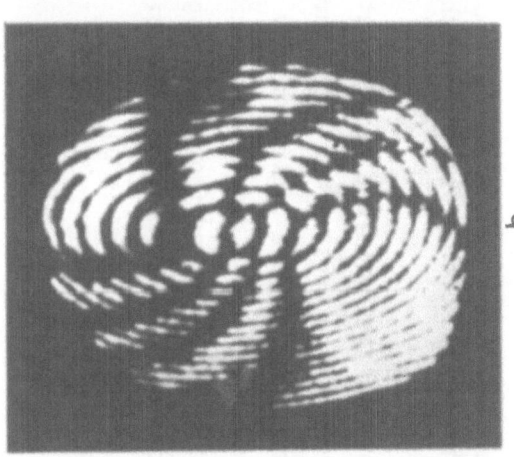

a

b

c

Fig.8: Differential moiré coherent evaluation with photorefractive crystal

a) interferogram of the two wavefronts, reconstructed from double exposed hologram, that corresponds to the two states defining the deformation of DM (Fig.7)

b) interferogram (moiré pattern) of reconstruction from a) and object wave corresponding to another deformation of DM

c) interferogram of reconstructed wavefronts which correspond to the first and last state of deformation

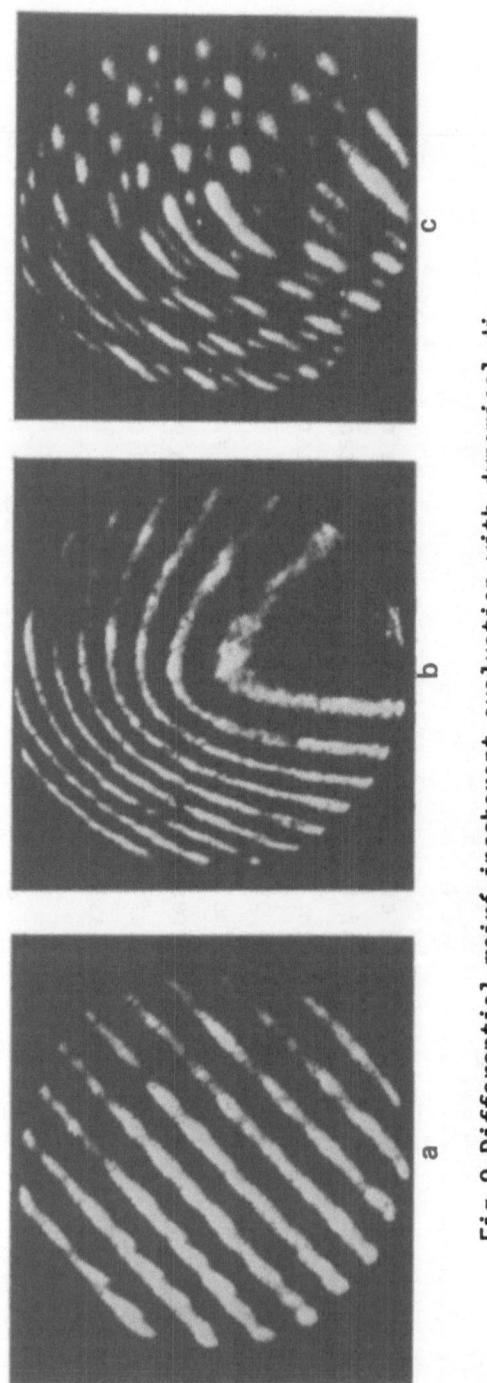

Fig.9 Differential moiré incoherent evaluation with dynamical digital memory

a) interferogram of the wavefront reconstructed from the hologram stored in the crystal and the deformed object
wavefront as stored in the digital memory

b) interferogram of the last deformed object wavefront and
the first one reconstructed from the hologram

c) moirégram observed in the monitor after digital subtraction
of the interferograms shown in a) and b)

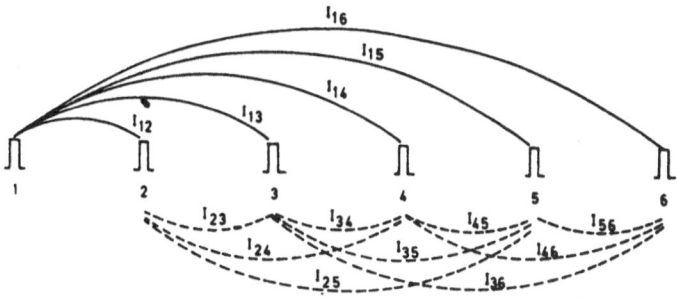

Fig.10: Generalization of differential moiré evaluation concept (10)

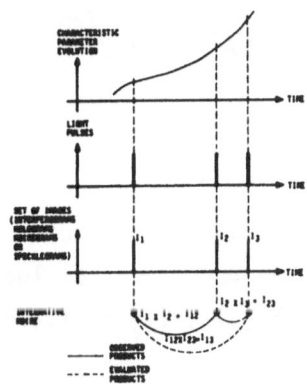

Fig.11: Integrative moiré evaluation principle (10)

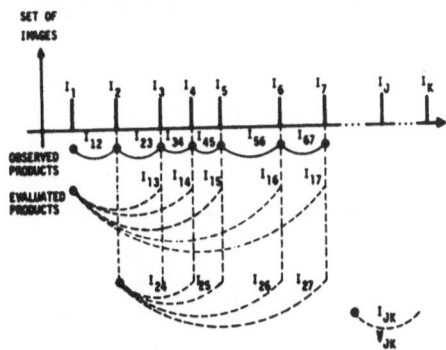

Fig.12: Generalization of the integrative moiré evaluation concept (10)

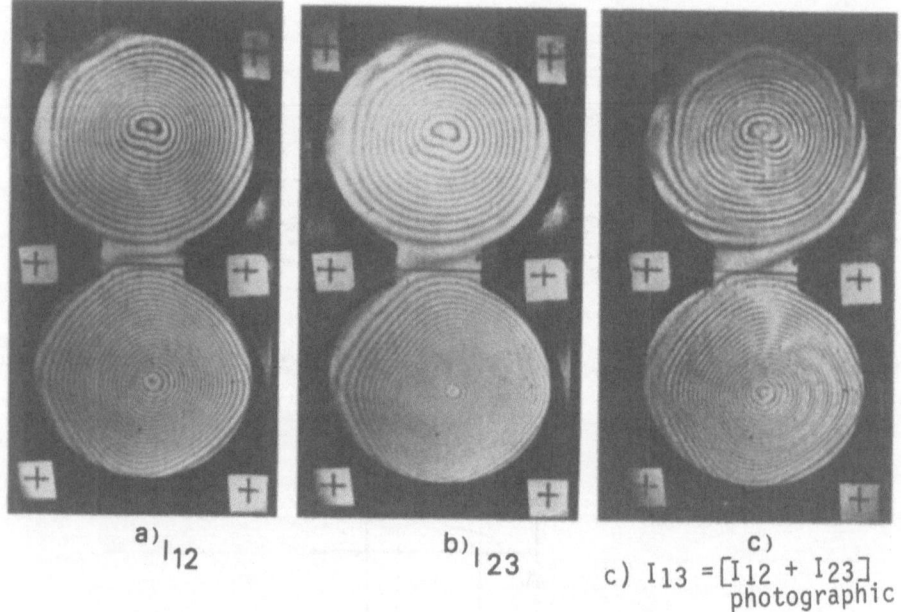

a) I_{12} b) I_{23} c) $I_{13} = [I_{12} + I_{23}]_{photographic}$

Fig.13: Integrative moiré evaluation (overlapped) for a couple of
circular membranes under central loading.
   a) according to Fig.11, I_{12} real time holographic interfe-
      rogram
   b) according to Fig.11, I_{23} real time holographic interfe-
      rogram
   c) double exposure on photographic film with I_{12} and I_{23}
      real-time holographic interferograms

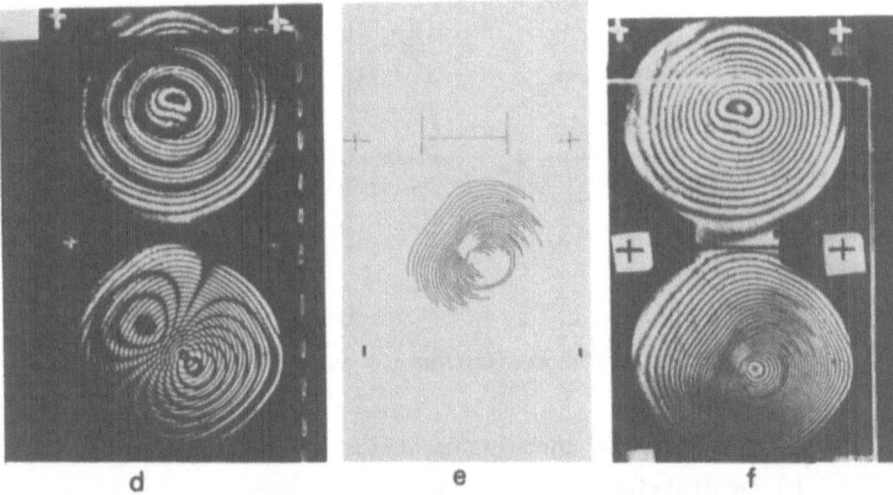

d   e   f

d) $I_{13} = \overline{[I_{12} \times I_{23}]}$
    overlapped

Fig.13: Integrative moiré evaluation(overlapped)for a couple of
    circular membranes under central loading.

    d) integrative moiré evaluation of I_{13}

    e) integrative moiré evaluation of I_{23} from photograph d)

    f) coincidence of results from e) and b)

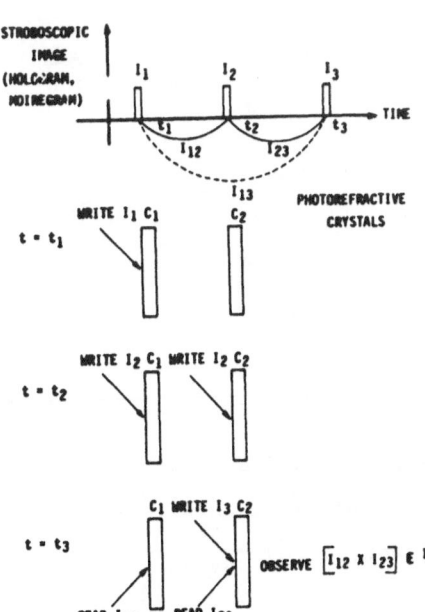

Fig.14: Transposition of the
    moiré integrative
    principle to an
    implementation with
    photorefractive
    crystals

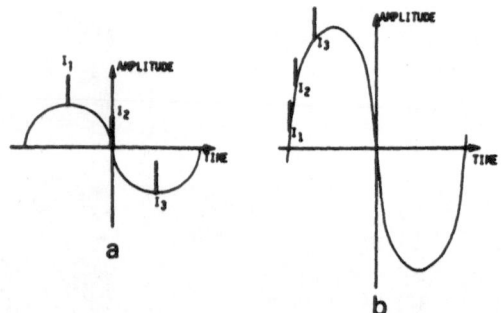

a

b

Fig.15: Integrative moiré evaluation to study vibrations of large
amplitude
a) intermediate correlating image at resting position
b) subdivision of amplitude

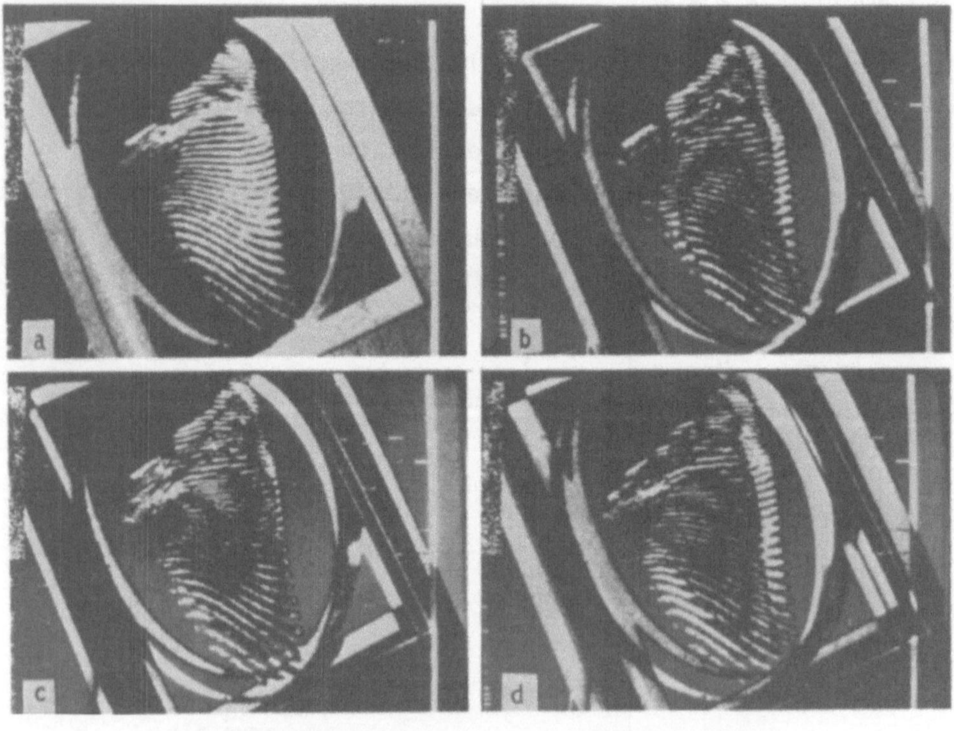

Fig.16: Spatial moiré derivation along a direction
a) fringe pattern b) $\frac{\partial}{\partial x}$ horizontal direction
c) $\frac{\partial}{\partial y}$ vertical direction d) $\frac{\partial^2}{\partial x\, \partial y}$
A dynamical digital memory was used (14)

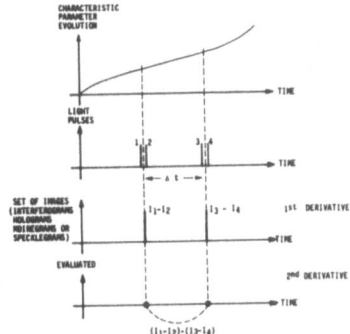

Fig.17: Principle of moiré derivation in time

ELECTRONIC PROCESSING FOR HOLOGRAPHIC INTERFEROMETRY

R. Dändliker and R. Thalmann

Institut de Microtechnique de l'Université,
CH-2000 Neuchâtel, Switzerland.

In many interferometric setups, heterodyne or quasi-heterodyne
techniques have become a powerful tool for high accuracy
interference fringe interpolation. In holographic interferometry,
heterodyning has been applied to real time as well as to double-
exposure holography. For double-exposure holography two reference
beams are required. During reconstruction a change of the relative
phase between the two references allows the shifting of the fringes
in the interferogram at will.Heterodyne holographic interferometry
offers high spatial resolution and interpolation up to 1/1000 of a
fringe.However,it requires sophisticated electronic equipment and
mechanical scanning of the image by photodetectors. Quasi-heterodyne techniques are more adequate for digitial processing and TV
detection. Two-reference-beam holography with reference sources
close together and video-electronic processing allows measurement
of the interference phase with an accuracy of 1/100 fringe at any
point in the TV image. This system combines effectively the
simplicity of standard double-exposure holography, video-electronic
processing, and the power of heterodyne holographic interferometry.

1. INTRODUCTION

Holographic interferometry has become a well known technique
to investigate deformations and vibrations of solid, three-
dimensional objects with optically rough surfaces. In classical
double-exposure holographic interferometry, the two wavefields are
recorded consecutively on the same hologram.The phase differences
due to the change of the optical path lengths between the two
reconstructed wavefields show up as intensity variations, the
so-called interference fringes, in the image of the object.

Soares, O.D.D. (ed), Optical Metrology
© *1987. Martinus Nijhoff Publishers, Dordrecht. Printed in the Netherlands.*

For many practical applications in deformation and strain measurements, the quantitative and automated determination of interference phase versus position is required. From the usual fringe pattern, however, quantitative information on the interference phase can only be obtained reliably of the minima and maxima of the interference fringes. Accurate interference phase measurement requires fringe interpolation, independently of fringe position and intensity variations in the reconstructed image. This is only possible by applying two-reference-beam holography [1,2], which allows one to vary the relative phase of the interfering wavefields during reconstruction.

Two kinds of methods to evaluate the interference phase are known. In heterodyne methods [3] the relative phase increases linearly in time and the interference phase is measured electronically at the beat frequency of the reconstructed wavefields. Quasi-heterodyne techniques change the relative phase stepwise, using at least three different values [4]. Compared with classical double-exposure holographic interferometry the necessity to use two-reference-beam holography unfortunately introduces some additional difficulties with unwanted reconstructions and repositioning of the hologram [2,5]. It has been shown recently [6] how, under certain conditions, these difficulties can be overcome.

2. INTERFEROMETRY AND FRINGE INTERPOLATION

Interferometry is used to transform phase differences of wavefields into detectable variations, i.e. interference fringes. Assuming that the two wavefields to be compared are given by

$$V_1(\vec{x}) = a_1(\vec{x}) \ \cos[\omega_1 t + \phi_1(\vec{x})],$$

$$V_2(\vec{x}) = a_2(\vec{x}) \ \cos[\omega_2 t + \phi_2(\vec{x}) + \phi_k], \tag{1}$$

where ϕ_k is an additional constant phase, the local intensity $I(\vec{x})$ of their superposition becomes

$$I(\vec{x}) = |V_1 + V_2|^2 = a(\vec{x}) \ \{1 + m(\vec{x}) \ \cos[\Delta\omega t + \phi(\vec{x}) + \phi_k]\}, \tag{2}$$

where $a(\vec{x})$ is the local mean intensity, $m(\vec{x})$ the fringe contrast, and $\phi(\vec{x}) = \phi_2(\vec{x}) - \phi_1(\vec{x})$ the phase difference of the two wavefields.

In classical interferometry, the optical frequencies ω_1 and ω_2 are identical ($\Delta\omega = \omega_2 - \omega_1 = 0$) and the interference phase $\phi(\vec{x})$ is transformed into interference fringes independent of time. Interpolation of the phase $\phi(\vec{x})$ between dark fringes is not accurately achievable because mean intensity $a(\vec{x})$ and fringe contrast $m(\vec{x})$ may change as well with position $\vec{x}$.

In heterodyne interferometry the two optical frequencies are chosen to differ by a small amount $\Delta\omega = \omega_2 - \omega_1$. The superposition intensity is then time dependent and the interference phase $\phi(\vec{x})$ is transformed into the phase of the beat frequency signal. As the beat frequency $\Delta\omega$ is chosen low enough (<100MHz) to be resolved by the optoelectronic detector employed, the interference phase can be measured with high accuracy, independently of the amplitudes a_1 and a_2, using an electronic phasemeter. This way, both the interpolation problem and the sign ambiguity of classical interferometry are solved. This method requires special equipment, such as acousto-optical modulators and phasemeters. The image is scanned mechanically by photodetectors to measure the interference phase $\phi(\vec{x})$ locally. Therefore the speed is relatively low ($\sim$1 sec per point) but the accuracy ($\Delta\phi \leq 0.3°$ or 1/1000 of a fringe) and the spatial resolution (> 10^6 resolvable points) are extremely high [1].

For moderate accuracy and spatial resolution, quasi-heterodyne techniques [4] have been developed, which allow electronic scanning of the image by photodiode-arrays (CCD) or TV-cameras, and use microprocessor controlled digital phase evaluation [7]. For this purpose, the linear increase of the relative phase ($\phi_t = \Delta\omega \cdot t$) due to the heterodyne frequency offset $\Delta\omega$ is replaced by a stepwise change of ϕ_k. To determine the interference phase $\phi(\vec{x})$, the interferogram is analyzed at least three times and the mutual phase is changed each time between two analyses by a well controlled phase-shift. The local intensities $I_k(\vec{x})$ are then given by

$$I_k(\vec{x}) = a(\vec{x}) \{1 + m(\vec{x}) \cos[\phi(\vec{x}) + \phi_k]\}, \qquad (3)$$

which correspond for k = 1, 2, 3 to a system of three equations with three unknown values: the mean intensity $a(\vec{x})$, the fringe contrast $m(\vec{x})$ and the interference phase $\phi(\vec{x})$. For the interference phase $\phi(\vec{x})$ one gets from Eqs.(3)

$$\tan \phi(\vec{x}) = \frac{(I_3-I_2)\cos\phi_1 + (I_1-I_3)\cos\phi_2 + (I_2-I_1)\cos\phi_3}{(I_3-I_2)\sin\phi_1 + (I_1-I_3)\sin\phi_2 + (I_2-I_1)\sin\phi_3} . \qquad (4)$$

The other unknown values are given by

$$a(\vec{x})\cdot m(\vec{x}) = (I_1 - I_2)/\{\cos[\phi(\vec{x}) + \phi_1] - \cos[\phi(\vec{x}) + \phi_2]\}, \qquad (5)$$

$$a(\vec{x}) = I_1 - a(\vec{x}) \, m(\vec{x}) \cos[\phi(\vec{x}) + \phi_1]. \qquad (6)$$

For both heterodyne and quasi-heterodyne fringe evaluation, independent access to the two wavefields is necessary in order to control their relative phase. For holographic interferometry, this implies independent recording of each wavefield by using different reference sources [1,2]. The frequency offset $\Delta\omega$ or the change of mutual phase ϕ_k can be introduced by acting on the respective reference beams.

3. TWO-REFERENCE-BEAM HOLOGRAPHIC INTERFEROMETRY

The optical arrangements for two-reference-beam holographic interferometry are sketched in Fig.1(a) for well separated reference sources and in Fig.1(b) for reference sources close together. Except for the two references, this setup is the same as for classical holographic interferometry. In both arrangements, the object is illuminated by a point source, and an imaging system permits observation of the interferogram on a screen.Consecutively, the first object state O_1 is recorded by reference R_1 and the second object state O_2 by reference R_2 on the same hologram plate.

As described in Ref.2, two-reference-beam holography requires special attention to the multiplicity of the reconstructed images and the influence of misalignment of the hologram with respect to the reference beams. Illuminating the hologram with both reference beams R_1 and R_2 yields eight reconstructions. Four of them are conjugate reconstructions. The other ones are the two desired primary self-reconstructions ($R_1R^*O_1$ and $R_2R^*O_2$), which give rise to the interference pattern, and the two undesired cross-reconstructions ($R_2R^*O_1$ and $R_1R^*O_2$). The direction of propagation of the various reconstructed waves depends on the geometry of the optical setup. The primary reconstructions are shown for both cases in Fig.2(a) and Fig.2(b).

To avoid disturbing overlap of the different reconstructions [Fig.2(a)] the two reference sources must be chosen on the same side of the object,with a mutual separation just larger than the angular size of the object in the corresponding direction [Fig.1(a)]. However,the consequence of a large separation of the reference sources is high sensitivity to repositioning errors, as described hereafter. Therefore, reference sources close together [Fig.1(b)] would be prefered if, under certain conditions, overlapping [Fig.2(b)] of the cross-reconstructions could be tolerated without undue loss of accuracy for the interference phase measurement.

The sensitivity of two-reference-beam holographic interferometry to repositioning occurs because the propagation of the two reconstructed wavefields are differently affected [1,2]. For small changes of the hologram position from recording to reconstruction the resulting additional phase difference $\psi(\vec{x}_H)$ in the hologram plane between the reconstructed waves corresponding to the objects O_1 and O_2 is given by

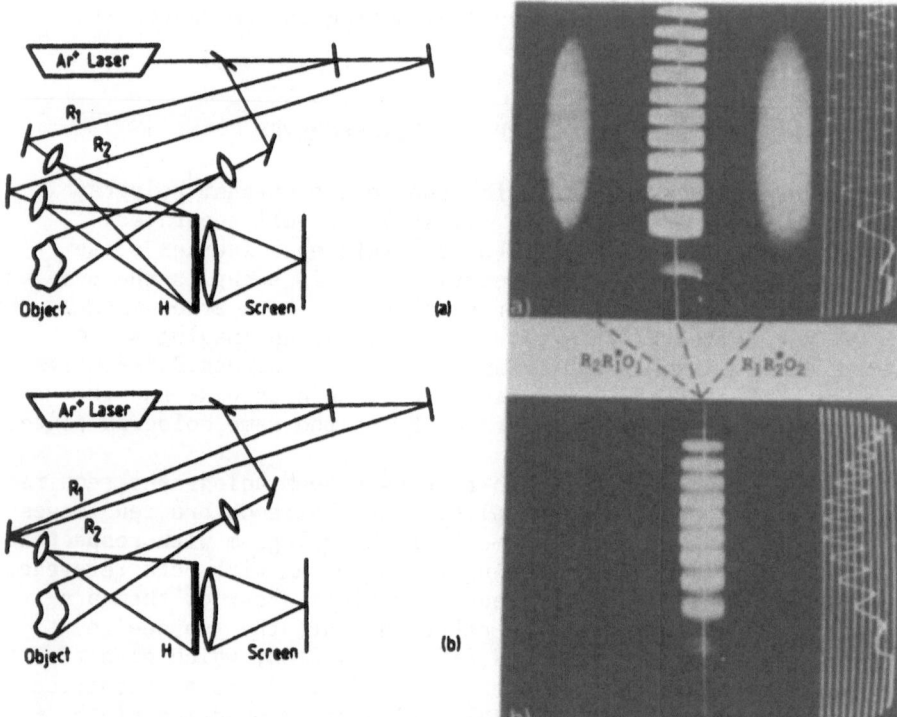

Fig.1. Setup for two-reference-beam holographic interferometry, (a) well separated references,

(b) references close together.

Fig.2. Reconstructed images and intensity profiles on TV screen, (a) separated cross-reconstructions $R_2R\ddagger O_1$ and $R_1R\ddagger O_2$, (b) overlapping reconstructions.

$$\psi(\vec{x}_H) = [(\vec{k}_1 - \vec{k}_2) \times \vec{w}] \cdot \vec{x}_H, \tag{7}$$

where $\vec{w} = (\Delta\alpha, \Delta\beta, \Delta\gamma)$ is the rotation vector for small hologram rotations $\Delta\alpha$, $\Delta\beta$, $\Delta\gamma$ around the x, y, z axes, respectively, and $\vec{x}_H$ are the coordinates in the hologram plane. Note that a pure translation of the hologram causes only a constant phase shift and can be ignored. It is seen from Eq.(7) that a rotation of the hologram introduces a linear phase deviation across the hologram plane which depends only on the difference vector $\Delta\vec{k} = \vec{k}_1 - \vec{k}_2$ of the two references. This means that the repositioning sensitivity is much smaller for reference sources close together. The effect of a linear phase deviation in the hologram plane results mainly in a mutual transverse shift between the two desired reconstructions in the image plane [1,2], given by

$$u_I = (d_I/k)|\vec{grad}_H \psi|$$

$$= (d_I/k)[(\Delta k_y \Delta \gamma - \Delta k_z \Delta \beta)^2 + (\Delta k_z \Delta \alpha - \Delta k_x \Delta \gamma)^2]^{1/2} , \tag{8}$$

where d_I is the distance from the lens to the image plane and $\vec{grad}_H \psi$ is the component of $\vec{grad} \psi$ in the hologram plane.

The magnitude of the interference term depends essentially on the autocorrelation $C_h(u_I)$ of the impulse response function of the imaging system, which defines also the speckle size in the image. The interference fringes are only visible as long as the mutual shift of the speckle patterns is smaller than the speckle size. For a circular aperture of diameter D, the fringe contrast $\gamma(u_I)$ is thus given by the well known Airy function

$$\gamma(u_I) = C_h(u_I) = 2J_1(\pi D u_I / \lambda d_I) / (\pi D u_I / \lambda d_I), \tag{9}$$

where $J_1(.)$ is the first order Bessel function. The fringe contrast decreases with transverse shift u_I in terms of the diffraction limited optical resolution $(\Delta x)_I \doteq \lambda d_I / D$.

In Fig.3 theoretical and experimental results for the fringe contrast versus hologram rotation $\Delta \gamma$ are presented. The two reference waves had an angular separation of $\Delta k_y/k = 0.19$, which is necessary to separate the reconstructions [Fig.1(a) and 2(a)] in a typical holographic setup. A one-to-one image of the reconstructed object was formed by an objective of $f = 300$ mm ($d_I = 600$ mm) and of $D = 9$ mm (f/32) effective aperture. The allowed repositioning error for a reduction of the fringe contrast γ to 0.5 as obtained from Eqs.(8) and (9) is found to be only $\Delta \gamma = 0.013°$ (Fig.3).

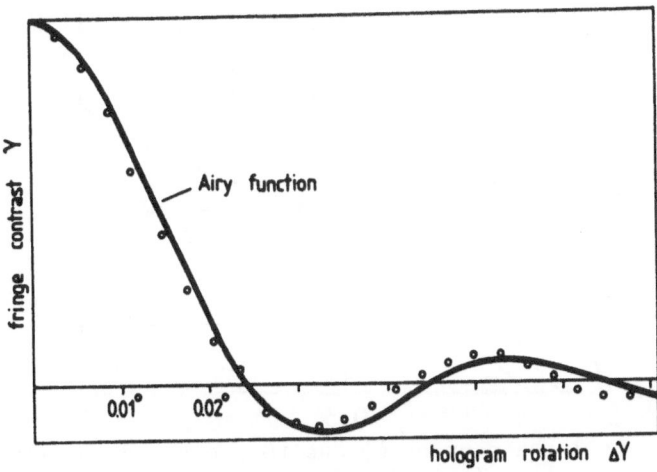

Fig.3. Fringe contrast γ versus angular misalignement $\Delta \gamma$ of the hologram for well separated references ($\Delta k_y/k = 0.19$).

432

On the other hand, as long as the lateral separation of the cross-reconstructions $R_2R_1^*O_1$ and $R_1R_2^*O_2$ is larger than the speckle size $(\Delta x)_I$, the speckle patterns of the desired and undesired images are uncorrelated and their superposition does not produce macroscopic interference. The required minimum angular separation between the two reference sources is given by the diffraction limit of the resolution imposed by the circular aperture D of the imaging lens. In other words, the two reference sources must produce several interference fringes across the lens aperture. The sensitivity to misalignment in such a setup [Fig.1(b)] is reduced to less than $\Delta\gamma = 10°$, compared to the 0.01° for the well separated cross-reconstructions.

However, the overlapping with the uncorrelated cross-reconstructions reduces the overall fringe contrast by a factor of two, as can be seen by comparing Figs.2(a) and 2(b), and introduces a statistical error to the interference phase. This error can be adequately reduced by spatial averaging using a detection area which covers many speckles [5]. The theoretical values of the statistical error $\delta\phi$ due to loss of fringe contrast by speckle decorrelation [Eq.(9), Fig.3] and overlapping reconstructions are presented in Fig.4 as a function of the normalized transverse displacement $u_I D/\lambda d_I$ in the image. In both cases the phase error depends essentially on the number N of independent speckles or

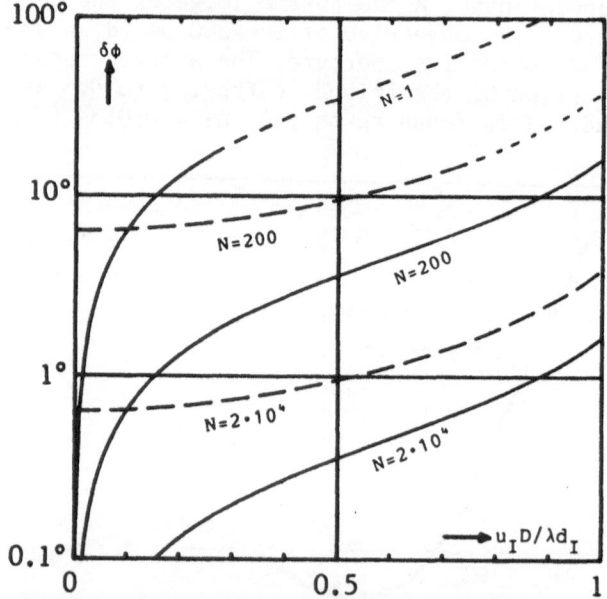

Fig.4. Statistical phase error $\delta\phi$ due to speckle decorrelation by transverse displacement $(u_I D/\lambda d_I)$ for well separated (solid) and overlapping (dashed) cross-reconstructions. N is the number of speckles within the detector area [Eq.(10)].

correlation cells within the detector area A_D, and decreases with $(1 + N)^{-\frac{1}{2}}$. For practical purposes N can be estimated from the F-number and the focal length f of the imaging lens through the relation

$$N = A_D \pi (f/2F\lambda d_I)^2. \tag{10}$$

The solid and the dashed curves in Fig.4 correspond to well separated and to overlapping reconstructions, respectively. The minimum error for $u_I = 0$ becomes zero in the former case, whereas in the latter it has a finite value given by

$$\delta\phi^2 = 5/2(1 + N). \tag{11}$$

Additional errors of systematic nature may occur due to the fact that the hologram will usually not be exactly in the pupil-plane of the imaging lens, or due to nonlinear cross-talk in the hologram reconstruction [8]. The latter can be avoided by carefully removing from the recorded scene spurious reference light sources, reference light sources, such as those from optical components or objects which remain unchanged between the two exposures.

4. HETERODYNE HOLOGRAPHIC INTERFEROMETRY

A typical setup for double-exposure heterodyne holographic interferometry is shown in Fig.5. The frequency $\Delta\omega/2\pi$ shift of about 100 kHz is realized by two commercially available acousto-optical modulators (AOM) in cascade, to give opposite frequency shifts. During recording, both modulators are driven with 40 MHz, so the the net shift is zero. During reconstruction, one modulator is driven with 40 MHz and the other one with 40.1 MHz, so that the net shift is the desired beat frequency of 100 kHz. The interference phase $\phi(\vec{x})$ is obtained by scanning the image of the object with a photodetector D_1 and measuring the phase of the beat frequency with respect to the reference phase obtained from a second detector D_2.

In practice an array of three detectors is used to determine the two phase-differences $\Delta\phi_x$ and $\Delta\phi_y$ in the orthogonal directions x and y, rather than the interference phase $\phi(x,y)$ itself. The latter can be easily and accurately calculated by summation of the measured differences along a given path. All electronic amplifiers and filters in the signal path should be designed carefully to avoid phase distortion which could reduce the accuracy of the phase measurement [9]. The detector array can be realized by the ends of optical fibers or fiber bundles, which feed the light to three photomultiplier tubes. The beat frequency signals at 100 kHz are filtered with a bandwidth of 10 kHz and the amplitudes are kept constant, independent of the intensity variations across the image, by a feedback control of the

434

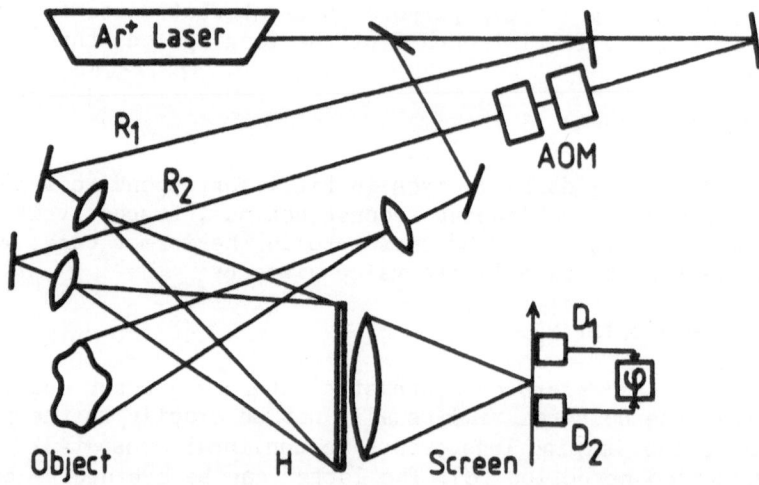

Fig.5. Setup for heterodyne holgraphic interferometry using two acousto-optical modulators (AOM) for the frequency-offset.

photo-multiplier voltages. The phase differences $\Delta\phi_x$ and $\Delta\phi_y$ are measured with two zero-crossing phasemeters, which interpolate the phase angle to 0.1° and count also multiples of 360°, which corresponds to the fringe number [9]. The detector array is mounted on a stepmotor driven stage to scan the image. Scanning and data-acquisition is automated and computer controlled. The measuring time for one position, including displacement of the detector array, is about one second.

The overall accuracy and reproducibility of heterodyne holographic interferometry depend mainly on the specification of the phasemeter, the signal-to-noise-ratio (SNR) of the detector signals and the mechanical stability of the optical setup [10]. The ultimate limits due to loss of fringe contrast by speckle decor-relation and overlapping reconstructions are discussed in the preceding section. Zero-crossing phase meters require for proper operation, i.e. to avoid multiple zero crossings, a SNR of at least 20 dB and a noise bandwidth of less than the signal frequency. The noise introduces a phase error $\delta\phi$ due to the fluctuations of the zero crossings. This phase error is found to be

$$\delta\phi^2 = 1/(SNR \cdot M) = T/(\tau \cdot SNR) \tag{12}$$

where $M = \tau/T$ is the number of zero crossings observed during the integration time τ of the phase meter. This means that a single measurement ($M = 1$) with SNR = 20 dB yields $\delta\phi = 6°$. This is reduced to $\delta\phi = 0.06°$ for $\tau = 100$ ms and a frequency of 100 kHz ($T = 10^{-5}$s). The SNR of the detector signals can be estimated from the holographic setup, the hologram effiency, and the laser power

[10]. Typically one gets with a 500 mW single-frequency Ar-laser at 514 nm, a bleached hologram of 25% average efficiency, an object size of 0.1 m^2 in the image and a detector diameter of 1 mm (area 10^{-6} m^2) a SNR of 40 dB.

The capability of heterodyne holographic interferometry is demonstrated by the experimental results given in Fig.6 [1]. The bending of a cantilever, clamped at the base and loaded at the end, was measured and compared with theory. The second derivative d^2u_z/dx^2 of the measured normal displacement $u_z(x)$ should follow a straight line, proportional to the bending moment M(x). The measurements were taken at intervals of Δx = 3 mm and the comparison with theory indicates an accuracy for the interference phase of $\delta\phi$ = 0.3°, corresponding to δu_z = 0.2 nm.

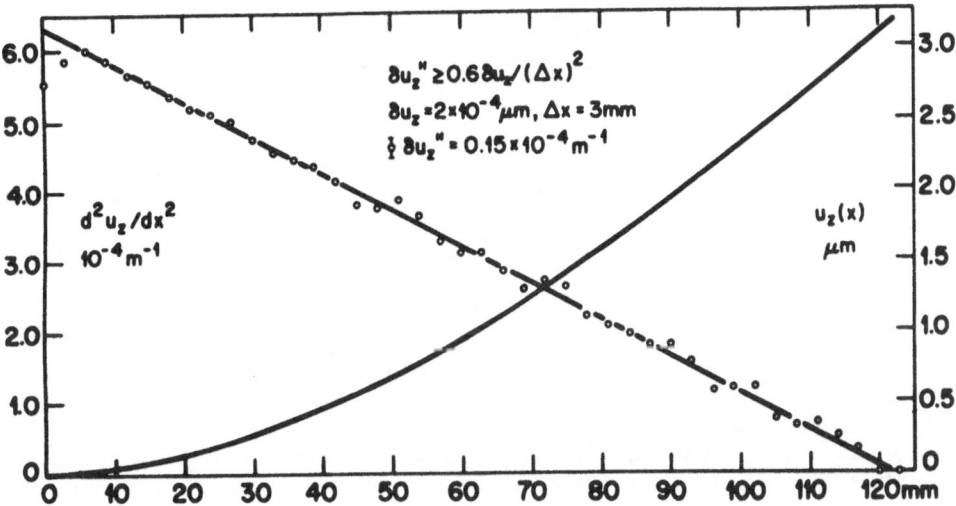

Fig.6. Bending of a cantilever measured by heterodyne holographic interferometry (Accuracy of phase measurement $\delta\phi$ = 0.3°).

5. QUASI-HETERODYNE INTERFEROGRAM PROCESSING

A typical setup for quasi-heterodyne holographic interferometry with the two reference sources close together is shown in Fig.7. Data acquisition and processing is carried out with a video-system and a micro-computer. The interferogram, projected on a ground glass screen, is observed with a TV-camera. A silicon target vidicon or CCD-array camera should be used to guarantee linear response for the detected signal. Three hologram reconstructions with different relative phases ϕ_k of the reference beams are investigated successively. These phases are adjusted carefully with a piezo mounted mirror to become ϕ_1 = 0°, ϕ_2 = 120°, ϕ_3 = 240°.

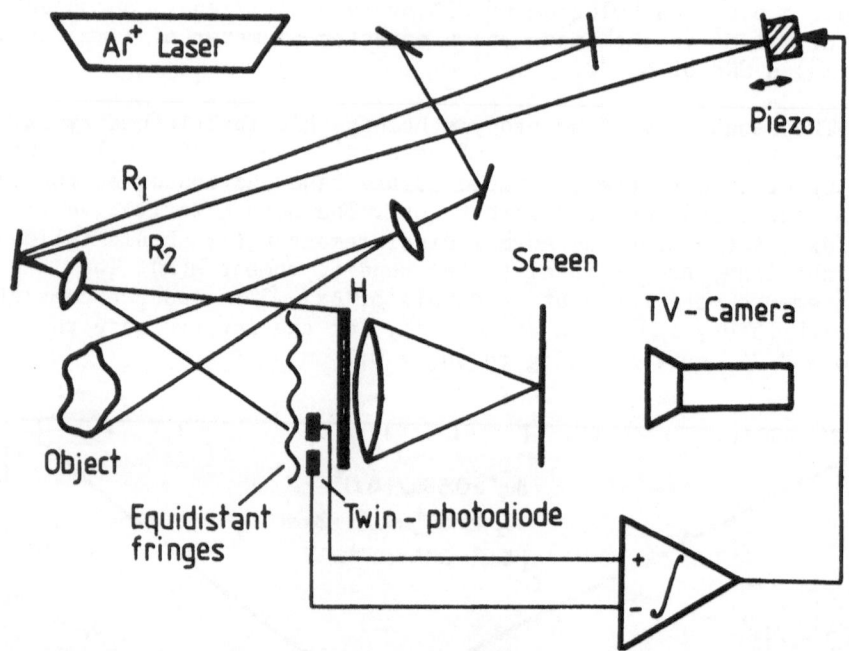

Fig.7.Setup for quasi-heterodyne holographic interferometry with fringe stabilization and TV observation.

In the case of two reference sources close together, the relative phase of the two reference beams can be detected and controlled very effectively by a twin-photodiode placed in the hologram plane, just outside the field used for the reconstruction, as shown in Fig.7. This is possible easily and accurately since the two reference beams produce on the hologram a pattern of parallel and equidistant interference fringes of typically 2 to 5 mm separation. The twin-photodiode detects the fringe maximum and an integrating feed back loop to the piezo electric mirror keeps the fringes stable. The different phases ϕ_k for the three reconstructions can be established very accurately by moving the twin-photodiode in steps of 1/3 of the fringe separation. With this stabilization device fluctuations $\delta\phi_k$ of less than 1° for a time constant of 10 ms have been achieved.

To estimate the accuracy of the phase measurement one has to distinguish between systematical and statistical sources of errors. For the statistical error $\delta\phi$ the expression

$$\hat{\delta\phi}^2 = (2/3 \ m^2)(\hat{\delta}I/\overline{I})^2 + (\delta\phi_k)^2/2 \tag{13}$$

can be derived from Eq.(4). It depends on the relative fluctuations $\delta I/\overline{I}$ of the intensities measured by the TV-camera, and on the statistical fluctuations $\delta\phi_k$ of the relative phases between the reference beams. A systematic error is introduced by wrong phase shifts ϕ_k. It gives rise to a periodic error in $\phi(\vec{x})$. For three measurements with phase shifts ϕ_k in steps of 120°, the maximum error of $\phi(\vec{x})$ is of the same order as the errors for the ϕ_k. It is important to point out that the absolute value of the ϕ_k must be guaranteed with a precision equal to that desired of the phase $\phi(\vec{x})$.

To determine the statistical errors (variance) of the phase measurement, the same object was recorded simultaneously with the two reference beams to give a two-reference-beam hologram of an undeformed object [6]. This object was illuminated with a 600 mW single frequency Ar-laser at 514 nm. The intensity ratios between object illumination and reference R_1 and R_2 can be adjusted independently by two variable beam splitters to get, first, optimum recording conditions, and afterwards maximum power in the recon- structing reference waves. The double-exposure hologram was recorded on a Millimask plate (50 x 50 mm). The interferogram was reconstructed on a ground glass screen, at unit magnification, through an objective having 300mm focal length and aperture f/5.6, and observed by a silicon target Ultricon TV-camera. A video-analyzer extracts an analog signal along a straight line through the object (fringe profile). This signal is converted into digital and the micro- computer takes the average of three successive scans as input.

The average electrical noise ratio $\delta I/\bar{I}$ of the TV-camera was determined experimentally to be about 1.7%. Averaging over three independent scans reduces this value by a factor $1/\sqrt{3}$. One can there- fore expect a phase error of $\delta\phi = 1.2°$, calculated from Eq.(13) for a fringe contrast of m = 0.4. From the measured phase along the undeformed object [Fig.8], the statistical phase error was deter- mined to be $\delta\phi = 1.6°$. This agrees quite well with the expected value. Note that the phase error is slightly larger at both ends of the object, due to the lower mean intensity $\bar{I}$, which yields a reduction of the TV signal-to-noise ratio.

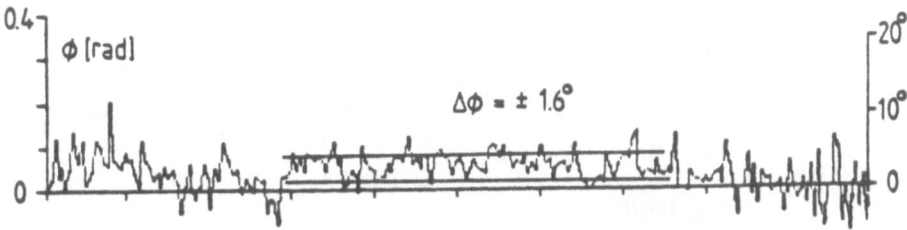

Fig.8. Phase variations (statistical error) measured by quasi- heterodyne holographic interferometry for an undeformed object.

In conclusion, quasi-heterodyne evaluation of holographic interferograms offers an accuracy of 1/100 fringe by using video-electronic processing. The required two-reference-beam holography can be operated as easily as classical double-exposure holography by using two reference beams close together (about 0.01° angular separation). With this arrangement, the misalignment sensitivity of two-reference-beam holography is drastically reduced; and even wavelength changes between recording and reconstruction (e.g. ruby laser for pulsed recording and He-Ne laser for recon-struction)can be tolerated.Moreover,the relative phase of the two reference waves during reconstruction can be stabilized and controlled very easily and accurately. The described system effectively combines the simplicity of standard double-exposure holography, video-electronic processing, and the power of heterodyne holographic interferometry.

The described setup can be applied to real-time, double-exposure and double-pulse holography. Using two different illumi-nation beams S_1 and S_2 with each reference beam, holographic contouring can be made in exactly the same experimental setup (Fig.9). Experimental results for the holographic contouring of a cube are shown in Fig.10. This allows quantitative determination of displacement vector components on the surface of arbitrarily shaped objects. The high accuracy fringe interpolation offers also the possibility for automated determination of 3-D displacement vector fields from one and the same hologram plate [11].

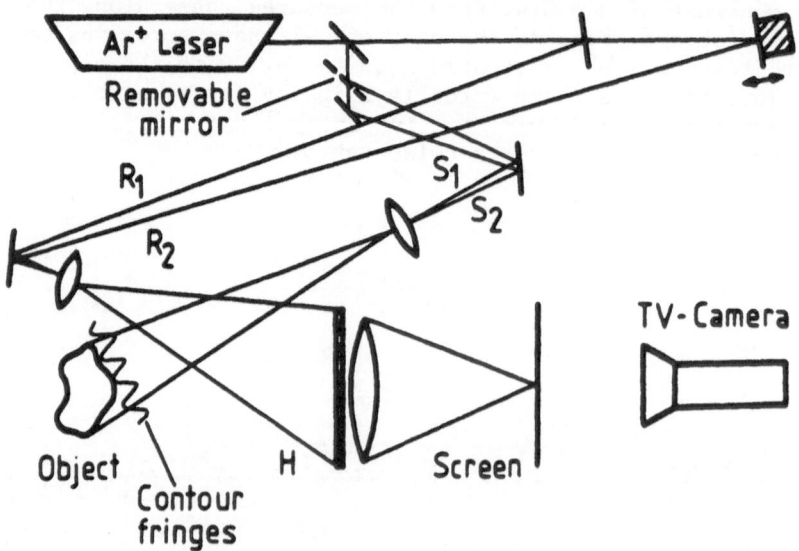

Fig.9. Setup for holographic contouring by changing the illumina-tion direction (S_1,S_2) and quasi-heterodyne processing.

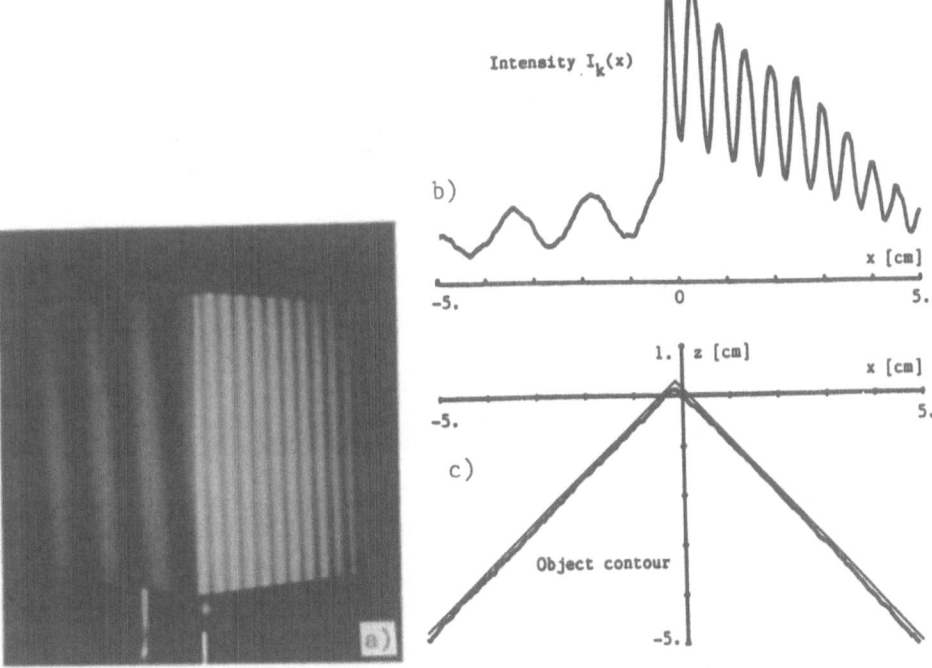

Fig.10. Holographic contouring of a cube: (a) object with contour fringes, (b) intensity profile across the object, (c) object contour evaluated by quasi-heterodyne processing.

REFERENCES

1. Dändliker, R., Heterodyne Holographic Interferometry, in Progress in Optics, vol XVII, E. Wolf, ed. (North Holland, Amsterdam, 1980), pp.1–84.
2. Dändliker, R., E. Marom, F. M. Mottier, Two-Reference-Beam Holographic Interferometry, J.Opt.Soc.Am. 66 (1976) 23–30.
3. Crane, R., New Developments in Interferometry. V. Interference Phase Measurement, Appl.Opt. 8 (1969) 538–542.
4. Wyant, J. C., Interferometric Optical Metrology: Basic Principles and New Systems, Laser Focus, May 1982, pp.65–71.
5. Ref.1, pp.40–44.
6. Dändliker, R., R. Thalmann, J.-F. Willemin, Fringe Interpolation by Two-Reference-Beam Holographic Interferometry: Reducing Sensitivity to Hologram Misalignement, Opt.Commun. 42 (1982) 301–306.
7. Hariharan, P., B. F. Oreb, N. Brown, A Digitial Phase-Measurement System for Real-Time Holographic Interferometry, Opt.Commun. 41 (1982) 393–396.

8. Ref.1, pp.33-40.
9. Mastner, J. and V. Masek, Electronic Instrumentation for Heterodyne Holographic Interferometry, Rev.Sci.Instrum. 51 (1980) 926-931.
10. Ref.1, pp.54-60.
11. Dändliker, R. and R. Thalmann, Determination of 3-D Displacement and Strain by Holographic Interferometry for Non-plane Objects, SPIE vol.398 (1983) pp.11-16.

HOLOGRAPHY IN MEDICINE AND BIOLOGY
- STATE OF THE ART AND
THE PROBLEM OF INCREASING MILITARIZATION -

G. von Bally

Medical Acoustics and Biophysics Laboratory, ENT-Clinic, University of Muenster, Kardinal-von-Galen-Ring 10, D-4400 Muenster, Federal Republic of Germany

INTRODUCTION

Although commonly regarded as part of physical optics holography has turned out to be a widespread, interdisciplinary field of science. This is not only understandable from the fact that the principle of holograhy is not restricted to optical waves but can be applied to any wave phenomenon. Moreover, it can be explained by understanding holography in more general terms e.g. complex spatial filtering [1] or by demonstrating the mathematical similarity to communication theory on the basis of Fourier transform [2]. Accordingly, applications of that basic principle have been attempted in a variety of sciences. Thus, today not only physicists and engineers but also biological and medical scientists are exploring the potentials of holographic methods in their special field of work. Since most of the underlying physical principles are explained in detail in other contributions to this Institute this article is confined to applications of holography in biomedical sciences. Because of the great number of contributions and the variety of applications [3,4,5,6,7,8], in this review the investigations can only be mentioned briefly and the survey has to be confined to some examples. Thus, concerning the topics of this Institute it seems to be appropriate to emphasize optical holograhic methods, especially the applications of holographic interferometry.

As in all fields of optics and laser metrology, a review of biomedical applications of holography would be incomplete if military developments and their utilization were not mentioned. As will be demonstrated by selected examples the increasing interlacing of

science with the military does not stop at domains that traditionally are regarded as exclusively oriented to human welfare like
biomedical research [9]. Thus, the consequences - even in such
highly specialized fields like biomedical applications of
holography - will be discussed.

THREE-DIMENSIONAL IMAGING

Biomedical sciences are involved in research on biological processes and their interactions in living organisms, so that the knowledge of the spatial structure - in particular the microstructure -
of the object under consideration is of importance. Since three-
dimensional imaging and the additional advantage of a large field
of depth are basic features of holography, this technique has been
proposed for applications in biological and medical teaching [10].
Imaging by reconstruction from appropriate holograms may be of help
in cases where suitable subjects are not at disposal. Technical
difficulties and requirements of expensive equipment - compared to
conventional slide projection - may be the reason for the rare applications of this possibility. Therefore it would be desirable to
record holograms, which can be reconstructed in white light [11].
Yet, holographic three-dimensional imaging has been applied in ophthalmological research (fundus holography) [12,13,14,15] with the
advantage that - contrary to conventional fundus photography - any
layer of interest within the eye may be investigated in the reconstructed image.

A general problem in biomedical applications of holographic methods, particularly in clinical applications in-vivo without an-
esthesia, is the requirement of interferometric stability of the
experimental set-up and the object during exposure. This can be
solved satisfactorily only by utilizing a Q-switched ruby laser
[16]. Using this technique safety requirements have to be followed
carefully, especially in ophthalmological applications [12,13].
Due to the short exposure times the use of Q-switched ruby lasers
allows holographic imaging of fast moving objects. Therefore, it
can be applied e.g. to investigations of chemical reactions, the
distribution and size of droplets, jets or the function of spray
nozzles. Thus, in-line holographic arrangements have been used for
the assessment of aerosols and pollution [17]. Such investigations
are especially important e.g. for the development of sprays which
should optimally penetrate to the lung like anti-asthmatic sprays
or for optimizing the distribution of insecticides and pesticides
in agriculture. But there have been also proposals for applications in the defoliation-actions in Vietnam [18].

This demonstrates that the problem of the dualistic applicability of the same technique, even the same apparatus, for military and civil usage is present even in such specialized fields like bio-medical applications of holography. It is made obvious by this example that possible misuse is inherent in scientific and techno-logical achievements. Thus, it has to be a selfevident duty of each scientist and engineer to consider the consequences of his work with the same engagement and self criticism as applied to the solution of a particular scientific or technical problem.

HOLOGRAPHIC ENDOSCOPY

Most biological processes take place in the interior of living or-ganisms. Thus, for endoscopic investigations special optical de-vices have been developed. Conventional endoscopic photography suffers from the limited field of depth. Holographic methods may help to solve this problem. In developing a "holoendoscope" it has to be taken into account that external recording of holograms re-quires the reflected object wave to be led from the object space to the photographic plate by lenses and mirrors. This results par-ticularly in phase aberrations and a small entrance pupil limiting the parallax. Therefore an endoscope has been constructed includ-ing a holographic recording device inserted in the head of the in-strument [19]. The use of mirrors and lenses results necessarily in a rigid instrument. Thus, the introduction of fiber optics has considerably improved the performance of various endoscopes. In holograhic endoscopy multimode fiber bundles can only be used for object beam guidance due to phase distortions. Effects of image transmission in such a fiber bundle and movements of the object and the fiber bundle itself on the image quality, especially the inter-ference fringe visibility (ref. chapter "Double Exposure Holo-graphy") have been investigated [20], demonstrating that the use of pulsed ruby lasers is possible. Using a multimode fiber bundle for the illuminating object beam and a monomode fiber for reference beam guidance, holographic endoscopy is feasible in spite of the limited power transmission capability of a single monomode fiber. Thus, the development of an easy to handle and flexible set-up for a hand-held holoendoscopic camera became possible [21], which prov-ed to be usable for clinical applications (ref. chapter "Double Exposure Holography"). Investigations using a singlemode fiber in both the reference and object beam path and a multimode fiber bundle for image transmission are described in reference [22]. Re-duction of speckle noise caused by the small entrance pupil of en-doscopic optics has been gained by using a gradient index rod lens as illumination and imaging guide, simultaneously. Parasitic light reflexes from the endfaces of the gradient index rod lens could be suppressed in the reconstructed image by appropriate shaping and holographic subtraction [23].

HOLOGRAPHY WITH NON-VISIBLE WAVES

In addition to three-dimensional imaging of surface and shape of a biological object, 3-D display of its internal structure is also of interest. Due to the high absorption coefficient of biological materials, recording of holograms of internal structures using electromagnetical waves within the visible spectral region is usually not possible. On the other hand, certain non-visible waves are commonly used in biology and medicine because of their potentiality to penetrate tissue. Non-visible waves which can be generated coherently are e.g. micro-, infrared, acoustical waves and, recently, X-rays. Thus, there is the possibility of recording in-line and off-axis holograms, but with difficulties and unsolved problems in detection and transformation into the visible spectrum at reconstruction.

Microwave Holography

The capability of microwaves to penetrate optically opaque dielectrics can be used to locate internal anomalies by holographic methods [24]. Experimental results such as the detection of metallic objects like concealed weapons through clothes [25] may indicate potential biomedical applications.

Infrared Holography

As to the author's knowledge at the time there are only suggestions for the use of coherent infrared waves for holographic three-dimensional imaging of cancer of skin and breast [10]. The advantage would be contrast enhancement based on the different absorption of infrared radiation by normal and cancerous tissue.

Acoustical Holography

As far as biomedical applications of holography with non-visible waves are concerned most extensive investigations have been carried out in the field of ultrasonic holography. The interest in the use of ultrasonic waves is based on its capability to image soft tissue structures without - contrary to e.g. X-ray exposure - the risk of radiation damage. It is beyond the scope of this short survey to describe the different techniques for holographic ultrasonic imaging. Those particularly interested in biomedical applications of acoustical holography may consult comprehensive literature [10,26,27,28,29].

X-Ray Holography

Two-dimensional X-ray imaging is the most commonly used non-visible wave technique in medical diagnostic of internal structures within the human body. Thus, for a long time it has been hoped in biomedical research to develop a three-dimensional X-ray imaging technique using holography. Such a technique is expected to render possible recording of 3-d images of molecules, pinpointing cancer cells, and developing a safer and more effective X-ray therapy.

Basic ideas to overcome problems of finding an appropriate high resolving recording medium and a coherent X-ray source were preposed long ago [30]. However, the solution and (expensive) technical realization was reached only after military interest was raised. Today X-ray lasers emitting at short wavelengths are developed for military purposes but information is not available from the scientific literature [31], although detailed considerations on technical realization and expected results of e.g. X-ray biomicroholography are published [32].

HOLOGRAPHIC MULTIPLEXING

Holographic multiplexing can be a solution to the problem of three-dimensional imaging without a coherent X-ray source by combining holographically several two-dimensional X-ray pictures recorded from different views [33,34]. The advantage compared to conventional tomography is the possibility of analysing any layer of interest in the image reconstructed from the synthetic hologram, although it has been generated only by a limited number of radiographs. The type of recording process of the two-dimensional images used for holographic multiplexing is obviously of no importance. It could be an ultrasonic-B-scan record [10], electron micrograph [10], or simply two-dimensional photographs [11].

HOLOGRAPHIC MICROSCOPY

In studying microorganisms or microstructures of biological specimens microscopy is an important tool in biomedical sciences. Some properties of holography can be used with advantage in microscopic imaging, in particular the large field of depth and the two step principle of recording and reconstructing. This enables microscopic investigations to be carried out without preparing sections [35], or without focusing to a certain layer during recording e.g. in exobiology [36], or the method of analysis to be chosen afterwards (dark field, phase contrast, or interference microscopy) [37]. Basically there are two methods for holographic microscopic imaging. Firstly, there are so called holographic lensless techniques, which, however, lead to considerable aberrations caused by the enlargement of the expanding object and reference beam. Yet, pulsed ruby lasers can be applied since no cemented lens systems are used. Secondly, the already microscopically enlarged image can be used as object for holographic recording [10]. In this case the use of a groundglass leads to diffuse but speckled object illumination, while the use of a point source results in a non-uniform illumination. A good compromise seems to be found by using four light beams, entering the optical system with the aperture angle [38]. A regular interference pattern that remains superimposed on the reconstructed image may be eliminated e.g. by holographic spatial filtering techniques.

HOLOGRAPHIC SPATIAL FILTERING

The possibility of a "posteriori" image deblurring of photographs unintentionally blurred by motion, improper focusing, imperfect instruments etc., using a "holographic Fourier-transform division filter" [39] has also led to image improvements in electron microscopy [40]. A well known result of this application is an electron microscopic picture of the double-helical structure of a fd virus [41].

Holographic spatial filtering techniques can also be applied to non-coherent waves e.g. X- and Gamma-rays as used in radiology and nuclear medicine. Instead of taking serial two-dimensional radiographs for three-dimensional image by holographic multiplexing, the object can be projected simultaneously from different views, thus producing a coded image e.g. on a film. In these methods - known as coded source and coded aperture techniques [33,42] - on-axis and off-axis Fresnel-zone plates, or discrete point distributions may be used for the coding process. The latter may be an array of holes (Gamma-ray imaging), or a distribution of radiation sources

(X-ray imaging). Using a non-redundant distribution of the sources to optimize the signal-to-noise ratio, decoding can be provided during reconstruction in laser light e.g. by means of a Fourier-transform hologram of this distribution [43]. At present this latter technique ("flashing tomosynthesis") is under study with the aim of displaying layers of moving objects like the pulsating heart or fast flowing contrast media, among other objectives.

PATTERN RECOGNITION

The possibility to "recognize" wavefronts using holographic spatial filtering techniques gives rise to a variety of suggestions for biomedical applications of this feature of holography. Cell identification - especially differentiation between normal and cancerous cells - may be an important example for clinical use of holographic pattern recognition techniques [1,44]. Automatic recognition and counting of diatoms (algae) as a measure of water pollution is under study, using averaged filters [45]. Because of the great interindividual variety of the shape of biological specimens as well as the problem of orientation and size variance the generation of appropriate filters is sophisticated. New optical transforms particulary suited for scale, positional and rotational invariant correlations without loss in signal-to-noise ratio can be applied advantageously to holographic pattern recognition [46].

Spatial filters can also be generated by digital computers [47]. Biomedical applications of this technique e.g. in image enhancement and pattern recognition are mentioned in [48,49].

Another important domain in biomedical research is the study of alterations of form and structure as well as movements of biological objects. Utilizing holographic methods this can be realized either by succesive recording of single holograms of the process under consideration (cineholography) or by means of holographic interferometry.

CINEHOLOGRAPHY

Cineholography has been used for microscopic investigations of living marine plankton organisms [50]. Holograms have been taken by stroboscopic illumination using a pulsed Argon laser synchronized to the recording sequence of a camera. Because of the large field of depth it was possible to investigate the microscopic subjects as they moved freely in the object space. A similar technique has been applied to studies of Bends decompression sickness (divers sickness) which is caused by bubbles forming in the blood vessels. Generation of tiny gas bubbles within the vessels of a living hamster's cheeck pouch has been investigated by appropriately changing the air-pressure within a hyperbaric chamber [51].

HOLOGRAPHIC INTERFEROMETRY

As known by its applications to non-destructive testing, holographic interferometry provides the possibility of a three-dimensional, non-contactive, high resolving analysis of alterations either in shape, structure, and position of the object under test. These changes are characterized by interference fringes macroscopically visible in the reconstructed image. In the following, examples of biomedical applications are presented, classified by the different most commonly used techniques of holographic interferometry.

Time-Averaged Holography

In order to generate a macroscopically visible interference fringe pattern by time-averaged holography the object has to move periodically. Therefore, this technique is used for vibration analysis. A biological object with periodic movements within the range of displacement resolution of holographic interferometry, is the tympanic membrane. Thus, time-averaged holography has been used for the analysis of the vibration pattern of tympanic membranes in cats and human temporal bones to determine the role of the tympanic membrane in sound transmission by the middle ear [52,53]. Vibration analysis of the round window in cats [54] and of the human ossicular chain [55,56] has been carried out using this technique. Contrast enhancement of the interference fringes was achieved in the latter experiments by means of a phase modulated reference wave to shift high fringe orders to lower ones. In this way a small tilting movement of the stapes could be detected, besides the expected piston like oscillation. Phase modulation of the reference beam also renders possible phase mapping and increase in amplitude resolution to the order of 1 nm when using time-averaged holography which is otherwise insensitive to the phase of vibration.

Real-Time Holography

Biomedical applications of real-time holography are complicated by the requirement of precise repositioning of the reference hologram and the rapid changes of biological specimens, even in in-vitro experiments [57]. Therefore these investigations are restricted to experiments on models or objects, e.g. teeth or macerated bones, not suffering from uncontrollable alterations.

Since there are possibly adverse influences on osteosynthesis by unphysiological load, e.g. after implantation of prostheses or fixation by plates after fractures, the mechanical properties of bones have to be known. Comparative holographic investigations have been carried out in real-time on the human femur in-vitro before and after implantation to optimize hip joint prostheses [58,59], as well as on the human tibia after fracture fixation by compression plates [60]. Real-time holography has been used to study the function of the human ankle joint and the leg-foot complex [61], as well as the thermal expansion of human teeth and dental materials [62].

Biological membranes may vibrate unsymmetrically around the resting position. Time-averaged holography cannot detect such unsymmetry of oscillations because it does not provide (vibration-)phase information. Real-time holography combined with synchronized stroboscopic illumination proved to be capable of investigating arbitrary vibrations by Fourier analysis and synthesis [63] by damping mechanically one half-wave of the oscillation on models of the tympanic membrane.

Double-Exposure Holography

Concerning its basic principle double-exposure holography can be regarded as part of holographic multiplexing, a technique used, among other things, for three-dimensional imaging, as mentioned previously. Thus, there is a capability of this technique, known as "contour mapping", related to three-dimensional imaging rather than to the analysis of vibrations, deformations, or structure changes. Isocontour lines, generated by superposition of two holograms of the same object on one photographic plate, can be used to measure the three-dimensional contour of an object. For that purpose each hologram has to be recorded e.g. with a different wavelength or slightly changed angle of the reference or object beam. The first method has been used to generate depth contour lines of the eye [12,13] and it is suggested to determine in this way the curvature of the sclera or front corneal surface for the production of well fitting contact lenses [57]. A combination of contour mapping and real-time holography is described to measure the wear of knee prostheses [64]. Similar techniques have been used to investigate the wear of hip joint prostheses, dental materials, and prosthetic mitral heart valves [65].

For the analysis of vibration and deformation double-exposure holograms are taken by time selective holographic recording of two phases of vibration or states of deformation, resp., on the same photographic plate. Concerning the use of CW-lasers in the domain of biomedical applications it is similar to that of real-time holography, e.g. in-vitro experiments on the biomechanics of the locomotor system, particularly the investigation of the function of the tibia/fibula system [61,66], the pelvis [67], studies on the deformations under load of hip joint prostheses/femur systems [68], and human vertebrae [69,70]. A review of orthopedic applications of holographic interferometry can be found in [71]. Double-exposure holography has been used in experimental dentistry to investigate deformations of teeth, jaws, prosthodontic appliances, and skulls [72,73,74]. Dental applications of holographic interferometry are comprehensively reviewed e.g. in [75,76]. The possibility to measure the growth of seeds and plants by means of double-exposure holography has also been considered [77]. Using an extremely simple, easy to handle, and inexpensive set-up such an application of double-exposure holography can become helpful in on-spot experiments for the enhancement of food production in third world countries [78].

Using a Q-switched ruby laser for the recording of double-exposure holograms - besides periodic vibrations - fast, non-periodic processes can be studied. This possibility was discovered firstly during military oriented investigations of bullets in flight using a ruby laser which accidentally emitted two short pulses. The reconstructed images showed fringe patterns according to the propagation of the shock wave [79]. An interesting application of this method, which has been used in biomedical research, is the study of transient processes. Thus, movements of tympanic membranes subjected to acoustic impulses have been investigated in in-vitro experiments on guinea pigs by superposition of a hologram recorded at rest and a second hologram taken on the same holographic plate at a certain time after the acoustic event [80]. The laser pulses were separated by a time interval of about one minute for technical reasons, which prevents in-vivo applications. The aim of these experiments was the study of generation of lesions of the tympanic membrane caused by acoustic impulses such as bursts emanating from weapons [81].

Releasing two laser pulses within one flashlamp pulse of a Q-switched ruby laser system the same technique has been applied to the same object but for the development of clinical diagnostics in audiology. As pathological changes of the mechanical properties of the middle ear have an influence on the vibratory pattern of the tympanic membrane, a vibration analysis may provide the possibility of a differential diagnosis of dysfunctions without opening the tympanic cavity. After model and in-vitro experiments, results of which have demonstrated the capability of double-pulsed holography to detect unsymmetric oscillations, a special closed acoustic system for simultaneous application of sound and holographic recording of the tympanic membrane vibration through the intact outer ear canal has been used on patients [82]. For clinical routine applications it would be desirable to have an easy to handle, flexible holographic-endoscopic arrangement. Thus, a small hand held holootoscopic camera has been developed using fiber optics [21] (ref. chapter "Holographic Endoscopy"). Successful in-vivo experiments to investigate motions of teeth and bridge-work to optimize the design of prosthodontic appliances have been carried out on patients, releasing the laser pulses at certain masticatory force levels [83,84]. Human chest motions have been investigated in-vivo during inhalation with the aim of lung diagnosis [85], as well as by triggering the laser pulses in relation to heart action to test the possibility for detection of heart diseases by double-pulsed holography [86]. In order to study the function of the human vocal organ holographic vibration analysis has been carried out in-vivo on the frontal part of the human neck [87].

Electronic Speckle Pattern Interferometry

Speckles usually regarded as an inevitable disadvantage in coherent optics can be used for holographic interferometrical purposes, e.g. in combination with videotechnical means, as in electronic speckle pattern interferometry (ESPI). Basically, this is an in-line holographic technique using a video target as recording medium, which renders possible a quasi real-time display of speckle interferograms according to the TV frame rate [88]. Examples of biomedical applications of this technique are vibration analysis of the human tympanic membrane [89], ossicular chain [55], basilar membrane [90], and of the human skull [91].

Rigid Body Motions

Distortions of the interference fringe pattern in the reconstructed image caused by uncontrolled rigid body motions are a practical problem in holographic interferometry, particularly in biomedical applications. The effect of such motions can be compensated to some extent in real-time holography by appropriate adjustment of the illuminating beam [58,92]. In double-exposure holography this problem may be solved by means of a "posteriori" applied Moire-technique [93] and by sandwich holographic interferometry. The latter allows compensation of rigid body motions by appropriate alignment of the illuminating beam during reconstruction, as well as the determination of the sign of the deformation vector [94]. A modified technique has been developed in order to use a Q-switched, double pulsed ruby laser for sandwich holograhic interferometry [95], which allows in-vivo applications.

CONCLUSIONS

Although only some of the numerous examples have been mentioned, this review demonstrates that already early after the development of lasers, by which holography became a practical tool, the different holographic techniques have been used extensively in bio-medical sciences. Today holography has established its place in biomedical research. Unfortunately, clinical applications are still rare and none of the holographic techniques has been used really routinely in clinical diagnostics, up to now. Thus, before holography in the biomedical field goes out of the research laboratories a lot of work still has to be done within the laboratories in interdisciplinary cooperation between physicists, engineers and biological and medical experts.

On the other hand, the given examples of military applications demonstrate that the problems caused by an increasing interlacing of science with military interests are present even in such specialized fields like biomedical applications of holography, though this is commonly not anticipated:

o Dualistic applicability of techniques

Nearly all techniques can be used as well in the civil as in the military domain.

o Dualistic applicability of results of civil research

Not only the techniques themselves but also the results of their applications even in research fields commonly regarded as purely civil like biomedical sciences can be used for military purposes.

o Preference of military demands

In spite of a need for human welfare and medical care, which may have been expressed first, many technical developments - if not most - have been or will be used first in the military domain or exclusively developed for it.

o Preference of military budgets

   Among other things the preference of developments for military
   use is caused by the unbalanced increase of military and civil
   research budgets in favour of the arms race.

o Influence on researchers

   In turn, the preference of military demands and military budgets
   leads to situations, in which scientists - although originally
   working with different intentions - propose already by themselves military applications of their ideas, in order to get the
   funds for its realization.

It is obvious that the results of scientific and engineering research are at least one important cause of the arms race. This
competition for destructive power does not only distract scientific
and engineering resources from contributing to the solution of the
major social issues like hunger, overpopulation, insufficient
education and medical care, but rather increases the threat, man-
kind has to face [96]. Thus, no scientist or engineer can deny his
share of responsibility, including myself, since, as demonstrated
in this review, even results of holographic investigations in the
biomedical field can be and are used for military purposes.

Critical discussions of the consequences of our work as scientists
and engineers are a self evident part of our science and not only
of other fields like social sciences or politics. Therefore, this
important topic has to be part of our scientific meetings, including NATO Advanced Study Institutes like this, in order to raise and
increase consciousness about that problem, since "consequence research" and public and free exchange of information is an equally
important part of our social responsibility, as is our scientific
work itself.

454

REFERENCES

[1] Felleppa, E.J.: Biomedical applications of holography, Physics Today 22, 25 (1969).
[2] Leith, E.N. et al.: Reconstructed wavefronts and communication theory, J. Opt. Soc. Amer. 52, 1123 (1962).
[3] Greguss, P. (ed.): Holography in Medicine, IPC Science and Technol. Press (1975).
[4] Hoke, M. and G. von Bally (eds.): Proc. Symp. 1976 Spec. Res. Area 88 and Int. Conf. on Electrocochleography and Holography in Medicine, Muenster (1976).
[5] Marom, E., Friesem, A.A. and Wiener, E. (eds.): Proc. Int. Conf. Appl. Hol. and Opt. Data Process., Pergamon Press (1977).
[6] von Bally, G. (ed.): Holography in Medicine and Biology, Springer-Series in Optical Sciences, Springer-Verlag, Heidelberg, Berlin, New York, Vol. 18 (1979).
[7] Shankar, P.M. et. al.: Applications of Coherent Optics and Holography in Biomedical Engineering, IEEE Transactions on Biomedical Engineering 29, 8-15 (1982).
[8] von Bally, G. and P. Greguss (eds.): Optics in Biomedical Sciences, Springer-Series in Optical Sciences, Springer-Verlag, Heidelberg, Berlin, New York, Vol. 31 (1982).
[9] von Bally, G.: Remarks of the chairman: scientists, scientific societies, and military research, in: D. Vukicevic (ed.): Holographic Data Nondestructive Testing, SPIE 370, 26 (1983).
[10] Greguss, P.: Thoughts on the future of holograhy in biology and medicine, Optics and Laser Technol. 253 (1975).
[11] Tsujiuchi, J.: Holograhic stereograms as a tool of nondestructive testing, SPIE 370, 17 (1983).
[12] Vaughan, K.D. et al.: Holography of the eye: a critical review, in: M.L. Wolbarsht (ed.): Laser applications in medicine and biology, Plenum Press (1974), 77 pp.
[13] Calkins, J.L.: Fundus camera holography, see [3], 85 pp.
[14] Tokuda, A.R. et al.: Development of a holocamera for 3-D microscopy of the unanesthetized human eye, J. Opt. Soc. Am. 68, 1382 (1978).
[15] Ohzu, H. et al.: Application of holography in opthalmology, see [6], pp. 133.
[16] Ansley, D.A.: Techniques for pulsed laser holography of people, Appl. Opt., 9, 815 (1970).
[17] Bexon, R. et al.: In-line holography and the assessment of aerosols, Optics and Laser Technol. 8, 161 (1976).
[18] Bals, E.J.: The principles of and new developments in ultra low volume spraying, Proc. 5th Br. Insectic. Fungic. Conf. 189 (1969).
[19] Hadbawnik, D.: Holographische Endoskopie, Optik 45, 21 (1976).

[20] Yonemura, M. et al.: Endoscopic hologram interferometry using fiber optics, Appl. Opt. 20, 1664 (1981).

[21] von Bally, G.: Otoscopic investigations by holographic interferometry: a fiber endoscopic approach using a pulsed ruby laser system, see [8], pp. 110.

[22] Dudderar, T.D. et al.: Remote vibration measurement by time averaged holographic interferometry, Proc. Vth Int. Cong. Exp. Mech., Montreal, 362 (1984).

[23] von Bally, G. et al.: Gradient-index optical systems in holographic endoscopy, Appl. Opt. 23, 1725 (1984).

[24] Tricoles, G. et al.: Microwave holography: applications and techniques, Proc. IEEE. 65, 108 (1977).

[25] Farhat, N.H. et al.: Millimeter wave imaging of concealed weapons, Proc. IEEE 59, 1383 (1971).

[26] Proc. Int. Symp. Acoust. Holography, Plenum Press (1967) et seq.

[27] Greguss, P.: Optical evaluation of ultrasonic scattering in animal tissue, Ann. New York Acad. Sci. 267, 312 (1976).

[28] Hildebrand, B.P. et al.: An introduction to acoustical holography, Plenum Press (1972).

[29] Waidelich, W. et al.: Methoden der akustischen Holographie, in: Medizinische Physik in Forschung und Praxis, De Gruyter (1976), 146 pp.

[30] Caulfield, H. et al.: The applications of holography Wiley-Interscience, New York (1970).

[31] New harmonic technique opens up extreme UV, Laser and Applications 40 (1983).

[32] Solem, J.G.: X-ray biomicroholography, Opt. Eng. 23, 193 (1984).

[33] Groh, G.: Tomosynthesis and coded aperture imaging: new approaches to three-dimensional imaging in diagnostic radiography, Proc.R.Soc. Lond. B. 195, 299 (1977).

[34] Sugimura, K. et al.: Clinical application of multiplex holography, SPIE 370, 20 (1983).

[35] Greguss, P.: Laser as a probe in biomedical research, in: Waidelich, W. (ed.): Laser 75 Optoelectronics Conference Proc., Munich, (1975), pp. 155.

[36] van Ligten, R.F.: Holographic microscopy in exobiology, see [3], 44 pp.

[37] Ellis, G.: Holomicrography: transformation of image during reconstruction a posteriori, Science 154, 1195 (1966).

[38] Haendler, E. et al.: Contribution to experimental holographic microscopy, see [3], pp. 51.

[39] Stroke, G.W. et al.: Image improvement and three-dimensional reconstruction using holographic image processing, Proc. IEEE 65, 39 (1977).

[40] Stroke, G.W. et al.: Image improvement in high-resolution electron microscopy using holographic image deconvolution, Optik 41, 319 (1974).

[41] Stroke, G.W.: Optical computing, IEEE Spec. 9, 24 (1972).

[42] Barett, H.H. et al.: Fresnel zone plate imaging in radiology and nuclear medicine, Opt. Eng. 12, 8 (1973).

[43] Weiss, H. et al.: Coded aperture imaging with X-rays (flashing tomosynthesis) Opt. Acta 24, 305 (1977).

[44] Caulfield, H.J.: The applications of coherent optical image processing in medicine and biology, see [3], pp. 39.

[45] Almeida, S. et al.: Water pollution monitoring using matched spatial filtering, Appl. Opt. 15, 510 (1976).

[46] Casasent, D. et al.: New optical transforms for pattern recognition. Proc. IEEE 65, 77 (1977).

[47] Lohmann, A.W. et al.: Computer generated spatial filters for coherent optical data processing, Appl. Opt. 7, 651 (1968).

[48] Stroke, G.W. et al.: Holographic image restoration using Fourier spectrum analysis of blurred photographs in computer-aided synthesis of Wiener filters, Phys. Lett. 51A, 383 (1975).

[49] Huang, Th.S.: Computer holography and its possible applications to medical diagnosis, see [3], pp. 36.

[50] Knox, G. et al.: Holographic motion picture microscopy, Proc. Roy. Soc. Lond.B. 174, 115 (1969).

[51] van der Haagen, G.A.: Ein Mikroskop mit holographischer 16-mm-Filmaufzeichnung, Laser 2, (1970).

[52] Khanna, S.M. et al.: Tympanic membrane vibrations in cats studied by time-averaged holography, J. Acoust. Soc. Amer. 51, 1904 (1972).

[53] Tonndorf, J. et al.: Tympanic membrane vibrations in human cadaver ears studied by time-averaged holography, J. Acoust. Soc. Amer. 52, 1221 (1972).

[54] Khanna, S.M. et al.: The vibratory pattern of the round window in cats, J. Acoust. Soc. Amer. 50, 1475 (1971).

[55] Gundersen, T. et al.: Holographic vibration analysis of the ossicular chain, Acta Otolaryngol. 82, 16 (1976).

[56] Hogmoen, K. et al.: Holographic investigation of stapes foot plate measurements, Acustica 37, 198 (1977).

[57] Greguss, P.: Holograhic interferometry in biomedical sciences, Optics and Laser Technol. 8, 153 (1976).

[58] Haeusler, G. et al.: Holograhische Deformationsmessungen zur Optimierung von Hueftgelenksimplantaten, see [4], pp. 349.

[59] Hanser, U.: Anwendung der holographischen Interferometrie in der experimentellen Orthopaedie, see [4], pp. 343.

[60] Hardinge, K. et al.: A preliminary study of fracture fixation using holographic interferometry, see [4], pp. 307.

[61] Vukicevic, D. et al.: Holographic investigation of mechanical characteristics of the complex leg-foot in conditions of lesion and reconstruction, see [6], pp. 34.

[62] Kinder, J. et al.: Holographische Untersuchungen des thermischen Verhaltens von Schmelz, Dentin und ausgewaehlten Dentalstoffen, see [4], pp. 301.

[63] Sieger, C. et al.: Measurement of vibration waveforms using temporally modulated holography, see [6], pp. 247.

[64] Atkinson, J.T. et al.: Measurement of the area of real contact between, and wear of, articulating surfaces using holographic interferometry, see [5], pp. 289.

[65] Lalor, M. et al.: Holographic studies of wear in implant materials and devices, see [6], pp. 20.

[66] Wagner, J. et al.: Application de l'interferomètrie holographique à l'etude du complexe tibio-pèronier chargè axialement, Acta Orthop. Belgica 41, 24 (1975).

[67] Vukicevic, D. et al.: Holographic investigations of the human pelvis, see [8], pp. 138.

[68] Hanser, U.: Quantitative evaluation of holographic deformation investigations in experimental orthopedics, see [6], pp. 27.

[69] Wesendahl, Th. et al.: Untersuchung des Verformungsverhaltens menschlicher Wirbelkoerper mittels holographischer Interferometrie, Laser u. Elektrooptik 1, 37 (1977).

[70] Piwernetz, K. et al.: Elastomechnical properties of trabecular bone from the human vertebral body, see [6], pp. 15.

[71] Piwernetz, K. et al.: Holography in orthopedics, see [6], pp. 7.

[72] Pryputniewicz, R. et al.: Determination of arbitrary tooth displacements, J. Dent. Res. 57, 663 (1978).

[73] Dirtoft, I.: Holographic measurement of deformation in complete upper dentures - clinical application, see [8], pp. 100.

[74] Pavlin, P. et al.: Strain distribution in the facial skeleton arising from orthodontic appliance activity, see [6], pp. 177.

[75] Bjelkhagen, H.: Holography in dentistry, see [6], pp. 157.

[76] Dirtoft, I.: Dental Holography, SPIE 370, 108 (1983).

[77] Hinsch, K.: Einsatzmoeglichkeiten kohaerent optischer Methoden in der Bioindikation, Angew. Botanik 55, 179 (1981).

[78] Lunazzi, J. et al.: A simple set-up for using holograhic interferometry in studies on seeds, see [6], pp. 77.

[79] Brooks, R.E. et al.: (9A9) Pulsed laser holograms, IEEE QE-2, 275 (1966).

[80] Dancer, A.L. et al.: Holographic interferometry applied to the investigation of tympanic-membrane displacements in guinea pig ears subjected to acoustic impulses, J. Acoust. Soc. Amer. 58, 223 (1975).

[81] Smigielsky, P. et al.: Application de l'interfèromètrie holographique a l'ètude des dèformations du tympan du cobay sous l' effet de bruits de durèe brève, Nouv. Rev. Optique 6, 49 (1975).

[82] von Bally, G.: Otological investigations in living man using holographic interferometry, see [6], pp. 198.

[83] Wedendal, P. et al.: Holography in dentistry, in: M.L. Wolbarsht (ed.), Laser applications in medicine and biology, Plenum Press, (1977) pp. 221.

[84] Pryputniewicz, R.: Holographic determination of rigid body

motions, and application of the method to orthodontics, Appl. Opt. <u>18</u>, 1442 (1979).

[85] Zivi, S.M. et al.: Chest motion visualized by holographic interferometry, Med. Res. Eng. <u>9</u>, 5 (1970).

[86] Bjelkhagen, H.: Development of hologram interferometry, in particular pulsed sandwich holography, for engineering uses as well as applications within medicine and odontology, Dissertation, Stockholm (1978).

[87] Pawluczyk, R. et al.: Holographic vibration analysis of the frontal part of the human neck during singing, see [8], pp. 131.

[88] Løkberg, O.: Speckle techniques for use in biology and medicine, see [8], pp. 144.

[89] Løkberg, O. et al.: Use of ESPI to measure the vibration of the human eardrum in-vivo and other biological movements, see [6], pp. 212.

[90] Løkberg, O. et al.: Bio-medical applications of ESPI, see [8], pp. 154.

[91] von Bally, G. et al.: Potentials of holographic vibration analysis of the human skull, Arch. Otorhinolaryngol. Suppl. II (1984) (in press).

[92] Ferrano, G. et al.: Compensation of rigid body motions in holographic interferometry, see [6], pp. 258.

[93] Piwernetz, K.: A posteriori compensation for rigid body motion in holographic interferometry by means of a Moirè-technique, Optica Acta 24, 201 (1977).

[94] Abramson, N.: Sandwich holography and its applicability to biomedical investigations, see [6], pp. 235.

[95] Bjelkhagen, H.: Pulsed sandwich holography, Appl. Opt. <u>16</u>, 172 (1977).

[96] Statement of European Physicists, Europhysics News <u>13</u>, 2 (1982).

HOLOGRAPHIC CINEMATOGRAPHY WITH THE HELP OF A PULSE YAG LASER

P. Smigielski, H. Fagot, F. Albe

Institut Franco-Allemand de Recherches de Saint-Louis (ISL)
12, rue de l'Industrie 68301 Saint-Louis, France

ABSTRACT

For many years opticists have tried to achieve 3D-cinemato-graphy with the help of holography. The term "cineholography" has been introduced in 1965. At that time the method consisted in superimposing on the same photographic plate various holograms recorded at different times. The image separation was achieved by rotating either the plate in its plane or the reference beam during both the recording and reconstructing processes. The number of views was limited by the principle itself, but a high repetition rate (100 kHz) has been obtained with a pulsed ruby laser.

Then other experiments have been conducted by different authors, mainly in the Soviet Unions and in the United States of America, by using the principle of classical cinematography in which the image separation is obtained by translating the film. With this method we have recorded the first French holographic movies on 35 mm Agfa films with the help of a pulsed YAG laser built in our laboratory. This frequency doubled laser ($\lambda = 0.532$ μm) delivers pulses of 20 ns with an energy of 30 mJ at a repetition rate of 24 Hz.

The experimental arrangements are described and some images of diffuse moving objects are presented. The volume of the record-ing scene is greater than one cubic meter. The coherence length of the laser is higher than one meter and remains steady during the recording process. Results are discussed and an outlook on the future is given with special respect to high repetition rate techniques.

INTRODUCTION

For a number of years great efforts have been made to perform three-dimensional cinematography on the basis of holography. The term "cineholography" was introduced in 1965 [1]. The cineholographic method is characterized in that various holograms recorded at successive time instants are superimposed on the same photographic plate. The images are separated by rotating the plate in its plane or by varying the inclination of the reference beam in both the recording and reconstructing processes. The total number of views $_sN_s$ is strongly limited by the principle inherent in this technique ($N = 2\pi/\theta$, θ thereby denoting the angle at which the specimen is seen from the hologram) whereas the brightness of the reconstructed images diminishes according to $1/N^2$ [2]. However this method allows very high repetition rates (100 KHz) required for investigating ultra-rapid phase objects [3]. Another technique has been used in Great Britain for recording holograms of objects transilluminated at high repetition rates (10 to 20 kHz). Here the photographic plate is immobile and the various holograms are separated on the photographic plate (and not superimposed like in the previous case) with the aid of a mechano-optical deflection system [4]. In this method a particular half-scattering plate is used to generate simultaneously both the object beam and reference beam such that the size of the scene to be recorded is somewhat limited.

Comparable repetition rates (10 to 20 kHz) are attained in separating the elementary holograms by the rotation of the photographic plate or by the rotation of a disk placed ahead of the plate and provided with a series of apertures [5]. Once more the holograms recorded are those of transparent objects. Still higher repetition rates can be attained (GHz range) such that plasmas can be investigated with an experimental device which does not include any mobile component. The incident laser beam is split into ten parallel parts which are shifted in space by means of a set of prisms and shifted in time with the aid of optical delay lines [6].

Simultaneously experiments have also been conducted by other authors, chiefly in the Soviet Union [7] and in the United States of America [8]. These experiments were based on the principle of conventional cinematography in which the images are separated by translating the film. The holograms of objects which scatter light by reflection are recorded on 70-mm-movies at a low repetition rate (maximum rate equal to 24 Hz). The method used in the Soviet Union allows the image to be observed by several persons via a holographic screen. The source of light used for the recording process is a ruby laser, and thus it is not possible to attain the cinematographic repetition rate (24 Hz) unless many ruby lasers are used. The experiments performed in the United States attained

repetition rate of 20 Hz with the help of a YAGlaser, but the scene
recorded is very small (small object rotating at a speed of
1 r.p.s.)because of the limited power output of the laser(7 mJ).

At the French-German Research Institute we have designed and
built a frequency-doubled YAG laser ($\lambda = 0.532$ µm) which is operated
with a single amplifier. It delivers pulses of 20 ns duration with
a minimum energy of 30 mJ at a repetition rate of 24 Hz. The coherent length of the laser is in excess òf 1 meter. Furthermore
this length remains sufficiently stable during the entire recording
process. Thus scenes occupying a sensitive volume in the m^3-range
can be recorded on a 35-mm-film of the Agfa 10E56 type. It is the
purpose of this paper to describe the first experiments performed
in this field of interest. In another series of experiments conducted in co-operation between the French-German Research Institute
and the "Laboratoire d'Expérimentation dans les Arts Cinégraphiques
(LEAC) of the University of Paris 8 (C.EIZYKMANN, G.FIHMAN) pictures
were also recorded on a 70-mm-movie of the Agfa 8E56 type. These
experiments will be reported in a paper to be published later.

REMARK

The sequences recorded at a high repetition rate are seen to
be concerned with phase objects (transparent objects) illuminated
with either a directed beam of light or a diffused beam of light.
In contrast the experiments conducted at the repetition rate of
conventional moving picture recording refer to opaque objects
which scatter the beam of light in the direction of the hologram.
In the latter case a very great coherence length of the laser is
required and the allowable translation of the object during exposure must be kept very small. Furthermore the energy modulated by
the object is relatively low in the 2nd case (scattered reflection)
such that high-powered lasers have to be used. This is perhaps the
reason why only a few papers have been published so far in which
the cineholographic recording of three-dimensional objects with
scattered reflection is described.

EXPERIMENTAL ARRANGEMENT (Fig. 1)

A conventional holographic arrangement was used for the experiments described herein. A prism (PR) allows the incident beam of
light to be split into three parts. The transmitted beam of light
serves to illuminate the scene under investigation via a divergent
lens (L_1). The beam of light reflected from one of the two faces
of the prism allows formation of the reference beam using lens
(L_R).The light beam reflected from the other face of the prism is
used for checking the coherence of the laser with the aid of the
FP3 Fabry-Pérot interferometer.

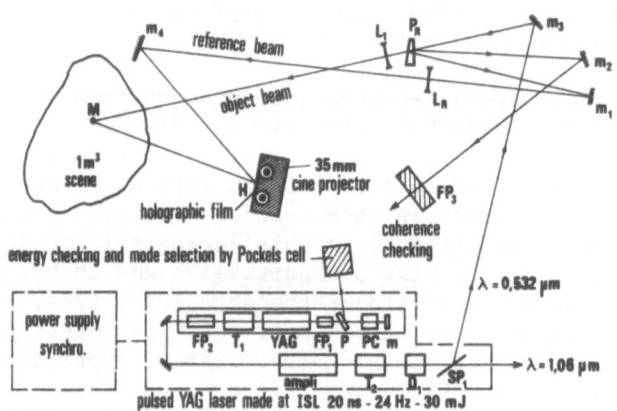

Fig. 1: Experimental set-up used for cineholography

The YAG laser used is a transverse single-mode laser (TEM$_{nn}$-mode).The required coherence which must be stable in time is achieved with two thermostatically controled Fabry-Pérot etalons (FP$_1$ and FP$_2$)and via the gradual opening of the electronically controlled Pockels cell (PC) [9].A telescopic resonator (T$_1$)allows for a better output of the laser energy and for an appropriate correction of the thermal lens effects generated as the YAG laser rod is heated [10].The light reflected from the polarizer (P) which is connected with the electro-optical shutter (Pockels cell PC),is used for both the checking of the pulse energy and selection of the longitudinal modes by cell (PC).Only a single amplifier is needed.The telescope (T$_2$)allows an adequate energy density to be obtained on the frequency-doubling crystal (D$_1$).

A 35-mm-film projector "HAHN II" has been especially fitted to the recording of the holographic pictures. Stops on the film were not compulsory, however, because of the very short exposure time (20 ns): a continuous film speed of 3 m/s was considered to be acceptable.

Each elementary hologram is 25 mm wide and 18 mm high. The height of the holograms can be reduced without difficulties to a few millimeters. The holographic film is of the Agfa 10E56 type. Its mean grain size (resolution: 3000 lines/mm) is too large with respect to the wavelength of the light beam (0.532 μm). Therefore the acceptable angle between the object beam and reference beam is limited and a noise is generated in the reconstructed pictures (Fig. 2). This noise had disappeared as the pictures were observed dynamically at the repetition rate of 24 Hz and the quality of the

reconstruction can therefore be regarded as acceptable. Remark:
for the experiments conducted later, we have selected a film of the
Agfa 8E56 type, the resolution of which exceeds 10 000 lines/mm
and allows high quality holograms to be recorded independently of
the angle formed between the object beam and reference beam.
However this film suffers from the following drawback. Its sensitivity is much lower (by a factor 20) than that of the 10E56 film
such that the volume of the scene to be recorded is reduced. With
the energy output of 30 mJ available in the green range, it is
nevertheless possible to take holograms of scenes occupying a volume
of approximately one m^3.

Several sequences of a mean duration equalling 15 s have been
recorded on a film of 6 m in length (300 pictures, 18 mm high, for
each hologram).

RECONSTRUCTION (Fig. 2)

The laser used for the reconstructing process was an argon
laser[*]) operated in the green range at a wavelength of 0.5145 μm
which approaches that of the recording laser (0.532 μm) such that
the geometric aberrations due to the difference between the above
mentioned wavelengths are relatively unimportant. More perturbing,
however, are the aberrations due to the possible curvature of the
film. This problem can be minimized by using the projector for the
reconstructing process. But if a reconstruction set-up is desired
which is operated independently of the recording device, a film-
transport (continuous or intermittent)must be developed in which
the curvature of the film is exactly the same as that used in the
recording process.

Taking into account the useful width of the film (35 mm), the
event can be observed with one eye only. For a comfortable viewing
with both eyes, the useful width of the film should attain about
100 mm, taking thereby into account that the distance between
pupils is approximately 64 mm.

In order to illustrate the results achieved, we have photographed a few pictures taken from the holographic film of an
electric miniature train which hits a wall consisting of stacked
parallelepipedes. From the figures 2a, b, c, d, e, f, the train is
seen to advance through the wall which does not fall in completely.

––––––––––––

*) The reconstruction can also be achieved with the aid of a YAG
laser, but the pulses delivered at a repetition rate of 20 Hz
have disturbing effects on the eye (scintillation). Smooth
operation is possible at 50 Hz, but the motion of the scene is
getting accelerated.

464

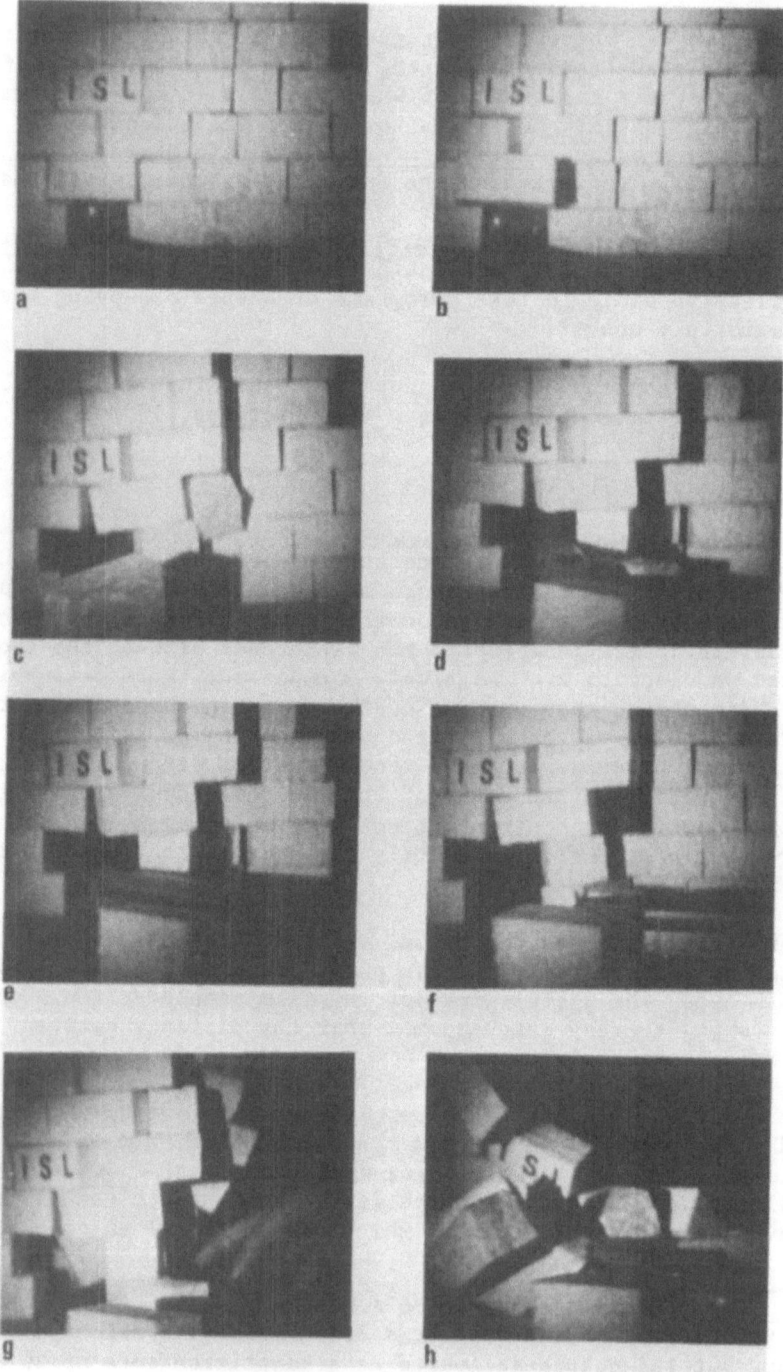

Fig. 2

The wall has been made unstable manually (Fig. 2g). At the next passage of the train, the wall has completely collapsed (Fig. 2h).

PERSPECTIVES

The results achieved are far from being extraordinary. But the experiments performed made it possible to conceive an improved holographic film which is characterized as follows: 70-mm-size (or more) on the Agfa 8E56 film; each elementary hologram is only a few millimeters wide; improved performance of the laser; use of a 70-mm-projector or development of a system allowing for a continuous run of the film (here enlarged sizes can be used ranging from 100 to 130 mm);development of a well adapted device for the reconstruction processes. Finally there is a possibility of taking secondary holograms from the original film such that the event can be observed simultaneously by several persons (for instance, by developing a holographic screen similar to that reported by Komar [7]).

The YAG laser is capable of delivering two pulses separated by a very short time interval (time interval between 0 and 100 μs or in excess of 3 ms) at a maximum repetition rate of 50 Hz (repetition rate presently attained) such that holographic interferometry can be applied to the following fields of interest: non-destructive testing of materials, variation of the density field in fluid flows, investigation of vibrations, ... [11]. The double pulse regime at high repetition rates can also be used for investigating the velocity field in a sheet of fluid [12].

For applications at high repetition rates (5 to 10 kHz) it is envisaged to use a copper vapor laser ($\lambda = 0.5106$ μm) presenting a good coherence in time and delivering an energy per pulse (several mJ) which is compatible with the sensitivity of holographic films.

The problem of the image separation on an immobile film can be solved, for instance, by using an acoustic-optical deflection system. It is obvious, however, that the number of images will be strongly limited in this case.

REFERENCES

1. H. Paques, P. Smigielski, Cinéholographie, C.R. Acad. Sc. Paris
   260, 6562 (1965)
2. H. Royer, P. Smigielski, Expositions multiples sur un holo-
   gramme. Qualités des images. Applications. Optica Acta, vol. 17,
   No 22, 97.105 (1970)
3. P. Smigielski, A. Hirth, New holographic studies of high speed
   phenomena, 9th Int. congress on High-Speed Photography, Denver,
   Col. (1970)
4. J.W.C. Gates, R.G.N. Hall, I.N. Ross, High-speed recording of
   transilluminated events, 9th Int. congress on High-Speed Photography, Denver, Col. (1970)
5. K.J. Ebeling, Investigation of cavitation bubble dynamics by
   high-speed ruby laser and ion laser holocinematography,
   1st European conference on optics applied to metrology,
   Strasbourg (1977)
6. M. Novaro, Caméra holographique ultra-rapide, 10ème congrès
   int. de Photographie Ultra-Rapide, Nice (1972)
7. V.G. Komar, Principle of the holographic cinematography,
   1st European conference on optics applied to metrology, Strasbourg (1977)
8. A.J. Decker, Holographic cinematography of time-varying reflecting and time-varying phase objects using a Nd:YAG laser.
   Optics letters, vol. 7, No 3, March (1982)
9. Y.K. Park, R.L. Byer, Electronic linewidth narrowing method for
   single axial mode operation of Q-switched Nd:YAG lasers. Optics
   communications, vol. 37, No. 6, 15 June (1981)
10. D.C Hanna, C.G. Sawyers, M.A. Yuratich, Telescopic resonators
    for large-volume TEM_{00}-mode operation. Optical and Quantum
    Electronics 13, 493-507 (1981)
11. P. Smigielski, H. Fagot, A. Hirth, L'utilisation du laser YAG
    pulsé en contrôle non destructif par holographie, 5ème congrès
    international sur les méthodes de contrôle non destructif,
    Bordeaux, juin (1983)
12. T. Ferrari, J. Saulnier, H. Fagot, P. Smigielski, H. Haertig,
    Instantaneous velocity fluid measurement in a sheet of fluid
    by means of chronophotography in coherent light: improvement
    of the method, 10th Int. congress on Instrumentation in Aero-
    space Simulation Facilities (ICIASF 83), Saint-Louis, France,
    sept. (1983).

PHASE CONJUGATION METROLOGY

Silverio P. Almeida and Luis M. Bernardo*

Virginia Polytechnic Institution and State University
Department of Physics
Blacksburg, Virginia 24061

I. INTRODUCTION

In this paper we present an introductory discussion on nonlinear optical phase conjugation (NOPC) properties and how they
might be used in metrology. Quantitative details and descriptions
of the various phase conjugation (PC) setups possible are cited
in the references (1). Among the different types of nonlinear
media available we shall describe PC results on ruby; the one
used in our study on speckle noise reduction.

We also present applications of two PC results from the
literature which could have a direct use in metrology. The first
application is the one by Bernardo (2) and involves PC speckle
reduction on a resolution chart while the second one discussed is
on projection lithography by phase conjugation by Levenson and
Co-workers (3).

II. NONLINEAR OPTICAL PHASE CONJUGATION

A. Properties of phase conjugation

Shown in fig. 1(a) and (b) is a comparison of what happens
when a ray of light strikes an ordinary mirror (OM) and a NOPC
mirror (NOPCM). The law of reflection is nicely obeyed in the
case of the OM (a). But what has happened when the ray strikes a
NOPCM? The NOPCM has the property that it causes ray $\vec{K}_{incident}$
to exactly reverse both its direction and its phase, i.e. ($\vec{K}_{out}$ =
$-\vec{K}_{inc.}$ and $e^{-ikz} = e^{ikz}$).

468

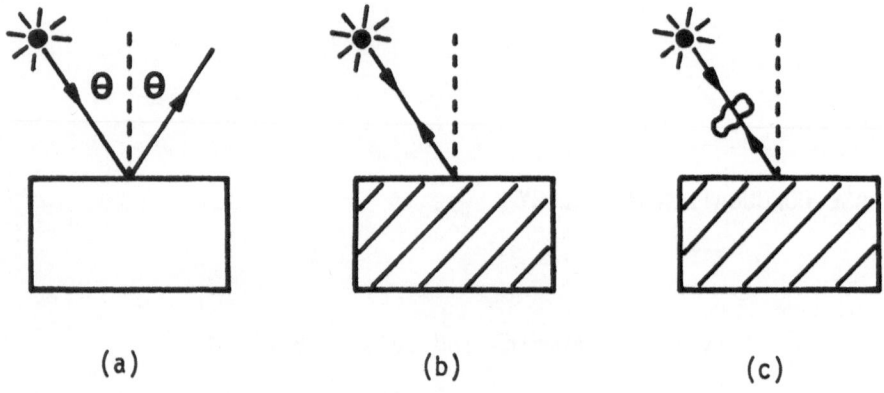

(a)   (b)   (c)

Fig. 1 (a) Reflection from an ordinary mirror. (b) Reflection
from a nonlinear optical phase conjugation mirror. (c) Same as
(b) but with a random phase object inserted in beam path.

The NOPCM will, therefore, take a converging beam, phase
conjugate it and have it reflected upon itself as a diverging
beam. In a similar fashion an incident diverging beam will be
reflected along its path as a converging one. A consequence of
the NOPCM is that if one were to intercept the beam in fig. 1(b)
with a phase distoring material, the phase conjugate beam (PCB)
would still reflect back to its origin as shown in Fig. 1(c).
This aberration correction (adaptive optics) due to the NOPC
medium can play an inportant role in improving metrological
signals by removing the phase distortions.

B. Phase conjugation (PC)

Phase conjugation refers to any process in which a wave

$$\vec{E}_4(\vec{r},t) = Re[\vec{A}_4(\vec{r})\ e^{i\omega t}] \tag{1}$$

is generated from an incident wave

$$\vec{E}_3(\vec{r},t) = Re[\vec{A}_3(\vec{r})\ e^{i\omega t}] \tag{2}$$

and the relation between their amplitudes is

$$\vec{A}_4(\vec{r}) = R \, \vec{A}_3^*(\vec{r}), \tag{3}$$

where R is any constant; Re stands for "real part of" and * for "complex conjugate". For simplicity we consider the case of plane waves. If we write

$$A_3 = \psi(r) \, e^{-ikz} \tag{4}$$

and because of the identity

$$\vec{E}_4(\vec{r},t) = Re[R \, \psi^*(\vec{r}) \, e^{ikz} \, e^{i\omega t}] = Re[R \, \psi^*(\vec{r}) \cdot e^{ikz} \, {}^{i\omega t}]*$$

$$= Re[R \, \psi(\vec{r}) \, e^{-ikz} \, e^{-i\omega t}] = Re[R \, \vec{A}_3(\vec{r}) \, e^{-i\omega t}], \tag{5}$$

we can say that $\vec{E}_4$ is the time reversal of $R \, \vec{E}_3$, i.e. $\vec{E}_4$ is equal to $R \, \vec{E}_3$ after t becomes -t. Phase conjugation and time reversal are therefore often used with the same meaning. The above discussion is also valid for non-plane waves if the superposition principle applies.

Optical phase conjugation can be obtained, in differential time, by holography (4) and in real-time by adaptive optical techniques (5) or several non-linear optical mechanisms; namely, 3-wave mixing, (6) stimulated backscattering (7) (such as Brillouin (SRS), (8) Raman (SRS), (9) Rayleigh (10) or other origins (11) and 4-wave mixing (FWM) (1,12). The 4-wave mixing mechanism is one of the most widely used and perhaps the easiest to implement.

C. Four-wave mixing (FWM)

FWM refers to the interaction of four waves in a non-linear medium. The term "Degenerate Four-Wave Mixing" (DFWM) is used when the waves have all the same frequency. The geometrical configuration for the interaction is shown in fig. 2, where two counter-propagating pump waves are used with a probe beam at some angle, is sometimes called "backward DFWM" interaction. In this way, the process can be distinguished from the forward DFWM, (13) where only one pump with a probe generate a forward phase conjugate signal. This interaction corresponds to 3-wave mixing; but, as the pump acts as if it is formed by two forward pumps (or equivalently two pump photons take part in the reaction), it can be considered a FWM process. We use "DFWM" instead of "backward DFWM" for simplicity during the exposition, since this is the only configuration we consider.

The nonlinear nature of DFWM implies that the response of the optical medium to the light fields is not linear, i.e. the

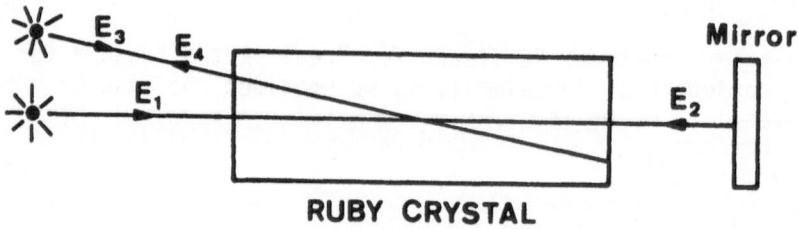

Fig. 2 E_1 and E_2 are pump beams. E_3 the probe beam. E_4 the conjugate beam generated in the nonlinear medium ruby.

polarization $\vec{P}(\vec{r},t)$ of the medium has terms which are nonlinear functions of the electric field $\vec{E}(\vec{r},t)$. Because four waves are involved, the first important term of the nonlinear polarization, when expanded in powers of $\vec{E}(\vec{r},t)$, is of third order (14)

$$\vec{p}^{NL}(r,t) = \chi^{(3)} \vdots \vec{E}\,\vec{E}\,\vec{E}, \tag{6}$$

where $\chi^{(3)}$ is the third order tensor susceptibility and

$$\vec{E} = \sum_i E_i(\omega), \ (i = 1,2,3,4), \tag{7}$$

is the field.

For isotropic media and when only the polarization of frequency ω is considered, the condensed expression eq. 6 reduces (15) to

$$\vec{p}^{NL} = \alpha(\vec{E}\cdot\vec{E}^*)\vec{E} + \gamma(\vec{E}\cdot\vec{E})\vec{E}^*. \tag{8}$$

The first term leads to the holographic analogy, (16,17) where each pump acts as the reading beam, generating the phase conjugate beam. The second term describes the oscillation at frequency 2ω of the nonlinear index of refraction which scatters one of the waves to generate the fourth, (18) parametric interaction. The coefficients α and γ can be made large by choosing the right non-linear medium: one-photon resonant medium for large α, two-photon resonant medium for large γ. Ruby behaves like a one-photon resonant medium and therefore a larger α is expected.

If multi-photon processes are important, odd terms of the polarization higher than the third power of $\vec{E}$ have to be considered, since their coefficients cannot be neglected.

The above considerations and general expressions for $\vec{p}^{NL}$ are valid for any nonlinear medium. The expressions for the suscept-ibility χ and for the parameters α and γ, however, result from the particular nonlinear mechanism involved. In our medium, ruby, the mechanism is resonant absorption and the expressions used are valid in general for those media.

III. PHASE CONJUGATION BY DFWM IN RUBY

As an example of PC by DFWM we describe the experimental setup used in the study of the resolution chart and whose results are presented in section IV-A. For this experiment we used a ruby crystal. The geometry of the crystal is shown in fig. 3. The phase conjugation setup used to perform the microscopic object (resolution chart) reconstruction is presented in fig. 4.

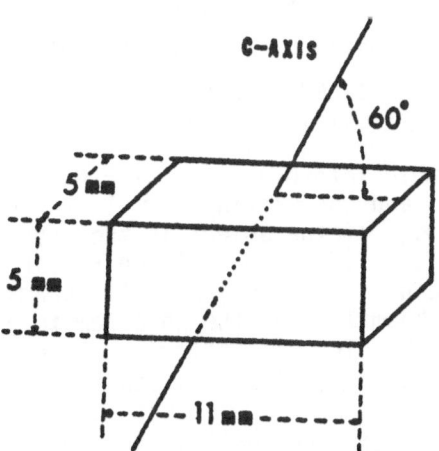

Fig. 3 Ruby crystal geometry. C-axis makes a 60 degree angle with non-polished surfaces 5 x 11 mm² and is parallel to polished surfaces 5 x 11 mm².

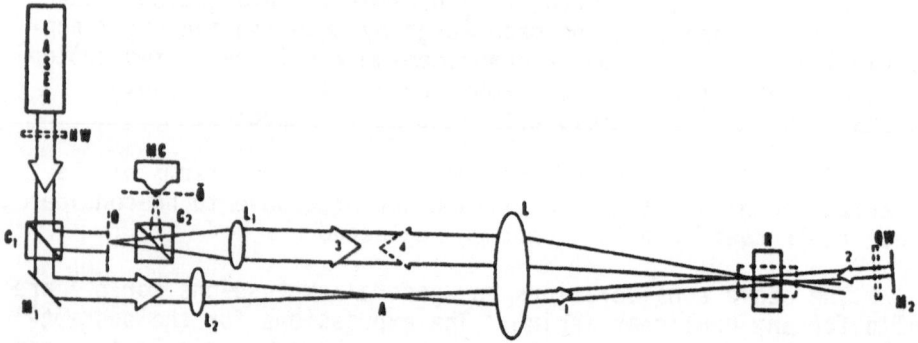

Fig. 4 Schematic of phase conjugation reconstruction setup to measure resolution chart. Both HW, QW are half and quarter wave plates; C_1 and C_2 cube beam splitters; M_1 and M_2 are mirrors; L, L_1 and L_2 lenses; MC camera or photodetector; O and $\bar{O}$ are the object and phase conjugate object plane.

The ruby crystal R is in the focal plane of lens L. L, L_1 and L_2 have the focal lengths F, F_1 and F_2; they are chosen according to the object and crystal dimensions in order to maximize the interaction length and power density. A magnified image of the object is obtained in the focal plane of L in the crystal. The pump beam 1 is collimated by lenses L_2 and L and its diameter can be adjusted by varying the ratio F_2/F_1. The other pump beam 2 is obtained by back-reflection of beam 1 on the mirror M_2 with reflectivity $R \simeq 0.9$.

In the holographic interpretation of FWM, the interference of the object beam 3 and the writing beam 1 produces a fringe modulation of the image of the object in the crystal. It is that grating that causes the diffraction of the reading beam, to generate the phase conjugate image, beam 4. Beams 2 and 3 also may interfere; then beam 1 reads the hologram to generate the phase conjugate signal. The last hologram can be experimentally eliminated by 90° rotation of the polarization of beam 2, using a λ/4 plate or by increasing the distance between the crystal and the mirror M_2 to be larger than half of the laser coherence length. In practice, the location of mirror M_2 far beyond the crystal has advantages, since the probe beam can be easily stopped after passing through the crystal. That way, all the noise, otherwise generated by the probe after being reflected in M_2, can be eliminated. The disadvantage is the reduction of the efficiency by half, since only one of the two possible gratings is formed. The presence of the reflected probe could also improve the efficiency, by the effect of multiple reflections (19).

Examples of the pump beam profile (using the laser wavelength of 514.5 nm) and the phase conjugate beam are shown in fig. 5 and 6. The beam was scanned by a vidicon, digitized (8-bits) and stored on disk by an on-line mini-computer. Fig. 5a, b show the computer plots of the raw data while fig. 6a, b show the same beam profiles as in 5a, b but after software programs have filtered out some of the noise (20).

IV. METROLOGICAL PC-TYPE APPLICATIONS

In this section we present two examples, among other possible ones, see ref. (1); to illustrate the potential use of phase conjugate techniques in metrology. The first example is a study of speckle noise removal to improve the PC resolution chart signal by Bernardo (2). Next we present results on projection photolithography by Levenson (3). Details of these examples are to be found as cited in the authors' original articles.

A. Speckle removal in PC of a resolution chart

We now consider the phase conjugate reconstruction of a microscopic object and its improvement by a speckle removal technique. The setup used in this experiment is shown in fig. 4. In that geometry, the parameters can be adjusted to get the maximum efficiency for fixed object dimensions. Simpler geometries with only a beam splitter between the object and the nonlinear medium have been used; (21,22) they show, however, some limitations. We point out some of the characteristics of our geometry:
  (a) Because of the image of the object to be reconstructed can be easily (de)magnified, the efficiency can be maximized for each particular object dimensions and available power.
  (b) Because a sharp image of the object, not a diffracted image (21) or its Fourier transform, (22) is in the nonlinear medium, the power spectrum of the object will not be distorted; i.e. no spatial filtering will occur.
  (c) Separation of the phase conjugate image and other disturbing signals from stray reflections is easily achieved. This is a consequence of the relatively large dimensions of our setup.
  (d) The ultimate limiting resolution in our geometry is, as in that of ref. (21), the wavelength of light.

The object is an Air Force chart transparency placed in plane O. Its phase conjugate image can be observed, through a microscope, in the so called phase conjugate plane Ō. The section of the chart we

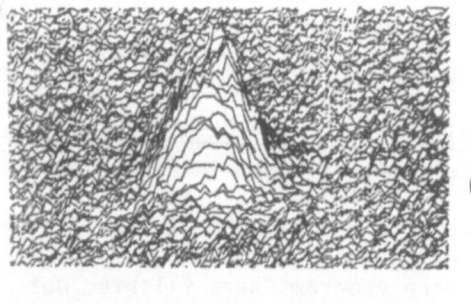

(a)

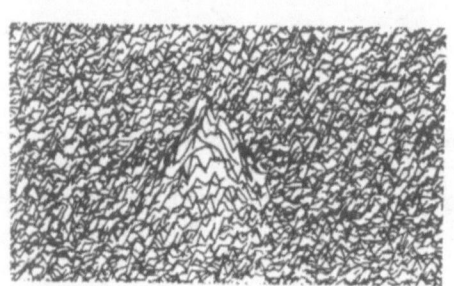

(b)

Fig. 5 (a) Unfiltered probe beam profile. (b) Unfiltered phase conjugated DFWM beam profile.

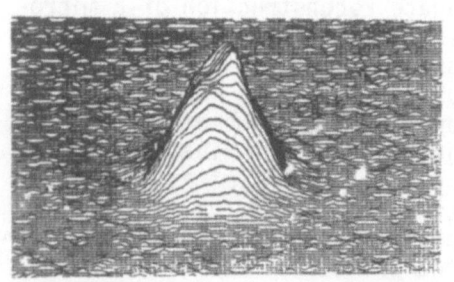

(a)

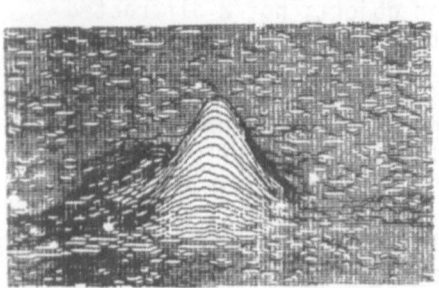

(b)

Fig. 6 (a) Filtered version of 5a using software noise reduction programs. (b) Filtered version of 5b.

selected is shown in fig. 7 the line spacing in the group 5-1 is $\simeq$ 15 μm. We have used the longitudinal orientation of the ruby crystal fig. 3 and λ = 514.5 nm, in order to get the maximum efficiency for a given power.

To have a good quality image we have to eliminate all the reflections coincident with the image and minimize the speckle coming from the optical components and the crystal itself. Stray reflections and the speckle associated with the optical components are easily eliminated in our geometry, because of its large dimensions. The speckle coming from the crystal itself is the major source of the noise. Its size, observed through the microscope, is comparable to the dimensions of some of the lines (see fig. 8a), limiting the resolution of the reconstruction. The origin of the speckle coming from the crystal is associated with the scattering of the beams in its bulk and surfaces. We can minimize the speckle by cancelling that coming from the scattering of beams 1 and 3, by using a λ/4 optical plate between the crystal and the mirror M_2. That way, the polarization of beam 2 is rotated by π/2. Then, we put an analyzer beyond plane $\bar{O}$ to cut down the polarizations of the beams 1 and 3. The only light observed through the microscope is that having normal polarization and, therefore, only the phase conjugate image and the speckle caused by beam 2 will be observed. Although improvements have been observed using that technique, the image is not speckle-free and the resolution limitations are not eliminated.

A much better improvement in resolution has been achieved, using a speckle averaging technique, different from others(23) but with similar effects. The speckle distribution in the image plane $\bar{O}$ can be changed by moving the crystal. If a multi-exposure photograph is made, moving the position of the crystal between the single exposures, an averaging of the speckle will result. For real-time observations, a continuous slow movement of the crystal will cause the same averaging effect, with a remarkable improvement of resolution. Figs. 8 a,b,c show the phase conjugate image: (a) single exposure, 1/8 sec., (b) multiple (7) exposure, 7 x 1/60 sec., (c) multiple (15) exposure, 15 x 1/125 sec.. Between each single exposure, the ruby was slightly and vertically shifted. A x 5, N.A = 0.10 microscope objective and a 10x ocular have been used to observe the image.

A carefull adjustment of the setup and of the position of the microscope shows that the image quality can be improved. Although in fig. 8 the maximum resolution is $\simeq$ 15 μm, we could resolve lines separated by $\simeq$ 5 μm, which is close to the diffraction-limited resolution associated with the microscope, λ/2 N.A. $\simeq$ 2.5 μm.

In conclusion, we have shown an experimental setup and a method to reconstruct a microscopic object, using DFWM in a ruby crystal.

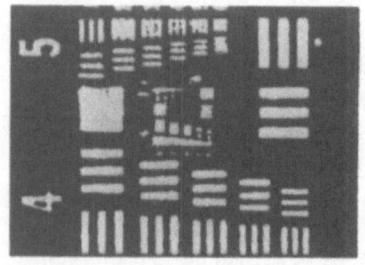

Fig. 7 Input object transparency placed at plane 0 of
fig. 4 to be reconstructed
by DFWM. Spatial frequency
of group 5-1 is 30 line pairs
per mm.

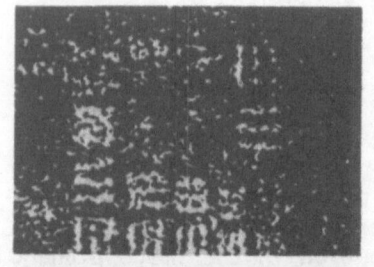

(a)

Fig. 8 Phase conjugate image
of fig. 7 object. (a) Single
exposure 1/8 sec. (b) Multiple
exposure 7 x 1/60 sec.
(c) Multiple exposure 15 x
1/125 sec.

(b)

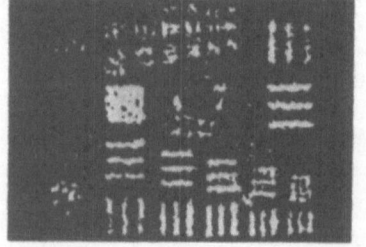

(c)

The maximum resolution is shown to be limited, after speckle averaging, essentially by the wavelength λ and the N.A. of the microscope. We have introduced a new speckle averaging technique appropriate to the characteristics of our setup geometry, which improves considerably the quality of the phase conjugate image.

B. PC Projection photolithography

Levenson and co-workers see ref. (3,10) have used DFWM techniques to project images with submicrometer resolution onto substrates coated with photoresist. The resolution obtained is shown to be better than 800 line pairs per millimeter when using a light source with a wavelength of 413. nm. In addition, the projected patterns for their setup shown in fig. 9 was not degraded by speckle. This lensless and contact free phase conjugate geometry shows great promise in visible illumination photolithography. Results of their DFWM phase conjugate are presented in fig. 10 which shows a pattern of five 0.75 micron lines whose spacing is 0.50 microns. These results represent the limiting resolution of this particular system. The authors' point out, however, that minor changes could result in an increased numerical aperture to about 0.60. This inturn would allow for the imaging of 0.50 micron lines with 0.50 micron gaps.

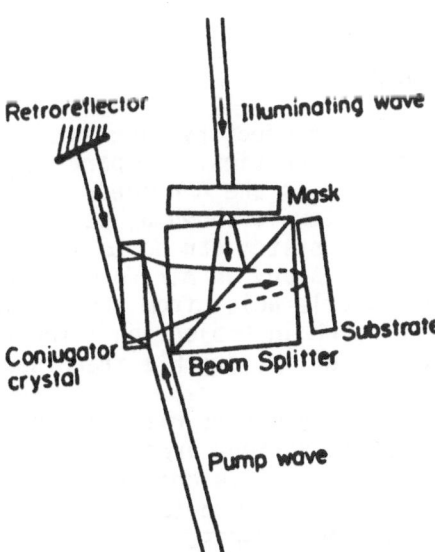

Fig. 9 Projection photolithography setup using DFWM. Conjugation crystal is LINbO$_3$. Both mask and photoresistcoated substrate were 0.5 mm from the beam splitter face. (After Levenson and co-workers, 1981. See ref. 3).

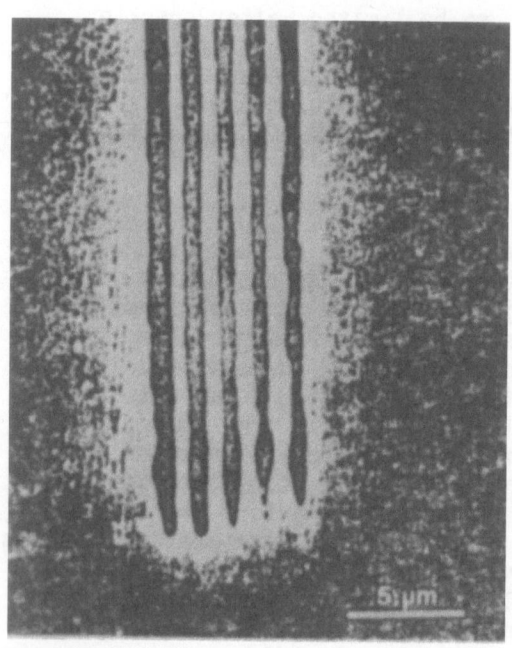

Fig. 10 Photolithography PC reconstruction of a pattern of five 0.75 micron lines with 0.50 micron spacing. Spatial frequency of pattern is 800 line pairs per mm in AZ1350B photoresist. Setup for this PC reconstruction is shown in fig. 9. (After Levenson and co-workers, 1981. See ref. 3). ·

V. SUMMARY AND CONCLUSION

We have presented an introductory discussion on nonlinear optical phase conjugation properties. In particular, we described a DFWM experiment using ruby as the nonlinear medium. The raw signal obtained in this system is shown as well as what a hybrid optical processing computer system can do to improve it. An example of DFWM is given to study an Air Force resolution chart. A method to remove the speckle noise from the PC image was presented. A second example of DFWM in projection lithography was presented and shows high resolution, speckle free results.

Methods such as these and others cited in the references can in some cases be extended to various areas in metrology. We believe that the real-time, adaptive optics benefits afforded by phase conjugation techniques will provide a powerful tool with which to perform optical measurements.

VI. ACKNOWLEDGEMENTS

We would like to thank Srisuda Puang-ngern for her technical assistance on this research. One of us (*) would also like to thank the Institute Nacional de Investigacão Cientifica, Lisbon, Portugal, for its financial support.

*Present address, Laboratorio de Fisica, Faculdade de Ciências, Universidade do Porto, P-4000 Porto, Portugal.

References

1. Fisher, Robert A. (editor), Optical Phase Conjugation (Academic Press, New York, 1983).
2. Bernardo, L. M. and S. P. Almeida, Appl. Opt. 22, 3926 (1983).
3. Levenson, M. D., K. M. Johnson, V. C. Hanchett and K. Chiang, Opt. Soc. Am. 71, 737-743 (1981).
4. See e.g. J. W. Goodman, Introduction to Fourier Optics (McGraw Hill Book Company, New York, 1968).
5. See special issue in Adative Optics, JOSA 67 (1977). Barret, H. H. and S. F. Jacobs, Opt. Lett. 4, 190 (1979). Orlov, V. K., Ya. Z. Virnik, S. P. Vorotilin, V. B. Gerasimov, Yu. A. Kalinin and A. Ya. Sogalovich, Sov. J. Quantum Elextron. 8, 799 (1978). See Nonlinear Optical Phase Conjugation issue, Opt. Eng. 21, March/April (1982). O'Meara, T. R. Opt. Eng. 21, 271 (1982). Jacobs, S. F. Opt. Eng. 21, 281 (1982).
6. Yariv, A. Appl. Phys. Lett. 28, 88 (1976); JOSA 66, 301 (1976); Opt. Comm. 21, 49 (1977). Avizonis, P. V. Appl. Phys. Lett. 31, 435 (1977).
7. Hellwarth, R. W. Opt. Eng. 21, 257 (1982); Optical Phase Conjugation, ed. R. A. Fisher (Academic Press, New York 1983). Zel'dovich, B. Ya., N. F. Pilipetskii and V. V. Shknuov, Sov. Phys. Usp. 25, 713 (1982); Optical Phase Conjugation, ed. R. A. Fisher (Academic Press, New York, 1983). Sokolovskaia, A. I., G. L. Brekhovskikh and A. D. Kudriavtseva, JOSA 73, 554 (1983). Mays, Jr., R. and R. J. Lysiak, Opt. Comm. 31, 89 (1979).
8. Hon, D. T. Opt. Eng. 21, 252 (1982). Wang, V. and C. R. Giulliano, Opt. Lett. 2, 4 (1978). Scott, A. M. Opt. Comm. 45, 127 (1983). Armandillo, E., D. Proch, Opt. Lett. 8, 523 (1983).
9. Sokolovskyaya, A. I., G. L. Brekhovskikh and A. D. Kudryavtseva, Opt. Comm. 24, 74 (1978). Izgorodin, V. M., S. B. Kormer, G. G. Kochemasov, V. D. Nikolaev and A. V. Pinegin, Sov. J. Quantum Electron. 12, 119 (1982). Kung, R. T. V. and J. H. Hammond, IEEE J. Quantum Electron. QE-18, 1306 (1982).

10. Mack, M. E., Phys. Rev. Lett. 22, 13 (1969).
    Zel'dovich, B. Ya. and T. V. Yakovleva, Sov. J. Quantum
    Electron. 10, 501 (1980).
    Chiang, K. and M. D. Levenson, Appl. Phys. B 29, 23
    (1982).
    Desai, R. C., M. D. Levenson and J. A. Barker, Phys. Rev.
    A 27, 1968 (1983).
    Levenson, M. D., J. Appl. Phus. 54, 4305 (1983).
11. Pepper, D. M. Opt. Eng. 21, 156 (1982).
    AuYeung, J. C. Optical Phase Conjugation, ed. R. A. Fisher
    (Academic Press, New York, 1983).
    Shtyrkov, E. I. Opt. Spectrosc. 45, 339 (1978).
12. Yariv, A. Physics of Quantum Electron (Addison-Wesley,
    Reading, Mass., 1978) Vol. 6, ch.6; IEEE J. Quantum
    Electron. QE-14, 650 (1978).
    Hellwarth, R. W. IEEE J. Quantum Electron. QE-15, 101
    (1979).
    Zel'dovich, B. Ya., N. F. Pilipetskii, V. V. Reful'skii
    and V. V. Shkunov, Sov. J. Quantum Electron. 8, 8 (1978).
    Scott, A. M. Opt. Comm. 45, 207 (1983).
    Jain, R. K. Opt. Eng. 21, 199 (1982).
13. Heer, C. V. and N. C. Griffen, Opt. Lett. 4, 239 (1978).
    Maruani, A. IEEE J. Quantum Electron. QE-16, 558 (1980).
    Smirl, A. L., T. F. Boggess and F. A. Hopf, Opt. Comm.
    34, 463 (1980).
14. Bloembergen, N. Nonlinear Optics (W. A. Benjamin, Inc.,
    New York, 1965).
15. Maker, P. D. and R. W. Terhune, Phys. Rev. A 137, 801
    (1965).
16. Yariv, A. IEEE J. Quantum Electron. QE-14, 650 (1978);
    Opt. Comm. 25, 23 (1978).
17. Stepanov, B. I., E. V. Ivakin, A. S. Rubanov, Sov. Phys.
    Dokl. 16, 44 (1971).
18. Steel, D. G. and J. F. Lam, Phys. Rev. Lett. 43, 1588
    (1979).
19. Zel'dovich, B. Ya. and T. V. Yakovleva, Soc. J. Quantum
    Electron. 11, 1144 (1981).
20. Bernardo, Luis M. Ph.D. Thesis, Virginia Polytechnic
    Institute and State University, Blacksburg, Virginia 24061.
    (1983).
21. Levenson, M. D., Opt. Lett. 5, 182 (1980).
22. Bloom, D. M. and G. C. Bjorklund, Appl. Phys. Lett. 31,
    592 (1977).
23. Goodman, J. W. JOSA 66, 1147 (1976).
    Kozma, A. and C. R. Christensen, JOSA 66, 1257 (1976).
    Martienssen, W. and S. Spiller, Phys. Lett. 24 A, 126
    (1967).
    Huignard, J. P., J. P. Herriau, L. Pichon and H. Marrakchi,
    Opt. Lett. 5, 436 (1980).

HOLOGRAPHIC METROLOGY AND NONDESTRUCTIVE TESTING - PAST AND FUTURE

Charles M. Vest

Department of Mechanical Engineering and Applied Mechanics
The University of Michigan
Ann Arbor, Michigan 48109, USA

1. INTRODUCTION

The objective of this paper is to review, in rather general terms, the status and future of holographic metrology with some particular emphasis on nondestructive testing and related techniques. This field of technology and its literature are now too vast to present a detailed review in a paper of modest size. Instead, I will attempt to present a very broad assessment by commenting in general terms on the process of applying holographic interferometry to a particular industrial or scientific problem.

Consideration of the process of applying holographic interferometry and related techniques has led me to categorize knowledge of holographic interferometry and nondestructive testing into six categories: applications, techniques, technologies, industrial and scientific uses, limitations and challenges. Some thoughts and assessments regarding each of these categories are summarized herein.

2. PROCESS AND APPLICATIONS

The process of applying holographic interferometry to any industrial or scientific problem consists of the following ten steps:

1. Problem definition and goals
2. Apparatus and setup
3. Loading the object
4. Recording the hologram
5. Developing the hologram

6. Reconstruction
7. Viewing and storing
8. Analysis of fringes
9. Physical interpretation
10. Use of information

This process in fact contains at least two feedback loops because physical interpretation generally feeds back to refinement of the apparatus and setup, and the use of information obtained from the holographic analysis usually feeds back to refinement of problem definition and goals.

The details of each of these ten process steps, as well as the requirements they place on the holographic system, technique and analysis of results, differ greatly depending on the particular application. Applications of holographic interferometry can be classified according to whether the object is <u>opaque</u> or <u>transparent</u>. In the former case, we measure or visualize surface displacement or deformation. In the latter case, we measure or visualize spatial variations of refractive index. In either case, one may require either quantitative or qualitative analysis. Obviously, applications such as strain measurement which require quantitative analysis, particularly in three-dimensions, require a greater sophistication of equipment and effort than do applications such as certain flow visualization and nondestructive testing situations, which require only qualitative analysis of fringe patterns. This will become apparent later in this paper.

Most, if not all, applications of holographic interferometry and related techniques fall into ten general categories:

1. Metrology
2. Solid mechanics
3. Biomedical
4. Nondestructive testing
5. Contour generation

6. Flow visualization
7. Flow measurement
8. Plasma diagnostics
9. Heat and mass transfer
10. Stress analysis

In each of the above applications the primary quantity displayed by and determined from holographic interferograms is a change of optical path length. However, this change of optical path length must be related to some physical property of interest. These physical properties vary widely among the ten application areas.

<u>Metrology</u> refers to a measurement of object motion or deformation in which the translation, rotation or global deformation is the primary quantity of interest. For example, figure 1 is a holographic interferogram of an aluminum cube which

Figure 1. Interferogram of an aluminum cube heated uniformly
between holographic exposures.

has been heated uniformly between holographic exposures. Patterns
of this type have been used to determine the thermal expansion
coefficient of materials [1]. Figure 2 illustrates an application
in solid mechanics. The object in this figure is a system
constructed of composite materials which is internally pressurized
to a high level. The fringes appearing on this object are
indicative of deformation under a small differential pressure about
a high mean pressure. From this interferogram qualitative and/or
quantitative information regarding the behavior of the object under
load can be inferred. In this case the information was used as
part of an iterative design/experiment process.

Although biomedical applications are specific instances of
metrology or mechanics, they form an interesting subgroup in their
own right. Figure 3 is an example of the use of holographic
interferometry to measure displacement of various points in the
facial bone structure of an animal as it bites an object.
Information such as this is scientifically useful and can be used
to design dental prosethetics.

Holographic nondestructive testing is a class of techniques by
which material flaws or inhomogeneities are detected through their
affect on surface deformation, as disclosed by optical holography,

484

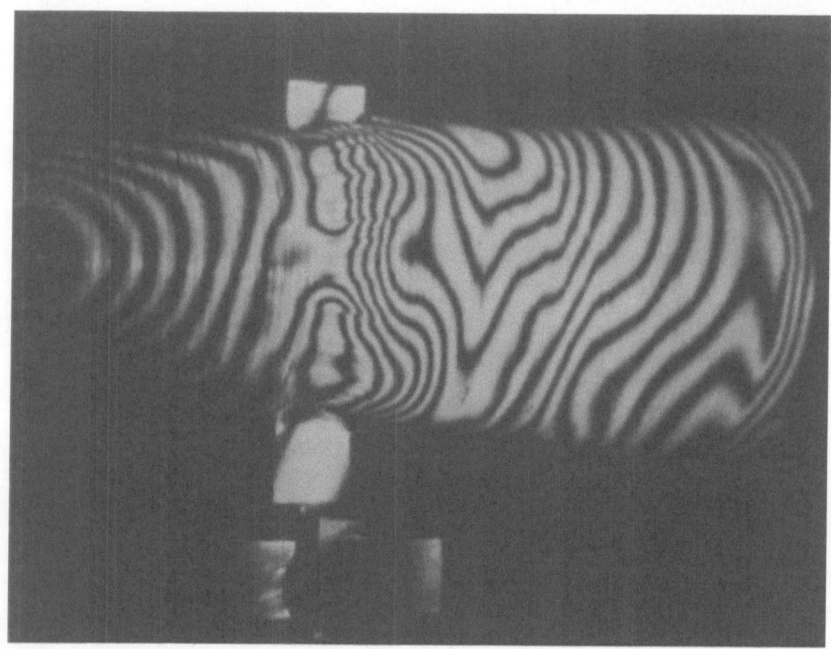

Figure 2. Composite structure pressurized to a high level and differentially pressurized between holographic exposure.

when the test object responds to some mechanical or thermal load. Again, this is a special case of application to metrology and solid mechanics in which the objective is the detection of some flaw within a manufactured system or component. An example is shown in figure 4 which is an interferometric fringe pattern due to differential pressurization of a honeycomb construction panel. In this case the surface bulges slightly due to the pressure difference, and the localized bulging above an area where a poor bond exists between the skin and core renders this defect immediately apparent to the observer. The status and future of holographic nondestructive testing are outlined in some detail in references 3 and 4, and additional comments about it are found below.

Contour generation refers to a series of techniques by which multiple wavelength holography, or an equivalent technique in which an object is submerged in a fluid whose refractive index can be changed, is used to generate topographical contours of an object surface [5, 6].

All of the above application areas involve opaque test objects. An equally important set of applications involves the study of transparent media for flow visualization, flow

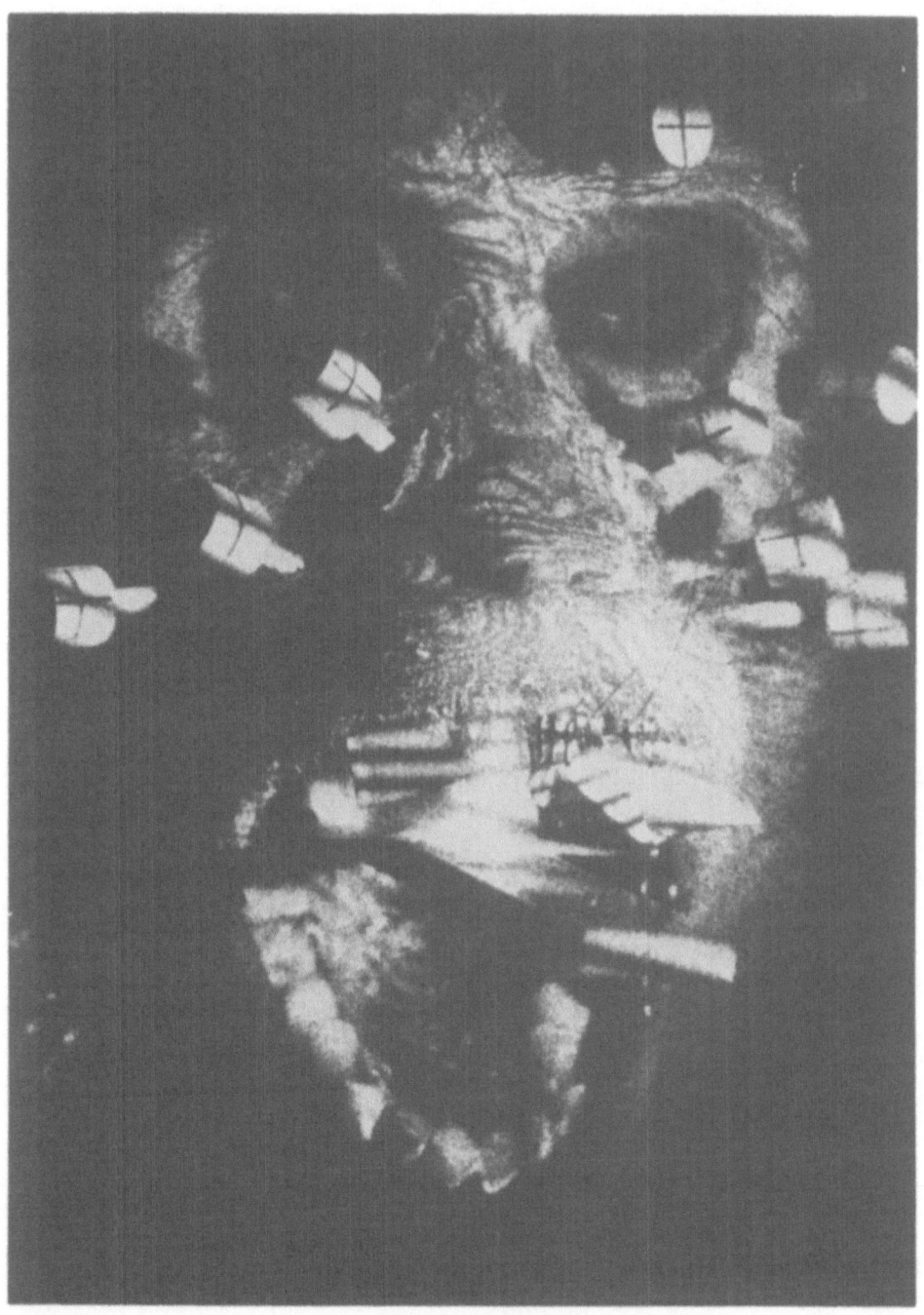

Figure 3. Interferogram showing motion of facial bones due to load on teeth (courtesy Dr. B. Goldin).

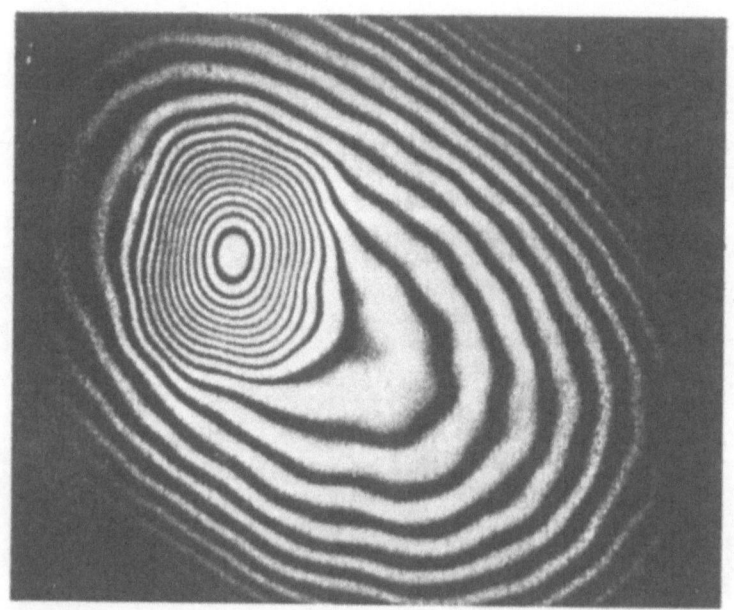

Figure 4. Disclosure of region of poor bond by differential vacuum of face of honeycomb panel.

measurement, plasma diagnostics, heat and mass transfer, and stress analysis. In this case the optical path length displayed or measured by holographic interferometry must be related to the refractive index variation within the test field. Refractive index in turn must be related to physical properties such as mass density, electron number density, temperature, species concentration, distribution of principal stresses, etc. In some cases, additional analysis can in turn provide measurements of rate quantities such as velocity and heat or mass transfer. Applications of holographic interferometry to the study of transparent media are summarized in a separate paper in these proceedings [7].

3. TECHNIQUES AND TECHNOLOGIES

Throughout most of its history there have been four basic techniques of holographic interferometry. These are two-exposure, real-time, time-average, and real-time/time-average. To these we might add a fifth, modulated-wave holographic interferometry. This latter category includes stroboscopic holograpy and various frequency or phase modulation techniques used in vibration measurement. These five fundamental techniques are all well known and will not be illustrated here.

Several other techniques have been developed or are currently under development for a variety of applications. Among these are:

1. Holographic moire
2. Heterodyne holography
3. Sandwich holography
4. Comparative holography
5. Moire analysis of interferograms

6. Coupling with stress analysis
7. Coupling with optical filtering
8. Coupling with speckle techniques

The above list may not be comprehensive, but it does give some idea of the variety of techniques developed to date. Holographic moire [8] is a strain measurement technique in which the object is illuminated by two object waves rather than one. This gives rise to a fringe pattern indicative of changes of in-plane displacement which therefore is directly useful for in-plane strain analysis.

Heterodyne holographic interferometry is an important technique which has been developed primarily by Dändliker and co-workers [9]. In this technique two separate reference waves are used when recording the hologram, one for each of two exposures. When the holographic image of the object is reconstructed an electrooptical device is used to introduce a constant temporal frequency difference between the two reconstruction waves. This causes the output of the hologram to have a temporal beat. The beat frequency is typically in the kHz or MHz range and therefore can be detected by photodiodes or photomultipliers and analyzed by electronic means. This enables the experimentalist to make very accurate measurements which are equivalent to interferometric measurements to accuracies as high as $\lambda/1000$. This technique is important for two reasons. First, it enables one to make very high accuracy measurements which are required for evaluation of derivatives necessary for determining strains, and second, because it effectively varies the range of sensitivity of holographic interferometry.

Sandwich holography, developed by Abramson [10] is a technique in which two holograms are recorded on separate plates, one before and one after deformation of the object. Before reconstruction these plates are physically placed together to form a sandwich. When this holographic sandwich is tilted and translated relative to its original position, it is possible to dynamically change the fringe pattern so as to perform important operations such as removal of rigid body components of object motion.

Comparative holography is a technique proposed by Neumann [11] which directly addresses an important problem in holographic nondestructive testing. It enables the experimenter to display the

<u>differences</u> of surface deformation of two nominally identical components in response to the same load. For example, one component may be a flaw-free master and the other a production component being inspected for flaws. Comparative holography produces a single interference pattern which is a display of this difference with an accuracy on the order of the wavelength of light.

<u>Moire</u> <u>analysis</u> <u>of</u> <u>interferograms</u> refers primarily to the formation of moire patterns using one or more interferograms to detect subtle indications of flaws or other localized deformations. It is well known that moire fringes formed by overlaying two nearly identical interference fringe patterns display contours of differences between the two patterns. Hence, if one pattern is that of a master object and the second is that of a test object, the moire difference pattern will indicate the presence of flaws or other differences between the two objects. In applications such as industrial nondestructive testing, it is unlikely that loading and placement can be easily reproduced with sufficient accuracy to make this technique work. However, small differences of loading can be compensated for by appropriate relative magnification of the images. Figure 5a illustrates this. It shows the moire pattern formed by superimposing, with corrective relative magnification, the fringe pattern formed by a uniformly pressurized aluminum disk and that formed by a nominally identical disk with two flaws (oval regions from which metal has been removed on the backside of the disk). The moire fringes clearly indicate the approximate size, shape and orientation of these flaws.

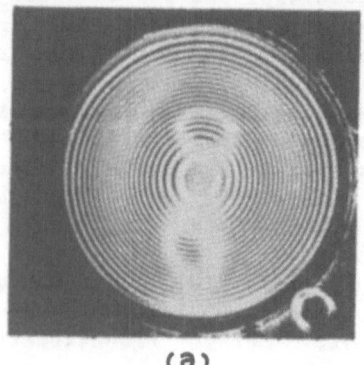 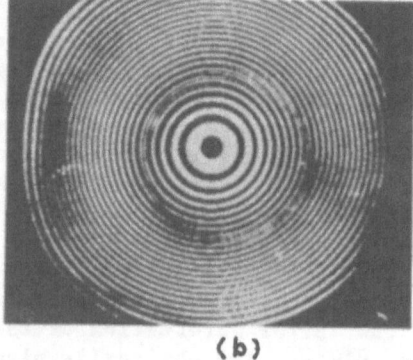

(a)　　　　　　　　　　　　(b)

Figure 5.　Moire analysis of flawed, pressurized metal plates. (a) Superposition of interferograms of flawed and unflawed plates under nearly identical loads. Small pressure differences are corrected for by magnification. (b) Computer-generated moire pattern formed by computationally superimposing two <u>identical</u> interferograms with a small relative magnification. The flaw is indicated on the left side.

At least in the case of basically radially symetrical deformation, it is possible to form a moire pattern indicative of flaws by superimposing two images of the same interference fringe pattern, i.e. no master object is used. This is illustrated by figure 5b. Figure 5b was generated by digitizing an image of a single interferogram, computing a magnified version of the same interferogram, and multiplying these two images together, after adding an appropriate bias brightness level [12]. Digital formation of moire patterns has the advantage of enabling the operator to impose various magnifications or other transformations as well as filtering operations to form a distinct image.

The concept of coupling holographic interferometry with stress analysis is likely to become increasingly important. For example, Dändliker has used the theory of elasticity to develop a scheme for extrapolation of strain and stress into a solid object from a knowledge of holographically measured surface deformation. In principle, and by numerical simulation, he has shown that all nine components of the deformation gradient required to calculate strains, rotations and tilts can be determined in a cone shaped region beneath the observed external surface. Although, as must be expected, this technique is susceptible to error amplification, it is important in that it addresses one of the major limitations of holographic interferometry [13]. An increasingly important application of holographic interferometry appears to be its interaction with modal analysis [14] and various finite element techniques [15] as part of code verification or iterative design/experimental procedures.

Optical filtering is more commonly associated with speckle metrology and interpretation of moire patterns than it is with holographic metrology. However, it is possible, particularly with transparent media, to apply optical or hybrid optical/digital processing to the interpretation of holograms, perhaps bypassing the two-exposure process. Figure 6 illustrates a method by which a single hologram of a wave front can be processed to determine its phase structure. This can be applied to windtunnel measurements and related applications. In this particular case a simple Fourier processor is used to differentiate the optical wave front. This output is then integrated by computer to obtain the desired phase information. It has been shown that by digitizing the outputs of this optical processor with the differentiating filter placed in three slightly different locations, most of the noise associated with coherent optical processing can be computationally eliminated [16].

Finally, it is well known, but rarely exploited, that a single hologram can be recorded in such a manner that out-of-plane deformation can be determined by interferometric fringe analysis and in-plane deformation can be determined by speckle photographic

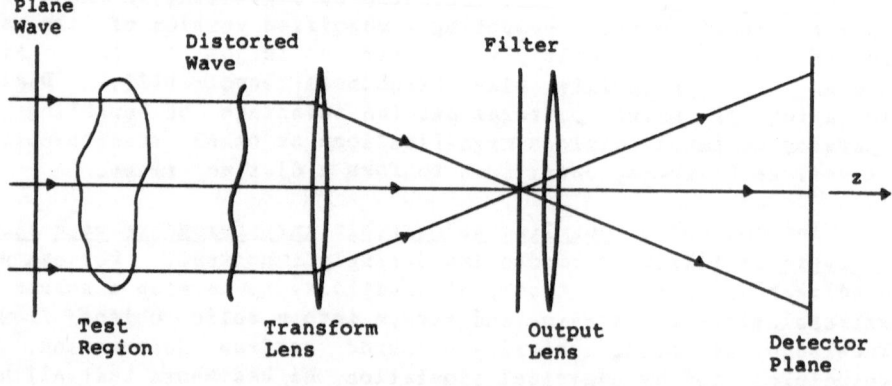

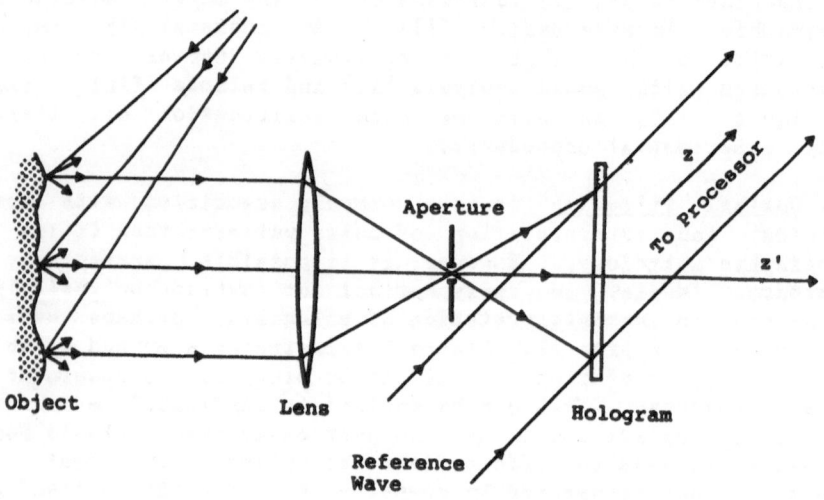

Figure 6. (a) Optical processor used as part of a hybrid system.
(b) Input to processor can be derived from a hologram.

techniques. Undoubtedly, other techniques exist or will be developed in time. This indicates that holographic interferometry remains a rich field which can usefully be combined with other techniques.

The future importance, particularly in industrial applications, of holographic metrology is strongly dependent on a number of technologies:

1. Recording media
2. Lasers
3. Fringe analysis systems
4. Video/detector arrays
5. Computers/microprocessors
6. Computer-generated holograms
7. Phase conjugation
8. Fiber optics

Although rapid development techniques and carefully engineered film drives render traditional silver halide recording systems surprisingly fast and convenient, rapidly-developed and erasable recording media would be extremely useful in many applications. The development and commercialization of photothermoplastic cameras has gone a long way to solve this problem. However, they suffer in some instances from relatively low sensitivity, limited spatial frequency range and modest aperture. Hence continued development of recording technology may be important.

Continued development of lasers is quite crucial for scientific and industrial applications of holographic interferometry. Continuous-wave lasers are highly developed, and good, reliable systems are available. Pulsed laser technology has made significant advances but still lacks some of the ruggedness and reliability required for many industrial applications. The broad wave length ranges and fine tuning capabilities of dye lasers may be of increasing importance in holographic metrology, particularly in areas such as holographic contour generation. Finally, ultrafast (picosecond and even femtosecond) lasers are now available in several laboratories. Such rapid laser pulses will have interesting uses in future applications of holographic metrology, as already has been demonstrated by Abramson's work in light-in-flight holography [17].

Although it is a view which is not universally shared, I believe that the future importance of holographic interferometry in both industrial and scientific application will strongly depend on the extent to which computers, microprocessors, video systems and solid-state detector arrays are used to develop sophisticated and well automated fringe analysis systems. This field is about to yield very important results. In areas like nondestructive testing Mitchell, et al [18] have applied pattern recognition techniques to search for fringe patterns characteristics indicative of flaws.

492

Tichenor, et al [19] have developed an algorithm which simply looks
at fringe density in various regions of an interferogram. This
can, and has, been used to develop a highly automated
nondestructive testing procedure. Several workers are currently
developing automated evaluation schemes which attempt to deal with
practical difficulties such as speckle noise, variation of image
intensity, and small regions in which fringes are discontinuous or
missing. Figure 7 illustrates the use of nonlinear regression
analysis to fit digitized intensity measurements from a noisy
holographic interferogram to a variation of the form

$$I(x) = b(x) + a(x)\cos[\Delta\phi(x)] \tag{1}$$

This technique accepts variations in background irradiance and
considerable speckle noise. It is useful for broad fringe patterns
which contain few fringes but from which one must obtain an
accurate interpolation of phase [20].

Among the interesting work currently being done on automated
interferogram evaluation is that of Becker, et al [21]. In their
algorithm the digitized intensity pattern is operated on to remove
background intensity variations, the contrast of fringes is
enhanced by a threshholding algorithm, a polygonal representation
of the fringes is established, fringe disconnections are removed,
fringe order numbers are assigned in a highly automated manner, and
the resulting fringe patterns are smoothed and represented by local
polynomial approximations.

Other interesting fringe analysis techniques are developed for
specific applications. For example, Goldberg [22] has recently
reported a sophisticated algorithm for determining components of

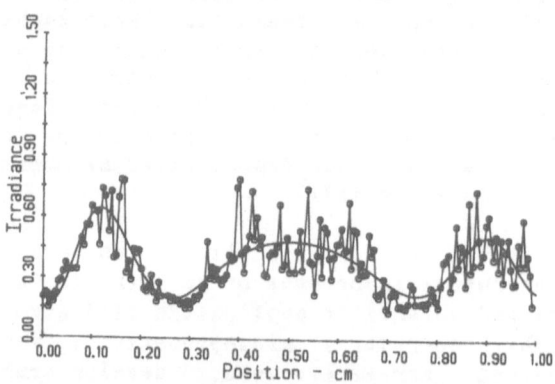

Figure 7. Nonlinear regression analysis of interference fringes.
Dots indicate digital irradiance data. The curve is a
fit by nonlinear regression analysis to Eq. (1).

surface strain on three-dimensional objects. Although this algorithm is based on the use of well known cubic spline techniques, the error amplification generally associated with taking derivatives of cubic splines passing through all data points is reduced by digitally filtering the Fourier spectrum of the representation of the strain function. This technique appears to be quite promising and is being developed for application to testing of large cast iron pipe.

The use of computer generated holograms for testing optical elements has been demonstrated, and computer-generated holographic filters may play a future role in processing of interferometric data of various kinds. The use of nonlinear recording materials based on phase conjugation makes possible a kind of real time holographic interferometry which may be of future importance.

Finally, it is obvious that fiberoptic technology will be of increasing importance in holographic metrology in two respects. First, object and reference waves can be transmitted around a system by fibers rather than by mirrors. This can greatly reduce the physical size, complexity and vibration problems of holographic systems. Second, and perhaps much more important, coherent fiberoptic bundles can be used to record holograms and interferograms in object regions and environments which normally would preclude optical access. Work on this problem is proceeding at ATT Bell Laboratories.

4. LIMITATIONS AND CHALLENGES

A number of fundamental or technical limitations of holographic metrology are implicit in the preceding discussion. Since these define many of the future challenges for workers in this area, I have tabulated several below:

1. Range of sensitivity
2. Mechanical stability requirements
3. Removal of rigid-body motion
4. Surface measurement only

5. Speckle noise
6. Optical access
7. Automation
8. Cost effectiveness
9. System reliability
10. Safety and acceptance

The significance of each of the factors noted above is obvious to anyone knowledgable in the field.

Based on the current status and limitations of holographic metrology, several general challenges to workers in the field can be identified. <u>Automation of fringe analysis</u> must progress substantially. Problems remain in interpolation and accuracy of analysis, but the most important challenge is to maximize the automation of global analysis of fringe patterns and assignment of order numbers. General work in <u>fringe interpretation</u> aimed at accurate strain analysis, removing rigid-body components of motion and detection of small defects still is very important. Further development of <u>rapid erasable recording media</u> is desirable. The ability to <u>vary the sensitivity of holographic metrology</u> is important to its future. One might note that techniques such as heterodyne interferometry, moire analysis and fringe interpolation effectively vary this range of sensitivity. The future of holographic nondestructive evaluation would be brightened greatly if significant advances in <u>quantitative flaw detection</u> are made. Finally, a general challenge remains the identification of industrial and scientific applications for which holography is a natural, cost effective and accurate technology.

5. SPECULATION AND OPINIONS ABOUT THE FUTURE

Holographic interferometry will remain a relatively specialized technique. It is not likely to develop into a black box immediately applicable to a wide variety of applications. It remains a challenge to identify industrial and scientific problems for which it is the best solution. However, I am confident that many such problems exist and should be addressed.

The operation of holographic systems will be simplified by the application of fiberoptics, nonlinear optical materials and computers. This will be particularly important in the development of routine industrial testing applications where theoperator is not a highly trained engineer or scientist.

Holographic metrology systems will become hybrid optical-digital or electronic systems. This effectively will vary the sensitivity of the technique. It no longer will be thought of as a modulo $\pi/2$ measurement technique. Of course, this will also enhance the ease with which data are interpreted by operators.

Artificial intelligence concepts should be applied to automation of fringe analysis. It is extremely difficult to completely remove the knowledgable human operator from interpretation of interference patterns. In specific application areas, it would seem that the use of expert systems and other artificial intelligence procedures in which the computer analysis algorithm πlearnsπ from human operators would be natural to apply to fringe analysis.

Finally, it is important that significant research and development projects be funded by industry and government and carried to completion. The amount of money which has been spent in this application area is probably small compared with the development of many other techniques over the years. I am confident that when sufficient funding and multidisciplinary effort is devoted to holographic and related optical metrology techniques they will be enormously successful. Examples of major projects which are successful, or well on their way to success, include holographic tire testing, helicopter rotor aerodynamic experiments, bubble chamber holography, holographic monotoring of experiments in space, and turbine blade dynamics studies. There appears to be growing interest in major programs for inspection of large composite structures.

REFERENCES

1. L. O. Heflinger, R. F. Wuerker and H. Spetzler, Thermal expansion coefficient measurement of diffusely reflecting samples by holographic interferometry, Rev. Sci. Instrum, 44, 629-633 (1979).

2. B. Goldin, Department of Orthodontics, University of Connecticut, private communication (1979).

3. G. Birnbaum and C. M. Vest, Holographic nondestructive evaluation: Status and future, International Advances in Nondestructive Testing, 9, 257-282 (1983).

4. C. M. Vest, Holographic NDE: Status and Future, Report No. NBS-GCR-81-318, Natl. Bur. Standards, Washington, DC, 20234 (1981).

5. J. R. Varner, Holographic and moire surface contouring, ch. 5 in Holographic Nondestructive Testing, R. K. Erf (Ed.), Academic Press, New York, 1974, pp. 106-147.

6. A. A. Friesem and V. Levy, Fringe formation in two-wavelength contour holography, Appl. Opt., 15, 3009-3020 (1976).

7. C. M. Vest, Optical metrology and computer tomography for measurement of temperature and density, These procedings (1984).

8. C. A. Sciammarella and J. A. Gilbert, A holographic-moire technique to obtain separate patterns for components of displacement, Exp. Mech., 16, 215-220 (1976).

9. R. Dandliker, Heterodyne holographic interferometry, in Progress in Optics, Vol. XVII, North-Holland, Amsterdam,

1980, pp. 3-84.

10. N. Abramson, The Making and Evaluation of Holograms, Academic Press, New York, 1981, pp. 304-312.

11. D. B. Neumann, Comparative Holography, Tech. Digest, Topical Meeting on Hologram Interferometry and Speckle Metrology, Opt. Soc. Am., 1980, pp. MB2-1 to MB2-4.

12. Xu Youren, C. M. Vest and E. J. Delp, Optical and digital moire detection of flaws applied to holographic nondestructive testing, Appl. Opt., 8, 452-454 (1983).

13. R. Dandliker, Extrapolation of strain and stress from holographically measured surface displacement, J. Appl. Mechs, 46, 581-586 (1979).

14. M. K. Rao, M. P. Zebrowski, and H. C. Crabb, Proc. Intl. Modal Analysis Conf., Orlando, Florida, February 6-9, 1984, pp. 408-414.

15. R. Pryputniewicz, Unification of laser holography with finite element methods, These proceedings (1984).

16. J. Prikryl and C. M. Vest, Hybrid processing for phase measurement in metrology and flow diagnostics, Appl. Opt., 22, 2844-2849 (1983).

17. N. Abramson, Light-in-flight recording by holography, Opt. Lett., 3, 121 (1978).

18. O. R. Mitchell, E. J. Delp and W. K. Caldwallindir, Proc. IEEE 5th Intl. Conf. on Pattern Recognition (1980).

19. D. A. Tichenor and V. P. Madsen, Opt. Engin., 18 469-472 (1979).

20. J. B. Schemm and C. M. Vest, Fringe pattern recognition and interpolation using nonlinear regression analysis, Appl. Opt., 22, 2850-2853 (1983).

21. F. Becker, G. E. A. Meier and H. Wegner, Automatic evaluation of interferograms, Proc. SPIE, 359, 386-393 (1982).

22. J. L. Goldberg, A new algorithm for determining components of surface strain on three-dimensional objects using two-exposure holographic interferometry, Proc. Intl. Conf. Exp. Mechs, Montreal, June 1984.

SPECKLE METHODS

SPECKLE METROLOGY

Karl A. Stetson

United Technologies Research Center
East Hartford, Connecticut, U.S.A. 06108

1 INTRODUCTION

When diffusely reflecting objects are illuminated by light, they reflect fields that scatter in all directions. If the light is coherent, e.g. from a laser, the overlapping fields interfere randomly to form patterns .called laser speckles. Being random, these patterns can only be characterized by their statistical properties, the most useful of which are (a) the probability distribution of spatial frequencies and (b) the probability distribution of irradiance.

In virtually all situations where speckles are observed, they result from the light having passed through some aperture. Usually it is the pupil of a lens, but it may also be the boundary of the illuminated region of the object itself. The aperture defines the probability distribution of spatial frequencies that are present in the speckle pattern by defining all the possible angles and directions between the interfering fields. The distribution of spatial frequencies is a function that can be physically observed. If a photographic transparency of a speckle pattern (which may be called a specklegram) is illuminated with a converging beam of coherent light, a halo will form where the beam comes to focus that consists of light diffracted away from the illumination direction by the speckle pattern. The field amplitude of the halo is the Fourier transform of the field amplitude transmitted by the transparency and displays the distribution of spatial frequencies in the transparency. Because of the coherence of the converging light beam, the halo field is randomly speckled and generally has the same statistical properties as the pattern itself. Fourier transform theory shows, however, that the halo envelope is the auto-convolution of the transmittance function of the aperture defining the original speckle pattern. From this relationship, it is simple to estimate the size and shapes of halo patterns associated with various aperture functions.

Specklegram halos serve an important function in speckle metrology. Double-exposure specklegrams of identical but displaced speckle patterns generate halos containing fringe patterns that can be used to measure the speckle displacements. The fringes can be derived rigorously from the shift theorem of Fourier transforms but also may be modeled by considering common speckles as point sources for Young's interference. Halo fringes may also be obtained by combining separate specklegrams either by direct overlay or by means of a beam splitter. This is useful in compensating for large translations between exposures.

Provided that there is no specular component to the field reflected by an object, the probability distribution of irradiance of a speckle pattern depends essentially upon whether the interfering fields have a common polarization. If they do, the distribution follows a negative exponential function of irradiance with zero irradiance being most probable. If not, it is proportional to irradiance times the negative exponential function of irradiance and zero becomes the least probable irradiance. This latter function describes the incoherent addition of two independent speckle patterns and is useful in characterizing double-exposure specklegrams and speckle correlation interferometry.

The objective of these lectures is to describe the use of laser speckles in metrology of object displacements and deformations. Object displacements can give rise to changes in the phase of speckled wavefronts and/or to speckle displacements. The first can be measured by detecting changes in speckle correlation. The second can be detected by measurement of halo fringes from double-exposure specklegrams. We shall begin with a discussion of the theory and applications of speckle correlation, because of its relative simplicity, and progress to the theory, measurement, and applications of speckle displacements.

2 SPECKLE CORRELATION

The first examples of interferometry by speckle correlation were reported by Leendertz (1) who made use of an apparatus similar to that of Fig. 1. This shows a set up similar to a Michelson interferometer except that the customary mirrors have been replaced

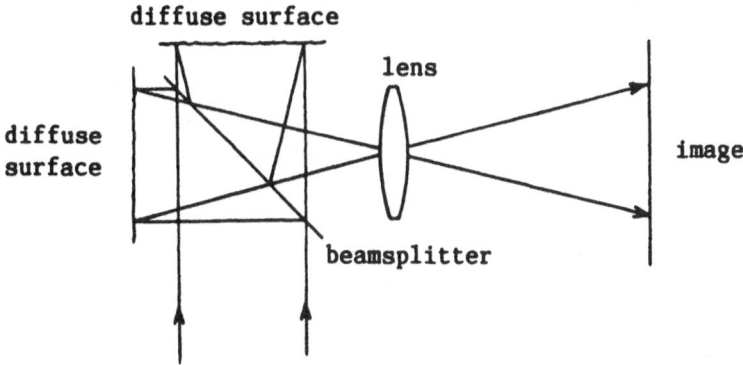

Fig. 1. A speckle-correlation interferometer.

by diffusely reflecting surfaces which are imaged by a lens in the output field. With a laser illuminating the system, the output fields will be sufficiently coherent to interfere even though the result may not be recognizable. The random speckles of the two fields are uncorrelated, and, because of the addition of the field amplitudes, their resulting interference pattern has the same statistics as either field alone.

Let the two fields be represented by the two functions S_1 and S_2 ao

(1) $S_1 = a_1(x,y)\exp[i\phi_1(x,y)]$, and $S_2 = a_2(x,y)\exp[i\phi_2(x,y)]$

where x and y are coordinates in the image plane, a_1 and a_2 are the field amplitudes, and ϕ_1 and ϕ_2 are phase functions of the two fields. Consider the irradiance of the sum of the two fields:

(2) $I = I_1 + I_2 + 2a_1a_2\cos(\phi_1 - \phi_2)$.

where I_1 and I_2 are the squares of the field amplitudes a_1 and a_2 respectively and represent the separate irradiance patterns of the two fields by themselves. The probability distribution of irradiance of either field, or their coherent sum, is a negative exponential function,(2)

(3) $p(I_1) = p(I_2) = p(I) = (1/I_{av})\exp(-I/I_{av})$,

where I is the irradiance and I_{av} is the average irradiance of the field under consideration. The probability density of irradiance for the incoherent sum of the two speckled fields (i.e. $I = I_1 + I_2$) is (2)

(4) $p(I_1 + I_2) = (4I/I_{av})\exp(-2I/I_{av})$

It is possible to distinguish between these two distributions by a number of methods which are generally based upon the fact that the speckle field due to addition of field amplitudes has higher contrast than the one due to the addition of field irradiances. With contrast defined as the standard deviation of irradiance divided by average irradiance, the ratio of contrasts for these two cases is $(2)^{1/2}$.(3) The secret to performing interferometry with speckles lay in finding a way to modulate the third term of Eq. (2).

A slight tilt of one of the diffuse reflectors in Fig. 1 will generate a slowly varying change in phase, ψ, across its corresponding image. This phase change will appear in the argument of the cosine function in Eq. (2). Consider two exposures (a) where the surface is tilted to a minus angle and (b) where it is tilted to a plus angle. The corresponding irradiance patterns, I_a and I_b, are

(5) $I_a = I_1 + I_2 + 2a_1a_2\cos(\phi_1 - \phi_2 - \psi)$, and

        $I_b = I_1 + I_2 + 2a_1a_2\cos(\phi_1 - \phi_2 + \psi)$.

When these patterns each successively expose a photograph for the time interval T, the resulting exposure, E, is

(6) $E = T(I_a + I_b) = 2T[I_1 + I_2 + 2a_1a_2\cos(\phi_1 - \phi_2)\cos\psi]$.

Clearly, when the final factor of $\cos\psi$ is zero, the exposure is equivalent to the intensity sum of two uncorrelated speckle patterns. When it is ± 1 the exposure is the amplitude sum of two patterns. When the total phase change $(=2\psi)$ between the two fields is plus or minus any nonzero multiple of 180°, the contrast will be 0.707. When the total phase change is zero or any multiple of 360°,

the contrast will be unity. By variations in contrast, therefore,
the double-exposure photograph encodes a fringe pattern that is
identical to what would have been obtained had the diffuse reflectors
been perfect mirrors.

The fringes can be extracted from double-exposure photographs
by a number of means. The simplest method employs photographic
transmittance. The average transmittance over a representative
number of speckles is the normalized integral of the transmittance-
versus-exposure function of the film times the probability density
function of the speckle irradiance. Because the high contrast
pattern has a higher probability of speckles with zero irradiance,
its average transmittance will be greater than that of the low
contrast pattern. The fringes can be seen, therefore, by simply
examining the photographic negative. Film with high photographic
contrast will enhance these fringes. The fringe contrast can also
be improved by superposing positive and negative transparencies,
or by various techniques of spatial frequency plane filtering(2).
The fringes may be observed concomitantly (in real-time) by reloca-
ting a negative transparency of the initial speckle pattern in
register with the pattern that exposed it. It will form a random
moiré mask for the illuminating field and generate a null in
the transmitted irradiance. Where decorrelation occurs, it results
in a noticable increase in transmitted light.

The system pictured in Fig. 1, which measures of out-of-plane
displacement, offers no advantage over hologram interferometry. Its
fringe contrast is so poor, in fact, that the system serves mainly
pedagogical interests. In the system shown in Fig. 2, however,
the object serves the function of the beam splitter and scatters
uncorrelated fields from two illumination beams toward the lens.
This system measures displacement along the direction of the bisector
of the two illumination directions according to the equation:

(7) $2\psi = (\underline{K}_2 - \underline{K}_1) \cdot \underline{L}$,

where $\underline{L}$ is the vectorial object displacement and $\underline{K}_2$ and $\underline{K}_1$ are the
propagation vectors of the two illumination beams. When a flat
surface is aligned parallel to $\underline{K}_2 - \underline{K}_1$, the resulting fringe spacing
in that direction is inversely proportional to strain along that
axis according to the equation:

(8) $\varepsilon = \lambda/2d_f \sin\theta$,

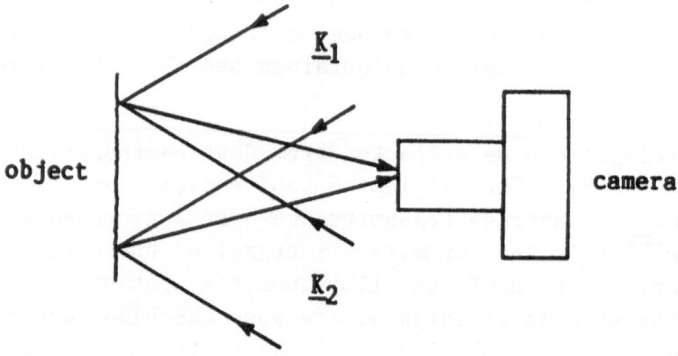

Fig. 2. A strain measurement system using speckle correlation.

where ε is strain, d_f is the fringe spacing, and θ is the half angle between the illumination beams. The ability of this system to measure strain directly, even in the presence of bulk displacements, (4) is often ample compenstion for low fringe contrast.

The system shown in Fig. 3 employs a beamsplitter to provide a speckle interference pattern between two sheared images of an object surface. Here the phase change that occurs between the two fields comes from the equation:

(9) $2\psi = (\underline{L}(x,y,z)-\underline{L}(x-\Delta x,y-\Delta y,z))\cdot(\underline{K}_2-\underline{K}_1)$,

where $\underline{K}_2$ and $\underline{K}_1$ are the propagation vectors for observation and illumination, and Δx and Δy are the amounts of image shear in the x and y directions. Let s lie along the direction of image shear and let $\underline{K} = \underline{K}_2-\underline{K}_1$. If the amount of shear is small enough to allow linear approximations, Eq. (9) may be simplified as

(10) $2\psi/\Delta s = \partial/\partial s(\underline{K}\cdot\underline{L})$.

Fringe orders are proportional to the derivative in the s direction of $\underline{K}\cdot\underline{L}$, which is the familiar fringe locus function of hologram interferometry. Fringe spacing in this system is inversely proportional to the second derivative of the fringe locus function and is useful in measurement of bending moments (5).

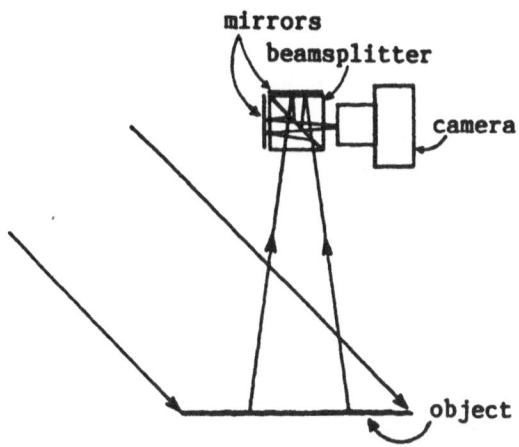

Fig. 3. A system for speckle shearing interferometry.

There are numerous systems for metrology that can be con-
trived using speckle correlation, of which the two above are merely
representative. For an extensive discussion of these systems the
reader must refer to available textbooks (6,7).

3 SPECKLE DISPLACEMENT

The displacement sensitivity of most speckle correlation
systems is similar to other interferometric systems because the
technique generally measures the phase changes of speckled fields.
In many cases, object displacements are far too large to be measured
conveniently by optical phase changes, and the technique of speckle
displacement measurement is preferable. In its simplest form, the
object is illuminated by laser light and imaged by a photographic
lens. A photographic plate is exposed by the image before and
after the object is loaded. The plate is developed and examined
point-by-point with a narrow, convergent beam of laser light to
generate halo fringes (See Fig. 4). The fringe spacings and orientations are measured and are related to the speckle displacements
by the equation

(11) $\underline{H} = (D/\lambda)(\hat{i}/d_{fx} + \hat{j}/d_{fy})$,

where $\underline{H}$ is the vectorial speckle displacement, D is the distance
from the speckle pattern recording (specklegram) to the halo, and
d_{fx} and d_{fy} are the fringe spacings in the x and y directions.

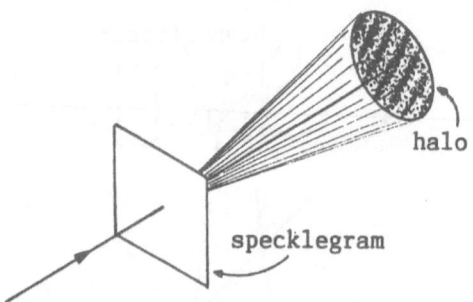

Fig. 4. A system for displaying specklegram halos.

  If the surface of the object is in focus, image speckles move as if they were attached to the surface, at least to first-order approximation. In this case the speckle displacements can be related to the object displacements by the image magnification and by the oblique projection derived by this author in the chapter on matrix methods dealing with spherical perspective. Object displacements are often less important than object strain, which may be determined by measuring the object displacements at nearby points and dividing differential displacement by the separation between points. Let $\underline{L}_o$ be the object displacement at the point of interest, $\underline{L}_1$ be the displacement at a point Δx to the side and $\underline{L}_2$ be the displacement at a point Δy above. The transpose of the surface strain-rotation matrix, $[f_s]$, may be approximated by the matrix

$$(12) \qquad [f_s]^T = \begin{bmatrix} (\underline{L}_1 - \underline{L}_o)/\Delta x \\ (\underline{L}_2 - \underline{L}_o)/\Delta y \end{bmatrix}.$$

  Image magnification plays no role in this type of strain measurement because strain is a dimensionless ratio of distances. Spherical perspective, however, can generate apparent isotropic strain if the object undergoes axial displacement. If, for simplicity, the image magnification is unity, the image displacement vector is

$$(13) \qquad \underline{L}_{im} = \hat{\imath}(L_x - L_z\, x/z) + \hat{\jmath}(L_y - L_z\, y/z),$$

where x and y are spatial coordinates on the object surface and z is the distance from the object to the lens. Consider the matric derivative of the image displacement.

(14) $\nabla \cdot L_{im} = [f_s]^T + [\varepsilon_{ap}]$,

where the apparent strain, $[\varepsilon_{ap}]$, is defined by

(15) $[\varepsilon_{ap}] = \begin{bmatrix} L_z/z & 0 \\ 0 & L_z/z \end{bmatrix}$.

The displacement toward the lens divided by the object-to-lens distance, L_z/z, appears as an isotropic expansion of the object surface. If the apparent strain is to be kept below 100 $\mu\varepsilon$, for example, the object-to-lens distance must be held constant to within 0.01%. For small working distances such as 100 to 200 mm, this is a stringent requirement, especially if high loads are applied to a test specimen. The problem can be solved by the use of a telecentric lens for small objects, or with a long working distance for large objects.

3.1 3-Dimensional Speckle Displacements

In the foregoing, it was tacitly assumed that the object surface was not only in focus but also had its normal aligned with the lens axis. This can be true only for a surface that is flat. Most surfaces of engineering interest are contoured and it is inevitable that problems of defocusing and obliquity be considered. Furthermore, a knowledge of speckle motions in three-dimensional space proves useful in both strain analysis and measuring surface rotations.

Speckles form in the neighborhood of the image plane of an optical system as a result of the lens aperture through which the random fields pass. When the object moves, the speckles move in three dimensions. To describe these movements, it is simpler to refer back to a set of virtual speckles near the object surface. These may be thought of as an observed speckle pattern of which the real speckles are an image. In this way, the issues of lens distortion, aberation, and perspective can be introduced a postiori.

The displacements of the virtual speckles surrounding the object surface have been analyzed.(8,9) Their displacements in three dimensions, $\underline{H}$, may be described by

(16) $\underline{H} = \underline{L} - D\underline{K}_{fobc}/k,$

where $\underline{L}$ is the object displacement, D is the line-of-sight distance
from the object surface to the point observed, k is $2\pi/\lambda$, and $\underline{K}_{fobc}/k$
is the angular slewing of the field surrounding the object. $\underline{K}_{fobc}$
also corresponds to a hypothetical telecentrically-observed fringe
vector for a double-exposure hologram recorded of the same object
displacement.

The physical interpretation of Eq. (16) is straightfoward.
When the object surface is in focus, D=0 and $\underline{H}=\underline{L}$ so that the displacement of the speckles equals the displacement of the object.
Otherwise, the slewing of the observed field generates an additional
speckle displacement that adds vectorially to the object displacement. This additional displacement is proportional to the distance
from the object, and it may be the dominant term if the defocusing
is severe. The field slewing is generated by surface rotations,
strain, and curvature of the illumination wavefront. If none of
these are present, the defocused speckles will move identically
with the focused speckles.

Now image these speckles with a lens system with unity
magnification. The image displacements are related to the object
field displacements by oblique projection, $[P_{mk}]$, onto the photographic plate, whose normal shall be $\hat{m}$.

(17) $\underline{H}_{im} = \underline{H}[P_{mk}] = \underline{L}[P_{mk}] - (D/k)\underline{K}_{fobc}[P_{mk}].$

Now substitute for the telecentrically observed fringe vector, $\underline{K}_{fobc}$,
which equals

(18) $\underline{K}_{fobc} = \{\underline{K}[f] - k\underline{L}_{i11}/R_{i11}\}[P_{kn}],$

where $\underline{K}$ is the sensitivity vector, [f] is the strain-rotation
matrix, $\underline{L}_{i11}$ is projection of the object displacement onto a plane
normal to the illumination, R_{i11} is the distance from the illumination source to the object, and $[P_{kn}]$ is the oblique projection
matrix between the object surface normal $\hat{n}$ and the observation
direction $\hat{k}_2$. Substituting Eq. (18) into Eq. (17) yields

(19) $\underline{H}_{im} = \underline{L}[P_{mk}] - (D/k)\{\underline{K}[f] - k\underline{L}_{i11}/R_{i11}\}[P_{kn}][P_{mk}].$

Equation (19) describes the motion of image speckles including the effects of camera perspective, defocus, and obliquity to the object surface.

There are numerous ways to utilize Eq. (19) to determine displacement and strain. First, consider the use of a number of specklegrams recorded from different observation directions to determine object displacement. If the point of interest is in focus in all specklegrams, the problem is identical to ordinary photogrammetry, which has been solved in the chapter on matrix methods. The result is

$$(20) \quad \underline{L} = [\Sigma_r [P_{mk}^r][P_{km}^r]]^{-1} \{\Sigma_r [P_{mk}^r] \underline{H}_{im}^r\},$$

where the superscript r denotes different observation directions.

Now suppose that because of object contours it is impossible to maintain it in focus over the entire field of view. Tandem specklegrams (i.e. plates held one in front of the other and exposed simultaneously) can be used to remove the effects of field slewing. Assume that the two amounts of defocusing, D_1 and D_2 are known. From one pair of tandem specklegrams two equations of the form of Eq. (19) may be written

$$(21) \quad \underline{H}_{im}^1 = \underline{L}[P_{mk}] - (D_1/k)\{\underline{K}[f] - k\underline{L}_{i11}/R_{i11}\}[P_{kn}][P_{mk}], \text{ and}$$

$$\underline{H}_{im}^2 = \underline{L}[P_{mk}] - (D_2/k)\{\underline{K}[f] - k\underline{L}_{i11}/R_{i11}\}[P_{kn}][P_{mk}].$$

These may be solved simultaneously to give

$$(22) \quad \underline{L}[P_{mk}] = (\underline{H}_{im}^1 D_2 - \underline{H}_{im}^2 D_1)/(D_2 - D_1).$$

Tandem specklegrams may also be used for strain analysis by eliminating the displacement terms from Eqs. (21). This gives

$$(23) \quad k\Delta\underline{H}_{im}/\Delta D = -\{\underline{K}[f] - k\underline{L}_{i11}/R_{i11}\}[P_{kn}][P_{mk}].$$

where $\Delta\underline{H}_{im} = \underline{H}_{im}^2 - \underline{H}_{im}^1$ and $\Delta D = D_2 - D_1$. Eq. (23) may now be multiplied from the right by $[P_k][P_n]$ and it is simple to prove that

$$(24) \quad [P_{kn}][P_{mk}][P_k][P_n] = [P_n]$$

This allows Eq. (23) to be rewritten as

$$(25) \qquad \underline{K}[f][P_n] = k\{\underline{L}_{i11}/R_{i11}\}[P_n] - k\{\Delta\underline{H}_{im}/\Delta D\}[P_k][P_n].$$

It is now possible to make the following redefinitions:

$\underline{k}_\sigma = \underline{K}/k$ is the normalized sensitivity vector
$[f_s] = [f][P_n]$ is the surface strain–rotation matrix
$\underline{\sigma}_{i11} = \underline{L}_{i11}/R_{i11}$ is the vectorial angle subtended by the object displacement relative to the illumination
$\underline{\sigma}_{sp} = \Delta\underline{H}_{im}/\Delta D$ is the vectorial angle subtended by the tandem pair of image speckle displacements

With these definitions, Eq. (25) may be rewritten as

$$(26) \qquad \underline{k}_\sigma[f_s] = \{\underline{\sigma}_{i11} - \underline{\sigma}_{sp}[P_k]\}[P_n].$$

The two projections in Eq. (26) correct for two effects. $[P_k]$ corrects for the obliquity of the photographic plate to the lens axis, and $[P_n]$ corrects for the obliquity of the object surface to the viewing direction. If data are available from three or more independent sensitivity vectors, it may be used to expand Eq. (26) into the matrix form,

$$(27) \qquad [k_\sigma][f_s] = \{[\sigma_{i11}] - [\sigma_{sp}][P_k]\}[P_n].$$

This may be solved for $[f_s]$ to give the least square error via the equation

$$(28) \qquad [f_s] = [[k_\sigma]^T][k_\sigma]]^{-1}[k_\sigma]^T\{[\sigma_{i11}] - [\sigma_{sp}][P_k]\}[P_n].$$

Equation (19) may also be used for strain analysis by the method of finite differences illustrated in Eq. (12). Speckle displacements are measured at neighboring locations and the incremental changes in displacement are divided by incremental changes in position to approximate the matric derivative of speckle displacement. (see discussion of Eq. 14) The speckle displacements in turn approximate the object surface displacements. Although this works well for normal observation of flat surfaces, examination of Eq. (19) reveals problems associated with surface contours and perspective. To clarify these problems, it is helpful to substitute the following relationship for $\underline{L}$:

(29) $\underline{L} = \underline{R}_o[f_s]^T + \underline{L}_T,$

where $\underline{R}_o$ is the space vector to points on the object surface and $\underline{L}_T$ is the bulk translation of the object. Because of perspective, $\underline{R}_o$ is the oblique projection of the three-dimensional space vector $\underline{R}$ along the viewing direction onto the object surface, i.e.

(30) $\underline{R}_o = \underline{R}[P_{nk}].$

Accordingly, Eq. (19) becomes

(31) $\underline{H}_{im} = \{\underline{R}[P_{nk}][f_s]^T + \underline{L}_T - D(\underline{k}_o[f] - \underline{L}_{i11}/R_{i11})[P_{kn}]\}[P_{mk}].$

In general, all parameters in Eq. (31) vary with x and y coordinates except the strain-rotation matrix, $[f_s]$, and the bulk displacement, L_T. If spherical perspective is eliminated by tele-centric or long radius viewing, the projection matrices may be considered constant. Similarly, collimated illumination can reduce the correction term for spherical illumination to zero. With these simplifications, the matric derivative of Eq. (31) becomes

(32) $[\nabla\otimes\underline{H}_{im}] = [P_{nk}][f_s]^T[P_{mk}] + [\nabla D\otimes\underline{k}_o][f_s][P_{kn}][P_{mk}].$

The first term of Eq. (32) describes the admixture of surface tilts with surface strains and rotations due to obliquity of the object surface and the specklegram plate to the viewing direction. Simply put, an inclined surface appears foreshortened to the viewer and changes in inclination increase or decrease that effect. The second term describes an additional source of admixture due to linear variations in defocusing. As stated previously, surface tilts cause the field reflected by the object to slew and this adds additional speckle motions to those generated by object translations. Inclination of the surface relative to the viewer causes a variation in defocusing of the surface, which, together with field slewing, causes linear variations in the speckle motions that generate apparent strain. Note that both of these effects persist even where the surface is in focus. In principle, it is possible to use data from two or more viewing directions to eliminate these admixtures.

4 CONCLUSION

This presentation has provided an introduction to the primary issues concerning speckle metrology. The wide range of application for speckle methods has given rise to numerous variations of the basic technology discussion of which must be omitted due to space limitations. The material provided, however, should give the reader the mathematical and conceptual background to understand the field as a whole.

References

1. Leendertz, J.A., Interferometric Displacement Measurement on Scattering Surfaces Utilizing Speckle Effect, J. Phys. E: Sci. Inst. 3 (1969) 214-218.
2. Ennos, A.E., Speckle Interferometry, in Progress in Optics vol. XVI, E. Wolf, Ed. (Amsterdam, North-Holland, 1978) 235-286.
3. Goodman, J.W., Statistical Properties of Laser Speckle Patterns, in Laser Speckle and Related Phenomena, J.C. Dainty, Ed. 2nd edition (Springer-Verlag, Berlin 1984) 9-75.
4. Stetson, K.A., Analysis of Double-Exposure Speckle Photography with Two-Beam Illumination, J. Opt. Soc. Amer. 64 (1974) 857-861.
5. Leendertz, J. A. and J.N. Butters, An Image-Shearing Speckle-Pattern Interferometer for Measuring Bending Moments, J. Phys. E: Sci. Inst. 6 (1973) 1107-1110.
6. Erf, R.K., Ed. Speckle Metrology (New York, Academic Press, 1978).
7. Jones, R. and C. Wykes, Holographic and Speckle Interferometry (Cambridge, Cambridge University Press, 1983).
8. Stetson, K.A., Problem of Defocusing in Speckle Photography, its Connection to Hologram Interferometry, and its Solution, J. Opt. Soc. Amer. 66 (1976) 1267-1271.
9. Jacquot, P. and P.K. Rastogi, Speckle Motions Induced by Rigid-Body Movements in Free Space Geometry: An Explicit Investigation and Extension to New Cases, Appl. Opt. 18 (1979) 2022-2032.

RECENT DEVELOPMENTS ON SPECKLE PATTERN CORRELATION BY PHOTOGRAPHICAL MEANS

J.J. Lunazzi and M. Muramatsu
Instituto de Fisica, UNICAMP, C.P. 6165, Campinas - SP - Brasil

1. INTRODUCTION

Correlation Techniques between a speckle pattern and its photographic negative are useful for measuring in real time displacements[1], vibrations[2] and the state of deformation of rigid objects during fatigue testing[3] [1]. A similar technique is to use the photographic record of the process to analyze the fringe visibility of the diffraction pattern. One aditional system is necessary in this case and some errors are introduced due to the halo effect. The capability of real-time observation is, however, lost; sometimes requiring the making of many photographic frames. The photodetection and subsequent electronic processing of the signal is in current use, although it does not achive the resolution and simplicity of the photographic systems.

We describe here three recent advances on the use of the photographic real-time technique:

- A masking technique[4] that permits one to obtain the absolute value of the correlation function independent of the photographic parameters.
- A derivative technique[5] that allows reduction of the exposure energy while also increasing the sensitivity.

- Its application to the testing of printed circuit
board soldering.

2. THEORETICAL FUNDAMENTALS

The photographic correlation of a speckle pattern requires
its registration on a photographic plate that remains in the original position; the transmited intensity being collected in a photodetector. The situation corresponds to that of Fig.1, the speckle pattern being generated by scattering of the laser light at the rough
surface of the object under study. We consider, for simplicity, a
one dimensional treatment. Assuming the plate displacement to be
x' the detected signal can be expressed by:

$$V(x') = V_0 - V_1 A(x') \qquad (1)$$

where V_0 is the signal corresponding to transmission through the
photographic background of the plate and V_1 represents the photographic parameters, the photodetector response and the exposure
energy. $A(x')$ is the normalized correlation function between the
illuminating speckle and the one that was registered. It takes the
value A_{max} when both patterns are at the original position, and
A_{min} when they are displaced to a large extent. Most authors[2]
have characterized the correlation process by using the modulation
of this signal, that is:

$$M_S = \frac{V_{max} - V_{min}}{V_{max}} \qquad (2)$$

we employed instead the definition:

$$M_C = \frac{A_{max} - A_{min}}{A_{max}} \qquad (3)$$

3. MASKING TECHNIQUE

The value of M_C can be obtained from the signal values by
combining expressions (3) and (1):

$$M_C = \frac{V_{max} - V_{min}}{V_0 - V_{min}} \qquad (4)$$

We have used the fact that M_C is equal to 0.5 for a fully developed speckle pattern in order to test our masking technique. Two
complementary masks were made, one (PM) consisted of many equally
distributed holes, and the other (NM) being the photographic contact
copy of the first. When exposing the plate the mask NM was interposed, giving a statistical ensemble of small unexposed areas over the
plate. The plate and the masks could always be repositioned by
means of a kinematic plate holder. We can then measure, within the
same set-up, the average transmission T_0 of the plate at the unexposed regions by means of mask PM. The value of V_0 for each plate
is then obtained as:

$$V_O = T_O \ V_N \qquad (5)$$

where V_N is the voltage signal obtained when only the mask NM is
present. By measuring V_{max}, V_{min} and V_n in a rapid sequence we
eliminated the problem of the laser intensity fluctuations. We
verified experimentally the theoretical ($M_C = 0.5$) within a 2%
precision, which does not constitute a practical limit. We also
observed that even large changes in the development time did not
affected M_C , within a 1% precision value. It is important to
mention that, for obtaining a true value for M_C it is necessary
to operate within the linear portion of the transmission response
curve of the plate, so that the mean energy of the speckle pattern
must not exceed one third of the limiting value for linear operation. The correlation values that are obtained in that way permits
the direct comparison of the results obtained in situations where
the photographic material or its processing are quite different.
We also obtain in this way an indicator of perfect repositioning
of a speckle pattern without requiring two precision translation
stages.

4.   THE DERIVATIVE TECHNIQUE
Sometimes it is necessary to expose a photographic plate at
a reduced energy level. In this case, the influence of the insta-

bility on the intensity of the laser beam affects enormously to the measurement error. It increases by a factor equal to M_s^{-1}, which can be, tipically, 7.7 for a reduction factor of five in the exposure energy. We obtained a solution to this problem that does not require the monitoring of the laser intensity values. It consisted of introducing a small oscillation on the speckle pattern or on the plate positions. The amplitude of the detected signal displays then the modulus of the derivate curve, which is influenced by the laser intensity fluctuations in the same proportion as in the full expore case; as an aditional result, we obtained an increasing of the sensitivity by a factor of 2.7 (see Fig.2) and a curve that does not need an initial displacement in order to operate linearly, when measuring displacements or vibrations which are limited in their movement to one side of the initial position.

5. THE TESTING OF PRINTED CIRCUIT BOARD SOLDERING

The testing of printed circuit board soldering has been performed previously by holographic correlation techniques. In one case[6] the residual deformation resultering after a thermal cycle reduced the original value of the maximum correlation signal that can be obtained, and this reduction represented the possibility of failure of the solder joint. In a second case[7], the reduction in correlation signal was obtained after heating uniformly the sample during a fixed time, having the same significance.

We have considered that the changes produced on the wave that is scattered by the surface of the solder joint after its deformation can also change the speckle pattern intensity through the interference of the light coming from the different regions of this surface. So that, in this case, the photographic correlation technique is also essentially interferometric, and offers a greater stability due to the fact that no reference beam becomes necessary. We have worked in the same manner than previous authors[6][7],

displacing the plate when it was necessary in order to follow the
position of the correlation peak. The only modification that was
introduced consisted of a very fine brushing of the metal surface,
sometimes necessary for obtaining a fine speckle pattern. The process of oxidation of the surface was then also observed and the
testing was performed in conditions were it was of minor significance, if any. The results showed a great similarity with those obtained by holographic correlation techniques. We can compare then very
well with the first case[6] by means of Fig.3 , where the loss in
correlation after each thermal cycling is represented by the lower
curve. In this case, we have tested a complete solder joint seven
days after brushing, and the thermal cycle can be said to be equal
to that of Jenkins[6].

The upper curve of the same figure represents the loss in
correlation for a similar solder joint were the contact lead was
not included, the test being performed one day after brushing. We
can conclude, in this way, that the loss in correlation signal
observed was due to a mechanical process during the thermal cycling
and not to an oxidation process.

ACKNOWLEDGMENTS
The financial support received in many ways from the National
Council of Research (CNP) and from the Foundation for Assistance
to Research of the São Paulo State (FAPESP) are gratefully acknowledged by the authors.

References:
(1) Groh, G. In "The Engineering Uses of Holography" Robertson ed.,
    Cambridge Press (1970) 483.
(2) Weigelt, G.P., Opt. Comm. 19,2 (1976) 223
(3) Maron, E., in "Holographic Nondestructive Testing", Editor
    R.K. Erf, Academic Press, 1984.
(4) Muramatsu, M., Lunazzi,J.J.,
    "Absolute speckle pattern correlation by photographic means",

Proc. of the ICO XIII Conference, Sapporo, Japan, August 1984.
(5) Muramatsu, M., Lunazzi, J.J. "Advantages of a derivative technique in performing speckle correlations", App.Opt. September 1984.
(6) Jenkins, R.W., and McIlwain, MC., "Holographic analysis of
printed circuit bcards", Mater. Evalvat. 29 (1971) 199
(7) Espy, P.N., "Testing of printed circuit board solder joints by
optical correlation", NASA TR R-449 (1975).

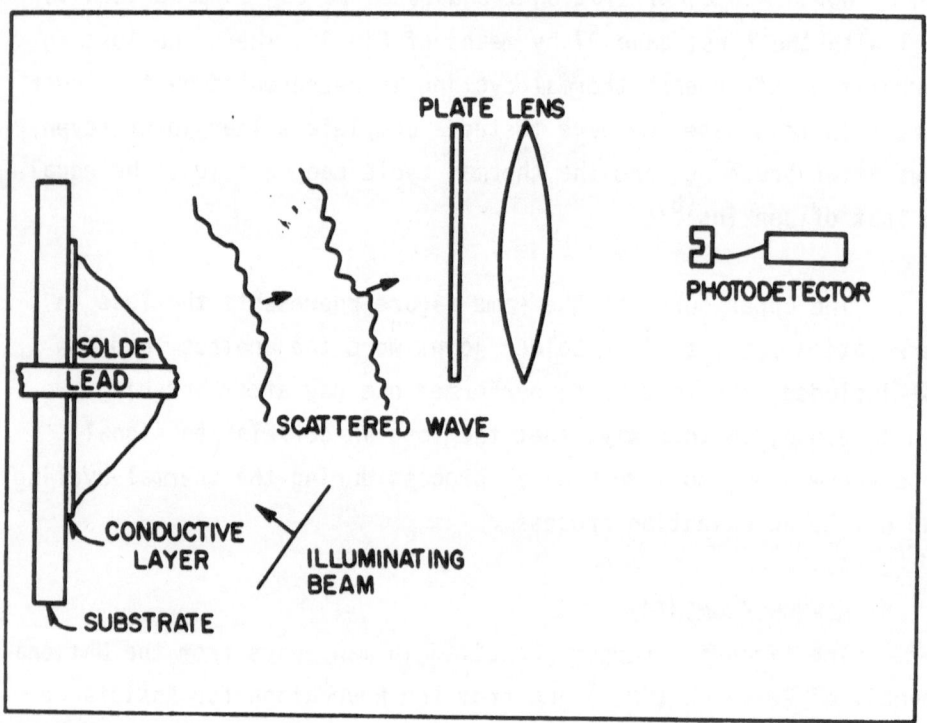

Fig. 1 : Schematic (out of scale) view of the
system, the object under study being
a typical soldering of a printed

circuit board.

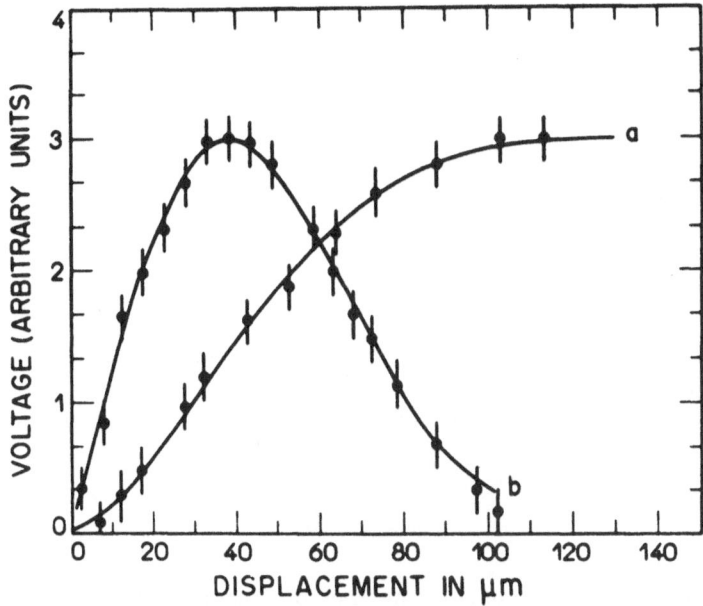

Fig. 2: a) Direct correlation curve.
b) The derivative curve.

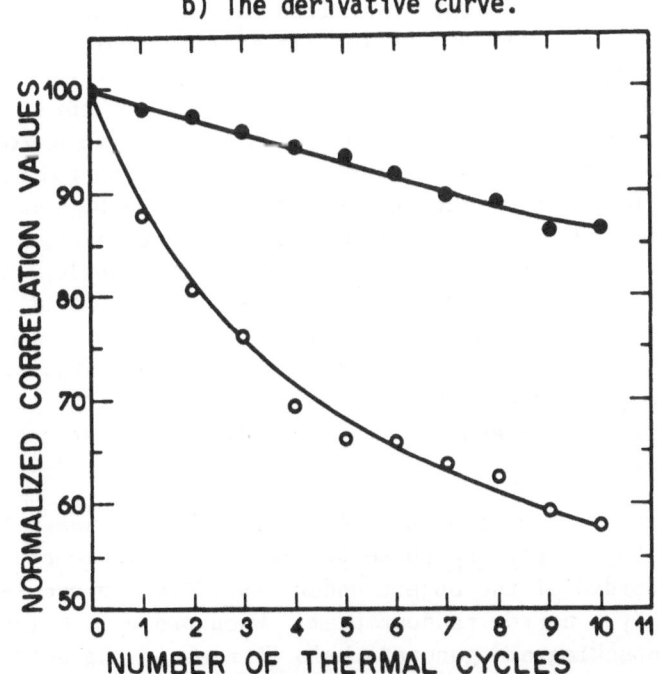

Fig. 3: The correlation values obtained in a
soldering after thermal cycling.

HETERODYNE SPECKLE PHOTOGRAMMETRY

Karl A. Stetson

United Technologies Research Center
East Hartford, Connecticut, U.S.A. 06108

1 INTRODUCTION

A number of problems beset the use of speckle photographs in photogrammetric measurement of displacements and strain. First of all, the range of measurement is limited by the number of fringes that may be read in a specklegram halo. It is not possible, in general, to read more than about twenty five fringes nor fewer than one, and, for good accuracy, a minimum number of two or three may be required. This restricts displacement measurements to a single order of magnitude, and in many engineering applications this range is too narrow to be useful. Second, strain must often be measured when it is accompanied by large, unknown, bulk displacements. Finally, systems that measure the spatial frequency of halo fringes(1) (and therefore displacement) are limited in accuracy to about 1%. When displacement is differentiated to yield strain, the measurement errors escalate and generate strain errors greater then 10%. A method of speckle photogrammetry is needed, therefore, that can ignore large displacements and detect extremely small displacements with high accuracy.

The technique of dual plate speckle photography was a significant step toward solving these problems.(2) Single-exposure specklegrams, recorded of the object under study, can be overlayed and interrogated by a narrow readout beam. When properly aligned, these overlayed specklegrams generate halo fringes in the same manner as double-exposure specklegrams. Overlayed specklegrams, however, have a separation between their emulsions which causes the halo fringes to appear circular, i.e. like Fresnel zones.

Separating the specklegram recordings of the object before and after deformation allows simple and easy compensation for large translations. The fringes that must be analysed for strain measurements, however, are now circular and quadratically spaced, and most automated readout systems are designed for straight, equally-spaced fringes. Although systems could be developed to deal with circular fringes, their accuracy would still be limited to about 1%. In the field of holography, the accuracy of fringe reading was greatly improved by the use of heterodyne interferometry.(3) The next logical step, therefore, was to use a heterodyne interferometer to evaluate halo fringes. The following sections will review the basic technology involved in such a system and discuss some of its applications.

2 HETERODYNE PHOTOCOMPARATOR

Heterodyne interferometry is accomplished by finding some means to continuously change the phase of one path of an interferometer relative to the other. When this is done, the output fringes of the interferometer scan across the field of view. A photodetector placed there will yield a sinusoidal signal whose phase may be measured electronically relative to some reference signal. This phase distribution as a function of the output coordinates will correspond to the fringe orders of the original pattern to within a constant.

The heterodyne readout of halo fringes is done most easily by combining the halos from two specklegrams in an interferometer as shown below in Fig. 1. This arrangement of six mirrors and two beamsplitters allows the two specklegrams to lie physically in the same plane and be equidistant from the output beamsplitter. The output beamsplitter is made from two identical etalons with the partially reflecting surface in the center. This allows the transmitted and reflected fields to pass through identical optical paths. The input laser beam passes through a rotating halfwave plate and a stationary quarterwave plate which create two orthogonally-polarized, colinear beams that are Doppler shifted by four times the rotation frequency. After passing through a lens that brings the beams to focus at the output plane, the two beams are separated by a polarizaton beamsplitter. The polarization of one beam is then rotated by 90° to match that of the other. The two specklegrams are

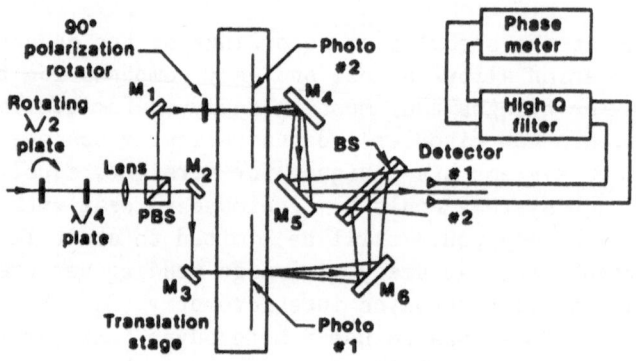

Fig. 1   Schematic Diagram of a Heterodyne Photocomparator.

mounted on a single translation stage with individual positioners
for horizontal and vertical translation and for in-plane rotation.
It is convenient to have the input beamsplitter and the mirror M_2
mounted on a common x,y translation stage as well as the output
beamsplitter and the mirror M_5.   These aid greatly in zeroing the
path lengths and centering the output beam on the input beam.

2.1 Alignment Procedures

The interferometer is aligned, initially with no specklegrams
in place and the halfwave plate stationary, so that the output beams
are parallel and colinear. Care must be taken to keep all beams in
a common plane and the two beams must be parallel as they pass
through the plane of the specklegrams. For final alignment, a
piece of ground glass may be placed in front of the input beamsplitter. The resulting fringes in the output plane may be adjusted
for best contrast and maximum broadness.

When the specklegrams are placed in the system, they may
be observed visually through the alternate side of the output
beamsplitter with a broad source of white light as an input. The
positioners may be used to align the details of one relative to
the other until the images match. This is usually sufficient to
obtain fringes in the output with laser beam as input. The positioners may then be adjusted for maximum contrast and broadness of
the fringes. The output translation stage may then be adjusted to
equalize the path length between the specklegrams and the output

beamsplitter which will remove any circular component of the fringe pattern. It will then be necessary to remove the specklegrams and zero the overall path length of the interferometer again.

2.2 Operation

For uniaxial strain measurement, two photodetectors are placed in the output field of the interferometer and aligned horizontally. The halfwave plate is set into rotation at a constant speed and the two detector outputs are fed through two matched high-Q filter circuits to an electronic counter. The counter is set to measure the time, t_1, between zero crossings of the two signals. The stage supporting the two specklegrams is translated a small amount, Δx, and the new time between zero crossings, t_2, is measured. To eliminate drift, it is helpful to return the stage to its original position and measure a third time, t_3. Let $\Delta t = 2t_2 - t_1 - t_3$. Strain in the horizontal direction may be computed by the formula

(1) $\epsilon = (\Delta t/\tau)(R\lambda/d\Delta x)$,

where τ is the period of the heterodyne signal, R is the distance from the specklegrams to the detector plane, d is the distance between detectors, and λ is the wavelength of light. $\Delta t/\tau$ is the phase change in radians between the signals. Residual fringes in the initial pattern will not affect the measurement; however, it is essential to gate the counter so that it does not add or subtract a period of the signal between measurement of t_1 and t_2. Slight rotations of the stage supporting the specklegrams will not affect the measurement since it will not change the horizontal spacing between the specklegrams. Vertical translation should not be used to measure vertical strain because slight rotations of the stage will affect the relative vertical positions of the specklegrams, owing of their lateral separation, and create apparent strain. This problem may be solved by translating the input beam rather than the plates themselves.

Two-dimensional strain can be measured by placing four detectors in the output plane, two horizontally spaced and two vertically spaced. The input beam may be sent through a thick etalon, tilting of which will translate the beam without changing its propagation direction. Phase changes between the horizontal pair of detectors and the vertical pair are measured for both horizontal and vertical beam displacements, resulting in four values. Let us denote these by a pair of subscripts, the first denoting the detector pair (x = horizontal and y = vertical) and the second denoting the beam deflection (x = horizontal and y = vertical). The components of the surface strain-rotation matrix are,

(2a) $\varepsilon_{xx} = (\Delta t_{xx}/\tau)(R\lambda/d\Delta x)$,

(2b) $\varepsilon_{xy}+\theta = (\Delta t_{xy}/\tau)(R\lambda/d\Delta y)$,

(2c) $\varepsilon_{yx}-\theta = (\Delta t_{yx}/\tau)(R\lambda/d\Delta x)$,

(2d) $\varepsilon_{yy} = (\Delta t_{yy}/\tau)(R\lambda/d\Delta y)$,

where ε_{xx} and ε_{yy} are linear strains in the x and y directions, $\varepsilon_{xy} = \varepsilon_{yx}$ is the shear strain, and θ is surface rotation.

Scanning the pair of specklegrams by translation of the input beam creates another source of apparent strain. This method actually compares, in effect, the two images of the specklegrams as seen through the output beamsplitter and the three associated mirrors. If the distance from the beamsplitter to one specklegram is not exactly the same as to the other, there will be an effective enlargement of the nearer one relative to the farther one in proportion to the ratio of the respective distances. This shows up in the output as an apparent isotropic strain. The position of the beamsplitter and mirror M_5 that minimizes this effect should be found empirically with a pair of identical specklegrams because it is not necessarily the position for maximum broadness of the halo fringes. Once known, this effect may be used to precompensate for thermal expansion of an object between a pair of specklegram recordings. Departures from flatness of the mirrors, however, can create an anamorphic distortion of the apparent specklegram images. This results in an orthotropic apparent strain that must be removed numerically from the results. With mirror flatness in the order of tenth wave, these may be kept less than ±100 µstrain.

This system can be straightfowardly adapted to computer control. The items requiring a computer interface are: (a) a tilting system for the etalon, (b) an x,z translation system for positioning the pair of specklegrams at the desired locations, (c) a switching network for the detectors, and (d) a counter for measuring the times between zero crossings. An inexpensive microcomputer will suffice to operate the equipment, log the data, and process it to obtain strain patterns. Once the system has been automated, the operator is required only to load the pairs of plates, align them, and change them after the data has been taken.

3 APPLICATIONS

One of the most attractive applications of heterodyne speckle photogrammetry is the measurement of steady-state strain on structures at high temperatures. At temperatures above 400°C, electrical strain gages face serious problems not only in maintaining good bonding to the structure, but also in large resistance changes versus temperature that depend upon the rate of temperature change. Because speckle photogrammetry does not require contact with the object surface and measures absolute changes in surface dimensions, it is immune to these problems. Only a stable surface is required from which the light may scatter to form speckles. Many refractory metals, such as Hasteloy X for example, form a self-protecting oxide layer that serves this purpose up to 850°C or more. It is possible, therefore, to make strain measurements by speckle photogrammetry where virtually no other means exists.

Speckle photogrammetry does face other problems, however, that are analogous to bonding and apparent strain. Speckle decorrelation is analogous to debonding because, if this happens, the speckles cannot be photogrammetrically compared and the ability to measure strain is lost. This can happen if, between specklegram recordings, the object moves toward the camera more than the focal depth of the lens. Using a lens system with a low numerical aperture reduces this problem. Although this reduces the subsequent halo size, and thus the potential measurement accuracy, the high accuracy of the heterodyne readout system can offset this. Strain accuracies of 10^{-5} can be obtained with apertures corresponding to f/10. Decorrelation can also occur if the object surface tilts more than half the angle subtended by the lens aperture, and this is improved if the numerical aperture of the lens is increased. Clearly, it is necessary to find a balance between these conflicting problems in any practical situation.

Anything that changes the apparent size of the recorded image can create apparent strain in speckle photogrammetry. First of all, this can happen if the object moves toward the camera lens as discussed in the section of these proceedings on Speckle Metrology. A telecentric lens may be required to eliminate this problem and it may need to be designed and built specifically for this purpose. Also, oblique observation of the surface causes it to

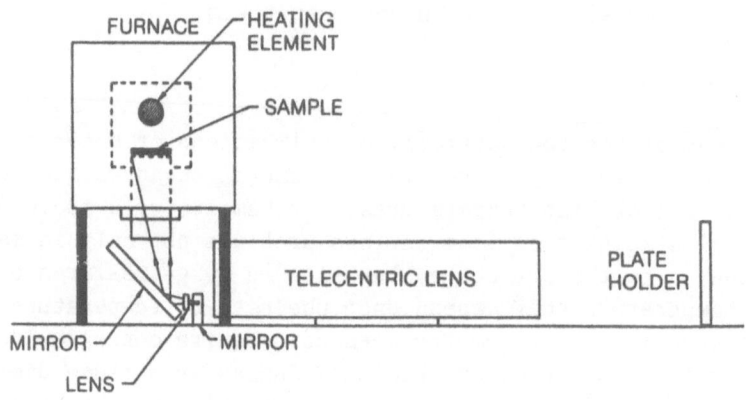

Fig. 2   Diagram of an Optical System for Recording Specklegrams of
a Sample in a Laboratory Oven.

appear foreshortened and this allows tilts to masquerade as strain.
(see also the section on Speckle Metrology) Next, the air or gas
through which the object is observed may not be homogeneous, particularly at high temperature and pressure. In these cases, turbulent
flow may create a pattern of lenslets that randomly magnify small
segments of the object surface, giving a random pattern of apparent
strain. At present, the only solution to that problem is to avoid
it. Finally, is it essential to know the surface temperature of
the object in order to separate thermal expansion from mechanical
strain. This may be obtained either via a matrix of thermocouples
or by infrared pyrometry.

In spite of the problems that face speckle photogrammetry
of high temperature objects, laboratory demonstrations have shown
it to work remarkably well.(4) Experiments have been performed on
samples of Hasteloy X heated in an oven as shown in Fig. 2. The
oven was open at the bottom and the sample sat on top of a ceramic
tube. A 45° mirror beneath the tube allowed illumination and
observation of the lower surface of the sample. A laser beam
entered the system perpendicular to the plane of the diagram, and
was directed by a small mirror through a small negative lens to
the larger mirror so that it illuminated the sample. The sample
was observed via a telecentric lens system, which formed an image
at the plane of a photographic plate. The vertical column of air
through which the sample was observed was hottest at the top and

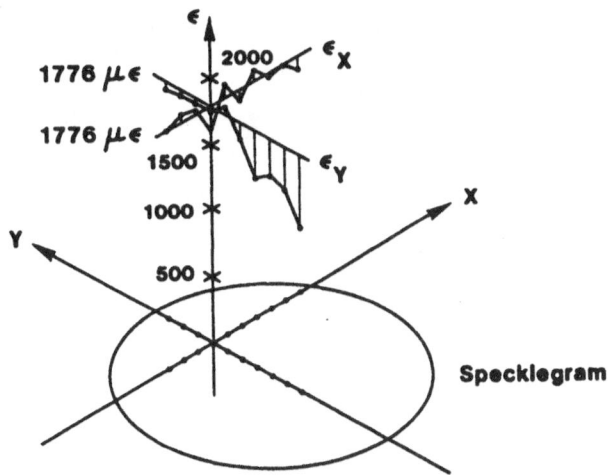

Fig. 3. Isometric Plot of a Thermal Strain Distribution. Two specklegrams of the sample at 883°C and 772°C were compared.

therefore free from convection. Essentially all apparent strains due to thermal gradients were eliminated by this configuration.

The initial experiment consisted of recording two speckle-grams at different temperatures to simulate an isotropic strain by means of thermal expansion. The chosen temperatures were 883°C and 772°C and the thermal expansion of Hasteloy X at those temper-atures should generate an apparent strain of 1.776×10^{-3}. Linear strains were measured at ten locations along each of two axes, which are indicated as x and y in Fig. 3. The sample is shown in isometric projection with the strains plotted in the z direction. A second set of axes, ε_x and ε_y, are drawn parallel to x and y at the height corresponding to 1.776×10^{-3} strain. The measured strains are plotted with respect to those axes above their corresponding locations.

The strains measured by heterodyne photogrammetry are within five to ten percent of the correct value except at the center of the sample. Analysis of the sample configuration in the oven indicated that the center of the sample may have been cooler than the edges (where the temperature was measured) so that this pattern of apparent strain may not be an error. The capacity of the technique to obtain strain patterns in two dimensions at high temperatures is demon-strated by these results.

528

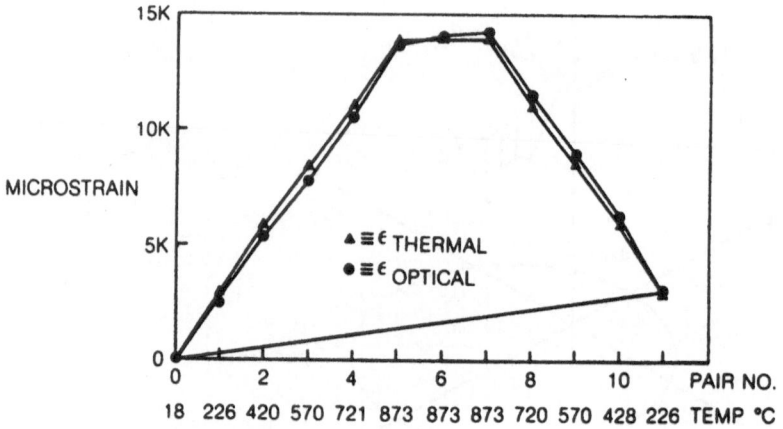

Fig. 4. Thermal Expansion History for an Unconstrained Sample of
Hastelloy X. Measurements via heterodyne photogrammetry are compared
with predictions from temperature measurements.

Further experiments were performed to demonstrate the
capacity of the technique to measure large excursions of strain by
a sequence of specklegram recordings. A sequence of 12 specklegrams
were recorded of the sample while its temperature was increased
stepwise from ambient to 873°C and back to 226°C. Linear strain
was measured at the center of the sample in the x and y directions
and the average was obtained in each case over a 5 mm span. Successive pairs of photographs were compared: #1 to #2, #2 to #3, etc.
and finally #12 to #1. In this way, any accumulated errors would
be revealed.

The results are plotted in Fig. 4. The x and y strains
are unresolved at this scale and both lie quite close to the predicted values. A systematic error is noticeable which is negative
during temperature increase and positive during decrease. This
was attributed to the fact that the thermocouple was not in direct
contact with the actual sample measured. Subsequent repetitions
of the experiment did not show this error when the thermocouples
were welded to the sample.

The experiment was repeated with a bimetalic system of a
Hastelloy X sample whose ends were welded to a block of alloy with
a slightly different expansion coefficient. Heating this system
induced mechanical strain into the Hastelloy X. The results are

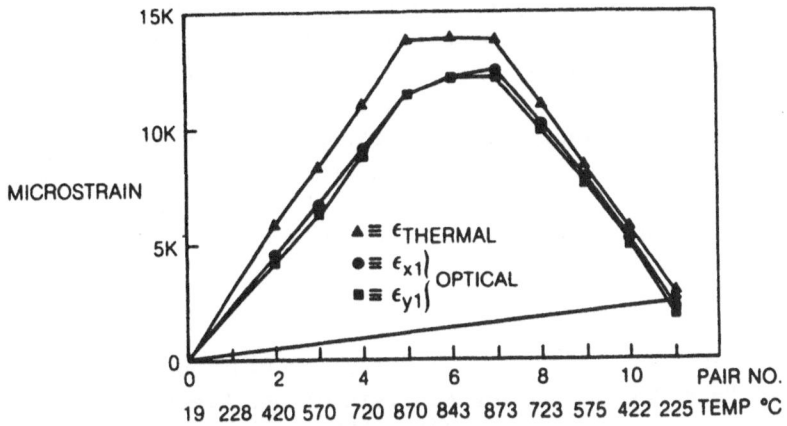

Fig. 5. Expansion History of Hastelloy X in a Bimetalic System.

shown in Fig. 5. Here the x and y strains are resolvable and both lie well below what was predicted due to thermal expansion.

Figure 6 shows a plots of the data from the two experiments with the thermal expansion removed. The results should show only the mechanical strain on the sample.

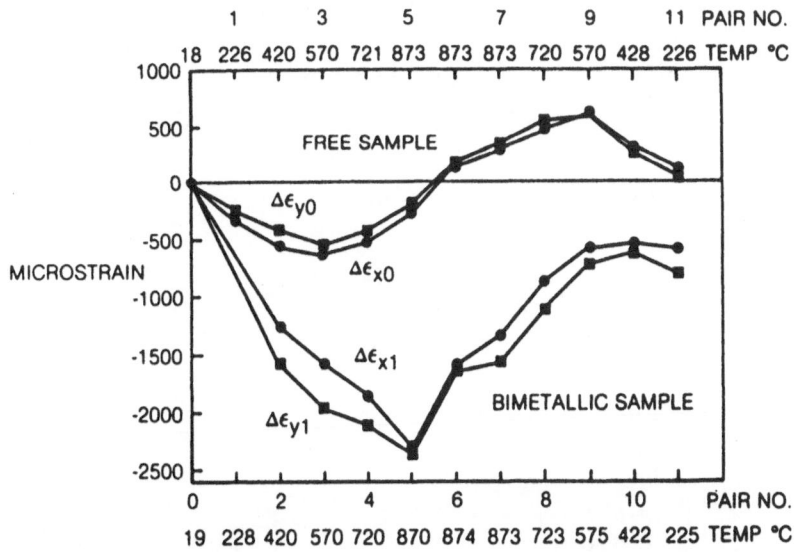

Fig. 6. Mechanical Strain of the Constrained and Unconstrained Samples.

4 CONCLUSION

The use of heterodyne interferometry to read fringe data from specklegram halos increases the accuracy with which this data may be obtained and make it easily available to numerical data processing systems. This results in a system for optical strain measurement that has good accuracy, extended range, and is tolerant to a considerable variety of environmental conditions. Steady-state strain distributions can be measured on object surfaces at temperatures higher than obtainable by electrical gages.

References

1. Kaufmann, G.H., A.E. Ennos, B. Gale, and D.J. Pugh. An Electro-Optical Read-Out System for Analysis of Speckle Photographs. J. Phys. E: Sci. Instrum., 13(1980)579-584.
2. Adams, F.D., and G.E. Maddux. Dual Plate Speckle Photography. AFFDL-TR-75-92 (Wright-Patterson Air Force Base, Ohio 45433).
3. Dändliker, R. Heterodyne Holographic Interferometry, in Progress in Optics XVII, E. Wolf, ed. (North-Holland, 1972), Chap. 1.
4. Stetson, K.A. The Use of Heterodyne Speckle Photogrammetry to Measure High-Temperature Strain Distributions. Proc. SPIE, 370(1983)46-55.

MEASURING MICROVIBRATIONS BY HETERODYNE SPECKLE INTERFEROMETRY

R. Dändliker and J.-F. Willemin

Institut de Microtechnique de l'Université,
CH-2000 Neuchâtel, Switzerland.

In-plane as well as out-of-plane displacements and vibrations of objects with diffusely scattering surfaces are measured in real time. Microvibrations with amplitudes down to 1 nm and frequencies up to 5 MHz were analyzed at a spatial resolution of 35 µm.

1. INTRODUCTION

For vibration analysis, the knowledge of amplitude and phase is essential to characterize the mechanical movement (which is supposed to be sinusoidal) completely. In particular, the behavior of the phase in the neighborhood of a minimum of the amplitude is important to verify the nodes of the vibration mode. The relative phase of the vibration components in plane and out of plane is also indispensable for the correct reconstruction of the three-dimensional movement of the object surface.

An opto-electronic system using real-time heterodyne speckle interferometry for microvibration analysis of objects with diffusely scattering surfaces has been developed [1-3]. With this system, in-plane as well as out-of-plane displacements and vibration amplitudes can be measured in the same optical arrangement. With appropriate electronics the vibration amplitude and the corresponding phase can be measured. Experimental results are shown to demonstrate the capabilities of this powerful tool for microvibration analysis.

Soares, O.D.D. (ed), Optical Metrology
© *1987. Martinus Nijhoff Publishers, Dordrecht.*

2. PRINCIPLE OF OPERATION

The optical arrrangement of the real-time heterodyne speckle interferometer is sketched in Fig.1. It is easily converted from in-plane [Fig.1(a)] to out-of-plane [Fig.1(b)] measurements.

For in-plane measurements, the diffusely scattering surface of the object is simultaneously illuminated by two focused laser beams arriving symmetrically with respect to the surface normal. The light scattered from the common spot of these two beams on the object surface is collected in an arbitrary direction and imaged onto the photodetector D_1. The corresponding aperture is chosen so that at least 100 speckles are picked up by the detector. This is necessary to assure sufficiently constant average optical power when the spot moves from one point to another on the surface. This setup is similar to those used in laser Doppler velocimetry [4] or in speckle interferometry of in-plane displacement [5].

Two acousto-optical modulators M_1 and M_2 (one in each beam) introduce the necessary frequency offset for the heterodyne detection. They are driven with frequencies ω_1 and ω_2, respectively, so that the net frequency shift between the two illuminating beams is $\Delta\omega = \omega_1 - \omega_2$. The driving frequencies $\omega_1/2\pi$ and $\omega_2/2\pi$ are independently adjustable by two frequency synthesizers in steps of 1 kHz between about 35 and 45 MHz to select any desired $\Delta\omega/2\pi$ up to about 10 MHz.

The output of detector D_1 is then given by

$$I_1(t) = a_1 + b_1 \cos(\Delta\omega t + \phi_x + \phi_1), \tag{1}$$

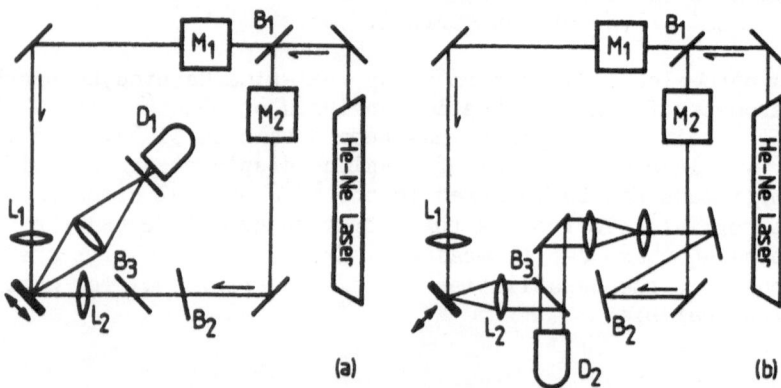

Fig.1. Optical arrangement for measuring (a) in-plane and (b) out-of-plane displacements.

where ϕ_x accounts for the in-plane displacement and ϕ_1 is constant for a fixed observation point but changes arbitrarily from point to point on the surface. The in-plane displacement u_x is obtained by measuring the change of the phase ϕ_x of the ac signal at the beat frequency $\Delta\omega$ and using the relation

$$\phi_x = (4\pi/\lambda) u_x \sin \alpha = \beta_x u_x, \tag{2}$$

where 2α is the mutual angle of the illumination directions [(Fig.1(a)] and λ is the laser wavelength.

For out-of-plane measurements [Fig.1(b)] the diffusely scattering surface of the object is illuminated only by one of the previously used laser beams, whereas the second focusing lens L_2 collects the light scattered off the surface, symmetrically to the surface normal. This scattered light is brought to interference with a plane reference wave by the beam splitter B_3 and observed by the photodetector D_2. The aperture is again chosen to contain at least about 100 speckles. This setup is similar to conventional speckle interferometry. However, to obtain a reasonably good average signal from all the statistically independent individual speckles, the average wave fronts of the scattered light and the reference wave have to be adjusted carefully. The frequency offset for the heterodyne detection is introduced in the same manner as previously by the two modulators M_1 and M_2. The output of detector D_2 is then given by

$$I_2(t) = a_2 + b_2 \cos(\Delta\omega t + \phi_z + \phi_2), \tag{3}$$

similar to Eq.(1).

The out-of-plane displacement u_z is again found by a phase measurement from

$$\phi_z = (4\pi/\lambda) u_z \cos \alpha = \beta_z u_z. \tag{4}$$

Changing from in-plane to out-of-plane measurements can be accomplished without moving any optical element simply by blocking appropriately the light paths not used in either system. Thus the measurements can be performed nearly simultaneously, and the observed point is guaranteed to be the same.

3. AMPLITUDE MEASUREMENT

In case of vibration analysis, the displacements are periodic functions in time of the form

$$u(t) = u_o \cos(\Omega t + \psi). \tag{5}$$

534

This leads to frequency-modulated output signals from detectors D_1 and D_2 with the carrier frequency $\Delta\omega$, the modulation frequency Ω and the modulation depth

$$\delta\omega = \Omega\phi_o = \Omega\beta u_o. \tag{6}$$

For both cases, the ac component of the photocurrent is thus given by

$$I(t) = b \cos[\Delta\omega t + \phi + \beta u_o \cos(\Omega t + \psi)], \tag{7}$$

where u_o, Ω and ψ are the amplitude, the frequency, and the phase of the vibration, respectively; ϕ is a constant phase for a fixed observation point but changes arbitrarily from point to point on the surface.

The corresponding rf spectra for different vibration amplitudes are shown in Fig.2. Such signals can be evaluated by the well-known frequency-modulation-demodulation techniques as long as $\delta\omega < \Delta\omega$. Small vibration amplitudes ($\beta u_o < 1$, i.e. $u_o \ll \lambda$) are readily found from the power ratio of the carrier frequency (P_o) and the first sideband (P_1) by inversion of the relation

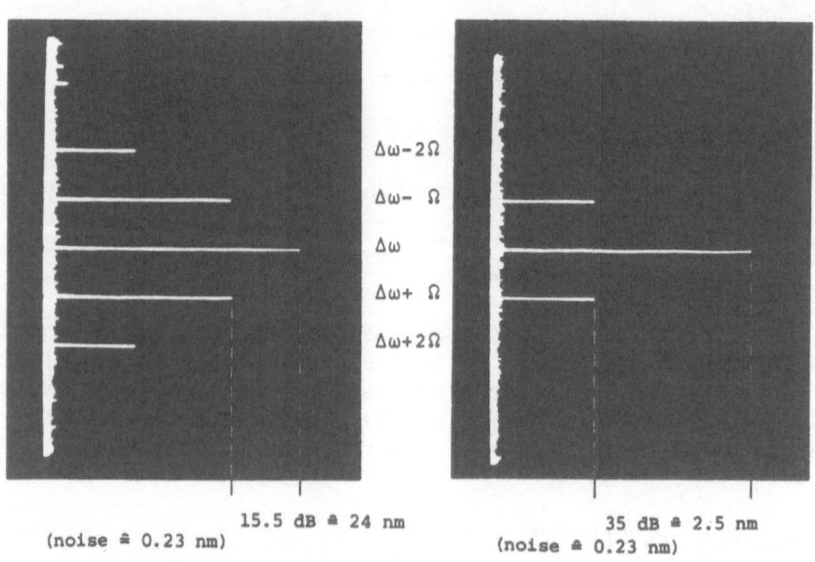

$\Delta\omega-2\Omega$

$\Delta\omega-\ \Omega$

$\Delta\omega$

$\Delta\omega+\ \Omega$

$\Delta\omega+2\Omega$

15.5 dB $\triangleq$ 24 nm   35 dB $\triangleq$ 2.5 nm
(noise $\triangleq$ 0.23 nm)   (noise $\triangleq$ 0.23 nm)

Fig.2. Two typical photocurrent spectra (log scale, 10 dB/div) for sinusoidal motion of the object. The central line corresponds to the beat frequency $\Delta\omega = \omega_1 - \omega_2$. ($\Delta\omega/2\pi = 5$ MHz). The sidebands are produced by the object vibration ($\Omega/2\pi = 1.1$ MHz).

$$J_1(\beta u_o)/J_0(\beta u_o) = (P_1/P_o)^{1/2}. \tag{8}$$

where J_0 and J_1 are Bessel functions. Practically, P_o and P_1 are measured with a spectrum analyzer, as shown in Fig.3. The sensitivity is essentially limited by the signal-to-noise ratio for the first sidebands. For shot-noise-limited detection, the minimum-detectable vibration amplitude is found to be

$$(u_o)_{min} = 2/\beta(SNR)^{1/2}, \tag{9}$$

where SNR is the signal-to-noise ratio for the carrier. For the experimental setup used, this limit falls below 0.14 nm (SNR > 60 dB) with a 2-mW He-Ne laser (λ = 633 nm), a detection bandwidth of 1 kHz, a spot size (spatial resolution) of d = 35 μm and an angle of α = 45°. The results of vibration measurements presented in Fig.2 are in good agreement with the predicted limits.

4. PHASE MEASUREMENT

With coherent or synchronous detection, it is possible to translate the spectrum of I [Fig.4(a)] to the origin [Fig.4(b)] so that the component at Ω, containing the wanted phase ψ, can be separated from the carrier. As shown in Fig.3, the detector signal

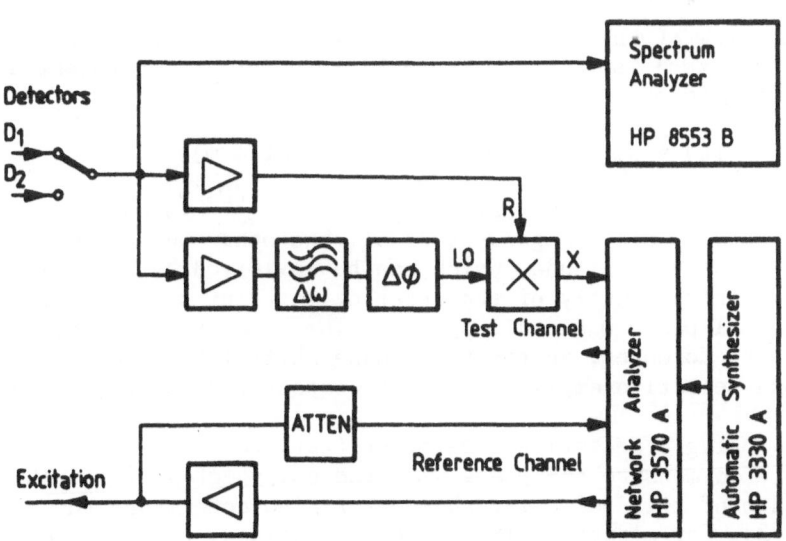

Fig.3. Block diagram of the electronic setup for amplitude and phase detection.

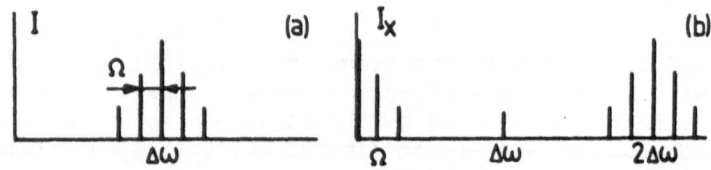

Fig.4. (a) Typical photocurrent spectrum for sinusiodal motion of the object (beat frequency $\Delta\omega$, vibration frequency Ω). (b) Spectrum of the mixer output signal.

I is multiplied after wideband amplification by a locally generated signal I_{LO}, which is phase synchronized to the carrier of I:

$$I_x(t) \propto [I(t) \times I_{LO}(t)]. \tag{10}$$

Practically, the local-oscillator (LO) signal is derived from the detector I by a selective amplifier tuned to the carrier frequency $\Delta\omega$. The bandwidth of this amplifier must be narrow enough ($<\Omega$) to block the other components of I. As will be seen below, I_{LO} must also be dephased by a controllable phase shift $\Delta\phi$. From Eq.(7) one finds that

$$I_{LO}(t) \propto J_o(\beta u_o) \cos(\Delta\omega t + \phi + \Delta\phi). \tag{11}$$

The spectrum of the resulting signal I_x (at the output of the balanced mixer) is given in Fig.4(b), and for its component at Ω one gets, from formulas (7), (10), and (11),

$$I_x^{(\Omega)}(t) \propto J_o(\beta u_o) J_1(\beta u_o) \sin\Delta\phi \cos(\Omega t + \psi). \tag{12}$$

Formula (12) shows that the phase of $I_x^{(\Omega)}$ is equal to the wanted phase ψ and independent of all the other phases. The phase $\Delta\phi$ appears in the amplitude, which can be maximized by adjusting $\Delta\phi$ near $\pi/2$ independently of the observation point because the unknown statistical phase ϕ has disappeared. The exact value of $\Delta\phi$ is not important. Moreover, an eventual phase shift introduced by the previous selective amplifier can be compensated for through $\Delta\phi$.

By using a selective phasemeter (network analyzer), it is possible to measure the phase ψ of the component Ω of I_x. The bandwidth of this selective phasemeter also determines the noise level [3]. For electrically excited vibrations, the phase of the exciting signal can be used as phase reference, as shown in Fig.3.

Some general remarks: (a) The phase of the vibration can also be measured with respect to the vibration at another point on the object surface; twin optics and a second detector-mixer unit are then required. (b) Another method to generate the LO signal consists of generating the carrier frequency $\Delta\omega$ electronically by mixing the driving frequencies of the two acousto-optical modulators. The disadvantage is that $\Delta\phi$ must be readjusted for every observation point to compensate for ϕ. Nevertheless, this method is useful for larger amplitudes, when the carrier in the detector signal vanishes, i.e., for $J_o(\beta u_o) = 0$. (c) Instead of a phasemeter, a lock-in amplifier (synchronous detector) can be employed as the phase-sensitive detector.

5. EXPERIMENTAL RESULTS

Amplitude and phase measurements were successfully verified with quartz resonators (32-kHz) having a known vibration mode for in-plane as well as out-of-plane movements (Fig.5). For reasonable signal-to-noise ratios of the first sideband P_1 [cf. Eq. (8), SNR(P_1) > 10 dB], the measurement accuracy is essentially given by the spectrum analyzer (±6%) for the amplitude and by the network analyzer (±4°) for the phase [6].

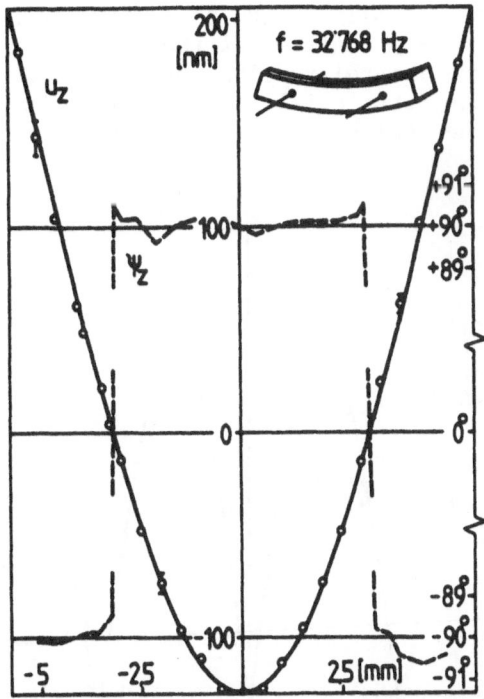

Fig.5. Fundamental bending mode (32 kHz) of a slab quartz resonator (length 11.6 mm). Solid line: theoretical amplitude with measured points (out-of-plane movement of top surface). Dashed line: measured phase (note the expanded scale).

538

The capability of this system was also demonstrated with a piezoelectric ceramic disk (diameter, 25.4 mm; thickness, 2 mm; full electrodes)[6].Figure 6 illustrates the excellent agreement between the theoretical mode shape of the third radial resonance (339 kHz) and the experimentally measured in-plane vibration amplitude along a diameter of the disk. For the second radial resonance (217 kHz) amplitude and phase profiles along a diameter have been measured for both in-plane (radial) and out-of-plane (axial) displacements (Fig.7). The in-plane distribution, Fig.7(a), corresponds to a pure radial mode, and the phase inversion denotes zero crossings of the vibration amplitude (nodal line). The complexity of the radial-axial coupling is well illustrated by comparing Figs.7(a) and 7(b). The observed behavior of the normal displacement (out-of-plane movement) is in good agreement with capacitively measured vibration amplitudes [7]. The superposition of the measured vibration components (Fig.7) allows one to reconstruct the three-dimensional movement of the object points, which in general is of elliptical shape, since one knows the relative phase of the components.

To get information about the resonance behavior of the piezo-electric transducer, amplitude and phase have been measured at the center of the disk as a function of the excitation frequency [6]. Figure 8 shows experimental results for the out-of-plane (axial) movement around the first axial resonance (approx 1 MHz). These results confirm the well known complex behavior of axial resonances

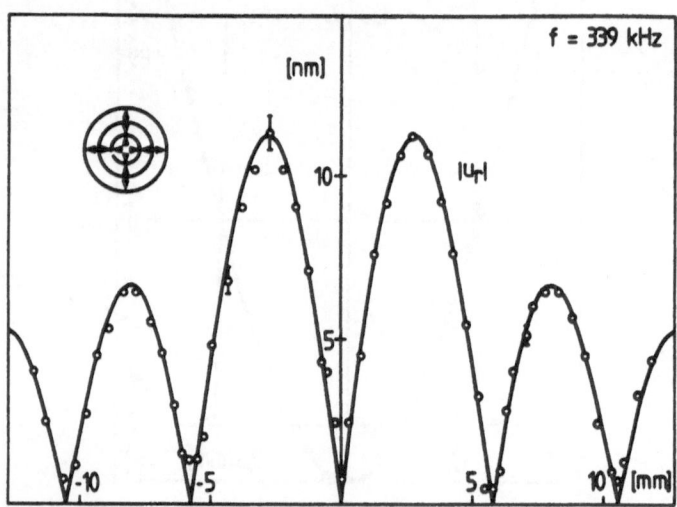

Fig.6. Comparison of the theoretical mode shape (solid line) and the in-plane amplitudes measured along a diameter of a piezoceramic disk excited at the third radial resonance (339 kHz).

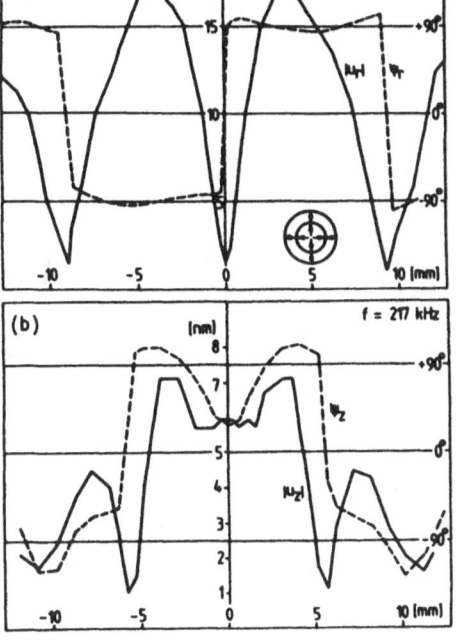

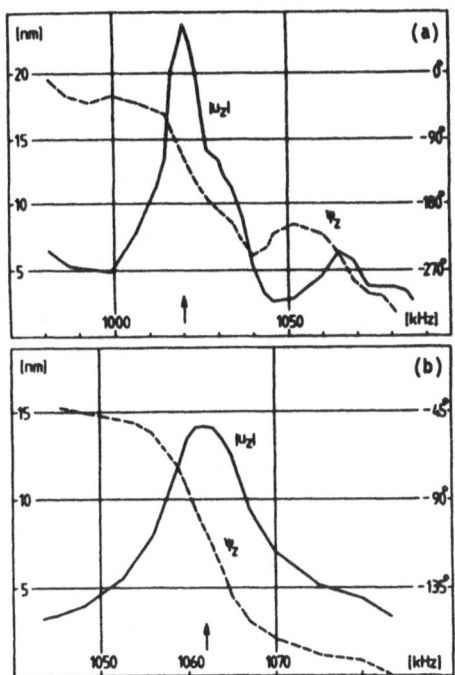

Fig.7. Amplitude (solid line) and phase (dashed line) for (a) in-plane and (b) out-of-plane displacement components measured along a diameter of a piezoce-ramic disk excited at the second radial resonance (217 kHz).

Fig.8. Amplitude (solid line) and phase (dashed line) of the out-of-plane movement measured at the center of piezoceramic disks as a function of excitation frequency around the first axial resonance. (a) full elec-trodes. (b) partial electrodes.

of piezoelectric disks with full electrodes, Fig.8(a), compared to disks with partial electrodes, [Fig.8(b)],which are designed for energy confinement.

6. FM-DEMODULATION AT LOW FREQUENCIES

FM-demodulation of the detector output, Eq. (7), would produce a signal which is proportional to the instantaneous speed of the mechanical movement versus time, as it is well known from laser Doppler velocimetry. For vibration frequencies in the audio-range (20 Hz to 20 kHz), such demodulators are readily available for carrier frequencies in the range of 87 MHz to 104 MHz in the form of standard FM-receivers. A suitable carrier frequency is easily obtained by using the optical setup shown in Fig.1 with the two acousto-optical modulators adjusted to give opposite frequency shifts of 44 MHz each, yielding an 88 MHz beat frequency. Note that

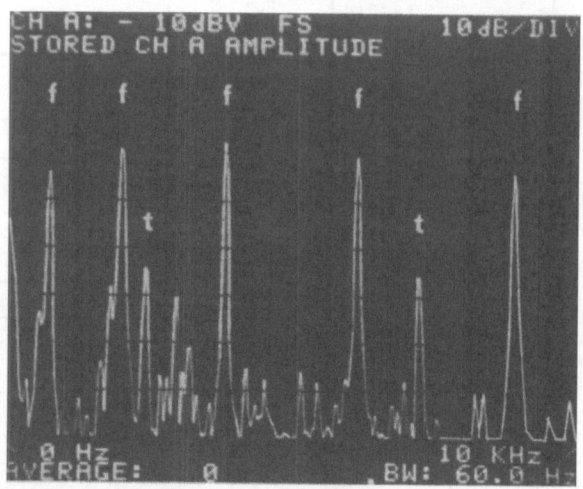

Fig.9. Vibration spectrum (0 - 10 kHz) of a clamped cantilever beam excited by shock. Bending modes (f): 720, 1980, 3800, 6160 and 8850 Hz. Torsional modes (t): 2380 and 7150 Hz.

FM-receivers are usually equipped with a deemphasis circuit which reduces the response above 3 KHz, so that the output becomes proportional to the displacement rather than to the velocity. The exact response of the FM-receiver over the whole audio-frequency range should be established first by calibration with a known input [6].

An example of experimental results obtained with a Revox A 76 FM-tuner followed by a FFT (fast Fourier transform) spectrum analyzer (HP 3582 A) is shown in Fig.9 [6]. It shows the vibration spectrum of a cantilever beam (steel: length 300 mm, cross-section 12.9 mm x 10.0 mm), clamped at one end and excited by shock. Five bending modes (f) and two torsional modes (t) are identified between 0 kHz and 10 kHz. The detection bandwidth was 60 Hz and the sensitivity for the amplitude measurement was still shot-noise limited, namely about 0.05 nm in this case.

REFERENCES

1. Dändliker, R. and J.-F. Willemin, Measuring Microvibrations by Heterodyne Speckle Interferometry, Proc.Soc.Photo-Opt. Instrum.Eng. vol. 236 (1980) 83-85.
2. Dändliker, R. and J.-F. Willemin, Measuring Microvibrations by Heterodyne Speckle Interferometry, Opt.Lett. 6 (1981) 165-167.
3. Willemin, J.-F. and R. Dändliker, Measuring Amplitude and Phase of Microvibrations by Hetrodyne Speckle Interferometry, Opt.Lett. 8 (1983) 102-104.

4. Leendertz, J.A. Interferometric Displacement Measurements on Scattering Surfaces Utilizing Speckle Effects, J.Phys.E 3 (1970) 214-218.
5. See e.g., Ennos, A.E. Speckle Interferometry, in Pogress in Optics, vol. XVI, E. Wolf, ed. (North-Holland, Amsterdam, 1978), pp.235-259.
6. Willemin, J.-F. Interférométrie Hétérodyne de Speckles: Application à la Mesure de Vibrations Mécaniques Microscopiques, Ph.D.thesis (University of Neuchâtel, 1983).
7. Lypacewicz, G. and L. Filipczynski, Vibration of Piezoelectric Ceramic Transducers Loaded Mechanically, Proc.Vibration Problems, 11 (1970) 283-299.

Electronic Speckle Pattern Interferometry

Ole J. Løkberg

Physics Department
The Norwegian Institute of Technology
N-7034   TRONDHEIM-NTH
NORWAY

1.   INTRODUCTION

Historically, the concept of replacing the photographic emulsion by video recording and display occured to several independent groups almost simultanously. Macovsky(1) in USA, Schwomma(2) in Austria and Köpf(3) in West-Germany considered the technique to be based on pure holography. The Lougborough group in England, headed by Butters and Leendertz(4) viewed the technique more as an off-spring of their speckle work and they also introduced the name ESPI as an abbreviation for electronic speckle pattern interferometry. Although the ESPI-name might be considered appropriately descriptive, its association with coarse, annoying speckles has undoubtedly hampered the general acceptance of the technique. Video holography might be a name with better sales appeal, but might be confused with techniques for direct transmission of three-dimensional holographic images(5).

The author will here follow the holographic approach, as

the equivalence between most ESPI-techniques and holographic
interferometry is obvious and most helpful for the basic
understanding (provided the holographic process is fairly well
understood). The presentation will be directed towards the practical
use of ESPI avoiding any lengthy mathematical developments. More
in-depth treatment of the technique can be found in the recent
book by Jones and Wykes (6). In addition there will be a chapter
on ESPI by Løkberg and Slettemoen(7) in a forthcoming volume
in the "Applied Optics and Optical Engineering" series.

2. THE WORKING PRINCIPLE OF ESPI

2.1. ESPI vs. hologram interferometry

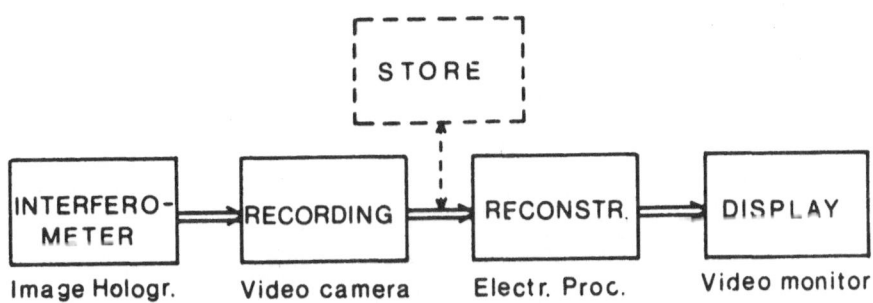

Fig. 1. The build-up of an ESPI

Fig. 1 represents the main building blocks in an ESPI set-
up. Briefly explained,the optical part of the ESPI consists
of an image holography set-up with an in-line reference wave.
The target of the video camera replaces the photographic film
as recording medium. The reconstruction step of holography is
performed by electronic processing on the video signal from the
camera. The reconstructed image is thereafter displayed on the
video monitor. An intermediate video store is used whenever
the difference between separate video frames is desired.

544

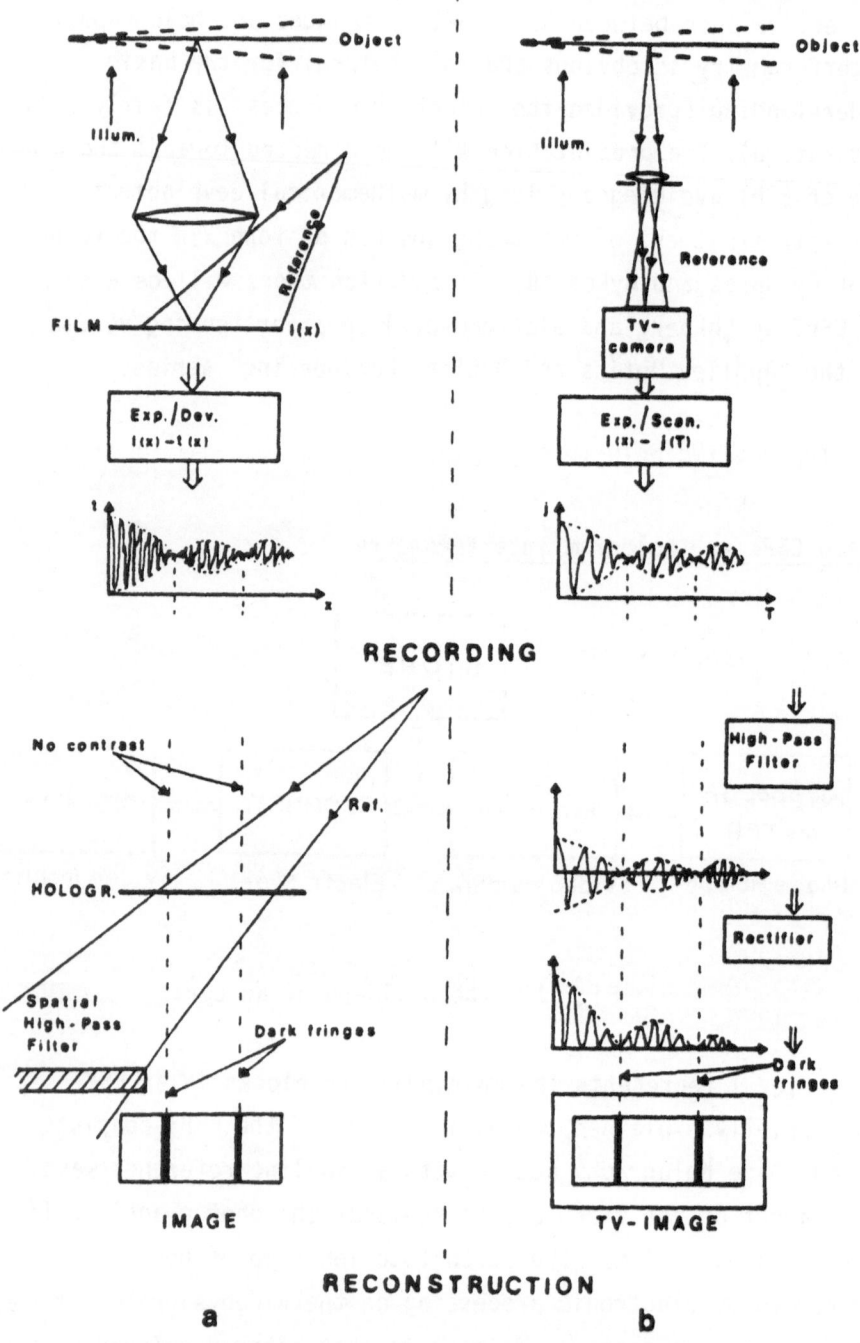

Fig. 2. The recording and reconstruction steps of (image) holography (a) and ESPI (b).

To give a better feeling for the contents and purpose of
the various blocks in fig. 1, we will in the following refer
to fig. 2, where the various steps of (image) hologram interferometry
and ESPI are lined up against each other.

The optical part of the recording step is essentially the same
for both techniques. However, the resolution of the video camera
is more than two decades below that of the commonly used holographic
film (30 1/mm vs. 3000 1/mm). The spatial frequency f of the
primary interference fringes (or the holographic information
carrier), is approximately proportional to the angle θ between the
interfering waves ($f \sim \theta/\lambda$, where λ is the laser wavelength).
In the holographic system we are free to use an off-axis
configuration with arbitrarily large θ. In ESPI we are forced
to use an in-line object-reference wave system where the aperture
of the lens in addition is closed down to keep the interfering
angle below 1°. (Note the close similarity between the optical
part of an ESPI and a visual speckle interferometer(8,9)).

The recording mechanism in holography is a film grain
exposure which after (chemical) development results in a
proportional amplitude (or phase) transmittance of the hologram.
In ESPI the exposure results in a charge distribution across
the target of the video camera, which is subsequently converted into
a proportional current variation by the TV scanning.

If the object has not moved during the exposure, both methods
record a high contrast (primary) interferogram. Any movement
of the object shifts the primary fringes resulting in a smoothing
of the integrated intensity. We get local contrast variations
related to the movement as shown on fig. 2 for an object vibrating
like a bar.

The purpose of the reconstruction in both techniques is
to convert the variation of contrast into a corresponding variation
of intensity which we will observe as secondary fringes.

In the optical reconstruction process the holographic images
are separated from the direct wave by diffraction due to the
complex grating representing the primary fringe pattern. The
diffraction process corresponds to spatial high pass filtering.
Finally we observe the intensity of the diffracted object wave
which corresponds to a square law rectification. The diffraction
efficiency is proportional to the fringe contrast of the primary
grating. Therefore we observe across the object image an intensity
variation which represents the secondary fringes of hologram
interferometry.

The ESPI reconstruction is a similar process except that
we act upon a video current in the time domain. As depicted
in fig. 1 the electronic high pass filter removes the DC-term,
while the full wave rectifier flips the polarity of the negative
part of the signal. The processed video signal is fed into the
video monitor where we see the object image covered with the
same bright and dark fringes as in hologram interferometry except
that the image has a coarser speckle structure as shown on fig. 3 a-b.

By direct comparison between the two recordings on fig. 3,
the spatial characteristics of the holographic image is superior
to the ESPI image. It is, however, the favourable temporal
properties which make ESPI an interesting tool for interferometric
measurements. By ESPI, 25-30 (European-American video standards)
holographic interferograms are recorded and reconstructed each
second. These interferograms which already are available as
electronic signals, may be analyzed in real-time or when required,
easily stored by video tape for later retrieval and analysis.

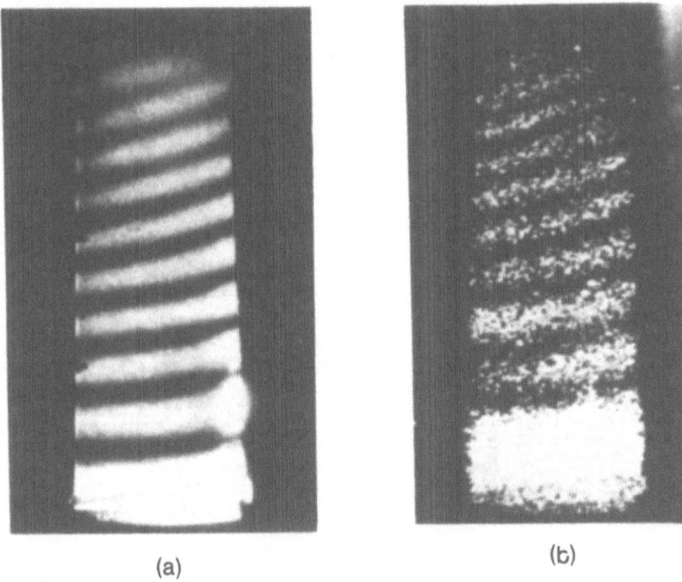

(a) (b)

Fig. 3. Holographic (a) and ESPI (b) recording of a vibrating
turbine blade

In addition, ESPI adds some new features to the holographic
techniques as complete video images, and not only the cross
interference part, can be subtracted from each other.

2.2. Basic ESPI Set-ups

All interferometric techniques are based on detecting the
phase change of one wavefront (simple or complex) relative to
another wavefront. The information in ESPI is therefore given
by the cross interference between the object- and reference wave.
As in hologram interferometry, the self interference or the
object/reference speckle pattern can be considered as optical
noise which should be filtered out or suppressed. (This is in
contrast to pure speckle techniques where modifications of the
self interference term are of interest).

We will now discuss how the basic ESPI set-up enhances the
cross interference information within the limited resolution

of the video system. Again for a more exhaustive discussion, the reader should refer to ref. 6 and 7.

Specular Reference ESPI

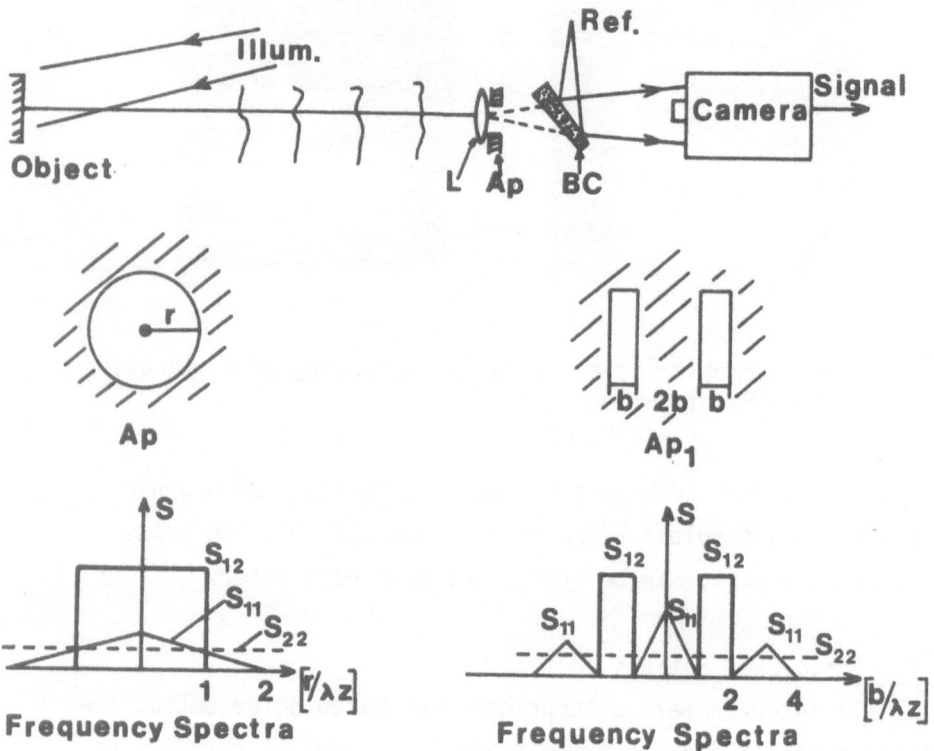

Fig. 4. Interferometric part of a specular reference ESPI with resulting frequency spectra.

This is the most commonly used set-up which also was used by all pioneering groups previously mentioned. A schematic of the interferometric part of this set-up is shown on fig. 4. Here the illuminated object scatters light diffusely and is imaged by a lens L with aperture Ap through the beam combiner BC down to the video camera. The spatially filtered reference wave is directed towards the object image by the beam combiner. (Note, for simplicity the beam combiner is represented by a glasswedge in fig. 4. More sophisticated solutions like e.g. a mirror-pinhole combination described in detail in ref. 6 usually produces a cleaner reference wave on the target).

The spatial frequency spectra for a circular aperture due to the interference between the object and reference wave is shown on the left graph on fig. 4. Here S is the DC-component mainly due to the reference wave intensity, while S_{22} represents light scattered by the reference wave. S_{11} is due to object self interference or simply the object speckle wave. Finally, S_{12} represents the information carrier or the cross interference between the object and reference wave. We see from the graph that the S_{11} and S_{22} terms will always overlap the S_{12}-term to produce background noise in the image. However, by adjusting the intensity ratio and the total intensity of the two waves we can enhance S_{12} relative to S_{11} and S_{22}. (A trick well known from holography). Another important parameter is the size of the aperture. Its diameter D determines the highest frequencies in S_{12} and S_{11}. If we suppose the resolution limit of the video target to be ν_c (1/mm), we find that no signal contributes to S_{12} at aperture diameters larger than:

$$D = 2\lambda z \nu_c \tag{1}$$

where: z is distance from exit pupil to video target
          and λ is the laser wavelength.

The optimal aperture diameter will be close to this value, but never larger as this increases S_{11}. If we suppose $\nu_c \sim 30$ 1/mm and $\lambda \sim 0.6$ μmm we find the effective f-number, F_{eff}, on the image side to be approximately 30. Note that this f-number is

referred to the image side, and does not necessarily represent
the actual f-number of the lens.

The video signal is thereafter filtered to pass an optimum
content of the cross interference spectrum. Usually a bandpass
filter will be used to remove the lower frequencies where S_{11}
is strongest ($\sim < 0.5$ MHz) while electronic noise may be reduced
by a filtering above 5 MHz.

Biedermann et al.(10) have replaced the circular aperture
by a double slit aperture which is equivalent to off-axis
holography. The resulting frequency spectra is shown on the
right graph on fig. 3. We see that by appropriate bandpass
filtering the self interference term S_{11} will be completely
suppressed. Thus this methods may give higher fringe contrast
than for a circular aperture. Due to the narrow band filtering,
however, we gain this contrast both at the expense of resolution
along the scan direction and at the expense of laser power(11).

Speckle Reference ESPI

This particular set-up was first described by Slettemoen(12)
and is drawn schematically in fig. 5 together with the resulting
frequency spectra. Here the (speckled) image fields from two
objects O_1 and O_2 overlap and interfere on the video target.
The two waves pass through different optical apertures which
makes it possible to separate the self interference terms S_{11}
and S_{22} from the cross term S_{12}. As shown by the front view
of the aperture element on fig. 5, light from object 1 is
transmitted through the transparent stripes (t), while light
from object 2 is reflected from the mirror stripes (r). The
transparent and reflective stripes are isolated by opaque
stripes (o) of the same width. The opaque stripes are introduced
to separate the different spectra as witnessed on the lower graphs

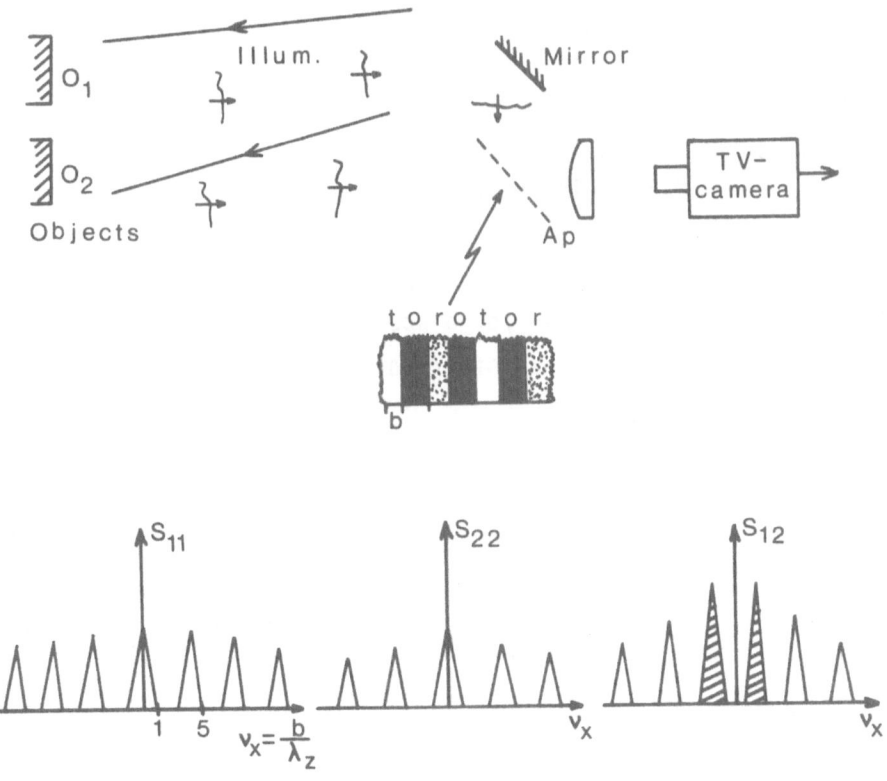

Fig. 5. Speckle reference ESPI with its resulting frequency spectra.

in fig. 5. We see that by using an electronic bandpass filter
which transmits the hatched parts of the cross interference
spectra S_{12} we completely exclude noise from the two self
interference spectra (as in double slit method).

We should note that this set-up (and the next one to be
described) does not have the same restrictions on the imaging
aperture as the previous set-ups. In a speckle reference ESPI
both waves are distributed throughout the entire exit pupil

plane. Therefore the camera can detect resolvable interference patterns from all parts of the pupil. The aperture may be opened until the saturation level of the video camera is reached. At low light levels the effective f-number may be decreased to 3-4 compared to about 30 for the specular reference ESPI. However, due to scattering losses in the reference object and the reduced bandwidth, the overall light gain compared to a specular reference ESPI is hardly noticable. Also the speckle structure is more coarse and contrasty(13) than for a specular reference ESPI. However, this set-up possesses other interesting properties to be discussed later.

Double illumination ESPI

The optical part of this set-up was first described by Leendertz(14) and later adapted to TV-camera registration(15).

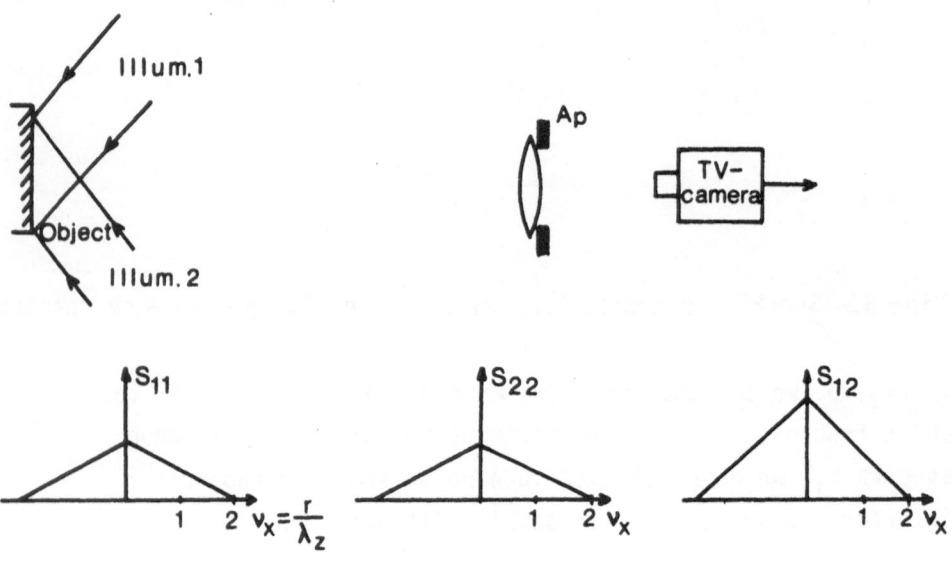

Fig. 6. Double illumination ESPI and its frequency spectra.

As shown on fig. 6 the object is (symmetrically) illuminated
by two (plane) waves. The object is imaged onto the video camera
without any separate reference wave being introduced. However,
if we want to use holographic terms we might consider the scattered
waves due to illumination waves 1 and 2 as the object- and reference
waves respectively. As both waves pass the same optical aperture
the self interference and the cross interference have identical
frequency spectrum. Consequently we can neither suppress the
self interference terms by adjusting the intensity ratio, nor
separate the terms by electronic filtering. We have to use a
video store to record the reference state of the object. This
stored image is subsequently subtracted from the camera images.
As the self interference or the object speckle remains unchanged
by small object movement, these noise terms subtract to zero.
The cross interference term varies, however, cyclically with
the phasechange to produce high contrast fringes across the
image.

2.3. ESPI hardware

So far we have discussed the formal structures of various
ESPI systems. We will now describe the ESPI-components which
are different from, or not commonly used in hologram interferometry
set-ups. We will refer to the specular reference ESPI depicted
on fig. 7.

The laser has so far been the conventional sources (He-Ne,
Ar and Rb). For industrial and medical work, a double pulsed
laser (e.g. a YAG) syncronized to the TV-scanning rate would be
extremely useful. The future cw-lasers in ESPI will undoubtedly
be laser diodes.

For shortened exposures and stroboscopic work we need chopping
or amplitude modulation - AM - of the laser beam. Usually accousto-
optics or electro-optics modulators are used as they are easily
syncronized to the video-scanning or the object excitation.

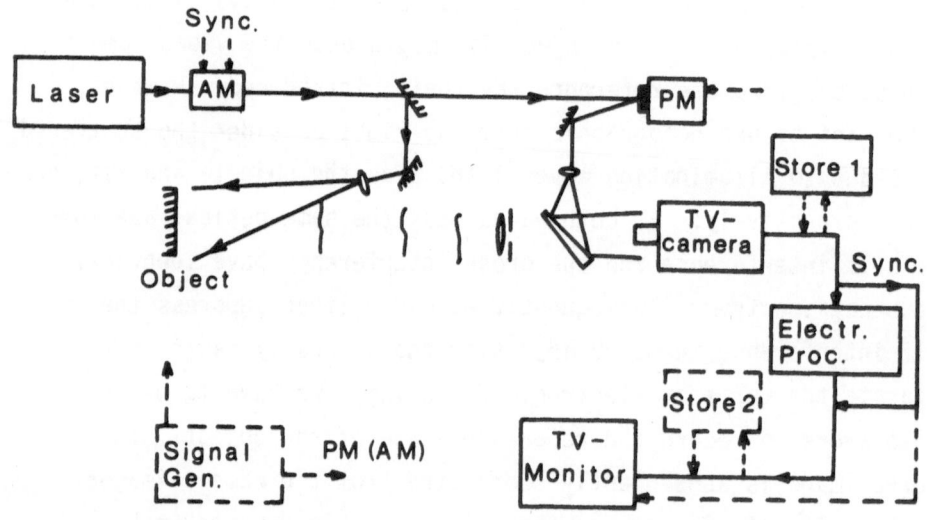

Fig. 7. An ESPI system.

In many applications we also want to introduce a controlled
phase shift or phase modulation - PM - into one of the branches
of the interferometer. In fig. 7 the PM is achieved by reflecting
the reference beam from a mirror which may be translated by either
piezoelectric or electromechanical means. If PM at high frequencies
(> 50 kHz) is desireable, the beam is passed through an electrooptical
crystal.

The optical components in the interferometer are similar
to those used in hologram interferometry. However, we again want
to stress the importance of a properly designed and adjusted
beamcombiner. A clean and uniform reference wave over the target
is essential in a specular reference ESPI. The reference wave
should be a spherical wave centered on the exit pupil of the
imaging lens system to obtain the lowest possible object-reference wave angles for a given lens aperture.

The recording medium, the TV-camera, is an essential part of an ESPI. Its main ingredient is the tube where different light sensitive materials are available as targets. The modern multi-alkali tubes like Newvicon and Chalnicon (Pasecon) are excellent performers both in terms of resolution and sensitivity. Ultricon tubes are interesting for work in the near infrared. At low light levels the SIT-tube has been cited as a possible candidate. However, its shorter dynamical range and higher electronic noise level makes the SIT-tube not much more sensitive than e.g. a Chalnicon for ESPI applications. For future ESPI, CCD and similar array cameras should be considered as their resolution increases towards acceptable levels. Especially for work in rough surroundings and close to strong magnetic fields these array cameras should prove invaluable. The camera should preferable scan in a 2:1 interlace and not randomly, which simplifies connection to e.g. digital video stores. Manual adjustment of the video amplification may also be a useful asset.

Continuing with fig. 7, storage 1 is used for storage and accurate comparison (subtraction) of TV-frames representing primary interferograms. Previously video disc, taperecorders(16) and video storage tubes(17) have been used, but digital video stores will undoubtedly be the ultimate solution. Note that for storage and comparison (addition) within a picture frame the TV-target itself does the job automatically as it accumulates charges between read-outs.

As explained earlier the electronic processing in ESPI consists of a bandpass filtering followed by a fullwave rectification. These electronic circuits may be conventionally designed with sufficient amplication. Storage 2 is a video tape recorder used to store the resulting fringe pattern for later replay and analysis (18, 19, 20).

The monitor should have a bright display with high contrast and low distortion.

3. ESPI APPLICATION TECHNIQUES

In this section we will describe techniques which produce fringe patterns related to changes in the object or to the object's topology. As some techniques have their exact counterparts in ordinary hologram interferometry they will be only briefly described. Other techniques which rely on the dynamical presentation of ESPI will be treated more in-depth. Note that by the term "movement" we include any change of the optical path length. Although displacement of solid objects are used to exemplify phase changes, also changes occuring in transmission objects (gases, liquids) may be measured.

We start with techniques which deal with measurements of rapid (or large) movements. In these cases sufficient phase changes take place within a TV-frame and no external store is needed. Slow movements or two-step changes where different frames have to be compared will be treated next. The common problems of all techniques, namely to determine the absolute magnitude and direction of the phase change, will be addressed at the end.

3.1. Damped oscillations

The decaying oscillations of an object excited by a sharp blow may be readily observed on the TV-monitor. Provided the object is lightly damped, the object's amplitude will be nearly constant during each TV-exposure. In this case the fringe function can be approximated by the common $J_o{}^2$-function associated with harmonic vibration to be treated next. Strangely enough this property of ESPI does not seem to have been applied to any serious research.

3.2. Harmonic vibrations

The <u>time average</u> method where we simply record the object

while it is vibrating is most commonly used. We write the
vibration $u_0(x,y,t)$ across the object's surface as:

$$u_0(x,y,t) = a_0(x,y)\cos[2\pi ft + \phi_0(x,y)] \tag{2}$$

where: $a_0(x,y)/\phi_0(x,y)$ represents the object's amplitude/phase
distribution and f is the vibration frequency.

We can use directly the results developed for hologram
interferometry to find the fringe pattern across the monitor
image. From e.g. ref. 21, we get:

$$I(x,y) = I_0(x,y)J_0^2(\frac{4\pi}{\lambda} a_0(x,y)) \tag{3}$$

where: $I_0(x,y)$ = intensity of interference image at rest
J_0 = Bessel function of first kind and zero order
and we assumed the illumination and observation
directions to be coincident with the movement vector

We should note the eq. 3) is exact only if the exposure
spans multiples or a great number of vibration cycles. At low
frequencies or by shortened exposures, the fringe function is
modified(19).

As in hologram interferometry the resulting fringe pattern
of eq. 3 is very easy to interpret in terms of amplitude
distribution. Constant amplitudes will be displayed as contours
of constant intensity or fringes. The center of the zero order
fringe, $J_0^2(o)$, represents the nodal line (or point) and is
displayed at full intensity. The bright fringes thereafter will
roughly represent areas of the object having a vibration amplitude
of an integer multiple of $\lambda/4$. The interpretation is further
aided by the fact that the intensity of the fringes decreases
with increasing amplitude.

By straight time average recordings we loose the phase
information as witnessed by eq. 2 and 3. Often the phase

distribution may be inferred by pre-knowledge. However, analysis of complex mode structures as encountered in e.g. loudspeaker cones is hardly possible without phase information. Time average ESPI has also restricted amplitude range. At the lower end, the first dark fringe represents an amplitude of about λ/5. The maximum upper values depend on the fringe pattern quality and complexity. On a simple pattern (like fig. 3b) more than 40 fringe orders have been detected, corresponding to about 10 λ.

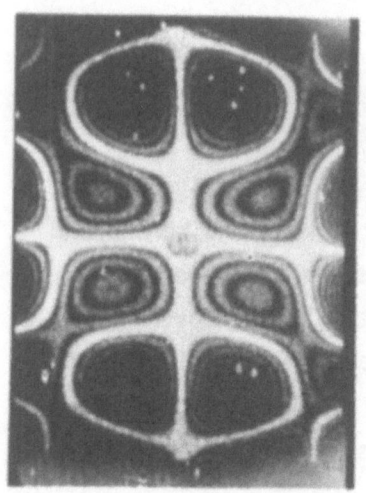 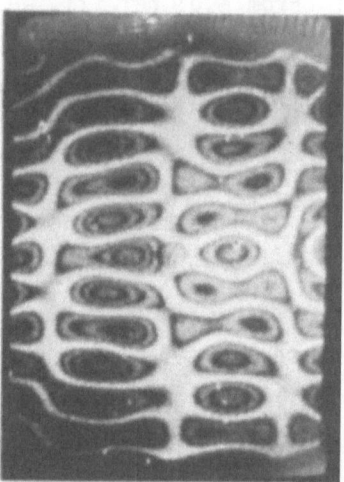

Fig. 8.   Time Average ESPI recordings.

On more complex patterns like the ones on fig. 8 a more realistic limit would be 8 fringe orders or 2 λ. Difficult optical surfaces, like biological specimens, cause volume scattering which limits the detection limit to 1-3 fringe orders even on simple patterns.

If we combine time average ESPI with phase modulation - PM - the method becomes phase sensitive and we get an extended measuring range. So far phase modulation only in its simplest form, i.e.

sinusoidal phase-modulation - SPM, has been used in ESPI. We
have already in sect. 2.3 described how SPM might be achieved
by, for example, reflection from a mirror in the reference path.
If we write the vibration of this mirror as:

$$u_R(x,y) = a_R \cos(2\pi ft + \theta_R) \tag{4}$$

the resulting intensity distribution $I_{mod}(x,y)$ of a vibrating
object may be expressed as(22):

$$I_{mod}(x,y) = I_0\{J_0^2 [\frac{4\pi}{\lambda} (a_0^2(x,y) + a_R^2 -$$

$$2a_0(x,y) \cdot a_R \cdot \cos(\phi_0(x,y) - \phi_R))]^{\frac{1}{2}}\} \tag{5}$$

Thus the fringe function itself remains unchanged, but its
argument now depends on the vectorial difference between the
two movements. To simplify the discussion of eq. 5 we may think
of SPM as a compensation principle: the bright zero order fringe,
$J_0^2(o)$, is observed wherever the amplitude $a_0(x,y)$ and phase
$\phi_0(x,y)$ of the object is compensated by the amplitude a_R and
phase ϕ_R of the induced SPM.

We use this principle to map the amplitude at levels out
of range for ordinary time average ESPI. The SPM is set at a
desired amplitude a_{R1}, while its phase is adjusted. The resulting
movement of the $J_0^2(o)$ fringe across the object represents and
traces the contour(s) of constant object amplitude $a_{o1} = a_{R1}$.
(Note, with classical ± vibrations only two reference phase
settings are necessary). Provided the reference mirror can be
sufficiently excited, the measuring range may be extended to
amplitudes corresponding to above 100 J_0^2-fringes(22). On
biological objects where the fringe quality is inferior, this
technique is most valuable also at lower amplitude levels(18).

To trace the contours of constant phase across the object(23),

the amplitude of the SPM is varied from zero to the maximum
object amplitude while its phase is kept at a constant value
ϕ_{R1}. The $J_0^2(o)$-fringe now traces the contours of constant
object phase $\phi_{o1} = \phi_{R1}$. In this way we can use SPM to get a
complete mapping of both the amplitude and the phase contours
across a vibrating surface as shown on fig. 9.

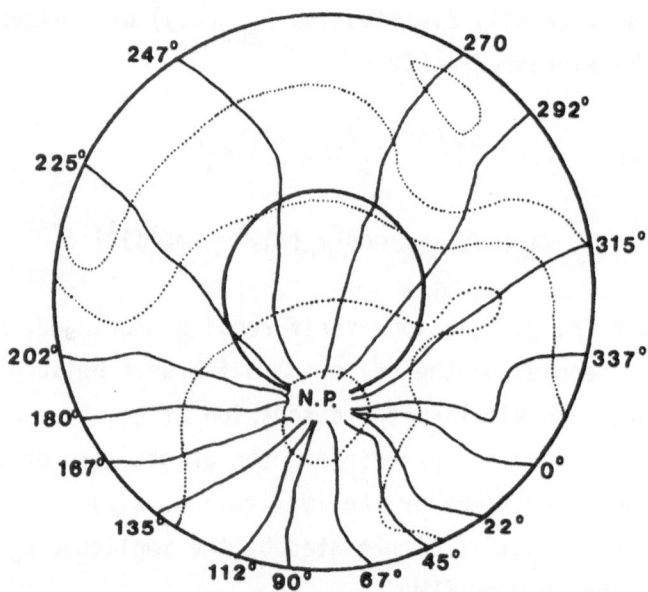

Fig. 9.   Amplitude (fully drawn) and phase (dotted contours of a
          loudspeaker cone (NP: nodal point).

Another problem which we often encounter in vibration
analysis is measurement of very small vibrations. This problem
is very efficiently solved by so-called dynamic SPM(24). Here
we modulate with an amplitude a_R which moves the working point
from zero sensitivity to the steep and linear part of the $J_0^2(o)$-
fringe. A small object amplitude a_o now causes the monitor
brightness to depend on the relative phase, $\phi_R - \phi_o$. The extremal
intensities, I_+ and I_-, occur when this relative phase is 0°
and 180°. To obtain a continously varying phase we modulate
with a SPM frequency slightly different (0.5 - 10 Hz) from the

object frequency. The resulting intensity variation, I_+ and I_- is directly proportional to the objects amplitude, while the timing of I_+ and I_- is directly given by the object phase. By visual observation of the dynamic brightness variation across the object image, amplitudes down to 1 nm have been detected. Visual interpretation of the pattern is possible from about 5 nm. Amplitudes down to 0.01 nm have been measured by lock-in detection of the monitor brightness variations. Note that these values have been reached without using any fringe stabilization as explained in ref. 24. We have also the possibility of storing the dynamic interferograms in a video tape recorder for later analysis. This last property is very useful in work with living or detoriating biological objects(18, 20).

In the SPM-methods described so far, the induced modulation has a constant value across the wavefront as the beam is reflected from a mirror with constant amplitude. An interesting extension of the technique is to have a spatially varying SPM(25). We obtain this by vibrating both objects O_1 and O_2 in the speckle reference ESPI on fig. 5, sect. 2.2. In this way we get a direct pointwise comparison between the vibration patterns of the two objects. Note that the difference fringe pattern which we observe may indicate an amplitude and/or a phase difference. The technique may also be used to separate the modes in a combination mode(25).

Stroboscopic techniques are also used to study harmonic vibrations(26). Unstable objects like e.g. the human ear-drum in-vivo have to be sampled by one double exposure each TV-frame. Stable objects can be studied by short double exposures synchronized to the object vibration. By these techniques we stop the object's motion twice each vibration cycle, usually half a cycle between subexposures. Formally the resulting intensity distribution $I_s(x,y)$ can be written as:

$$I_s(x,y) = I_0(x,y)\cos^2(\frac{4\pi}{\lambda} \Delta a_0(x,y)) \tag{6}$$

where: $\Delta a_o(x,y)$ is the object's movement between the pulses.

To map the vibration we have to vary the timing of the pulses relative to the vibration cycle and from the resulting patterns we determine the amplitude- and phase distribution. This works well for pure modes with 0°-180° phase values, but becomes time consuming for analysis of complex modes.

Due to the constant brightness of the higher order $\cos^2$-fringes of eq. 6, we are able to detect a high number of fringes. Under optimum conditions well above 100 ESPI fringes can be observed across the monitor image.

Finally, before leaving the harmonic vibration, we should note that none of the techniques described so far reveals asymmetries in the vibrations regarding to the object's rest position. To solve this problem we have to record the rest positions in a videostore and use one pulse each vibration cycle. By comparing the amplitude on symmetrical parts of the cycle the asymmetry can be determined.

3.3. General periodic vibrations

Both time average and stroboscopic techniques may contribute to a complete analysis of such vibrations, but no ESPI work has been reported in this field so far.

3.4. Non periodic movements

More arbitrary movements may be sampled by double exposures(19, 27). For analysis the interferograms should be stored by a video tape recorder to be later replayed by single-frame or in slow-motion. The technique should be really valuable for testing objects excited e.g. by noise. However, a double pulsed laser syncronized to the TV-rate is essential.

Another interesting area of application would be velocity mapping to supplement the conventional laser doppler velocity

measurements. By time average recordings of an object moving
at constant velocity, we will essentially observe fringes
indicating areas of no movement(17) or zero order fringes. The
higher fringes in the resulting fringe function will follow a
$(sinx/x)^2$-dependency which decreases rapidly. To measure
velocities different from zero, we note that interesting results
have been reported by combining linear phase modulation with
an ESPI(28-32). In literature the resulting system is called
a laser doppler imaging system. It displays either real time
1-dimensional velocity map or builds up 2-dimensional maps by
varying the phase modulation setting.

We will now proceed to describe techniques where a video store
is necessary as we wish to measure the phase changes over time
intervals longer than the TV-framing period. There are, as in
hologram interferometry, essentially two ways to use the storage:
Double exposure where the primary interferograms from two different
frames are subtracted or added by the external store to produce
a permanent display of the fringes.
Real time mode where the interferogram from a reference frame is
stored and subsequently subtracted or added from the frames
coming from the camera,to give a dynamic display of the phase
change.
    In general the latter technique is used during adjustment
and observation of the progression of an experiment, while a
permanent double exposure is useful for accurate measurement
on the fringe pattern. Both techniques give the same cos^2-fringe
pattern as in hologram interferometry(21) and already given by
eq. 6. We should note that by subtraction of frames, the zero
order fringe will be dark while addition gives a bright zero
order fringe.

3.5. General object movement

In general we are concerned with the same type of measurements or observation as in hologram interferometry - how does the object change due to force, heat, pressure, humidity or even time? The purpose of such measurements is apart from analysing the object behaviour, often to detect areas of excessive movements which indicate weak or faulty regions.

In practical work we follow the object's behaviour by the real time method. If necessary the fringe patterns are recorded by a video tape recorder. Whenever the fringe density gets too high we simply push a button on the video store and a new reference frame is recorded within 40 msec. The fringe build up can then be watched and recorded. In fact it is possible to record only the primary interferograms on the video tape recorder and afterwards combine these recordings in any succession by means of a digital store.

The acccuracy of the fringe analysis may be surprisingly high considering the coarsely speckled fringes. Even by visual inspection the accuracy may be better than $\lambda/25$ if an appropriate number of reference fringes is introduced across the monitor image. The maximum number of fringes we have resolved across the monitor image is about 120 which is close to the theoretical limit. In practice, however, we find 5 to 20 fringes, including the reference fringes, to give suitable fringe density.

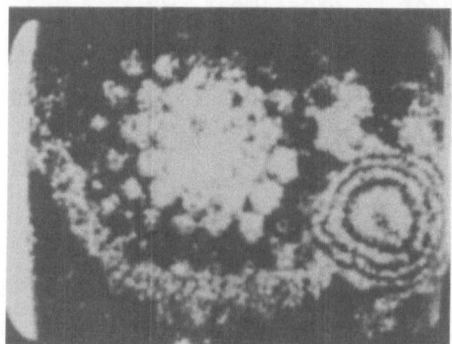

Fig. 10.   Interior flaw in a honeycomb panel.

Fig. 10 shows how ESPI is used to detect an interior weak area of a honeycomb structure.

3.6. Contouring

The holographic two wavelength contouring method(21) has been adapted to the specular reference ESPI by the Loughborough group(33-35). An extensive description of their work can be found in ref. 6.

In this technique the object is illuminated sequentially by the laser wavelengths λ_1 and λ_2. The two primary interferograms are subtracted by a video store, resulting in a contour map overlapping the object image. If we assume coincident illumination and observation, the contour spacing d will be given by:

$$d \sim \frac{\lambda_1 \lambda_2}{\lambda_2 - \lambda_2} \tag{7}$$

The interpretation of the contour map is mainly dependent on the geometrical shape of the illumination wave.
Absolute height contours are obtained if the illumination is a plane wave used in a telecentric observation system.
Relative height contours are obtained by use of more complex illumination geometries. By using for example spherical or cylindrical lenses in the illumination we may contour the object relative to the shape of spheres or cylinders. The Loughborough group has made holographic illumination elements which in principle can be adapted to any shape. In this way we can contour production objects relative to a prototype, which may be of great importance for example in turbine blade production. A general problem all these techniques share is that large aperture illuminators are needed.

3.7. Directional sensitivity

The ESPI has a directional sensitivity which is common to

all previous techniques described. If we assume the direction
of movement to be at angles θ_1 and θ_2 with the illumination and
observation respectively the angle factor will be:

$$\cos\theta_1 + \cos\theta_2 \tag{8}$$

In the eqs. (3,5,6) we have assumed this angle factor to
be 2, i.e. coincident illumination and observation in the movement
direction.

To determine the direction and magnitude of the movement
vector we need at least 3 independent measurement. In ESPI work
we usually vary the direction of illumination with a fixed
observation. In many specular reference ESPI systems the
illumination and observation directions very nearly coincide,
which gives directional sensitivity in the viewing direction.
The different sensitivity vectors are then obtained by viewing
the object from different positions.

In some cases we are interested in the in-plane movement
of plane or nearly plane object surfaces. The specular reference
ESPI will always have some sensitivity to displacements normal
to the object surface (or out of plane). To get pure in-plane
sensitivity of plane objects the double illumination ESPI previous
described in sect. 1.2. has to be used. By using a symmetrical
illumination with plane waves, we get pure in-plane sensitivity.
Double illumination ESPI has been used for strain measurement
by the Loughborough group(15,36).

3.8. Miscellaneous

We have already described in sect. 3.2. how speckle reference
ESPI may be used to directly compare the vibration patterns of
two similar objects. The same technique may of course also be
used to compare static deformations of objects. In fact the
potential application area is probably far greater here as mainly

one parameter, the magnitude of deformation, will be of importance. Apart from direct comparison between the deformations, this technique will also compensate for excessive movements.

Finally we should mention <u>shearing</u> ESPI where we observe the interference between the object (image) wave and the shifted or sheared object wave. Shearing ESPI is useful e.g. for calculating the mechanical strain of thin plates, as we reduce the order of mathematical derivation by one if differential shear in the x- and y-directions is used. Various shearing ESPI set-ups have been suggested. A Michelson interferometer where the light source is replaced by the illuminated object is most commonly used(37,38). In these cases an external video store has to be used to subtract images. Alternatively we may use a speckle reference ESPI for shearing(12,39), which gives shearing interferograms without external storage.

Shearing interferometers are very stable due to the common paths of the two waves and should be particularly applied for work in unstable surroundings.

4. SPECIFIC APPLICATIONS

In industry and industrially related research the most popular application has been vibration analysis. Turbine blades have been studied both under laboratory conditions(16,40) and on site(41). Another popular object has been the loudspeaker which has been studied in steady state(42,43,44) and excited by pulses(27). Transducers of various types have been studied and calibrated(45,46).

By double exposure recordings, delamination of skis and tires(40) and faults in brake shoes(47) have been revealed. By use of double illumination the strain field in metal and carbon fibre have been measured (48,49). The same technique with a pulsed laser has been used to measure strain in rotating flywheels(50). The opening of cracks in epoxy coatings(51) and metals(52) have been studied. Contouring of wheel caps and turbine blades has also been reported(49,53).

Another main area of application is the biomedical field where the author foresees a long row of applications which have hardly been touched upon yet. So far bioaccoustics has been the most active area of research. The vibrations of the human ear drum has been investigated in-vitro(54,55) and in-vivo(18,56). In-vitro measurements of the ossicular chain(57) and the basilar membrane(20) have also been reported. Finally, the natural movement of the human skin has been studied(19).

Although this list of applications includes most of the published material so far, we feel that it does not give a fair presentation of the real potential of the technique. Recent unpublished work indicates that possible ESPI projects would include most of the holographic work listed by Vest(58) plus some others.

5. AN EVALUATION OF ESPI

As a conclusion we will comment upon the negative and positive properties of the ESPI system where the comparison is naturally made against the current state of hologram interferometry.

The main drawback of ESPI is its coarsely speckled fringe pattern caused by using a low resolution recording device. However, the dynamic display of ESPI allows us to manipulate and shift the fringes which brings out far finer fringe details than observed on a static image. We should note that microscopic ESPI represents a special case as the object resolution will be comparable to holography. Finally, speckle averaging can be used to improve fringe patterns of vibrating objects (12).

At low light levels, the fixed maximum TV-exposure time of 40 msec. may be considered a drawback when the object stability is good. Increased exposures may be obtained to a certain degree by interupting the scan. However, with the available high sensitive

TV-tubes even small He-Ne lasers can be used to study most objects under normal conditions.

The cost of an (commercial) ESPI may seem high compared to a holographic system, especially if a digital video store is necessary. However, in this context we should also consider the greater experimental output of ESPI at almost nil running costs.

The sampling rate of ESPI (25 Hz) is high compared to ordinary holography, but this rate is too slow to study fast events like e.g. shock waves. An interesting solution may be to use Reticon arrays in one dimensional ESPI (59).

On the positive side the short exposure time and high sampling rate make the interferometer very stable. In fact by using shortened exposures the increased stability allows one to hold a vibrating object freely and still observe its vibration pattern(19). The speckle reference ESPI may also be made extremely stable even at normal exposures(39).

A properly designed ESPI is easy to adjust and use, especially when based upon a speckle reference wave where no spatial filter is used. The system works well in normal light conditions. The interferograms are easy to observe and interpret, especially if they are presented on a large TV-screen. These properties are useful when a group of people wants to discuss e.g. the vibration characteristics of the test object.

On the positive side we will again point out the possibility of recording an entire experiment on video tape for later analysis. The audio channel can be used to store reference signals or as a notebook.

The optical head, including the light source and detector of an ESPI may be very compactly constructed, which will be a favourable asset for on-site inspection. We should note that the video signal might be relayed by cable or wireless transmitted from the TV-camera to the observation station. More inaccessible areas may be reached by fiber optics(60).

The future acceptance of ESPI will probably depend on how well it will compete with holographic systems based upon the emerging self-developing, reusable recording media. If a detailed and once in a lifetime, three-dimensional recording of the behaviour of a large object is requested, ordinary holography is bound to emerge as the winner. It is unlikely, however, that holographic recording and display in the near future will obtain ESPI's combination of real-time capabilities, short exposure, high repetition rate, easy storage of primary interferograms and high sensitivity. ESPI should therefore be attractive for fast and troublefree testing and measurements in industry and research. Future work should stress the incorporation of computers in ESPI for automatic data acquisition and analysis, the first steps in this direction being reported in refs. (61-63).

REFERENCES

1. Macovski, A., Ramsey, D. and Schaefer, L.F. Appl. Opt. 10, 2722 (1971).
2. Schwomma, O. Österreichisches pat.nr. 298830 (1972).
3. Køpf, U. Messtechnik. 4, 105 (1972).
4. Butters, J.N. and Leendertz, J.A. J. Meas. Control. 4, 349 (1971).
5. Collier, R.J., Burckhardt, C.B. and Lin, L.H. "Optical Holography". Acad. Press. (1971).
6. Jones, R. and Wykes, C. "Holographic and Speckle Interferometry". Cambr. Univ. Press (1983).
7. Løkberg, O.J. and Slettemoen, G.A., to appear in "Applied Optics and Optical Engineering". (Eds. R. Shannon and J.C. Wyant), Acad. Press.
8. Archbold, E., Ennos, A.E. and Taylor, P.A. In "Optical Instruments and Techniques" (Ed. J. Home Dickson), pp. 265-275, Oriel Press. (1970).
9. Stetson, K.A. Opt. Laser Technol. 2, 179 (1970).

10. Biedermann, K., Ek, L. and Østlund, L. In "The Engineering Uses of Coherent Optics". (Ed. E.R. Robertson), pp. 219-222. Cambr. Univ. Press (1976).

11. Slettemoen, G.A. Opt. Acta 26, 313 (1979).

12. Slettemoen, G.A. Appl. Opt. 19, 616 (1980).

13. Slettemoen, G.A. J. Opt. Soc. Am. 71, 474 (1981).

14. Leendertz, J.A. J. Phys. E: Sci. Instrum. 3, 214

15. Butters, J.N. and Leendertz, J.A. J. Phys. E: Sci. Instrum., 4, 1 (1971).

16. Butters, J.N. In "The Engineering Uses of Coherent Optics" (Ed. E.R. Robertson), pp. 155-170, Cambr. Univ. Press (1976).

17. Løkberg, O.J., Holje, O.M. and Pedersen, H.M. Opt. Laser Technol. 8, 17 (1976).

18. Løkberg, O.J., Høgmoen, K. and Holje, O.M. Appl. Opt. 18, 763 (1979).

19. Løkberg, O.J. Appl. Opt. 18, 2377 (1979).

20. Neiswander, P. and Slettemoen, G.A. Appl. Opt. 20, 4271 (1981).

21. "Holographic Nondestructive Testing" (Ed. R.K. Erf). Acad. Press (1974).

22. Løkberg, O.J. and Høgmoen, K. J. Phys. E: Sci. Instr. 9, 847 (1976).

23. Løkberg, O.J. and Høgmoen, K. Appl. Opt. 15, 2701 (1976).

24. Høgmoen, K. and Løkberg, O.J. Appl. Opt. 16, 1869 (1977).

25. Løkberg, O.J. and Slettemoen, G.A. Appl. Opt. 20, 2630 (1981).

26. Pedersen, H.M., Løkberg, O.J. and Førre, B.M. Opt. Commun. 12, 421 (1974).

27. Cookson, T., Butters, J.N. and Pollard, H.C. Opt. Laser Technol. 10, 119 (1978).

28. Sato, T., Nakatani, Y. and Ueda, M. Appl. Opt. 13, 275 (1974).

29. Sato, T., Nakatani, Y. and Ueda, M. Appl. Opt. 15, 867 (1976).

30. Sato, T., Nakatani, Y. and Ueda, M. Appl. Opt. 16, 1263 (1977).

31. Yoshimura, T., Yamamoto, H. and Wakabayashi, N. Opt. Commun. 40, 10 (1981).

32. Yoshimura, T., Yamamoto, H. and Wakabayashi, N. Appl. Opt. 22, 2448 (1983).

33. Butters, J.N. and Leendertz, J.A. In "Proceedings of the Technical Program", Electro-Optics Conference, Kiver Commun. pp. 43-49 (1974).

34. Jones, R. and Butters, J.N. J. Phys. E: Sci. Instrum. 8, 231 (1975).

35. Denby, D., Quintanille, G.E. and Butters, J.N. In "The Engineering Uses of Coherent Optics" (Ed. E.R. Robertson), pp. 171-197. Cambr. Univ. Press (1976).

36. Denby, D. and Leendertz, J.A. J. Strain. Anal. 9, 17 (1974).

37. Leendertz, J.A. and Butters, J.N. J. Phys. E: Sci. Instr. 6, 1107 (1973).
38. Nakadate, S., Yatagai, T. and Saito, H. Appl. Opt. 19, 424 (1980).
39. Slettemoen, G.A. and Løkberg, O.J. Appl. Opt. 20, 3467 (1981).
40. Schwomma, O. Neuer Züricher Zeitung Beilage "Forschung und Technik" 257 (1975).
41. Løkberg, O.J. and Svenke, P. Opt. and Lasers in Engr. 2, 1 (1981).
42. Mc. Kehone, D. Electro-Optical System Design, 41 (1977).
43. Løkberg, O.J. Proc. SPIE, 215, 92 (1980).
44. Hurden, A.P.M. Opt. Laser Technol. 14, 21 (1982).
45. Jones, R. and Bergquist, B.D. In "The Engineering Uses of Coherent Optics". (Ed. E.R. Robertson), pp. 413-429. Cambr. Univ. Press (1976).
46. Koyuncu, B. Opt. and Lasers in Engr. 1, 21 (1980).
47. Butters, J.N. Opt. Laser Technol., 9, 117 (1977).
48. Jones, R. Opt. and Laser Technol. 8, 215 (1976).
49. Butters, J.N., Jones, R. and Wykes, C. In "Speckle Metrology" (Ed. R.K. Erf), pp. 111-157. Acad. Press (1979).
50. Preater, R.W.T. Proc. SPIE, 236, 58 (1980).
51. Richardsson, M.O.W., Al-Hassani, A.H.M. and Herbert, D.P. Trans. Inst. Metal Finish, 60, 84 (1982).
52. Løkberg, O.J. and Slettemoen, G.A. Proc. SPIE 398, 295 (1983).
53. Wykes, C., Jones, R. and Butters, J.N. Annal. CIRP, 27, 361 (1978).
54. Høgmoen, K. and Løkberg, O.J. In "The Engineering Uses of Coherent Optics". (Ed. E.R. Robertson), pp. 147-152. Cambr. Univ. Press (1976).
55. Burian, K., Fritze, S. and Schwomma, O. In "Symposium 1976 on Electro-Cochleography and Holography in Medicine", (Münster) A 29.
56. Løkberg, O.J., Høgmoen, K. and Gundersen, T. Acta Otolaryngol. 89, 37 (1980).
57. Løkberg, O.J. and Høgmoen, K. Proc. SPIE, 136, 222 (1977).
58. Vest, C.V. Proc. SPIE, 349, 186 (1982).
59. Ottonello, P. and Pontiggia, C. Opt. Commun. 30, 20 (1979).
60. Løkberg, O.J. and Krakhella, K. Opt. Commun. 38, 155 (1981).
61. Koyuncu, B. and Cookson, T.J. J. Phys. E. 13, 206 (1980).
62. Hurden, A.P.M. Opt. Laser Technol. 14, 21 (1982).
63. Dikici, A. Dr.ing.thesis "Automation of Electronic Speckle Interferometer", Norw. Inst. of Technol., Trondheim, Norway.

CONVECTIVE MASS TRANSFER COEFFICIENT MEASUREMENT BY HOLOGRAPHIC AND ELECTRONIC SPECKLE PATTERN INTERFEROMETRY

N. Macleod

University of Edinburgh (Chemical Engineering Department)

Engineering Importance of Mass Transfer Coefficient Measurement

Measurements of the rate of transfer of volatile or soluble material from a solid surface of specified shape to a turbulent stream of gas or liquid flowing over it, due to forced convection and diffusion, are of value:

a) In providing fundamental information about the transport properties of the fluid boundary layer, important to an understanding of all interfacial transport processes; and

b) Because such mass transfer situations can serve as 'cold models' for heat transfer situations in heat transfer equipment of geometrically similar form and dynamically similar flow characteristics. In particular, it is generally much easier to infer the overall and local heat transfer properties of a propsed new type of heat exchange surface (especially its liability to hot-spot formation at regions of low local convective transfer rate) from mass transfer experiments on an isothermal model at room temperature rather than from direct trials with heated surfaces. Established relations between mass and heat transfer processes can then be used to estimate tne required heat transfer behaviour from the mass transfer results.

Profilometric Technique for Mass Transfer Measurement

If the surface of the mass transferring model is coated with a volatile or soluble substance, the local flux of mass transfer to the experimental stream of gas or liquid can be deduced for any designated point on the surface by measuring local dimensional changes brought about by exposure to the fluid stream (at the same temperature as the model) for known times. From such flux measurements one can compute the corresponding values of local convective mass transfer coefficient, defined as the mass flux divided by tne difference of concentration of transferred substance at the interface and in the bulk of the stream. This quotient is a measure of diffusive and convective effects in the fluid boundary layer at the point considered. Neither its value at any particular point nor its pattern of variation over the surface can in the present state of knowledge be calculated a priori for most situations of engineering interest; but once it is known for a particular flow and surface configuration from mass transfer measurements, the thermal transport properties of the bounaary layer at corresponding locations in a dynamically similar system are known also. Corresponding values of the heat transfer coefficients that would be obtained in a geometrically similar prototype heat exchange system, in which solid surface and fluid were at different temperatures, can accordingly be obtained from well established 'analogy relations' based on the fundamental similarity of the differential equations describing the convective and diffusional transport of heat and mass within the boundary layer.

In the experimental method developed in Edinburgh [1], the mass transferring coating is of transparent silicone rubber, swollen to equilibrium before each experiment by immersion in a volatile or soluble ester or hydrocarbon swelling agent. The initial rate of transfer of swelling agent by evaporation or solution to the experimental gas or liquid stream can be shown to be identical to that of the pure swelling agent. This transfer rate, and hence transfer coefficient, is registered by measurements of coating shrinkage.

Application of Holographic Interferometry

Holographic interferometry provides an obvious means of registering such changes of mass-transferring coating thickness at every point within an entire field of view. This synoptic method of acquiring such data was first described from our laboratory more than a decade ago [1] and has been much developed since, both by us [2] and elsewhere [3,4]. The thin transferring coating can be treated as a phase object of nearly unchanging refracting index but varying thickness. It can be interrogated by the object beam of the holographic system by frontal viewing, with reflection of the beam from a diffusely reflecting substrate; or, as in the example shown here, by entry of the object beam into the rear of the coating via a transparent substrate (a 90° prism) followed by total internal reflection from the coating-fluid interface (Figs. 1,2 and 3).

By the internal reflection method, transfer coefficients at the interior wall of a closed duct carrying a turbulent fluid stream can be determined without disassembly of the apparatus, either by the frozen fringe method or, indeed, in real time, there being no intersection of the optical and fluid paths. Fig. 4 shows an (early) record of frozen fringes obtained using double exposure and total internal reflection methods, near the entrance of a rectangular duct carrying a stream of air at a Reynolds number not far above the turbulence transition value. It is doubtful whether the unexpected (and hitherto unreported) transverse variation of transfer coefficient revealed in such photographs could have been determined in any non-optical way.

Fringe Ordering from Double Exposure Records

Vibration has so far proved an obstacle to the accurate measurement of real-time fringes in these experiments and calculation of local transfer coefficients at the duct wall has had to be carried out from frozen fringe photographs. Absolute fringe orders are needed for these calculations and these have been obtained from sets of several (usually 3 or more) double exposed fringe photographs made at carefully measured time intervals for constant flow conditions - over which the rate of mass transfer and coating shrinkage remains constant.

A Formalism for the Ordering of Fringes by Fringe-Fraction Methods

Let the absolute fringe order (in general fractional) at a chosen point P on the mass-transferring surface after a run of duration t_1 be $n_1 + \nu_1$, where n_1 is the (integral) order number of the nearest fringe to P, and ν_1 is the fractional part of the fringe order at P. Then for a second experiment at the same flow conditions, of duration t_2, the fringe order $n_2 + \nu_2$ at P is related to $n_1 + \nu_1$ by

$$\frac{n_1 + \nu_1}{n_2 + \nu_2} = \frac{t_1}{t_2} \equiv \rho_{12} \tag{1}$$

ν_1, ν_2, t_1 and t_2 may be measured within certain limits of error, leaving n_1 and n_2 to be determined. The set of eqns. generated from a known set of t values has, of course, zero determinant; but the condition that n_1, n_2 etc. are integral (and odd for dark fringes) usefully restricts the indeterminacy of n value assignments. In particular, we find that if the error bounds on measurements of ρ and ν are $\pm m_1$ and $\pm m_2$ respectively, two successive odd integral values, n_1 and n_1' satisfying (1) are separated by the even integer 2q for ranges of ρ satisfying the condition:

$$\rho \notin [^p/q - \{n_1 m_1 + m_2 - m_1 q\}/q \;,\; ^p/q + \{n_1 m_1 + m_2 - m_1 q\}/q]$$

--- where p is an integer $\leqslant$ q $\tag{2}$

Relation (2) is expressed in graphical form in Fig.5 for the case $n_1 = 21$, $m_1 = \pm 1\%$, $m_2 = \pm 0.1$ fringe. For $\rho_{12} \equiv t_1/t_2 = 0.33$ (or 0.67) it is then evident that the same data would fit eqn. (1) with $2q = 6$, i.e. for n_1 values 15, 21 and 27. An additional experiment with $\rho_{13} \equiv t_1/t_3 = 0.16$ (or 0.84) would give $2q = 2$ (excluded by the first result) or 10. The intersection of the resulting sets of permitted n_1 values evaluates n_1 unambiguously, according to the roughest of independent checks.

The relationship (2) is the same whether ρ represents a ratio of times of observation in two different experiments or a ratio of the appropriate geometrical functions in a single experiment using two differently oriented object beams; or, in a single experiment, ρ may represent the ratio of two different wavelengths of illumination, λ_1/λ_2 . All these quantities enter into the fringe-spacing equations in the same way. When $\rho \equiv \lambda_1/\lambda_2$, m_1 can be taken as zero, since laser light wavelengths are so exactly known. This special case has been treated in a different, but equivalent, way by Tilford [5] . Our relation (2) then simplifies to:

$$\rho_1 \notin [\; (^p/q - {}^{m_2}/q) \; , \; (^p/q + {}^{m_2}/q) \;] \quad\quad (3)$$

Fig.6 representing (3) for this case of two-wavelength illumination with $m_2 = \pm 0.1$ fringe as before shows that fringing with the 514.5 nm line of the Argon ion laser and the 632.8 nm line of the He-Ne laser would give $2q = 10$ (irrespective of the nominal value of n_1), satisfying (1) for the 21st black fringe at apparent orders 11, 21 and 31 - an ambiguity easily resolved by external checks. The two strongest Argon lines, 488 and 514.5 nm, would not do this, giving $2q = 2$ and alternative apparent fringe orders 19, 21 and 23.

Application of Electronic Speckle Pattern Interferometry

An alternative method of obtaining fringe-order information in our mass transfer experiments, and an essential means of locating maxima and minima of mass transfer rate, is by real time observation. The mass transfer rate or coefficient at any point on the surface is then obtained simply by counting the rate at which fringes cross that point.

Real time observation is much facilitated by the use of electronic speckle pattern interferometry as an alternative to holography. We have accordingly for some time been using the commercial ESP1 instrument made by Messrs Vinten of Bury St. Edmunds, England, according to designs developed at Loughborough University. The mass transfer fringes appear directly on a television monitor and the entire course of an experiment can be stored on videotape and data-processed as required by an on-line computer without photographic processing (see Figs. 7-11).

The results shown are for mass transfer from a flat plate set normal to an impinging jet of air. When the jet-flow is laminar, the radial distribution of mass transfer coefficient, and hence of coating-recession rate, is monotonic decreasing with radial distance from the jet axis according to a well-verified relation based on boundary layer theory. For a turbulent jet, the mass-transfer rate shows maxima and minima in its variation with radial distance. The location and amplitude of these has been a matter of controversy; but we have found them readily observable directly with the ESP1 instrument as regions from which the fringes appear in real time to radiate (maxima) or upon which they appear to converge (minima). No satisfactory theory of spatial distribution of convective transfer rate is yet available for this case.

REFERENCES

1. An application of Holography to the Determination of Local Mass Transfer Coefficients.
   Kapur, D.N. and Macleod, N., Nature Physical Science, 1972, 237 (73), 57-59.

2. The determination of Local Mass Transfer Coefficients by Holographic Interferometry. I. General Principles: Their application and verification for mass transfer measurements at a flat plate exposed to a laminar round air-jet.
   Idem. Int. J. Ht. Mass Transfer, 1974, 17, 1151-1162.

3. Holographic Determination of Mass Transfer due to Impinging Square Jet.
   Masliyah, J.H. and Nguyen, T.T., Can. J. Chem. Eng., 1976, 54, 299-304.

4. Application of the Swollen Polymer Technique to the Study of Heat Transfer on Film Cooled Surfaces.
   Hay, N., Lampard, D. and Saluja, C.L., 7th Int. Heat Transf. Conf., Munich, 1982, 4, 503-508.

5. Analytical Procedure for Determining lengths from Fractional Fringes.
   Tilford, C.R., Applied Optics, 1977, 16 (7), 1857-1860.

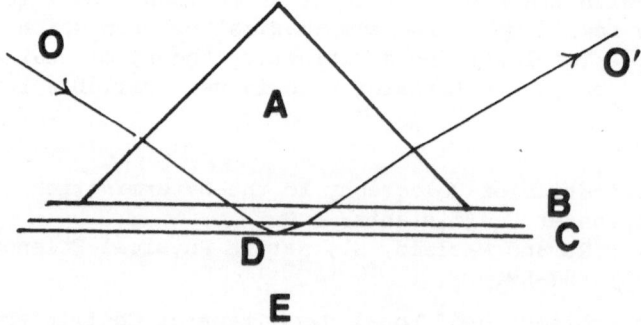

FIG. 1. BASIS OF TOTAL INTERNAL REFLECTION METHOD FOR INTERFEROMETRIC DETERMINATION OF COATING THICKNESS CHANGES.

A – Perspex Prism   B – Glass wall of duct
C – Mass transferring coating   D – Interface between fluid stream E and coating
O – Object beam in O' – Object beam out.

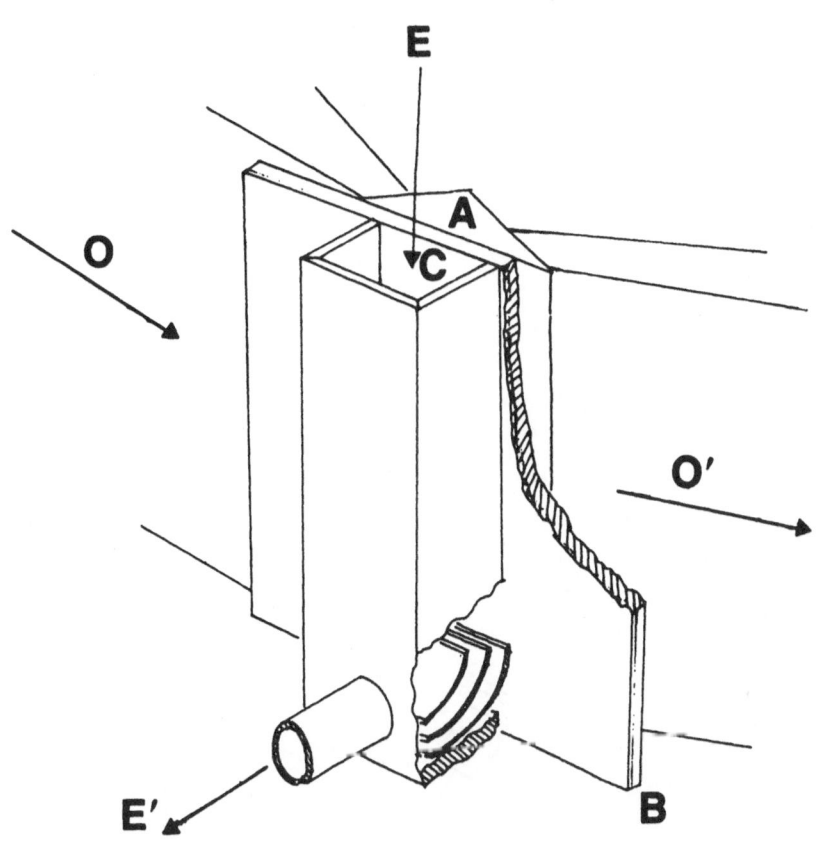

FIG. 2. ARRANGEMENT OF PRISM AND DUCT.

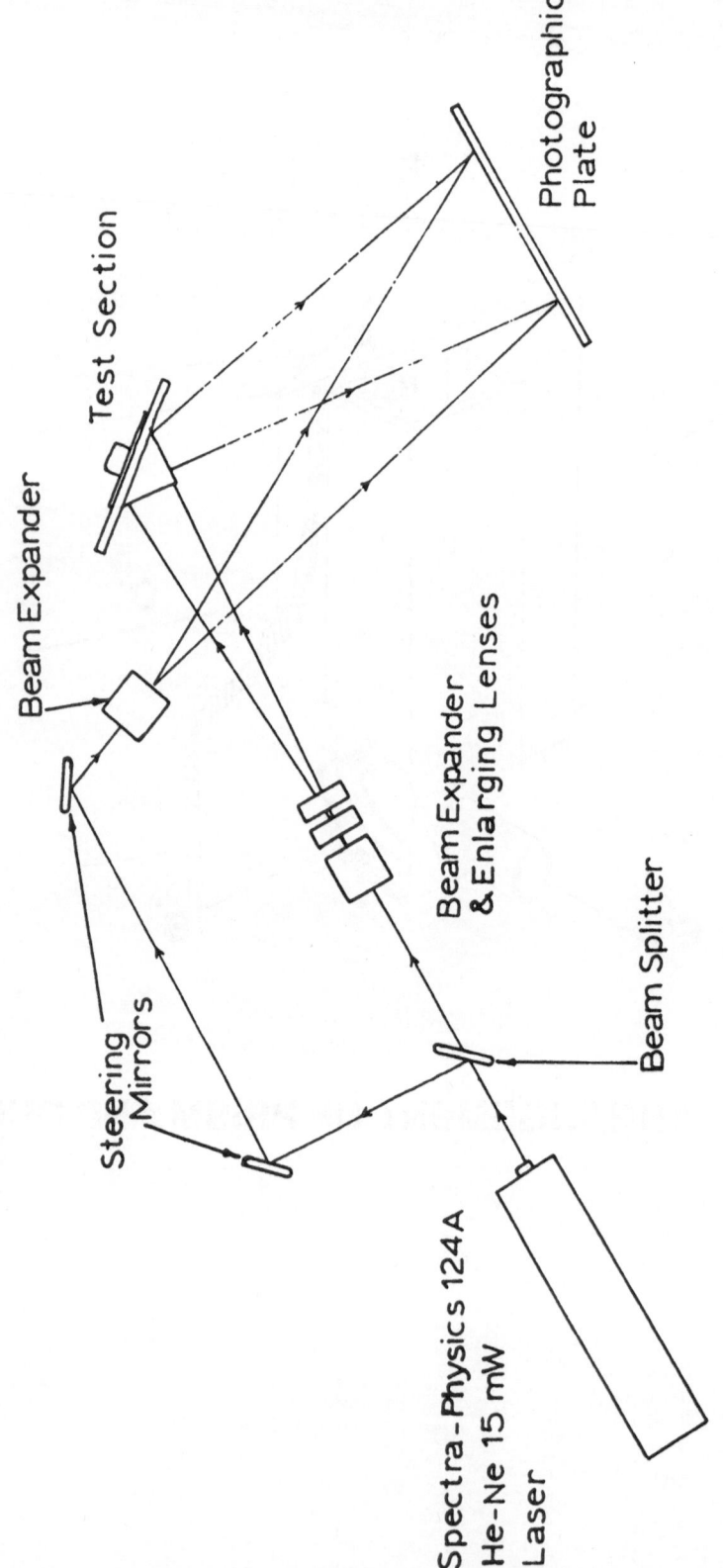

Photographic Plate

Test Section

Beam Expander

Beam Expander & Enlarging Lenses

Steering Mirrors

Beam Splitter

Spectra-Physics 124A He-Ne 15 mW Laser

Figure 3. : The Optical Circuit.

FLOW ↓ DIRECTION

**FIG. 4. MASS TRANSFER FRINGES
NEAR DUCT ENTRANCE.**

582

FIG. 5. FORBIDDEN e INTERVALS PLOTTED AGAINST q FOR $\hat{n} = 21$ WITH $m_1 = \pm 0.01$, $m_2 = \pm 0.1$

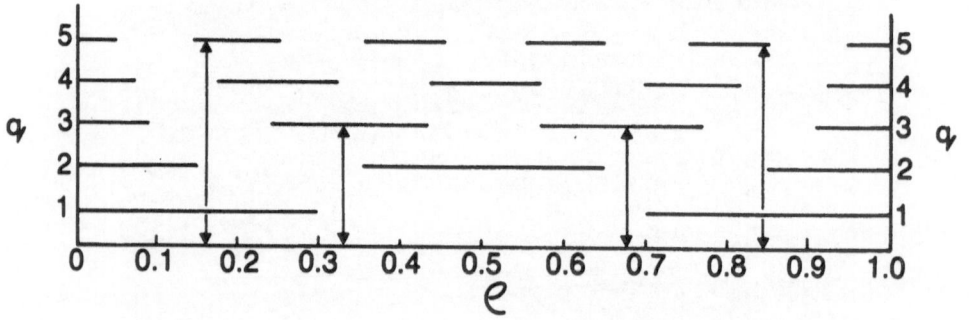

FIG. 6. FORBIDDEN e INTERVALS PLOTTED AGAINST q FOR $m_1 = 0$, $m_2 = \pm 0.1$

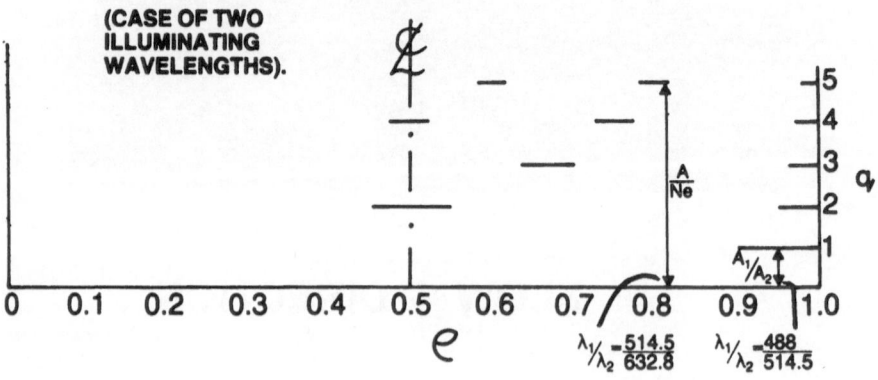

(CASE OF TWO ILLUMINATING WAVELENGTHS).

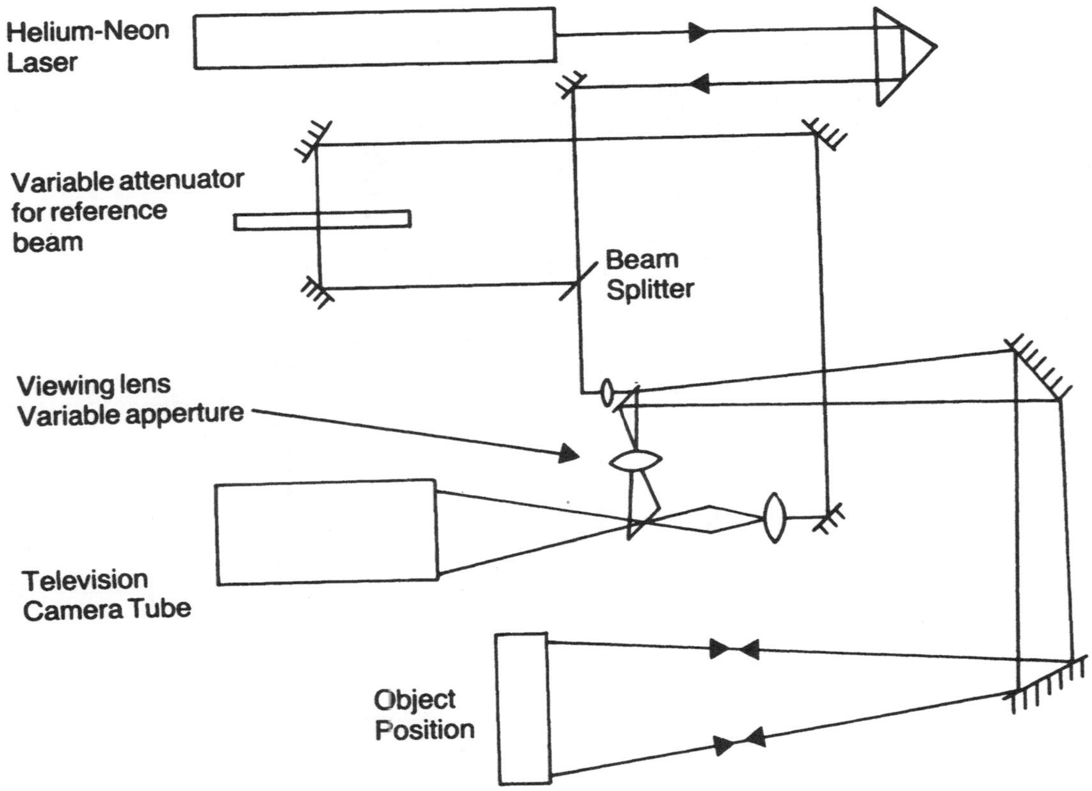
Helium-Neon
Laser
Variable attenuator
for reference
beam
Beam
Splitter
Viewing lens
Variable apperture
Television
Camera Tube
Object
Position

FIG. 7. OPTICAL SYSTEM OF SPECKLE INTERFEROMETER

FIG. 8. JET AND PLATE ARRANGEMENT

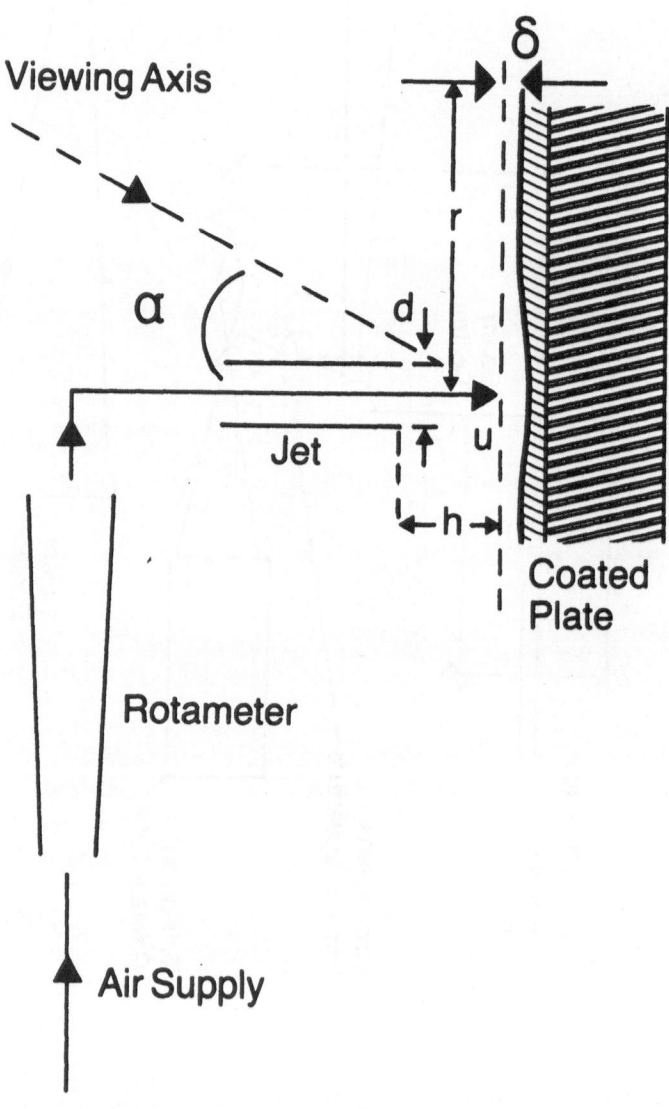

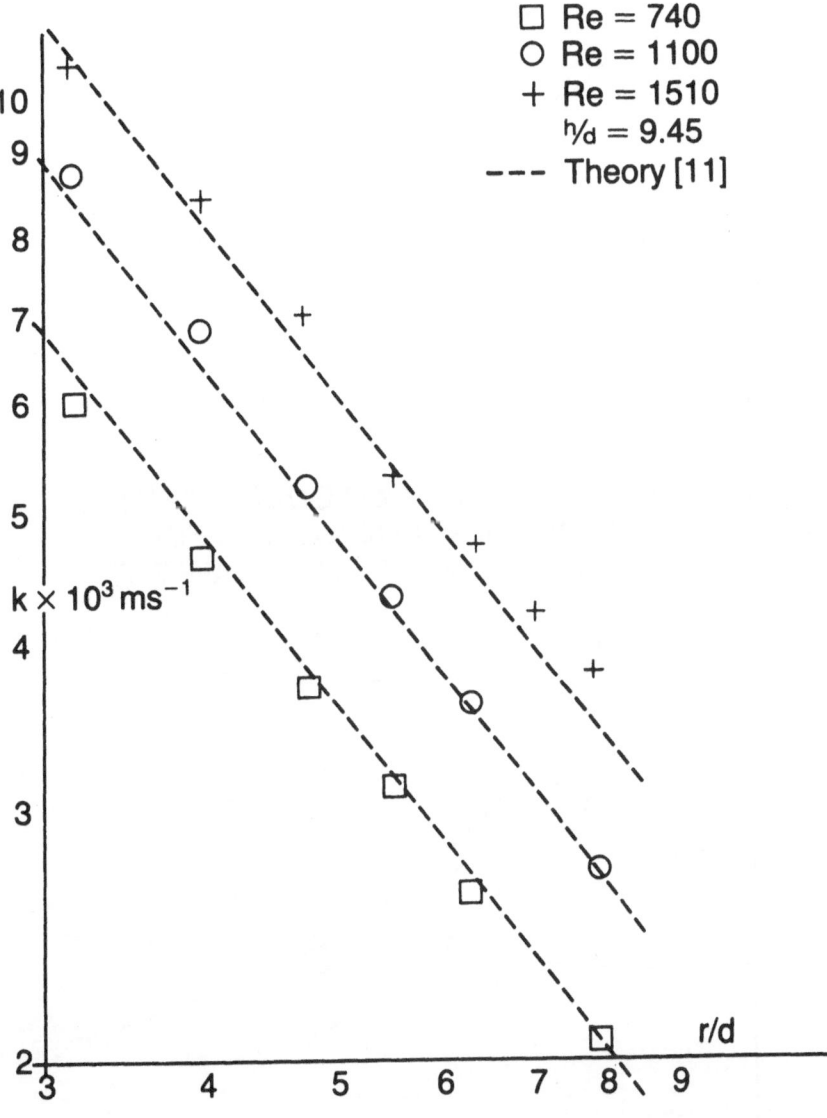

**FIG. 9. MASS TRANSFER COEFFICIENT
k VERSUS RADIAL LOCATION
r FOR LAMINAR JET
IMPINGEMENT.**

□ Re = 740
○ Re = 1100
+ Re = 1510
h/d = 9.45
--- Theory [11]

FIG. 10. MASS TRANSFER COEFFICIENT K VS. RADIAL LOCATION R FOR TURBULENT JET IMPINGEMENT.

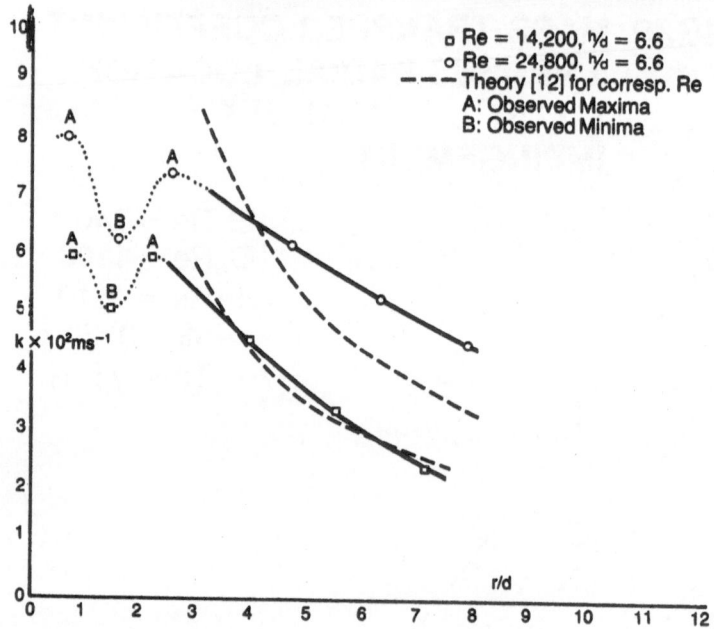

FIG. 11. MASS TRANSFER COEFFICIENT K VS. RADIAL LOCATION R FOR TURBULENT JET IMPINGEMENT.

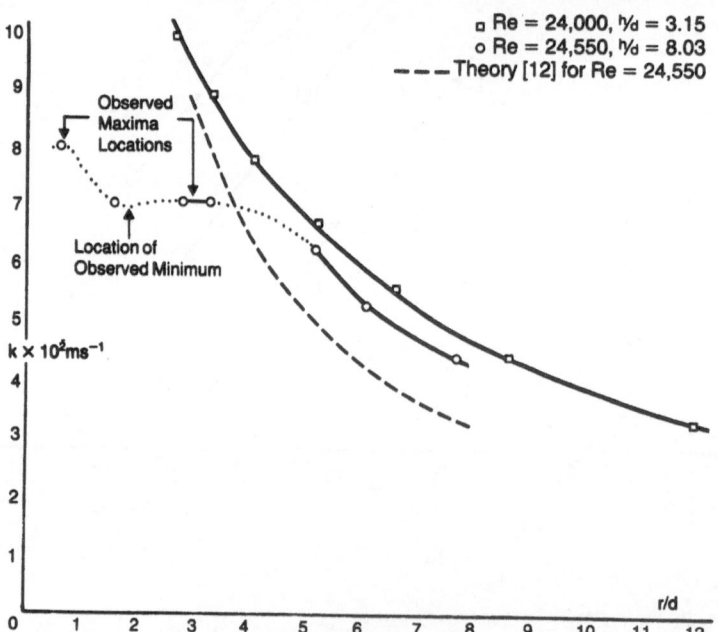

Improvements on Electronic Speckle Pattern Interferometry

O.D.D. Soares, A.L.V.S. Lage, H. Sakowski*

Centro de Física, Universidade do Porto, Portugal

Abstract

Fibre optics and synchronized amplitude and phase illuminating pulse modulation, controlled by microprocessor, are combined to improve performance and capabilities of measurements with E.S.P.I.. Cosmetic and resolution betterment of specklegrams is achieved while at the same time it is possible to design a portable and compact set-up, insensitive to vibrations and orientable towards the object. Novel features and concepts are introduced that further open the field of research with E.P.S.I. techniques.

----

* On leave from University of Münster, present address Polytecnicum of Koln.

Introduction

Speckle metrology (1,2) is receiving increasing interest in view of its engineering and biomedical applications. Major problems in outside laboratory implementation of speckle methods have been tackled with ingenious novelty using specialized techniques.

Electronic Speckle Pattern Interferometry (3) (E.P.S.I.) presents particular interest as an alternative to interferometric image holography at a lower spatial resolution because of its advantageous features, such as short exposure, real-time high sample video rate specklegrams, video display and analogical processing. Furthermore, dynamic studies (e.g. vibrations and transients) may be carried out without the need for a pulsed Laser.

However, a number of systematic difficulties have contributed to the slow rate of expansion in the use of the technique. The most frequent objection to the application of E.S.P.I. is that the appearance of the fringe patterns is too coarsely speckled and that there are a limited number of fringes in the field of observation. Phase ambiguity is also common to E.S.P.I. creating uncertainty in the interpretation of specklegrams, in particular, whenever identification of the signal of the direction of deformation is requested.

Nonetheless, E.S.P.I. is a powerful technique so that it is justifiable to search for solutions to the recognized difficulties of its application.

An introduction of optical fibres and pulse modulation both in amplitude and phase is presented with discussion of their features and advantages to practical implementation of the technique.

The objectives of the research are to develop instrumentation and methods for the analysis of transitory dynamical regimes.

E.S.P.I. with Optical Fibres (4)

Several authors have explored the advantages of introducing optical fibres to guide both reference and object beam in interferometric techniques.

A great variety of optical fibre interferometric sensors have also been developed. E. S. P. I. using optical fibres has already been proposed (5).

A novel approach to the use of optical fibre was evaluated, Fig. 1 in which the reference beam is guided by a monomode optical fibre passing through the imaging lens, along its optical axis and with the end face positioned at the section of effective minimum diameter within the imaging forming system, Fig.4. Alternatively, the positioning of the fibre end is made in a way that its numerical aperture and the image numerical aperture seen from the TV-target are identical and co-axial, Fig.5.

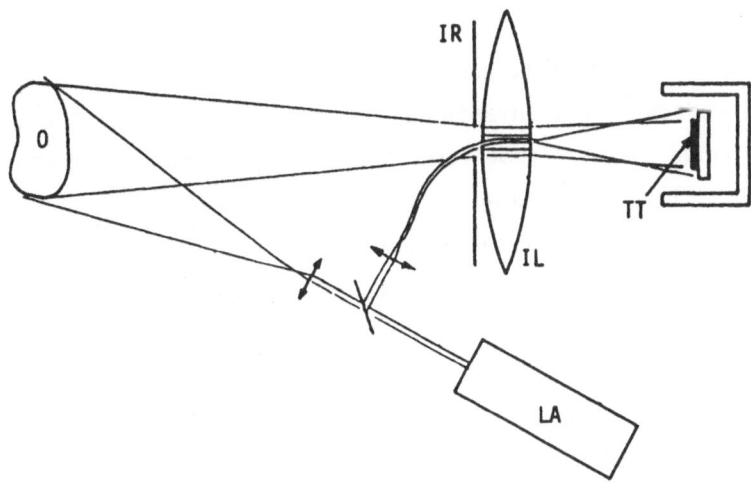

Fig.1: Concept design of ESPI with optical fibre guided reference beam through imaging lens (4).
(LA - laser, O - object, IL - imaging lens, IR - iris,
TT -TV-camera target)

The optical fibre may be just positioned before (reference beam then disturbed by artifacts on the lens) or after (availability of space may be limited) the imaging lens.

Main advantages of the concept are:

i) reference beam co-axial with imaging forming wavefront. Consequently, spatial frequency bandwidth is at minimum, resulting in optimum use of available resolution of recording target. Higher frequencies of the image may still be encoded with final resolution improvement, i.e. more fringes may be observed on the TV-monitor

ii) contrast improvement of correlation fringes is achieved by attenuating the zero order component and the lower frequencies of object image once the optical axis of imaging lens is used for optical fibre positioning.

iii) number of optical components is reduced and a simpler configuration is achieved. The distance from imaging lens to TV-camera target can be reduced with advantage. A large area of the object may be observed with the same Laser beam intensity and target sensitivity. Eventually, the imaging lens will be substituted by another imaging forming system (e.g. for microscopic objects imaging, image guiding for surfaces of difficult access, or a zooming system)

iv) clean reference beam is provided improving interference pattern quality

v) easy alignment and simultaneous adjustment of reference and object beam is obtained

vi) desensitization to parasitic vibrations results.

Optical fibres may also be used both for object illumination and image guiding. Fig.2 presents a design concept to obtain miniaturization, mobility, compactibility and portability of the entire system. The laser beam may be coupled with appropriate power division by recourse to a holographic coupler (6). Areas of difficult acess can be conveniently illuminated by appropriate guiding of illuminating beam and surface image. Fig.3 presents the use of a GRIN rod in a configuration designed for endoscopy.

Fig.4 compares results obtained from the study of membrane vibration with an E.S.P.I. set-up using beam-splitter, and implemented with monomode optical fibre with arrangement of Fig.1.

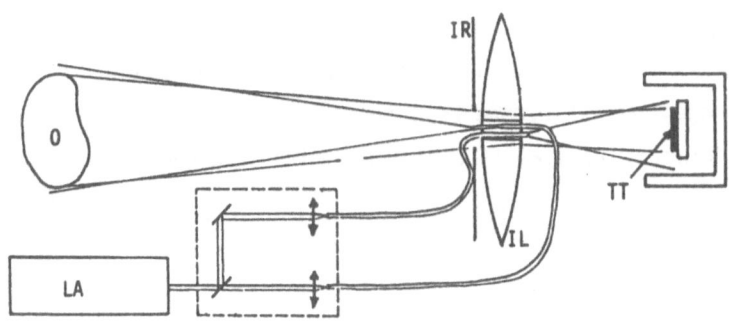

Fig.2: Compact design of an E.S.P.I. system using optical fibres (4). Ultimately, it would consist of a head comporting imaging lenses, optical fibre bundle with Laser beam coupler and adjustable iris diaphragm, to be mounted on a TV-camera

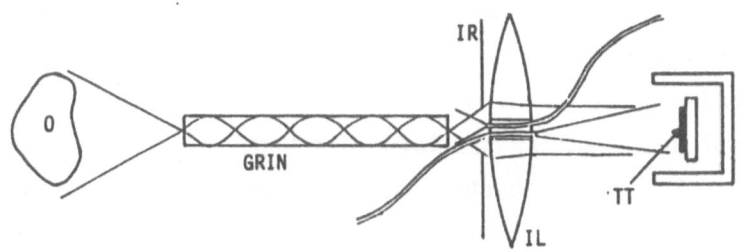

Fig.3: Image guiding with a GRIN rod on E.S.P.I. for endoscopic uses (4)

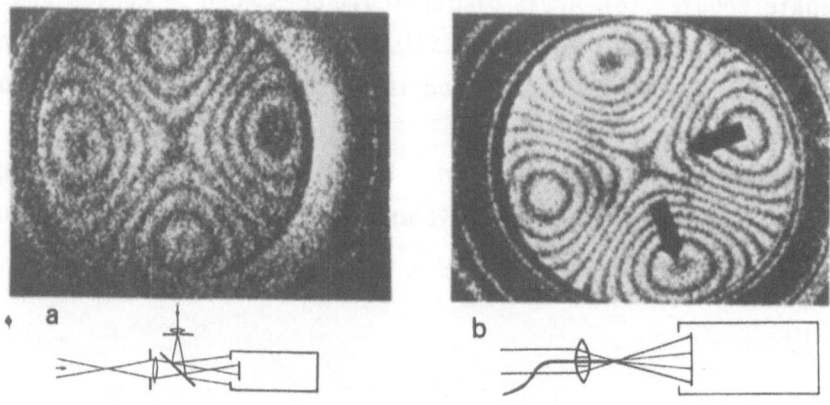

Fig.4: Time-average E.S.P.I. analysis of membrane vibration

    a) Object and reference beam combined by a beam splitter.

    b) Monomode optical fibre is used with arrangement of Fig.1.

E.S.P.I. with Pulse Modulation

A convenient method for dynamical studies is pulsed illumination synchronized with course of events.

There has been developed a microcomputer based system to pulse a CW Laser beam by electrooptic modulation both in amplitude and phase (7). The amplitude modulation is designed to create a sequence of pulses. The number of pulses, width, time interval between pulses, and synchronization of the pulse set with observed event can be fully programmed and controlled.

The control may be determined by a loaded programme and the input of data or dynamically adjusted both at the operator command or under sensing and control signals.

The phase modulation is used to vary the relative phase between pulses (object or reference beam) to obtain either a single or combined effect. The main effects are:

i) full exploration of the vibration cycle
ii) asymmetries identification at each half cycle of vibration
iii) phase measurement of vibrational mode spectrum
iv) resolving ambiguity on displacement direction
v) speckle averaging to optimize fringe quality
vi) stable background elimination on the specklegram

A general discussion being outside of the scope of the present text a number of examples of application will be discussed.

Fig.5 shows schematically the general lay-out of the laboratory set-up, and Fig. 6 presents specklegrams of resonant modes on a circular membrane.

When analysing vibrations a phase difference exists between the generated excitation signal and the mechanical forced vibration that is dependent on vibration frequency and mode. It is revelant to stroboscopic methods either to measure this phase difference or to localize the light pulses in relation to the effective state of vibration. Fig.7 shows specklegrams of forced vibrations at resonant frequency 1550 Hz on a metal plate (70x70 mm^2) exhibiting the phase delay between the effective object vibration and excitation sinusoidal signal. The double pulse set (pulse width 5 to 8 µs) is shifted continuously along the vibration excitation signal cycle up to the point where maximum deformation is observed, Fig.7 b).

The Fig.8 shows the evolution of phase delay with frequency.

The Fig.9 represents the modulus of relative deformation at frequency 1.5 KHz along a direction of observation.

To explore the vibration cycle one of the pulses remains locked (e.g. maximun deformation observed, corresponding to synchronism at an extreme of the amplitude) and the other pulse varies its position along the cycle,

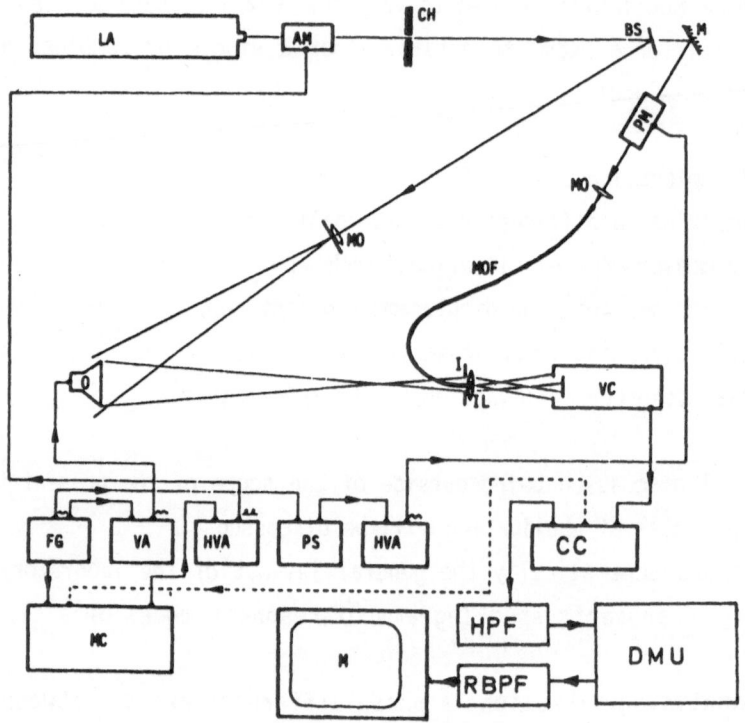

Fig.5: Laboratory arrangement for E.S.P.I.
(LA - Ar Laser, MOF - monomode optical fibre, IL - imaging lens, VC - high resolution video camera, FG - function generator, HVA - high voltage video amplifier, VA - video amplifier, PS - phase shifter, VC - video controller, CC -camera controller, DMU - dynamical memory unit, HPF- high-pass filter, RBPF - rectifier band-pass filter, M - monitor, MC-microcomputer, AM - electro-optic amplitude modulator, PM-electro-optic phase modulator, CH - chopper, M - mirror, BS - beam splitter, O-Object, MO - microscope objective, I-iris diaphragm)

Fig.10. The relative deformations along a direction are plotted in Fig.11. If one takes the successive deformation values along the vibration cycle, e.g. point A, Fig.11, the temporal law of effective deformation at the point is obtained, Fig.12, so that by Fourier analysis spectral components can be estimated.

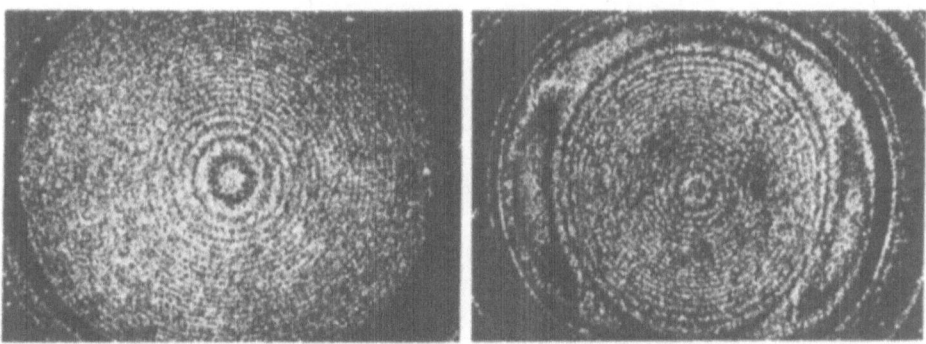

Fig.6: Stroboscopic twin pulse illumination specklegrams

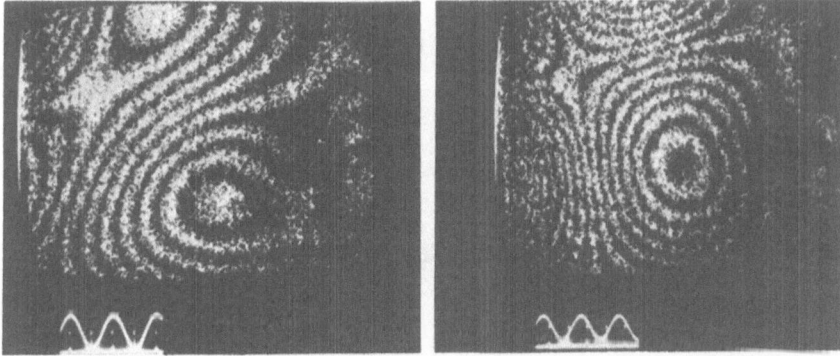

Fig.7: Double pulse stroboscopic E.S.P.I. for analysis of vibrations with mechanical excitation phase delay evaluation
   a) Double pulse set positioned at extreme amplitude of generated excitation signal.
   b) Localization of synchronized double pulse illumination with effective maximum amplitude of object vibration.

596

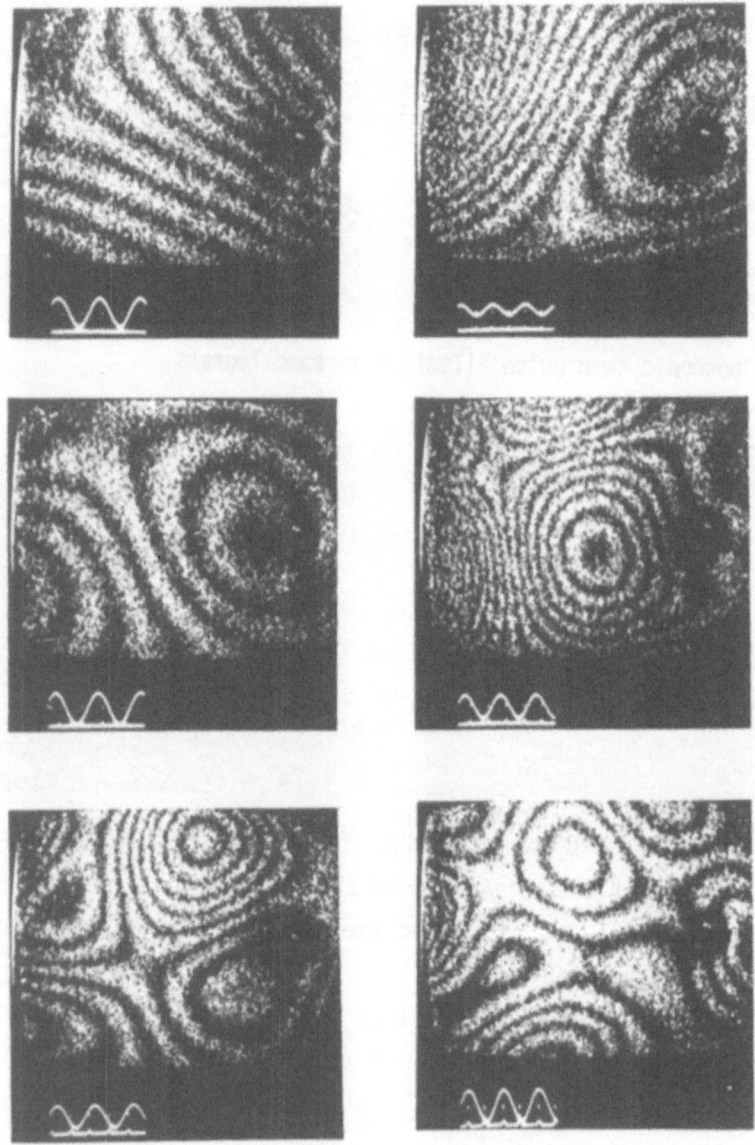

Fig.8: Mechanical excitation phase delay variation with frequency on a metal plate

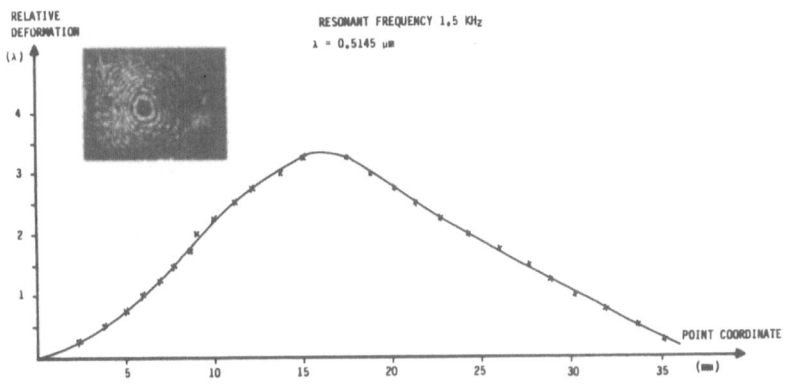

Fig.9: Relative deformation along an observed direction taken from the specklegram of Fig.7b (Stroboscopic E.S.P.I.)

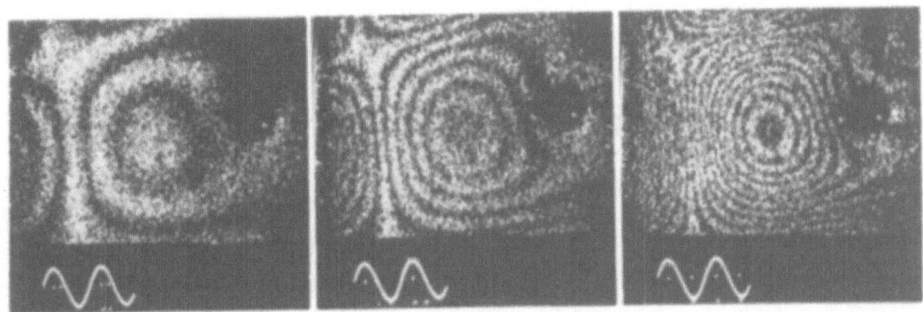

Fig.10: Exploration of vibration cycle by variation of the time interval between twin pulses.

The ambiguity on the direction of deformation can be resolved by introducing a known phase shift on reference beam with phase modulator (Fig.5) and observing the fringe movement as represented by the self-explanatory Fig.13.

Transient studies and parasitic vibration effect removal could be handled by providing the double pulse illumination set just at the begining of every video image. Alternatively, it is possible to start the target reading after double exposure on the target (CCD cameras may be used). This implies the solution of two problems: synchronization and achievement of high extinction rate of the pulse modulator (7). The microcomputer based system is capable of solving the synchronization

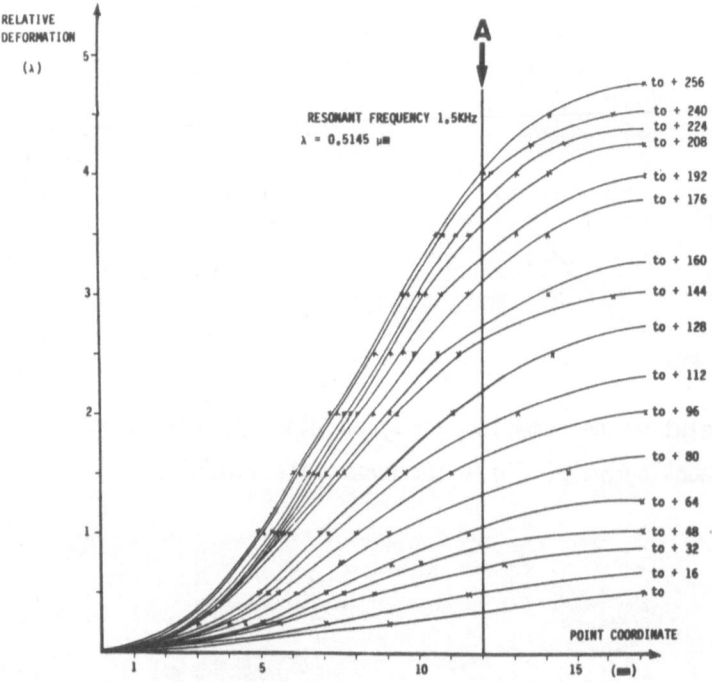

Fig.11: Relative deformation temporal evolution during the vibration cycle along an observed direction. Time interval between pulses was increased by 16 us (1/94 of the period)

problem, however improvements are still needed to increase the extinction rate of the amplitude modulator (7). Fig.14 shows the specklegram of resonant modes on a circular membrane with described tecnique. To eliminate the time average component due to improper extinction ratio of the amplitude modulator a chopper was introduced and openned only during the twin pulse set time interval, Fig.14b). (Another solution is to introduce on the reference beam a phaseshift from $-\pi/2$ to $+\pi/2$ distributed within the periodicity of the twin pulse set equally. Acousto-optic modulation can also be used).

Fringe quality in the observation of stationary events can be improved by speckle averaging, varying the relative phase between object and

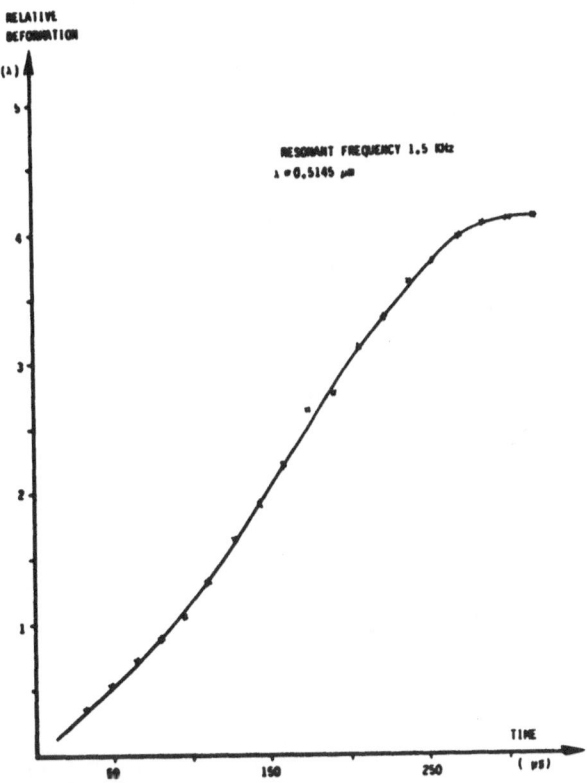

Fig.12: Relative deformation at one point during the vibration cycle
(Point A, Fig.11)

reference wavefront, and by changing the illumination of the object.

   Stable background elimination, including spurious speckle, is obtained
by reversing the phase of the reference wavefront during the illumination
by one of the twin pulses. If multiple vibrational cycles are integrated
on the target it is sufficient to alternate from 0 to π the relative
phase of the reference wavefront during the illumination of the double
pulse set in sequence.

600

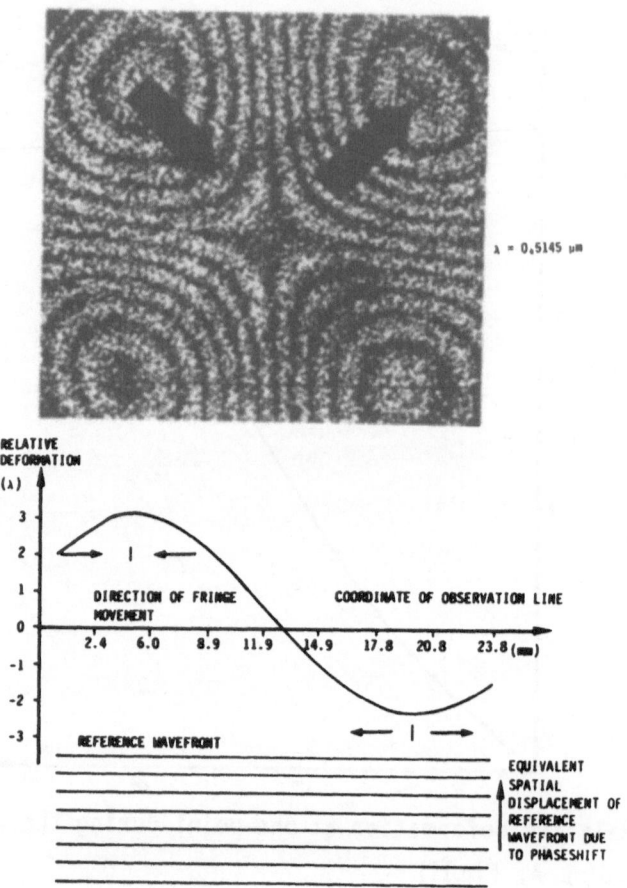

Fig.13: Elimination of signal ambiguity on the direction of deformation by fringe displacement, due to selected phase shift on the reference beam

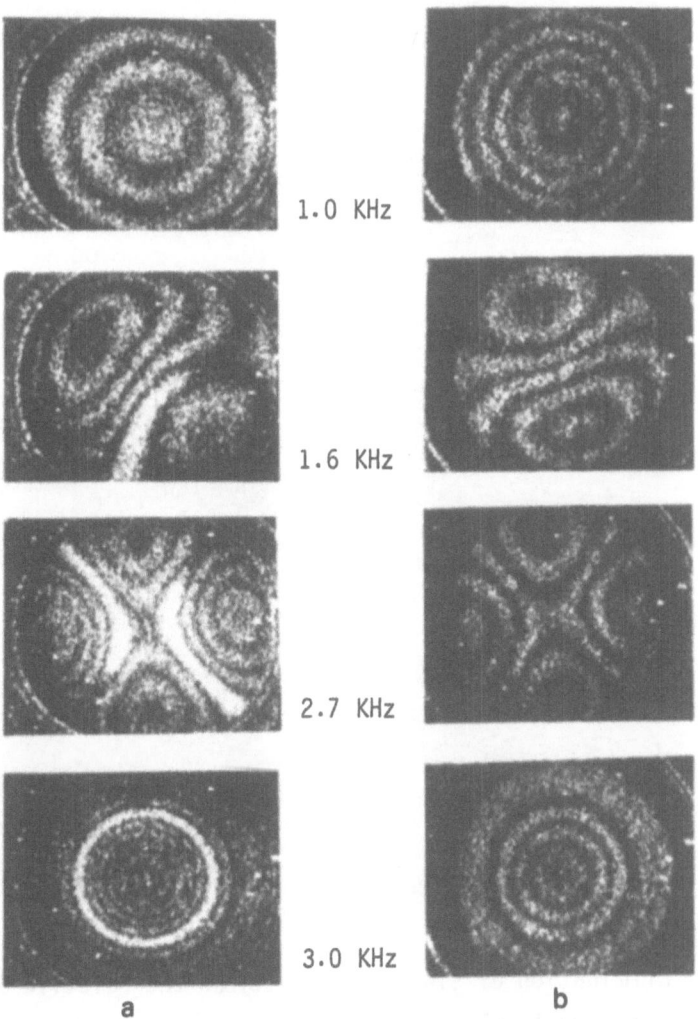

1.0 KHz

1.6 KHz

2.7 KHz

3.0 KHz

a   b

Fig.14: E.S.P.I. with double pulse illumination synchronized with video
image reading and excitation
a) The resonant modes on a circular membrane are shown but poor
extinction rate of amplitude modulator blurs the image.
b) A chopper is introduced to eliminate the time average component
by blocking the beam outside the time interval of the pulse
set.

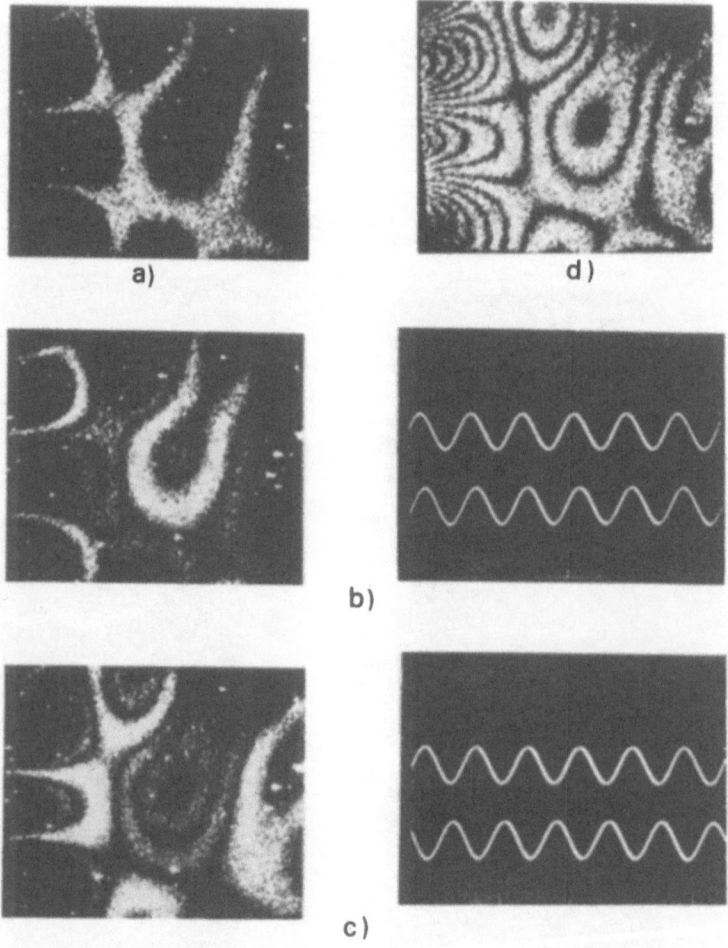

a)

d)

b)

c)

Fig.15: Elimination of signal ambiguity on the relative displacement direction in time average E.S.P.I. by phase shift with synchronous phase modulation of reference beam

a) Time average specklegram of metal plate at resonant frequency 2.9 KHz.

b) Time average specklegram with synchronous phase modulation of reference beam.

c) Same as b) but with a phase shift of π

d) Double pulse specklegram.

Discussion

Phase modulation of the reference beam wavefront can be used to resolve the phase ambiguity of resonant modes in time averaging E.S.P.I., Fig.,15. The phase modulator is run at the same frequency of the object vibration. If a phase shift of π is introduced in the reference wavefront the fringe pattern changes. Regions now in phase with reference wavefront will have a lower number of fringes, and complementally, regions in opposition of the phase will reveal an increase in the number of fringes.

Heterodyning proprieties of phase modulation of reference beam are currently being investigated.

The disadvantages of time average as compared with stroboscopic E.S.P.I. are well known:

i) elaborate extraction of data
ii) fringe contrast decay with crescent vibration amplitude.

The considered system requires a convenient image processing back-up and computer data handling for adequate data extraction. Furthermore, moiré evaluation techniques (8) have been considered of great pertinence to transient analysis so that a dynamical digital memory has been developed (9,10) and incorporated in the system.

Transposition of some of the results seem also to deserve investigation into real-time interferometric holography with photorefractive crystals.

Conclusion

Improvements in E.S.P.I. have been described where fibre optics, pulse modulation and phase modulation have been combined to realize a more versatile and higher resolution set-up available to industrial and biomedical applications.

It is believed that ultimately it would be possible to overcome the coarsely speckled fringe pattern appearance, to tailor the exposure time and dilute the high initial cost of the system by extending the simplicity and flexibility of its operation, improving its stability and covering a broad range of measurement to take maximum advantage of a system which has practically no running costs.

Acknowledgements

The reported results belong to a joint research-project with the University of Munster. Presented experimental results were mainly taken from the report of H. Sakowski, covering his stay at University of Porto. The authors greatly acknowledge the donation of equipment by the Deutsche Gesellschaft fur Technich Zusammenarbeit (G.T.Z.) and the support of scientific exchange by the Deutscher Akadmischer Austauschdienst (DAAD).

Contributions to the project development by A.O.S. Gomes, J.C.M. Santos and technical assistance by J.S. Fernandes and L. Vilaça were of great value.

Research funds from Centro de Física, INIC, were provided for the project.

References

1. Erf, R.K., Speckle Metrology, Academic Press, NY (1978)

2. Jones, R.; Wykes, C., Holographic and Speckle Interferometry Cambridge University Press, Cambridge (1983)

3. Lokberg, O.J., Electronic Speckle Pattern Interferometry Optical Metrology, Martinus Nijhoff (1985)

4. Soares, O.D.D.; Lage, A.L.V.S., Sistema Optico para Interferometria de Granitado Laser com Fibras Opticas, Patent Nº 80335, INPI, Portugal (1985)

5. Lokberg, O.J.; Krakhella, K., Electronic Speckle Pattern Interferometry Using Optical Fibres, Optical Communications 38 (1981), 155-158

6. Soares, O.D.D., Optical Coupling by Real Image Projection Holography, Recent Advances in Optical Physics, ICO - 10, Prage (1975), 537-546

7. Soares, O.D.D.; Lage, A.L.V.S., Controllable Synchronized Multipulse Illumination System for ESPI and Holography, Proc. SPIE vol. 427 (1983)

8. Soares, O.D.D.; Lage, A.L.V.S.; Bernardo, L.M.,Moiré Evaluation with Fringe Patterns of Interferograms, Holograms and Specklegrams, Optical Metrology, Martinus Nijhoff (1985)

9. Soares, O.D.D.; Lage, A.V.LS., Use of TV-Frame-Memory on Electronic Speckle Pattern Interferometry Applied to Orthopedics, Proc SPIE vol. 348 (1982), 838-844

10. Soares, O.D.D.; Lage, A.V.L.S.; Gomes, A.O.S.; Santos, J.C.M. Dynamical Digital Memory for Holography, Moiré and ESPI Optical Metrology, Martinus Nijhoff (1985)

SPECKLE REDUCTION IN FOUR-WAVE MIXING IMAGING WITH A Bi_{12} Si O_{20} CRYSTAL

L.M. Bernardo, O.D.D. Soares

Laboratório de Física, Faculdade de Ciências, U.P.
4000 Porto, Portugal

Speckle on the image plane of a four-wave mixing imaging system may represent an important source of noise. In most of the practical applications the signal/speckle-noise ratio (SNR) has to be increased by some kind of speckle reduction technique. (1,2)

There are, in general, two sources of noise which degrade the image quality: the background speckle, due to non-ideal quality and cleanness of the optical system, and the speckle intrinsic to the real conditions of image reconstruction. In this short communication, techniques for their reduction are studied.

Fig.1 shows the set-up used for image reconstruction by four-wave mixing. The recording medium is a Bi_{12} Si O_{20} (BSO) crystal, with dimensions 10x10x2 mm^3, in transverse configuration, without applied external electric field. The object is a specular square with dimensions 0.8x0.8 cm^2. Its phase conjugate image is received on a vidicon tube, V, and seen in a TV monitor, MO.

The background speckle is caused by reference beams 1,2 and the object beam 3 and originates at the different components of the set-up. Fig.2 shows background speckle, at the image plane, corresponding to different structures, sizes and intensities. It was found that the crystal, illuminated by reference beam 1, was the most important source of noise, due to the scattering on its surface as a result of dust and inhomogenities.

When a BSO crystal is used as recording medium, the polarization of the signal beam 4, is slightly elliptical with its minor axis rotated 90º from incident beams. The use of an analyser crossed with incident polarization will improve the SNR, by reducing the background speckle (3). If a $\lambda/4$ retardation plate is inserted in beam 2, the axes of tne elliptical polarization are rotated by 90º with the major axis normal to the polarization of beams 1 and 3. If the analyser is kept normal to the polarization of those beams, the image intensity is larger and SNR is further increased. Fig.3a shows the reconstructed image without both retardation plate and analyser. Figs. 4b, c show the image with retardation plate and analyser axis parallel (b) and normal (c) to the polarization plane of beams 1 and 3. Fig.3c presents a quite remarkable improvement on SNR, when compared with Fig. 3a,b. Ringing of the images is mainly attributed to the fact that the laser beam is not spatially filtered in view of power availability.

Once background speckle is eliminated, the intrinsic speckle is clearly distinguishable as in Fig. 3c. Two different averaging techniques to reduce that speckle have been tried. Both techniques are based on superposition of uncorrelated speckle patterns(4).The first one relies on the movement of the crystal, in its own plane, during the exposure time (5), the second one on moving the hologram fringes, inside the crystal. This is obtained by phase modulation of the reference beams that corresponds to a periodic (1-10 Hz was used) shear of hologram fringes with an amplitude of their spatial period. The results achieved by using those two techniques are shown in Figs. 4a,b, respectively. Their averaging effects for speckle reduction appear to be very similar. However, if the analyser is not used in the signal beam, moving the crystal would be much more effective, as both intrinsic and background speckle are averaged. Since the intrinsic speckle, on the image, can be averaged by relative movement of the hologram and the crystal, it probably results from scattering in the region of the crystal, where the hologram is registered. The scattering centers may be related with inhomogenities and impurities in the crystal bulk.

In conclusion, SNR can be increased significantly, in the image plane of a four-wave mixing system in a BSO crystal, by the use of a quarter wave retardation plate in the reading beam and analyser, to eliminate background speckle. By moving the crystal or the written hologram the intrinsic speckle can be reduced. Since the relative movement of the hologram and the crystal can decrease the four-wave mixing reflectivity (6), the response time of the recording medium and the exposure time are parameters that have to be taken into consideration, whenever the described techniques are to be implemented.

These preliminary results suggest the interest of further studies to descriminate the contributions for the speckle seen at the image.

The authors would like to thank Prof. S.P. Almeida for fruitful discussions and A.L. Lage for technical assistance. Donation of equipment by the Deutsch Gesselschaft für Technik Zusammenarbeit (GTZ) and a research grant from INIC made this research possible.

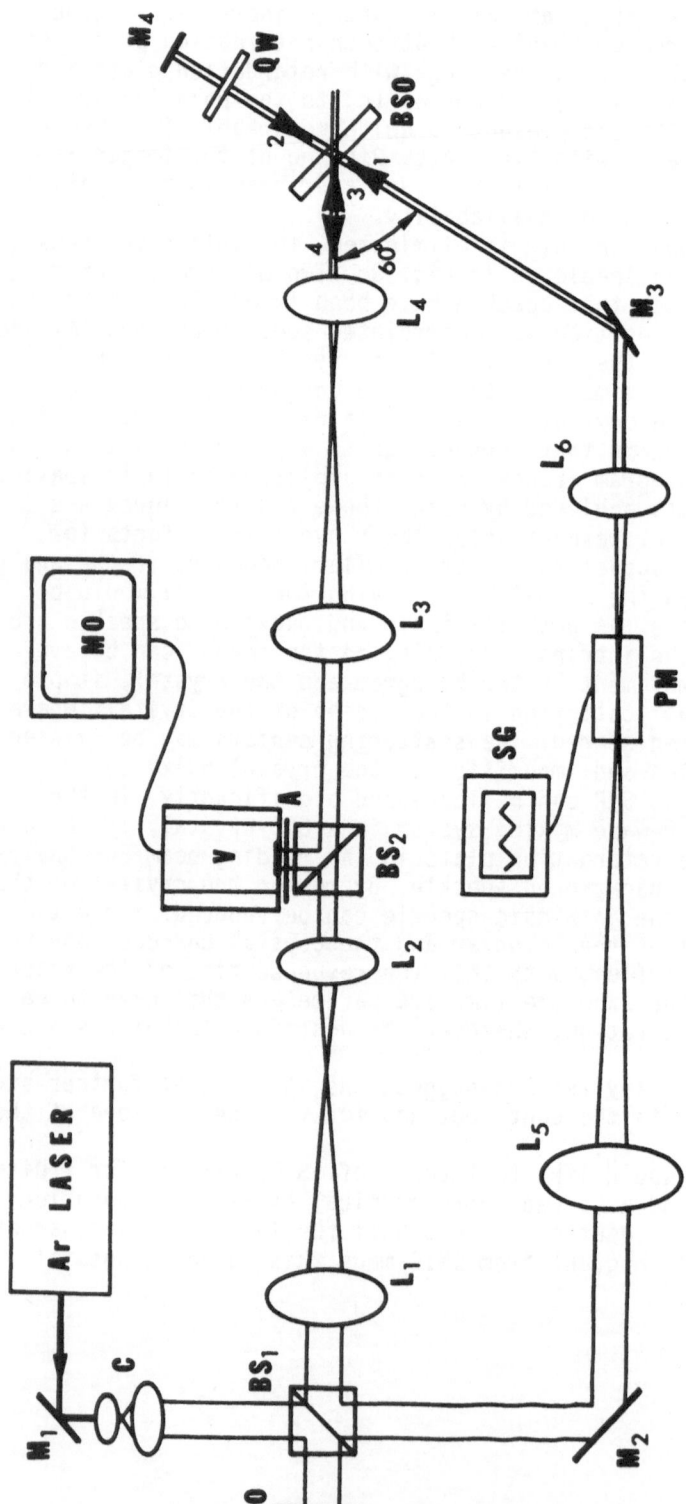

Fig.1 Scheme of the set-up for the four-wave mixing imaging system. (C, collimator; BS, beam spliters; L, lenses; M, mirrors; O, reflecting plane; QW, quarter wave retardation plate). The object is on the focal plane of L, whose focal distance is f_1. The ratios f_1/f_2, f_3/f_4 and f_5/f_6 are respectively 4, 5 and 8. The demagnified image of the object is formed in the BSO crystal at a distance f_4 from L_4. V is the vidicon tube and MO the TV-monitor; A is the analyser; PM, the phase modulator and SG its signal generator and driver.

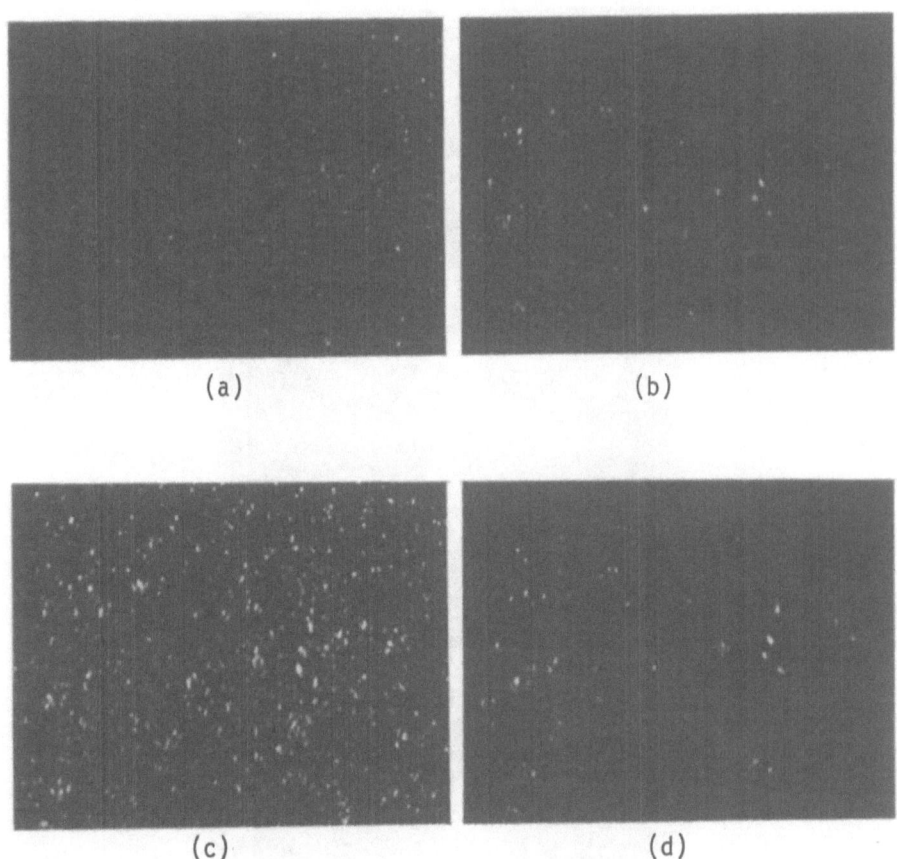

Fig.2 Speckle originated from the different beams and optical components. For all the beams present, a stop is introduced
between: (a) L_3 and BS_2, (b) BSO and L_4, (c) BS_2 and L_2, (d)
the same as in (c) with beam 2 also stopped.

610

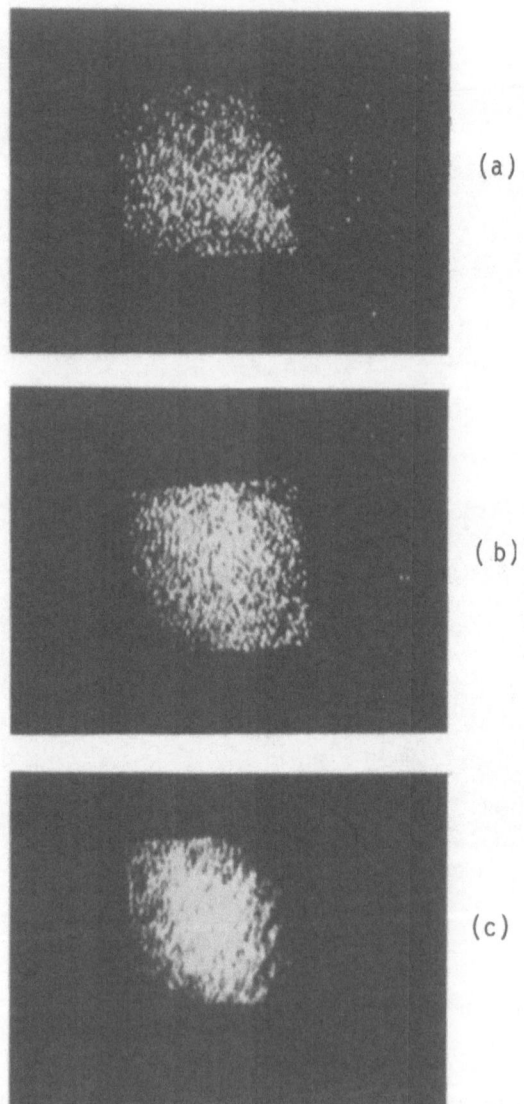

Fig.3 Phase conjugate reconstructed image as seen in the TV-monitor: (a) when beam 2 polarization is not changed outside the crystal and no analyser is inserted in the signal beam; rotation of the polarization of the beam 2 and insertion of analyser with its axis parallel (b) and normal (c) to the polarization of beam 1 and 3.

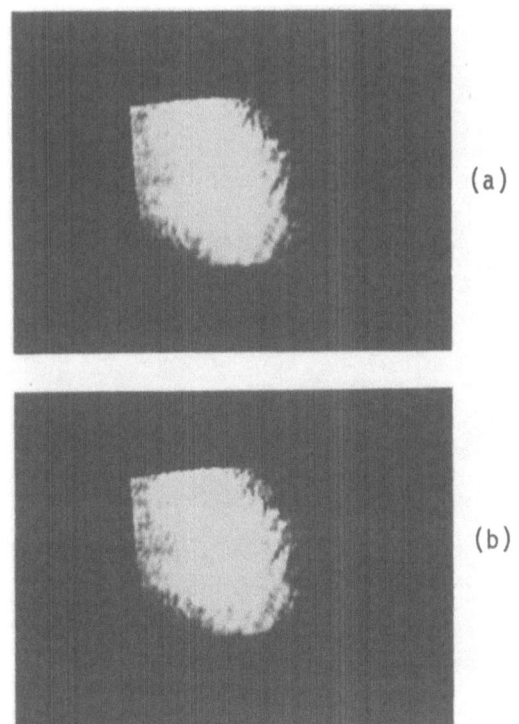

Fig.4 Time-averaged images under conditions of the Fig.3c, corresponding to the situations such that during the exposure time: (a) the crystal is moved, (b) the hologram is continously shifted in the crystal by phase modulation.

REFERENCES:

1. J.P. Huignard, J.P. Herriau, L. Pichon and A. Marrakchi, Opt. Lett. 5, 436 (1980).
2. M.D. Levenson, J. Appl. Phys. 54, 4305 (1983)
3. J.P. Herriau, J.P. Huignard and P. Aubourg, Appl. Opt. 17, 1851 (1978)
4. W. Martienssen and S. Spiller, Phys. Lett. A 24, 126 (1967)
5. L.M. Bernardo and S.P. Almeida, Appl. Opt. 22, 3926 (1983)
6. S.I. Stepanov, V.V. Kulinov, M.P. Petrov, Opt. Comm. 44, 19 (1982)

OPTICAL FIBRE METROLOGY

SINGLE-MODE FIBER OPTICAL SENSORS

R. Dändliker and A. Bertholds

Institut de Microtechnique de l'Université,
CH-2000 Neuchâtel, Switzerland.

Single-mode optical fibers can be used to carry coherent light in any kind of interferometric arrangement. External action on the fiber may change the phase of the light propagating through the fiber. Fiber optical sensors employ this type of interferometry, together with sophisticated opto-electronic detection, to measure different kinds of physical or chemical effects. Some basic considerations and some practical examples are presented.

1. INTRODUCTION

1.1. Single-mode Fibers

Single-mode silica fibers for visible (He-Ne) or near-infrared (GaAs) laser light have core diameters of typically 3 to 6 µm, sufficiently small to guide only the fundamental mode, which has a nearly Gaussian shaped intensity profile. The phase-velocitiy is described by the propagation constant β, which depends on the wavelength through the refractive index of the fiber material (material dispersion) and the waveguide effect (modal dispersion). However, this fundamental mode may still have two independent states of polarization.

Single-mode fibers are usually birefringent as a result of an (often unintentional) lack of circular symmetry in the core cross section. This and an associated stress anisotropy allows the fiber to support two nearly degenerate orthogonally polarized modes with a small phase-velocity difference. In a typical fiber the magnitude of this intrinsic birefringence is relatively small and can be severely modified by environmental factors such as pressure, twists

Soares, O.D.D. (ed), Optical Metrology
© *1987. Martinus Nijhoff Publishers, Dordrecht.*

616

and bends. Thus the overall fiber birefringence cannot be predetermined and the output polarization state is not maintained. In
so-called "polarization-maintaining" fibers the internal birefringence is increased to a level well above the environmental
effects. Thus a single linearly polarized mode can be selected and
sustained without coupling to its orthogonally polarized partner.
There is a clear need for both low- and high-birefringence fibers
in sensor and communications technology and both types of single-
mode fibers are currently under development [1].

1.2. Interferometry

There are basically three possibilities, shown in Fig.1, for
interferometric arrangements with single-mode fibers. First,
classical two-branch interferometers, such as Michelson or Mach-
Zehnder [Fig.1(a)], with two fibers, one in each arm. The fibers,
and therefore also the light paths, may take any desired shape, as
e.g. in hydrophones [2]. Second, counter-propagating waves in one
and the same fiber [Fig.1(b)], such as in optical gyroscopes [2,3].
Third, two orthogonally polarized waves in one and the same fiber
[Fig.1(c)], as used in the so-called polarimetric sensors, e.g. for
mechanical force (induced birefringence) or magnetic fields
(Faraday rotation) [3].

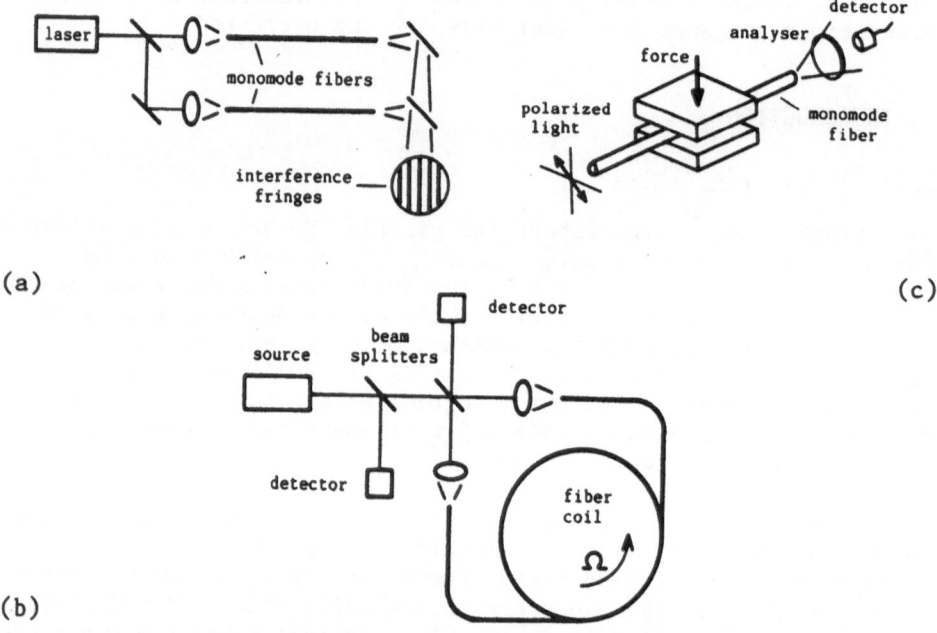

(a)

(c)

(b)

Fig.1. Interferometric arrangements with single-mode fibers.
(a) Mach-Zehnder interferometer, (b) Sagnac interferometer,
(c) polarimetric interferometer.

1.3. Detection Schemes

Different opto-electronic detection schemes are employed, depending on the desired sensitivity, accuracy, resolution, stability, dynamic range, speed, etc of the fiber optic sensor. Many of them are already known from conventional interferometry. However, preference is given to practical realizations which are compact and small to fit best the miniaturisation potential of fiber optics [2].

The following list gives the most important schemes with their main properties:
- fringe counting: low sensitivity, digitial resolution and large dynamic range;
- homodyne detection: high sensitivity, limited dynamic range, sensitivity drift;
- homodyne detection with phase compensation [4]: high sensitivity for ac signals, limited resolution and dynamic range;
- phase compensation: high resolution, speed and accuracy limited by the compensation element;
- sinusoidal phase modulation: high sensitivity, low dynamic range;
- heterodyne detection (linear phase modulation): high sensitivity, high resolution and large dynamic range;
- microprocessor based fringe interpolation and counting [5]: high accuracy and resolution, large dynamic range.

There is no single optimum detection system for all applications of fiber optic interferometric sensors. Each scheme has advantages and drawbacks which must be evaluated in the light of a particular application. Unfortunately, this evaluation is made more difficult because not all the detection systems have been investigated in sufficient detail. Some of the above-mentioned detection schemes will be described in connection with the following examples of single-mode fiber sensors.

2. FIBER-OPTIC ROTATION SENSORS

Fiber-optic rotation sensors [Fig.1(b)] are based, similar to the laser gyroscopes, on the Sagnac effect [2,6]. The two counter-propagating waves in the ring-shaped interferometer travel a slightly different distance, when the interferometer rotates with an angular velocity Ω about an axis normal to its plane. This leads to a mutual phase shift 2ϕ between the two waves. In the case of a fiber-optic gyroscope, this phase shift is found to be

$$2\phi = 8\pi NA\Omega/\lambda c = 2\pi D\ell\Omega/\lambda c, \tag{1}$$

where A is the area enclosed by the fiber ring, N is the number of turns of the fiber coil, λ and c are the vacuum wavelength and velocity of the light, respectively. The sensitivity of such a gyroscope increases proportionally with the number of turns N and thus also with the fiber length. The second part of Eq.(1) holds for a fiber of length ℓ wound on a circular coil of diameter D. A typical example of a device with 1 km fiber on a drum of 10 cm diameter and operating at a wavelength of λ = 850 nm produces a phase shift of only $2\phi = 1.2 \times 10^{-5}$ rad for a rotation rate of 1 °/h (earth's rotation = 15 °/h). The best experimental set-ups exibit sensitivity in the 0.1 °/h√Hz range in a laboratory environment.

The achievement of sufficient accuracy and stability of the minimum detectable rotation rate required the solution of numerous problems. For example, it is necessary to guarantee absolutely equal optical path lengths for the two counter-propagating waves, regardless of tolerances of components (reciprocity problem). The polarization of the two interfering beams must be adjusted and maintained for maximum contrast (polarization problem). Backscattered light from a highly coherent source may interfere with the signal and produce noise and drift (coherence problem). Moreover, suitable detection schemes must be chosen to get highest sensitivity and indication of the sense for low rotation rates. With the construction shown in Fig.2 several of these problems have been solved [7]. The use of a superluminescent diode as source solves the coherence problem, the polarizer in the common port for injection and detection assures the reciprocity, and the depolarizer inside the interferometer ring eliminates the polarization problem.

As detection scheme a sinusoidal phase modulation at low frequency was chosen in this case. Low frequency phase modulation can be readily realized by stretching the fiber periodically, e.g. with the help of a piezoelectric transducer, so as to elasto-optically introduce a sinusodially varying phase shift. If the phase modulator is positioned at one end of the fiber coil, as shown in Fig.2, and it produces a phase modulation at a frequency

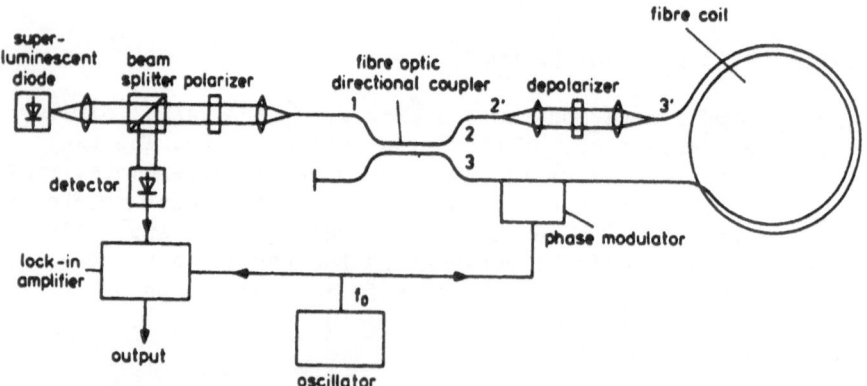

Fig.2. Experimental realization of a high-resolution fiber optical gyroscope [7].

f_o of amplitude ψ, then the detected output intensity becomes

$$I(t) = I_o \cos^2(\phi - \psi \sin\alpha \sin2\pi f_o t), \tag{2}$$

where $2\alpha = 2\pi f_o T$ and T is the transit time through the fiber coil. Expanding Eq.(2) into its frequency components shows that the amplitude of the fundamental frequency f_o contains the wanted phase in the appropriate form. Thus, after phase sensitive detection at f_o one gets a signal proportional to

$$I_f = I_o J_1(\eta) \sin2\phi, \tag{3}$$

where $J_1(\eta)$ is the first order Bessel function at $\eta = 2\psi\sin\alpha$. The amplitude ψ of the phase modulation is chosen so as to make approximately $\eta = 1.8$, and hence maximize the sensitivity. For small roatation rates ($2\phi \ll 1$) the signal I_f is proportional to Ω and indicates also the sense of the rotation.

3. MECHANICAL EFFECTS

Optical fibers are often subjected to mechanical effects, such as stretching, bending, twisting, pressing etc. Some of them may be used for sensors, others are perturbations in sensor or communication applications. In general, these mechanical effects introduce a change of the propagation constant (refractive index) and the polarization properties (birefringence, optical activity). For a theoretical description, the following three problems have to be solved successively [8-10]: First, the stress distribution in the fiber has to be calculated using the theory of elasticity. Second, the resulting changes of the index ellipsoid or the tensor of the dielectric constant have to be established through the photoelastic and piezooptic relations. Third, the changes of the propagation of light in the single-mode fiber can be determined by a coupled mode approach [8].

3.1. Photoelastic Relations

Light propagation in an anisotropic material is best described by the so-called indicatrix $(1/n^2)_{ij}$, which is identical to the inverse tensor B_{ij} of the dielectric constants ε_{ij}, namely

$$B_{ij} = \varepsilon_{ij}^{-1} = (1/n^2)_{ij}, \qquad (i,j = x,y,z). \tag{3}$$

For an originally isotropic material of refractive index n, small changes of B_{ij} and ε_{ij} are related approximately by

$$\Delta\varepsilon_{ij} = - n^4 \Delta B_{ij}. \tag{4}$$

Both, ΔB_{ij} and $\Delta \varepsilon_{ij}$, are symmetric 3×3 tensors, thus having not more than 6 independent components.

The change of B_{ij} due to mechanical strain S_{ij} is described by the so-called elastooptic relations. For convenience, the components of the symmetric tensors $\Delta \varepsilon_{ij}$, ΔB_{ij} and S_{ij} are usually represented in the form of 6-dimensional vectors $\Delta \varepsilon_k$, ΔB_k and S_k, respectively, using the convention

$$
\begin{aligned}
&k = 1 \quad 2 \quad 3 \quad 4 \quad 5 \quad 6 \\
&ij = xx \quad yy \quad zz \quad yz \quad zx \quad xy.
\end{aligned}
\tag{5}
$$

The elastooptic relations are then given by

$$
\Delta B_n = p_{nk} S_k,
\tag{6}
$$

where p_{nk} is a 6×6 matrix, describing the materials elastooptic properties. In the case of an isotropic material'symmetry requires that the strain-optic tensor p_{nk} is of the form

$$
p_{nk} = \begin{bmatrix}
p_{11} & p_{12} & p_{12} & 0 & 0 & 0 \\
p_{12} & p_{11} & p_{12} & 0 & 0 & 0 \\
p_{12} & p_{12} & p_{11} & 0 & 0 & 0 \\
0 & 0 & 0 & p_{44} & 0 & 0 \\
0 & 0 & 0 & 0 & p_{44} & 0 \\
0 & 0 & 0 & 0 & 0 & p_{44}
\end{bmatrix}, \quad p_{44} = (p_{11} - p_{12})/2.
\tag{7}
$$

For silica the numerical values are $p_{11} = 0.12$ and $p_{12} = 0.27$.

As an example, the effect of stretching a single-mode fiber will be discussed in the following. Applying an axial force to the fiber causes an elongation $\Delta L = \varepsilon_\ell L$, where L is the fiber length and ε_ℓ is the longitudinal strain. The components of the strain tensor are then given by

$$
S_1 = S_2 = -\nu \varepsilon_\ell, \quad S_3 = \varepsilon_\ell, \quad S_4 = S_5 = S_6 = 0,
\tag{8}
$$

where the Poisson's ratio ν describes the induced transverse strain under uniaxial stress in the z-direction. Using Eqs.(4), (6) and (7) one gets for the changes of the dielectric constants

$$
\begin{aligned}
&\Delta \varepsilon_1 = \Delta \varepsilon_2 = -n^4 \varepsilon_\ell [p_{12} - \nu(p_{11} + p_{12})], \\
&\Delta \varepsilon_3 = -n^4 \varepsilon_\ell [p_{11} - 2\nu p_{12}], \quad \Delta \varepsilon_4 = \Delta \varepsilon_5 = \Delta \varepsilon_6 = 0.
\end{aligned}
\tag{9}
$$

Since $\Delta\varepsilon_{ij}$ is diagonal and isotropic in the transverse (x,y)-directions [Eq.(9)], light propagating in the z-direction, which is the fiber axis, sees following Eq.(3) independently of its polarization a change of the refractive index equal to

$$\Delta n = (1/2n)\Delta\varepsilon_1 = -(n^3/2)\varepsilon_\ell[p_{12} - \nu(p_{11} + p_{12})]. \tag{10}$$

For silica one gets, with n = 1.458 and ν = 0.17, the numerical relation $\Delta n = -0.315\ \varepsilon_\ell$.

The total accumulated phase change ϕ at the end of a stretched fiber due to the elongation ΔL and the induced Δn becomes finally

$$\phi = k(n\Delta L + L\Delta n) = kL(n\varepsilon_\ell + \Delta n) = \alpha_o L\varepsilon_\ell, \tag{11}$$

with k = $2\pi/\lambda$. For silica fibers one gets at λ = 633 nm (He-Ne) a numerical value of α_o = 1.13×10^7 rad/m for the sensitivity.

3.2. Piezooptic Relations

In a linear elastic material strain and stress are related by Hooke's law. Using the contracted index representation, introduced by Eq.(5), the relation between the stress tensor σ_n and the strain tensor S_k is given by

$$\sigma_n = c_{nk}S_k, \quad \text{or} \quad S_k = c_{kn}^{-1}\sigma_n, \tag{12}$$

where c_{nk} is a 6x6 matrix, describing the material's elastic properties. In the case of an isotropic material, symmetry requires that the strain-stress tensor c_{nk} is of the form

$$c_{nk}^{-1} = (1/E)\begin{bmatrix} 1 & -\nu & -\nu & 0 & 0 & 0 \\ -\nu & 1 & -\nu & 0 & 0 & 0 \\ -\nu & -\nu & 1 & 0 & 0 & 0 \\ 0 & 0 & 0 & 1+\nu & 0 & 0 \\ 0 & 0 & 0 & 0 & 1+\nu & 0 \\ 0 & 0 & 0 & 0 & 0 & 1+\nu \end{bmatrix}, \tag{13}$$

where E is Young's modulus of elasticity and ν is Poisson's ratio. The numerical values for silica are E = 7.6×10^{10} N/m^2 and ν = 0.17.

Using Eqs.(6) and (12) one finds the piezooptic tensor q_{nk}, which relates the change of the indicatrix ΔB_n to the stress σ_k by

$$\Delta B_n = q_{nk}\sigma_k = p_{nm}c_{mk}^{-1}\sigma_k. \tag{14}$$

For an isotropic material the stress-optic tensor q_{nk} has exactly the same symmetries as the strain-optic tensor p_{nk}, shown in Eq.(7). From Eq.(14) the two relevant components are found to be

$$q_{11} = (p_{11} - 2\nu p_{12})/E, \quad q_{12} = [p_{12} - \nu(p_{11} + p_{12})]/E. \tag{15}$$

The corresponding values for silica are $q_{11} = 0.38 \times 10^{-12}$ m^2/N and $q_{12} = 2.68 \times 10^{-12}$ m^2/N.

As an example the effect of a transverse pressure acting on a single-mode fiber will be discussed [9,10]. Since the guiding core (≈ 5 μm) is very small compared to the outer diameter (≈ 125 μm) of the fiber, one may assume that the light wave sees a nearly constant stress field at the center of the circular fiber cross-section. In this case the components of the stress tensor are

$$\sigma_1 = F/\pi RL = \sigma_0, \quad \sigma_2 = -3\sigma_0, \quad \sigma_3 = \sigma_4 = \sigma_5 = \sigma_6 = 0, \tag{16}$$

where F/L is the transverse force per unit length acting on the fiber in the x-direction and R is the outer radius of the fiber, which is assumed to be of uniform silica material.

With the help of Eqs.(4) and (14) one finds for the change of the dielectric constants

$$\Delta\varepsilon_1 = -n^4\sigma_0(q_{11} - 3q_{12}), \quad \Delta\varepsilon_2 = -n^4\sigma_0(q_{12} - 3q_{11}),$$
$$\Delta\varepsilon_3 = 2n^4\sigma_0 q_{12}, \quad\quad\quad \Delta\varepsilon_4 = \Delta\varepsilon_5 = \Delta\varepsilon_6 = 0. \tag{17}$$

Since $\Delta\varepsilon_{ij}$ has different values in the transverse (x,y)-directions, the two corresponding linear polarizations see, following Eq.(3), different refractive indices. The fiber has become birefringent. The accumulated phase difference $\Delta\phi$ between the two polarizations is finally given by

$$\Delta\phi = kL(\Delta n_1 - \Delta n_2) = 2kn^3(q_{12} - q_{11})F/\pi R. \tag{18}$$

For a silica fiber of diameter $2R = 125$ μm and $\lambda = 800$ nm one calculates a sensitivity of $\Delta\phi/F = 0.57$ rad/N $= 32.7$ °/N.

4. HIGH-RESOLUTION ELONGATION MEASUREMENT

The deformation of single-mode fibers resulting from a longitudinally applied force has been measured experimentally by means of high resolution heterodyne interferometry and analysed theoretically using the second order theory of elasticity for an isotropic material under finite deformation [11].

A Mach-Zehnder interferometer with a He-Ne laser was employed (Fig.3), where the light beams in the two arms travel through two identical and equally disposed single-mode fibers. The optical phase-shift induced by the deformation of one of the fibers was

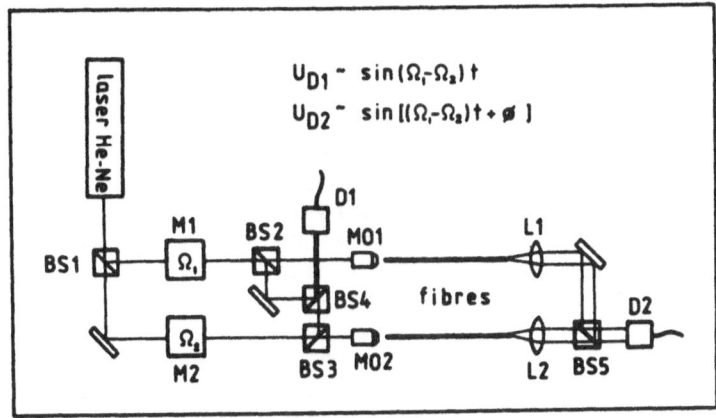

Fig.3. Heterodyne Mach-Zehnder interferometer.

measured with a precision of 1° (≈1/300 of a fringe) by means of heterodyne phase measurement. The two acousto-optical modulators M1 and M2 produce a beat-frequency of 100 kHz. The sinusoidal signals U_{D1} and U_{D2} , resulting from the interference of the two beams before and after travelling through the fibers, are connected to a phasemeter and a fringe counter.

The set-up for the elongation of the fiber by means of calibrated weights is shown schematically in Fig.4. The fiber is loaded via a T-shaped construction, which is vertically suspended by four cantilever blades. The spring constant of the suspension was sufficiently small so that the applied force F = Mg (M is the applied mass and g = 9.81 m/s^2 is the gravity constant) did not need to be corrected for. The uncoated fiber samples were fixed to the aluminium supports by small drops of epoxy. The resolution

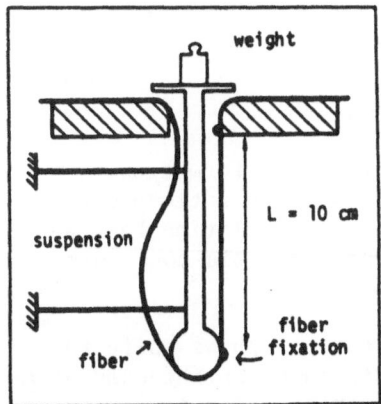

Fig.4. Set-up for stretching a fiber by calibrated weights.

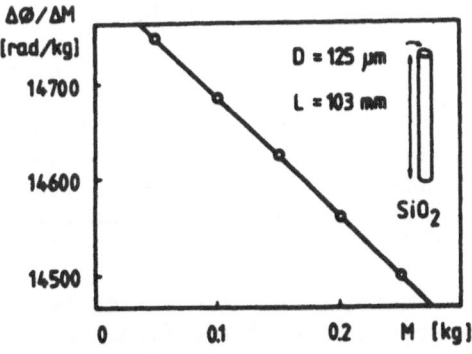

Fig.5. Measured sensitivity as a function of the bias load.

obtained was 1 mg and the maximum load applied to the fiber was 300 g, corresponding to a strain of 0.3%. Thus, the number of resolved points was 3×10^5.

To reduce the influence of relaxation effects noticed during the deformation measurements (probably caused by "creeping" of the epoxy at the fixation points) and the drift of the room temperature, the relation between the applied load M and the resulting optical phase-shift ϕ has been determined by differential measurements. Thus, the sensitivity $\Delta\phi/\Delta M$ (phase-shift induced by an additional load of 10 g) has been measured for different bias loads. A typical result is presented in Fig.5, where a linear decrease of the sensitivity is noticed. By integrating the straight line fitted to the experimental points, one gets

$$\phi = aM - (b/2)M^2, \tag{19}$$

with $a = 14810 \pm 10$ rad/kg and $b = 1240 \pm 8$ rad/kg^2.

The observed linear decrease of the sensitivity can be explained by analysing the theory of elasticity for an isotropic material under finite deformation. For a fiber under simple tension, second-order theory predicts a non-linear relation between the applied force and the resulting longitudinal and transverse strains [12], namely

$$\varepsilon_\ell = \varepsilon_o + \delta\varepsilon_o^2, \tag{20}$$

$$\varepsilon_t = -\nu\varepsilon_o + \beta\varepsilon_o^2, \tag{21}$$

where ε_o denotes the applied stress σ_o divided by the Young's modulus E_o, ν the Poisson's ratio and δ and β are two non-linearity constants. The stress σ_o is defined as the applied force per unit area of the unstrained fiber, so that

$$\varepsilon_o = \sigma_o/E_o = Mg/\pi R_o^2 E_o, \tag{22}$$

where R_o is the initial radius of the fiber.

Thus, for a fiber under simple tension, its deformation is characterized by the Young's modulus, the Poisson's ratio and the two constants δ and β of the fiber material. The relation between these four parameters and the five fundamental elastic constants of an isotropic material is described in Ref.12.

Further, the relation connecting the optical phase-shift to the deformation of the fiber has to be determined. In the case of a longitudinally strained fiber, the induced phase-shift ϕ results from changes both in fiber length and refractive index. By introducing the finite deformations from Eqs.(20) and (21) into the strain tensor and then evaluating the photoelastic effect one gets

$$\phi = knL\{\epsilon_\ell - (n^2/2)[S_3 p_{12} + S_1(p_{11} + p_{12})]\}, \tag{23}$$

where $k = 2\pi/\lambda$ denotes the free space wavenumber, n the refractive index of the core, L the fiber length and p_{11}, p_{12} are the components of the strain-optic tensor. $S_1 = S_2 = \epsilon_t + \nu^2\epsilon_o^2/2$ and $S_3 = \epsilon_\ell + \epsilon_o^2/2$ are the diagonal elements of the strain tensor.

For the silica parameters $n = 1.458$, $\nu = 0.17$, $p_{11} = 0.121$ and $p_{12} = 0.270$, and for $\lambda = 633$ nm , Eq.(23) becomes

$$\phi = \alpha_o L\epsilon_o[1 + \epsilon_o(0.91\delta - 0.53\beta - 0.19)], \tag{24}$$

where $\alpha_o = 1.13 \times 10^7$ rad/m is the phase sensitivity to elongation in linear theory [Eq.(11)].

If one compares the quadratic terms of Eqs.(19) and (24), one sees that measuring the phase-shift of light passing through the fiber does not provide enough information to determine δ and β independently, since both the longitudinal and transverse strains contribute to the photoelastic effect. It is therefore necessary to measure the longitudinal strain ϵ_ℓ separately as a function of the applied load in order to determine the values of the Young's modulus E_o and the constant δ, using Eqs.(20) and (22). Once E_o and δ are known, the remaining two parameters ν and β can be obtained from Eqs.(19), (22) and (24).

So far, only the phase-shift of the light passing through the fiber as a function of the applied load has been analysed experimentally. These measurements have been carried out for five different samples of the same silica fiber (Lightwave Technologies, type F1506B, outer diameter 125 μm). By comparing Eq.(24) with the experimental results, and taking $\nu = 0.17$, a Young's modulus of $E_o = 6.28 \times 10^{10}$ N/m^2 an average value for the relation connecting δ and β of

$$0.91\delta - 0.53\beta = -2.7 \pm 0.4 \tag{25}$$

has been calculated.

Let us assume that the non-linear behaviour of the longitudinal and transverse strains are the same, that is, $\epsilon_t = -\nu\epsilon_\ell$. Then the left side of Eq.(25) becomes equal to δ, and consequently, the average value of δ is -2.7, with a standard deviation of 0.4. This value compares favourably with previously reported ones, namely -3.0 ± 0.9 [13] and -2.4 ± 0.7 [14], measured by different techniques.

5. MEASURING FORCE BY INDUCED BIREFRINGENCE

An important class of fiber optical sensors are the polarimetric sensors [Fig.1(c)]. An example of an experimental arrangement is shown in Fig.6, where the birefringence induced in a single-mode optical fiber by a transverse force [Eq.(18)] is measured. The linear polarization of the input light is orientated at 45° with respect to the direction of the applied force. By placing appropriate optical elements (polarizer, quarter-wave plate) in front of the photodetectors, two signals in quadrature (90° phase-shift) can be obtained. This is necessary for bi-directional fringe counting and allows fringe interpolation with phase independent accuracy. Further, to eliminate the effect of intensity fluctuations, an additional reference detector is employed to monitor the total intensity.

In practice, the detector signals as a function of the interference phase ϕ are given by

$$I_1 = a_1 [1 + m_1\cos\phi], \quad I = a_2 [1 + m_2\cos(\phi - \psi)], \tag{26}$$

where a_1 and a_2 are the DC-levels, m_1 and m_2 the modulation depths and ψ is the relative phase-shift of about 90°. To determine the interference phase accurately, the exact values of these parameters have to be known or to be controlled. Traditionally, they are adjusted to predetermined values ($a_1 = a_2$, $m_1 = m_2$, $\psi = 90°$) by means of optical alignement and electrical adjustement. However, these parameters may change with time due to thermal effects, mechanical and optical missalignement, electronical drift, etc.,

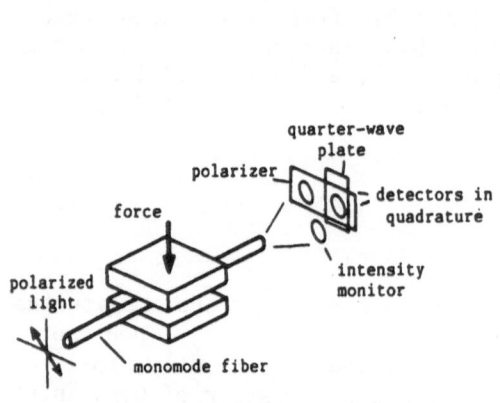

Fig.6. Experimental arrangement to measure force induced birefringence with high resolution.

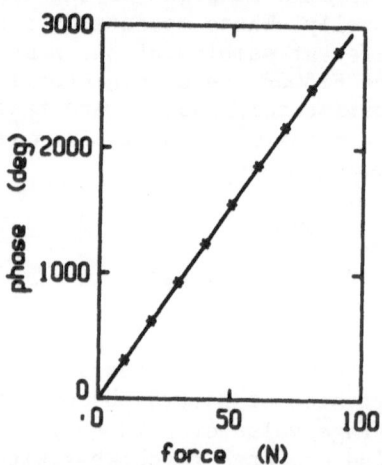

Fig.7. Measured phase between the two linear polarizations versus force (silica fiber, diameter 125 µm, λ = 800 nm).

thus limiting the long-term accuracy. The influence of the most likely and most rapid changes, namely the intensity fluctuations (source power fluctuations, sensor losses), is eliminated immediately by normalizing the signals I_1 and I_2 to the reference intensity I_r.

A typical experimental result for a transverse force of maximum 100 N acting on a silica fiber of 125 μm diameter and about 60 cm length, using a laser diode source at 800 nm, is presented in Fig.7. The measured sensitivity of 32°/N corresponds very well to the theoretical value calculated with Eq.(18). The linearity and the accuracy has been verified to be at least 0.02 N, which corresponds in this case to a resolution of about one part in 5000.

A microprocessor based approach to the problem of high resolution phase determination has the advantage that all hardware adjustments, electronic, optical and mechanical, are replaced by adaptation of numerical parameters in software [5]. The microprocessor calculates the value of the interference phase from the simultaneously detected and A/D converted signal levels I_1, I_2 and I_r, using the relation

$$\tan\phi = (\{m_1[(I_2/a_2 I_r)-1]/m_2[(I_1/a_1 I_r)-1]\} - \cos\psi)/\sin\psi. \quad (27)$$

Through a calibration procedure, the actual values of the parameters a_1, a_2, m_1, m_2 and ψ, which are assumed to change only slowly, are automatically determined, stored, and updated from time to time. Finally, the number of periods counted (by means of a bi-directional counter) and the interpolated phase are appropriately added to get the total accumulated phase.

To determine accurately the interference phase from the detected signal levels, it is important to sample them simultaneously. The number of required bits for the A/D conversion depends on the desired precision. For example, using Eq.(27) with the ideal values of the calibration parameters ($m_1 = m_2 = 1$, $\psi = 90°$), 8 bits are required for a phase resolution of about 1°. As explained earlier, the values of the parameters a_i, m_i and ψ, which may vary with time, are determined by a calibration procedure. The overall precision of the phase measurement system depends therefore on how accurately their values are calculated and on how frequently the calibration is performed.

To calculate these parameters, the values of the three output signals I_1, I_1 and I_r, are sampled simultaneously at six particular positions within the same period. These positions should not differ too much from the six ideal positions shown in Fig.8. Six of the 18 sampled values are intensity references (see Fig.8). These are used to normalize the values of the other two detectors. The resulting 12 values form a set of 12 equations with 6 unknown phases corres-

628

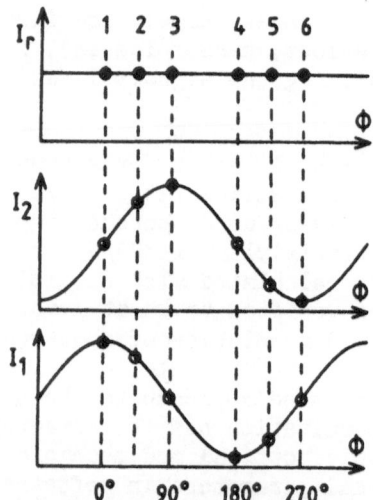

Fig.8. Detector signals as a function of phase ϕ, with the (ideal) sampling positions numbered from 1 to 6.

ponding to the exact positions of the samples and the 5 calibration parameters to be determined, giving a total of 11 unknown parameters. This set of equations is solved by means of iteration, using an algorithm which benefits from the fact that the approximmate values of the sample positions are known, namely the 6 ideal positions shown in Fig.8.

Note that, to enable this sampling procedure to work, the signals at the output of the sensor must exhibit phase changes of at least one period, preferably during a short time interval to assure accuracy. Fortunately, this is what usually happens in most sensor applications, where, during normal operation, several periods are passed through. If this is not the case, a special phase modulator may be added for this purpose.

A phase measuring system of this kind has been tested experimentally by simulating a polarimetric sensor using a Soleil-Babinet compensator. The reproducibility of the calibration procedure and the interpolation accuracy has been measured to correspond to at least 1/1000 of a period.

REFERENCES

1. Payne, D. N., A. J. Barlow, J. J. Ramskov Hansen, Development of Low- and High-Birefringence Optical Fibers, IEEE J.Quatum El. QE-18 (1982) 477-487.

2. Giallorenzi, T. G., et al., Optical Fiber Sensor Technology, IEEE J. Quantum El. QE-18 (1982) 626-664.
3. Ulrich, R., Fiber-Optic Sensors, Proc.Journée d'Electronique 1982 (Presses Polytechniques Romandes, Lausanne, 1982) pp.73-87.
4. Dandridge, A., and A. B. Tveten, Phase Compensation in Interferometric Fiber-Optic Sensors, Opt.Lett. 7 (1982) 279-281.
5. Dändliker, R., and A. Bertholds, Microprocessor Based Phase Determination for High Resolution Optical Sensors, to be published in Proc.Journée d'Electronique 1984 (Presses Polytechniques Romandes, Lausanne, 1984) pp.239-245.
6. Ezekiel, S., and H. J. Arditty, Fiber-Optic Rotation Sensors Tutorial Review, in Optical Sciences No. 32 (Springer, Berlin, 1982) pp.2-27.
7. Böhm, K., et al., Low-Drift Fiber Gyro Using a Superluminescent Diode, Electron.Lett. (1981) 352-353.
8. Ulrich, R., and Simon A., Polarization Optics of Twisted Single-Mode Fibers, Appl.Opt. 18 (1979) 2241-2251.
9. Sakai, J., and T. Kimura, Birefringence and Polarization Chracteristics of Single-Mode Optical Fibers under Elastic Deformations, IEEE J.Quantum El. QE-17 (1981) 1041-1051.
10. Rashleigh, S. C., Origins and Control of Polarization Effects in Single-Mode Fibers, J.Lightwave Tech. LT-1 (1983) 312-331.
11. Bertholds, A., and R. Dändliker, Interferometric Investigation of Monomode Fibers under Longitudinal Stress, to be published in Proc.OFS'84 (VDE-Verlag, Berlin, 1984).
12. Murnaghan, F., Finite Deformation of an Elastic Solid (Wiley & Sons, New York, 1951) pp.1-118.
13. Mallinder, F., and B. Proctor, Elastic Constants as a Function of Large Tensile Strain, Phys.Chem.Glasses, 5 (1964) 91-103.
14. Powell, B., and M. Skove, Measurement of Higher-Order Elastic Constants, Using Finite Deformations, Phys.Rev. 174 (1968) 977-983.

DESIGN OF A FIBER OPTIC HYDROPHONE*

E. Giese   and E. O. Schulz-DuBois
Honeywell-Elac-Nautik GmbH.   Institute of Applied Physics
2300 Kiel, F. R. Germany   University of Kiel, F. R. Germany

Abstract. As an example of modern sensor technology based on mono-mode glass fibers, the design of a hydrophone is described. In order to obtain practically useful operating characteristics, some development effort was necessary as regards sensitivity of the sensor head, the problem of maintaining the state of polarisation, the development of rugged fiber optic couplers and some means of frequency control to maintain the optimum sensitivity in the presence of drifts.

1. INTRODUCTION

During the past two decades, unprecedented progress has been made in the development of glass fibers. Attenuation rates, previously of the order of 1000 dB/km, were reduced to about 1 dB/km. The progress is chiefly due to improved chemical purity, notably due to the elimination of OH radical and divalent iron impurities in the quartz raw material. The availability of low-loss fibers made it possible to design long distance fiber communication systems. If, for example, an amplifier for regenerating the signals is required as soon as the input signal is attenuated by 60 dB, the repeater spacing was an unreasonable 60 m with the high-loss fiber, but is a very attractive 60 km with the new low-loss fiber. As is well known, the development and installation of fiber optical communication systems is under way and will continue for the next decades. Continued development and quantity production of fibers is almost exclusively for communication purposes.

---

* A hydrophone is a microphone for underwater use

The attractive properties of these fibers prompted scientists and engineers to look for other applications. Most other uses of fibers may be summarized under the heading of fiber sensors. An incomplete list of physical parameters which may be measured by means of optical fibers, includes mechanical strain, magnetic field, acceleration, electrical current, rotation rate with respect to the inertial reference frame, and sound pressure. Most of those sensors are still under development. An excellent survey of this field is given in two recent books [1, 2].

The object of this lecture is to report on a particular development, that of a fiber optic hydrophone. The emphasis will be on a number of technical problems and their solutions realized in our laboratory. For brevity, possible alternative solutions will not be discussed. The reader interested in fiber optical techniques will find a discussion of technical problems which will occur in one way or another in all fiber optical applications.

In the lecture, the hydrophone system and its function will first be surveyed. Then a number of technical details will be discussed. It is assumed that the reader has a basic understanding of glass fibers and of light propagating in them. In the following, the term glass fiber is always used for a monomode fiber whose core diameter is typically 5 μm. Note that most current fiber communications projects employ multimode fiber with a typical core diameter of 50 μm.

2. THE FIBER HYDROPHONE SYSTEM

A block diagram of the system is shown in Fig. 1. The pressure sensitive part, the sensor head, represents the glass fiber version of an optical Mach-Zehnder interferometer. It is connected to the remainder of the system by three fibers shown as dashed lines in the figure. The connecting fibers may be made very long without degrading the acoustical sensitivity of the system. This feature obviously is very helpful because it allows flexible layouts with sensor heads at locations far removed from the electronics.

The half-transparent 45° mirrors of conventional Mach-Zehnder interferometers are replaced by integrated fiber couplers similar to 3 dB couplers used in microwave waveguide technology. The four arms of a coupler are mutually related in the same way as the 0° and 90° beam directions on either side of a 45° half-transparent mirror. The first coupler is used as a beam splitter, the other one superposes light from both branches of the interferometer. In both coupler arms, the superposition is additive and subtractive, respectively, with respect to light amplitudes so that the

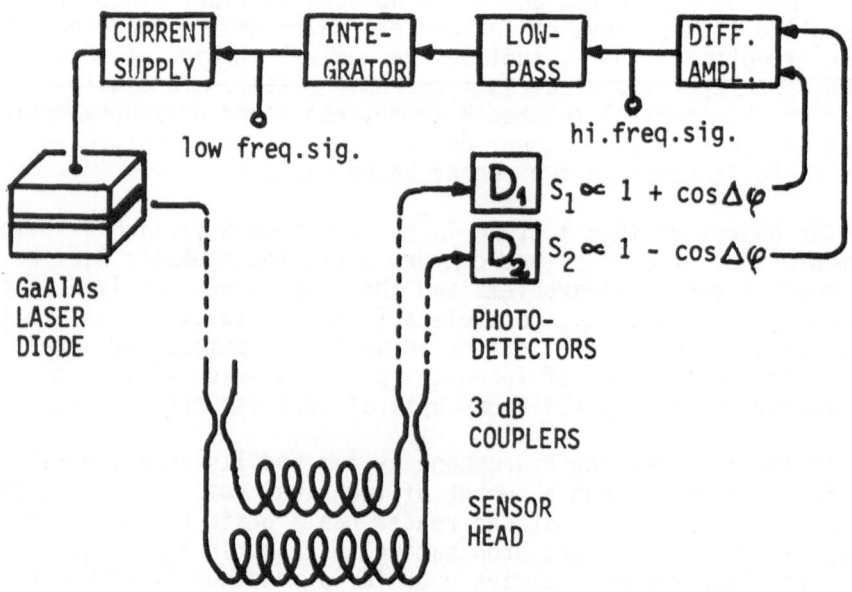

Fig. 1   Schematic layout of fiber optic hydrophone system

effects of interference can be monitored by the photodetectors D1 and D2. Both branches of the interferometer consist of several meters of fiber which are coiled and embedded in a mechanically soft epoxy tubing. The geometrical arrangement of both tubes in the sensor head is such that as a result of increased pressure the diameter of one of the tubes shrinks, that of the other increases. As a result the embedded fiber in the first tube shortens, the other one lengthens. The change of length leads to a change of optical phase of opposite sign for light passing these fibers. In a sense, the fibers act as strain gauges where length changes cause phase changes. Upon superposition of both light fields through the second coupler and photo detection, the pressure effect finally results in opposite intensity changes at the detectors. After forming the difference, an electric signal proportional to the instantaneous sound pressure is available for further processing.

As shown in Fig. 1, the light source is a semiconductor laser. It is small, it requires supply voltage and current in a range that is compatible with other semiconductor electronics, and power and, to a small degree, output wavelength may be controled by current changes. In the system, the supply current for the laser diode is controled by a negative feedback system for purposes of fre-

quency stabilization. This will be explained below in paragraph
3.4. The entire section 3 is devoted to a discussion of several
technical problems and their solution.

3. DETAILED TECHNICAL PROBLEMS

3.1 Sensitivity

Quartz glass, the basic material of fibers, is a fairly hard material so that its mechanical dimensions are not changed very much
when the fiber is exposed to isotropic pressure or longitudinal
stress. The induced associated phase changes $\Delta\varphi$ are for isotropic
pressure Δp

$$\frac{\Delta\varphi}{l\ \Delta p} = -4.1\quad 10^{-5}\ \text{rad/m Pa} \tag{1a}$$

and for longitudinal forces F

$$\frac{\Delta\varphi\ A}{l\ F} = -18.4\ 10^{-5}\ \text{rad m/N}\ . \tag{1b}$$

Here l is length and A is cross section. Both quantities (1a, b)
have the same dimension since 1 Pascal = 1 Newton/m^2. Under isotropic pressure, all linear dimensions shrink by the same percentage. But decreased length leads to less phase shift, decreased
cross section leads to larger index of refraction and more phase
shift. These effects tend to cancel so that the resulting phase
shift is small. Under longitudinal stress, the fiber lengthens and
the cross section decreases. The phase shifts associated with both
changes have the same sign, leading to a larger total phase shift.

In absolute terms, these phase shifts are rather small. A
change by the atmospheric pressure, $\Delta p = 10^5$ Pa, is needed in order to get a phase shift of more than π in one meter of fiber.
This should be compared with the acoustical environment of the
oceans where typical sound pressure fluctuations are around
$\Delta p \approx 10^{-5}$ Pa. Obviously it is reasonable to push the sensitivity
of a hydrophone to this limit of resolution.

A simple way to enhance the sensitivity of the fiber is by
coating it with a soft plastic material. In fact, fibers are always manufactured with a protective plastic coating. If the plastic coating has a cross section small compared to that of the
fiber, the elastic properties of the combination essentially will
be those of the fiber. On the other hand, if the cross section of
the coating is much larger, then the elastic properties

essentially will be those of the plastic. In this way the phase changes induced in the fiber may be 10 to 80 times larger with plastic coating.

A better geometry yet is circular tubing made out of plastic with coiled fiber embedded in the wall. Suppose the inside of the tube is maintained at a reference pressure. Sound pressure applied to the outside of the tube then leads to a much larger change in diameter than it would for a solid cylinder. It is not difficult to enhance the acoustical sensitivity of the fiber by a factor 10^3 in this way compared to the value for bare fiber.

The design of the sensor head is illustrated in Fig. 2. It

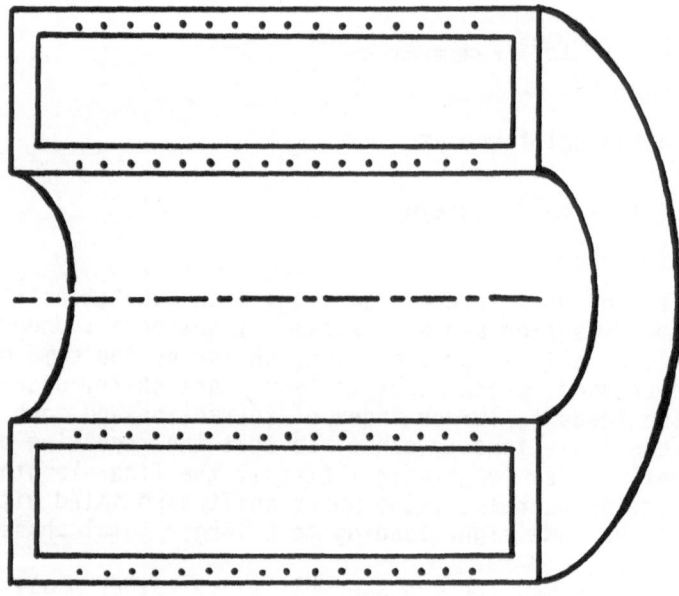

Fig. 2   Sensor head

comprises two tubes with embedded fiber and a hollow annular space between them. Upon application of pressure to the medium surrounding the head, the outer tube diameter shrinks and that of the inner one increases. Thus the associated phase changes in both coils have opposite sign. The interferometer is like a push-pull circuit where both phase changes combine so as to yield an enhanced signal. This design has another beneficial effect. The fiber shows a rather large temperature sensitivity. The phase change due to a temperature change of 1 K is about the same as that by a pressure change of 10^5 Pa, referring to bare fiber. Thus it is advisable to have both fiber coils as closely as possible at the same temperature, otherwise the drift in temperature between the two could mask any sound pressure effects. In the concentric sensor design both fiber coils are exposed to practically the same temperature, hence drifts due to temperature unbalance should be small and slow.

3.2 Polarization

As is well known, interference between two optical waves occurs only to the extent that both have the same polarization. Conventional monomode fiber ideally has a rotationally symmetric cross section so that it should offer the same propagation conditions for the two possible polarization states of the propagating mode. In practice, however, the fiber core is less than perfectly circular, it usually was twisted during manufacture, and coiling and bending or squeezing lead to birefringeance, that is to different propagation constants for two appropriately chosen orthogonal polarizations. As a result, there is no way of knowing what the state of polarization will be at the fiber output if a certain polarization is applied to the input. What is more, the output polarization changes with time, presumably due to room temperature drifts or ambient acoustic noise. In laboratory practice, these changes are slow, with an upper cutoff frequency of 10 Hz. It should be mentioned, however, that as long as the fiber may be considered lossless, two orthogonal polarizations at the input will remain orthogonal during propagation and past the output although one cannot predict what they will be.

One way to obtain a prescribed polarization at the output is by a negative feedback system [3]. The output intensity at the desired polarization is sensed and compared to a reference level. In the case of deviations, an error signal is derived and a proportional current actuates an electromagnetic fiber squeezer. This unit, essentially a relay, introduces the right amount of birefringeance into the fiber near the input in order to have the correct intensity of the desired polarization at the output. In the hydrophone interferometer, one would need two such feedback systems. This does not seem to be a desirable solution.

The other way would be to make the interferometer out of polarization maintaining fiber. As pointed out, for example, in the paper by Kaminov [2], the term polarization maintaining is somewhat misleading. It is a fiber of oval core cross section. As a result the two orthogonal polarizations have different propagation

636

constants, β_x and β_y. One defines the beat length B, an easily measurable quantity, by

$$(\beta_x - \beta_y)\ B = 2\pi\ .\tag{2}$$

For polarization maintaining fiber, B is of the order 1 mm. For ideally circular fiber, B would be infinite; fibers manufactured with the intent of making them circular, have beat lengths in the centimeter range. The polarization maintaining fiber does its job only under the following conditions: One has to launch only one of the eigen polarizations at the input, and all bends, kinks or other disturbances in the fiber should be so gradual that, in a spectral sense, all these disturbances are associated with wavelengths much longer than B.

Experiments with polarization maintaining fiber are on the way and as yet there is no reason to doubt that satisfactory sensor heads can be made with their help. However, more practice is needed in the manufacture of integrated couplers using these fibers.

3.3 Couplers

The beam splitter in a conventional interferometer setup, Fig. 3,

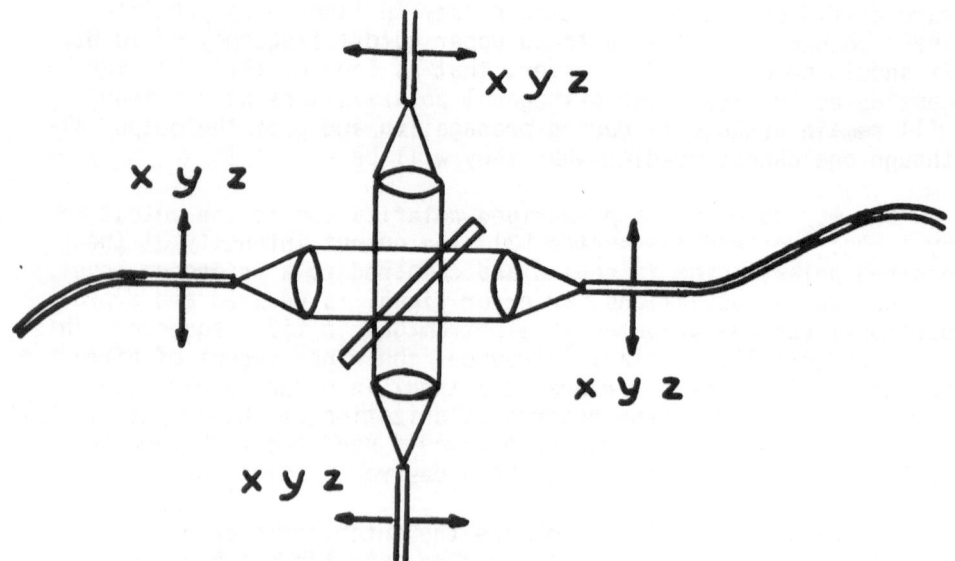

Fig. 3   Conventional beam splitter, requiring one semitransparent mirror, 4 microscope objectives and 4 x-, y-, z-positioners

usually is a semitransparent mirror at 45⁰ orientation. If the light propagation in the four possible directions should be inside glass fibers, one needs four microscope objectives which match the nearly plane wave near the mirror to the fiber input. The focussed light should hit the fiber core to within a tolerance of about 1 μm in all three dimensions. Thus one needs, additionally, four high-precision mechanical x-, y-, z-positioning stages. Needless to say that this setup is highly susceptible to vibrations and, therefore, should be mounted on a shock proof table.

The same optical function is realized in an integrated coupler as shown in Fig. 4. To readers familiar with microwave or high

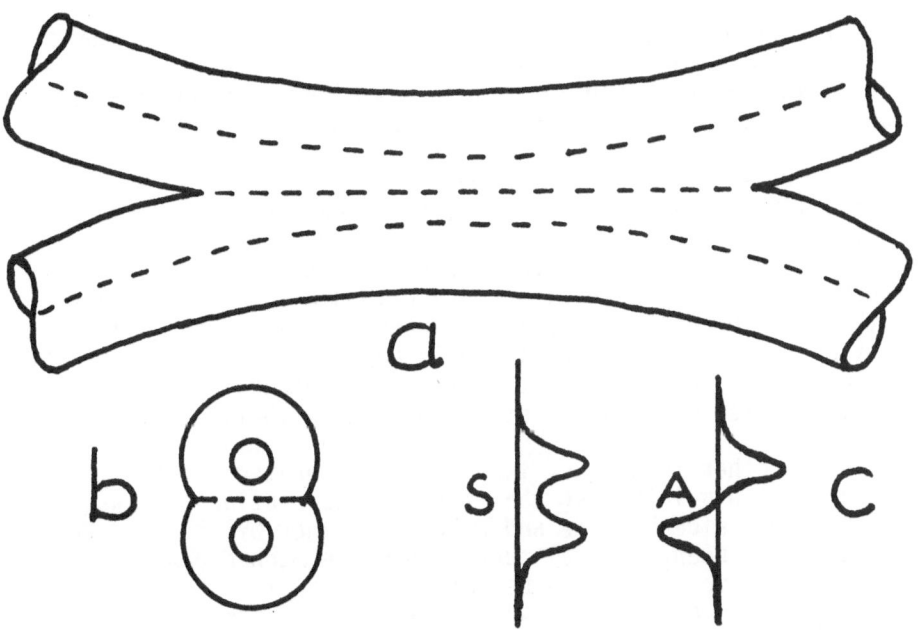

Fig. 4   Fiber optic coupler; a. Side view; b. Cross section; c. Eigen modes of propagation

frequency techniques it may be mentioned that this coupler parallels the function and the design of couplers used in these frequency ranges.

The function of the coupler is readily explained in terms of eigenmodes of propagation. Eigenmodes are characterized by the fact that they remain invariant under propagation. In the central section of the coupler, the cores of two fibers run parallel and close to one another. Here the eigenmodes show a symmetric (S) or antisymmetric (A) field configuration, respectively, and they differ in their propagation constants. When used as a beam splitter, there is a finite amplitude a in one of the input fibers, but zero in the other. In terms of s and A eigenmodes, the input is a linear superposition

$$\left\{ \begin{matrix} a \\ 0 \end{matrix} \right\} = \frac{a}{2} A + \frac{a}{2} s .$$ (3)

A 3 dB-coupler is realized if over the length of the coupler the s and A modes develop a 90° phase shift,

$$\frac{a}{2} A + \frac{ia}{2} s = \frac{a}{2} \left\{ \begin{matrix} 1 + i \\ 1 - i \end{matrix} \right\} ,$$ (4)

so that both outputs have the same amplitude, but 90° of phase between them. Smaller degrees of coupling, for example 10 dB, require a shorter coupling length. In a similar fashion, the function of the coupler for beam recombination may be described.

In our laboratory we experimented with two approaches of coupler fabrication. In one technique, the fiber is fastened to a curved metal block in which a slit receives the fiber, and a flat surface is ground on block and fiber close to the core. Two such blocks with their fibers attached are then placed on top of one another and the attained degree of coupling is checked with laser light while the relative position is adjusted. There remains the problem of finding a method to fasten both blocks in such a way that the degree of coupling stays the same with time.

In the other technique, two fibers are heated by an electric arc and pulled a little so that their cross section shrinks. They are then held side by side and welded together by the arc. Up to now, this has been a manual operation where much depends on the skill of the operator. A number of useful couplers has been made in this way, however. After welding they are fastened by epoxy to a small glass plate, resulting in a small, rugged device whose optical performance is insensitive to shock and vibration. It seems that much better results can be achieved by perfecting and automating this process.

3.4 Frequency stabilization

The output of both detectors D1 and D2 in Fig. 1 depends on the phase change $\Delta\varphi$ in the interferometer. As shown in Fig. 5

$$i_1 = a + b \cos \Delta\varphi$$

$$i_2 = c - d \cos \Delta\varphi ,$$ (5)

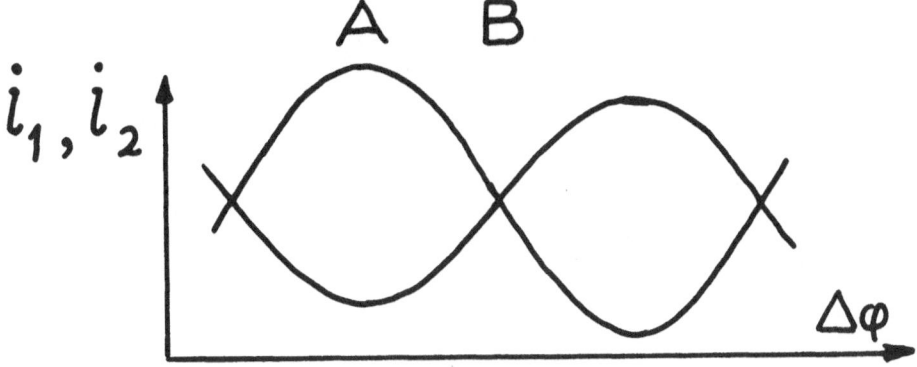

Fig. 5   Output current of detectors 1 and 2 versus differential phase shift $\Delta\varphi$; A operating point of zero sensitivity, B operating point of maximum sensitivity, to be stabilized by feedback system

where, for an ideally adjusted system, the constants a, b, c, d are all alike whereas in practice they are close to one another. If the interferometer operating condition (see A in Fig. 5) is near $\Delta\varphi$ = 0 mod 2π, then small changes of $\Delta\varphi$ do not lead to any detectable signal, at least not in first order. Similarly, a slow drift of $\Delta\varphi$ leads to changing sensitivity. Clearly this should be avoided. One should find means to stabilize the operating conditions of the interferometer such that $\Delta\varphi$ = $\pi/2$ mod 2π (see B in Fig. 5).

This is achieved by the negative feedback system shown in Fig. 1. In a difference amplifier both signals are combined such that the result is proportional to cos Δ . This requires somewhat different amplification levels for both channels to make effectively a = c . The difference signal is low pass filtered and fed to an integrator. Its output controls the laser diode current. By current control, the emitted laser wavelength of our GaAlAs laser diode changes by

$$\Delta\lambda/\Delta i = 5.8 \ 10^{-2} \ \text{Å/mA} \ . \tag{6}$$

A change of wavelength leads to the desired change in phaseshift $\Delta\varphi$ if both interferometer branches are unequal in length. A length difference of 10 cm proved adequate for this purpose. When the feedback system locks, $\Delta\varphi$ is maintained at $\pi/2$ mod 2π as slow changes are concerned. Fast changes of $\Delta\varphi$, for example above 100 Hz, are still present at the difference amplifier output. For high frequencies this is the hydrophone output. Low frequency output is available at the integrator output although there slow drifts are also present. It is a problem of electronic filtering to separate periodic low-frequency sound from other effects such as thermal drift.

In summary, the proposed hydrophone system offers the sensitivity required in the oceanic environment, it has a favorable layout with electronic parts separated from the fiber optic sensor head and only three fibers connecting the two, it has a built-in compensation for temperature and other drift, and with some further development it should be possible to produce practical units. The problems discussed here are similar to those encountered in the development of other high-sensitivity monomode fiber sensors.

References

1. A. B. Sharma, S. J. Halme, and M. M. Butusov,
    Optical fiber systems and their components,
    Springer Series in Optical Sciences vol. 24 (1980)
2. S. Ezechiel and H. J. Arditty, eds.,
    Fiber-optic rotation sensors and related technologies,
    Springer Series in Optical Sciences vol. 32 (1982)
3. E. Giese, K. Schätzel and E. O. Schulz-DuBois
    Optics Letters 7, 337 (1982)

FIBER OPTIC SENSORS AS DOSIMETERS FOR IONIZING RADIATION

G. Sehrig, W. Gaebler

Hahn-Meitner-Institut für Kernforschung Berlin GmbH

The starting point of our investigations was the impact of ioniz-ing radiation emerging from natural sources on fiber optic wave-guides. Some of them showed a great sensitivity resulting in a darkening of the glass when exposed to ionizing radiation. The idea was to use this effect (unwanted in optical communication-systems) for a new type of radiation sensor or dosimeter /1/ /15/. Optical fibers are made of highly pure synthetic quartz or fused silica which can be doped (e.g. with Ge, B, P, F) to achieve the required refractive index-profile, or of silica- and leadglass.

Some basic processes induced by radiation in glass are shown in fig. 1. Irradiation generates new centers which enable transitions of electrons from the valence-band to the center (1.). Thus optic-al absorption is increased at a wavelength corresponding to ener-gy E_a /3/. Alternatively these centers can be filled by UV-stimu-lation and the electrons can fall down into deeper centers emit-ting light of energy E_e (radiophotoluminescence, 2.).

Electrons lifted to the conduction-band during irradiation can be trapped in shallow centers where they are stored (3.). Heat-ing up the sample enables them to reach the conduction-band from which they fall down into deep traps under emission of light at energy E_e (thermoluminescence, 3.). If the electrons fall down immediately after generation the process is called cathodolumi-nescence (4.).

Two of these processes were investigated for sensor applications (fig. 2).

The thermoluminescence-sensor can be used for integrating measurements with destructive readout. The fiber can either be cut into pieces which are heated in a small heater or a special fiber with a metal coating can be heated locally by electric current. Both methods allow a spatially resolved readout. Our recent investigations concentrated on transmission sensors which use the radiation induced optical attenuation to measure the dose.

The so-called backscatter-sensor allows one to detect the amount of radiation-induced attenuation (with a limited dynamic range of 23 dB) and the distance from the end of the fiber at which the event occurred (distance range is up to 10 km with 1 % resolution) by using the Optical-Time-Domain-Reflectometry-technique. The OTDR-technique uses a short ligthpulse from a semiconductor-laser which is launched into the fiber. A small amount of light is scattered back along the fiber due to Raleigh-scattering and is detected by a receiver. The logarithmic light-intensity vs time gives the attenuation along the length of the fiber and clearly shows parts of increased attenuation due to radiation exposition.

Transmission-sensors (fig. 2) measuring the integral radiation-induced attenuation make it possible to construct simple dosimeters using standard fiber optic equipment. Some design considerations and test results of the sensor properties of commercial fibers are discussed below.

The detection range for optical attenuation measurements on fibers (fig. 4) is determined by two factors: 1. the minimum detectable power of optical receivers which depends on the type of detector used (Si-/Ge-detector, cooled/uncooled, large/small area) and 2. the maximum output power of commercial sources coupled into the fiber. The difference between these powers gives the detection range at a certain wavelength. Fig. 4 shows typical data of commercially available receivers and sources and the corresponding measurement range for wavelengths from 633 to 1550 nm. The dynamic range of the laboratory set-up used for the investigations is up to 100 dB at 633 nm.

The sensor consists of a fiber of known length terminated by standard fiber optic connectors. The fiber-length determines the sensitivity of the sensor because at a given specific sensitivity S_B (dB/kmSv) the increase in attenuation is larger if the exposed fiber-length is larger. Furthermore the length determines the maximum dose that can be measured (given by the maximum detectable attenuation) and the error limit of the measured dose. The error limit is given by the relationship of the attenuation-reproducibility of the connectors to the radiation induced attenuation of the fiber.

Fig. 5 shows some results of recent investigations on commercial fibers for communication applications.

The two most important parameters determining the detectable dose and the fading properties of the sensor are the glass composition and the wavelength of measurement. The specific sensitivity of induced attenuation is a strongly varying function of the wavelength of the transmitted light /2 - 4/. By changing this wavelength, the dose measurement range can be varied by a factor of up to 100 for a given fiber.

Fibers made of fused silica have a very low specific sensitivity and negligible fading when stored in the dark at room temperature. On the other hand their attenuation saturates at relatively low values so that the range of measurement is limited.

With increasing doping level the specific sensitivity of fibers increases as does the fading. We assume that the fading properties are mainly determined by the amount of residual impurities introduced in the fiber during production.

Leadglass fibers (which are not used anymore today) have a very high specific sensitivity but very bad fading properties. The best linearity combined with low fading of some percent per month occur in Ge- or Ge/B/P-doped fibers which allow a doserange of 1 Sv to 30 kSv to be measured when changing the wavelength from 633 to 1550 nm. These results are achieved with a typical sensorlength of 15 m which can easily be concentrated in a close volume as a coil. Other advantages are the possibility of a batch calibration of some km of a sensor fiber, its safety against electromagnetic disturbance and the great resistivity against corrosive chemicals.

References

1. W.Gaebler, G.Sehrig; Fiber Optic Sensors in Environmental Surveillance; 6th Int. Conf. IRPA, Berlin May 1984
2. E.J.Friebele; Optical Fiber Waveguides in Radiation Environments; Opt.Eng. 18.6(1979) 552-561
3. G.H.Sigel jr.; Fiber Transmission Losses in High Radiation Fields; Proc. IEEE 68(1980) 1236-1240
4. W.Gaebler, G.Sulz,D.Bräunig; Radiation Effects Testing of Optical Fiber Waveguides; PHOTON´83, Paris May 1983
5. W.Gaebler; Characteristics of Fiber Optic Radiation Detectors; PHOTON´83, Paris May 1983

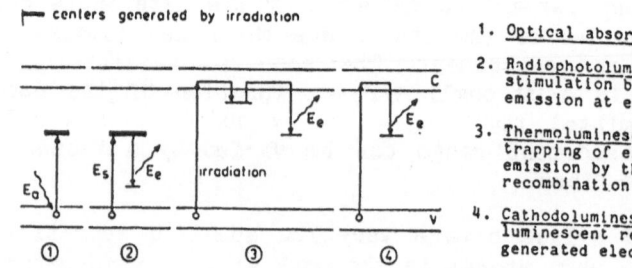

centers generated by irradiation

1. Optical absorption at energy E_a

2. Radiophotoluminescence:
   stimulation by UV-light at energy E_s,
   emission at energy E_e.

3. Thermoluminescence:
   trapping of electrons during irradiation,
   emission by thermal activation,
   recombination with E_e emission.

4. Cathodoluminescence:
   luminescent recombination of irradiation
   generated electrons.

Fig.1 Radiation Induced Processes in Glass

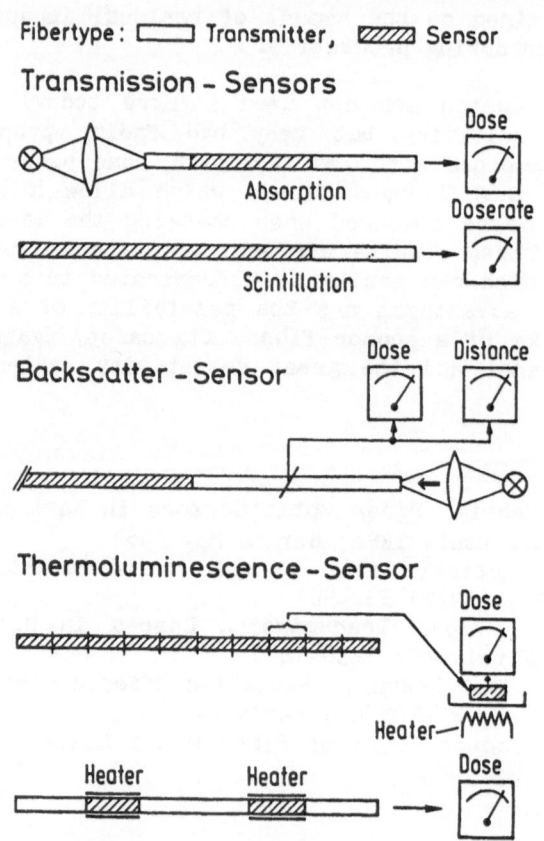

Fibertype: ▭ Transmitter, ▨ Sensor

Transmission – Sensors

Absorption → Dose

Scintillation → Doserate

Backscatter – Sensor

Dose Distance

Thermoluminescence – Sensor

Dose

Heater—

Heater Heater Dose

Fig.2 Fiber Optic Sensors for Ionizing Radiation

RADIATION INDUCED ATTENUATION

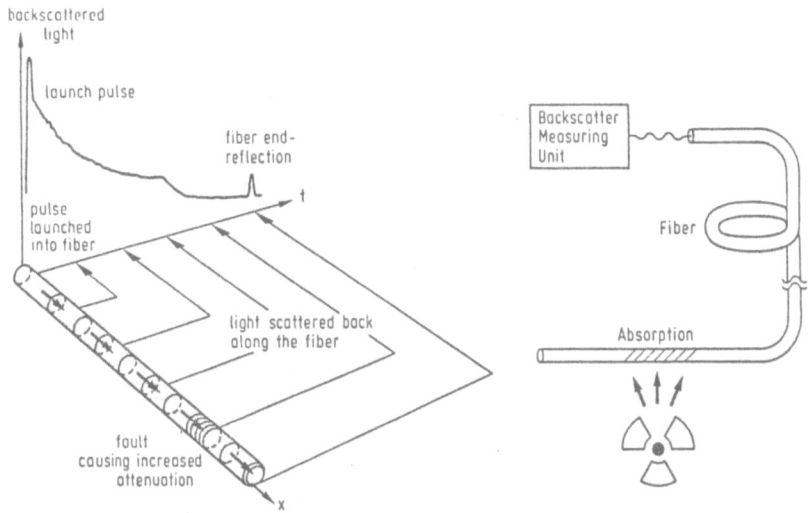

Fig.3. Optical-Time-Domain-Reflectometry

USABLE DETECTION RANGE FOR OPTICAL ATTENUATION MEASUREMENTS

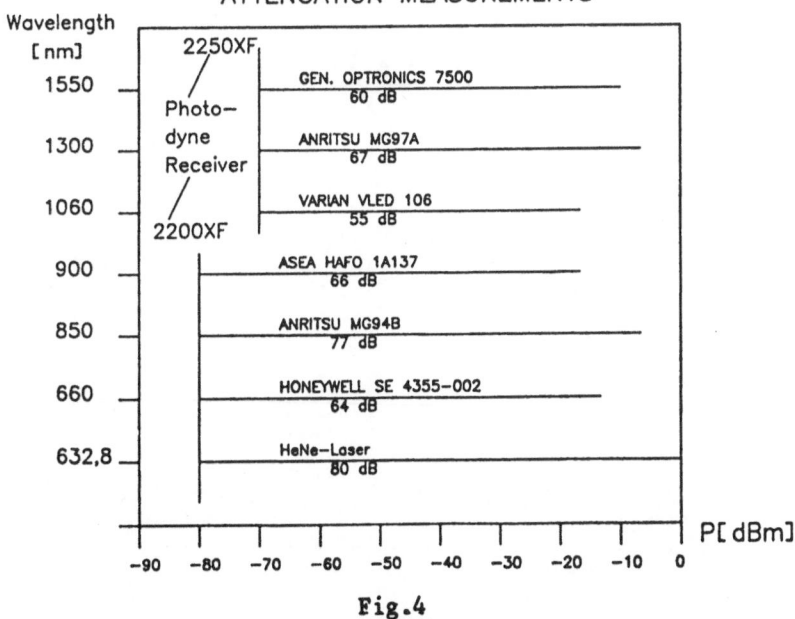

Fig.4

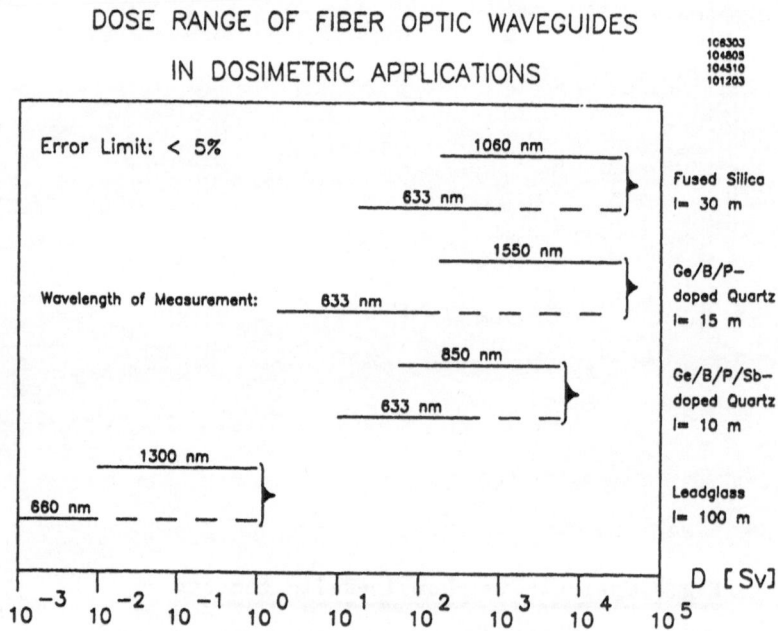

Fig.5

PICOSECOND PULSE METROLOGY

PICOSECOND METROLOGY

Claude FROEHLY
Laboratoire d'Optique (U.A. C.N.R.S. n°356),
123, rue A. Thomas, F.87060-LIMOGES Cédex (France)

1 - PICOSECOND OPTICAL PULSES

1.1. General features

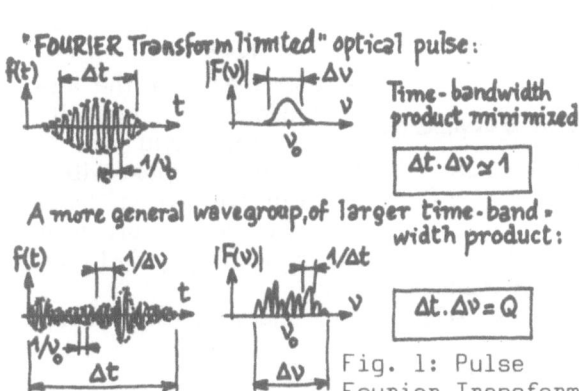

Fig. 1: Pulse Fourier Transform

f(t) and F(ν), respectively the temporal and spectral field distributions, are FOURIER transformed from each other.

IMPORTANT :
F(t) and F(ν) exhibit a **fine structure** of correlation length $1/\Delta\nu$ and $1/\Delta t$ respectively :

the total number of temporal samples needed for full description of a wavetrain of duration Δt and bandwidth $\Delta\nu$ is $Q \simeq \Delta t.\Delta\nu$

The Q samples of a wavegroup can be really observed if
. Radiation contains enough energy per sample
. Observation devices are able to resolve each of these samples temporally or spectrally.

1.2. When are picosecond pulses observable ?

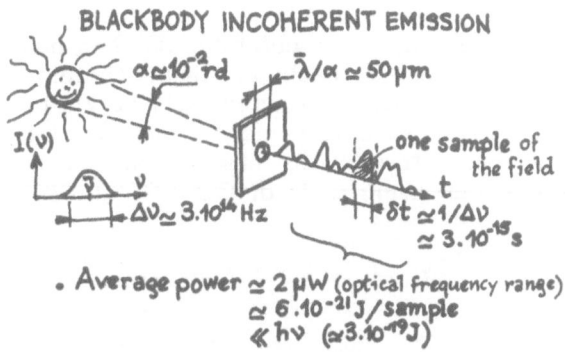

Fig. 2: Incoherent Emission

First example,
where no individual detection of the samples (= short subpulses) is possible, because the average sample energy is lower than the photon energy.

Thus measurements of the instantaneous power are not possible ;

measurements concern averaged moments of the field (average power and power spectrum). Such a radiation is TEMPORALLY INCOHERENT.

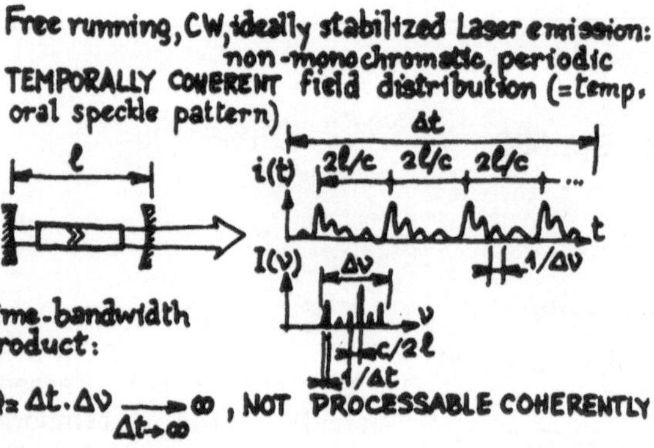

Second example

Free running, CW, ideally stabilized Laser emission:
non-monochromatic, periodic
TEMPORALLY COHERENT field distribution (= temporal speckle pattern)

Time-bandwidth product:

$$Q = \Delta t \cdot \Delta \nu \xrightarrow[\Delta t \to \infty]{} \infty \; , \; \text{NOT PROCESSABLE COHERENTLY}$$

Fig. 3: Laser Emission

TEM_{00} He-Ne laser beam of average power 1mW, frequency bandwidth 1 GHz (correlation time 10^{-9}s): average energy per sample $\simeq 10^{-9} \times 10^{-9}$ J, $\simeq 10^{-12}$ J, that is $N \simeq 3.10^6$ photons. Measurements of the instantaneous intensity i of this field are in principle possible within a relative uncertainty $\Delta i/i$ of the order of $N^{-1/2} \simeq 10^{-3}$: such a radiation has to be considered nearly TEMPORALLY COHERENT, **even if non monochromatic.**

But COHERENT OBSERVATION (i.e. phase measurements) would need both
- spectral resolution higher than $1/\Delta t$,
- temporal resolution higher than $1/\Delta \nu = 10^{-9}$s, which is not possible over long Δt (seconds or more ...), as the huge number Q of samples exceeds the capacity of the analyzing channel.

Thus CW laser emission, though possessing rather high intrinsic coherence, has always to be analysed statistically because the number of samples to be processed simultaneously is too large (a situation similar to statistical analysis of a coherent speckle pattern).

Third example

Pulsed laser, of 1 mJ energy, 1 ns duration, 10 GHz bandwidth : $Q = 14$ 10 independent samples, each of them containing $N \simeq 3.3 \times 10^{14}$ photons, so that the field uncertainty is very small ($\Delta i/i \simeq 5.10^{-8}$) ; this radiation possesses a very high degree of intrinsic coherence ; and coherent (i.e. deterministic) observation and processing of each sample are possible, using tools of COHERENT PICOSECOND OPTICS.

1.3. Short pulse generation by mode locked laser emission

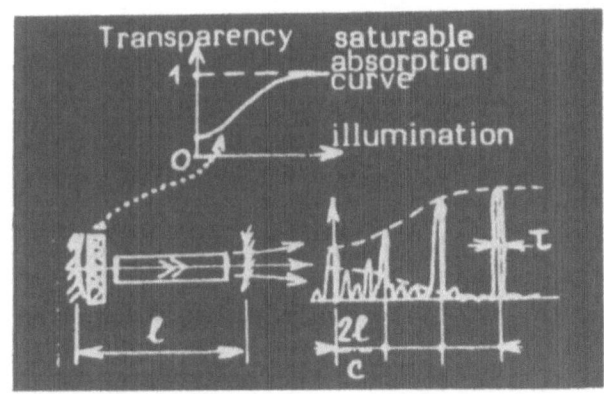

Fig. 4: Intracavity Mode Locking

Inserting both an optical amplifier (pumped material) and a nonlinear, bleachable transparency in a resonant cavity selects and enhances a periodical sequence of peaks, starting from noise, of length τ reciprocal to the frequency bandwidth of the amplifying material.

Pulse lengths of a few picoseconds were produced by applying this simple technique to self-mode-locking of glass/Nd lasers.

Generation of subpicosecond pulses requires more sophisticated principles, such as the arrangement shown on figure 5 : the cw mode locked YAg/Nd laser generates a periodic pulse train at wavelength λ_1 = 1.06 μm ;

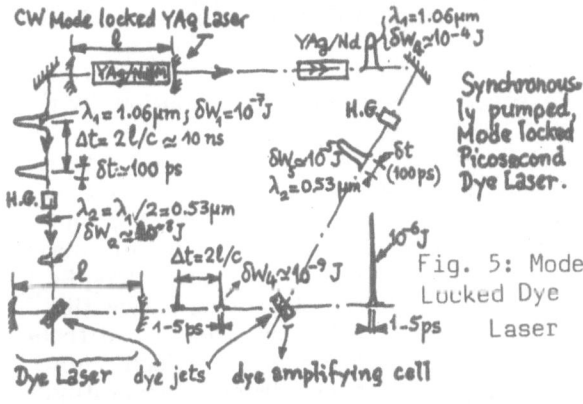

Fig. 5: Mode Locked Dye Laser

harmonic generation through nonlinear materials converts a part of this wavetrain into the wavelength λ_a = 0.53 μm suitable for pumping a dye, for instance Rhodamine 6G, at the period of the round trips in the dye laser cavity. The strong modulation of gain and losses of the dye laser results in periodical selection and enhancement of pulses, with generation of pico to subpicosecond cw pulse trains.

Of course, the very careful adjustments and the high stability needed by exact synchronous pumping limit the present use of subpicosecond sources to Laboratory experiments.

652

1.4. Two classical observation techniques

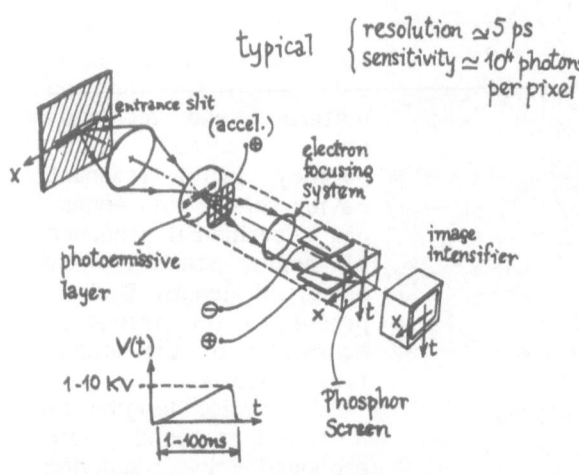

typical $\begin{cases} \text{resolution} \simeq 5\,ps \\ \text{sensitivity} \simeq 10^4\,photons \\ \text{per pixel} \end{cases}$

Fig. 6: Electronic Streak Camera

The electronic streak camera involves : entrance slit illuminated by the radiation to be observed ; photocathode converting the slit image into an electron source ; accelerating electrode ; imaging electrodes ; deflecting electrodes supplied by a linear tension ramp ; phosphor screen followed by an image intensifier.

The tension variation V(t) deflects the electronic image of the slit across the phosphor screen. The distribution of the screen brightness is an image of the space-time intensity distribution i(x,t) in the observed pulse structure.

Streak cameras are powerful tools in picosecond optics but their high price and limited temporal resolution restricts their use.

In the subpicosecond domain, only second order and higher order intensity correlations via nonlinear optics provide information on pulse lengths, structures and intensities. The most widespread technique is the "non collinear harmonic generation", where a harmonic generating crystal performs the product of a pulse and its delayed replica. As the delay time exceeds the pulse length, the harmonic intensity falls down to zero : pulse lengths will be deduced from measurements of path differences between the two arms of an interferometer.

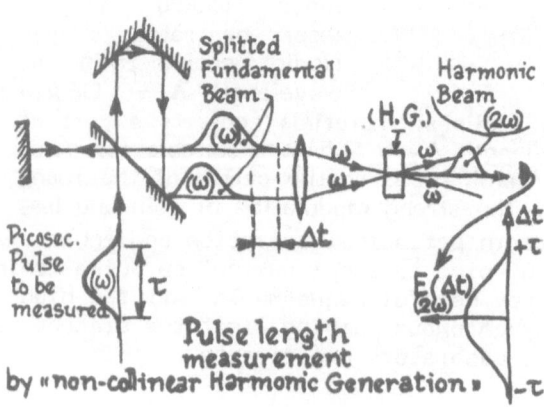

Fig. 7: Pulse Length Evaluation

Nevertheless this simple experiment provides very ambiguous information on the pulse shapes. More accurate details on the pulse structures will be extracted from third order correlation experiments described in further sections (self-phase modulation of pulses in silica fibers).

2 - COHERENT LINEAR OPTICS WITH PICOSECOND PULSES

2.1. Pulse stretching and filtering by apertures

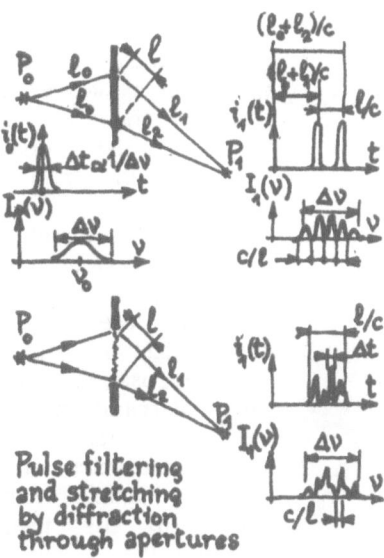

Pulse filtering and stretching by diffraction through apertures

Fig. 8: Pulse Processing

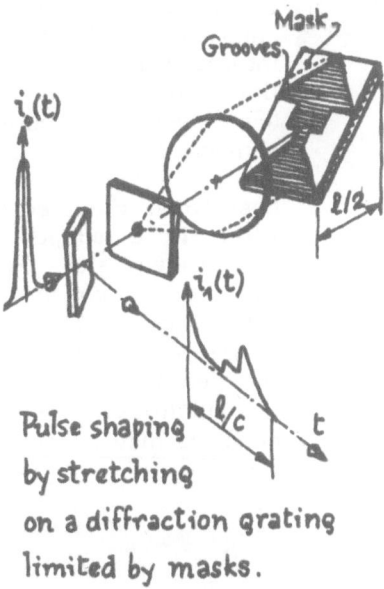

Pulse shaping by stretching on a diffraction grating limited by masks.

Fig. 9: Pulse Stretching

As shown in the figure opposite diffraction through apertures perturbs both temporal and spectral structures of pulses if the path differences involved by diffraction are larger than the initial pulse length.

Then we define a temporal impulse response - or its FOURIER transform the frequency transfer function - of the aperture, describing the linear filtering of the frequency spectrum of the transmitted radiation. Any incident radiation will be temporally convolved with the temporal impulse response and spectrally modulated by the frequency transfer function.

Of particular interest is the case of diffraction through periodic gratings. It generates wavetrains of length l/c, l being the "optical depth" of the grating. By masking the grooves partially it was shown to be easy to produce various temporal shapes over the length l/c starting from FOURIER transform limited shorter pulses.

Diffraction gratings are also known as angular dispersers of temporal frequencies. Great care has to be

654

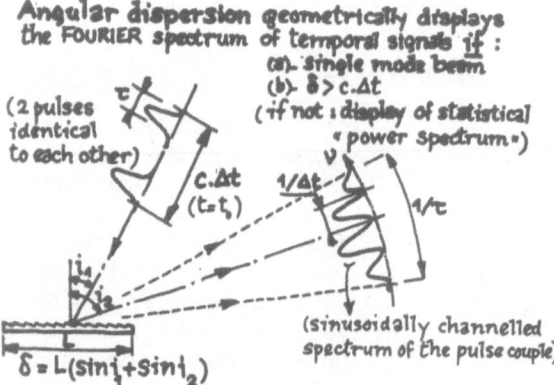

Fig. 10: Chanelled Spectrum

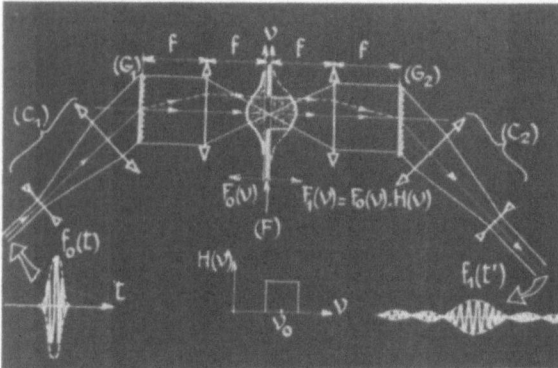

Fig. 11: Double Dispersion

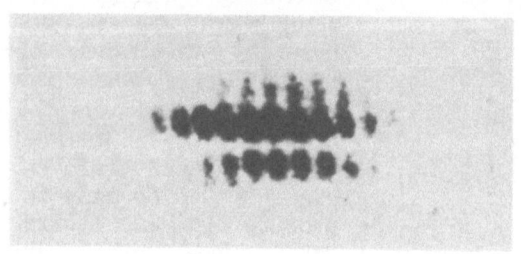

Fig. 12: Cosine Intensity Modulations

taken that the optical depth of the grating exceeds the whole length of the analysed vibration, if it is wanted to display the true coherent spectrum of the field (here, the sinusoidally channelled spectrum of a split pulse).

The double dispersion device represented in figure 11 requires the condition above to be fulfilled. Then picosecond temporal distributions may be synthesized simply by interposition of amplitude and/or phase filters in the frequency plane (F).

In figure 12 will be reproduced a streak camera recording of the cosine intensity modulation resulting from interposition of two thin slits in the plane (F). The next figure 13 demonstrates the $(sint/t)^2$ shape of the output signal when the frequency filter is a single wide rectangular slit (situation skizzed on the drawing, figure 11). In the same way filtering of the phase rather than the amplitude of the frequencies may be achieved by placing transparent optical pieces (lenses, prisms, or more complicated variations of optical thickness) in the FOURIER plane (F), instead of selecting frequencies by opaque screens.

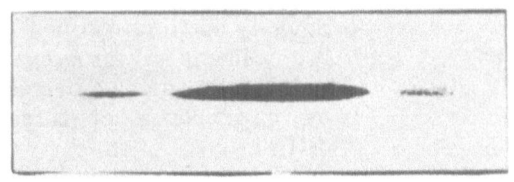

Fig. 13: $(\sin t/t)^2$ Output Signal

2.2. Linear dispersion of pulses

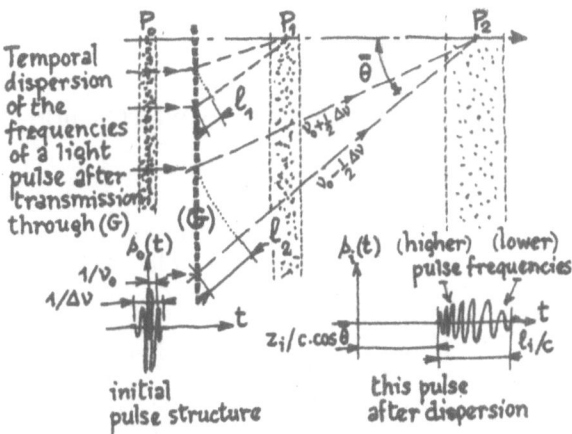

Fig. 14: Frequency Filtering

After a dispersive propagation length the

This a particular case of frequency filtering by a filter inducing parabolic phase modulation of the spectrum.

Propagation behind diffraction grating (G) does it, in the approximation of narrow bandwidths needed for angular dispersion to be considered nearly linear with respect to the optical frequency (ν) various spectral components of the pulse meet the observation point P successively. This temporal "chirping" of the frequencies increases with the propagation length up to the limit settled by the grating size.

In dispersive materials (for instance single mode optical fibers, as in section 3) the amount of chirping will be limited by light scattering.

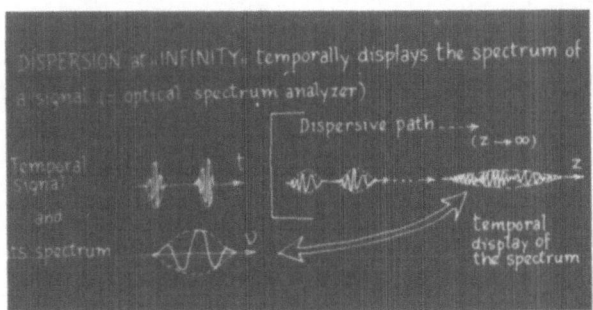

Fig. 15: Spectrum Analyser by Dispersion

Grating dispersion does not keep the single mode structure of a parallel light beam. Figure 16 (a) depicts an experimental

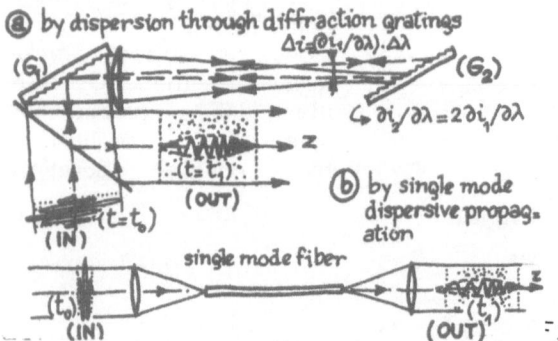

"Chirping" of pulses carried by single mode Laser beams

(a) *by dispersion through diffraction gratings*

$\Delta i = (\partial i_1/\partial \lambda) \cdot \Delta\lambda$

(G_1) (G_2)

$\hookrightarrow \partial i_2/\partial\lambda = 2\partial i_1/\partial\lambda$

z

(OUT) $(t=t_1)$

$(t=t_0)$

(IN)

(b) *by single mode dispersive propagation*

single mode fiber

(t_0) (t_1) z

(IN)   (OUT)

device performing near-ly linear dispersion of single mode beams by a sequence of three diffractions on two gratings (G_1) and (G_2).

These dispersive lines are of constant use in optical pulse compression.

Fig. 16: Linear Dispersion

2.3. Two dimensional representation of dispersion ; temporal YOUNG's experiment, FRESNEL-KIRCHOFF integral and HUYGENS principle.

Two independent temporal variables characterize the linear dispersion of a narrow bandwidth field distribution : they are the group time $t_g = z/v_g$ (v_g = group velocity) and the variation $\Delta t(t_g)$ of this time with frequency responsible for pulse spreading and chirping.

It suggests the geometrical representation of dispersion in the two-dimensional frame $t_g, \Delta t$.

The existence of a carrier frequency under the pulse envelope tilts the "temporal beam" axis by an angle θ_0 with respect to the horizontal coordinate t_g. This angle may be related directly to the dispersion magnitude, usually expressed in picoseconds per meter of propagation length and per nanometer of spectral bandwidth, that is nearly proportional to the temporal stretching rate $(\Delta t/t_g)(\Delta\nu)^{-1}$.

In the rectangular frame t, $t_g = z/v_g$, dispersive propagation of « FOURIER Transform limited » pulses looks like diffraction of « diffraction limited » beams.

$c \cdot \Delta t_0$ $c \cdot \Delta t_1$ $c \cdot \Delta t_2$

λ_0 z

z_1

z_2

λ_0/c Δt $(t_g)_1 = z_1/v_g$ $(t_g)_2 = z_2/v_g$ t_g

Λ Δt_1 θ_0

Λ Δt_2

$\Delta\nu = 1/\Delta t_0$

$\tan\theta_0(1+(\tan\theta_0)^2) = (c/\lambda_0)(\Delta t/t_g)(1/\Delta\nu)$

$\Lambda = (\lambda_0/c)\sin\theta_0$ = «Temporal wavelength»

Fig. 17: Dispersion Representation

Then the definition of a temporal wavelength completes the full analogy of pulse dispersion with two-dimensional beam diffraction;

Such a representation is accessible physically by using a streak camera, the entrance slit of which will be placed along the

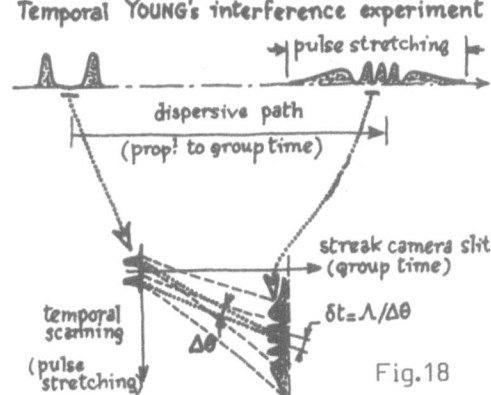

Fig. 18: Young Interference

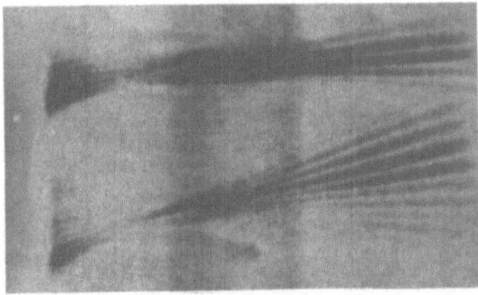

Fig. 19: Two Pulse Interference

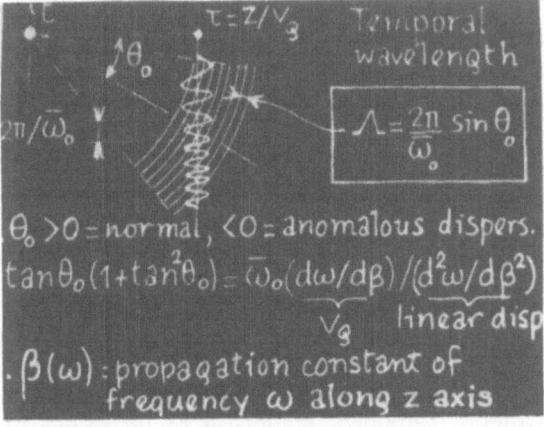

Fig. 20: Huygens Wavelets

propagation axis z : the group time t_g is proportional to this coordinate ; the temporal evolution at each point z will be displayed vertically by the electronic slit deflection.

In this experiment dispersion of a pulse pair looks like interference of two monochromatic beams issuing from a pair of YOUNG's slits.

On figure 19 two successive pulse pairs were recorded on the same photograph. Unfortunately overexposition of the region of maximum intensity could not be prevented, due to the low dynamical range of the camera : thus pulse splitting at zero dispersion is not very clear.

These images demonstrate the physical reality of the temporal HUYGENS wavelets emitted from DIRAC impulses (figure 20)with a temporal wavelength determined by the pulse dispersion and its average carrier frequency.

The temporal evolution of field distributions

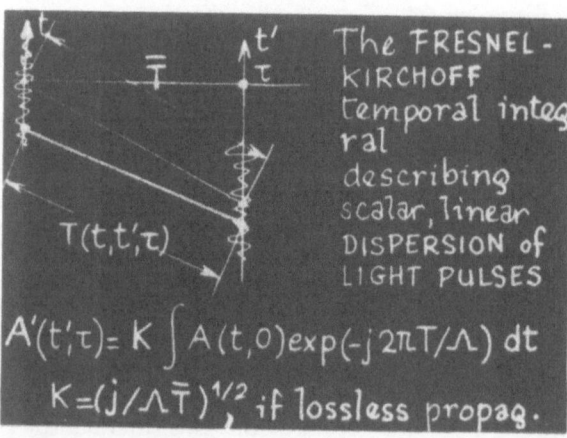

Fig. 21: Fresnel-Kirchoff Diffraction

between two successive dispersive steps may be calculated using the two-dimensional form of the scalar FRESNEL-KIRCHOFF diffraction integral (figure 21).

It proceeds from the above considerations that optical pulse analysis may be grounded on the basis of usual coherent FOURIER optics, the concepts of which keep their entire validity in the domain of TEMPORALLY COHERENT OPTICS.

Let's emphasize the fact, that temporal coherence **does not refer at all to field monochromaticity** but, on the contrary, to the deterministic structure of **polychromatic** light fields allowing their processing by the tools of coherent linear optics : amplitude and phase filtering and modulation, interferometry, holography ...

It's now the place to recall the principles of temporal FOURIER holography.

2.4. Temporal FOURIER Holography

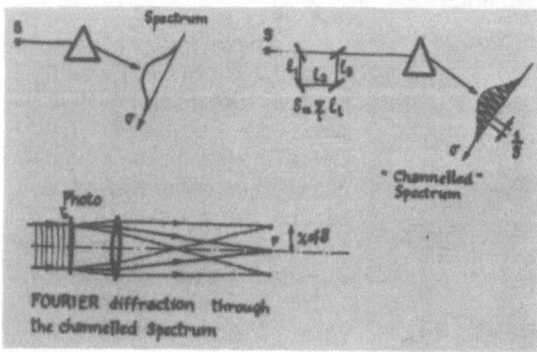

Fig. 22: Temporal Fourier Holography

The photographic recording of the spectrum of a pulse pair is a field of sinusoidal fringes ("channelled spectrum"). Far field diffraction of a monochromatic laser beam through the channelled spectrogram generates two bright spots symmetrical to each other with respect to the image focus (F) of the FOURIER transforming lens. This is the elementary two-step process involved in temporal FOURIER Holography : the "Object" temporal structure to be recorded

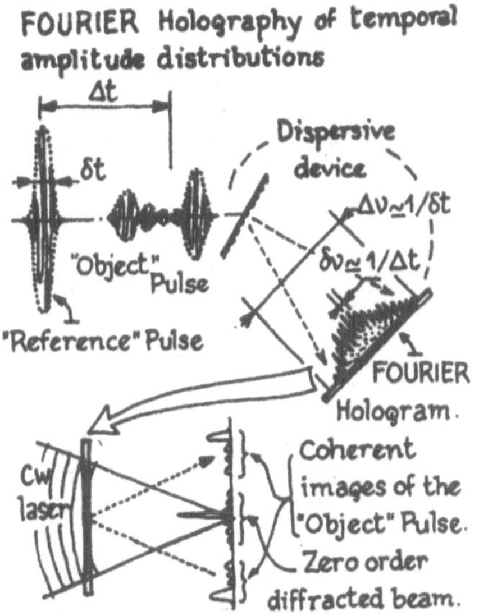

FOURIER Holography of temporal amplitude distributions

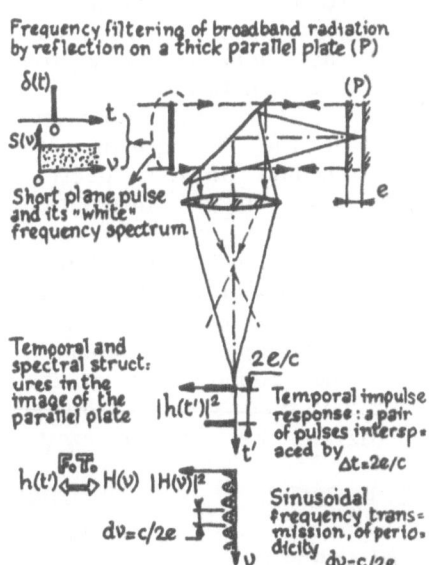

Fig. 23: Reconstruction Fourier Hologram

will be accompanied by a much shorter, FOURIER transform limited "reference" signal possesing at least the same frequency bandwidth as the "object". The interpulse delay Δt has to be choosen larger than the object length. The generalized channelled spectrum available at the spectroscope output will behave as a FOURIER hologram when placed in the converging beam of a monochromatic laser : the diffraction pattern in the image focal plane consists of three spots, two of them (figure 23) are complex conjugate coherent images of the "temporal object".

This natural conversion of picosecond structures into coherent images underlines the complete symmetry among coherent optical operations in both spatial and temporal domains.

2.5. "Path difference metrology" using broadband radiation, spectral analysis and temporal holography.

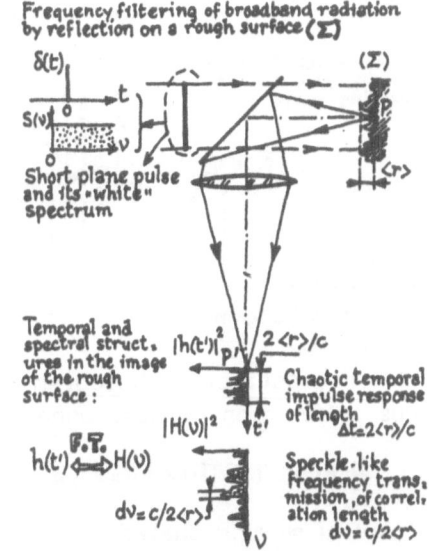

Fig. 24: Metrology of Flat Surface

Fig. 25: Metrology of Rough Surface

From the figures 24 and 25 it can be seen that the optical thickness 2e of a parallel plate, or 2<r> of a rough surface may be measured by observation of the spectral correlation length of the reflected or scattered radiation. This length is reciprocal to the temporal response length 2e/c or 2 <r>/c.

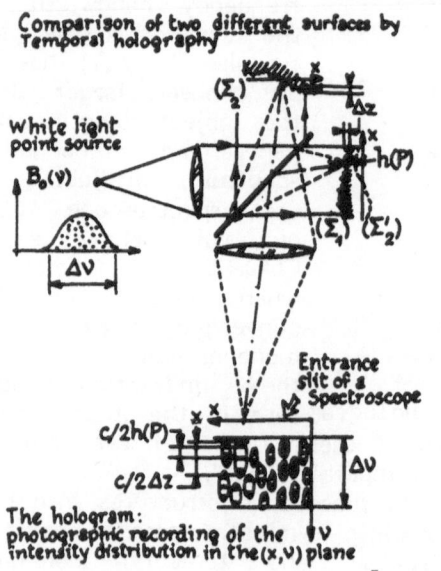

Fig. 26: Speckle Reference Holography

Holographic imaging of the distance h(P) of the rough surfaces (Σ₁) and (Σ'₂)

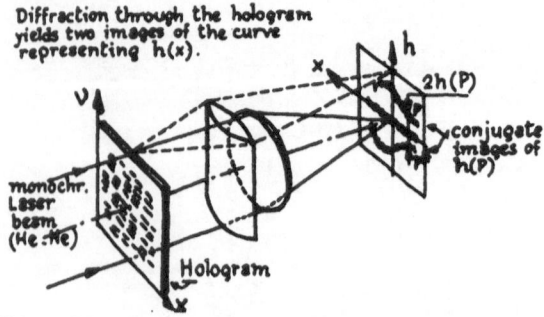

Fig. 27: Reconstruction Hologram Fig. 26.

Such analyses provide statistical information about average optical path differences. More accurate observations of surface shapes, including the unusual comparison of two **different rough surfaces** are possible using a temporal version of holography with speckled reference beam, as in figures 26 and 27 : spectral analysis of the field selected by the entrance slit of the spectroscope displays a two-dimensional speckle pattern. Speckle cells exhibit a thin spectral cosine modulation of periodicity reciprocal to the local depth h(P) separating the two surface images through the beamsplitter.

FOURIER analysis of the photographic recording of the spectrogram needs diffraction of a monochromatic sphero cylindrical laser beam. The images of the curve h(P) are speckled lines, of thickness proportional to the roughness z of the observed surfaces. Image outlines suffer from optical uncertainty (dz) of the order of $c/\Delta\nu$.

3 - TEMPORAL METROLOGY OF OPTICAL FIBERS

3.1. Spectral speckle analysis of intermodal delays

Straightforward spectral analysis of frequency filtering through fibers allows measurements of perturbations

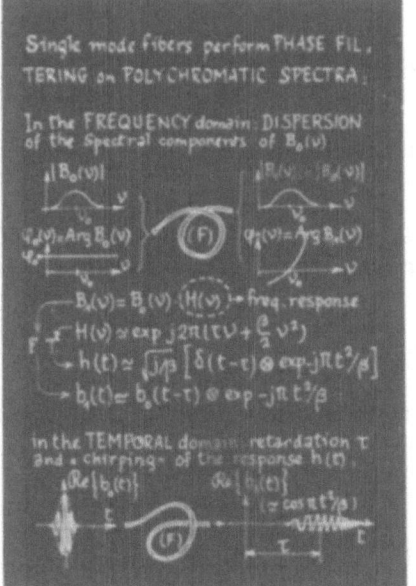

In the figure:

Single mode fibers perform PHASE FILTERING on POLYCHROMATIC SPECTRA

In the FREQUENCY domain: DISPERSION of the Spectral components of $B_e(\nu)$

$|B_e(\nu)|$ $|B_s(\nu)| = |B_e(\nu)|$

$\varphi_e(\nu) = Arg\, B_e(\nu)$ $\varphi_s(\nu) = Arg\, B_s(\nu)$

$B_s(\nu) = B_e(\nu) \cdot (H(\nu)) \leftarrow$ freq. response

$H(\nu) \propto \exp\, j2\pi(\tau\nu + \frac{\beta}{2}\nu^2)$

$h(t) \simeq \sqrt{j/\beta}\,[\delta(t-\tau) \otimes \exp\text{-}j\pi t^2/\beta]$

$b_s(t) \simeq b_e(t-\tau) \otimes \exp\text{-}j\pi t^2/\beta$

In the TEMPORAL domain retardation τ and a «chirping» of the response $h(t)$

$Re\{b_e(t)\}$ $Re\{b_s(t)\}$ $(\propto \cos \pi t^2/\beta)$

Fig. 28: Monomode Fiber Dispersion

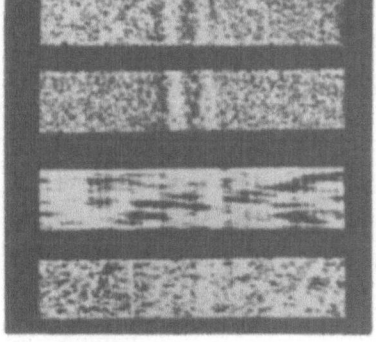

In the figure:

MULTIMODE fibers perform AMPLITUDE and PHASE filtering on POLYCHROMATIC radiation

STEP INDEX Fiber

NEARLY EQUALIZED f.

$\Delta t, \Delta t'$, lengths of the responses $b_s(t)$, $b'_s(t)$, are equal to the time differences between the shorter and the longer paths in the fiber cores.

Fig.29 Multimode Fiber Response

1 nm λ

occasioned to pulses after transmission through optical waveguides.

This section restricts itself to the **optically linear distortions** occuring at low powers of the electromagnetic field (section 4 will consider a few nonlinear distortions).

Figure 28 again illustrates the effect of linear dispersion involved in single mode fiber propagation : it consists of parabolic modulation of the spectral phase distribution, or linear "chirping" of the temporal response. The spectral energy will be transmitted without attenuation assuming lossless propagation. This is no longer the case in multimode fibers : temporal and frequency responses look like temporal and spectral speckle patterns, as the multimode waveguide behaves as a multipath interferometer : each excited mode corresponds to a particular geometrical path. The maximum intermodal delay Δt may be deduced immediately from analysis of the fiber spectral output : in figure 30 four different fibers modulate the injected broad and smooth spectrum by speckle patterns of various correlation lengths. The temporal frequency being displayed horizontally, it appears clear that fiber (c) causes much less pulse dispersion than the three others : intermodal delays

(a)

(b)

(c)

(d)

Fig.30: Fiber Intermodal Delay

are spread over a time interval of about O.5 ps, instead of about 4 ps for fiber (d) and 8 ps for both the others.

3.2. Temporal holographic imaging of pulse dispersion;

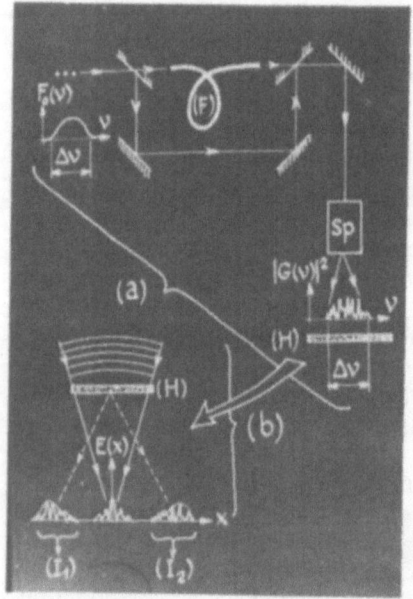

Fig. 31: Pulse Dispersion Holography

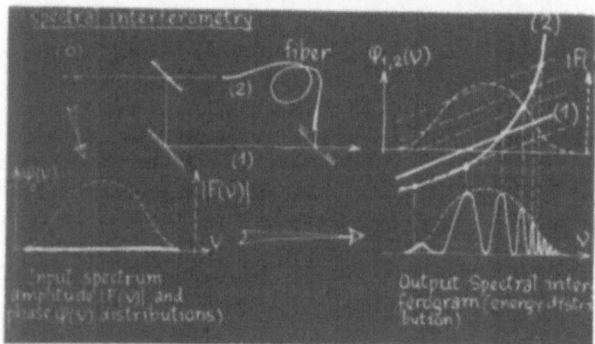

Fig. 32: Newton Rings Analogy

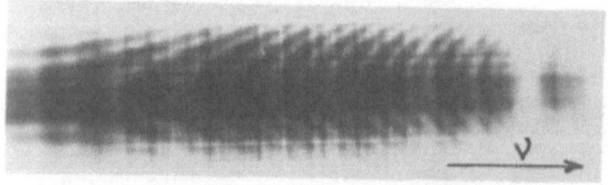

Fig. 33: Interferogram Relating to Fig. 32

The holographic set-up (MACH-ZEHNDER interferometer, single or multimode fiber (F), spectroscope (Sp)) has to be illuminated by a spatially coherent beam possessing a broad and smooth spectrum. The photographic recording of the output spectrum generates a hologram (H) of the temporal response of the fiber.

in the **first case of single mode fibers**, the hologram looks like NEWTON's fringes of a cylindrical lens ; indeed it results from interference of two spectral wavesurfaces $\varphi_1(\nu), \varphi_2(\nu)$ of different curvatures, as explained in figure 32 ; figure 33 reproduces an enlarged view of such an interferogram. The hyperbolicity of the fringe pattern comes from some undesired geometrical curvature of the interfering wavefronts.

Shining the spectrogram by a monochromatic laser beam allows observation of two coherent images of the fiber response in the FOURIER plane of the diffraction device (figure 34 (b)).

The experiment here described was performed on a 10 cm long silica fiber transmitting the radiation issued from a Rhodamine 6 G dye laser.

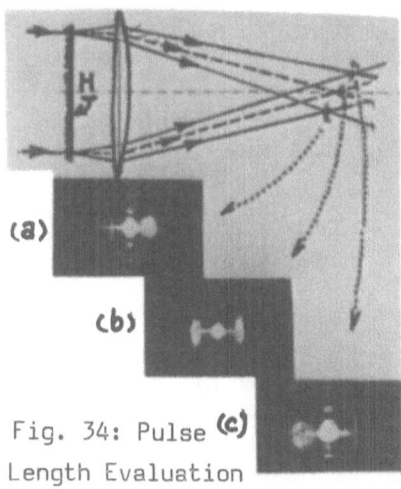

(a)

(b)

Fig. 34: Pulse (c)
Length Evaluation

The output **pulse length,** deduced from spot widths of figure 34 (b), was found equal to about 0.3 ps.

The output **pulse phase distortion** due to linear dispersion of silica may be deduced immediately from the amount of defocusing needed for either spot to concentrate into a thin vertical line (on the left of image (a), on the right of image (c)). This proves that the images (b) consist of elementary cylindrical wavelets, the curvature of which translate the temporal pulse chirping in the geometrical space. The dispersive phase shift was found to be of the order of $10 \times 2\pi$ radians, in agreement with the measured pulse stretching.

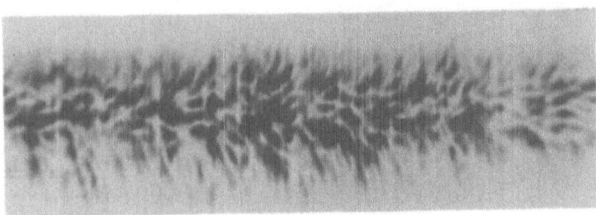

Fig. 35: Multimode Fiber Hologram

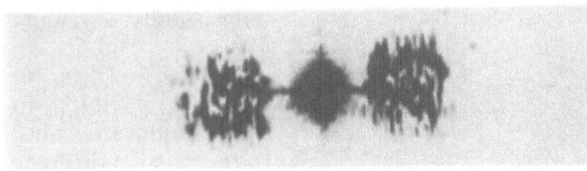

Fig. 36: Reconstruction Relating to Fig. 35

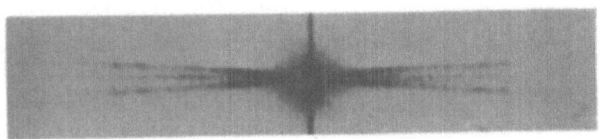

Fig. 37: Step-Index Fiber Hologram
- Far Field

In the **second case** of **multimode fibers** the speckle-like holographic intensity distribution (fig.35) comes from random interference of the various excited modes.

The reconstructed image of the temporal response also exhibits a sharp space-time structure (figure 36).

Let's note the interesting observation of figure 37, where the temporal hologram was recorded in the **far field** of a **step index fiber.** Then the horizontal coordinate is proportional to time, whereas the vertical one represents the angular tilt of rays with respect to the fiber axis : the temporal delay of oblique rays appears as a rapidly growing (parabolic) function of the tilt angle.

3.3. Subpicosecond display of modes by temporal holography combined with "Spectrographic imaging".

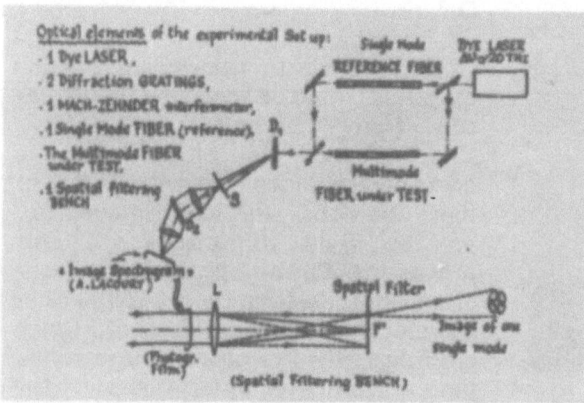

Fig. 38: Spectrographic Imaging

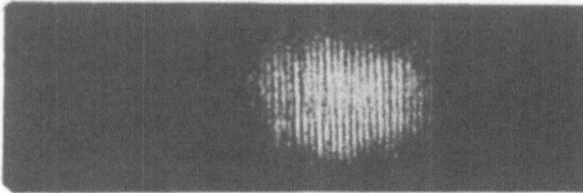

Fig. 39: The "Image Hologram" ...

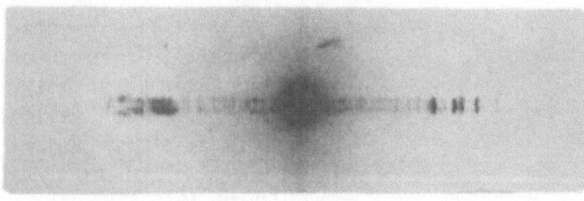

Fig. 40: ... and Its Diffracted Far Field

Replacing simple spectrographic analysis by the double dispersion filtering device introduced ten years ago by A. LACOURT (Besançon University) greatly improves the device performances : this allows the temporal display of the modal patterns excited in the tested multimode fiber ; the hologram will now be made of the superposition of many "image holograms" of the modes, each of them being modulated by a carrier frequency proportional to the considered modal delay (figure 39).

The diffraction far field of this hologram contains a discrete sequence of bright spots (figure 40, on the right). The horizontal abscissa of each spot is proportional to its group delay time. Its coherent struc-

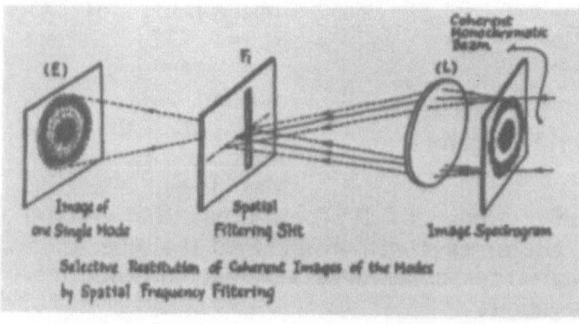

Fig. 41: Mode Restitution with Filtering

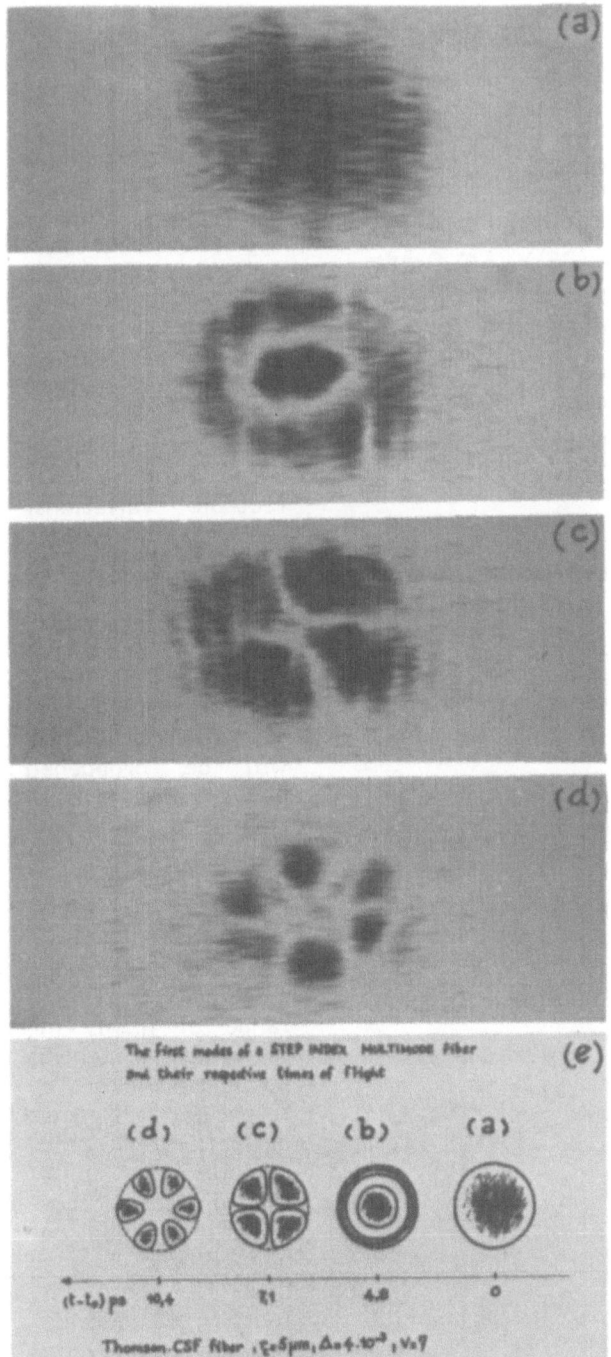

Fig. 42: Modal Patterns

ture contains the whole information needed for reconstruction of the corresponding modal pattern : as shown by figure 41 successive selection of each bright spot by the filtering slit gives successive images of the first excited modes. The modal images of figure 42 are very noisy ; but the signal to noise ratio would be increased by orders of magnitude if a smoother power spectrum could replace the irregular dye laser spectrum used in this experiment.

4 - EXAMPLES OF NON LINEAR PICOSECOND OPTICS

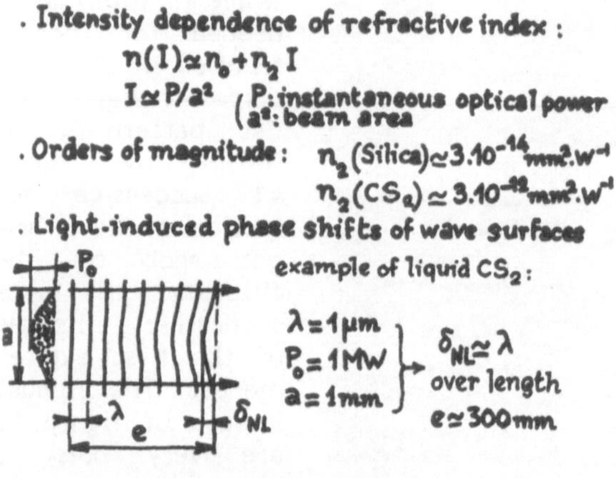

. Intensity dependence of refractive index :

$$n(I) \propto n_o + n_2 I$$

$$I \propto P/a^2 \quad \left(\begin{array}{l}P:\text{instantaneous optical power}\\ a^2:\text{beam area}\end{array}\right)$$

. Orders of magnitude :

$$n_2(\text{Silica}) \simeq 3.10^{-14} \text{mm}^2.\text{w}^{-1}$$

$$n_2(\text{CS}_2) \simeq 3.10^{-12} \text{mm}^2.\text{w}^{-1}$$

. Light-induced phase shifts of wave surfaces

example of liquid CS_2 :

$$\left.\begin{array}{l}\lambda = 1\mu m\\ P_o = 1MW\\ a = 1mm\end{array}\right\} \begin{array}{l}\delta_{NL} \simeq \lambda\\ \text{over length}\\ e \simeq 300mm\end{array}$$

Fig. 43: Non-linear Effect

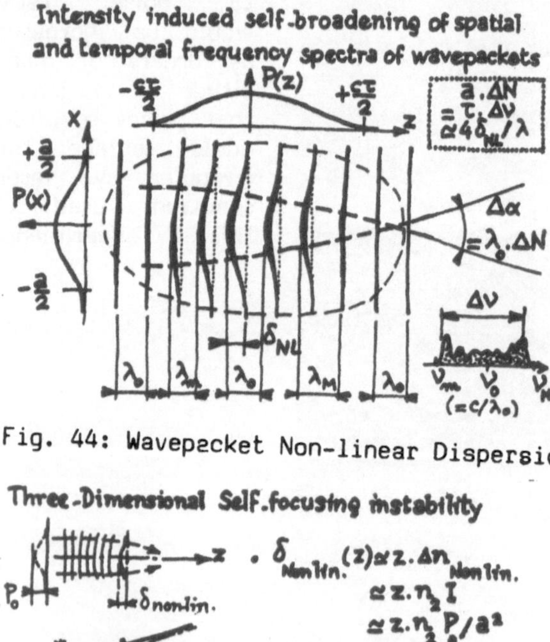

Intensity induced self-broadening of spatial and temporal frequency spectra of wavepackets

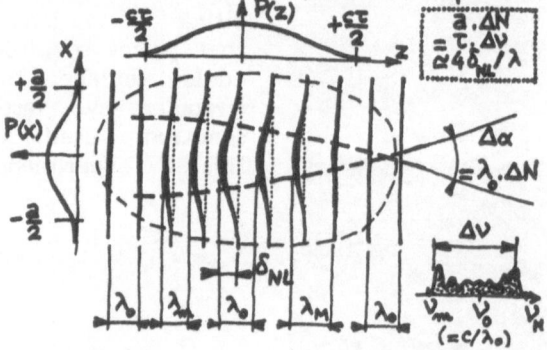

$$\begin{array}{l}a.\Delta N\\ = \tau.\Delta\nu\\ \simeq 4\delta_{NL}/\lambda\end{array}$$

$$\Delta\alpha = \lambda_o.\Delta N$$

$$\Delta\nu$$

$$\left(=c/\lambda_o\right)$$

Fig. 44: Wavepacket Non-linear Dispersion

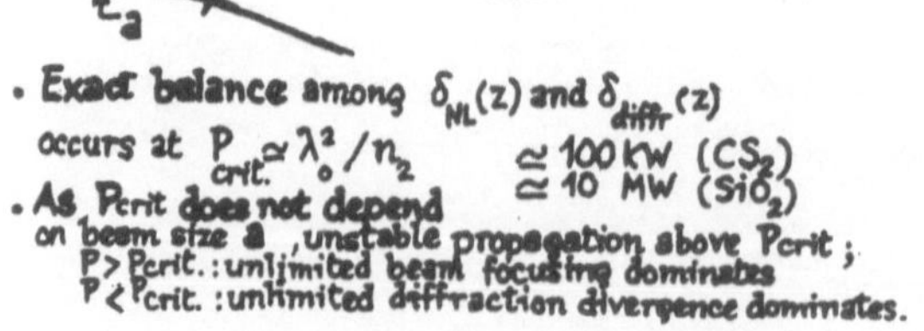

Three-Dimensional Self-focusing instability

. $\delta_{Nonlin.}(z) \propto z.\Delta n_{Nonlin.}$
$\simeq z.n_2 I$
$\simeq z.n_2 P_o/a^2$

. $\delta_{diffr.}(z) \simeq z\alpha^2 \simeq z(\lambda/a)^2$

. Exact balance among $\delta_{NL}(z)$ and $\delta_{diffr.}(z)$
occurs at $P_{crit.} \simeq \lambda_o^2/n_2 \quad \simeq 100\,kW \ (CS_2)$
$\simeq 10\,MW \ (SiO_2)$

. As P_{crit} does not depend
on beam size a, unstable propagation above P_{crit} ;
$P > P_{crit.}$: unlimited beam focusing dominates
$P < P_{crit.}$: unlimited diffraction divergence dominates.

Fig. 45: Self-focusing Instability

Picosecond optical pulses, even of weak energy, carry high peak powers : microjoule energy and picosecond pulse length correspond to megawatt peak power. At these levels of optical illumination the refractive index of transparent materials exhibits detectable variations.

The effects of these variations concern the spatial as well as the temporal structure of the pulses : their spatial and temporal spectral widths will be broadened by a ratio proportional to the amplitude of the light induced non-linear phase shift.

Self phase and self frequency modulation offer a range of highly interesting potential applications to "opto-optical" beam deflection, pulse compression, picosecond laser stabilization, optical bistability... A good deal of these effects were considered in early non linear optical stu-

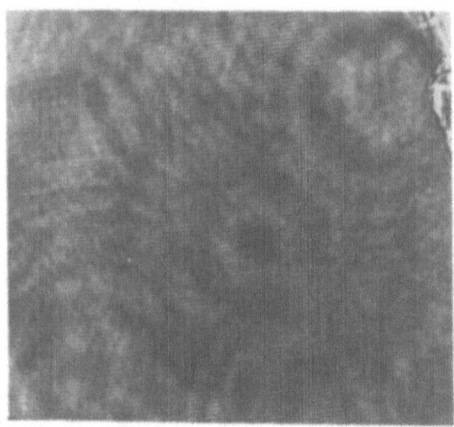

Fig. 46: Laser Beam Pattern

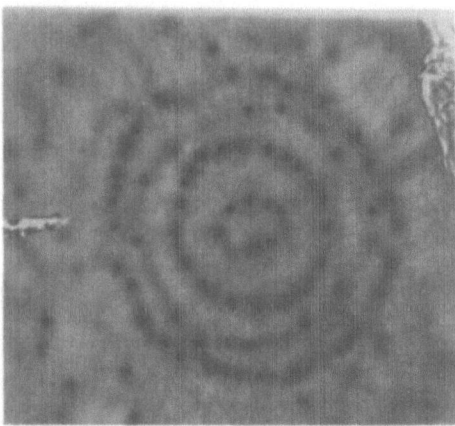

Fig. 47: Example of Self-focusing
Instability

dies. Unfortunately none of them, with the exception of the famous "Kerr oprical gate", was able to give usable and reproducible results. Failures in convincing experimental demonstrations were generally caused by beam and pulse instabilities : as soon as the nonlinear phase shifts integrated over the material thickness attaining a threshold very close to 2π radians, the laser beam breaks up suddenly into discrete and bright spots ; figures 46 and 47 show the cross section of the same laser beam at pulse powers respectively lower and higher than this threshold. The power carried by each spot is close to the self focusing critical power $P_{crit.}$ defined in figure 45. The geometrical location of spots is not predictable, but it appears from figures 46 and 47 that their distribution will be ruled by the intensity gradients of the coherent noise always present in laser beams.

It is clear that such chaotic random structures prevents any deterministic use of self phase modulation of light in applications to coherent optics.

Great progress was made when R. STOLEN and CHINLON LIN, at Bell Laboratories (1978), initiated single mode nonlinear propagation along single mode silica fibers : surprisingly the strong inten-

668

Self-phase and frequency modulations of
picosecond optical pulses through NON DISPERSIVE
single mode guided propagation

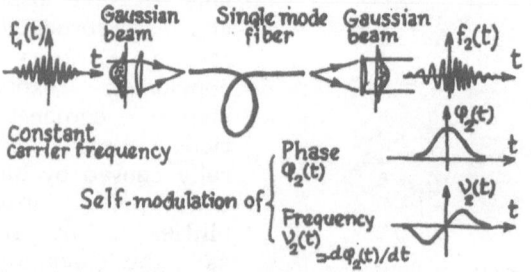

Fig. 48: Non-dispersive Self-modulation

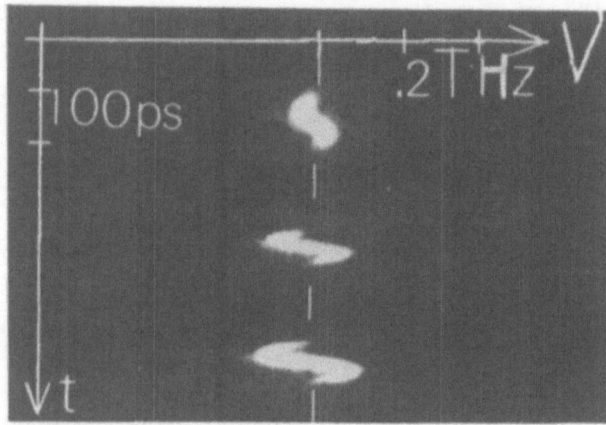

Fig. 49: Self-modulation Effect

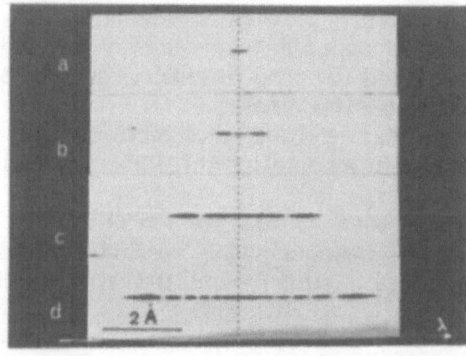

Fig. 50: Spectral Modulation

sity gradients in the fiber core did not generate self focusing. This is due to the weakness of the nonlinear refractive index variation with respect to the fiber core/cladding index gradient : the nonlinear phase shift perturbs the modal structure in a quite negligible way.

Under these original conditions of stable propagation of reproducible, intensity proportional self-frequency modulation does occur effectively ;

figure 49 shows streak camera recordings of the instantaneous frequency of the harmonic pulse of 100 ps YAg/Nd pulse (initial bandwidth 10^{10} Hz) after transmission over 3 m of silica single mode fiber at three different input powers.

The maximum spectral broadening about 4.10^{11} Hz, was observed at 700 W peak power, the pulse energy being as low as 10^{-7} J. The maximum nonlinear phase shift may be deduced from time-integrated spectral analysis using a spectroscope of spectral resolution higher than the reciprocal of the pulse length : in figure 50 (a) to (d) the ratio to π radians of the phase shift will be determined either by counting the number of maximums or minimums of the spectral modulation, or by measuring the half of the spectral broadening ratio.

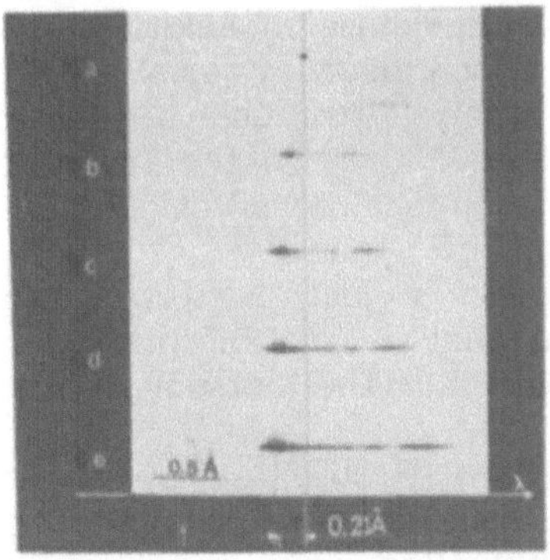

The spectral pattern figure 50 d corresponds to a maximum phase shift close to 14 π radians.

Fig. 51: Asymmetrical Spectral Modulation

Figure 51 represents the spectral distributions observed at increasing powers in the case of asymmetrical envelope shapes (nearly linear ramp) : on the left of the central vertical dotted line an intense narrow spectral peak showing a frequency shift proportional to the ramp slope and, consequently, to the peak power of the pulse. This was an experiment of **continously adjustable frequency translation** of a laser line performed at the Limoges University in 1979 ; the weaker parts of the spectrum (on the right hand) are due to the trailing edge of the ramp, contain a minor part of the pulse energy and play the part of a noise. The accessible frequency range is equal to the product of the normalized phase shift -as defined above- by the initial laser linewidth ; for instance in the case of figure 51 the 5.10^{-3} nm wide laser line was up-shifted by 21.10^{-3} nm after accumulation of about $5 \times 2\pi$ radians of nonlinear phase modulation.

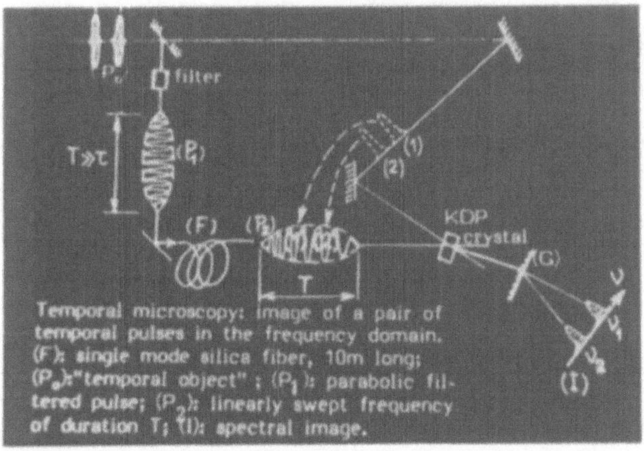

Then it would be of major interest to cumulate large phase modulations over very long, low loss optical fibers, thus providing new tunable laser lines in a very simple way.

Availability of wide self broadening of frequencies would also give the

Temporal microscopy: image of a pair of temporal pulses in the frequency domain. (F): single mode silica fiber, 10m long; (P$_0$):"temporal object" ; (P$_1$): parabolic filtered pulse; (P$_2$): linearly swept frequency of duration T; (I): spectral image.

Fig. 52: Temporal Microscopy

"temporal microscope" (figure 52) a very high resolving power ; this device demonstrated at Limoges University in 1983 performs the same operations on temporal field distributions as magnifying glasses do on geometric images : they multiply the "object" structure- here a pair (P_o) of pulses- by a parabolic phase modulation - the "temporal lens", of length T - inside a nonlinear material (KDP crystal in our experiment) ; a FOURIER analysis through a spectroscope (G) images the pulse pair as two lines ν_1 , ν_2 in the frequency domain, in an analogous way to the final image observation through a magnifying lens in the angular (FOURIER) space.

The temporal resolution is just reciprocal to the self frequency modulation bandwidth, that is, to the maximum nonlinear phase difference between the top and the bottom of the parabolic pulse (P_2).

Unfortunately it is presently observed by all the workers in fiber nonlinear optics that the maximum nonlinear phase shift attainable in a stable, well controlled way, never exceeds about $20 \times 2\pi$ radians. What does occur beyond this limit ? Figure 53

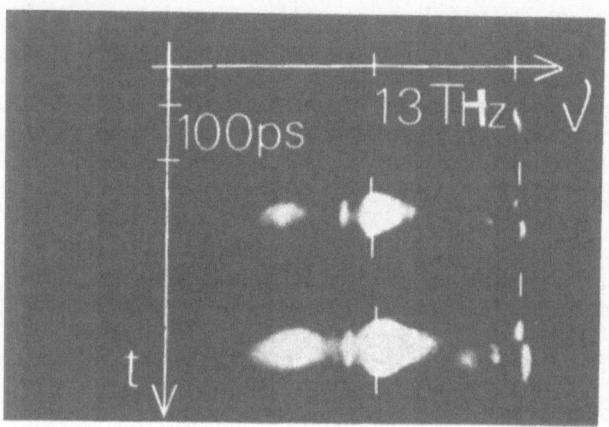

shows it : a streak camera recording of the spectral/temporal evolution demonstrates the existence of a well defined threshold of the product Power density x fiber length above which very broad Raman "lines" exhibit sudden growth : that is the "stimulated Raman Scattering", that presently settles unpassed barriers to the useful range of coherent self frequency modulation of short light pulses.

Fig. 53: Limit on Non-linear Phase Shift

More about picosecond optics

General

- Ultrashort light pulses ; S.L. SHAPIRO Ed. - Springer Verlag, Berlin, 1977.

- Ultrafast Phenomena IV, D.H. AUSTON, K.B. EISENTHAL Ed. Springer Verlag 1984.

FOURIER approach and processing

- Shaping and Analysis of picosecond light pulses : C. FROEHLY, B. COLOMBEAU, M. VAMPOUILLE, Progress in Optics - Vol. XX, E. Wolf Ed., North Holland Publ., 1983).

- B. COLOMBEAU, thesis, Limoges Univers. 1983 (in French, available from author on request).

Application to coherent guided propagation

- Temporally coherent Fiber Optics; C. FROEHLY, in "New directions in guided wave and coherent Optics", vol.1 and 2, NATO.ASI series E, n°72 - D.B. Ostrowsky and E. Spitz Ed., Martinus Nijhoff Publ., 1984.

Optical SOLITONS: a short presentation.

Claude FROEHLY
Laboratoire d'Optique (U.A. C.N.R.S. n°356),
123, rue A. Thomas, F.87060-LIMOGES Cédex (France)

Solitons

They are a very wide class of stable solutions of nonlinear propagation equations. The first Soliton officially recognized in Physics was the famous Gravity Wave in water observed by Scott Russel (read for instance "the Soliton and its History", in "Solitons", Topics in Current Physics, R.K. Bullough, P.J. Caudrey, Springer Verlag 1980) during a horse ride along a canal. The experimental conditions were quite similar to these having generated Swimming Pool Solitons last summer at our Optical Metrology Institute...

Optical Solitons

They are stable coherent optical waves propagating through nonlinear materials, for instance in presence of intensity-dependent transparency or refractive index. Their existence was discussed theoretically in the early ages of Nonlinear Optics (intensive works on the so-called "self-induced filamentation" of pulsed Laser beams).

Here we will only consider Soliton propagation due to self-phase modulation of light waves by the dependence of the refractive index with respect to the wave intensity.

The pioneering experiment on optical solitons was performed in a single mode Silica fiber by L.F. Mollenauer, R.H. Stolen and J.P. Gordon at the Bell Laboratories (Experimental observation of picosecond pulse narrowing and solitons in optical fibers, Phys. Rev. Lett. 45, 1095, 1980.)

This experiment evidenced a set of surprisingly simple particular situations where dispersion of picosecond pulses along a fiber cancelled because of its exact compensation by the frequency shifts resulting from the light-induced refractive index variations of the fiber core. Fiber-guided Soliton propagation was shown to occur at well-defined pulse powers and temporal shapes.

Thanks to the close parallelism of pulse dispersion with monochromatic beam diffraction "Soliton beams" should also exist, resulting from exact cancellation of diffraction divergence by the self-induced nonlinear convergence of the beam. Such stable "self-trapping" of quasi-monochromatic parallel beams was effectively demonstrated experimentally at the Limoges University in 1984 (in publication).

Both classes of solitons just considered are solutions of the same **Cubic Nonlinear Schrödinger Equation** written either in the temporal or in the geometrical domain according as the matter concerns soliton pulses or beams respectively.

Thus we will restrict the next presentation of the conditions required by soliton propagation to the geometrical case of beam self-trapping, the temporal self-confining of soliton pulses being explained by the same analysis after replacement of the usual wavelength by the "temporal wavelength" depending on the dispersion of the propagation line.

SOLITON BEAM: two-dimensional, self-induced single mode waveguide generated by compensation of DIFFRACTION by SELF-FOCUSING

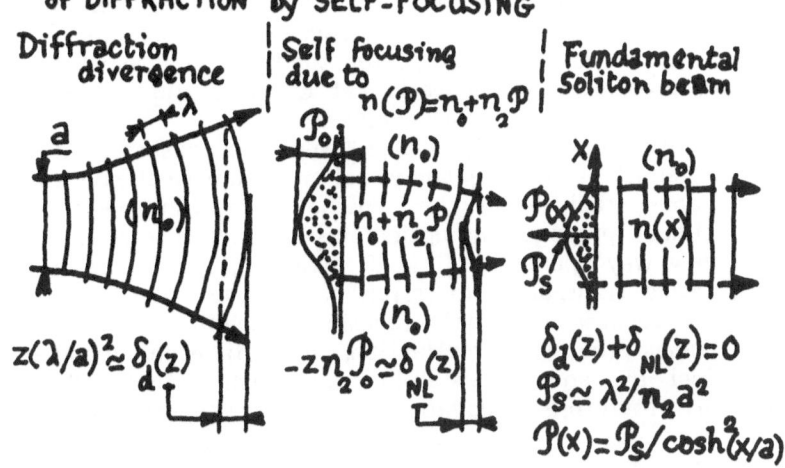

On the contrary to the unstable self-focusing beam distortion arising in three-dimensional propagation (see section 4), stable equilibrium exists in a two-dimensional propagation space, provided that beam shape and power satisfy to the "Soliton conditions" given on the figure above. This case represents "bright solitons". "Dark solitons", of hyperbolic tangent shape, were also theoretically considered when the sign of either the nonlinear coefficient n_2 or the wavelength is changed. No experimental observation of dark soliton beams was performed up to the present time to our knowledge. Dark or bright field distributions satisfy to the Schrödinger equation with a nonlinear term in the cubic power of the beam amplitude:

$$j.df_z(x)/dz = a.\partial^2 f_z(x)/\partial x^2 + b.\left|f_z(x)\right|^2.f_z(x),$$

where $a = \lambda/4\pi n_0$, $b = 2\pi n_2/\lambda$, $\lambda = n_0 \cdot \lambda$.

Demonstrating this equation is easy, using elementary Fourier analysis: it only consists to calculate the total perturbation experienced by the amplitude distribution $f_z(x)$ over a propagation length dz by summing the effects of an elementary diffraction step and an elementary nonlinear phase shift (A. Barthelemy, C. Froehly: Fourier analysis of geometrical and temporal solitons; J.of China Institute of Telecommunications, vol.6, n°1, 1985).

"First order" soliton solutions may be found in the form of an inhomogeneous, propagation invariant plane wave : $f_z(x) = f_0(x) \times \exp(-j\beta z)$. They are either of the sech(x) type or of the tanh(x) type, depending on the sign of the coefficient b.

Generation of bright soliton pulses along single mode fibers requires positive dispersion $dv_g/d\nu$ of the group velocity v_g with respect to the optical frequency (ν). It made the experiment rather difficult, needing the use of a special "color center" mode locked picosecond Laser emitting at 1.55 micrometer wavelength. Dark soliton pulses could be expected in the whole range of normal dipersion of silica fibers, i.e. at wavelengths shorter than 1.33 micrometers. Experiments are presently in preparation at Limoges and Brussels Universities, concerning these "negative pulses".

The future of "Soliton Optics" seems promising as it is an "islet of knowledge", admitting analytical and numerical studies, exhibiting welcome stability in the moving and often disappointing ocean of nonlinear propagation, optical instabilities, catastrophes and chaos. This new way opens the field of efficient, lossless Laser beam interactions inside transparent materials, which could never be overcome previously because of the self-focusing instability. "Opto-optical" beam and pulse processing (deflection, magnification, modulation, shaping; optical bi- and multistability and many other all-optical operations) are therefore becoming much more realistic than only three years ago, since plane Laser waves are now able to cumulate nonlinear phase shifts without being converted immediately into chaotic noise similar to the pattern of   figure 47 !

Note by the Editor:

Participants at the Optical Metrology Institute excite solitons (1).

Reference:

1. Olsen, J.; Smith, H.; and Scott, A.C.; Solitons in a Wave Tank, Am. J. Phys. 52 (1984), 826-830

PHOTOELASTICITY

PHOTOELASTICITY

J.F.Silva Gomes

University of Oporto, Department of Mechanical Engineering

1 FUNDAMENTALS OF THE THEORY OF ELASTICITY

1.1 Introduction

Photoelasticity is an experimental technique for stress and strain analysis. It is based upon a unique property of certain transparent noncrystalline materials that are optically isotropic when free of stresses, but which become optically anisotropic and display characteristics similar to cristals when they are stressed.

Photoelastic analysis is widely used for engineering problems in which stress or strain information is required for extended regions of structures or mechanical components. The experimental technique results in measurements of either stress or strain, from which the other quantity must be inferred. The necessary relationships amoung the stresses and the strains and between stress and strain are given by the theories of elasticity and plasticity. This short introduction to *Photoelasticity* is a review of the most useful concepts in the field of two- and three-dimensional elasticity, and covers the quantities and the equations most valuable in the photoelastic analysis.

1.2 The Concept of Stress

Fig. 1 represents a body in equilibrium under the action of *external forces* P_1, P_2, P_3,..., P_7 and *body forces* of intensity F per unit of volume of the body. To study the internal forces produced in the body, let us divide it by a plane s-s into two parts A and B, and consider one of these parts, say part A. This part is in equilibrium under the action of the external forces and body

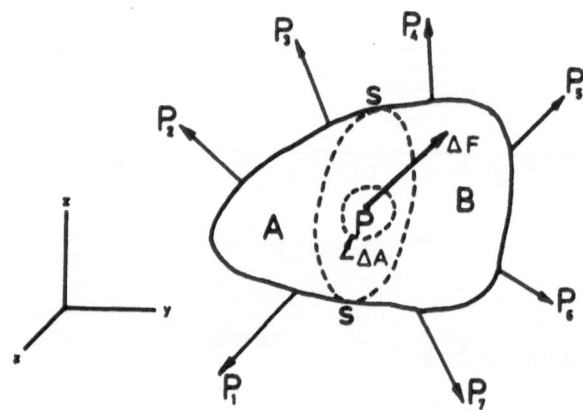

Fig.1 - Solid body subjécted to external forces

forces acting on part A, and the internal forces, continuously
distributed over the plane s-s₃ and representing the action of the
material of part B on the material of part A. The magnitudes of the
latter forces are usually defined at each point by their intensity,
i.e, by the amount of force per unit area of the plane on which
they act:

$$S = \lim_{\Delta A \to 0} \frac{\Delta F}{\Delta A}$$

where S is the *stress* at point P, ΔA is a small area around P, and
ΔF is the resultant of forces acting on ΔA.

In the general case of Fig.1, the direction of the stress S
is inclined to the area ΔA, on which it acts, and it can be resolved
into two components, as illustrated in Fig.2: a *normal stress* σ,
perpendicular to the area ΔA, and a *shear stress* τ, acting in the
plane of the area ΔA. Thus the general state of stress at a point
in the body may be expressed by a combination of normal and shear
stresses, whose magnitudes depend on the orientation of the plane
of the section.

In many cases, the external and body forces define a plane
which contains all of the stresses. This condition is known as *two-
-dimensional*, or *bi-axial*, or *plane stress*, and most of the stress
analysis problems are of this type. If the plane of the stresses
is taken as the x y - plane, a general two-dimensional stress state
at a point may be represented by the stresses acting on the rectangular element shown in Fig.3. Each of the faces is subjected to a
combination of normal stress, (either *tensile*, which is considered
positive, or *compressive*, considered negative), and shear stress.

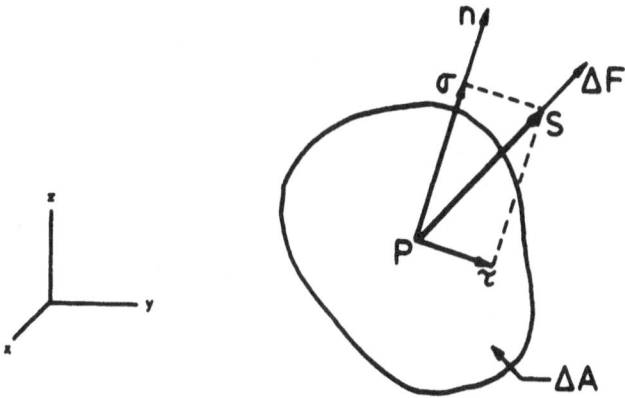

Fig.2 - Normal and shear stress

Usually the notation is a follows:

σ_x - *normal stress* on a plane perpendicular to the x-axis
σ_y - *normal stress* on a plane perpendicular to the y-axis
τ_{xy} - *shear stress*; the first subscript refers to the axis perpendicular to the plane on which the stress acts, the
second subscript refers to the direction of the stress in
that plane. It can be shown that, for reasons of equilibrium, $\tau_{yx} = \tau_{xy}$.

As the orientation of the axis is changed, the magnitudes of
stress components will change. In Fig.4 is shown a set of axes x'y',
which is rotated through an angle θ from xy. Equilibrium of forces
acting on the triangular elements 0 A B and 0 B C give the following

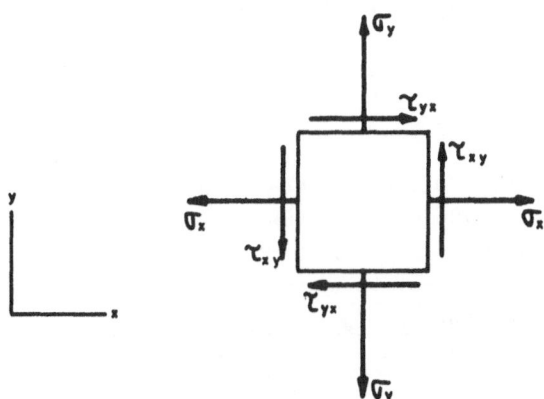

Fig.3 - Two-dimensional element showing stress components

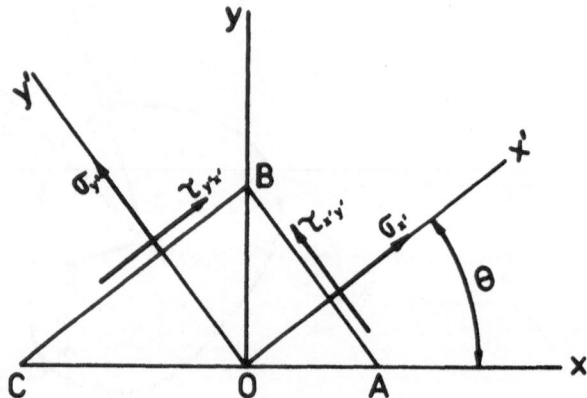

Fig.4 - Rotation of two-dimensional cartesian axes

relations between the stress components:

$$\sigma_{x'} = \frac{\sigma_x + \sigma_y}{2} + \frac{\sigma_x - \sigma_y}{2} \cos(2\theta) + \tau_{xy}\sin(2\theta)$$

$$\sigma_{y'} = \frac{\sigma_x + \sigma_y}{2} - \frac{\sigma_x - \sigma_y}{2} \cos(2\theta) - \tau_{xy}\sin(2\theta) \qquad (1)$$

$$\tau_{x'y'} = \tau_{xy}\cos(2\theta) - \frac{\sigma_x - \sigma_y}{2} \sin(2\theta)$$

Manipulation of these equations shows that an orientation of x'y' can be found which makes the shear stress $\tau_{x'y'}$ zero, and the corresponding normal stresses on the planes perpendicular to these axes are the maximum and the minimum which exist at the point under examination. This occurs for an angle θ_p given by

$$\tan(2\theta_p) = \frac{2\tau_{xy}}{\sigma_x - \sigma_y} \qquad (2)$$

These axes are called *principal stress axes* and the corresponding normal stresses, σ_1 and σ_2, are the *principal stresses* at that point. They are given by

$$\sigma_{1,2} = \frac{\sigma_x + \sigma_y}{2} \pm \sqrt{\left(\frac{\sigma_x - \sigma_y}{2}\right)^2 + \tau_{xy}^2} \qquad (3)$$

1.3 Mohr's Circle for Stress

If the principal stress directions are taken as the x and y axes, τ_{xy} vanishes in equations (1) and it is obtained, for the plane perpendicular to x':

$$\sigma = \frac{\sigma_1 + \sigma_2}{2} + \frac{\sigma_1 - \sigma_2}{2} \cos(2\theta)$$

$$\tau = \frac{1}{2} (\sigma_2 - \sigma_1) \sin(2\theta)$$

(4)

The normal and shearing components given by equations (4) are the coordinates of point D of the circle shown in Fig.5-a. For the construction of this circle, the τ axis is taken positive in the upward direction and the shear stresses considered as positive when they give a couple in the clockwise direction, as on the sides bc and ad of the element abcd (Fig.5-b). As the angle θ, in Fig.4, varies from 0 to $\pi/2$, the point D of Fig.5-a moves from A to B, so that the upper half-circle represents the stress variation for all values of θ within these limits. The lower half-circle gives stresses for $0>\theta>-\pi/2$. Prolonging the radius CD to the point D_1, Fig.5-a, i.e., if the angle $\pi+2\theta$ is taken instead of 2θ, the stresses on the plane normal to AB, in Fig.4, are obtained. This shows that the shear stresses on two mutually perpendicular planes are numerically equal. As for normal stresses, it is seen from Fig.5, that their sum remains constant when the angle θ changes. The maximum shear stress is given in Fig.5 by the maximum ordinate of the circle, and it is equal to

$$\tau_{max} = \frac{\sigma_1 - \sigma_2}{2}$$

(5)

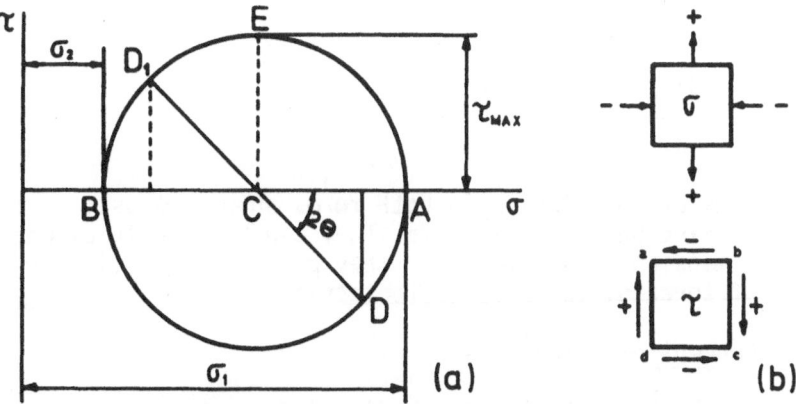

Fig.5 - Mohr's stress circle in two-dimensions

The circle above can also be used for determining the principal stress directions. If x and y are not the principal stress axes, and the stress components σ_x, σ_y and τ_{xy} are known, the two points such as D and D_1, in Fig.5, can be plotted. This gives the diameter DD_1 of the circle. Making the corresponding circle, points A and B are obtained, giving the magnitudes of principal stresses and of the angle 2θ defining the orientation of the principal axes. This graphical representation of the state of stress is known as *Mohr's stress circle* and it provides a powerful aid to visualization of the stress relationships at a point.

1.4 Stress Equations of Equilibrium

If consideration is given to the variation of stresses over a two-dimensional stress field, the conditions of equilibrium across a small element lead to a pair of differential equations known as the *equations of equilibrium*. Referred to a rectangular set of axes, xy, they are as follows:

$$\frac{\partial \sigma_x}{\partial x} + \frac{\partial \tau_{xy}}{\partial y} + X = 0$$

$$\frac{\partial \sigma_y}{\partial y} + \frac{\partial \tau_{xy}}{\partial x} + Y = 0$$

(6)

where X and Y are the components of the body force intensity in the x and y directions. In many problems, body forces such as weight may be neglected, in comparison with the effects of the transmitted stresses, and the above equations may be simplified further to give

$$\frac{\partial \sigma_x}{\partial x} + \frac{\partial \tau_{xy}}{\partial y} = 0$$

$$\frac{\partial \sigma_y}{\partial y} + \frac{\partial \tau_{xy}}{\partial x} = 0$$

(7)

The equations of equilibrium in both forms shown are used extensively in reducing photoelastic data. They can be readily extended to three dimensions, in which form they provide the basis for the solution of three-dimensional photoelastic problems.

1.5 The Concept of Strain

It is well known that solid bodies deform when subjected to load. For elastic materials, the amount of deformation is proportional to the applied loads. In discussing deformation of an elastic body, it will be assumed that there are enough constraints to

prevent motion as a rigid body, so that no displacements of particles of the body are possible without deformation. Only small deformations such as commonly occur in engineering structures will be
considered.

Consider an infinitesimal rectangular element, $\ell_x \times \ell_y$, (Fig.
6-a), subjected to normal stresses σ_x and σ_y. The applied stresses
cause variations of $\Delta\ell_x$ and $\Delta\ell_y$ in lengths of segments OA and OB,
both sides remaining perpendicular to each other. The *unit elongation* or *strain* at point 0, in the x and y directions are:

$$\varepsilon_x = \frac{\Delta\ell_x}{\ell_x} = \frac{\partial u}{\partial x}$$

$$\varepsilon_y = \frac{\Delta\ell_y}{\ell_y} = \frac{\partial v}{\partial y}$$

(8)

where u and v are the displacements of the point 0 in the x and y
directions. As for the shear stress τ_{xy}, Fig.6-b, it causes the
initially right angle AOB between the two sides OA and OB to dimi-
nish by the amount:

$$\gamma_{xy} = \frac{\partial u}{\partial y} + \frac{\partial v}{\partial x}$$

(9)

which represents the *shear strain* between the directions Ox and Oy.

The state of strain, defined relative to x and y - axes as the
combination of ε_x, ε_y and γ_{xy}, may also be expressed with reference
to another set x'y' rotated relative to xy, as in Fig.4. The assump-
tion that strains are small enough that only first order effects

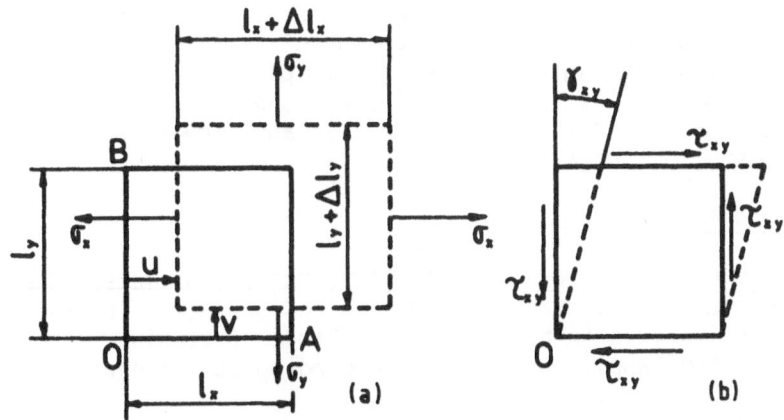

Fig.6 - Normal and shearing strain

need to be considered, (and this is a satisfactory approximation to a majority of engineering structures), leads to the following relationships:

$$\epsilon_{x'} = \frac{\epsilon_x + \epsilon_y}{2} + \frac{\epsilon_x - \epsilon_y}{2} \cos(2\theta) + \frac{\gamma_{xy}}{2} \sin(2\theta)$$

$$\sigma_{y'} = \frac{\epsilon_x + \epsilon_y}{2} - \frac{\epsilon_x - \epsilon_y}{2} \cos(2\theta) - \frac{\gamma_{xy}}{2} \sin(2\theta) \tag{10}$$

$$\gamma_{x'y'} = \gamma_{xy} \cos(2\theta) - (\epsilon_x - \epsilon_y) \sin(2\theta)$$

where θ is the angle measured from the x-axis to the x'-axis.

Comparison of equations (10) above with equations (1) for stresses, reveals an identity of form. It is observed that the equations for stress transformation are converted into strain relationships by replacing σ with ϵ and τ with $\gamma/2$. By analogy with stresses, the *principal strain directions*, (where $\gamma_{x'y'} = 0$), are found by the equation,

$$\tan(2\theta_p) = \frac{\gamma_{xy}}{\epsilon_x - \epsilon_y} \tag{11}$$

and the *principal strains*, ϵ_1 and ϵ_2, by

$$\epsilon_{1,2} = \frac{\epsilon_x + \epsilon_y}{2} \pm \sqrt{\left(\frac{\epsilon_x - \epsilon_y}{2}\right)^2 + \left(\frac{\gamma_{xy}}{2}\right)^2} \tag{12}$$

It is also apparent that a *Mohr's circle for strain* may be drawn in a similar manner as for stresses. In Mohr's circle for strain, the normal strains are plotted in the horizontal axis, positive to the right. For a positive shear strain, $\gamma_{xy}/2$ is plotted at a distance $\gamma_{xy}/2$ below the ϵ line, (point A, Fig. 7), and γ_{yx} at a distance $\gamma_{xy}/2$ above the ϵ line, (point B, Fig. 7); and vice-versa when the shear strain γ_{xy} is negative.

1.6 The Hooke's Law

Experimental observations show that, for a tensile specimen made of homogeneous, isotropic material, and subjected to uniaxial stress along its axis (x-axis), the magnitude of strain ϵ_x is proportional to the applied stress σ_x, according to the equation

$$\epsilon_x = \frac{\sigma_x}{E} \tag{13}$$

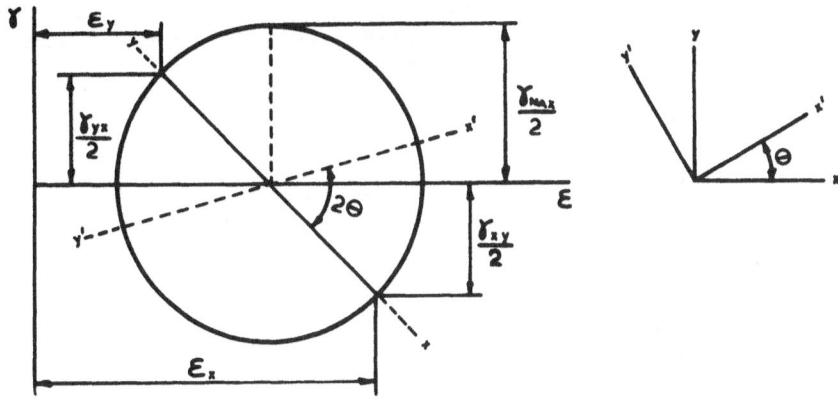

Fig.7 - Mohr's circle for strain

where E is the *Young's modulus* of the material. This extension in the axial direction is accompanied by lateral contractions proportional in magnitude to the strain ε_x:

$$\varepsilon_y = \varepsilon_z = - \nu \varepsilon_x = - \nu \frac{\sigma_x}{E} \tag{14}$$

The constant of proportionality, ν, is known as the *Poisson's ratio.* Thus the generalized expressions for *Hooke's law* may be assembled by superposition of those effects:

$$\varepsilon_x = \frac{\sigma_x}{E} - \frac{\nu}{E} (\sigma_y + \sigma_z)$$

$$\varepsilon_y = \frac{\sigma_y}{E} - \frac{\nu}{E} (\sigma_z + \sigma_x) \tag{15}$$

$$\varepsilon_z = \frac{\sigma_z}{E} - \frac{\nu}{E} (\sigma_x + \sigma_y)$$

These equations may be solved for stresses to give three equations of the form,

$$\sigma_x = \frac{E}{(1+\nu)(1-2\nu)} [(1-\nu) \varepsilon_x + \nu \varepsilon_y)]$$

$$\sigma_y = \frac{E}{(1+\nu)(1-2\nu)} [(1-\nu) \varepsilon_y + \nu \varepsilon_x)] \tag{16}$$

$$\sigma_z = \frac{\nu E}{(1+\nu)(1-2\nu)} (\varepsilon_x + \varepsilon_y)$$

For the common problem of determining the state of strain on the free surface of a body, where the stress normal to the surface, say σ_z, is zero, equations above reduce to,

$$\varepsilon_x = \frac{1}{E} (\sigma_x - \nu\sigma_y)$$

$$\varepsilon_y = \frac{1}{E} (\sigma_y - \nu\sigma_x) \tag{17}$$

$$\varepsilon_z = -\frac{\nu}{E} (\sigma_x + \sigma_y)$$

or, solved for stresses,

$$\sigma_x = \frac{E}{(1-\nu^2)} (\varepsilon_x + \nu\varepsilon_y)$$

$$\sigma_y = \frac{E}{(1-\nu^2)} (\varepsilon_y + \nu\varepsilon_x) \tag{18}$$

Shear stresses and strains are also related linearly as follows:

$$\tau_{xy} = G \gamma_{xy}$$

$$\tau_{yz} = G \gamma_{yz} \tag{19}$$

$$\tau_{zx} = G \gamma_{zx}$$

where G is the *shear modulus*. Because of the inter-relations involved in equations (1), (10), (17) and (19), the shear modulus and the Young's modulus are related by

$$G = \frac{E}{2(1+\nu)} \tag{20}$$

1.7 Three-Dimensional Relationships

For some problems, when the body shape or loading is too complex to be successfully modeled in a single plane, the two-dimensional relationships given above are inadequate, and must be replaced by the more complete three-dimensional stress and strain equations. The rectangular element should be replaced by a cartesian cube-shaped element, Fig.8, on each face of which act a component of normal stress and two of shear stress, identified by subscripts in accord with the notation described earlier. Consideration of the equilibrium of forces acting on that element leads to the three-dimensional equations of equilibrium,

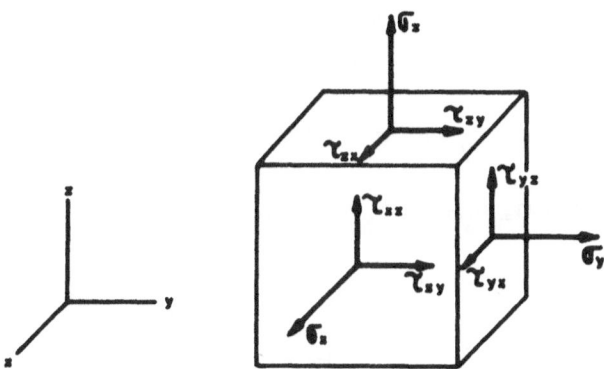

Fig.8 - Three-dimensional element showing stress components

$$\frac{\partial\sigma_x}{\partial x} + \frac{\partial\tau_{xy}}{\partial y} + \frac{\partial\tau_{yz}}{\partial z} + X = 0$$

$$\frac{\partial\sigma_y}{\partial y} + \frac{\partial\tau_{yz}}{\partial z} + \frac{\partial\tau_{zx}}{\partial x} + Y = 0 \qquad (21)$$

$$\frac{\partial\sigma_z}{\partial z} + \frac{\partial\tau_{zx}}{\partial x} + \frac{\partial\tau_{xy}}{\partial y} + Z = 0$$

where X,Y,Z, are the components of the body force intensity. It could also be shown that, for reasons of equilibrium,

$$\tau_{xy} = \tau_{yx} \; ; \; \tau_{yz} = \tau_{zy} \; ; \; \tau_{zx} = \tau_{xz} \qquad (22)$$

It becomes apparent, then, that a three-dimensional state of stress may be specified in terms of six components which, in cartesian coordinates, are the normal stresses σ_x, σ_y, σ_z, and shear stresses τ_{xy}, τ_{yz}, τ_{zx}.

For two sets of cartesian coordinate axes, whose relative orientation is described in terms of the direction cosines between them (Table - 1), the equations for transformation of stresses are as

TABLE 1 - Direction cosines between two sets of cartesian axes

	x	y	z
x'	ℓ_1	m_1	n_1
y'	ℓ_2	m_2	n_2
z'	ℓ_3	m_3	n_3

follows:

$$\sigma_x' = \ell_1^2\sigma_x + m_1^2\sigma_y + n_1^2\sigma_z + 2\ell_1 m_1 \tau_{xy} + 2m_1 n_1 \tau_{yz} + 2n_1 \ell_1 \tau_{zx}$$

$$\sigma_{y'} = \ell_2^2\sigma_x + m_2^2\sigma_y + n_2^2\sigma_z + 2\ell_2 m_2 \tau_{xy} + 2m_2 n_2 \tau_{yz} + 2n_2 \ell_2 \tau_{zx}$$

$$\sigma_{z'} = \ell_3^2\sigma_x + m_3^2\sigma_y + n_3^2\sigma_z + 2\ell_3 m_3 \tau_{xy} + 2m_3 n_3 \tau_{yz} + 2n_3 \ell_3 \tau_{zx}$$

$$\tau_{x'y'} = \ell_1\ell_2\sigma_x + m_1 m_2\sigma_y + n_1 n_2\sigma_z + (\ell_1 m_2 + \ell_2 m_1)\,\tau_{xy}$$

$$\qquad + (m_1 n_2 + m_2 n_1)\tau_{yz} + (n_1 \ell_2 + n_2 \ell_1)\,\tau_{zx} \qquad\qquad (23)$$

$$\tau_{y'z} = \ell_2\ell_3\sigma_x + m_2 m_3\sigma_y + n_2 n_3\sigma_z + (\ell_2 m_3 + \ell_3 m_2)\,\tau_{xy}$$

$$\qquad + (m_2 n_3 + m_3 n_2)\tau_{yz} + (n_2 \ell_3 + n_3 \ell_2)\tau_{zx}$$

$$\tau_{z'x'} = \ell_3\ell_1\sigma_x + m_3 m_1\sigma_y + n_3 n_1\sigma_z + (\ell_3 m_1 + \ell_1 m_3)\,\tau_{xy}$$

$$\qquad + (m_3 n_1 + m_1 n_3)\tau_{yz} + (n_3 \ell_1 + n_1 \ell_3)\tau_{zx}$$

As in the plane-stress case, an orientation of x', y', z' can be found for which the shear stresses vanish, and $\sigma_{x'}$, $\sigma_{y'}$ and $\sigma_{z'}$ are the *principal stresses* at the point. It is usual to adopt for the principal stresses the notation $\sigma_1 > \sigma_2 > \sigma_3$, acting on the *principal planes* 1, 2 and 3. The principal stresses are found as the three roots of the cubic equation

$$\sigma_i^3 - (\sigma_x + \sigma_y + \sigma_z)\sigma_i^2 + (\sigma_x\sigma_y + \sigma_y\sigma_z + \sigma_z\sigma_x - \tau_{xy}^2 - \tau_{yz}^2 - \tau_{zx}^2)\,\sigma_i$$

$$\qquad - (\sigma_x\sigma_y\sigma_z - \sigma_x\tau_{yz}^2 - \sigma_y\tau_{zx}^2 - \sigma_z\tau_{xy}^2 + 2\tau_{xy}\tau_{yz}\tau_{zx}) = 0 \qquad (24)$$

and the directions of the *principal axes* may be found by the simultaneous solution of the following three equations:

$$\ell\,(\sigma_x - \sigma_i) + m\tau_{xy} + n\tau_{zx} = 0$$

$$\ell\tau_{xy} + m\,(\sigma_y - \sigma_i) + n\tau_{yz} = 0 \qquad\qquad (25)$$

$$\ell\tau_{zx} + m\tau_{yz} + n(\sigma_z - \sigma_i) = 0$$

with the relationship

$$\ell^2 + m^2 + n^2 = 1$$

where ℓ, m and n are the direction cosines of the principal stress axis, i, relative to x, y and z.

Observation of any of the principal planes from the positive end of its normal reveals in it a set of stress components related to orientation in that plane in the same manner as given by the Mohr's circle for two-dimensional state of stress. The complete stress state is represented by three Mohr's circles, Fig.9, one for each of the principal planes.

A three dimensional state of strain is defined by six components ϵ_x, ϵ_y, ϵ_z, γ_{xy}, γ_{xz}, γ_{zx}, and, as pointed out earlier, small-strain theory leads to transformation equations having the same form of the stress equations by substitution of normal strain ϵ, for normal stress, σ, and one-half of shear strains, $\gamma/2$, for shear stress, τ. Again, there are three *principal strains*, ϵ_1, ϵ_2 and ϵ_3, along three principal directions, 1, 2 and 3, for which the shear strains are zero.

Another aspect important to stress analysis is the result of the continuity of the elastic material. As the body is strained, contiguous elements must change shape so as to maintain contact with each other. This condition leads to six partial differential equations in the six strain components, know as the *equations of compatibility*:

$$\frac{\partial^2 \gamma_{xy}}{\partial x\, \partial y} = \frac{\partial^2 \epsilon_x}{\partial y^2} + \frac{\partial^2 \epsilon_y}{\partial x^2}$$

$$\frac{\partial^2 \gamma_{yz}}{\partial y\, \partial z} = \frac{\partial^2 \epsilon_y}{\partial z^2} + \frac{\partial^2 \epsilon_z}{\partial y^2}$$

$$\frac{\partial^2 \gamma_{zx}}{\partial z\, \partial x} = \frac{\partial^2 \epsilon_z}{\partial x^2} + \frac{\partial^2 \epsilon_x}{\partial z^2}$$

$$2\frac{\partial^2 \epsilon_x}{\partial y \partial z} = \frac{\partial}{\partial x}\left(- \frac{\partial \gamma_{yz}}{\partial x} + \frac{\partial \gamma_{zx}}{\partial y} + \frac{\partial \gamma_{xy}}{\partial z}\right) \qquad (26)$$

$$2\frac{\partial^2 \epsilon_y}{\partial z \partial x} = \frac{\partial}{\partial y}\left(\frac{\partial \gamma_{yz}}{\partial x} - \frac{\partial \gamma_{zx}}{\partial y} + \frac{\partial \gamma_{xy}}{\partial z}\right)$$

$$2\frac{\partial^2 \epsilon_z}{\partial x \partial y} = \frac{\partial}{\partial z}\left(\frac{\partial \gamma_{yz}}{\partial x} + \frac{\partial \gamma_{zx}}{\partial y} - \frac{\partial \gamma_{xy}}{\partial z}\right)$$

690

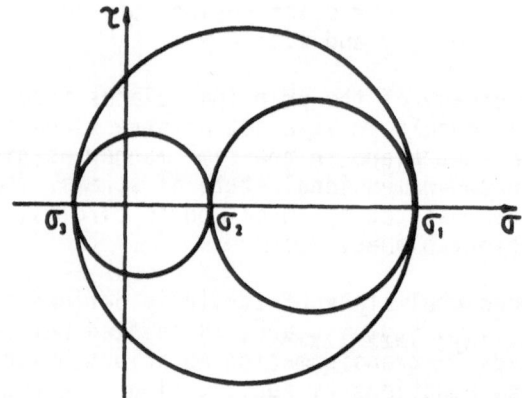

*Fig.*9 - Mohr's stress circles in three-dimensions

Because of the Hooke's law relation between stress and strain, the compatibility equations may also by expressed in terms of the stresses:

$$\nabla^2\sigma_x + \frac{1}{1+\nu}\frac{\partial^2}{\partial x^2}I_1 = -\frac{\nu}{1-\nu}\left(\frac{\partial X}{\partial x}+\frac{\partial Y}{\partial y}+\frac{\partial Z}{\partial z}\right)- 2\frac{\partial X}{\partial x}$$

$$\nabla^2\sigma_y + \frac{1}{1+\nu}\frac{\partial^2}{\partial y^2}I_1 = -\frac{\nu}{1-\nu}\left(\frac{\partial X}{\partial x}+\frac{\partial Y}{\partial y}+\frac{\partial Z}{\partial z}\right)- 2\frac{\partial Y}{\partial y}$$

$$\nabla^2\sigma_z + \frac{1}{1+\nu}\frac{\partial^2}{\partial z^2}I_1 = -\frac{\nu}{1-\nu}\left(\frac{\partial X}{\partial x}+\frac{\partial Y}{\partial y}+\frac{\partial Z}{\partial z}\right)- 2\frac{\partial Z}{\partial z}$$

$$\nabla^2\tau_{xy} + \frac{1}{1+\nu}\frac{\partial^2}{\partial x\partial y}I_1 = -\left(\frac{\partial X}{\partial y}+\frac{\partial Y}{\partial x}\right)$$

$$\nabla^2\tau_{yz} + \frac{1}{1+\nu}\frac{\partial^2}{\partial y\partial z}I_1 = -\left(\frac{\partial Y}{\partial z}+\frac{\partial Z}{\partial y}\right)$$

$$\nabla^2\tau_{zx} + \frac{1}{1+\nu}\frac{\partial^2}{\partial z\partial x}I_1 = -\left(\frac{\partial X}{\partial z}+\frac{\partial Z}{\partial x}\right)$$

(27)

where ∇^2 is the operator $(\partial^2/\partial x^2+\partial^2/\partial y^2+\partial^2/\partial z^2)$, I_1 is the first invariant of stress, $I_1=\sigma_x+\sigma_y+\sigma_z$, and X, Y, Z are the body force components in the x, y and z directions.

1.8 Secondary Principal Stresses

If a set of cartesian axes is rotated about one of the axes, say the z-axis, through an angle θ, (Fig.10), it will produce changes in σ_x, σ_y and τ_{xy}, which are described by equations (23). Since the z and z' axes coincide, those equations reduce to:

$$\sigma_{x'} = \frac{\sigma_x + \sigma_y}{2} + \frac{\sigma_x - \sigma_y}{2} \cos(2\theta) + \tau_{xy} \sin(2\theta)$$

$$\sigma_{y'} = \frac{\sigma_x + \sigma_y}{2} - \frac{\sigma_x - \sigma_y}{2} \cos(2\theta) - \tau_{xy} \sin(2\theta) \tag{28}$$

$$\tau_{x'y'} = \tau_{xy} \cos(2\theta) - \frac{\sigma_x - \sigma_y}{2} \sin(2\theta)$$

which are of the same form as equations (1) for the two-dimensional case of stress. This means that the stress state in the x' y' plane can be represented by a Mohr's circle like that of Fig.5, and that a pair of planes may be found, orthogonal to each other and to the z-plane, on which $\tau_{x'y'}$ vanishes and the normal stresses $\sigma_{x'}$ and $\sigma_{y'}$ are maximum and mimimum. Such planes are known as *secondary principal stresses*. These concepts of secondary principal planes and stresses are of importance in the solution of some problemes by two- or three-dimensional photoelastic methods.
dimensional photoelastic methods.

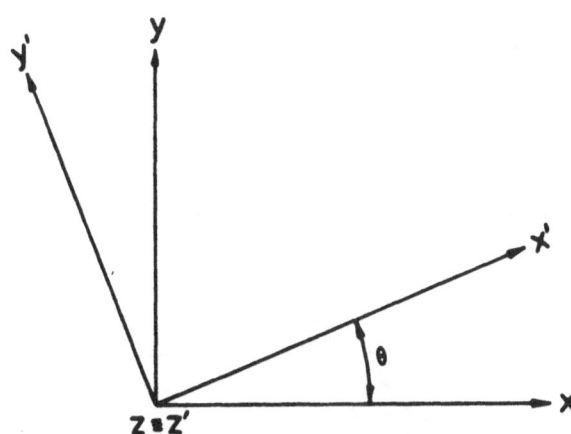

Fig.10 - Rotation of xy-plane about z-axis

2   INTRODUCTION TO PHOTOELASTICITY

2.1 The Nature of Light

According to the principles of the electromagnetic theory of light, proposed by Maxwell in 1864, a beam of light can be represented by the electric or magnetic field vector - *the light vector* - which is always at right angles to the direction of the wave propagation, and lies in the wave-front plane. This simple concept of the light vector has been successfully applied in many problems of light propagation, including effects of polarization optics dealing with the passage of polarized light through a number of birefringent or optical rotating media. The state of polarization of the light beam is completely determined if, at each position along the path of propagation, the direction and the magnitude of the light vector are known.

By considering the usual case of plane harmonic waves, the magnitude of the light vector or one of its components, will be described by

$$A = a \, \sin \frac{2\pi}{\lambda} \, (z - ct) \tag{29}$$

where a is the amplitude, z the coordinate along the axis of propagation, c the velocity of propagation, and t the time. A graphical representation of the magnitude of the light vector, at two different times, is illustrated in Fig.11. The length from peak to peak on the sinusoidal magnitude graph of the light vector is defined as the *wavelength* λ. The time required for the passage of two successive peaks at some fixed values of z is defined as the *period* T

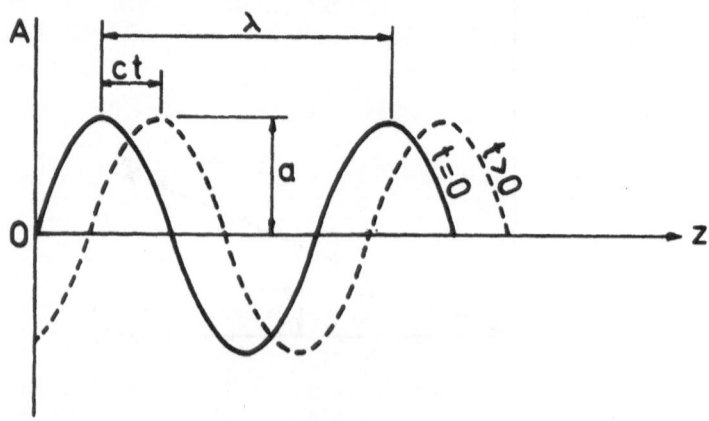

Fig.11 - Graphical representation of the magnitude of light vector

which is given by,

$$T = \frac{\lambda}{c} \tag{30}$$

The *frequency* of the light vector, or one of its components, is defined by the number of oscillations per second. Thus, the frequency f is given by

$$f = \frac{1}{T} = \frac{c}{\lambda} \tag{31}$$

and the *angular frequency* ω by

$$\omega = \frac{2\pi}{T} = 2\pi f \tag{32}$$

Two waves having the same frequency but different phases and amplitudes are shown in Fig.12. These two waves can be expressed as

$$A_1 = a \sin \frac{2\pi}{\lambda} (z + \delta_1 - ct)$$
$$\tag{33}$$
$$A_2 = b \sin \frac{2\pi}{\lambda} (z + \delta_2 - ct)$$

where δ_1 is the initial phase of A_1 and δ_2 the initial phase of A_2. The distance $\delta = \delta_2 - \delta_1$ between two consecutive peaks of the waves is the *linear phase difference* between waves. The difference between the phase angles of the two waves is called the *angular phase difference* Δ and related to the linear phase difference δ by

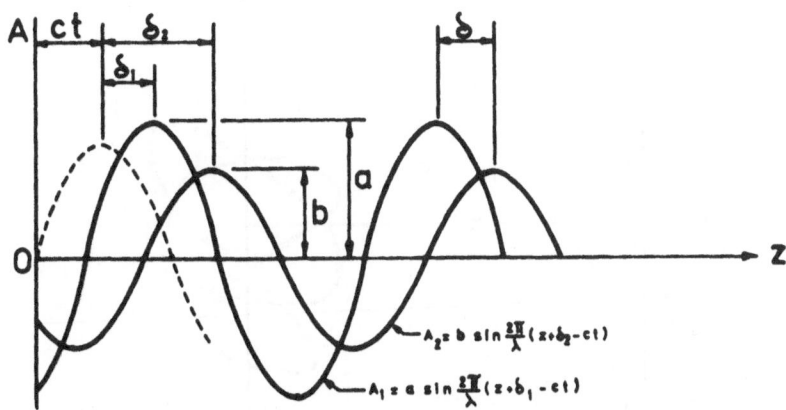

Fig.12 - Sketch showing the phase difference between two waves

694

the equation

$$\Delta = \frac{2\pi}{\lambda}\ \delta \qquad\qquad (34)$$

The colour of light which the human eye recognizes is determined by the frequency of the components of the light vector. The colours in the visible spectrum range from deep red, (390×10^{12} Hz), to a deep violet, (770×10^{12} Hz). When the light vector is composed of vibrations A_1, A_2, A_3,...., etc, which all have the same frequency, it is said to be *monochromatic*, of one colour depending upon the value of the frequency. The colours that can be observed by the human eye are, in the order of decreasing of frequency, violet, indigo, blue, green, yellow, orange and red. When the components of the light vector, A_1, A_2, A_3,...., etc are of different frequencies, the colours are mixed and the the human eye reccords this mixture as *white light*.

2.2 Ordinary and Polarized Light

According to the Maxwells electromagnetic theory, *ordinary* or *unpolarized* light can be described by the electric or magnetic vector, which moves in space irregularly and does not show any preferred directional properties. This is illustrated in Fig.13-a, in which the direction of the optical ray is perpendicular to the plane of the paper, and the light vector OA moves at random in that plane. The tip of the light vector describes a complicated curve as shown in Fig.13-b.

When some kind of order is introduced into the irregular motion of the light vector, the resulting light is said to be *polarized*. The end point of the light vector in polarization moves along well-

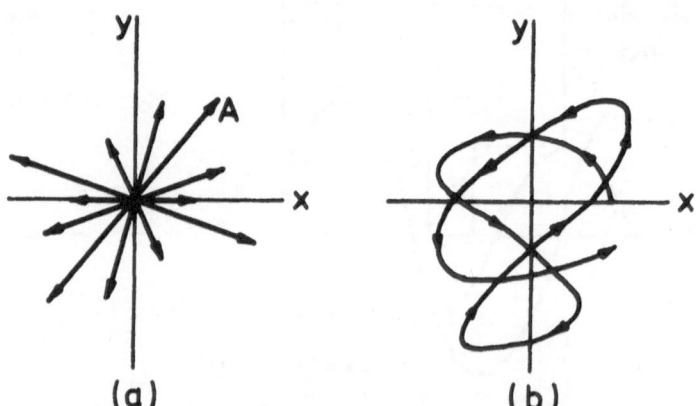

(a)   (b)

Fig.13 - Chaotic motion of the light vector in ordinary light

defined simple curves, in a definite direction. There are several forms of polarized light, according to the particular type of curve along which the tip of the light vector moves:

(i) - *Linear Polarized Light.* A polarized light is said to be linear or plane polarized when the light vector is confined to a single plane parallel to the direction of propagation and known as the *plane of polarization.* In this case, the light vector does not change in direction with time, but only in magnitude. Such type of light, or each of its components, can be represented by a sinusoidal curve which shows the magnitude of the wave train at a single instant, as it would be seen by an observer looking along the direction normal to the plane of polarization, Fig.14-a. The linear type of polarized light includes a one-parameter family, each form of linear polarization being defined by the inclination ψ of the light vector to the horizontal, Fig.14-b. Ordinary or common light may be considered as being made up of an infinite number of plane-polarizad components whose planes of polarization have every conceivable orientation.

(ii) - *Circular Polarized Light.* This type of polarized light is obtained when the tip of the light vector describes a circular helix, as the light propagates along the z-axis, as shown in Fig. 15-a. Thus, the magnitude of the light vector remains constant, while its inclination varies continuously between 0 and 2π. Two different types of circular polarization can be distinguished: The right-circular polarization and the left-circular polarization. The right-circular polarization is defined by the clockwise motion of the light vector, as it is seen by an observer looking along the propagation direction toward the light source, whereas the counter-clockwise motion corresponds to left-circular polarization Fig.15-b.

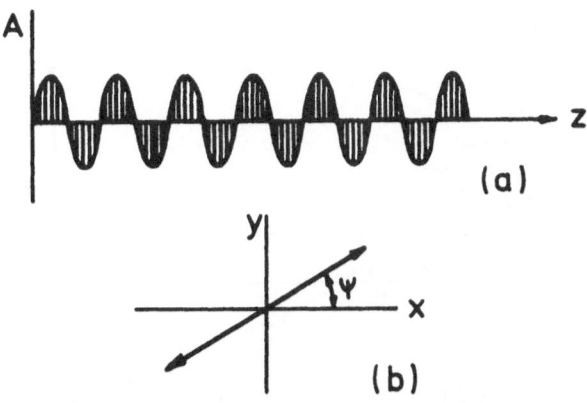

Fig.14 - Plane-polarized light

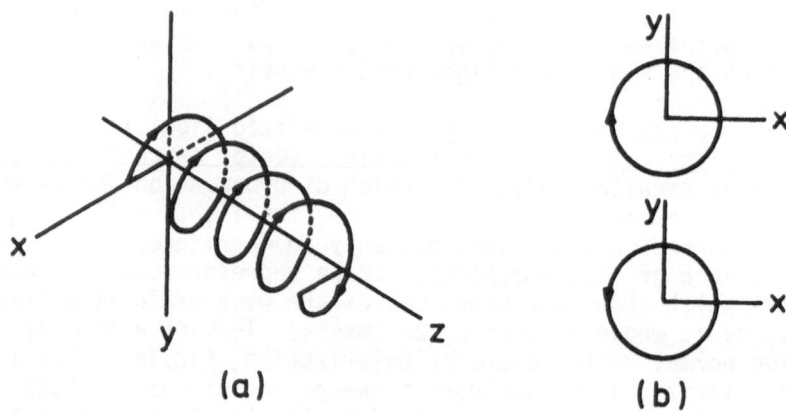

Fig.15 - Circular-polarized light

(iii) - *Elliptical Polarized Light*. This is the most general form of polarized light and includes both linear and circular polarized light as special cases. In the elliptical type of polarized light, the tip of the light vector describes an elliptical helix as the light propagates along the z-axis, as illustrated in Fig.16. Three different parameters must be used to define completely the elliptical polarization, i.e., the ratio b/a of the semi-axes of the ellipse, called the ellipticity; the inclination ψ of the major semi-axis of the ellipse with respect to the x-axis, called the azimuth; and the direction of motion of the light vector (clockwise or counterclockwise). Linear and circular forms of polarization correspond to elliptical polarized light for values of ellipticity (b/a) equal to 0 and 1, respectively.

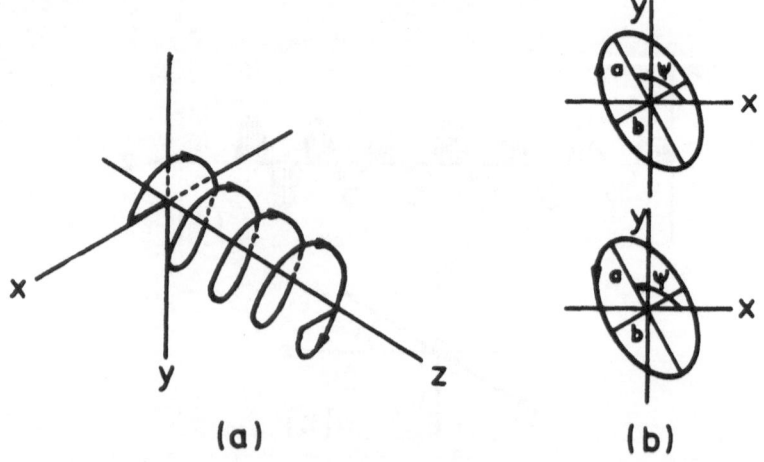

Fig.16 - Elliptical-polarized light

2.3 Plane Polarizers

A plane or linear polarizer is an optical element that divides the incident light beam into two orthogonal components, absorbing one of them. When a light vector passes through a plane polarizer, this optical element absorbs that component which is perpendicular to the axis of polarization, and transmit the parallel component, as illustrated in Fig.17. Thus, a linear polarizer whose axis corresponds to the vertical transmits, without any loss of intensity, all of the vertically polarized light, and extinguishes all of the horizontally polarized light.

If the plane polarizer is positioned at some point along the z-axis, the equation for the magnitude of the sinusoidal component of the light vector may be written as

$$A = a \sin \left(\frac{2\pi}{\lambda} c t\right)$$

which can be reduced to

$$A = a \sin (\omega t) \tag{35}$$

The absorbed and transmitted components of the light vector are

$$A_a = a \sin (\omega t) \ \sin\theta \tag{36}$$

$$A_t = a \sin (\omega t) \ \cos\theta \tag{37}$$

where θ is the angle between the axis of polarization and the light vector.

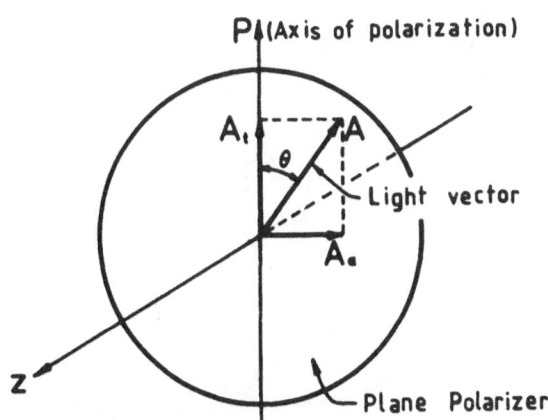

Fig.17 - Transmission of light through a plane-polarizer

Plane polarized light can be produced by a number of different methods, which includes reflected light at 57° angle of incidence, a glass pile, and Nicol prism. However, a sheet of Polaroid material is by far the most popular and generally used device for polarizing light in photoelastic polariscopes. The Polaroid filters have the advantage of providing a large field of very well polarized light at a relatively low cost.

2.4 Wave Plates

Plates made of certain crystalline materials, such as mica, have the ability of resolving the incident light vector A into two orthogonal components, A_1 and A_2, Fig.18, transmitting these components at velocities c_1 and c_2, respectively. This phenomenon is referred to as *double refraction* or *birefringence*, and plates which possess such a characteristic are called *wave, retardation* or *double refracting plates*. Directions 1 and 2 define the two principal axes of the plate, to which correspond refraction indices n_1 and n_2, respectively. Since $c_1 > c_2$, axis 1 is often called the *fast axis*, and axis 2 the *slow axis*.

As the two components A_1 and A_2 are transmitted with different velocities, they will emerge from the plate at different times. Consequently, one component is retarded timewise relative to the other component, Fig.19. This retardation can be expressed in terms of a difference in phase Δ between the two waves, which is dependent upon the thickness h of the plate, the wavelength of the light λ, and the properties of the plate material, as described by $(n_1 - n_2)$:

$$\Delta = \frac{2\pi h}{\lambda} (n_1 - n_2) \tag{38}$$

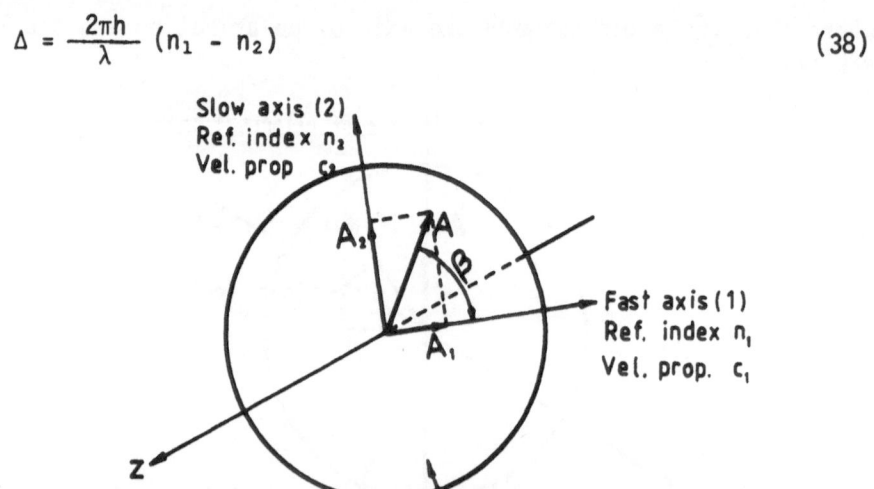

Fig.18 - Transmission of light through a wave-plate

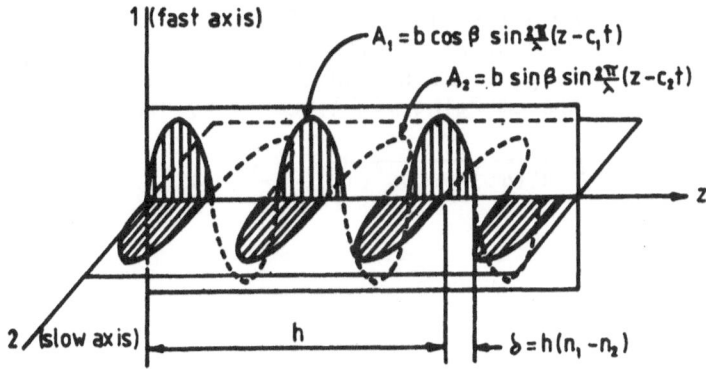

Fig.19 - The fast and slow components of the light vector in a
wave-plate

Upon emergence from the wave plate, the two components of light
are described by the equations

$$A'_1 = a \cos\beta \sin(\omega t + \Delta) \tag{39}$$

$$A'_2 = a \cos\beta \sin(\omega t)$$

where β is the inclination of the incident light vector relative to
the fast axis of the wave plate. The magnitude of the light vector
produced by these two components may be expressed as

$$A' = \sqrt{A'^2_1 + A'^2_2} = a \sqrt{\sin^2(\omega t + \Delta)\cos^2\beta + \sin^2(\omega t)\sin^2\beta} \tag{40}$$

Also, the angle γ that the emerging light vector makes with the
fast axis is given by

$$\tan \gamma = \frac{A'_2}{A'_1} = \frac{\sin(\omega t)}{\sin(\omega t + \Delta)} \tan\beta \tag{41}$$

which shows that the wave plate produces a rotation of the light
vector.

2.5 The Quarter-Wave Plate. Circular Polarized Light

When the double refracting plate is designed to give an
angular retardation of $\Delta = \pi/2$, it is called a *quarter-wave plate*.
From equation (38) it follows that the thickness of a quarter-wave
plate depends on the wavelength of the light being used. That is,
a quarter-wave plate suitable for one wavelength of monochromatic
light will not be suitable for a different wavelength.

700

The quarter-wave plate is an essential element in the production of circular polarized light. In fact, consider a beam of plane polarized light incident perpendicularly to a quarter-wave plate, as illustrated in Fig.20. If the angle β is set equal to π/4, then the magnitude of the emerging light vector is given by

$$A' = \frac{\sqrt{2}}{2} a\sqrt{\sin^2(\omega t) + \cos^2(\omega t)} = \frac{\sqrt{2}}{2} a \qquad (42)$$

and its inclination by

$$\tan\gamma = \tan(\omega t)$$

or

$$\gamma = \omega t \qquad (43)$$

Equation (42) shows that the magnitude of the emerging light vector has a constant value independent of time. And equation (43) indicates that the angle of emergence increases linearly with time. Hence the tip of the light vector sweeps out a circle. As the light propagates along the z-axis, this circle is opened up a circular helix corresponding to the circular polarized light.

2.6 Temporary Double Refraction. The Stress-Optic Law

The phenomenon of double refraction may also occur in certain isotropic materials, such as plastics, when subjected to a strain or stress. This condition is temporary, however, and disappears when the strains are removed. The discovery of *temporary double refraction* under strain was made in 1816, by Sir David Brewster, and makes the beginning of the *Science of Photoelasticity*. The stresses

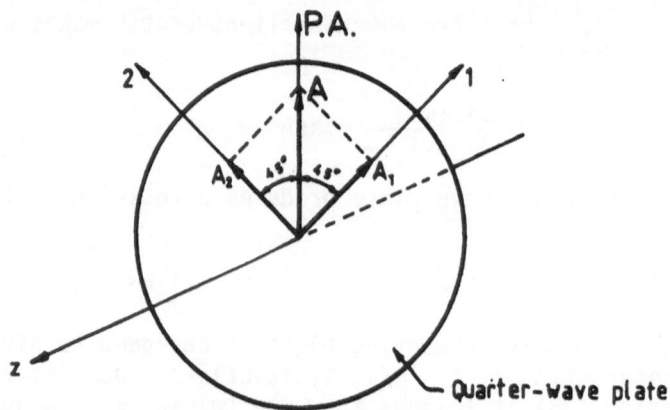

Fig.20 - The use of a quarter-wave plate to produce circular polarized light

and strains produce physical deformations which completely alter the initial isotropic character of the body, with reference to light. The principal axes of stress at any point of the model correspond to the fast and slow axes of the local double refracting plate.

The relations between temporary double refraction and the stresses, which are believed to be their cause, rest upon experimental evidence. They can be reduced to a simple equation of the form

$$\Delta = \frac{2 \pi h K}{\lambda} (\sigma_1 - \sigma_2) \tag{44}$$

where Δ is the relative retardation between the fast and the slow components of the light vector, h is the thickness of the plate, K is the stress optic coefficient of the photoelastic material, λ is the wavelength of the light, and σ_1 and σ_2 are the principal stresses at the point of interest. Equation (44) is the classical description of the stress-optic law for photoelastic materials, and it shows that the relative retardation Δ is linearly proportional to the difference of the principal stresses $(\sigma_1 - \sigma_2)$. Also, the relative retardation is linearly proportional to the plate thickness h and inversely proportional to the wavelength of the light passing through the plate.

In photoelastic practice, it is more convenient to rewrite equation (44) in form of

$$(\sigma_1 - \sigma_2) = \frac{N f_\sigma}{h} \tag{45}$$

where $N = \dfrac{\Delta}{2\pi}$, the *fringe order*, is the relative retardation in terms of a complete cycle of retardation, 2π (dimensionless)

$f_\sigma = \dfrac{\lambda}{K}$, is the *material fringe value* ($N \times m^{-1}$) for a given wavelength of light.

Thus, the principal stress difference $(\sigma_1 - \sigma_2)$ in a two-dimensional model can be determined if the relative retardation N can be measured, and if the material fringe value f_σ is known, or obtained by calibration. The fringe order N at each point in the photoelastic model can be measured by observing the model in the *Polariscope*.

The photoelastic behaviour can also be described in terms of strains. In fact, for a material exhibiting a perfectly linear elastic behaviour, the stress-strain relations for a two-dimensional state of stress are:

$$\varepsilon_1 = \frac{1}{E} (\sigma_1 - \nu\sigma_2)$$

$$\varepsilon_2 = \frac{1}{E} (\sigma_2 - \nu\sigma_1) \tag{46}$$

Thus

$$\sigma_1 - \sigma_2 = \frac{E}{1-\nu} (\varepsilon_1 - \varepsilon_2) \tag{47}$$

or, by substitution in equation (45),

$$\frac{N f_\sigma}{h} = \frac{E}{1-\nu} (\varepsilon_1 - \varepsilon_2)$$

Hence

$$\varepsilon_1 - \varepsilon_2 = \frac{N f_\varepsilon}{h} \tag{48}$$

where $f_\varepsilon = \frac{1+\nu}{E} f_\sigma$ is the *material fringe value* in terms of strain.

3   PHOTOELASTIC METHODS OF ANALYSIS

3.1 Introduction

The device or optical system most frequently employed to produce the necessary polarized beam of light to interpret the photoelastic effect in terms of stress (or strain) is called a *polariscope*. It may take a variety of different forms, depending on the desired use. However, in general, the polariscope consists of a light source, a polarizing device called the *polarizer*, the photoelastic model and a second polarizing device known as the *analyzer*. In addition, there may be a system of lenses, a viewing screen, and other accessories for convenient visual observation or photographic recording.

3.2. Plane Polariscope

The *plane polariscope* is the simplest optical system used in photoelastic analysis, for it consists of only two linear polarizers and a light source arranged in the manner illustrated in Fig. 21. In this type of polariscope the axes of the polarizer and analyzer are always crossed. With this arrangement, no light is transmitted through the analyzer, thus producing a *dark field*. The photoelastic model is then inserted between the two crossed elements and viewed through the analyzer.

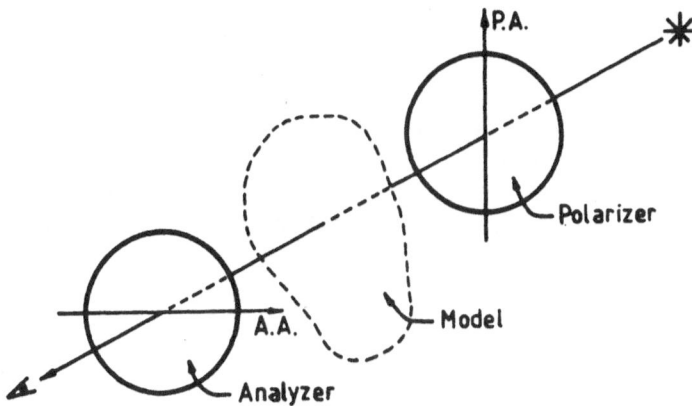

Fig.21 - Arrangement of the optical elements in a plane polariscope

Consider the case of a plane stressed photoelastic model, positioned perpendicularly to the axis of the polariscope. The light emerging from the polarizer is plane polarized, and represented by a light vector vibrating in the vertical plane according to the law

$$A = k \sin(\omega t) \tag{49}$$

At the point of interest, the stressed model has one plane of principal stress σ_1 at some angle α to the axis of polarization of the polarizer, Fig.22, and the other plane of principal stress σ_2 normal to σ_1. On reaching the model, the light vector is split into two components which vibrate in the two principal planes of σ_1 and σ_2, so producing A_1 and A_2, Fig.22,

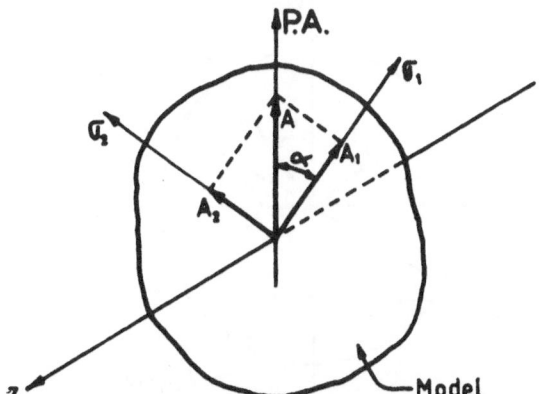

Fig.22 - Resolution of the light vector along σ_1 and σ_2 in a plane-
-polariscope

$$A_1 = k \sin(\omega t) \cos \alpha$$

$$A_2 = k \sin(\omega t) \sin \alpha \qquad (50)$$

These two components of the light vector propagate through the stressed model at different velocities and, consequently, they are out of phase when they emerge from the other side of the model. According to equation (45), the relative phase difference between the two components is given by

$$\Delta = 2\pi N = \frac{h}{f_\sigma} (\sigma_1 - \sigma_2) \qquad (51)$$

At the analyzer, the horizontal components of A'_1 and A'_2 are taken in the polarization axis, giving two vectores A''_1 and A''_2, Fig.23:

$$A''_1 = \frac{k}{2} \sin(2\alpha) \sin(\omega t + \theta)$$

$$A''_2 = \frac{k}{2} \sin(2\alpha) \sin(\omega t + \theta - \Delta) \qquad (52)$$

where θ is the phase of the fast component A'_1 at time t=0. As components A''_1 and A''_2 are vibrating in the same plane, their resultant vector can be obtained by simple addtion:

$$A = \frac{k}{2} \sin(2\alpha)[\sin(\omega t+\theta) - \sin(\omega t+\theta-\Delta)]$$

or

$$A = k \sin(2\alpha) \sin\frac{\Delta}{2} \cos(\omega t+\theta- \frac{\Delta}{2}) \qquad (53)$$

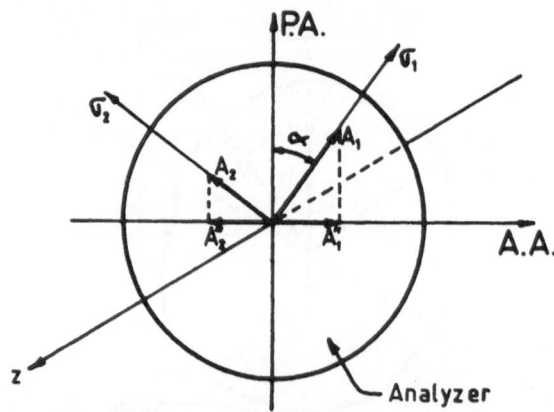

Fig.23 - Components of the light vector transmitted through the analyzer of a plane polariscope

For given conditions, the angle α and the relative retardation Δ are constant, so that equation (53) is of the form

$$A = C \cos(\omega t + \gamma)$$

Thus, the motion is simple harmonic, with the term $C = k \sin(2\alpha) \sin(\frac{\Delta}{2})$ giving the maximum amplitude. On the other hand, the intensity of the light, as recorded by the eye, is proportional to the square of the corresponding light vector amplitude. Hence, the intensity I of the emerging light, at that particular point in the model, is given by

$$I = k^2 \sin^2(2\alpha) \sin^2\left(\frac{\Delta}{2}\right) \tag{54}$$

or, in terms of the fringe order N, as

$$I = k^2 \sin^2(2\alpha) \sin^2(\pi N) \tag{55}$$

with k^2 represents the intensity of light emerging from the polarizer. By utilizing equation (45), the relative retardation N can be expressed in terms of model stress, to give the intensity of the emerging light as

$$I = k^2 \sin^2(2\alpha) \sin^2\left[\frac{\pi h (\sigma_1 - \sigma_2)}{f_\sigma}\right] \tag{56}$$

Equations (54)-(56) show that the intensity of the light emerging from the analyzer in the plane polariscope is a function of the angle α and the retardation Δ and, hence, it is influenced by the directions of the principal stresses and by the difference between the two principal stresses at the given point in the model. A dark spot will be observed on the model's image for every point at which the intensity of light vanishes. This occurs when

$$\sin(2\alpha) \sin\left[\frac{\pi h(\sigma_1 - \sigma_2)}{f_\sigma}\right] = 0 \tag{57}$$

Such dark points are, in general, linked together to form loci representing one of two conditions, namely: (i) - loci of constant principal stress direction called *isoclinic fringes* (when $\alpha = 0$ or $\alpha = \pi/2$); or (ii) - loci of constant difference $(\sigma_1 - \sigma_2)$ between the principal stresses, and referred to a *isochromatic fringes*, (for those cases in which $N = (\sigma_1 - \sigma_2)h/f_\sigma = 0,1,2,\ldots$ etc). Unfortunately these two fringe patterns are superimposed, as shown in the example of Fig.24, and their separation requires special techniques which will be described later.

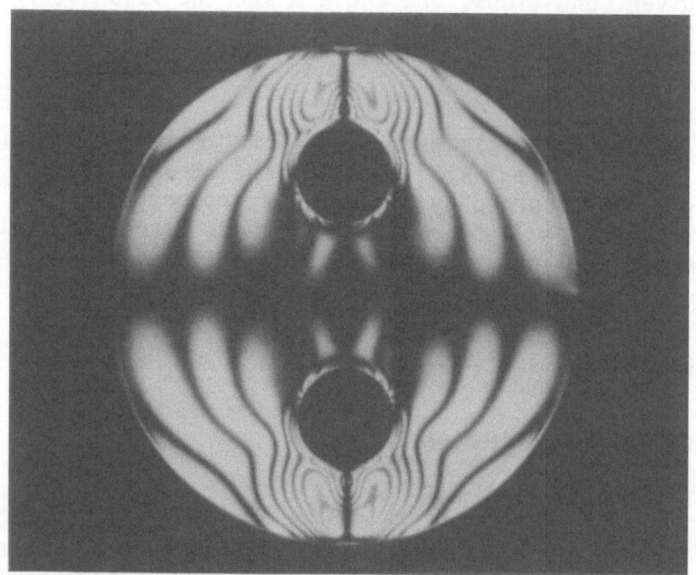

Fig.24 - Superimposed isochromatic and isoclinic fringe pattern in the photoelastic model

The photograph presented in Fig.24 shows de fringes as bands which have considerable width. Also, a direct visual examination of the fringe patterns obtained in the polariscope will show again that the fringes are bands and not lines. The fringe width is due to the recording characteristics of either the eye or photographic film. If the intensity of the light emerging from the analyzer were recorded by using a suitable photoelectric cell, the fringes would show a minimum intensity at some point near their centre, which will coincide with the exact extinction line.

3.3 Circular Polariscope

The *circular polariscope* employs circular polarized light; consequently, the photoelastic apparatus contains four optical elements and a light source, which is illustrated in Fig.25. The first element following the light source is the *polarizer*, as in the plane polariscope. The second element is a *quarter-wave plate* set at an angle $\beta = \pi/4$ to the axis of polarization, and it converts the plane polarized light into circular polarized light. The second *quarter-wave plate* is set with its fast axis parallel to the slow axis of the first quarter-wave plate, thus reconverting the circular polarized light into plane polarized light, which is again vibrating in the vertical plane. The last element is the *analyzer*, which can be orientated with its axis of polarization either perpendicular or parallel to the axis of the polarizer,

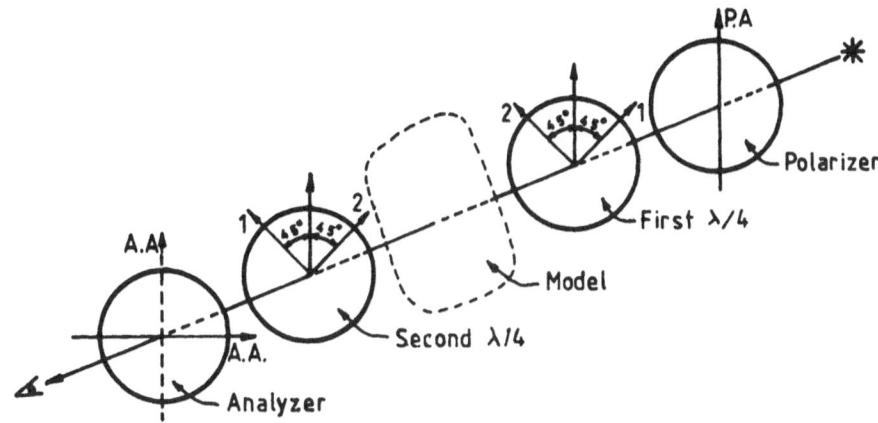

Fig.25 - Arrangement of the optical elements in a circular polariscope.

giving the standard *dark field* or *light field* arrangements of the
circular polariscope, respectively.

When a plane stressed photoelastic model is inserted between
the two quarter-wave plates of a circular polariscope, (dark field
arrangement, for example), the optical effects differ somewhat from
those observed in the plane polariscope. According to equation (39)
the light emerging from the first quarter-wave plate is described
by its fast and slow components given by

$$A'_1 = \frac{\sqrt{2}}{2} k \sin(\omega t + \pi/2) = \frac{\sqrt{2}}{2} k \cos(\omega t)$$

$$A'_2 = \frac{\sqrt{2}}{2} k \sin(\omega t)$$

(58)

These two components will propagate in the air at the same velocity
and they enter the model as illustrated in Fig.26.

The stressed photoelastic model has one plane of principal
stress, σ_1, at some angle α to the axis of the polarizer, and the
other principal plane, σ_2, normal to σ_1. Each component A'_1 and A'_2
is then resolved into the direction of the principal axes σ_1 and
σ_2, giving

$$A''_1 = \frac{\sqrt{2}}{2} k \cos(\omega t + \alpha - \pi/4)$$

$$A''_2 = \frac{\sqrt{2}}{2} k \sin(\omega t + \alpha - \pi/4)$$

(59)

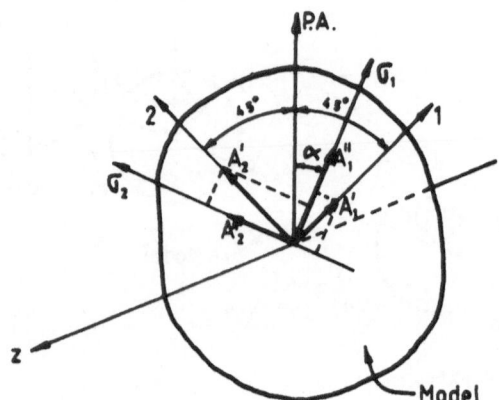

Fig.26 - Resolution of the light vector along σ_1 and σ_2 in the circular polariscope

According to equation (45), the relative phase difference between those two components after emerging from the photoelastic model is given by

$$\Delta = 2\pi N = \frac{h}{f_\sigma}(\sigma_1 - \sigma_2) \tag{60}$$

Thus,

$$A'''_1 = \frac{\sqrt{2}}{2} k \cos(\omega t + \alpha - \pi/4)$$

$$A'''_2 = \frac{\sqrt{2}}{2} k \sin(\omega t + \alpha - \pi/4 - \Delta) \tag{61}$$

Components A'''_1 and A'''_2 enter the second quarter-wave plate according to the diagram shown in Fig.27. The components associated with the fast and slow axes of the second quarter-wave plate are

$$A^{IV}_1 = \frac{\sqrt{2}}{2} k \left[\cos(\omega t + \alpha - \frac{\pi}{4})\sin(\frac{\pi}{4} - \alpha) + \sin(\omega t + \alpha - \frac{\pi}{4} - \Delta)\right.$$
$$\left.\cos(\frac{\pi}{4} - \alpha)\right] \tag{62}$$

$$A^{IV}_2 = \frac{\sqrt{2}}{2} k \left[\cos(\omega t + \alpha - \frac{\pi}{4})\cos(\frac{\pi}{4} - \alpha) - \sin(\omega t + \alpha - \frac{\pi}{4} - \Delta)\right.$$
$$\left.\sin(\frac{\pi}{4} - \alpha)\right]$$

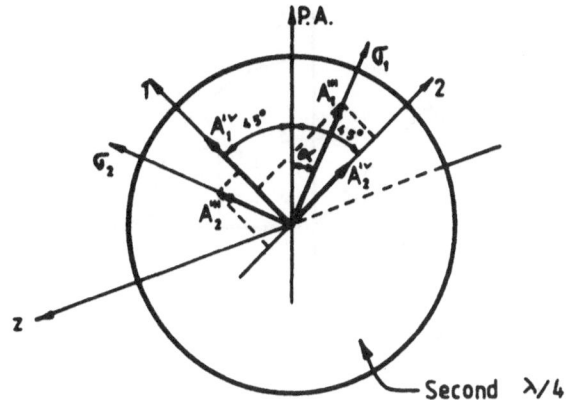

*Fig.*27 - Components of the light vectors as they enter the second
quarter wave plate of a circular polariscope

A relative phase shift of $\Delta=\pi/2$ is then imposed between components
A'_{v_1} and A'_{v_2}. Thus, the wave emerging from the second quarter-wave
plate can be expressed by

$$A^v_1 = \frac{\sqrt{2}}{2} k[\cos(\omega t + \alpha - \frac{\pi}{4})\sin(\frac{\pi}{4} - \alpha) + \sin(\omega t + \alpha - \frac{\pi}{4} - \Delta)$$

$$\cos(\frac{\pi}{4} - \alpha)] \qquad (63)$$

$$A^v_2 = \frac{\sqrt{2}}{2} k[\sin(\omega t + \alpha - \frac{\pi}{4})\cos(\frac{\pi}{4} - \alpha) + \cos(\omega t + \alpha - \frac{\pi}{4} - \Delta)$$

$$\sin(\frac{\pi}{4} - \alpha)]$$

On reaching the analyzer, Fig.28, the vertical components of A^v_1 and
A^v_2 are absorbed, while the horizontal components are transmitted,
to give,

$$A = k \sin(\frac{\Delta}{2}) \sin(\omega t + 2\alpha - \frac{\Delta}{2}) \qquad (64)$$

The intensity of the light emerging from a circular polariscope,
(dark filed arrangement), is then given by

$$I = K \sin^2(\frac{\Delta}{2}) = K \sin^2(\pi N) \qquad (65)$$

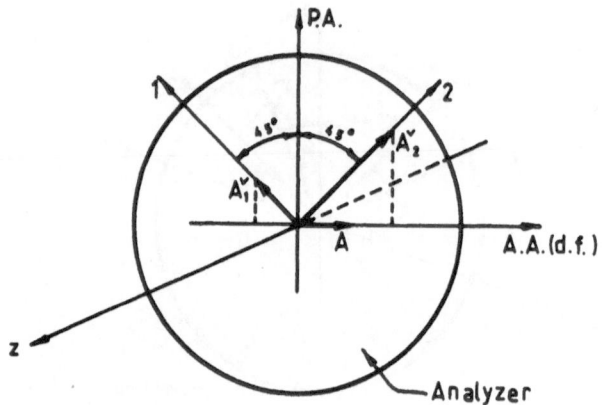

Fig.28 - Transmission of the light vector through the analyzer of a circular polariscope (dark field)

Equation (65) shows that the intensity of the light beam emerging from the circular polariscope is a function only of the principal stress difference $(\sigma_1 - \sigma_2)$. The isoclinic fringes appearing in the plane polariscope have been eliminated. The extinction fringe patterns obtained are associated with points where $\frac{\Delta}{2} = n\pi$, for $n = 0, 1, 2, 3, \ldots$, and they correspond to the integer isochromatic

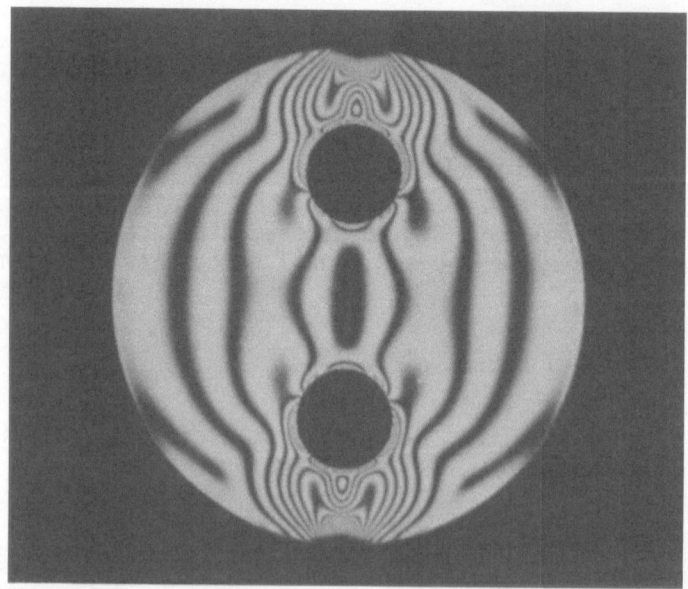

Fig.29 - Isochromatic fringe pattern obtained in the circular polariscope (dark field)

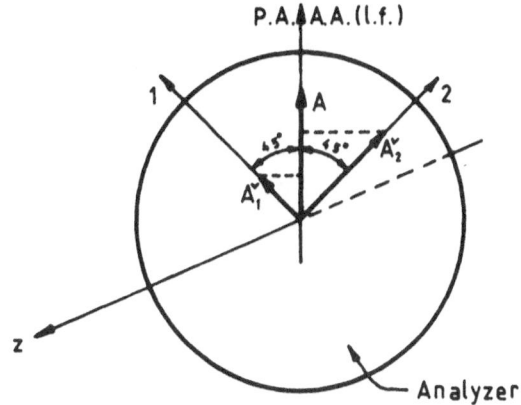

P.A.A.A.A.(l.f.)

Analyzer

Fig.30 - Transmission of the light vector through the analyzer of
a circular polariscope (light field)

fringe order N=0,1,2,3..., respectively. An example of this fringe
pattern is shown in Fig.29.

If the axis of the analyzer is set parallel to the polarizer
axis, (light field arrangement), Fig.30, the intensity of the

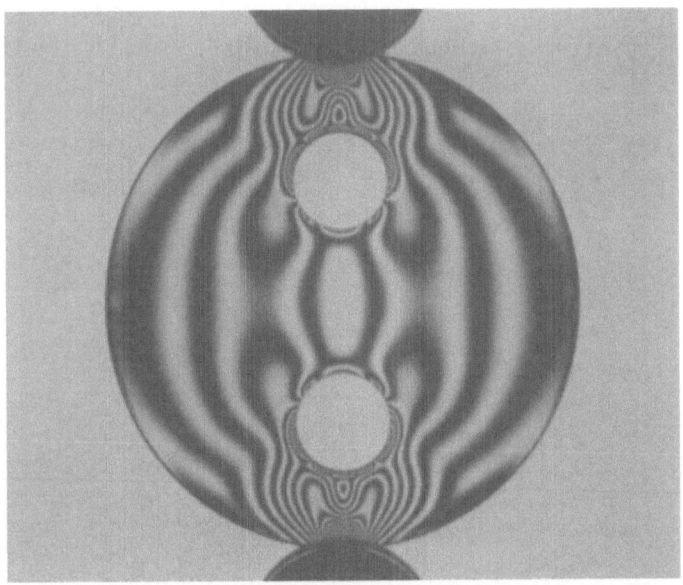

Fig.31 - Isochromatic fringe pattern obtained in the circular polariscope (light field)

712

light beam emerging from the polariscope can be shown to be given by

$$I = K \cos^2 \frac{\Delta}{2} \tag{66}$$

which shows that extinction will occur when

$$\frac{\Delta}{2} = \frac{1+2n}{2} \pi \quad \text{for, } n = 0,1,2,3,\ldots$$

or

$$N = \frac{\Delta}{2\pi} = \frac{1}{2} + n$$

corresponding to the isochromatic fringes of half order, i.e., $N = \frac{1}{2}$, $\frac{3}{2}$, $\frac{5}{2}$, ... An example of a light field isochromatic fringe pattern is shown in Fig.31.

A photograph of a commercially available 430 mm diameter polariscope is shown in Fig.32.

Fig. 32 - Photograph of a 430 mm diameter polariscope at University of Oporto, Dept. Mech. Eng.

3.4   Reflection Polariscope

A reflection type of polariscope is required for examination of birefringent coatings on metal or any other non-transparent prototypes. Optically, a reflection polariscope corresponds to the conventional transmission polariscope, which has been folded at the point where the photoelastic model is inserted, Fig.33. In the case of the reflection polariscope, the model is commonly replaced by an actual test object or structure, which has been coated with a photo-elastic plastic bonded in place on the surface with a reflective adhesive.

When the specimen is loaded, the surface displacements of the specimen at the interface are transmitted to the birefringent coating, if the bond is adequate. As the coating responds to these transmitted displacements, stresses and an associated amount of birefringence are induced. Observation of the coating by means of a reflection polariscope gives a fringe pattern, which is related to the surface strains of the specimen by the usual equation,

$$\epsilon_1 - \epsilon_2 = \frac{N f_\epsilon}{2 h} \tag{67}$$

where N is the fringe order, f_ϵ is the coating fringe value, and h is the thickness of the coating. A photograph of a commercially available reflection polariscope is shown in Fig.34.

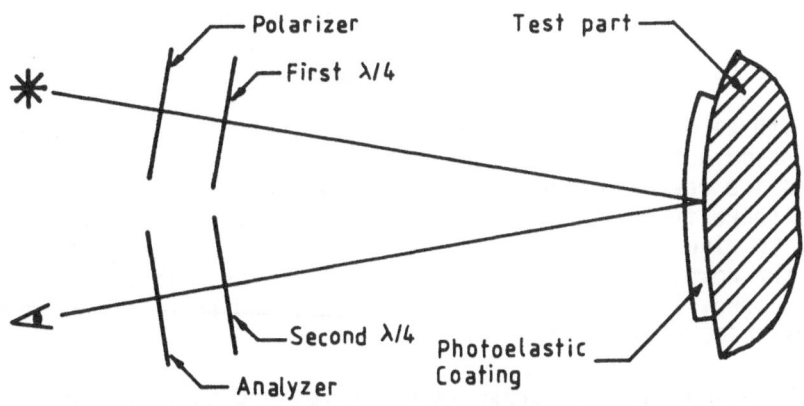

Fig.33 - Arrangement of the optical elements in a reflection pola-riscope.

Fig. 34 - Photograph of a reflection polariscope at University of Oporto, Dept. Mech. Eng.

3.5 Photoelastic Effect Using White Light

With monochromatic light, it has been shown that, by looking through the analyzer of a crossed polariscope, the surface of a stressed model appears to be covered with a number of black and monochromatic fringes. If white light is used, the model will appear to be covered with a series of brilliantly coloured bands of different colours. This appearance is due to the varying relative retardations at different points of the model, which cause each colour to be extinguished in turn, according to its wavelength.

With zero relative retardation, all light is extinguished in the crossed polariscope. As the retardation in increased, the colours are restored to produce first a grey tone, and then white light. Thereafter, the two components of the violet become out of phase, to give a residual colour of white minus violet, which produces the complementary yellow. The blue is the next colour to be extinguished, to produce an orange colour. And so on, through the various colours of the spectrum.

In estimating the stress value corresponding to a particular colour, the experimentalist must rely upon his subjective colour sense, thereby introducing a personal element with inevitable large errors. This is due to the fact that, although the colour varies with $(\sigma_1 - \sigma_2)$, the changes in the shades cannot be clearly recognized by the human eye for a rather wide range of stress. However the errors in estimating the stress field from a coloured fringe pattern can be considerably reduced by using the colour which is known as the *tint of passage*. During the transition from red to blue, the complementary colour is a dull purple which is very sensitive to small changes in $(\sigma_1 - \sigma_2)$. This colour is called *tint of passage*, and a very small change in stress causes this colour to change into red or blue.

3.6 Compensation Methods

The process of stress measurement by photoelasticity consists of, first determining the fringe order at any point of interest, and then multiplying the observed fringe order by an appropriate constant, $(\sigma_1 - \sigma_2 = N \times \frac{f\sigma}{h})$, to obtain the principal stresses difference at the point of interest of the test object. Thus, the primary function of any polariscope is the measurement of fringe orders.

It has been shown that, by employing the circular polariscope in both dark field and light field arrangements, the isochromatic fringe order at any point can be determined, at best, to the nearest 1/2 order. However, this limitation is readily overcome, and accurate measurements of fractional fringe order can be made at any arbitrary point in the field of view by *compensation methods*. The Babinet-Soleil method and the Tardy method will be described here. They are the two methods of compensation most commonly used in photoelasticity.

(i) - *The Babinet-Soleil or Null-Ballance Method*. The Babinet--Soleil method of compensation operates on the principle of introducing into the light path of the circular polariscope, (dark field arrangement),a calibrated variable birefringence of opposite sign to that induced in the photoelastic material by the stress field. The optical effect of the superposition of these two sources of birefringence is presented in Fig.35. When the opposite-sign variable birefringence is ajusted to precisely match the magnitude of the stress-induced birefringence, complete cancellation will occur and the combined fringe order goes to zero. The condition of zero birefringence is easily recognized because it produces a black fringe in the isochromatic pattern where, before introducing the compensator, a coloured fringe existed.

The construction details of the Babinet-Soleil compensator are given in Fig.36. It employs a pair of linearly birefringent plates arranged in a tandem so that the total birefringence introduced into the light path is proportional to the displacement of one plate with respect to the other. The birefringence exhibited by the compensator can be controlled by adjusting the micrometer screw that commands the displacement of the movable plate. With

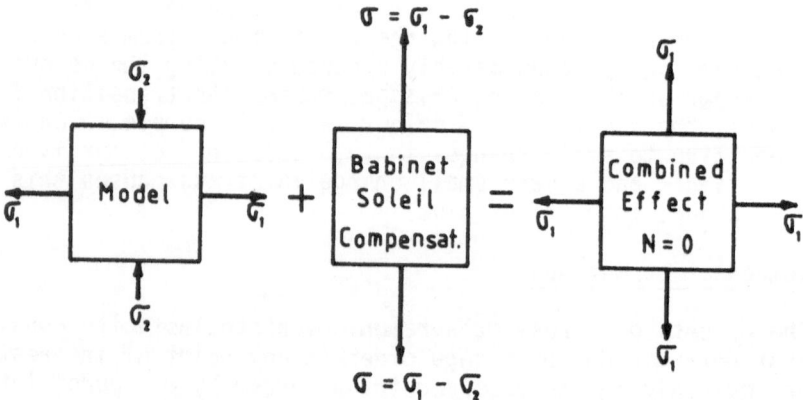

Fig. 35 - Superposition of birefringence effects exhibited by the model and Babinet-Soleil compensator.

this method of compensation, determination of N can be made with at least two-and possibly three-decimal point precision.

(ii) - *The Tardy Method.* The Tardy method of compensation is a fast, simple and accurate technique for measuring fractional fringe orders of isochromatic patterns. To employ the Tardy method, the optical elements of the polariscope should be aligned according to the standard dark-field circular polariscope arrangement, the axis of the polarizer being parallel to the principal stress σ_1 at the point of interest ($\alpha = 0$). The light vector emerging from the second quarter-wave plate is given by its components, Fig.37,

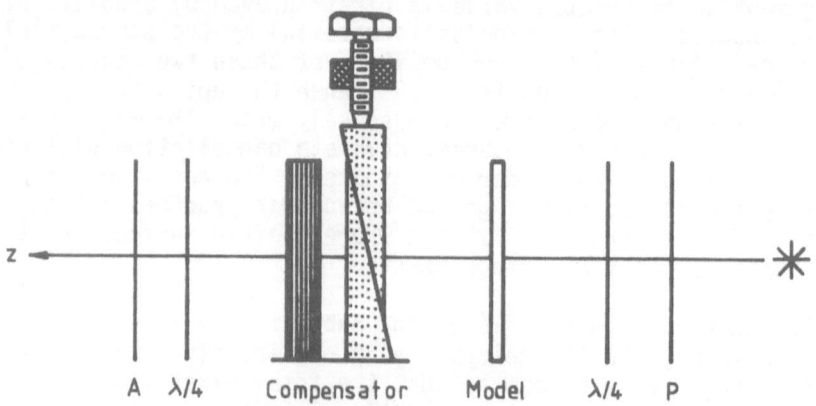

Fig. 36 - Positioning and construction details of the Babinet-Soleil compensator.

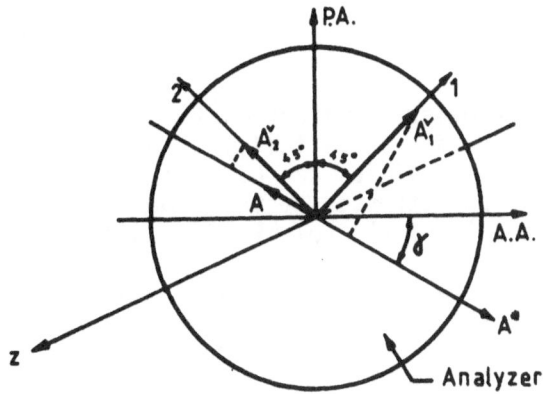

Fig.37 - Rotation of the analyzer in the Tardy method of compensation.

$$A_1^v = \frac{\sqrt{2}}{2} k \cos(\omega t)[\cos(-\frac{\Delta}{2}) - \sin(-\frac{\Delta}{2})]$$

$$A_2^v = \frac{\sqrt{2}}{2} k \cos(\omega t)[\cos(-\frac{\Delta}{2}) + \sin(-\frac{\Delta}{2})]$$

(68)

Now, if the analyzer is rotated through an angle γ, (position A*,
Fig.37), the magnitude of the light vector emerging from the analyzer is given by

$$A = A_2^v \cos(-\frac{\pi}{4} + \gamma) - A_1^v \cos(-\frac{\pi}{4} - \gamma) =$$

$$= \frac{k}{2} \cos(\omega t)[(\cos \frac{\Delta}{2} + \sin \frac{\Delta}{2})(\cos\gamma - \sin\gamma) - (\cos \frac{\Delta}{2} - \sin\frac{\Delta}{2})$$

$$(\cos\gamma + \sin\gamma)]$$

(69)

and extinction of light will take place when

$$(\cos \frac{\Delta}{2} + \sin \frac{\Delta}{2})(\cos\gamma - \sin\gamma) - (\cos \frac{\Delta}{2} - \sin \frac{\Delta}{2})$$

$$(\cos\gamma + \sin\gamma) = 0$$

(70)

Values of Δ satisfaying equation (70) are given by

$$\Delta = 2\gamma \pm 2n\pi \qquad , \text{ for } n = 0,1,2,3,\ldots.$$

The fringe order at the point of interest is then,

$$N = \frac{\Delta}{2\pi} = n \pm \frac{\gamma}{\pi} \qquad\qquad (71)$$

The procedure for measuring fractional fringe orders by the Tardy method of compensation is as follows. A plane polariscope is first employed so that isoclinics can be used to establish the directions of the principal stresses at the point of interest, as shown in Fig.38. The axis of the polarizer is then aligned with a principal stress direction, ($\alpha = 0$ or $\pi/2$), and the other elements of the polariscope are oriented to produce a standard dark-field circular polariscope. The analyzer is then rotated cockwise until extinction occurs at the point of interest. If, when the analyzer is rotated clockwise, the lower fringe (n in Fig.38) moves to the test point, the total fringe order at that point is,

$$N = n + \frac{\gamma}{\pi} \qquad \text{(N is positive)} \qquad\qquad (72)$$

If the higher order fringe (n + 1, Fig.38) moves to the test point, the total fringe order is,

$$N = -(n + 1 - \frac{\gamma}{\pi}) \text{ (N is negative)} \qquad\qquad (73)$$

The accuracy of the Tardy method depends upon the quality of the quarter-wave plates employed in the polariscope, but fringe orders accurate to two decimal points are often obtained.

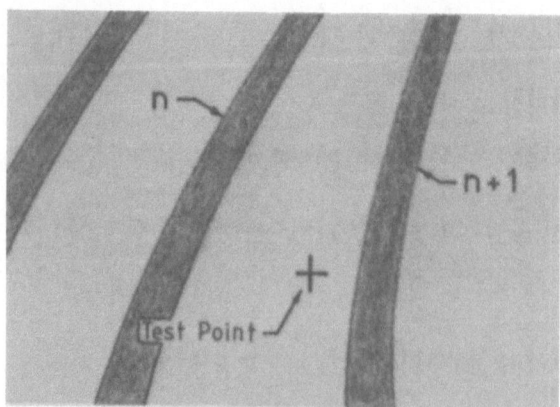

$Fig.38$ - Using the Tardy method for fractional fringe order measurements

3.7 The Oblique-Incidence Method for Separation of Principal Stresses

It has been shown that the principal stress difference $(\sigma_1-\sigma_2)$ in the photoelastic model can be determined directly by analysis of the isochromatic fringe pattern obtained in the polariscope. Also, at free boundaries, the principal stress normal to the boundary is zero and, therefore, the isochromatic data yield directly the value of the other principal stress. In some instances, however, such as at an interior region of the body, it may be necessary to separate the principal stresses, i.e., to determine the individual principal stress magnitudes, σ_1 and σ_2. The information for accomplishing this resides within the model, but the procedure for doing so requires the use of supplementary data, or the employment of numerical methods. Numerous techniques have been develloped to provide the supplementary information for the complete stress solution at interior points of the photoelastic model. In this section, the *oblique-incidence method* will be described, which employs a second or additional measurement of fringe order, but with the light ray traversing the model along an oblique line with respect to the model surface.

Suppose the separate principal stresses at some point A within a two-dimensional model are required as in Fig.39. Examination in a plane polariscope will reveal the isoclinics, and by suitable rotation of the polarizer and analyzer, or of the model by itself, an isoclinic can be made to pass through the point A, and so to give the directions of the principal stresses. The model is now examined by oblique incidence in the circular polariscope, where the light passes through the model at some angle θ, as shown in Fig.40. The isochromatic fringe pattern produced is related to the

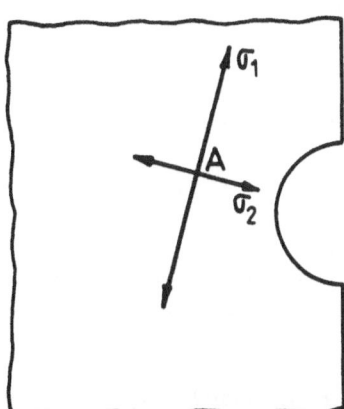

Fig. 39 - Principal stress directions at an interior point of the model.

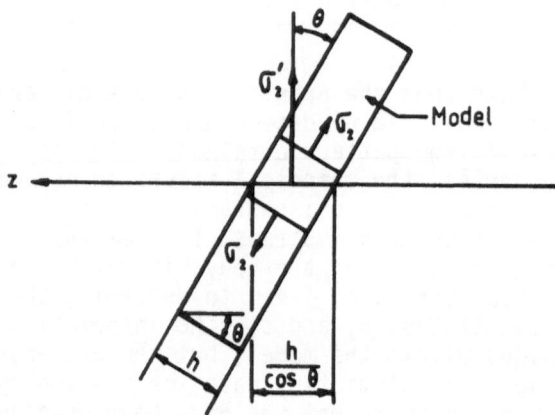

Fig.40 - Examination by the oblique incidence method

thickness h/cosθ traversed by the light and to the secondary principal stresses, σ_1 and σ_2, lying in the plane normal to the axis of
the light, according to the equation

$$\sigma_1 - \sigma_2' = \frac{f_\sigma N_\theta}{h/\cos\theta} \tag{74}$$

where N_θ is the fringe order associated with the oblique incidence
fringe pattern, and

$$\sigma'_2 = \sigma_2 \cos^2\theta \tag{75}$$

Hence,

$$\sigma_1 - \sigma_2 \cos\theta = \frac{f_\sigma N_\theta}{h/\cos\theta} \tag{76}$$

But, from the normal incidence fringe pattern,

$$\sigma_1 - \sigma_2 = \frac{N_0 f_\sigma}{h} \tag{77}$$

where N_0 is the fringe order at the point of interest, for normal
incidence. Thus, solving equations (76) and (77) for σ_1 and σ_2,

$$\sigma_1 = \frac{f_\sigma}{h} \frac{\cos\theta}{\sin^2\theta} (N_\theta - N_0 \cos\theta)$$

$$\sigma_2 = \frac{f_\sigma}{h} \frac{1}{\sin^2\theta} (N_\theta \cos\theta - N_0) \tag{78}$$

which give the separate principal stresses σ_1 and σ_2, obtained from the isochromatic fringe orders, N_o and N_θ, corresponding to normal and oblique incidences, respectively.

3.8 Three-Dimensional Photoelasticity

So far, only two-dimensional techniques have been discussed for the analysis of stresses occuring in the plane of the model, effectively of uniform thickness. In many practical problems, however, the shape of the structure or mechanical component is too complicated to permit a useful analysis to be made on a simplified two-dimensional model. Thus, the need arises for three-dimensional techniques. These are based on the construction and loading of a three-dimensional model, interior planes of which are analyzed photoelastically by using either *frozen-stress* or *scattered-light methods*.

The frozen-stress techniques are based upon another property of certain photoelastic materials, by which it is possible to lock model deformations, and the associated optical response, into a loaded three-dimensional model. A model made from such a material is heated up to its stress annealing temperature, stressed, and then cooled slowly, with the loads still applied. At room temperature, the loads are removed, but a major portion of the deformation and the associated photoelastic effects remain locked in the model. Once the stress-freezing process is completed, the model is sliced and analyzed in the polariscope as an ordinary two-dimensional photoelastic model. The mechanical and optical anisotropy remains permanently fixed in the model and, with suitable materials, isochromatic fringe patterns remains unchanged after several years of storage.

With the scattered-light method, interior stress information can be obtained without stress-freezing or slicing of the model. The phenomenon of scattering of light is atributed to the absorption of the light energy by the particles of the medium, which subsequently are put to a vibrating motion, and re-emit the incident light. To illustrate this scattering phenomenon, consider a beam of ordinary light propagating in the z-direction, and vibrating in the xy-plane, as shown in Fig.41. The incident light will scatter at every point and will produce a secondary source of plane polarized light which propagates radially outward from the source. It is possible to use this polarization produced by scattering of the light within the photoelastic model in place of either the polarizer or the analyzer in a photoelastic polariscope. The situation is as illustrated in Fig.42. From two incidences across lines PQ and P'Q', it is possible to measure the secondary principal stress difference $(\sigma'_1 - \sigma'_2)$ over a centralized plane of observation with thickness Δh. In practice, a sheet of light is used to illu-

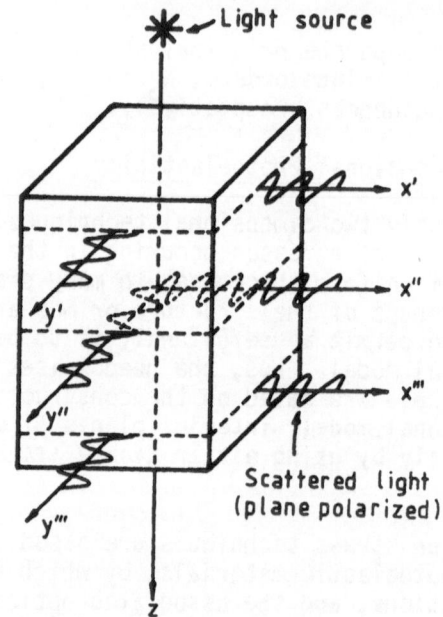

Fig. 41 - The scattered-light phenomenon

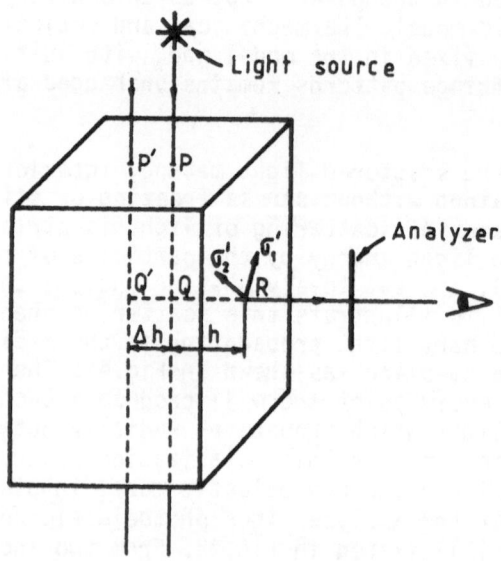

Fig. 42 - Polarization produced by scattering used as polarizer

minate a plane in the model rather than a line, as indicated above. The use of a sheet of light permits the determination of a fringe pattern over the whole field of the plane.

The scattered-light method has the advantage over the frozen--stress technique that it is a non-destructive and, therefore, all of the disadvantages that result from locking-in the stresses in the three-dimensional photoelastic model are removed. Thus, in addition to the fact that the model does not need to be cut, the laws of transition from model to prototype are more appropriately applied in the scattered-light method, in which the model is tested at room temperature. Therefore, Poisson's ratio is approximately equal to that of the prototype, whereas in the frozen-stress method, because of the heating of the model, the value of the Poisson's ratio approaches 0.5, i.e., the value for all rubbery polymers, which, therefore, differs greatly from Poisson's ratio of the prototype.

BIBLIOGRAPHY

1. Timoshenko, S.P., and J.N. Goodier. Theory of Elasticity (McGraw-Hill Book Company, New York 1951)
2. Frocht, M.M.. Photoelasticity, Vol. I, Vol. II (Wiley and Sons, New York 1941, 1948)
3. Jessop, H.T. and F.C. Harris. Photoelasticity. Principles and Methods (Cleaver-Hume Press, London 1949)
4. Hetényi, M.. Handbook of Experimental Stress Analysis (Wiley and Sons, New York 1950)
5. Heywood, R.B.. Photoelasticity for Designers (Pergamon Press, London 1969)
6. Dally, J.W. and W.F. Riley. Experimental Stress Analysis (McGraw-Hill Book Company, New York 1965)
7. Holister, G.S.. Experimental Stress Analysis (Cambridge University Press, Cambridge 1967)
8. Coker, E.G. and L.N.G. Filon. A Treatise on Photoelasticity, 2nd Edition (Cambridge University Press, Cambridge 1957)
9. Durelli, A.J. and W.F. Riley. Introduction to Photomechanics (Prentice-Hall, Englewood Cliffs, New York 1965)
10. Kuske, A. and G. Robertson. Photoelastic Stress Analysis (Wiley and Sons, New York 1974)
11. Theocaris, P.S. and E.E. Gdoutos. Matrix Theory of Photoelasticity (Springer-Verlag Berlin Heidelberg New York 1979)

NON-DESTRUCTIVE 3D PHOTOELASTICITY

R. Desailly

Laboratoire de Mécanique des Solides
Equipe de Recherche Associée au CNRS
40, avenue du Recteur Pineau
86022 Poitiers Cedex

Among the methods of experimental analysis photoelasticity plays a major role in the investigation of a three-dimensional model. Users of this technique are faced with two main problems : the difficulty of performing precise measurements in a three dimensional medium and the propagation of polarized light in anisotropic medium, (more precisely the effect of the rotation of the principal axes on optical phenomena). The first difficulty is often overcome by the classical technique of freezing and slicing. However this method has certain disadvantages such as time consumption, destructiveness and the problem due to a Poisson coefficient equal to .5. The second difficulty is considered but with some important hypotheses in the more classical techniques using scattered light. We first consider the practical need of taking into account the effect of rotation of the principal axes using an adequate optical theory even in the case of the analysis of a thin slice.

After a brief review of the classical techniques of scattered light photoelasticity we describe two examples of the more recent non destructive methods both working in more general conditions. We discuss their applications.

1. THE CURRENT SCHEMES

Let us first settle the basic hypothesis. In order to proceed with the necessary precision on the propagation direction for a light wave in a photoelastic medium we will assume this medium to be isotropic (indeed the current photoelastic materials are slightly anisotropic). It follows that for a ray of light entering along the

$\vec{z}$ direction the planes of vibrations for the component waves (x, y) are orthogonal to $\vec{z}$. For a photoelastic medium one shows that the principal directions for the index of light refraction tensor and the tensor of mechanical stress coïncide and that the change of index is related to the stress in the following form (in coordinate system associated with principal directions).

$$n'-n_0 = C_1 \sigma' + C_2 (\sigma''+\sigma_z)$$

$$n''-n_0 = C_1 \sigma'' + C_2 (\sigma'+\sigma_z)$$

n' and n" designate the principal secondary indices in the wave-plane (x, y) and σ', σ'' are the corresponding secondary principal stresses ; C_1, C_2 are independent constant for the photoelastic material.

In current three-dimensional photoelasticity approaches it is assumed that the direction of secondary principal stresses and their values rest constant following the thickness dz of a slice having its parallel faces normal to $\vec{z}$. The foregoing assumption permits to consider such an element as a birefringent plate characterized by the two parameters
- the angle $\alpha = (x, \sigma')$
- the angular birefringence $\phi = \dfrac{2\pi\delta}{\lambda}$, $\delta = dx(n'-n'') = C(\sigma'-\sigma'')dz$
(C being a photoelastic constant).

2. PROPAGATION OF LIGHT THROUGH PHOTOELASTIC MEDIUM

The problem of propagation of light waves was studied by many authors (1), (2), (3), (4). Here, the essential concepts from Aben's works will be introduced.

Aben, in 1966, considered the basic equations of three-dimensional photoelastic and analysed the matrix representation of the solutions of those equations. He showed that when rotation of principal axes was present, there were always two pairs of perpendicular conjugate "characteristic directions" (Fig. 2). He distinguished the primary characteristic directions at the entrance of light (Δ'_e, Δ''_e), and the secondary directions (Δ'_s, Δ''_s) for the light emergence from the medium. The light linearly polarised at the entrance along one of the primary directions emerges linearly polarised along the conjugate secondary direction. We will denote by R the angle determined by two such directions $R = (\Delta'_e, \Delta'_s)$, and by α^* the angle (x, Δ'_e). Aben has thus generalized the concept of "isoclinic" of the plane photoelasticity to the general case when a rotation of the secondary principal directions takes place (5). The characteristic directions are generally different from the secondary principal directions of the stress tensor (or those of the principal indices) at the entrance (σ'_e, σ''_e) and at the emergence (σ'_s, σ''_s) from the medium. We denote $\alpha_e = (\sigma'_e, \sigma'_s)$. Thus we conclude that the medium under consideration can be represented in general as an "elementary birefringent" (having the axes (Δ'_e, Δ''_e) located by $\alpha^* = (x, \Delta'_e)$ and characterized by an angular birefringence ϕ^*) followed or preceded by a rotatory power R.

2.1 Particular hypothese for a thin slice

For the case where $d\alpha/dz$ and $\sigma'-\sigma''$ are constant through a thickness, important conclusions follow (Aben) :
- the bisecting lines for the angles formed by two associated "characteristic directions" coincide with these for the angles formed by the associated secondary principal directions at the entrance and at the emergence (Fig. 2).
Remark : the bisectors mentioned correspond to the secondary principal directions (mechanical or optical) at mid thickness ; so their directions are defined by the angles $\pm R/2$ from the characteristic directions,
- the phase difference ϕ^* characteristic of the medium traversed by the light wave along the two characteristic orthogonal directions is generally different from the angular birefringence

$$\phi = \frac{2\pi e}{\lambda} (n'-n'')$$

which would result in the absence of rotation R,
- the quantities R, ϕ^*, ϕ and α_0 obey the following relationships

$$tg\ R = \frac{tg\ \alpha_0 - \frac{\alpha_0}{X} tg\ X}{1 + \frac{\alpha_0}{X} tg\ \alpha_0\ tg\ X}$$

$$cos\ \phi^* = 1 - \frac{\phi^2}{2X^2} sin^2 X$$

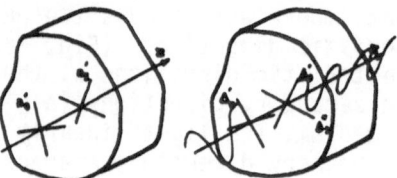

$$\alpha_o = (n'_e, n'_s)$$

$\Delta'.\Delta'$: characteristic directions

$$R = (\Delta'_e, \Delta'_s)$$

Fig. 1 : Rotation of the secondary principal directions

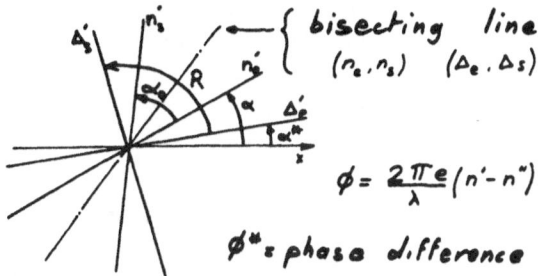

Fig. 2 : Secondary principal directions - Characteristic directions

where $X = \dfrac{\sqrt{\phi^2 + 4\,\alpha_0^2}}{2}$.

Note that for R = 0, $\alpha = \alpha_0$ and $\phi = \phi^*$; one finds again the classical scheme of homogeneous element representing a slice dissected from a model. Using the expressions above abacuses have been drawn (5) (6) which give ϕ and α_0 versus physical quantities ϕ^* and R. One can thus determine the difference of secondary principal stresses $\sigma'-\sigma''$ and secondary principal stress directions (through

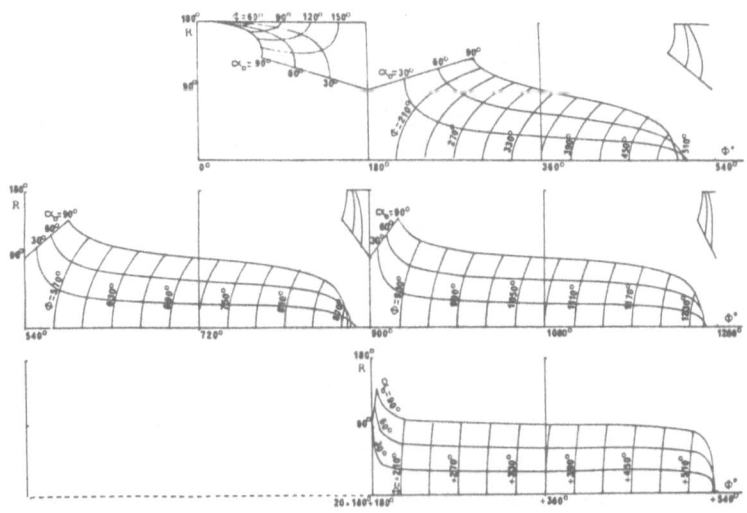

Fig. 3 : Mechanical quantities (ϕ and α_0) versus physical quantities (ϕ^* and R)

the angle α) if the angle α* describing the orientation of characteristic directions at the entrance is known.

The abacuses provide evidence for the following particular situations :
- φ* makes φ increase if φ > π. One finds here the result of Drucker and Mindlin (1) who indicated that "the rotation of secondary principal directions multiplied the number of isochromatic fringes",
- φ* makes φ reduced if φ > π and in the neighbourhood of this value the error becomes important α_0 = 30° ⟹ φ* = 140°,
- for the ratio α_0/ϕ small one obtains φ = φ* and there is coïncidence of the secondary principal directions and the characteristic ones.

3. ANALYSIS OF A THIN SLICE

We analyse the effects of a classical analysis of a slice regarded as an elementary birefringent when rotation of secondary direction is not negligeable. The slice is considered here as being extracted from a stress-frozen photoelastic model.

3.1 Point-wise analysis

To measure the parameters of a thin slice one can ellipsometry or a compensation technique which, employing the circular incident light, permits to determine φ* ; the error made in measuring φ* instead of φ can reach 40° when φ averages 180° for a rotation of α_0 = 30°. The advantage in precision of the most recent punctual techniques is thus subordinated to the necessity of carrying out a supplementary measurement which is that of the rotation power of the slice.

3.2 Whole field analysis with a plane polariscope

Here the analysis with a light-field polariscope is presented as it corresponds to the whole-field method by optical slicing. One can proceed with the parallel study for a dark-field polariscope (7).

Let us consider a slice or sheet put together within a plane (rectiline) polariscope permitting a light-field pattern to be formed. The slice is assumed to act as an elementary birefringent and a rotatory power. The slice we are speaking about can be optically sliced within a model or obtained by the stress-freezing and cutting classical technique. Let I_0 designate uniform light-field illumination and x the polarizing axis of the polarizer. We have for the light intensity.

$$I = I_0 \left(\cos^2 R - \sin 2 \ \alpha^* \sin^2 (\alpha^*+R) \sin^2 \frac{\phi^*}{2}\right)$$

The extremum values for intensity distribution correspond to :

$$\alpha^* = - \frac{R}{2} + k \frac{\pi}{4} \qquad k = 0, 1, 2, \ldots$$

Two cases can be distinguished. The first one corresponds to

$$\alpha^* = - \frac{R}{2} + k \frac{\pi}{2}$$

and the intensity expressed by

$$I_{max} = I_0 \left(1 - \sin^2 R \cos^2 \frac{\phi^*}{2}\right)$$

attains its maximum. The second case correspond to

$$\alpha^* = - \frac{R}{2} + \frac{\pi}{4} + k \frac{\pi}{2}$$

and the intensity given by

$$I_{min} = I_0 \left(\cos^2 R \cos^2 \frac{\phi^*}{2}\right)$$

attains a minimum.

In the framework of Aben's theory, we can express the rotation power R and the dephasing ϕ^* as function of the rotation of the secondary main axis α_0 and the angular birefringence $\phi = \frac{2\pi C e (\sigma' - \sigma'')}{\lambda}$ where C is the Brewster's constant, e is the thickness of the slice and λ is the wave length. We have traced for different values of α_0, and as function of ϕ the extreme curves of I as well as an interme-

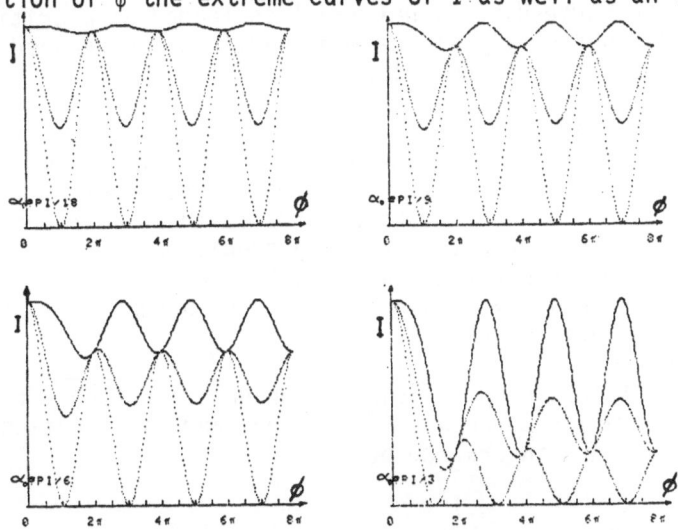

Fig. 4 - Evolution of the intensity for different values of α_0

diate curve corresponding to $\alpha = \frac{R}{2} + \frac{\pi}{8} + k\frac{\pi}{4}$. Let us note that for low values of $\alpha_0 (< 20°)$ the maximum of the intensity has a much reduced modulation. It characterizes then an isoclinic zone, this one occuring for $\alpha = \frac{R}{2} + k\frac{\pi}{2}$; it enables us to distinguish the orientation of the secondary main axis in the median plane of the slice when their rotation is uniform. The representative curve of the intensity when $\alpha = \frac{R}{2} + \frac{\pi}{4}$ characterizes the isochromatic fringes. We can notice that for $\alpha_0 < 30°$ the minima and the maxima occur for $\phi = k\pi$ which means that their localization are practically independent of the rotation of the secondary main axis. When the rotation α_0 is important for example for $\alpha_0 = 60°$ the intensity always includes a modulation, and as a result the isoclinic zone disappears.

In conclusion, when we limit ourselves to rotations of the secondary main axis inferior to 30° on the thickness of the slice, the isoclinic zone characterizes the orientation of the principal directions in the median plane and the maxima and minima of the isochromatic fringes define the points where the angular birefringence ϕ is a multiple of 180°. The rectilinear analysis between two parallel or crossed polarizers of a slice endowed with a rotation power or its equivalent by means of our method is thus a suitable technique provided the rotation of the secondary main axes does not exceed 30°.

4. THREE-DIMENSIONAL PHOTOELASTICITY STATE OF AFFAIRS

4.1 Stress freezing method

In this most commonly used technique the stresses are "frozen in" when the three-dimensional model is loaded at a critical temperature (9). Subsquently slices and subslices are removed from the model and analysed in the polariscope.

This technique presents some disadvantages :
- the birefringences are three to four times lower than those which would have resulted for the same deformations at the ambient temperature. One can introduce eventually the higher deformations but that might disturbe the boundary conditions ;
- the model-prototype similarity is not assured exactly because the Poisson's ratio effective value is .5 instead of .3 to .35 for most of structural metals.

Stress freezing and slicing techniques are used effectively in fracture mechanics to obtain flow shapes and distributions of stress-intensity factors in cracked bodies.

4.2 Scattered light techniques

The objection to loading the model at an elevated temperature and destroying it is overcome by the use of scattered light-photoelasticity.

Weller has the merit of having first applied scattered light to the non destructive study of three dimensional photoelastic models. He proposed in 1939 to employ scattering as either polariser or analyser (10) but later seems to have used only the second alternative (11). He used a plane-polarised parallel beam of light and observed a model slice contained within this beam from directions in a plane normal to the beam axis. When a polarised light beam penetrates a photoelastic model, the form of the light changes along the beam and depends on the relative entry point onwards. To determine this difference, Weller proposed to follow the light changes by analysing the light scatter. His implicit assumption was in our opinion as follows : the principal directions maintain their orientation along the light beam and, more generaly, the amplitudes of the light vibrations projected into the principal directions remain cosntant even if these directions rotate slightly. This full-field method seems attractive in its simplicity but requires large relative retardations, since many fringes must appear along the beam. It remains valid only in so far as the mentioned assumption is satisfied. Drucker and Mindlin (1) and later Aben (4) have shown that the assumption is true only on condition that the ratio α_0/ϕ of the rotation of the secondary principal axes to the phase retardation produced by the medium remains small. A simple method easily made automatic was later proposed by Cheng and permitted study of cross-sectional stresses in a bar under combined torsion and tension. It is similar in principle to Weller's method and subject to the same limitations.

Robert and Guillemet (13) in 1963 proposed a different method related to the idealisation used in stress freezing and slicing of a photoelastic medium : the part of the model traversed by the light beam is represented by a finite number of birefringent element, each corresponding to a thin slice normal to the beam. The authors for the first time used Poincaré's theorem (14) allowing them to replace the action of a finite number of birefringent element by a single retarding element followed by a rotatory power. They formulated relations for synthesis of birefringents : knowledge of the retardation and of the optical axes associated with two 'thick media", which differ from a birefringent, leads to the characteristics of the retarding element independently of the rotary power. In order to determine the characteristics of the thick medium, the authors developed a method using the scatter phenomenon as an analyser together with the technique of a rotating analyser. The method involves rotation of the model about the axis of observation and is valid only if the depolarisation coefficient of the scattered radiation remains

constant during the rotation. It also relies exclusively on finding
an extremum of the sinusoïdal amplitude, which is not generally
precise if only for reasons of intensity fluctuations occuring in
scattered light. Doubtless for these various reasons the authors
limit presentation of results to finding the position of points
where the angle of retardation associated with the thick medium is
$(k\frac{\pi}{2})$ and hence where the minimum of the modulated amplitude is zero.
Besides, as far as we know, the method has been used only for the
particular case of constant principal directions along the scattered
beam.

Employing the same idealisation of the thick medium as a finite
number of birefringents along the light path, Gross-Peterson (15) in
1973 advocated the use of a compensator for direct determination of
the optical characteristics of any thin slice inside the photoelastic model. In principle this method cancels out the retardation
associated with the thick medium from the model boundary to any
given internal point and then determines by a new compensation the
optical axes and relative retardation in the thin slice immediately
behind that point. The interest of this method lies in the precision
with which the orientation and the retardation adjustement of the
compensator can be obtained. However, the orientation of a main beam
of rotating linearly polarised light does not allow the author to
propose a systematic experimental procedure based on precise criteria. Hence the method can be used easily only when the retarding
element associated with the thick medium or with the thin slice is
a quarter-wave plate.

This brief survey of non-destructive photoelastic methods using
scattered light indicates that the problem of three-dimensional model
analysis has been attacked in two fundamentally different ways :
    - the first leads to "Weller type" point or full-field methods
which are generally rather simple to perform but have limited application because results are usable only for small ratios of axis
rotation to retardation,
    - the second idealises the photoelastic model as a sequence of
birefringent laminae each representing a thin slice of the model,
and leads in essence to point methods offering two options : use of
formulae for synthesising birefringents or compensation of the retarding effect associated with the thick medium.

4.3 Integrated photoelasticity

Non destructive stress analysis can also be conducted in some
simple cases by transmission photoelasticity by analysing the integrated effects along the path of light through a three dimensional
model. But this technique needs different experimental conditions
to obtain enough information for solving the algorithms which give
the stress distribution (4).

5. NEW SOLUTIONS

These scattered light techniques take into account the more exact representation of a photoelastic slice discussed previously.

5.1 Point wise techniques

The method described here has been proposed by Brillaud and Lagarde (17) (6). It follows Gross-Peterson's procedure of compensating the thick medium so that the combination of thick medium and compensator is equivalent to a rotatory power. It may be noticed that Robert and Royer (18) have contemporaneously proposed a similar method without involving a compensator. The characteristics of a thin slice are obtained by synthesis of birefringents.

5.1.1 Principle of an ellipsometer with linear detection

In scattered light the principal obstacle of the measurement of the characteristics of the light ellipse is the presence of the depolarised part of the scattered radiation. The modulation created by a rotating analyser (19) begins to solve this problem since it makes it possible to have a precise measurement (by a criterion of zero sine amplitude) in the case of the circular light ; however it remains insufficient in the general case. Brillaud et Lagarde associate a revolving quarter wave plate with the turning analyser in order to make us completely free from the depolarisation of the light. The choice of a signal of reference joined to the quarter wave plate allows, by varying the rotation of the latter, to carry out a linear detection of the parameters of the light ellipse. The light vibration propagates in the U direction and after passing through a revolving quarter wave plate and an analyser turning at the angular speed Ω, the light signal is received on a photomultiplier. The ellipticity of the light is characterized by the angle I such that $|tg\ I|$ may be equal to the ratio of the length of the short axis to that of the long axis of the ellipse. The angle 2I is positive when the light vector describes the ellipse in the positive sense (right vibration) and negative in the contrary case (left vibration). The long axis of the ellipse, the fast axis of the quarter wave plate and the analysis direction are shown respectively by the axes OX, OX_1 et Oa(t).

The expression of the light intensity transmitted by the rotating analyser is :

$$E = \frac{E_0}{2} (1 + \mathcal{A} \cos (2\,\Omega t + \psi)$$

with $\mathcal{A} \cos \psi = \cos 2I \cos 2\beta$

$\mathcal{A} \sin \psi = \sin 2I$

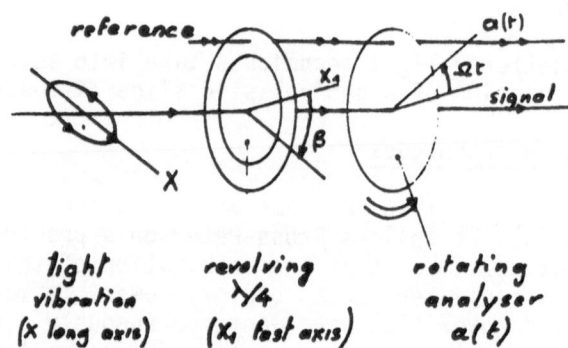

Fig. 5 - Schematic arrangement of the ellipsometer

E is intensity of the elliptic vibration.

The quantity is positive or zero, the angle ψ belongs to the interval $(-\pi, +\pi)$. A reference signal related to the rotation of the large axis of the plate $\lambda/4$ is chosen.

From a rectilinear vibration rotated to coïncide with the fast axis of the quarter wave plate, the light intensity transmitted by the rotating analyser is written :

$$e = \frac{e_0}{2} (1 + \cos 2 \Omega t)$$

ψ is the difference of phase between the alternative components of $\bar{e}$ and e.

When the scattering phenomenon is used as a polarizer, the only state of polarization that we can consider at the entrance into the thick medium is a rectilinear polarization having a known orientation. That leads us to consider the following set up :
- the incident rectilinear vibration is analysed by the ellipsometer ;

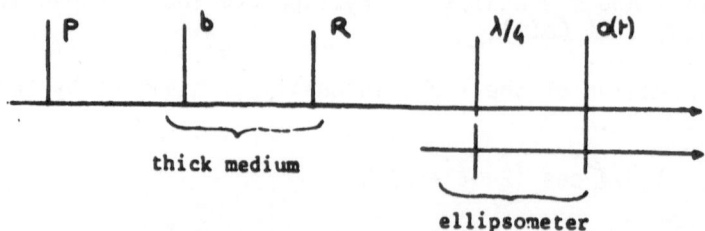

Fig. 6 - Experimental set-up

- after propagating through the thick medium that we characterize by a birefringent with a fast axis b, with birefringence $\delta = \frac{\phi}{2\pi}$ followed by a rotatory power R. Let p and x respectively be the directions of the rectilinear polarization and of the long axis of the light ellipse issuing from the thick medium. The angles $\theta = (b, p)$ et $\chi = (b, x)$ are orientated by the light propagation sense. In order to make the presentation of the method easier it is interesting to write the parameters 2I and β in terms of θ, R, ϕ and the angle $\alpha = (b, x)$, the expression of the signal issued by the ellipsometer becomes :

$$\mathcal{A} \cos \psi = \cos 2\beta \quad \cos 2I = \cos 2\theta \cos 2(\alpha - R) +$$

$$\sin 2\theta \cos \phi \sin 2(\alpha - R)$$

$$\mathcal{A} \sin \psi = \sin 2I = \sin 2\theta \sin \phi$$

We are proposing to determine the parameters of the thick medium by choosing the following way :
- first the polarization direction is orientated in such a manner to get a zero difference of phase, so it coïncides necessarily with one of the birefringent axes ($\theta = 0$ mod $\frac{\pi}{2}$). The vibration issued from the thick medium is rectilinear and makes an angle equal to R with the polarization direction. We are looking for the two orientations of the quarter wave plate making the sine amplitude zero ; the fast axis of the plate makes an angle $\pm \pi/4 + R$ with the polarization direction (obviously maintained in the previous position). Observing that these two orientations can be easily identified by following the evolution of the phase-shift while passing from one to the other, we determine the rotatory power mod π.
- Finally the fast axis of the quarter wave plate is kept making an angle of $\frac{\pi}{4} + R$ with the polarization direction, and the latter is shifted at $+\frac{\pi}{4}$, then the phase shift ψ is equal to ϕ or $-\phi$ depending on whether θ is initially zero or $\frac{\pi}{2}$.

5.1.2 Three-dimensional photoelasticity

By using the scattering phenomenon as a polarizer and the method of linear detection previously described Brillaud and Lagarde propose the principle of 3D photoelasticimeter able to determine the characteristics of a thin slice considered in its most general schema. The photoelastic model is placed in a immersion tank with index liquid ; a laser beam propagating in the U direction penetrates into the model and the observation of the radiation, scattered by a point M, is effected in the U direction. The principal beam, staying subject to passing through the point M can be revolved in a perpendicular plan towards the direction of the observation. Using the law of Rayleigh at the point M, the light scattered in the U direction, is rectilinear polarized following a direction orthogonal to the plan (u, U).

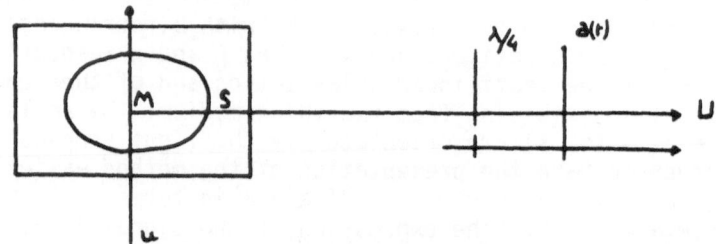

Fig. 7 - Experimental set-up

Brillaud and Lagarde propose to compensate first the thick medium M$_2$S and then to determine the characteristics of a slice M$_1$M$_2$ resulting from a small translation of the principal beam.

The method requires two steps :

. 1st step : compensation of the medium M$_2$S the axis and the rotary power M$_2$S through which the scattered light propagates are determined by the orientation of the polarization direction giving a zero phase and by the orientation of the fast axis of the quarter wave plate giving a zero sine amplitude, the compensator is placed "at R" from the axes of the birefringent (orientation easily deduced from that of the quarter wave plate when the sine amplitude is zero). Then, the polarization direction is rotated to make an angle $\frac{\pi}{4}$ from the axes of the birefringent and the birefringence of the compensator is adjusted in order to obtain a rectilinear vibration at the exit of the compensator.

. 2nd step : Measurements of the characteristics of a thin slice : the point M$_1$ located on the same observation axis close to the point M$_2$ is then illuminated. Thus a thin sheet is added to the rotatory power R of the set thick medium-compensator.

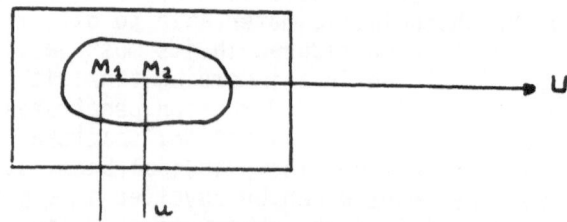

Fig. 8 - Location of a thin slice

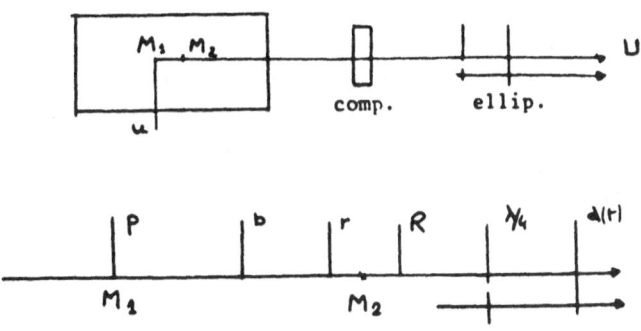

Fig. 9 - Schematic arrangement of the photoelasticimeter

This thin sheet can be representated generally by a birefringent whose fast axis is b, and birefringence is $\delta = \frac{\lambda \phi}{2\pi}$ followed by a rotatory power r. Then we have on the path of the scattered light the following elements (fig. 9).

The orientation of the axis b, the value ϕ of the birefringence and the rotatory power R + r, are then directly determined by the ellipsometer.

5.1.3 Applications

This technique has been successfully applied :
- to the measurement of stress concentrations in turbine blade (6) ;
- to the determination of the stress tensor in bars loaded in torsion (6) ;
- to the determination of the speed gradient distribution in a birefringent flow of milling yellow dye (20).

5.2 Whole field method

We have exploited the new possibilities created by the laser beams associated with the properties of diffused light to elaborate a new method of three-dimensional photoelasticity called "optical slicing" (21) (22) (23). It carries out, without destroying the model, the rectilinear analysis of a given slice optically singled out by two plane parallel laser beams. We recall briefly the principles of this method (7).

5.2.1 Basic idea

A slice of the photoelastic model is singled out by two parallel plane beams. The two dimensional field of scattered light is analysed along a direction perpendicular to the plane of the two illuminated sections. At its origin, the beam is rectilinearly polarized along a direction which is perpendicular to the directions of incidence and analysis (Rayleigh's law (fig. 10)).

The possibilities of interference between the beams from the first and second illuminated sections of the model depend on the birefringence of the singled out slice. If the birefringence of the latter is equal to zero or if one of its main directions is parallel to the scattered light direction of polarization, the polarization of the light issued from the first section is not modified by passing through the slice. The two beams have the same polarization, so they can interfere. On the other hand, if the angular birefrigence is an odd multiple of π or if one of its main directions forms an angle of $\pi/4$ with the direction of polarization, the beam issued from the first section becomes perpendicular to the beam issued from the second section and cannot interfere with it. More generally, it has been shown that the correlation factor of beams issued from each section could be rendered by the following expression.

$$\gamma = \sqrt{1 - \sin^2 2\alpha \, \sin^2 \frac{\phi}{2}}$$

where α is the angle formed by one of the main directions of the slice and the direction of polarization of the scattered light and ϕ is the angular birefringence of the slice. It must be noticed that this expression is not dependent on the state of the existing medium between the slice and the boundary of the model.

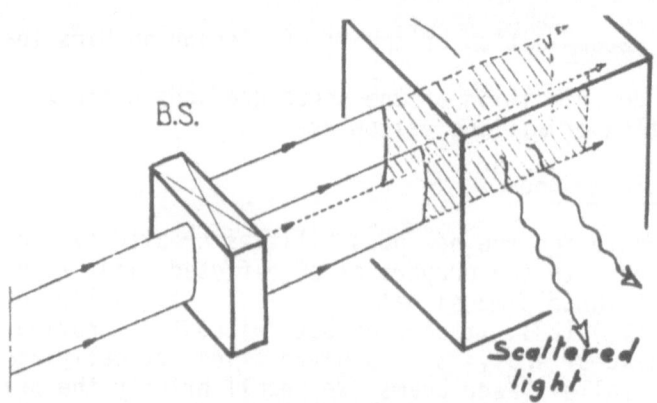

Fig. 10 : Optical slicing

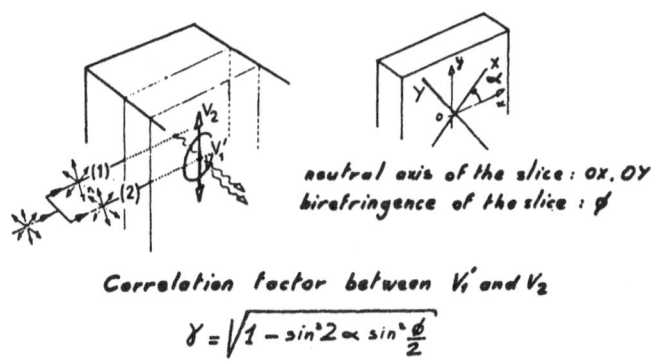

neutral axis of the slice: OX, OY
birefringence of the slice : φ

Correlation factor between V_1' and V_2

$$\gamma = \sqrt{1 - \sin^2 2\alpha \sin^2 \frac{\phi}{2}}$$

Fig. 11 : Correlation factor

It can be seen that the expression of the correlation factor is similar to that of the illumination I obtained in the course of the analysis of an identical slice (obtained for example by freezing and slicing) between two parallel polarizers (fig. 11).

$$I = 1 - \sin^2 2\alpha \sin^2 \frac{\phi}{2} \,.$$

We use the properties of polychromatic light to have a global visualization of the isochromatic and isoclinic fringes of the slice

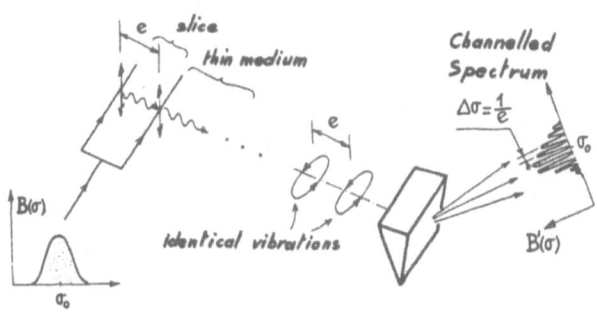

Fig. 12 : Interference observation in polychromatic light (case of MAX. interference)

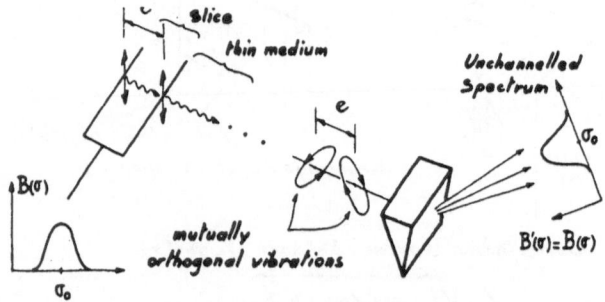

Fig. 13 : Interference observation in polychromatic
light (case of MIN. interference)

under investigation (22) (23). The two sections are illuminated by a
polychromatic beam. In the case of maximal interference the observation reveals two sources of polychromatic light, one delayed in
relation to the other. Analysis of the resulting beam with the held
of a spectrocopic device produces a channelled spectrum.

In order to obtain information simultaneously on the whole surface of the slice under investigation, we project the image of the
two sections illuminated by polychromatic light through the spectros-

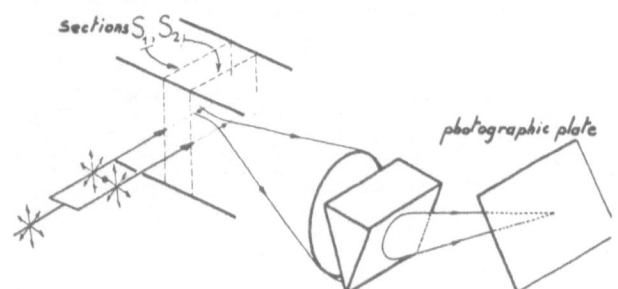

Fig. 14 : Spectrally resolved imaging of the slice

cope. For the areas of the image in which the correlation factor is maximal random patterns whose grains exhibit a sinusoïdal modulation and for those in which this factor is non-existent a random pattern whose grains are just stretched along the direction of analysis of the spectroscope. The discovery of the image areas exhibiting a modulation, that is to say the visualization of isochromatic and isoclinic fringes of the negative is made through pass-band optical filtering of the negative.

5.2.2 Experimental set-up

The polychromatic source of light is a dye-laser (rhodamine 6G) of a 600 mW output power. The optical spectral width is obtained by scanning by means of a Lyot filter monitored by a cam device. The experimental set-up giving the plane light beams consists of :
- a half wave plate allowing an orientation of the rectilinear polarization of the incident beam,
- a lens of 10 mm in focal length
- a Wollaston prism which gives two orthogonally polarized beams (vertical and horizontal),
- a half wave plate whose function is to realign the rectilinear polarization of the two beams at a $\pi/4$ angle to the vertical axis,
- a Babinet compensator placed according the position described in reference (7),
- a cylindrical lens which allows the focalisation of the two beams along two parallel lines.

The intensities of the two beams are balanced by rotating the first half-wave plate. The gap between the two light beams is fixed by the translation of the Wollaston prism. The translation of the

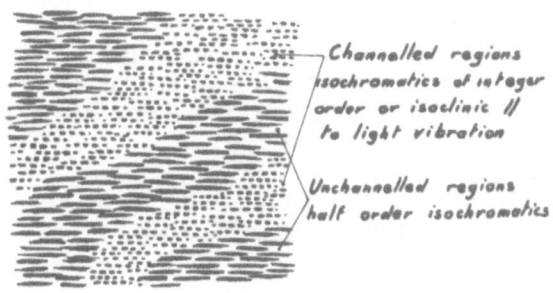

Fig. 15 : Spectrally resolved image

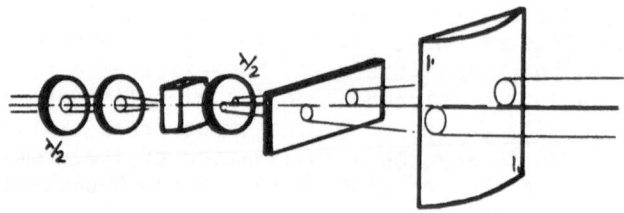

Fig. 16 : Illumination set-up

cylindrical lens allows a focalization of the beams at the level of
the model. The size of the measurement field depends on the distance
along which the thickness of the beam may be considered as constant :
the Fresnel length ; it depends on the aperture of the two beams
and has the value :

$$\rho = \frac{4\lambda}{\beta^2}$$

Example : if a measurement field of 100 mm in length according to
the direction of the beams is to be obtaine, then the thickness of
each will be :

$$s = \frac{\lambda}{\beta} = \sqrt{\frac{\rho\lambda}{4}} = 1,22.10^{-1} \text{ mm}$$

that is to say the thickness of all the beams is .25mm over a length
of 100 mm if the angle β is equal to $5,10^{-3}$ rd. This thickness is
reduced to .17 mm over a length of 50 mm for an angle equal to
7.10^{-3} rd.

The spectral analysis is carried out by a diffraction pattern
whose resolution must be sufficient to resolve a channelled spectrum
whose pitch is fixed according to the thickness of the slice. We
have selected a diffraction grating of 70 mm including 600 lines per
millimeter, which fixes its resolution power to 15 Å. The interval
between channels has been determined so as to correspond to a
scanning of 30 Å. It is fixed by a spatial delay of 12 mm between
the two diffused waves.

Taking into account the refraction index n of the medium this
delay is obtained with a thickness e of 7.5 mm :

$$e = \frac{g}{n} , \quad n = 1.6$$

While studying slices of different thickness, we are obliged
to place on the path of one or the other of the incident beams glass
panels in order to re-establish this gap. The thickness h of this
panel depends on the following relation.

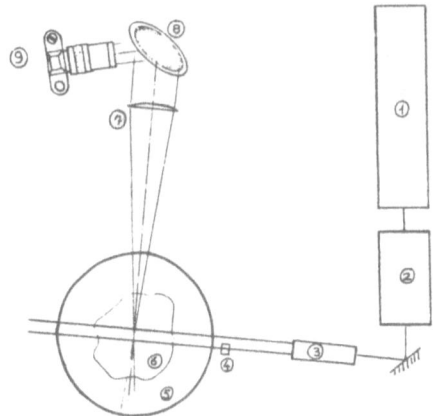

Fig. 17 : Experimental set-up

$g = n'h + n e$ where $n' = 1.5$ is the glass refractive index

we have imposed a scanning of 3 Å so as to obtain 10 channels in each grain of the image structure. In order not to introduce any distortion in the image, the diffraction grating is placed so as to work in autocollimation. To achieve that, the lines are horizontally oriented and the grating is tilted so that the incident and diffrac-ted beams are also on horizontal plane.

In order to obtain a uniform illumination picture, the grating should not introduce any dimming variations in function of the pola-rization of the incident light. To that effect, the value of the reflection angle in the horizontal plane has been optimized so that the partial polarization introduced by the diffraction on the gra-ting is compensated by the one corresponding to the reflection on the grating. The image is formed with a lens of 1 mm in focal length and with an objective of 300 mm of focal length after reflection on a diffraction grating 70 mm wide including 600 lines per mm. It is recorded on a Kodak film SO.253. The exposure time is of the order of 10 s for a incident power of 400 mW.

One should remember that a complete study of a photoelastic model requires a device for the positioning and the marking of the six degrees of freedom of a body (3 translations and 3 rotations). It is easy to design this sort of device when one has all the necessary room around the body by using shaft connections. In the framework of our study, we have however to leave room for the solid angles occupied by the incident and diffused beams. The mechanical device would be too sophisticated ; so we have solved the problem by using a spherical immersion tank. A glass sphere of 500 mm in diameter is supported by an hydrostatic water bearing allowing an easy orientation of the tank model set. In the connection between the tank and the loading system of the model a set of three translation plates is incorporated. The rotary connection by spherical tank allows us to adopt the system axis which is most suitable for our study. This is particularly true in the case of study of crack problems ; in that case it is sufficient to draw on the tank the circle corresponding to its plane. This type of identifying marking is easily carried out with a laser He-Ne whose beam materializes the direction of observation. A more general three-dimensional study involves more complex tracing of parallels and meridians. The positioning-precision taking into account the radius of the sphere is of one degree.

5.2.3 Rotation of the secondary principal directions

We examine here the influence of the rotation of the secondary principal lines on the validity of the results we obtain. It is to be remenbered that the method is characterized by a complete independence from the effects of the birefringents constituting the model other than the investigated slice. So the only effect to be considered is the rotation effect in the thickness of the slice. In the case of a rotation of the secondary principal directions, the single-out slice can no longer be considered as a single birefringent but as the combination of a birefringent B* and of a rotation power. Let (X_1, Y_1) be its characteristic directions when entering the slice and (X'_1, Y_1) those when leaving the slice (the rotation power is then equivalent to $R = (X_1, X'_1)$). Let ϕ^* be the angular birefringence of B* ; the direction of polarization of diffused light X is marked by the angle $\alpha = (X, X_1)$.

A simple calculation (7) gives the square of the following correlation factor.

$$\gamma^2 = \cos^2 R - \sin 2\alpha \sin^2 (\alpha + R) \sin^2 \frac{\phi^*}{2}$$

Let us note that the square of this correlation factor has the same expression as that of the intensity obtained by the analysis of this same slice between two parallel polarisers. The results of the analysis of § 3.2 can be directly transposed to this technique.

5.2.4 Some experimental results

We have studied with this method a bar of square section loaded in torsion. We have carried out the optical slicing of slices tilted at an $\pi/4$ angle to the axis for different directions of analysis. The studied slices being tilted at an angle $\pi/4$ to the axis the birefringence as well as the orientation of the secondary principal directions evolve along the direction of observation between the two sections bounding the slice.

We have determined in the framework of linear fracture mechanics the value of the characteristics parameters K_I and σ_{on} for a semi-elliptical crack loaded in opening mode in a bar in tension. The results have been compared with numerical investigations (7).

CONCLUSION

Using the optical theory of light propagation in anisotropic medium due to Aben we recall that a birefringent medium can be represented as a combined set composed of an elementary birefrigent and a rotatory power. This set is characterized by three optical parameters.

The assumtion of uniformity, for the difference of secondary principal stresses and for the rate of rotation of their directions throughout the thickness of a thin slice, allows the determination of two parameters related to mechanical state.

When the ratio of the rotation to the retardation is small the representation of the light propagation can be simplified and the classical techniques of scattered light are reliable.

In the more general case the examination of a slice from a stress frozen model in a classical plane polariscope leads to a correct evaluation of the isoclinic and isochromatic fringes. This is a rather comforting result which explains the reliability of this technique favoured by engineers. On the contrary the point-wise methods are not in general suitable for this purpose.

The more exact representation of a photoelastic slice discussed here is being taken into account by the two recent non destructive techniques described.

The optical slicing technique is more suitable to problems which present high stress gradient. The point-wise method with linear detection is particularly appropriate for the measurement of weak values of birefringence.

REFERENCES

1. Drucker D. et Mindlin R. "Stress Analysis by Three-Dimensional Photoelastic methods". J. Appli. Phys., Vol. 11, 1940.
2. Mindlin R. et Goodman L. "The Optical Equations of Three-Dimensional Photoelasticity". J. Appl. Phys., 20, 1949.

3. Lee L. "Effects of Rotation of Principal Stresses on Photoelastic Retardation". Exp. Mechanics, 4, 10, 1964.
4. Aben H. "Optical Phenomena in Photoelastic Models by the Rotation of Principal Axes". Exp. Mechanics, 6, 1, 1966.
5. Aben H. "Integrated Photoelasticity". Mac Graw Hill, 1979.
6. Brillaud J. "Mesure des Paramètres Caractéristiques en Milieu Photoélastique Tridimensionnel. Réalisation d'un Photoélasticimètre Automatique. Applications". Thèse d'Etat, n° 385, Poitiers, 1984.
7. Desailly R. "Méthode Non-Destructive de Découpage Optique en Photoélasticimétrie Tridimensionnelle. Application à la Mécanique de la Rupture". Thèse d'Etat, n° 336, 1981.
8. Hickson V.M. "Errors in Stress Determination at the Free Boundaries of "Frozen Stress" Photoelastic Models". Brit, J. Appl. Phys. Vol. 3, n° 6, p. 176-181, 1952.
9 Oppel G.U. "Forschungs Gebiete Ing.". 7, 1936, 240-8, NACA T.M., 824, 1937.
10. Weller R. "Journal of Applied Physics", t. 10, 4, p. 226, 1939.
11. Weller R. "Journal of Applied Physics", t. 12, 8, p. 610, 1941.
12. Cheng Y.F. "A Dual Observation Method for Determining Photoelastic Parameters in Scattered Light". Experimental Mechanics, Vol. 7, n° 3, p. 140-144, 1967.
13. Robert A. et Guillemet E. "Nouvelle Méthode d'Utilisation de la Lumière Diffusée en Photoélasticimétrie à Trois Dimensions". R.F.M., 1963, n° 5/6, p. 147-157.
14. Poincaré H. "Théorie Mathématique de la Lumière". Gauthiers-Villars, 1889.
15. Gross-Petersen "A Compensation Method in Scattered Light Photoelasticity". I.U.T.A.M. Symposium 'The Photoelastic Effect and its Applications'. Sept. 1975, Springer Verlag.
16. Brillaud J. et Lagarde A. "Ellipsométrie en Lumière Diffusée et son Application à la Détermination des Caractéristiques Optiques d'une Tranche Mince en Photoélasticimétrie Tridimensionnelle". Symposium I.U.T.A.M. 'Optical Methods in Mechanics of Solids' Poitiers, Septembre 1979, Ed. A. Lagarde, (Sijthoff Noordhoff).
17. Brillaud J. et Lagarde A. "Punctual Determination of Stress Deviator in Three Dimensional Photoelasticity". Communication présentée au 14e Congrès International I.U.T.A.M. Delft, sept. 1976.
18. Robert A. et Royer J. "Principe de Mesure des Biréfringences à l'Intérieur d'un Solide Transparent en Vue de son Application à la Photoélasticimétrie Tridimensionnelle". C.R.A.S., t. 281, Série B, 1975.
19. Sapaly J. "Contribution à l'Etude de la Photo-Extensométrie Statique et Dynamique (Thèse, Paris, 1961).
20. Monnet P. "Contribution à l'Etude des Lois de Comportement Rhéo-Optique de Certains Fluides Biréfringents. Application à un Ecoulement Non Viscométrique de Solutions d'Alphonogelb". Thèse de 3e cycle, Poitiers, 1980.

21. Desailly R. "Visualization of Isoclinics and Isochromatics in a Birefringent Slice Optically Singled out in a Three Dimensional Model". Optics Communications, Vol. 19, n° 1, Octobre 1976.
22. Froehly C. et Desailly R. "Polychromatic Speckle Technique for Three-Dimensional Non-Destructive Photoelasticity". Optics Communications, Vol. 21, n° 2, Mai 1977.
23. Desailly R. et Froehly C. "Whole Field Method in Three Dimensional Photoelasticimetry : Improvement in Contrast Fringes". Communication au Symposium I.U.T.A.M. de Poitiers 'Optical Methods in Mechanics of Solids', 10-14 Septembre 1979, Sijthoff & Noordhoff, 1931.
24. Desailly R. "Méthode Non-Destructive de Découpage Optique en Photoélasticimétrie Tridimensionnelle. Application à la Mécanique de de la Rupture". Thèse d'Etat, Poitiers, n° 336, 1981.

FURTHER TOPICS

OPTICAL METROLOGY AT LOUGHBOROUGH UNIVERSITY

P.C. Montgomery, A.C. Rowland.

Department of Mechanical Engineering,
University of Technology, Loughborough U.K.

ABSTRACT

The main objective of the laser group in the Mechanical Engineering
Department at Loughborough University is to develop methods of
extracting data by remote means that can be used by industry.
Three such whole field techniques are Electronic Speckle Pattern
Interferometry (ESPI), Holographic Interferometry and Grid
Projection Moire. Recent publications of the group are described,
including optimisation of ESPI systems, applications of ESPI, the
development of a portable holocamera and a Moire technique for
producing numerical data of the profiles of turbine blades.
Present work being carried out is also mentioned.

INTRODUCTION

The last fifteen to twenty years has seen rapid growth in remote
sensing using transducers in acoustic and electro-magnetic fields.
At Loughborough there is great interest in using visible light as
a means of extracting data from engineering components. This
data is typically in the form of an interferogram displayed on a
T.V. monitor, indicating the absolute shape of a component or
the change in shape under static or dynamic loading. A trained
operator can use this qualitatively to investigate vibration modes,
defects in fabricated structures or areas likely to fail due to
high strain conditions. By computer processing of the video signal,
numerical results can be obtained for 3D coordinate measurement
and stress analysis, which is useful throughout design, production
and quality control in industry.

Soares, O.D.D. (ed), Optical Metrology
© *1987. Martinus Nijhoff Publishers, Dordrecht.*

The group started, under the leadership of the late Professor
J.N. Butters, with work in holographic interferometry and has
spent much of its subsequent effort in developing Electronic
Speckle Pattern Interferometry, a form of T.V. image plane
holography. In recent years, grid projection Moire has been
used in shape measurement. This paper gives details of recent
publications in these areas and current work showing the group's
aim of making optical measurement techniques suitable for the
industrial environment.

DEVELOPMENT OF ESPI

Electronic Speckle Pattern Interferometry can be thought of in
terms of focussed image holography, in which a video camera
producing a video signal that is electronically processed
replaces the emulsion that is chemically processed. The video
camera detects the intensity of phase referenced speckles, from
an object illuminated with laser light, yielding phase information
(1) and forming a monochrome interferogram (Fig. 1).

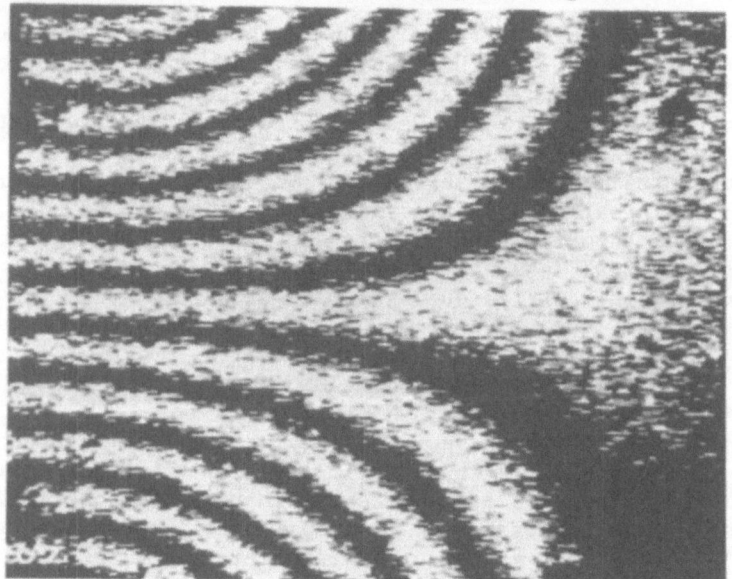

Fig 1. A typical ESPI Interferogram photographed from a T.V.
monitor showing torsion in a steel plate clamped on one
side.

As in holography, this has wavelength sensitivity but has reduced
contrast due to speckle noise. The great advantage of the
technique is that the interferogram is updated every   1/25

second, the frame rate of the camera. The video signal is also readily digitised and processed by computer. Because ESPI can operate under room lighting conditions, with sufficient vibration isolation (which is far less stringent than in holography) it is compatible with many industrial environments. Much has been published in the field of ESPI since the development work of Leendertz and Butters (2, 3). The theoretical and practical aspects of the technique have been extensively studied at Loughborough by Jones and Wykes, covered in their book, "Holographic and Speckle Interferometry" (4).

Recent work at Loughborough has included the theoretical basis for the design and optimisation of Electronic Speckle Pattern Interferometers (5) improving the fringe contrast (6) and smoothing the fringe patterns by computer processing (Fig. 2, ref. 7). At present, work is being carried out on optimising the electronic processing of the video signal using digital storage methods. The layout of the optics in the interferometer is also being improved for more efficient use of light, stability and ease of operation. With careful design, fringes have been obtained on an area of 1 m^2 using only 15 mW of HeNe light. Steps are being taken to find a company who are prepared to manufacture ESPI systems under licence from the British Technology Group (BTG) who hold the patents.

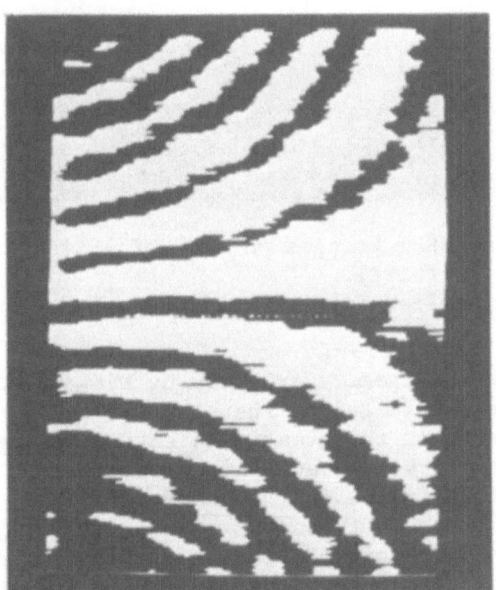

Fig 2. Fourier Transform Filtering and Contrast Enhancement of an ESPI Interferogram (courtesy of Mr. P.O. Varman and Dr. C. Wykes).

© Elsevier Applied Science Publishers Ltd., England 1984.

APPLICATIONS OF ESPI

Applications of ESPI have been extensively reported (1, 3, 4).
Wykes and Jones (8) cover further details on calibrating
measurements and converting fringe data to surface shape information
in two wavelength ESPI contouring. ESPI has been sucessfully
used looking through a microscope at areas of 1mm^2 for out-of-
plane deformation (Fig. 3) by Herbert (9) and applied to the
investigation of the plastic zone in front of a crack tip induced
in epoxy resins (10).

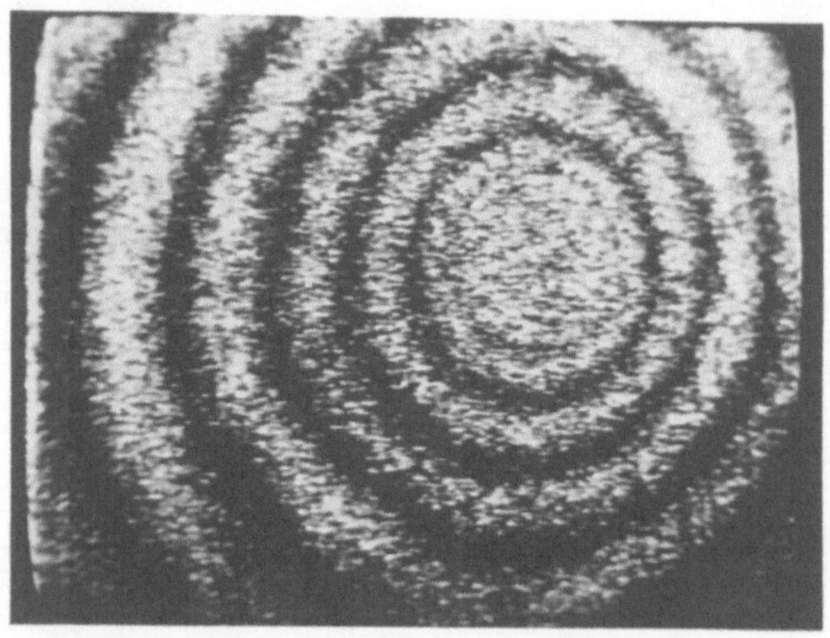

Fig 3. Out-of-Plane ESPI Fringes on 1mm^2 of a steel disc (courtesy
        of Mr. D.P. Herbert).
        ©Elsevier Applied Science Publishers Ltd., England 1984.

Bergquist, Tyrer and Montgomery are proving ESPI to be a useful
tool in the vehicle manufacturing industry through work with
British Leyland, using it to investigate engine resonance modes,
weld joints and general stress analysis problems. It is similarly
being used on problems in the fuel industry.

Holownia and Rowland are using ESPI to measure the dynamic bulk
modulus of elastomers (to be published). Other industrial
problems being investigated are NDT of glued laminates and
vibration modes of turbine blades and loudspeakers.

HOLOGRAPHIC INTERFEROMETRY

Holographic interferometry is being used as a measurement tool in student projects in the Mechanical Engineering Department as well as for industrial applications requiring higher contrast fringe patterns than those obtained by ESPI. For such studies Rowley (11) has developed a holocamera using a converted SLR camera (fig. 4). By using an incoherent fibre optic bundle to provide a reference beam, a focussed image hologram is produced on 35 mm format film. This is a convenient design for a low cost portable system for making good quality holograms viewable in white light. (Fig. 5).

Fig 4. Holocamera Using Fibre Optic Reference Beam (courtesy of Dr. D. Rowley).

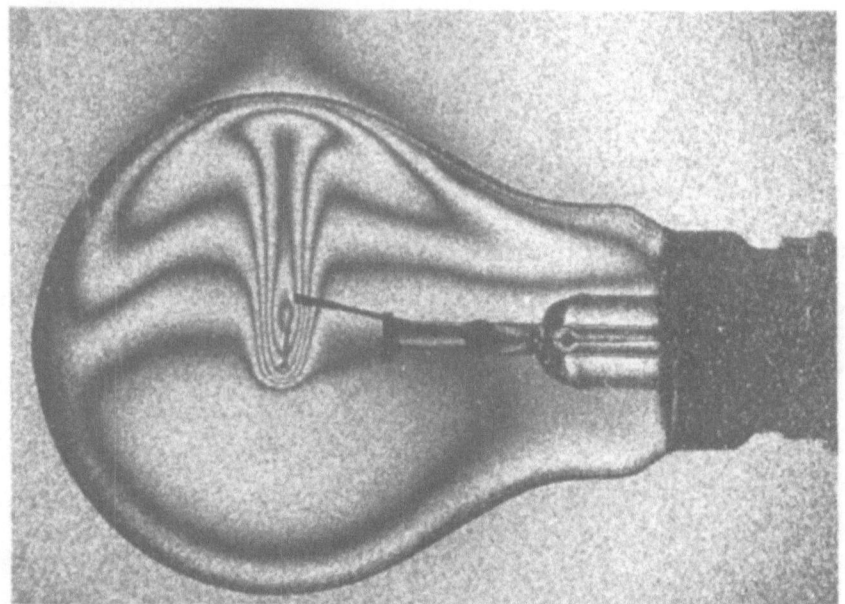

Fig. 5. Double Exposure Hologram of Convection Currents in a Lightbulb Taken With the Holocamera (courtesy of Dr. D. Rowley.

In collaboration with the Central Electricity Generating Board (CEGB), Bergquist and Henry are attempting to make accurate measurements of fuel rod dimensions from projected image holograms made inside a nuclear reactor core.

Another area of great industrial interest is the use of a pulsed laser, on site, for carrying out self excited studies on large pieces of machinery. Moving out of the laboratory environment to a remote location to make double pulsed holograms has been successfully carried out by Tyrer on a 1m diameter feedpipe and a 3m diameter turbo compressor driven by an RB2-11 gas turbine engine.

GRID PROJECTION MOIRE

While ESP1 provides measurements in the region of 0.1 to 20 μm, grid projection Moire is less sensitive, generally in the region of 25 μm up to many centimetres. Varman (12) has successfully used Moire for producing numerical data of the profile of a turbine blade using a computer and video store (Fig. 6).

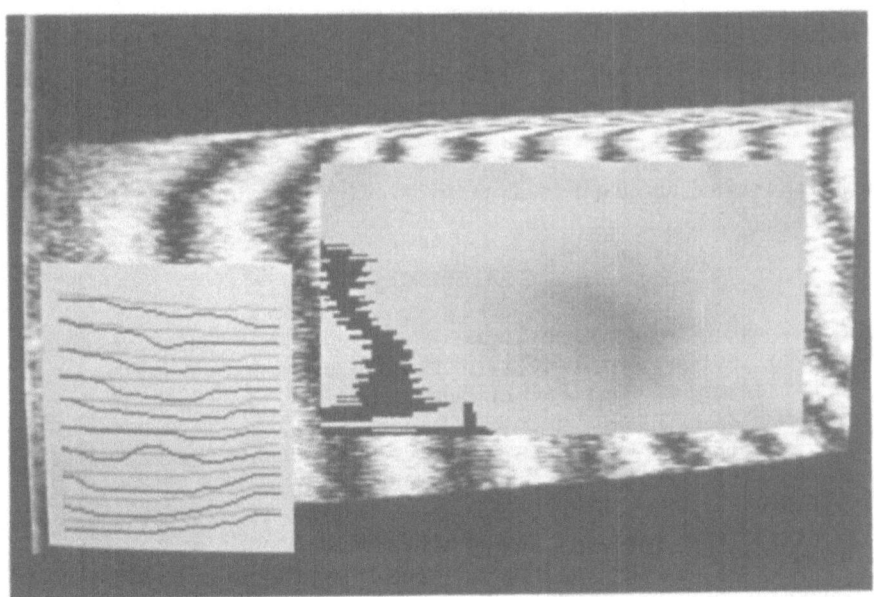

Fig 6. Visual Display of the Numerical data of the difference in
Profile of Two Turbine Blades (courtesy of Mr. P.O. Varman).

In collaboration with Wykes (7) Varman has also investigated the
use of computer processing for smoothing Moire fringes.

Work has been initiated on the development of an automatic shape
measurement system by Montgomery using white light grid projection
for looking at objects of 1m diameter and larger.

CONCLUSIONS

The aim of the laser group at Loughborough University is to
develop non-contacting optical measurement techniques for use in
Industry.

Three fields of particular interest are Electronic Speckle
Pattern Interferometry (ESPI), Holographic Interferometry, and
Grid Projection Moire. Work within these areas consists of
optimising system designs, proving applications, automating
analysis of interferograms and tailoring optical systems for use
in specific industrial situations.

ACKNOWLEDGEMENTS

We would like to thank the SERC for the financial support of the laser group's work and for the continued encouragement of Mr. B.D. Bergquist and Mr. J. Tyrer.

Thanks are also due to Mr. K. Topley, for the photographs.

REFERENCES

1. C. Wykes, "Use of Electronic Speckle Pattern Interferometry (ESPI) in the measurement of static and dynamic surface Displacements", Optical Engineering, Vol. 3, 1982, pp 400-406.
2. J.A. Leendertz, "Interferometric Displacement Measurement on Scattering Surfaces Utilizing Speckle Effect", J. Phys. E: Scientific Instruments, Vol. 3, 1970, pp 214-218.
3. J.N. Butters and J.A. Leendertz, "Holographic and Video Techniques Applied to Engineering Measurement ", Journal of Measurement and Control, Vol. 4, 1971, pp 344-350.
4. R. Jones and C. Wykes, "Holographic and Speckle Interferometry", (Cambridge, Cambridge University Press, 1983).
5. R. Jones and C. Wykes, "General Parameters for the Design and Optimization of Electronic Speckle Pattern Interferometers", Optica Acta, Vol, 28, 1981, pp 949-972.
6. C. Wykes, J.N. Butters and R. Jones, "Fringe Contrast in Electronic Speckle Pattern Interferometry", Appl. Optics, Vol. 20, 1981, pp 720-721.
7. P. Varman and C. Wykes, "Smoothing of Speckle and Moire Fringes by Computer Processing", Optics and Lasers in Engineering Vol. 20, 1981, pp 720-721.
8. C. Wykes and R. Jones, "Advances in Optical Metrology of Complex Objects", SPIE, Vol. 369 Max Born, 1983.
9. D.P. Herbert, "Inspection of Out-of-Plane Surface Movements Over Small Areas Using Electronic Speckle Pattern Interferometry", Optics and Lasers in Engineering, Vol. 4, 1983, pp.229-239.
10. M.O.W. Richardson, A.H.M. Al-Hassani and D.P. Herbert, "Epoxy Powder Coating Fracture Analysis by Electronic Speckle Pattern Interferometry", Trans. Inst. Metal Finishing, Vol. 60, 1982, p. 57.
11. D.M. Rowley, "The Use of a Fibre-Optic Reference Beam in a Focussed Image Holographic Interferometer," Optics and Laser Technology, August 1983, pp. 194-198.
12. P.O. Varman, "A Moire System for Producing Numerical Data of the Profile of a Turbine Blade Using a Computer and Video Store", Optics and Lasers in Engineering, Vol. 5, 1984, pp 41-58.

CLOSING REMARKS

CONCLUSIONS and FUTURE EVENTS

Charles M. Vest and O.D.D. Soares*

Department of Mechanical Engineering and Applied Mechanics
The University of Michigan, Ann Arbor
Michigan 48109, USA
*Centro de Fisica, Universidade do Porto
4000 Porto, Portugal

CONCLUSIONS

Under the directorship of Professor O.D.D. Soares an International Advanced Study Institute on Optical Metrology was successfully
held on 16-27 July 1984, in Viana do Castelo, Portugal.

Optical Metrology deals with the use of optical effects, its
interaction with matter and propagation of light to measure physical
quantities. The field therefore, is quite broad although emphasis
is primarily on mechanical topics. Properties of light include amplitude, phase, speed, frequency, state of polarization, direction of
propagation, coherence properties, diffraction and image formation.
Quantities which can be measured or inferred from changes in such
properties of light due to interaction with a test object include
size, shape, distance, velocity, stress, strain, rotation, alignment,
temperature, pressure, species concentration and statistics of surface
profile. All of the above properties of light and measured quantities
were dealt with by various lecturers during the Institute. The majo-

rity of optical processes discussed were based on linear effects, although there is an increasing trend toward the application of non-linear materials and effects.

The size scale of objects measured by Institute participants ranged from large civil engineering structures such as bridges and buildings to submicrometer particles. Accuracies of the various techniques ranged from centimeters to measurements made to one part in 10^{12}. The physical bases of measurement techniques ranged from elementary laser beam deflection, based on Snell's law, to nonlinear four-wave mixing. Applications ranged from heavy engineering problems such as the testing of railroad track clips to studies of mode coupling in optical fibers. The speeds and time scales ranged from rotational speeds of a few degrees per year to phenomena at optical frequencies and speeds.

The domains of activity in optical metrology discussed at the Institute ranged from engineering applications of holographic interferometry, to image synthesis and moiré techniques through the physics of phase conjugation and picosecond phenomena. The various talks displayed a continuum of research from physics through to engineering practice.

A number of themes were prominent in the lecturers presented. In some cases lectures dealt with the development of techniques to solve particular industrial problems. In others, interesting physical phenomena were detailed and the ways in which appropriate industrial or scientific applications were being sought. The coupling of optical methods to video, electronic and, perhaps most important of all, digital processing is a very important current theme. Attempts to develop the maximum possible accuracy and measurement information from known techniques were emphasized by several lecturers. The desirability, and indeed necessity, for cooperation among mechanical engineers, physicists and electronics/computer engineers was apparent and was the key to many of the most interesting applications discussed.

The Institute was timely in that some fields of optical metrology have matured into industrial practice while others are just being incubated. Hence there was a range of points of view and experience to share. It also was apparent that in a variety of countries, universities and other departments, activities in optical metrology or major subsets of it are currently under development.

It was generally agreed among participants of the Institute that coverage of the main topics had been achieved satisfactorily, though of course, it was impossible to cover the entire field. Even so, a wide range of tutorials, reviews and state-of-the-art presentations were programmed throughout. The Institute could not be all inclusive and some important topics such as testing of optical elements, spectroscopy, etc., were not treated. The emphasis was, quite properly, heavily on lasers, but certainly was not limited in this manner since a variety of white light techniques were also discussed. The poster sessions were found to be particularly useful in broadening the scope of the Institute and providing a mechanism for larger numbers of scientists and engineers to make their work known.

There was a strong theme of industrial relevance in the presentations. Many examples of techniques already applied to practical industrial problems were presented. At the same time, lecturers were quite willing to point out the pitfalls and the additional development needed for further progress on industrial application of individual techniques.

It was clear that the Institute had done much to foster interaction among participants from various nations. Each returned home with an increased awareness of the available technology and ideas for further research and applications. Publication of the proceedings of the Institute will reinforce this and help to spread the ideas and techniques discussed.

Finally, it was quite appropriate that this Institute, devoted to international cooperation in science and technology of Optical Metrology was held in Portugal, a nation which played such an important historical role in the development of navigation, trade and communication over vast distances.

Charles M. Vest

FUTURE EVENTS

Experimentalists understand full well the difficulties involved in obtaining reliable observations, and how rather unsafe and risky it is to make predictions.

However, my actual dilemma is even greater since the last lectures seemed to have made it clear that with the shortest pulsed excitation and interaction with systems, most information can be collected. This is known from linear system theory but has been examined and studied for non-linear systems.

Accordingly, the arrival of the "fentosecond area" could produce the inference that my statement should have finished by now!

Possibly it may be of some interest to expand the time scale and recall some of the interesting ideas, reflections and prospects projected into the near future that occupied discussions.

The relevance of the theme and its interdisciplinary character resulted in the strong recommendation that the organization of a course with similar scope should be considered at least every 3 years in order to continue fostering the interaction between University and Industry, and to keep pace with the fast moving expansion in the field of research, as well as to nucleate new efforts.

There are many pertinent reasons for organizing an Institute, the most obvious of which is the general realization that the value of participation extends beyond a conference scope. The critical in--depth treatment of specifically selected topics by acknowledged world experts through a well coordinated programme provides an excellent atmosphere in which to reflect on current research and development work. Furthermore, it frequently leads to a reorientation of activities or points the way to important follow up studies. In addition, it offers an ideal opportunity for identifying those research topics which are most likely to polarize multi-national teams for cooperative research and technological development.

Due to the importance of metrology studies and the high demand for education and training the conclusion has been reached that strong steps should be taken to determine the requirements and structure of a useful curriculum which would identify the major educational modules, inclusive of importance to other disciplines, in order to produce a good text book and to rationalize choice of research areas and laboratory equipment. The first step appears to be the preparation of a survey to be carried out at various Universities, Laboratories and Industries in Europe and the U.S.A., including Canada, to establish whether any Optical Metrology curriculum already exists.

The Institute was designed to induce cultural communication by bringing together individuals with different orientations and backgrounds, involved in research and education, and those concerned with practical applications, academics and experts from industry, to promote cross-fertilization of ideas via interdisciplinarity, to encourage industry to play an even stronger role, and to develop future cooperation between participants. So far the success of the Institute has been the direct result of the work of all its active members both

lecturers and other participants so that they deserve congratulations
for their work but it is mainly in the time ahead that it will be
judged how fruitfull, in relevant results, these endeavours have
been. Hopefully, another Institite, in some years time, will evaluate
the response to the already foreseen challenges.

O.D.D. Soares

REMARK:
HOLOGRAPHY AND RELATIVITY

Nils Abramson

Industrial Metrology, Dept. of Production Engineering, The Royal
Institute of Technology, S-100 44 Stockholm 70, Sweden

Everything we see depends on our point of view. Three dimensional
objects, perhaps aside from spheres, all look different if the
observation angle is changed. In our ordinary world we are so used
to this phenomenon that we automatically make the necessary transformations and corrections to form the "true image" in our brain.

Image forming processes are usually explained either by the concept
of rays of light, or waves of light. The spherical wave is the
most fundamental idea of all optics, while rays can be formed by
the cooperative effect of many waves (e.g. Youngs fringes). However,
even these basic elements of optics, the spherical wavefronts,
have an apparent shape depending on the point of view.

Let us imagine a person standing in a stationary space full of
scattering particles like smoke and even larger objects as e.g.
ping-pong-balls fixed to wires. The person emits one single short
pulse of light e.g. 10 picoseconds (3 mm) long. This pulse expands
in all directions in the form of a sphere. Ten nanoseconds (3 m)
afterwards he makes a short (picosecond) observation. Certainly he
will find himself in the center of a spherical luminous shell of
light having an apparent radius of 1,5 m. Only those particles and
those ping-pong-balls will be seen illuminated that are at such a
distance that the total time of flight for the light will be ten
nanoseconds.

Let us now repeat the experiment but this time using two persons.
One (A) emits the picosecond lightpulse, the other (B) makes the
picosecond observation ten nanoseconds later. If the distance separating A and B is four meters, what will B see? Will he see a lumi-

nous sphere around A with a radius of 1,5 meter? Certainly not! He will see nothing, because light has not yet reached him. The sphere can never be seen from outside.

Let B wait 10 nanoseconds more. What will he see after the delay of 20 nanoseconds (6 m)? He will find himself surrounded by a luminous ellipsoid having A and B as focalpoints. Only those particles and ping-pong-balls will be illuminated that are at such a distance that the total time of flight for the light from A to B will be twenty nanoseconds. If B repeats his observations every ten picoseconds he will find himself surrounded by a set of concentric ellipsoids all having A and B as focalpoints. The closest ellipsoids will have a high ellipticity, the more distant ones will become more and more spheroidal. The separation of adjacent ellipsoidal shells will be a function of k = 1/cos α, where α is half the angle separating illumination (A) and observation (B). These are the ellipsoids of the holo-diagram.

Thus we understand that the ellipsoids of the holo-diagram <u>are</u> the spherical wavefronts from A <u>as seen</u> from B. When the apearance of even the most basic concept in optics, the spherical wavefronts, depend on the point of view (B), it is not surprising that this is also the case for interferometric measurements. Thus in holographic interferometry the displacement can be calculated as d = n · k · 0,5λ, where d is displacement normal to ellipsoids, n is number of interference fringes, k is 1/cos α, α is half the angle separating A and B, and finally λ is the wavelength. This formula, which also works for conventional interferometers at oblique incidence, shows that "k" is a desensitizing factor, or that "k · λ" functions <u>as if</u> there was a redshift caused by α.

Thus we have demonstrated that an observer who does not know about the angle α would make errorneous measurements. If he believes that α is zero, when it is not, he will be using light with a "redshift" of which he is ignorent. Thus, the result of his measurement will be that distances (or displacements) appear shorter than they really are. The mistake he makes is caused by the fact that he is not aware that his observation sphere has become an observation ellipsoid. In practice, however, it is rather unlikely that he does not know that A and B are separated. If there were no other means, he could look for shadows on the objects.

Let us now again change the experimental situation so that it becomes dynamic instead of static. Let this time A and B be represented by the same person who is traveling at high speed, so that he moves from A to B during the twenty nanoseconds separating the emittance and the one single observation, of the picosecond pulse. If he does not know that he is traveling, or if he is ignorant of influences of his travel, he will again mistake his observation ellipsoid for

an observation sphere. Nothing in this situation will differ from
the static situation already discussed, except in the dynamic case
the lightrays are aberrated by his velocity in such a way that it
is impossible to him to observe the separation of A and B. Thus
again he makes errors in his measurement. Because of the angle 2α
between observation and illumination there is a correction factor
which is identical to the k-value of the conventional static holo-
diagram.

Let the speed of light be c, the velocity of the observer be v. In
that case the k-value anywhere on the Y-axis of the holo-diagram
will be:

$$k = \frac{1}{\cos \alpha} = \frac{c}{\sqrt{c^2 - v^2}} = \frac{1}{\sqrt{1 - \left(\frac{v}{c}\right)^2}}$$

This k-value is identical to the apparent redshift (the Transverse
Doppler Ratio of the relativity theory) which has to be corrected
for when interferometric measurements (or any other measurements)
are made on a system at high velocity in relation to the observer.
Without this correction object dimensions along the ellipsoids
appear contracted (the Lorentz Contraction) by the factor k of the
conventional holo-diagram.

Thus we have shown how interferometric measurements are influenced
the same way by a static separation as by a dynamic separation of
the points A and B of the holo-diagram. In the former case the
separation of course depends on the position of the observer in
relation to the object. In the later case the separation depends
on the velocity of the observer in relation to the object.

Einstein's theory of relativity is physically based on one-dimensional measurements using a conventional interferometer (the Michelson-
Morley experiment). Now that holographic interferometry enables us
to make three-dimensional measurements there exists new possibilities to solve problems in the relativity theory. Certainly much of
the knowledge now applied to fringe interpretation could be used
to interpret image distortions caused by ultrahigh velocities of
the recorded object. The advent of pico- and femtoseconds laser-
pulses will revolutionize high-speed photography and make compen-
sations for relativistic effects necessary. This is especially the
case when light-in-flight recordings by holography are used because
they represent four dimensions: the three dimensions found in ordinary holograms, plus the time domain.

After these reflections about the importance of the observers point of view I want, as a student and as a lecturer, thank professor Soares and his collaborators for making this Advanced Study Institute in Optical Metrology such a great success both from a social and from a scientific point of view.

SUBJECT INDEX

774